Springer Optimization ar

Volume 171

Aims and Scope

Optimization has continued to expand in all directions at an astonishing rate. New algorithmic and theoretical techniques are continually developing and the diffusion into other disciplines is proceeding at a rapid pace, with a spot light on machine learning, artificial intelligence, and quantum computing. Our knowledge of all aspects of the field has grown even more profound. At the same time, one of the most striking trends in optimization is the constantly increasing emphasis on the interdisciplinary nature of the field. Optimization has been a basic tool in areas not limited to applied mathematics, engineering, medicine, economics, computer science, operations research, and other sciences.

The series **Springer Optimization and Its Applications (SOIA)** aims to publish state-of-the-art expository works (monographs, contributed volumes, textbooks, handbooks) that focus on theory, methods, and applications of optimization. Topics covered include, but are not limited to, nonlinear optimization, combinatorial optimization, continuous optimization, stochastic optimization, Bayesian optimization, optimal control, discrete optimization, multi-objective optimization, and more. New to the series portfolio include Works at the intersection of optimization and machine learning, artificial intelligence, and quantum computing.

Volumes from this series are indexed by Web of Science, zbMATH, Mathematical Reviews, and SCOPUS.

More information about this series at http://www.springer.com/series/7393

Kok Lay Teo • Bin Li • Changjun Yu
Volker Rehbock

Applied and Computational Optimal Control

A Control Parametrization Approach

Springer

Kok Lay Teo
School of Mathematical Sciences
Sunway University
Selangor Darul Ehsan, Malaysia

Bin Li
College of Electrical Engineering
Sichuan University
Chengdu, China

Changjun Yu
College of Sciences
Shanghai University
Shanghai, China

Volker Rehbock
School of Electrical Engineering
Computing and Mathematical Sciences
Curtin University
Perth, WA, Australia

ISSN 1931-6828 ISSN 1931-6836 (electronic)
Springer Optimization and Its Applications
ISBN 978-3-030-69915-4 ISBN 978-3-030-69913-0 (eBook)
https://doi.org/10.1007/978-3-030-69913-0

Mathematics Subject Classification: 49M25, 34H05, 93C10

This Springer imprint is published by the registered company Springer Nature Switzerland AG
The registered company address is: Gewerbestrasse 11, 6330 Cham, Switzerland

Preface

For an optimal control problem, one seeks to optimize a performance measure subject to a set of dynamic, and possibly algebraic, constraints. The dynamic constraints may be expressed by a set of differential equations (ordinary or partial) or a set of difference equations. These equations may be deterministic or stochastic in nature. Based on the theoretical foundation laid by many great mathematicians of our time, optimal control has developed into a well-established research area. It has attracted the interests of many top researchers and practitioners working in several, often unrelated, disciplines, such as economics, management science, environmental management, forestry, agriculture, defence, core engineering (civil, chemical, electrical and mechanical), biology and social sciences. There is a large volume of papers, textbooks and research monographs dealing with theoretical as well as practical aspects of optimal control available in the literature.

For many practical real-life optimal control problems, the underlying dynamical systems are often large scale and highly complex. These optimal control problems are also subject to rigid algebraic constraints arising naturally due to practical limitations and engineering specifications. Thus, it is often not possible to obtain their analytical solutions. Therefore, we can only depend on computational methods for solving these real-world problems. For this reason, many successful computational methods have been developed to solve many different classes of optimal control problems with various types of constraints.

The computational methods developed based on the control parametrization form a specific family among these computational methods. Similar to the book titled, "A Unified Computational Approach to Optimal Control Problems", by Teo, Goh and Wong, the focus of this book is on this family of computational methods. The book by Teo, Goh and Wong was published in 1991, but it has been out of print since 1996. Hence, it contains only those results obtained prior to 1991. Many new theoretical results, new computational methods and new applications have been obtained and published since 1991. For this reason, we have been motivated to write this new book. To ensure that the book is self-contained, essential fundamental results from the 1991 book are included in this new book. A revised version of basic results on unconstrained and constrained optimization problems, and optimization problems subject to continuous inequality constraints are also included.

A tremendous proliferation of results published based on control parametrization has appeared in the literature after 1991. To keep the size of the book reasonable, we restricted ourselves to a discussion of those results obtained by the authors and their past and present collaborators and students. This choice ensures that the results presented in this book could be organized to form a unified computational approach to solve various real-world practical optimal control problems. These computational methods are supported by rigorous convergence analysis, easily programmable and adaptable to existing efficient optimization software packages.

We do not claim that this family of computational methods is necessarily superior to others found in the literature. Direct (Runge-Kutta) discretization of optimal control problems or pseudospectral techniques are two examples of methods that have been intensively studied by many researchers. A brief review of these techniques is included in Section 1.3.5.

This book can serve as a reference for researchers and students working in the areas of optimal control theory and its applications, and for professionals using optimal control to solve their problems. It is noted that many scientists, engineers and practitioners may not be thoroughly familiar with optimal control theory. Thus, the optimal control software MISER, which was developed based on the control parametrization technique, can help them to apply optimal control theory as a tool to solve their problems. We wish to emphasize that the aim of this book is to furnish a rigorous and detailed exposition of the concept of control parametrization and the time scaling transformation to develop new theory and new computational methods for solving various optimal control problems numerically and in a unified fashion. Based on the knowledge gained from this book, research scientists or engineers can develop new theory and new computational methods to solve other complex problems that are not being covered in this book.

The background required to understand the computational methods presented in this book, and their application to solve practical problems, is advanced calculus. However, to analyse the convergence properties of these computational methods, some results in real and functional analysis are also required. For the convenience of the reader, these mathematical concepts and facts are stated without proofs in Appendix A.1. Engineers and applied scientists should be able to follow the proofs of the convergence theorems with the aid of the results presented in Appendix A.1. For global optimization, a filled function method is presented in Appendix A.2. Some basic concepts and results on probability theory are discussed in Appendix A.3.

Chapter 1 introduces the reader to some essential concepts of optimal control theory. It also contains examples drawn from many fields of engineering and science. This ends with a brief survey of the existing computational techniques for solving optimal control problems. In Chapter 2, some fundamental results for unconstrained optimization problems are discussed. Chapter 3 contains basic results on constrained optimization problems. Chapter 4 considers optimization problems subject to continuous inequality constraints. Three computational methods are developed, two are based on the constraint transcription technique used in conjunction with a local smoothing method and the third one is developed based on the exact penalty function method. These results are important because after control parametrization, an optimal control problem is reduced to an optimal parameter selection problem, which can be viewed as an optimization problem.

Chapter 5 contains some fundamental results on discrete time optimal control problems. It contains discrete time minimum principle and dynamic programming technique, and computational methods for solving discrete time optimal control problems. Chapter 6 contains some essential results in optimal control theory. They are for those readers who are not familiar with optimal control theory. Chapter 7 is devoted to the derivations of gradient formulae for various kinds of optimal parameter selection problems, including optimal parameter selection problems with the heights and the switching times of the piecewise constant control being taken as decision variables; optimal parameter selection problems with discrete valued control; optimal parameter selection problems of switched systems; time-delay optimal parameter selection problem; and optimal control problems with multiple characteristic time points. With these gradient formulae, the respective optimal parameter selection problems can be solved as mathematical programming problems.

Chapter 8 considers optimal control problems in canonical form. The concept of control parametrization is introduced and applied to these canonically constrained optimal control problems. Gradient-based computational methods are derived. They are supported by a rigorous convergence analysis. A time scaling transform is also introduced to supplement the control parametrization technique for solving these canonically constrained optimal control problems. Chapter 9 considers a class of optimal control problems subject to continuous inequality constraints as well as terminal inequality constraints on the state and/or control variables. The constraint transcription method and the exact penalty function method are used to derive respective computational methods for solving these optimal control problems subject to continuous inequality constraints. Chapter 10 aims to develop computational methods for solving three classes of optimal control problems—time-lag optimal control problems, state-dependent switched time-delay optimal control problems and min-max optimal control problems.

In Chapter 11, we introduce two approaches to constructing suboptimal feedback controls for constrained optimal control problems. The first approach is known as the neighbouring extremals approach, while the second approach is to construct an optimal PID control for a class of optimal control problems subject to continuous inequality constraints and terminal equality constraint.

In Chapter 12, we consider two classes of stochastic dynamic optimization problems. The first one is a combined optimal parameter selection and optimal control problem in which the dynamical system is governed by linear Ito stochastic differential equation involving a Wiener process. Both the control and system parameter vectors may, however, appear nonlinearly in the system dynamics. The cost functional is taken as an expected value of a quadratic function of the state vector, where the weighting matrices are time invariant but are allowed to be nonlinear in both the control and system parameter. Furthermore, certain realistic features such as probabilistic constraints on the state vector may also be included. Another problem considered in Chapter 12 is a partially observed linear stochastic control problem described by three sets of stochastic differential equations: one for the system to be controlled, one for the observer (measurement) channel and one for the control channel driven by the observed process. The noise processes perturbing the system and observer dynamics are vector-valued Poisson processes. For both of these stochastic dynamic optimization problems, we show that they are equivalent to two respective deterministic dynamic optimization problems. These equivalent deterministic dynamic optimization problems are further transformed into special cases of the form considered in Chapter 9.

It is our pleasure to express gratitude to many of our colleagues and collaborators, and to those PhD students, Postdoctoral Fellows and Visiting Research Fellows of the first author listed as follows: Changzhi Wu, Zhiguo Feng, Joseph Lee, Kar Hung Wong, Ryan Loxton, Qun Lin, Rui Li, Chongyang Liu, Zhaohua Gong and Canghua Jiang. They have made great contributions to the book, and many of the results presented in this book are from various joint papers co-authored with them. Further details are indicated in the respective chapters.

We wish to thank Yanqing Liu and Zhaohua Gong for recalculating the examples in Chapters 8 and 9. Zhaohua Gong has redrawn the figures for the examples being solved in Chapters 8–11. Also, we wish to thank Xiaoyi Guan, Gaoqi Liu, Yanqing Liu, Shuxuan Su, Di Wu, Lei Yuan and Xi Zhu for their help in LaTeX.

Our thanks also go to Professor Panos M. Pardalos, the Editor of the Book Series, *Springer Optimization and Its Applications*, for his encouragement, leading to the publication of the book in his book series. We thank the reviewers for their constructive and detailed comments and suggestions.

We also wish to express our appreciation to the staff of Springer, especially to Elizabeth Loew, for their expert collaboration. Last but not least, our most sincere thanks go to our families for their support, patience and understanding. They have done much to improve the work, but any shortcomings are totally ours.

School of Mathematical Sciences, Sunway University Selangor Darul Ehsan, Malaysia — Kok Lay Teo

College of Electrical Engineering, Sichuan University Chengdu, China — Bin Li

College of Sciences, Shanghai University Shanghai, China — Changjun Yu

School of Electrical Engineering, Computing and Mathematical Sciences, Curtin University Perth, WA, Australia — Volker Rehbock

August 20, 2020

Contents

Chapter 1
Introduction

1.1 Optimal Control Problems

Broadly speaking, an optimal control problem seeks to optimize a performance measure subject to a set of dynamic constraints. The dynamic constraints may constitute a set of differential equations (ordinary or partial) or a set of difference equations. These equations may be deterministic or stochastic in nature. Furthermore, optimal control problems are often subject to constraints on the state and/or control. These constraints arise due to engineering regulations or design specifications, such as the specified product quality or the constraint on the safety requirement [55]. Optimal control problems subject to constraints on the state and/or control are referred to as constrained optimal control problems.

Optimal control has applications in almost every area of science and engineering, including aquaculture operation [28], cancer chemotherapy (see, for example, [180, 181]), switched power converters (see, for example, [93, 167]), spacecraft control (see, for example, [82, 100, 101, 111, 137, 255, 311, 312]), ship steering [132], underwater vehicles [45], process control (see, for example, [25, 41, 46, 139, 156–160, 173–176, 271, 272, 291–293, 302]), core engineering (see, for example [35, 42, 79, 80, 102, 105, 123]), optimal control of automobiles or trains (see, for example, [95–97, 128, 310]), and crystallization processes (see, for example, [194, 214]) and management sciences (see, for example, [70, 94, 133, 135, 136, 226, 233–236]).

The famous Pontryagin maximum principle (see, for example, [3, 4, 29, 40, 64, 69, 206]) is a set of first order necessary conditions of optimality for a constrained optimal control problem. It provides the means to solve many practical problems in various disciplines. In particular, researchers in economics started to model many of their problems in an optimal control context and were able to solve them using the maximum principle (see, for example, [74, 110, 226]). Markov-Dubins path is the shortest planar curve

K. L. Teo et al., *Applied and Computational Optimal Control*, Springer Optimization and Its Applications 171,
https://doi.org/10.1007/978-3-030-69913-0_1

joining two points with prescribed tangents, where a specified bound is imposed on its curvature. An elegantly simple solution was obtained by Dubins in 1957—a selection of at most three arcs are concatenated, each of which is either a circular arc of maximum (prescribed) curvature or a straight line. The Markov-Dubins problem is reformulated as an optimal control problem in various papers, and Pontryagin maximum principle is used to obtain the same results as those obtained by Dubins. In [114], under the same reformulation of the Markov-Dubins problems, the maximum principle is applied to derive Dubins result again. The new insights are: abnormal control solutions do exist; these solutions can be characterized as a concatenation of at most two circular arcs; they are also solutions of the normal problem; and any feasible path of the types mentioned in Dubins result satisfies the Pontryagin maximum principle. A numerical method for computing Markov-Dubins path is proposed. Dynamic Programming Principle developed by Bellman [18–20] has been used to determine optimal feedback controls for a range of practical optimal control problems. However, its application typically requires the solution of a highly nonlinear partial differential equation known as the Hamilton-Jacobi-Bellman (HJB) equation. Some numerical methods for solving this HJB equation for low dimensional problems are available in the literature (see, for example, [2, 6, 98, 99, 188, 208, 273, 274, 303–309]).

Practical problems, however, are usually too complex to be solved analytically using either Pontryagin maximum principle or dynamic programming. Thus, many numerical solution techniques have been developed and implemented on computers. These numerical solution techniques coupled with modern computing power are able to solve a wide range of highly complex optimal control problems.

There are several survey articles and books in the literature on optimal control computation. See, for example, [48, 69, 148, 210, 215]. For optimal control computational methods based on Euler discretization, see, for example, [36, 49, 50, 85, 113, 115, 178, 267].

In this book, our attention is centred on optimal control problems involving systems of ordinary differential equations with as well as without time delayed arguments and systems of difference equations. Note that these classes of problems have many practical applications across a wide range of disciplines such as those mentioned above.

1.2 Illustrative Examples

Example 1.2.1 (The Student Problem) Various versions of this problem have appeared in the literature over the past few decades (see, for example, [43, 120, 202, 209]). Consider a lazy and forgetful student who wishes to pass an examination with a minimum expenditure of effort. Assume that the student's knowledge level at any time t during the semester is a reflection of his/her performance if an examination was given at that time. Also, suppose that the rate of knowledge intake is a linear function of the work rate and that the student is constantly forgetting a proportion of what he/she already knows. Let $w(t)$ denote the rate of work done at time t, let $k(t)$ denote the knowledge level at time t and let T be the total time of the semester in weeks. Furthermore, let $c > 0$ be the forgetfulness factor, let $b > 0$ be a constant that determines how efficiently the student's work is being converted to knowledge, let $\bar{w}$ be an upper bound on the work rate, let k_0 be the initial knowledge level and let k_T be the desired final knowledge level (assumed to be the minimum level required to pass the examination). Then the problem may be stated as follows:

$$\text{Minimize} \;\; g(w) = \int_0^T w(t)\,dt$$

subject to

$$\begin{aligned} \frac{dk(t)}{dt} &= bw(t) - ck(t) \\ k(0) &= k_0 \\ k(T) &= k_T \end{aligned}$$

and $0 \leq w(t) \leq \bar{w}$ for all $0 \leq t < T$. Although this is a rather trivial example of an optimal control problem, it serves well to illustrate the basic class of optimal control problems we want to consider: an objective functional, $g(w)$, is to be minimized subject to a dynamical system governing the behavior of the state variable, $k(t)$, and subject to constraints. We need to find a control function $w(t)$, $t \in [0, T)$, subject to given bounds, which will optimize the objective functional. A somewhat more realistic version of the student problem, which assumes some ambition in the student, may be stated as follows:

$$\text{Maximize} \;\; g(w) = ak(T) - \int_0^T \left(\alpha_1 w(t) + \alpha_2 (w(t))^2\right) dt$$

subject to

$$\begin{aligned} \frac{dk(t)}{dt} &= (b_1 + b_2 k(t))w(t) - ck(t) \\ k(0) &= k_0 \\ k(T) &\geq k_T, \end{aligned}$$

and $0 \leq w(t) \leq \bar{w}$ for all $0 \leq t < T$. This version of the problem assumes that the student has some interest in maximizing his/her examination mark, is more averse to high rates of work and can gain knowledge more readily when he/she already has a high level of knowledge. Both problems stated in

Example 1.2.1 can be solved readily by using Pontryagin maximum principle or dynamic programming.

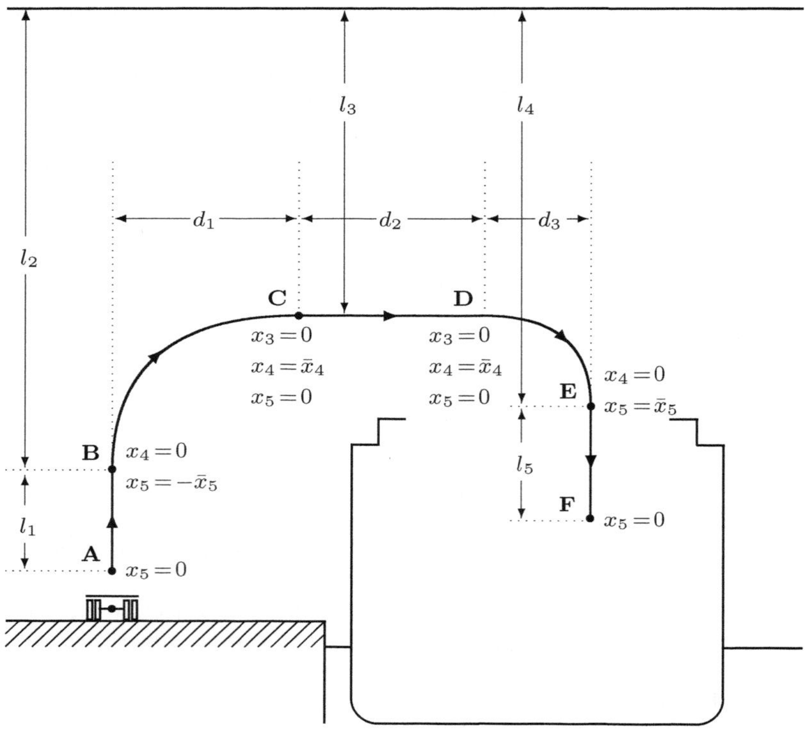

Fig. 1.2.1: Movement of the container

Example 1.2.2 (Optimal Control of Container Cranes) This problem originally appeared in [219]. Consider the task of loading a container from a truck waiting on a wharf onto a ship tied up to the wharf, see Figure 1.2.1. The mechanics of the crane are schematically shown in Figure 1.2.2. The crane is driven by two motors. The trolley motor (left) controls the horizontal position of the crane trolley, while the hoist motor (top) effectively controls the rope length of the crane. J_1 and J_2 denote the total moment of inertia of the trolley and hoist motors, respectively, including their associated components (reduction gears, brake, drum etc.). Similarly, b_1 and b_2 denote the drum radii of the trolley and hoist motors, respectively. Furthermore, we let $\theta_1(t)$ and $\theta_2(t)$ denote the angles of rotation of the respective motors in radians at time t. The underlying control variables are the driving torques

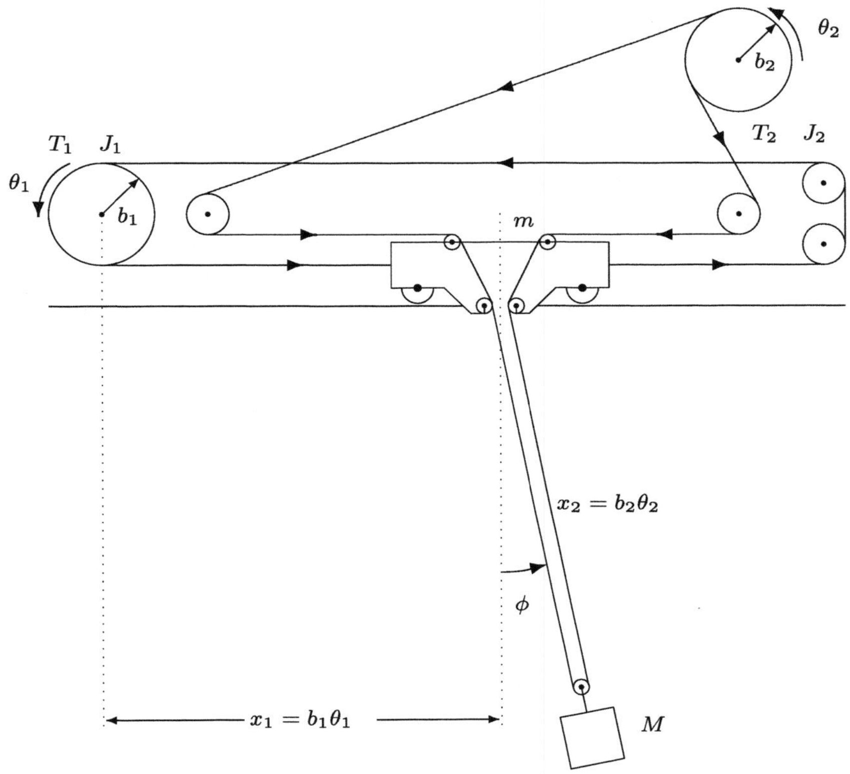

Fig. 1.2.2: Schematic of the container crane

generated by the trolley and hoist motors, denoted by $T_1(t)$ and $T_2(t)$, respectively. Finally, M denotes the total mass of the container and attached equipment, m is the total mass of the crane trolley and the operator's cab, $\phi(t)$ is the load swing angle and g denotes acceleration due to gravity.
We define

$$\begin{aligned}
x_1(t) &= b_1\theta_1(t) \text{ (horizontal position of trolley)},\\
x_2(t) &= b_2\theta_2(t) \text{ (rope length)},\\
x_3(t) &= \phi(t) \text{(swing angle of the load)},\\
x_4(t) &= \frac{dx_1(t)}{dt},\\
x_5(t) &= \frac{dx_2(t)}{dt},\\
x_6(t) &= \frac{dx_3(t)}{dt},
\end{aligned}$$

$$v_1(t) = \frac{b_1 T_1(t)}{J_1 + mb_1^2},$$

$$v_2(t) = \frac{b_2(T_2(t) + Mb_2 g)}{J_2 + Mb_2^2},$$

$$\delta_1 = \frac{Mb_1^2}{J_1 + mb_1^2} \text{ and } \delta_2 = \frac{Mb_2^2}{J_2 + Mb_2^2}.$$

Assuming that the load swing angle is small in magnitude, that the load can be regarded as a point and that frictional torques can be neglected, the dynamics of the container crane are given by Sakawa and Shindo [219]

$$\frac{dx_1(t)}{dt} = x_4(t) \tag{1.2.1}$$

$$\frac{dx_2(t)}{dt} = x_5(t) \tag{1.2.2}$$

$$\frac{dx_3(t)}{dt} = x_6(t) \tag{1.2.3}$$

$$\frac{dx_4(t)}{dt} = v_1(t) - \delta_1 x_3(t) v_2(t) + \delta_1 g x_3(t) \tag{1.2.4}$$

$$\frac{dx_5(t)}{dt} = -\delta_2 x_3(t) v_1(t) + v_2(t) \tag{1.2.5}$$

$$\frac{dx_6(t)}{dt} = -\frac{1}{x_2(t)}[v_1(t) - \delta_1 x_3(t) v_2(t) + (1+\delta_1) g x_3(t) + 2x_5(t) x_6(t)]. \tag{1.2.6}$$

Note that v_1 and v_2 are the effective control variables in these dynamics, and they are subject to the following bounds:

$$|v_1(t)| \leq \bar{v}_1, \quad \forall t, \tag{1.2.7}$$

$$\underline{v}_2 \leq v_2(t) \leq \bar{v}_2, \quad \forall t, \tag{1.2.8}$$

where $\bar{v}_1$, $\underline{v}_2$ and $\bar{v}_2$ are defined in terms of the maximum torques of the trolley drive and hoist motors, respectively (see [219]). Due to safety requirements, the following bounds on the state variables are imposed:

$$|x_4(t)| \leq \bar{x}_4, \quad \forall t, \tag{1.2.9}$$

$$|x_5(t)| \leq \bar{x}_5, \quad \forall t. \tag{1.2.10}$$

The movement of the container is generally divided into 5 distinct sections. The first section (from A to B in Figure 1.2.1) constitutes just a simple vertical lift off the truck. Note that x_4 and x_5 are equal to zero at the start of this section. At the end, the container is moving at the maximum allowed vertical velocity, so $x_4 = 0$ and $x_5 = -\bar{x}_5$. These conditions carry over as the starting values for the next section of the path, from B to C. Unlike the previous section for which the optimal control can be determined analytically, moving the container from B to C is a non-trivial optimal control task that forms the basis of the problem we present here. Note that the container must arrive at C with $x_3 = 0$ (no swing), $x_4 = \bar{x}_4$ (maximum allowed horizontal velocity) and $x_5 = 0$ (zero vertical velocity). Finally, the sections from C to D and E to F again constitute trivial problems, while the section from D to E has a similar complexity to that from B to C. Consider the section from B to C over the time horizon $[0, T]$. We impose the following initial conditions and terminal state constraints on the problem:

$$\begin{aligned} &x_1(0) = 0,\ x_2(0) = l_2,\ x_3(0) = 0, \\ &x_4(0) = 0,\ x_5(0) = -\bar{x}_5,\ x_6(0) = 0, \end{aligned} \tag{1.2.11}$$

$$\begin{aligned} &x_1(T) = d_1,\ x_2(T) = l_3,\ x_3(T) = 0, \\ &x_4(T) = \bar{x}_4,\ x_5(T) = 0,\ x_6(T) = 0, \end{aligned} \tag{1.2.12}$$

with the various constants illustrated in Figure 1.2.1.

The objective functional to be minimized subject to the dynamics and constraints given in (1.2.1–1.2.12) is

$$g(\mathbf{v}) = \frac{1}{2} \int_0^T \left(w_1(x_3(t))^2 + w_2(x_6(t))^2 \right) dt, \tag{1.2.13}$$

where w_1 and w_2 are given weights. This objective functional can be regarded as a measure of the total amount of swing experienced by the load.

Alternatively, since the speed of the operation is clearly an important issue [219], one may want to minimize

$$\bar{g}(\mathbf{v}) = \int_0^T 1\, dt = T \tag{1.2.14}$$

subject to (1.2.1) and (1.2.12) and subject to the additional swing constraint

$$\int_0^T \left(w_1(x_3(t))^2 + w_2(x_6(t))^2 \right) dt \leq S_{\max}, \tag{1.2.15}$$

where S_{max} is a given parameter.

Example 1.2.3 (Optimal Production of Penicillin) The production of penicillin takes place in a fed-batch fermentation process where the feed rate of the substrate is the main control variable [200]. The aim is to find the substrate feed rate that will optimize the final amount of penicillin. The problem may be stated as follows. Maximize

$$g(u) = P(T) \tag{1.2.16}$$

subject to

$$\begin{aligned}
\frac{dX(t)}{dt} =& \mu(X(t), S(t), V(t))X(t) && (1.2.17)\\
\frac{dP(t)}{dt} =& \pi(X(t), S(t), V(t))X(t) - kP(t) && (1.2.18)\\
\frac{dS(t)}{dt} =& -\sigma(X(t), S(t), V(t))X(t) + s_F u(t) && (1.2.19)\\
\frac{dV(t)}{dt} =& u(t) && (1.2.20)\\
X(0) =& 10.5 && (1.2.21)\\
P(0) =& 0 && (1.2.22)\\
S(0) =& 0 && (1.2.23)\\
V(0) =& 7 && (1.2.24)\\
V(T) =& 10 && (1.2.25)\\
0 \leq u(t) \leq& u_{\max}, \forall t \in [0, T). && (1.2.26)
\end{aligned}$$

Here, μ, π and σ are the growth rate functions given by

$$\begin{aligned}
\mu(X, S, V) =& \frac{\mu_{\max} S}{\mu_1 X + S} && (1.2.27)\\
\pi(X, S, V) =& \frac{\pi_{\max} SV}{\pi_1 V^2 + S(V + \pi_2 S)} && (1.2.28)\\
\sigma(X, S, V) =& \frac{\mu}{\sigma_1} + \frac{\pi}{\sigma_2} + \frac{\sigma_3 S}{\pi_1 V + S}, && (1.2.29)
\end{aligned}$$

where the $\mu_{\max}$, $\pi_{\max}$, μ_1, π_1, σ_1, σ_2 and σ_3 are given constants. Here, X represents the biomass in the reactor, P is the amount of product, S is the amount of substrate and V is the total volume of fluid in the reactor. This is a challenging problem even for numerical algorithms, due to the rigid behavior of the specific growth rate functions (see [200] and the references cited therein).

Example 1.2.4 (Optimal Driving Strategy for a Train) The following system of differential equations is a model of the dynamics of a train on a level track similar to that discussed in [95]:

$$\begin{aligned}
\frac{dx_1}{dt} =& x_2\\
\frac{dx_2}{dt} =& \varphi(x_2)\, u_1 + \zeta_2\, u_2 + \rho(x_2),
\end{aligned}$$

where x_1 is the distance along the track, x_2 is the speed of the train, u_1 is the fuel setting and u_2 models the deceleration applied to the train by the brakes. The function

$$\varphi(x_2) = \begin{cases} \zeta_1/x_2, & \text{if } x_2 \geq \zeta_3 + \zeta_4, \\ \zeta_1/\zeta_3 + \eta_1 (x_2 - (\zeta_3 - \zeta_4))^2 \\ +\eta_2 (x_2 - (\zeta_3 - \zeta_4))^3, & \text{if } \zeta_3 - \zeta_4 \leq x_2 < \zeta_3 + \zeta_4, \\ \zeta_1/\zeta_3, & \text{if } x_2 < \zeta_3 - \zeta_4, \end{cases}$$

where

$$\eta_1 = \zeta_1 \left\{ \left\{ \frac{1}{\zeta_3 + \zeta_4} - \frac{1}{\zeta_3} \right\} \frac{3}{4\,\zeta_4^2} + \frac{1}{2\,\zeta_4\,(\zeta_3 + \zeta_4)^2} \right\}$$

and

$$\eta_2 = \zeta_1 \left\{ - \left\{ \frac{1}{\zeta_3 + \zeta_4} - \frac{1}{\zeta_3} \right\} \frac{1}{4\,\zeta_4^3} - \frac{1}{4\,\zeta_4^2 (\zeta_3 + \zeta_4)^2} \right\}$$

represent the tractive effort of the locomotive. A somewhat simpler form of φ was used in [95], but the form used here models the actual data (see Figure 1 of [95]) more accurately. The function ρ is the resistive acceleration due to friction, given by $\rho(x_2) = \zeta_5 + \zeta_6\, x_2 + \zeta_7\, x_2^2$. ζ_i, $i = 1, \ldots, 7$, are constants with given values $\zeta_1 = 1.5$, $\zeta_2 = 1$, $\zeta_3 = 1.4$, $\zeta_4 = 0.1$, $\zeta_5 = -0.015$, $\zeta_6 = -0.00003$ and $\zeta_7 = -0.000006$. Also, $x_1(0) = 0$, $x_2(0) = 0$, $x_1(1500) = 18000$ and $x_2(1500) = 0$, i.e., the train starts from the origin at rest and comes to rest again 18000m away at $t_f = 1500$. The train is not allowed to move backwards, so we require $x_2(t) \geq 0$ for all $t \in [0, 1500]$.

The control $\begin{bmatrix} u_1 \\ u_2 \end{bmatrix}$ is restricted to the discrete set $\mathbf{U} = \left\{ \begin{bmatrix} 1 \\ 0 \end{bmatrix}, \begin{bmatrix} 0 \\ 0 \end{bmatrix}, \begin{bmatrix} 0 \\ -1 \end{bmatrix} \right\}$, so that the train is either powered by the engine, coasting or being slowed by the brakes. Note that power and brakes cannot be applied simultaneously. The objective is to minimize the fuel used on the journey, i.e., to minimize

$$J_0(\mathbf{u}) = \int_0^{1500} u_1 \, dt.$$

More realistic versions of the problem include multiple fuel settings and speed limit constraints, see [126].

Example 1.2.5 (Optimal Control of Crystallization) Crystallization from solution is a purification and separation technique of great economic importance to the chemical industry. The quality of the crystallization product and the efficiency of downstream product recovery processes are primarily determined by the crystal size distribution (CSD). Here, we deal with the precipitation of aluminium trihydroxide ($Al(OH)_3$) from supersaturated sodium aluminate solutions. As part of the Bayer process, it represents an important step in the production of aluminium. In industry practice, the precipitation of aluminium trihydroxide is carried out in a continuous manner using a cascade of crystallizers. For the purpose of a theoretical analysis, however, the complete process can be approximated by a single batch cooling crystallizer.

Following the approach detailed in [94], we discretize the solid particle distribution into m distinct size intervals $[L_i, L_{i+1}]$, $i = 1, \ldots, m$, where size is a measure of the diameter of a particle. A large range of particle sizes can be captured if L_i are defined by $L_{i+1} = rL_i$, $i = 1, \ldots, m$, where $r = \sqrt[3]{2}$ and $L_1 = 3.7 \times 10^{-6}$ metres. Let N_i denote the number of particles in the i-th size interval, and let C be the concentration of the solution. The rate of change of each N_i consists of 3 terms that reflect the effects of nucleation, crystal growth and agglomeration. The nucleation of new particles is assumed to occur only for the first size interval. The dynamics of the number of crystals in individual size intervals as well as the solute concentration are then described as follows:

$$\frac{dN_1}{dt} = \underbrace{\frac{2G}{L_1(1+r)}\left[\left(1-\frac{r^2}{r^2-1}\right)N_1 - \frac{r}{r^2-1}N_2\right]}_{\text{growth}} + \underbrace{B_u}_{\text{nucleation}} - \underbrace{\beta N_1 \sum_{j=1}^{m} N_j}_{\text{agglomeration}} \quad (1.2.30)$$

$$\frac{dN_i}{dt} = \underbrace{\frac{2G}{L_i(1+r)}\left[\frac{r}{r^2-1}N_{i-1} + N_i - \frac{r}{r^2-1}N_{i+1}\right]}_{\text{growth}}$$
$$+ \underbrace{\beta\left[N_{i-1}\sum_{j=1}^{i-2} 2^{j-i+1}N_j + \frac{1}{2}(N_{i-1})^2 - N_i\sum_{j=1}^{i-1}2^{j-i}N_j - N_i\sum_{j=i}^{m}N_j\right]}_{\text{agglomeration}},$$
$$i = 2, \ldots, m-1, \quad (1.2.31)$$

$$\frac{dN_m}{dt} = \underbrace{\frac{2G}{L_m(1+r)}\left[\frac{r}{r^2-1}N_{m-1} + N_m\right]}_{\text{growth}}$$
$$+ \underbrace{\beta\left[N_{m-1}\sum_{j=1}^{m-2} 2^{j-m+1}N_j + \frac{1}{2}(N_{m-1})^2 - N_m\sum_{j=1}^{N-1}2^{j-m}N_j - (N_m)^2\right]}_{\text{agglomeration}},$$
$$(1.2.32)$$

$$\frac{dC}{dt} = \frac{-3k_v\rho_s}{\varepsilon}G\sum_{i=1}^{m}N_iS_i^2 - \frac{\rho_s}{\varepsilon}k_vS_1^3B_u. \quad (1.2.33)$$

We consider $m = 25$ size intervals, S_i, $i = 1, \ldots, m$, denotes the average particle size in the interval $[L_i, L_{i+1}]$, equation (1.2.33) models the rate of change of the concentration of the solution if the change of volume is assumed negligible, G is a measure of growth, B_u denotes the rate of nucleation in the first size interval, β is known as the agglomeration kernel (a measure of the frequency of collisions between particles), assumed to be independent

of the particle size here, $k_v = 0.5$ is a volume shape factor, $\varepsilon = 0.8$ and $\rho_s = 2420\text{kg/m}^3$ is the density of the resulting solid.
Furthermore, letting T denote the temperature of the solution in degrees Kelvin, we can define the solubility as a function of temperature and caustic concentration [194], i.e.,

$$C^*_{\mathrm{Al_2O_3}} = C_{\mathrm{Na_2O}} e^{6.21 - \frac{2486.7}{T} + \frac{1.0875 C_{\mathrm{Na_2O}}}{T}},$$

where $C_{\mathrm{Na_2O}} = 100\,\text{kg/m}^3$. The supersaturation, defined as $\Delta C = C - C^*$, is the main driving force for the three processes of nucleation, growth and agglomeration. The growth is modeled by $G = k_g(\Delta C)^2$, where, assuming $C_{\mathrm{Na_2O}} = 100\,\text{kg/m}^3$ as before, $k_g = 6.2135e^{-\frac{7600}{T}}$. The dependence of the agglomeration kernel is modeled by $\beta = k_a(\Delta C)^4$, where $k_a = 6.8972 \times 10^{-21}T - 2.29 \times 10^{-18}$. Finally, the dependence of nucleation on ΔC and T is suitably modeled by

$$B_u = k_n(\Delta C)^{0.8} \left(k_s \sum_{i=1}^{m} N_i S_i^2 \right)^{1.7},$$

where $k_s = \pi$ is a surface shape factor and k_n is an empirical coefficient depending on temperature. It is required that all functions involved in the dynamics are continuously differentiable. The $(\Delta C)^{0.8}$ term appearing in the equations above does not satisfy this assumption as $\Delta C \to 0$. Hence, we replace this term by a smooth cubic approximation for small values of ΔC, i.e.,

$$B_u = k_n f_c(\Delta C) \left(k_s \sum_{i=1}^{m} N_i S_i^2 \right)^{1.7},$$

where

$$f_c(\Delta C) = \begin{cases} (\Delta C)^{0.8}, & \text{if } \Delta C > 1, \\ -1.2(\Delta C)^3 + 2.2(\Delta C)^2, & \text{if } 0 \le \Delta C < 1. \end{cases}$$

Furthermore, it has been shown experimentally that nucleation decreases markedly at temperatures above 70 °C and does not occur beyond 80 °C. For temperatures below 70 °C, we take $k_n = 9.8 \times 10^{22} e^{\frac{-10407.265}{T}}$, while $k_n = 0$ for temperatures above 80 °C. In between, we use a smooth cubic interpolation for k_n, i.e.,

$$k_n(T) = \begin{cases} 9.8 \times 10^{22} e^{\frac{-10407.265}{T}}, & \text{if } T \le 343.2\ ^\circ\text{K}, \\ (0.002c_1 + 0.01c_2)(T - 353.2)^3 \\ \quad +(0.03c_1 + 0.1c_2)(T - 353.2)^2, & \text{if } T \le 353.2\ ^\circ\text{K}, \\ 0, & \text{if } T > 353.2\ ^\circ\text{K}, \end{cases}$$

where $c_1 = 52673.694$ and $c_2 = 4654.0948$ and T is measured in degrees Kelvin. These dynamics are active over $[0, t_f]$, where t_f may be fixed or variable.
A narrow size distribution, i.e., one with a small variance, is usually desired, along with a large final mean crystal size. Thus, the aim is to maximize

$$J(T) = -\ln\left(\sqrt{\frac{M_5}{M_4^4} - \frac{1}{M_3 M_4^2}}\right),$$

where $M_3 = \sum_{i=1}^{N} N_i(t_f)S_i^3$, $M_4 = \sum_{i=1}^{N} N_i(t_f)S_i^4$ and $M_5 = \sum_{i=1}^{N} N_i(t_f)S_i^5$. This is equivalent to maximizing the ratio of the mean crystal size over the variance in the crystal size. Other versions of the problem, where seed crystals are added to the solution throughout the process, can also be readily formulated [214].

1.3 Computational Algorithms

Application of the well-known Pontryagin maximum principle can yield solutions to many optimal control problems by analytic means. This is particularly true in the field of economics, where many interesting problems have been solved this way, often yielding important practical principles (see, for example, [21, 110, 226]). Many other theoretical results regarding various classes of optimal control problems may be found in the literature (see, for example, [3–5, 9, 33, 40, 69, 83, 88–90, 121, 130, 149, 250, 253, 276]). However, most practical problems arising in engineering and science have a high degree of complexity and may not be able to find solutions by analytic means. Hence, many computational algorithms have been developed to calculate numerical solutions of optimal control problems. Numerical optimal control computation methods can be roughly divided into two categories: indirect methods and direct methods. In an indirect method, the maximum principle is used to determine the form of the optimal control in terms of state and costate variables. Then, it gives rise to a multiple-point boundary-value problem. Its solution is obtained through solving the multiple-point boundary-value problem. In a direct method, the optimal control problem is approximated as a nonlinear programming problem (NLP) through the discretization of the control and/or state over the time interval. The NLP is then solved using optimization techniques. Several survey articles and books on optimal control computation can be found in the open literature (see, for example, [48, 69, 148, 210, 215]). For optimal control computational methods based on Euler discretization, see, for example, [36, 49, 50, 85, 113, 115, 178, 258].

In what follows, we provide an overview of several numerical solution methods.

1.3.1 Dynamic Programming and Iterative Dynamic Programming

The Iterative Dynamic Programming (IDP) technique is a computational variation of the dynamic programming principle. The technique was initially developed in [173] and then refined in [175] to improve the computational efficiency. The method uses a grid structure for discretizing both the state variables and the control variables. The grid of the state defines accessible points in the state trajectory, and the grid of the controls defines admissible control values. The method typically starts with a coarse grid over a large region of the state space. Successive refinements of the grid are then implemented around the optimal trajectory until a satisfactory control policy is obtained. Initial development employed piecewise constant control, and this was later extended to piecewise linear continuous control policies [174]. Constraints are handled by a penalty function approach. The IDP technique has been successfully applied to a wide range of optimal control problems particularly in the field of chemical engineering [175]. According to [175], advantages of the IDP technique include its robustness, and, as a gradient-free method, its ability to steer away from local optima. It is also noted in [175] that the method involves numerous algorithmic parameters that can be difficult to tune for a novice user. These include the region contraction factor, the number of allowable values for controls, the number of grid points, the initial region size and the restoration factor.

1.3.2 Leapfrog Algorithm and STC algorithm

The algorithm developed in [112] is known as the leapfrog algorithm. In this algorithm, an initially feasible trajectory is given and subdivided over the time horizon. In each subinterval a piecewise-optimal trajectory is obtained. The junctions of these sub-trajectories are then updated through a scheme of midpoint maps. Under some broad assumptions, the sequence of trajectories is shown to converge to a trajectory that satisfies the Maximum Principle. In [117], the switching time computation (STC) method is incorporated in a time-optimal bang- bang control (TOBC) algorithm [111] for solving a class of optimal control problems governed by nonlinear dynamics system with single input. In this method, a concatenation of constant-input arcs is used to move from an initial point to the target, and an optimization procedure is utilized to find the necessary time lengths of the arcs. The difficulties of the STC method in finding the necessary arc time lengths are discussed. The gradients with respect to the switching time variables are calculated in a manner that avoids the need for costate variables, and this can be a computational advantage in some problems. Derivation of these gradients is given in Section 7.4.1, where the limitations of this approach are also discussed. In [111], the time-optimal switching (TOS) algorithm is developed for solving a class of time optimal switching control problems involving nonlinear systems with a single control input. In this algorithm, the problem is formulated in the arc times space, where arc times are the durations of the arcs. A feasible switching control is found using the STC method [117] to move from an initial point to a target point with a given number of switching. The cost is expressed as the summation of the arc times. Then, by using a constrained optimization technique, a minimum-time switching control solution is obtained. In [186], a numerical scheme is developed for constructing optimal bang-bang controls. Then, the second order sufficient conditions developed in [185] are used to check numerically whether the controls obtained are optimal.

1.3.3 Control Parametrization

The control parametrization method (see, for example, [36, 63, 69, 89, 143, 145, 148, 151, 153, 154, 160–162, 164, 166, 169–171, 181, 215, 229, 230, 238, 244, 249, 253, 255, 260, 284, 288, 294, 298, 300, 301, 311]) relies on the discretization of the control variables using a finite set of parameters. This is most commonly done by partitioning the time horizon of a given problem into several subintervals such that each control can be approximated by a piecewise constant function that is consistent with the corresponding partition. The approximating piecewise constant function can be defined in terms of a finite set of parameters, known as control parameters. Upon such an approximation, an optimal control problem becomes a finite dimensional optimal parameter selection problem. In real world, optimal control problems are often subject to constraints on the state and/or control. These constraints can be point constraints and/or continuous inequality constraints. The point constraints are expressed as functions of the states at the end point or some intermediate interior points of the time horizon. These point constraints can be handled without much difficulty. However, for the continuous inequality constraints, they are expressed as functions of the states and/or controls over the entire time horizon and hence are very difficult to handle. Through the control parametrization, a continuous inequality constrained optimal control problem is approximated as a continuous inequality constrained optimal parameter selection problem, which can be viewed as a semi-infinite programming (SIP) problem involving dynamic system. A popular approach to deal with the continuous inequality constraints on the state and control is known as the constraint transcription (see, for example, [76, 103, 135, 136, 148, 245, 246, 249, 253, 259]). Details will be given in later chapters. Another effective method to handle continuous inequality constraints is the exact penalty functions method (see, for example, [134, 300, 301]). It is also discussed in detail in later chapter. After the use of the constraint transcription method or the exact penalty function method, the continuous inequality constrained optimal parameter selection problem becomes an optimal parameter selection problem subject to constraints in the form of the objective functional, and these constraints are called canonical constraints. Each of these optimal parameter selection problems with canonical constraints can be regarded as a mathematical programming problem, and its solution is to be obtained by constrained optimization techniques. The control parametrization technique has been used in conjunction with the constraint transcription or the exact penalty function extensively in the literature (see, for example, [104, 134, 138, 145, 162, 164, 168, 171, 180, 181, 214, 215, 236, 244, 245, 249, 254, 294]). In [148], a survey and recent developments of the technique are presented. The technique has been proven to be very efficient in solving a wide range of optimal control problems. In particular, several computational algorithms to deal with a variety of different classes of problems together with a sound theoretical convergence analysis are reported in the literature (see, for example, [230, 237, 240, 245, 246, 248, 249, 253, 260, 279–281]). Under some mild assumptions, convergence of the sequence of approximate optimal costs obtained from a corresponding sequence of partition refinements of the time horizon to the optimal cost of the original optimal control problem has been demonstrated. Furthermore, the solution obtained for each approximate optimal control problem, which is regarded as a constrained optimization problem, is such that the KKT conditions are satisfied. However, there is no proof of the convergence of the approximate optimal controls to true optimal control. Therefore, the approximate optimal control obtained is likely to be not identically the same as the true optimal control, but the difference in the approximate optimal cost and the true optimal cost is insignificant. This is sufficient in real-world applications. In the next section, full discretization schemes based on Runge-Kutta discretization of the optimal control problems will be briefly discussed. These full discretization schemes can solve some optimal control problems such that the controls obtained can be verified to satisfy the optimality conditions.

Finally, note that the standard control parametrization approach assumes a fixed partition for the piecewise constant (or polynomial) approximation of the control. In many practical problems, it is desirable to allow the knot points of the partition to be variable as well. For this, the Control Parametrization Enhancing Transform (CPET) is introduced in the literature (see, for example, [125, 126, 138, 215, 256]). It is now called the time scaling transformation to better reflect the actual meaning of the transformation. It is now widely used in the literature, such as [106–108, 142, 144, 148, 150, 151, 162, 165–171, 311]. The time scaling transformation can be used to convert problems with variable knot points for the control into equivalent problems where the control is defined on a fixed partition once more. The transformed problem can then be readily solved by the standard control parametrization approach. Details of the transformation and many of its applications are described in later chapters.

In [63], an algorithm is developed for solving constrained optimal control problems. Through the control parametrization, a constrained optimal control problem is approximated by a SIP problem. The algorithm proposed seeks to locate a feasible point such that the KKT conditions to a specified tolerance are achieved. Based on the right hand restriction method proposed in [195] for standard SIP, the proposed algorithm solves the path constrained optimal control problem iteratively through the approximation of the path constrained optimal control problem by restricting the right hand side of the path constraint to a finite number of time points. Then, the approximate optimization problem with finitely many constraints is solved such that local optimality conditions are satisfied at each iteration. The established algorithm will find a feasible point in a finite number of iterations such that the first order KKT conditions are satisfied to a specified accuracy.

1.3.4 Collocation Methods

For a direct local collocation method, the state and control are approximated using a specified functional form. The time interval $[t_0, T]$ is partitioned into N subintervals $[t_{i-1}, t_i], i = 1, \ldots, N$, where $t_N = T$. Since the state is required to be continuous across intervals, the following condition is imposed for each $i = 1, \ldots, N$:

$$x(t_i^-) = x(t_i^+), i = 2, \ldots, N-1,$$

where $x(t_i^-) = \lim_{t \uparrow t_i} x(t)$ and $x(t_i^+) = \lim_{t \downarrow t_i} x(t)$. Two types of discretization schemes are normally used in the development of algorithms for solving optimal control problems: (i) Runge-Kutta methods and (ii) orthogonal collocation methods. Runge-Kutta discretization schemes are normally in implicit form. This is because they have better stability properties than those of explicit methods. In [212], an algorithm is developed to solve optimal control problems based on orthogonal collocation method, where Legendre-Gauss points, which are chosen as discretized points, are used together with cubic splines over each subinterval. In [52], Lagrange polynomials are used, instead of cubic spline. Note that the application of a direct local collocation to an optimal control problem will give rise to a nonlinear programming problem of very high dimension containing thousands to tens of thousands of variables and a similar number of constraints. However, the nonlinear programming problem will tend to be very sparse with many of the derivatives of the constraint Jacobian being zero. Thus, it can be solved efficiently using nonlinear programming solvers.

A pseudospectral is a global orthogonal collocation method. It approximates the state using a global polynomial, and the collocation is carried out at appropriately chosen discretized points. Typically, the basis functions used are Chebyshev or Lagrange polynomials. For local collocation, the degree of the polynomial is fixed while the number of meshes is varied. On the other hand, for a pseudospectral method, the number of meshes is fixed while the degree of the polynomial is varied. Pseudospectral method is originally developed to solve problems in computational fluid dynamics [39].

Psudostectral methods have been used to develop numerical algorithms for solving optimal control problems, where appropriate discretized points (with the basis functions being Lagrange polynomials) are to be chosen. For example, the Legendre-Gauss-Lobatto points or Chebysheve-Gauss-Lobatto points [52] are chosen as the discretized points for the Gauss-Lobatto pseudospectral method, the Legendre-Gauss points [212] are used as discretized points in the Gauss pseudospectral method, and for the Radau pseudospectral method [101], Legendre- Gauss-Radau Points are used as discretized points.

1.3.5 Full Parametrization

Full parametrization (discretization) is a popular approach for solving optimal control problems, where an optimal control problem is discretized as a finite dimensional optimization problem by using Euler, midpoint, trapezoid or, in general, Runge-Kutta discretization schemes. See, for example, [13, 36, 49, 50, 85, 113, 115, 178, 196]. Among these discretization schemes, Euler discretization scheme is the simplest but most popular one, for which the optimality conditions can be expressed easily. In [36], two discretization schemes—full discretization and control discretization—are developed to approximate optimal control problems as nonlinear constrained optimization problems. Then, SQP optimization method is utilized to find an optimal (local) control. In addition, SQP optimization method is used again to check numerically whether the obtained control satisfies the second order sufficient conditions, and the post-optimal calculation of the adjoint variables. The Inexact Restoration (IR) method is an iterative finite dimensional optimization method developed in [26, 179, 182], which is an extension of the gradient restoration methods proposed in [191–193]. Each IR iteration consists of two phases—the feasibility phase and optimality phase. They are solved separately as two separate subproblems. It has been shown that if feasibility is improved, while the magnitude of the update in control variables is kept small, then the magnitude of the update in state variables is also small. A local convergence analysis and an associated algorithm for the IR method are given in [26]. This method has been applied to the discretization of optimal control problems in [13, 113, 115]. Particularly, it is applied to Euler discretization of state and control of a constrained optimal control problem in [13], where the convergence of the discretized (finite dimensional optimization) problem to an approximate solution using the Inexact Restoration method, and the convergence of the approximate solution to a solution of the original continuous time optimal control problem are established. A practical algorithm is developed for the IR method in [85]. By using the modeling language AMPL [182], the adapted version of the algorithm is coded for constrained optimal control problems, where the optimization software Ipopt [186] is used. The convergence of the solution of the Euler discretized problem to a continuous time solution of the original constrained optimal control problem is also established in [116], where the time derivative of the pure state constraints is adjoined to the Hamiltonian function [185]. This approach is known as the indirect adjoining approach. The adjoint (or costate) variables so obtained differ from those obtained by using the direct adjoining approach. Four discretization methods for ordinary differential equations and differential algebraic equations are proposed in [69], which are discretized by one-step method, backward differentiation formula, linearized implicit Runge-Kutta method and automatic step-size selection. For the discretization of optimal control problems, the approaches being proposed are full discretization, reduced discretization and control discretization. The convergence results of Euler discretization and Runge-Kutta discretization are obtained in [69], where real-time control, model predictive control and mixed-integer optimal control are also covered. In [267], a class of terminal optimal control problems involving linear systems is considered. For Runge-Kutta direct discretizations of these terminal optimal control problems, error estimates are obtained. If certain sufficient conditions for structural stability are satisfied, the estimate is of first order; otherwise, the estimate is of fractional order.

1.4 Optimal Control Software Packages

Several general-purpose software packages are available in the literature for solving constrained optimal control problems.

Recursive Integration Optimal Trajectory Solver (RIOTS) [224] is a collection of programs for solving optimal control problems, designed as a MATLAB [183] toolbox. The method underlying the program is the representation of controls by finite dimensional B-splines to discretize the optimal control problems. In this sense, it is an example of the control parametrization approach. The integration of the system dynamics is carried out using fixed step-size Runge-Kutta integration. The use of Runge-Kutta method to numerically integrate the system dynamics leads to approximations of the optimal control problem. It is shown in [225] that there exists a class of higher order explicit Runge-Kutta methods that provide consistent approximations to the original problem, where consistency is defined according the theory introduced in [203]. Consequently, it is guaranteed that stationary points of the approximating problems converge to stationary points of the original problem and that global solutions (or strict local solutions) of the approximating problems converge to global (or local) solutions of the original problems as the step-size of the Runge-Kutta method is decreased. Hence an optimal control problem is solved through solving a sequence of approximating discrete time optimal control problems. The software can solve a large class of finite time optimal control problems involving path and terminal time constraints, control bounds, variable initial conditions and problems with integral as well as endpoint cost functionals. It also has a special feature for dealing with singular optimal control problems. It is mentioned in [225] that RIOTS comes with some limitations on the type of problems it can effectively solve. Among those, it has difficulty in solving problems with inequality state constraints that require a very high level of discretization. Another disadvantage is associated with the consistent approximations, which require that the approximating problems be defined on finite dimensional subspaces of the control space to which Runge-Kutta methods can be extended. The selection of the control subspaces affects both the accuracy of numerical integration and the accuracy of the approximate solutions to the original problem.

A general-purpose MATLAB software program called GPOPS-II [211] is now available for solving multiple-phase optimal control problems using variable-order Gaussian quadrature collocation methods. In this software, a Legendre-Gauss-Radau quadrature orthogonal collocation method is used to approximate the continuous time optimal control problem by a large sparse nonlinear programming (NLP) problem. Then, an adaptive mesh refinement scheme is utilized to determine the number of mesh intervals and the degree of the approximating polynomial within each mesh interval such that a specified accuracy is achieved. The optimization solver with the software is NLP solver, which can either be quasi-Newton or Newton solver. The derivatives of the functions involved in the optimal control problem, which are required by the NLP solver, are calculated using sparse finite differencing.

The optimal control software package MISER3.3 [104] is an implementation of the control parametrization technique. It has considerable flexibility in the range of features that can be handled. A large variety of constraints is catered for by allowing a general canonical constraint formulation as well as several special types of constraints. In particular, the algorithm of [103] is applied to handle continuous time inequality constraints on the state, and it has been incorporated in the software. Note that an approach developed in [259] and the exact penalty function approach developed in [299, 301, 305] can also be used as alternative means of handling continuous time inequality constraints. MISER3.3 has been successfully used to solve a significantly large number of practical optimal control problems. See, for example, those mentioned above in this chapter and the relevant references cited in these papers. MISER3.3 is written in the FORTRAN programming language as well as in the MATLAB enviroment. Recently, a new version known as Visual MISER [294] was developed with the Visual FORTRAN compiler within the Microsoft Visual Studio environment. It provides an easy to use interface while retaining the computational efficiency of the MISER3.3 software.

NUDOCCCS (NUmerical Discretization method for Optimal Control problems with Constraints in Controls and States) [34] is, like MISER3.3, a FORTRAN-based package aimed at solving a quite general class of optimal control problems. The underlying ODE system is integrated by an implicit Runga-Kutta scheme, and NUDOCCCS solves the resulting nonlinear problem with a sequential quadratic programming method. The package has been successfully used to solve a wide range of practical optimal control problems. See [35] for an example.

There are other software packages for solving optimal control problems. See, for example, [1].

Chapter 2
Unconstrained Optimization Techniques

2.1 Introduction

The numerical algorithms to be developed for optimal control computation in this book are based on the control parametrization technique in conjunction with a novel time scaling transform. Essentially, an optimal control problem with its control functions being approximated by an appropriate linear combination of spline functions is reduced to an optimal parameter selection problem. Thus, the determination of optimal control function is reduced to the selection of optimal parameters representing the control. Although the constraint of the dynamical system still exists, the problem may, after the parametrization, be viewed as an implicit mathematical programming problem. The solution to the optimal control problem may thus be obtained through solving a sequence of resulting mathematical programming problems, although the computational procedure is much more involved. Thus, a basic understanding of the fundamental concepts, theory and methods of mathematical programming is required.

To begin, we point out that the notation used in this chapter is applicable only to this chapter and Chapters 3 and 4. For example, the n-vector $\boldsymbol{x}$ used in this chapter, Chapter 3 and Chapter 4 should not be confused with the state vectors in other chapters. Also, the vector norm is denoted by $\|\cdot\|$ in these three chapters, but by $|\cdot|$ in later chapters. In this book, the gradient of a function is assumed to be a column vector.

There are already many excellent books on nonlinear optimization. For example, see [10, 22, 61, 172, 187, 199, 232, 275]. In this chapter, we summarize some essential concepts and results in nonlinear unconstrained optimization. It is based on the lecture notes on optimization prepared and used by the authors. These lecture notes have also been used by their colleagues. In addition to these lecture notes, this chapter includes also some important

K. L. Teo et al., *Applied and Computational Optimal Control*, Springer Optimization and Its Applications 171,
https://doi.org/10.1007/978-3-030-69913-0_2

results from the references listed at the beginning of the paragraph and from [60, 87, 204, 207, 220–222].

Unlike optimal control problems, mathematical programming problems are static in nature. The general constrained mathematical programming problem is to find an $\boldsymbol{x} \in \mathbb{R}^n$ to minimize the objective function

$$f(\boldsymbol{x}) \tag{2.1.1}$$

subject to the constraints

$$h_i(\boldsymbol{x}) = 0, \quad i = 1, \ldots, m, \tag{2.1.2}$$

$$h_i(\boldsymbol{x}) \leq 0, \quad i = m+1, \ldots, m+r, \tag{2.1.3}$$

where f and h_i, $i = 1, \ldots, m+r$, are continuously differentiable functions. Let Ω be the set which consists of all $\boldsymbol{x} \in \mathbb{R}^n$ such that (2.1.2) and (2.1.3) are satisfied. This set is called the *feasible set.*

2.2 Basic Concepts

For completeness, we shall first present some important concepts and results in unconstrained optimization techniques. Some basic theory and algorithms for constrained optimization will be given in Chapter 3. The unconstrained optimization problem is to choose an $\boldsymbol{x} = [x_1, \ldots, x_n]^\top \in \mathbb{R}^n$ to minimize an objective function $f(\boldsymbol{x})$. It is a special case of the general problem in Section 2.1, where the feasible region is the entire space $\mathbb{R}^n$.

Definition 2.2.1 *The point $\boldsymbol{x}^* \in \mathbb{R}^n$ is said to be a* global minimum (minimizer) *if*

$$f(\boldsymbol{x}^*) \leq f(\boldsymbol{x}), \text{ for all } \boldsymbol{x} \in \mathbb{R}^n.$$

Definition 2.2.2 *The point $\boldsymbol{x}^* \in \mathbb{R}^n$ is said to be the* strict global minimum (minimizer) *if*

$$f(\boldsymbol{x}^*) < f(\boldsymbol{x}), \text{ for all } \boldsymbol{x} \in \mathbb{R}^n \backslash \{\boldsymbol{x}^*\}.$$

Definition 2.2.3 *The point $\boldsymbol{x}^* \in \mathbb{R}^n$ is said to be a* local minimum (minimizer) *if there exists an $\varepsilon > 0$ such that*

$$f(\boldsymbol{x}^*) \leq f(\boldsymbol{x}), \text{ for all } \boldsymbol{x} \in \mathcal{N}_\varepsilon(\boldsymbol{x}^*),$$

where $\mathcal{N}_\varepsilon(\boldsymbol{x}^*) = \{\boldsymbol{x} \in \mathbb{R}^n : \|\boldsymbol{x} - \boldsymbol{x}^*\| < \varepsilon\}$ is an ε-neighbourhood of $\boldsymbol{x}^*$.

Definition 2.2.4 *The point $\boldsymbol{x}^* \in \mathbb{R}^n$ is said to be a* strict local minimum (minimizer) *if there exists an $\varepsilon > 0$ such that*

$$f(\boldsymbol{x}^*) < f(\boldsymbol{x}), \text{ for all } \boldsymbol{x} \in \mathcal{N}_\varepsilon(\boldsymbol{x}^*) \backslash \{\boldsymbol{x}^*\}.$$

Let $f(\boldsymbol{x})$ be a continuously differentiable function defined on an open set $X \subset \mathbb{R}^n$. Then, by Taylor's Theorem, for any two points $\boldsymbol{x}$ and $\boldsymbol{y} = \boldsymbol{x} + \boldsymbol{s}$ in X, there exists a θ, $0 \le \theta \le 1$, such that

$$f(\boldsymbol{y}) = f(\boldsymbol{x}) + (\boldsymbol{g}(\theta \boldsymbol{x} + (1-\theta)\boldsymbol{y}))^\top \boldsymbol{s}, \tag{2.2.1}$$

where

$$\boldsymbol{g}(\boldsymbol{\xi}) = (\nabla_{\boldsymbol{\xi}} f(\boldsymbol{\xi}))^\top = (\nabla f(\boldsymbol{\xi}))^\top = \left[\frac{\partial f(\boldsymbol{\xi})}{\partial \xi_1}, \frac{\partial f(\boldsymbol{\xi})}{\partial \xi_2}, \ldots, \frac{\partial f(\boldsymbol{\xi})}{\partial \xi_n}\right]^\top,$$

$\boldsymbol{\xi} = [\xi_1, \xi_2, \ldots, \xi_n]^\top$, $\boldsymbol{s} = [s_1, s_2, \ldots, s_n]^\top$ and the superscript "$\top$" denotes the transpose.

The formula (2.2.1) can be generalized to functions which are twice continuously differentiable as follows. Let $f(\boldsymbol{x})$ be a twice continuously differentiable function defined on an open set $X \subset \mathbb{R}^n$. Then, for any two points $\boldsymbol{x}$ and $\boldsymbol{y} = \boldsymbol{x} + \boldsymbol{s}$ in X, there exists a θ, $0 \le \theta \le 1$, such that

$$f(\boldsymbol{y}) = f(\boldsymbol{x}) + (\boldsymbol{g}(\boldsymbol{x}))^\top \boldsymbol{s} + \frac{1}{2}\boldsymbol{s}^\top G(\theta \boldsymbol{x} + (1-\theta)\boldsymbol{y})\boldsymbol{s}, \tag{2.2.2}$$

where $G(\boldsymbol{\xi})$ denotes the Hessian of the function f defined by

$$\begin{aligned} G &= \nabla_{\boldsymbol{x}\boldsymbol{x}} f(\boldsymbol{x}) = \nabla^2_{\boldsymbol{x}} f(\boldsymbol{x}) = (\nabla_{\boldsymbol{x}})^\top \nabla_{\boldsymbol{x}} f(\boldsymbol{x}) \\ &= \begin{bmatrix} \dfrac{\partial^2 f}{\partial x_1^2} & \dfrac{\partial^2 f}{\partial x_1 \partial x_2} & \cdots & \dfrac{\partial^2 f}{\partial x_1 \partial x_n} \\ \dfrac{\partial^2 f}{\partial x_2 \partial x_1} & \dfrac{\partial^2 f}{\partial^2 x_2} & \cdots & \dfrac{\partial^2 f}{\partial x_2 \partial x_n} \\ \vdots & \vdots & \ddots & \vdots \\ \dfrac{\partial^2 f}{\partial x_n \partial x_1} & \dfrac{\partial^2 f}{\partial x_n \partial x_2} & \cdots & \dfrac{\partial^2 f}{\partial x_n^2} \end{bmatrix}. \end{aligned}$$

If $f \in C^2$, then its Hessian is symmetric. A symmetric matrix A is said to be *positive definite* if $\boldsymbol{x}^\top A \boldsymbol{x} > 0$ for all $\boldsymbol{x} \neq \boldsymbol{0}$. Similarly, A is said to be *positive semi-definite* if $\boldsymbol{x}^\top A \boldsymbol{x} \ge 0$ for all $\boldsymbol{x} \in \mathbb{R}^n$.

The point $\boldsymbol{x}^0 \in \mathbb{R}^n$ is said to be a *stationary point* if

$$\boldsymbol{g}(\boldsymbol{x}^0) = \left(\nabla f(\boldsymbol{x}^0)\right)^\top = \boldsymbol{0}. \tag{2.2.3}$$

Note that (2.2.3) is equivalent to

$$\frac{\partial f(\boldsymbol{x}^0)}{\partial x_j} = \left.\frac{\partial f(\boldsymbol{x})}{\partial x_j}\right|_{\boldsymbol{x}=\boldsymbol{x}^0} = 0, \quad j = 1, 2, \ldots, n. \tag{2.2.4}$$

Theorem 2.2.1 *(Necessary Condition for Local Minima) If $\boldsymbol{x}^0$ is a local minimum, then $\nabla f(\boldsymbol{x}^0) = \boldsymbol{0}$.*

Proof. Suppose that $\nabla f(\boldsymbol{x}^0) \neq \mathbf{0}$. Choose $\boldsymbol{s} = -\left(\nabla f(\boldsymbol{x}^0)\right)^\top$. Then for any sufficiently small $\alpha > 0$, it follows from Taylor's theorem that

$$\begin{aligned} f(\boldsymbol{x}^0 + \alpha \boldsymbol{s}) &= f(\boldsymbol{x}^0) + \alpha \nabla f(\boldsymbol{x}^0)\boldsymbol{s} + o(\alpha) \\ &= f(\boldsymbol{x}^0) - \alpha \left\|\nabla f(\boldsymbol{x}^0)\right\|^2 + o(\alpha) \\ &< f(\boldsymbol{x}^0). \end{aligned}$$

This is a contradiction to the fact that $\boldsymbol{x}^0$ is a local minimum.

Theorem 2.2.2 *(Necessary Condition for Local Minima) Let $\boldsymbol{x}^0$ be a solution to (2.2.3). If $\boldsymbol{x}^0$ is a local minimum, then the Hessian $G(\boldsymbol{x}^0)$ of the function f evaluated at $\boldsymbol{x} = \boldsymbol{x}^0$ is positive semi-definite.*

Proof. For $\alpha > 0$, define $\boldsymbol{x}(\alpha) = \boldsymbol{x}^0 + \alpha \boldsymbol{s}$ and $\beta(\alpha) = f(\boldsymbol{x}(\alpha))$, where $\boldsymbol{s} \in \mathbb{R}^n$ is arbitrary. Clearly, $\left.\frac{d\beta(\alpha)}{d\alpha}\right|_{\alpha=0} = (\nabla f(\boldsymbol{x}(\alpha))|_{\alpha=0})\boldsymbol{s} = 0$. By Taylor's theorem, we have $\beta(\alpha) - \beta(0) = \frac{1}{2}\frac{d^2\beta(0)}{d\alpha^2}\alpha^2 + o(\alpha^2)$, where $\lim_{\alpha\to 0} \frac{o(\alpha^2)}{\alpha^2} = 0$. If $\frac{d^2\beta(0)}{d\alpha^2} < 0$, then the right hand side of the above equation is negative when $\alpha > 0$ is sufficiently small. This contradicts the fact that x^0 is a local minimum. Thus,

$$\frac{d^2\beta(0)}{d\alpha^2} = \boldsymbol{s}^\top G(\boldsymbol{x}^0)\boldsymbol{s} \geq 0$$

and the result follows.

Theorem 2.2.3 *(Sufficient Condition for Local Minima) Let $\boldsymbol{x}^0$ be a solution to (2.2.3). If the Hessian, $G(\boldsymbol{x}^0)$, of the function f evaluated at $\boldsymbol{x}^0$ is positive definite, then $\boldsymbol{x}^0$ is a strict local minimum.*

Proof. Since $G(\boldsymbol{x}^0)$ is positive definite, $\boldsymbol{s}^\top G(\boldsymbol{x}^0)\boldsymbol{s} > 0$ for all $\boldsymbol{s} \neq \mathbf{0}$. For any unit vector $\boldsymbol{s} \in \mathbb{R}^n$ and any sufficiently small $\alpha > 0$, by Taylor's Theorem, we have

$$f(\boldsymbol{x}^0 + \alpha \boldsymbol{s}) - f(\boldsymbol{x}^0) = \frac{1}{2}\alpha^2 \boldsymbol{s}^\top G(\boldsymbol{x}^0)\boldsymbol{s} + o(\alpha^2).$$

For small values of $\alpha > 0$, the first term on the right of the last equation dominates the second. Thus, it follows that $f(\boldsymbol{x}^0 + \alpha \boldsymbol{s}) - f(\boldsymbol{x}^0) > 0$ provided $\alpha > 0$ is sufficiently small. Since the direction of $\boldsymbol{s}$ is arbitrary, this shows that $\boldsymbol{x}^0$ is a strict local minimum and the proof is complete.

2.3 Gradient Methods

Consider the unconstrained optimization problem introduced in the previous section. An algorithm which generates a sequence of points $\{\boldsymbol{x}^{(k)}\}$ such that

$$f\left(\boldsymbol{x}^{(k+1)}\right) < f\left(\boldsymbol{x}^{(k)}\right) \tag{2.3.1}$$

for all $k \geq 0$, is referred to as a *descent algorithm* (i.e., the objective function value is reduced at each iteration).

Define

$$\boldsymbol{g}^{(k)} = \left(\nabla f\left(\boldsymbol{x}^{(k)}\right)\right)^{\top} \quad \text{and} \quad G^{(k)} = \nabla^2 f\left(\boldsymbol{x}^{(k)}\right). \tag{2.3.2}$$

Let $\boldsymbol{s}^{(k)}$ be a given vector in $\mathbb{R}^n$. Consider the function $f(\boldsymbol{x})$ along the line

$$\boldsymbol{x}(\alpha) = \boldsymbol{x}^{(k)} + \alpha \boldsymbol{s}^{(k)},$$

for $\alpha \geq 0$. Clearly, $f(\boldsymbol{x}(\alpha))$ may be regarded as a function of α alone. The slope of the function is

$$\frac{df\left(\boldsymbol{x}(\alpha)\right)}{d\alpha} = \frac{d}{d\alpha} f\left(\boldsymbol{x}^{(k)} + \alpha \boldsymbol{s}^{(k)}\right) = \nabla f\left(\boldsymbol{x}^{(k)} + \alpha \boldsymbol{s}^{(k)}\right) \boldsymbol{s}^{(k)}.$$

At $\alpha = 0$, the slope is

$$\left.\frac{df\left(\boldsymbol{x}(\alpha)\right)}{d\alpha}\right|_{\alpha=0} = \nabla f\left(\boldsymbol{x}^{(k)}\right) \boldsymbol{s}^{(k)} = \left(\boldsymbol{g}^{(k)}\right)^{\top} \boldsymbol{s}^{(k)}.$$

If a direction $\boldsymbol{s}^{(k)}$ is such that

$$\left(\boldsymbol{g}^{(k)}\right)^{\top} \boldsymbol{s}^{(k)} < 0, \tag{2.3.3}$$

then $\boldsymbol{s}^{(k)}$ is called a *descent direction* of the objective function at $\boldsymbol{x}^{(k)}$. The objective function value is reduced along this direction for all sufficiently small $\alpha > 0$.

The following is a general structure for the descent algorithm that we will consider:

Algorithm 2.3.1
Step 1. Choose $\boldsymbol{x}^{(0)}$ *and set* $k = 0$.
Step 2. Determine a search direction $\boldsymbol{s}^{(k)}$.
Step 3. Check for convergence.
Step 4. Find α_k *that minimizes* $f\left(\boldsymbol{x}^{(k)} + \alpha \boldsymbol{s}^{(k)}\right)$ *with respect to* α.
Step 5. Set $\boldsymbol{x}^{(k+1)} = \boldsymbol{x}^{(k)} + \alpha_k \boldsymbol{s}^{(k)}$, *and set* $k := k + 1$. *Go to Step 2.*

Remark 2.3.1

(i) Different descent methods arise from different ways of generating the search direction $\boldsymbol{s}^{(k)}$.

(ii) Step 4 is a one-dimensional optimization problem and is referred to as a line search. *Here, the line search is idealized. In practice, an exact line search is impossible.*

Note that if α_k minimizes $f\left(\boldsymbol{x}^{(k)} + \alpha \boldsymbol{s}^{(k)}\right)$, then

$$\left.\frac{df\left(\boldsymbol{x}^{(k)}+\alpha \boldsymbol{s}^{(k)}\right)}{d\alpha}\right|_{\alpha=\alpha_k}=0.$$

Thus, a necessary condition for an exact line search is

$$\frac{df\left(\boldsymbol{x}^{(k)}+\alpha_k \boldsymbol{s}^{(k)}\right)}{d\alpha}=\nabla f\left(\boldsymbol{x}^{(k)}+\alpha^{(k)}\boldsymbol{s}^{(k)}\right)\boldsymbol{s}^{(k)}=\left(\boldsymbol{g}^{(k+1)}\right)^{\top}\boldsymbol{s}^{(k)}=0. \tag{2.3.4}$$

2.4 Steepest Descent Method

Note that $-\boldsymbol{g}^{(k)}$ is the direction in which the objective function value decreases most rapidly at $\boldsymbol{x}^{(k)}$. By choosing $\boldsymbol{s}^{(k)}=-\boldsymbol{g}^{(k)}$ in Algorithm 2.3.1, we obtain the *steepest descent method.* This method is the simplest one among all gradient-based unconstrained optimization methods. It requires only the objective function value and its gradient. Moreover, with the steepest descent method, we have global convergence (i.e., the method converges from any starting point) as we will now demonstrate.

Assume that there exists a point $\boldsymbol{x}^*\in\mathbb{R}^n$ such that $\nabla f(\boldsymbol{x}^*)=\boldsymbol{0}$. Furthermore, we suppose that $\nabla f(\boldsymbol{x})\neq\boldsymbol{0}$ if $\boldsymbol{x}\neq\boldsymbol{x}^*$. Then, the steepest descent method constructs a sequence $\{\boldsymbol{x}^{(k)}\}$ with

$$f\left(\boldsymbol{x}^{(k+1)}\right)=\min_{0\le\alpha<\infty} f\left(\boldsymbol{x}^{(k)}-\alpha\boldsymbol{g}^{(k)}\right)\le f\left(\boldsymbol{x}^{(k)}\right),\quad \textit{for all } k\ge 0.$$

Since $f(\boldsymbol{x}^*)\le f\left(\boldsymbol{x}^{(k)}\right)\le f(\boldsymbol{x}^{(0)})$ for all $k\ge 0$, it is clear that $\left\{f\left(\boldsymbol{x}^{(k)}\right)\right\}$ is a bounded monotone sequence. This implies that $\left\{f\left(\boldsymbol{x}^{(k)}\right)\right\}$ is convergent for any initial point $\boldsymbol{x}^{(0)}$. In other words, $\left\{f\left(\boldsymbol{x}^{(k)}\right)\right\}$ is globally convergent to $f(\boldsymbol{x}^*)$ However, it should be noted that the convergence of the steepest descent method can be very slow. The convergence rate of this method is given in the following theorem:

Theorem 2.4.1 *Suppose that $f(\boldsymbol{x})$ defined on $\mathbb{R}^n$ has continuous second order partial derivatives, and has a local minimum at $\boldsymbol{x}^*$. If $\{\boldsymbol{x}^{(k)}\}$ is a sequence generated by the steepest descent method that converges to $\boldsymbol{x}^*$, then*

$$\left\|f\left(\boldsymbol{x}^{(k+1)}\right)-f(\boldsymbol{x}^*)\right\|\le\left(\frac{r-1}{r+1}\right)^2\left\|\boldsymbol{x}^{(k)}-\boldsymbol{x}^*\right\|,$$

where r is the condition number of the Hessian $G^*=\nabla^2 f(\boldsymbol{x}^*)$. Note that the condition number is given by $r=A/a$, where A and a are, respectively, the largest and the smallest eigenvalues of G^*.

2.5 Newton's Method

Newton's method is based on the quadratic approximation of the objective function obtained by truncating the Taylor series expansion of $f(\boldsymbol{x})$ about $\boldsymbol{x}^{(k)}$. That is, for $\boldsymbol{\delta} \in \mathbb{R}^n$, the objective function $f(\boldsymbol{x}^{(k)} + \boldsymbol{\delta})$ is approximated by the following quadratic function:

$$q^{(k)}(\boldsymbol{\delta}) = f^{(k)} + \left(\boldsymbol{g}^{(k)}\right)^\top \boldsymbol{\delta} + \frac{1}{2}\boldsymbol{\delta}^\top G^{(k)}\boldsymbol{\delta}, \tag{2.5.1}$$

where $f^{(k)} = f\left(\boldsymbol{x}^{(k)}\right)$. The next iterate $\boldsymbol{x}^{(k+1)}$ is chosen as the minimizer of this quadratic approximation. That is, we choose $\boldsymbol{x}^{(k+1)} = \boldsymbol{x}^{(k)} + \boldsymbol{\delta}^{(k)}$, where $\boldsymbol{\delta}^{(k)}$ is the solution of

$$\nabla q^{(k)}(\boldsymbol{\delta}) = \mathbf{0}. \tag{2.5.2}$$

If $G^{(k)}$ is positive definite, then

$$\boldsymbol{\delta}^{(k)} = -\left(G^{(k)}\right)^{-1} \boldsymbol{g}^{(k)}. \tag{2.5.3}$$

Remark 2.5.1

(i) Newton's method requires the information on $f^{(k)}$, $\boldsymbol{g}^{(k)}$ and $G^{(k)}$, i.e., function values and first and second order partial derivatives.

(ii) The basic Newton's method does not involve a line search. The choice of $\boldsymbol{\delta}^{(k)}$ ensures that the minimum of the quadratic approximation is achieved.

(iii) Assuming G^* is positive definite, Newton's method has good local convergence if the starting point is sufficiently close to $\boldsymbol{x}^*$.

(iv) Choosing $\boldsymbol{\delta}^{(k)}$ as the solution of (2.5.2) *is* only *appropriate and* well-defined if *the quadratic approximation has a minimum, i.e., $G^{(k)}$ is positive definite. This may not be the case if $\boldsymbol{x}^{(k)}$ is remote from $\boldsymbol{x}^*$, where $\boldsymbol{x}^*$ is a local minimum.*

Algorithm 2.5.1 *(Newton's Method)*
Step 1. Choose $\boldsymbol{x}^{(0)}$ and set $k = 0$.
Step 2. If $\boldsymbol{g}^{(k)} = \mathbf{0}$, stop.
Step 3. Solve $G^{(k)}\boldsymbol{\delta} = -\boldsymbol{g}^{(k)}$ for $\boldsymbol{\delta} = \boldsymbol{\delta}^{(k)}$.
Step 4. Set $\boldsymbol{x}^{(k+1)} = \boldsymbol{x}^{(k)} + \boldsymbol{\delta}^{(k)}$.
Step 5. Set $k := k + 1$. Go to Step 2.

We will now examine the convergence rate of Newton's method. From Algorithm 2.5.1, we have, at the k-th iterate,

$$\boldsymbol{x}^{(k+1)} = \boldsymbol{x}^{(k)} - \left(G^{(k)}\right)^{-1} \boldsymbol{g}^{(k)}. \tag{2.5.4}$$

Suppose that $\boldsymbol{x}^*$ is a point such that $\boldsymbol{g}(\boldsymbol{x}^*) = \mathbf{0}$ and that $G(\boldsymbol{x}^*)$ is positive definite. Then, it follows from Taylor's theorem that

$$\begin{aligned}\boldsymbol{g}^{(k)} = \boldsymbol{g}\left(\boldsymbol{x}^{(k)}\right) &= \boldsymbol{g}(\boldsymbol{x}^*) + G\left(\boldsymbol{x}^{(k)}\right)\left(\boldsymbol{x}^{(k)} - \boldsymbol{x}^*\right) + O\left(\left\|\boldsymbol{x}^{(k)} - \boldsymbol{x}^*\right\|^2\right) \\ &= G^{(k)}\left(\boldsymbol{x}^{(k)} - \boldsymbol{x}^*\right) + O\left(\left\|\boldsymbol{x}^{(k)} - \boldsymbol{x}^*\right\|^2\right), \end{aligned} \tag{2.5.5}$$

where the last term is understood to be vector valued and

$$\lim_{\|\boldsymbol{\xi}\|\to 0} \frac{\|O(\|\boldsymbol{\xi}\|)\|}{\|\boldsymbol{\xi}\|} = c,$$

for some constant $c > 0$. Let $k = k_0$. If $\boldsymbol{x}^{(k_0)}$ is sufficiently close to $\boldsymbol{x}^*$, then we may assume that $\{\boldsymbol{x}^{(k)}\}$ is in a neighbourhood of $\boldsymbol{x}^*$ for all $k \geq k_0$. Since f is twice continuously differentiable and $G(\boldsymbol{x}^*)$ is positive definite, we can find a constant c_1 such that $\left\|\left(G\left(\boldsymbol{x}^{(k)}\right)\right)^{-1}\right\| \leq c_1$ for all $k \geq k_0$. Multiplying both sides of (2.5.5) by $\left(G^{(k)}\right)^{-1}$ yields

$$\left(G^{(k)}\right)^{-1}\boldsymbol{g}^{(k)} = \boldsymbol{x}^{(k)} - \boldsymbol{x}^* + O\left(\left\|\boldsymbol{x}^{(k)} - \boldsymbol{x}^*\right\|^2\right). \tag{2.5.6}$$

From Step 3 of Algorithm 2.5.1, the left hand side of (2.5.6) may be replaced by $-\boldsymbol{\delta}^{(k)}$. Since $\boldsymbol{\delta}^{(k)} = \boldsymbol{x}^{(k+1)} - \boldsymbol{x}^{(k)}$, we have

$$-\left(\boldsymbol{x}^{(k+1)} - \boldsymbol{x}^{(k)}\right) = \boldsymbol{x}^{(k)} - \boldsymbol{x}^* + O\left(\left\|\boldsymbol{x}^{(k)} - \boldsymbol{x}^*\right\|^2\right).$$

Simplifying, we thus find that

$$-\left(\boldsymbol{x}^{(k+1)} - \boldsymbol{x}^*\right) = O\left(\left\|\boldsymbol{x}^{(k)} - \boldsymbol{x}^*\right\|^2\right).$$

By the definition of $O\left(\left\|\boldsymbol{x}^{(k)} - \boldsymbol{x}^*\right\|^2\right)$, there exists a constant c_2 such that

$$\left\|\boldsymbol{x}^{(k+1)} - \boldsymbol{x}^*\right\| \leq c_2\left\|\boldsymbol{x}^{(k)} - \boldsymbol{x}^*\right\|^2. \tag{2.5.7}$$

Therefore, if $\boldsymbol{x}^{(k)}$ is close to $\boldsymbol{x}^*$, $\{\boldsymbol{x}^{(k)}\}$ converges to $\boldsymbol{x}^*$ at a rate of at least *second order.*

2.6 Modifications to Newton's Method

Assume that $G^{(k)}$ has eigenvalues $\lambda_1^k < \lambda_2^k < \ldots < \lambda_n^k$. Choose ε_k such that

$$\varepsilon_k + \lambda_1^k > 0.$$

Obviously, if $\lambda_1^k > 0$, we choose $\varepsilon_k = 0$. Consider the matrix $\varepsilon_k I + G^{(k)}$. Clearly, it has eigenvalues

$$0 < \varepsilon_k + \lambda_1^k < \varepsilon_k + \lambda_2^k < \cdots < \varepsilon_k + \lambda_n^k.$$

Since they are all positive, $\varepsilon_k I + G^{(k)}$ is positive definite. Thus, we can construct a modified Newton's method:

$$\boldsymbol{x}^{(k+1)} = \boldsymbol{x}^{(k)} - \alpha_k \left(\varepsilon_k I + G^{(k)}\right)^{-1} \boldsymbol{g}^{(k)}, \qquad k = 0, 1, \ldots \tag{2.6.1}$$

Remark 2.6.1 *When ε_k is sufficiently large (on the order of 10^4), the term $\varepsilon_k I$ dominates $\varepsilon_k I + G_k$ and $(\varepsilon_k I + G_k)^{-1} \approx [\varepsilon_k I]^{-1} = \frac{1}{\varepsilon_k} I$. Thus, the search direction is*

$$-\left(\varepsilon_k I + G^{(k)}\right)^{-1} \boldsymbol{g}^{(k)} \approx -\frac{1}{\varepsilon_k} \boldsymbol{g}^{(k)}$$

which is the steepest descent direction. When ε_k is small, $\varepsilon_k I + G^{(k)} \approx G^{(k)}$ and the method is similar to Newton's method.

Let us describe this modified Newton's method, which is called *Marquardt's method*, as follows.

Algorithm 2.6.1

Step 1. *Choose $\boldsymbol{x}^{(0)}$, $\varepsilon_0 > 0$ (in the order of 10^4), c_1 (where $0 < c_1 < 1$), c_2 (where $c_2 > 1$) and $\varepsilon > 0$ (in the order of 10^{-2}). Set $k = 0$.*

Step 2. *Compute $\boldsymbol{g}^{(k)}$.*

Step 3. *If $\left\|\boldsymbol{g}^{(k)}\right\| \le \varepsilon$, stop. $\boldsymbol{x}^{(k)}$ is taken as a local minimum. Otherwise, go to Step 4.*

Step 4. *Find α_k^* such that $f\left(\boldsymbol{x}^{(k)} - \alpha_k \left[\varepsilon_k I + G^{(k)}\right]^{-1} \boldsymbol{g}^{(k)}\right)$ is minimized. Compute $\boldsymbol{x}^{(k+1)}$ according to*

$$\boldsymbol{x}^{(k+1)} = \boldsymbol{x}^{(k)} - \alpha_k^* \left[\varepsilon_k I + G^{(k)}\right]^{-1} \boldsymbol{g}^{(k)}. \tag{2.6.2}$$

Step 5. *Compare the values of $f\left(\boldsymbol{x}^{(k+1)}\right)$ and $f\left(\boldsymbol{x}^{(k)}\right)$. If $f\left(\boldsymbol{x}^{(k+1)}\right) < f\left(\boldsymbol{x}^{(k)}\right)$, go to Step 6. If $f\left(\boldsymbol{x}^{(k+1)}\right) \ge f\left(\boldsymbol{x}^{(k)}\right)$, go to Step 7.*

Step 6. *Set $\varepsilon_{k+1} = c_1 \varepsilon_k$, $k := k + 1$, and go to Step 2 followed by Step 3.*

Step 7. *Set $\varepsilon_{k+1} = c_2 \varepsilon_k$, $k := k + 1$, and go to Step 2 followed by Step 4.*

Remark 2.6.2 *In Marquardt's method, the value of ε_k is taken to be large at the beginning and then is reduced to zero gradually as the iterative process progresses. Thus, as the value of ε_k decreases from a large value to zero, the characteristics of the method change from those of a steepest descent method to those of Newton's method. Hence, the method takes advantage of both the steepest descent method (global convergence) and the Newton's method (fast convergence when near a local minimum).*

2.7 Line Search

The steepest descent method and, in fact, most gradient-based methods require a line search. The choice of α^* that *exactly* minimizes $f(\boldsymbol{x}+\alpha\boldsymbol{s})$ is called the exact line search. The exact line search condition is given by (2.3.4). However, an exact line search condition is difficult to implement using a computer. Thus, we must resort to an approximate line search.

If $\boldsymbol{s}$ is a descent direction at $\boldsymbol{x}$, as described by (2.3.3), then $f(\boldsymbol{x}+\alpha\boldsymbol{s}) < f(\boldsymbol{x})$ for all $\alpha > 0$ sufficiently small. Thus, we could replace finding the exact minimizer of $f(\boldsymbol{x}+\alpha\boldsymbol{s})$ by finding any α such that $f(\boldsymbol{x}+\alpha\boldsymbol{s}) < f(\boldsymbol{x})$. However, if α is chosen too small, we may not get to the minimum of f. We need at least linear decrease in function value to guarantee convergence. If α is chosen too large, then $\boldsymbol{s}$ may no longer be a descent direction.

An approximate minimizer $\bar{\alpha}$ of $f(\boldsymbol{x}+\alpha\boldsymbol{s})$ must be chosen such that the following conditions are satisfied:

(1) Sufficient function decrease:

$$f(\boldsymbol{x}+\bar{\alpha}\boldsymbol{s}) \le f(\boldsymbol{x}) + \rho\bar{\alpha}\boldsymbol{s}^\top \left(\nabla f(\boldsymbol{x})\right)^\top . \tag{2.7.1}$$

(2) Sufficient slope improvement:

$$|\nabla f(\boldsymbol{x}+\bar{\alpha}\boldsymbol{s})\boldsymbol{s}| \le -\delta\boldsymbol{s}^\top \left(\nabla f(\boldsymbol{x})\right)^\top . \tag{2.7.2}$$

Remark 2.7.1 *ρ and δ are constants satisfying $0 < \rho < \delta < 1$. If $\delta = 0$, then $\bar{\alpha} = \alpha^*$.*

For illustration, let us look at

$$h(\alpha) = f(\boldsymbol{x}+\alpha\boldsymbol{s}).$$

If we choose $\rho = \bar{\rho}$, then $\bar{\alpha} \le \alpha$. Furthermore,

$$\left|\frac{dh(\bar{\alpha})}{d\alpha}\right| \le -\delta\frac{dh(0)}{d\alpha} \Rightarrow \bar{\alpha} \in [b,c].$$

An acceptable approximate minimizer is any point $\bar{\alpha}$ satisfying both of these conditions, i.e., any point in $[b,c]$ in the diagram of Figure 2.7.1.

Remark 2.7.2

(i) To ensure the existence of a point satisfying both these conditions, we need $\rho \le \delta$.

(ii) As $\delta \to 0$, line search becomes more accurate. Typical values of δ are chosen as follows: $\delta = 0.9$—a weak line search, i.e., not very accurate line search; and $\delta = 0.1$—a strong line search , i.e., fairly accurate line search.

(iii) ρ is typically taken to be quite small, e.g., 0.01, where $0 < \rho < \delta < 1$.

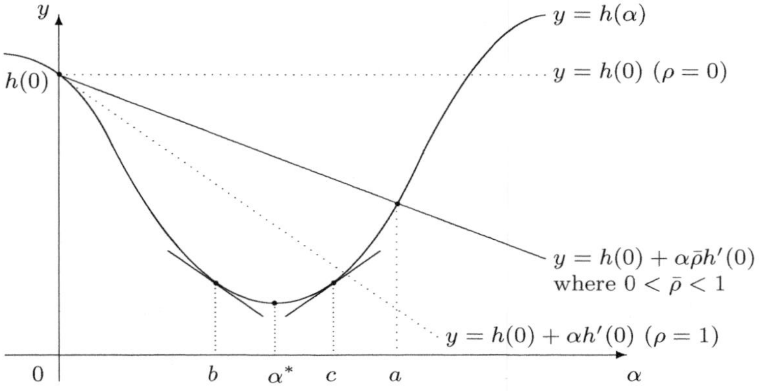

Fig. 2.7.1: The choices of α and ρ

For the steepest descent method with approximate line search, we have the following convergence result, which can be found in any book on optimization. See, for example, [172].

Theorem 2.7.1 *Let $f \in C^2$ and $\{\boldsymbol{x}^{(k)}\}$ be a sequence of points generated by the steepest descent method using an approximate line search. Then, one of the following conditions is to be satisfied:*

(i) $\boldsymbol{g}^{(k)} = \boldsymbol{0}$ for some k; or
(ii) $\boldsymbol{g}^{(k)} \to \boldsymbol{0}$ as $k \to \infty$; or
(iii) $f^{(k)} \to -\infty$ (there is no finite minimum).

2.8 Conjugate Gradient Methods

Conjugate gradient methods are originally developed for the minimization of a quadratic objective function, and then extended for general unconstrained minimization problems. Let us consider the following quadratic objective function:

$$f(\boldsymbol{x}) = \frac{1}{2}\boldsymbol{x}^\top G\boldsymbol{x} + \boldsymbol{c}^\top \boldsymbol{x} \tag{2.8.1}$$

over $\mathbb{R}^n$, where G is a symmetric positive definite matrix.

Definition 2.8.1 *Given a symmetric matrix G, two vectors $\boldsymbol{d}^{(1)}$ and $\boldsymbol{d}^{(2)}$ are said to be G-conjugate, if $\langle \boldsymbol{d}^{(1)}, G\boldsymbol{d}^{(2)} \rangle = \left(\boldsymbol{d}^{(1)}\right)^\top G\boldsymbol{d}^{(2)} = 0$, where $\langle \cdot, \cdot \rangle$ denotes the inner product.*

Theorem 2.8.1 *If G is positive definite and the set of non-zero vectors $\boldsymbol{d}^{(i)}$, $i = 0, 1, 2, \ldots, k$, are G-conjugate, then they are linearly independent.*

Proof. Suppose they satisfy

$$\alpha_0 \boldsymbol{d}^{(0)} + \alpha_1 \boldsymbol{d}^{(1)} + \cdots + \alpha_k \boldsymbol{d}^{(k)} = \mathbf{0} \tag{2.8.2}$$

for a set of constants α_i, $i = 0, 1, 2, \ldots, k$. Taking the inner product with $G\boldsymbol{d}^{(i)}$, we have

$$\alpha_0 \left(\boldsymbol{d}^{(0)}\right)^\top G\boldsymbol{d}^{(i)} + \alpha_1 \left(\boldsymbol{d}^{(1)}\right)^\top G\boldsymbol{d}^{(i)} + \cdots + \alpha_k \left(\boldsymbol{d}^{(k)}\right)^\top G\boldsymbol{d}^{(i)} = 0. \tag{2.8.3}$$

From the conjugacy, we have $\left(\boldsymbol{d}^{(i)}\right)^\top G\boldsymbol{d}^{(j)} = 0$ for all $i \neq j$. Thus, it follows from (2.8.3) that $\alpha_i \left(\boldsymbol{d}^{(i)}\right)^\top G\boldsymbol{d}^{(i)} = 0$. Since G is positive definite, we have $\left(\boldsymbol{d}^{(i)}\right)^\top G\boldsymbol{d}^{(i)} > 0$, and hence $\alpha_i = 0$. This is true for all $i = 0, 1, \ldots, k$. Thus, $\boldsymbol{d}^{(i)}$, $i = 0, 1, \ldots, k$, are linearly independent.

Theorem 2.8.2 *(Principal Theorem of Conjugate Direction Method) Consider a quadratic function given by* (2.8.1). *For an arbitrary $\boldsymbol{x}^{(0)} \in \mathbb{R}^n$, let $\boldsymbol{x}^{(i)}$, $i = 1, 2, \ldots, k$, with $k \leq n-1$ be generated by a conjugate direction method, where the search directions $\boldsymbol{s}^{(0)}, \ldots, \boldsymbol{s}^{(n-1)}$ are G-conjugate, and where, for each $i \leq n-1$, $\boldsymbol{x}^{(i+1)}$ is chosen as*

$$\boldsymbol{x}^{(i+1)} = \boldsymbol{x}^{(i)} + \alpha_i^* \boldsymbol{s}^{(i)}, \tag{2.8.4}$$

where

$$\alpha_i^* = \underset{\alpha \geq 0}{\operatorname{argmin}} \left\{ f\left(\boldsymbol{x}^{(i)} + \alpha \boldsymbol{s}^{(i)}\right)\right\}. \tag{2.8.5}$$

Then, the method terminates in at most n exact line searches. Furthermore, for each $i \leq n$, $\boldsymbol{x}^{(i)}$ is the minimizer of the function $f(\boldsymbol{x})$ given by (2.8.1) *on the affine subspace.*

$$U_i = \boldsymbol{x}^{(0)} + span\left\{\boldsymbol{s}^{(0)}, \boldsymbol{s}^{(1)}, \ldots, \boldsymbol{s}^{(i-1)}\right\}, \tag{2.8.6}$$

where $span\left\{\boldsymbol{s}^{(0)}, \boldsymbol{s}^{(1)}, \ldots, \boldsymbol{s}^{(i-1)}\right\}$ denotes the linear subspace spanned by $\boldsymbol{s}^{(0)}$, $\boldsymbol{s}^{(1)}$, ..., $\boldsymbol{s}^{(i-1)}$.

Proof. For all $i \leq n-1$, recall the notation

$$\boldsymbol{g}^{(i)} = \left(\nabla f\left(\boldsymbol{x}^{(i)}\right)\right)^\top$$

from (2.3.2). We have

$$\begin{aligned}
\boldsymbol{g}^{(i+1)} - \boldsymbol{g}^{(i)} &= \left(G\boldsymbol{x}^{(i+1)} + \boldsymbol{c}\right) - \left(G\boldsymbol{x}^{(i)} + \boldsymbol{c}\right) \\
&= G\left(\boldsymbol{x}^{(i+1)} - \boldsymbol{x}^{(i)}\right) \\
&= G\left(\boldsymbol{x}^{(i)} + \alpha_i^* \boldsymbol{s}^{(i)} - \boldsymbol{x}^{(i)}\right)
\end{aligned}$$

$$=\alpha_i^* G\boldsymbol{s}^{(i)}. \tag{2.8.7}$$

For $j < i$, note that we may write

$$\begin{aligned}\boldsymbol{g}^{(i+1)} =&\boldsymbol{g}^{(j+1)} + \left(\boldsymbol{g}^{(j+2)} - \boldsymbol{g}^{(j+1)}\right) + \ldots + \left(\boldsymbol{g}^{(i+1)} - \boldsymbol{g}^{(i)}\right) \\ =&\boldsymbol{g}^{(j+1)} + \sum_{k=j+1}^{i} \left(\boldsymbol{g}^{(k+1)} - \boldsymbol{g}^{(k)}\right). \end{aligned} \tag{2.8.8}$$

Using (2.8.8) and (2.8.7), it follows that

$$\begin{aligned}\left(\boldsymbol{g}^{(i+1)}\right)^\top \boldsymbol{s}^{(j)} &= \left(\boldsymbol{g}^{(j+1)}\right)^\top \boldsymbol{s}^{(j)} + \sum_{k=j+1}^{i} \left(\boldsymbol{g}^{(k+1)} - \boldsymbol{g}^{(k)}\right)^\top \boldsymbol{s}^{(j)} \\ &= \left(\boldsymbol{g}^{(j+1)}\right)^\top \boldsymbol{s}^{(j)} + \sum_{k=j+1}^{i} \alpha_k \left(\boldsymbol{s}^{(k)}\right)^\top G\boldsymbol{s}^{(j)} \\ &=0, \end{aligned} \tag{2.8.9}$$

where the last equality is based on the exact line search condition (2.3.4) and the G-conjugacy of $\{\boldsymbol{s}^{(0)}, \ldots, \boldsymbol{s}^{(i)}\}$. Now, consider the case when $j = i$. Again, by the exact line search condition (2.3.4), we have

$$\left(\boldsymbol{g}^{(i+1)}\right)^\top \boldsymbol{s}^{(i)} = 0. \tag{2.8.10}$$

Since (2.8.9) and (2.8.10) hold for all $i \leq n - 1$ and since Theorem 2.8.1 insures that $\{\boldsymbol{s}^{(0)}, \boldsymbol{s}^{(1)}, \ldots, \boldsymbol{s}^{(n-1)}\}$ is linearly independent, it follows from those two equations that

$$\boldsymbol{g}^{(n)} = \boldsymbol{0}.$$

Hence $\boldsymbol{x}^{(n)}$ is the minimizer of $f(\boldsymbol{x})$ given by (2.8.1) and consequently the search must terminate in at most n exact line searches.

We now move on to prove the second part of the theorem. We need to show that for each $i \leq n$, $\boldsymbol{x}^{(i)}$ is the minimizer of $f(\boldsymbol{x})$ on the set U_i defined by (2.8.6). We shall use an induction argument. The statement is true for $i = 1$, since the exact line search (2.8.4)–(2.8.5) is clearly over U_1 in this case. Suppose the statement is true for some $j < n$, i.e., $\boldsymbol{x}^{(j)}$ is optimal on U_j. Then we want to show that

$$\boldsymbol{x}^{(j+1)} = \boldsymbol{x}^{(j)} + \alpha_j^* \boldsymbol{s}^{(j)}$$

is optimal on $U_{j+1} = \boldsymbol{x}^{(0)} + \operatorname{span}\{\boldsymbol{s}^{(0)}, \boldsymbol{s}^{(1)}, \ldots, \boldsymbol{s}^{(j)}\}$, where, by the exact line search condition (2.3.4), $\boldsymbol{g}^{(j+1)}\boldsymbol{s}^{(j)} = 0$, i.e.,

$$\left(G\left(\boldsymbol{x}^{(j)} + \alpha_j^* \boldsymbol{s}^{(j)}\right) + \boldsymbol{c}\right)^\top = 0,$$

which, in turn, requires

$$\alpha_j^* = -\frac{(G\boldsymbol{x}^{(j)} + \boldsymbol{c})^\top \boldsymbol{s}^{(j)}}{(\boldsymbol{s}^{(j)})^\top G \boldsymbol{s}^{(j)}}. \tag{2.8.11}$$

To do so, consider the task of finding $\boldsymbol{x} = \boldsymbol{y} + \alpha \boldsymbol{s}^{(j)} \in U_{j+1}$ to minimize $f(\boldsymbol{x})$, where $\boldsymbol{y} \in U_j$ and $\alpha \geq 0$ need to be determined. We have

$$\begin{aligned}
f(\boldsymbol{x}) =& f\left(\boldsymbol{y} + \alpha \boldsymbol{s}^{(j)}\right) \\
=& \frac{1}{2}\left(\boldsymbol{y} + \alpha \boldsymbol{s}^{(j)}\right)^\top G \left(\boldsymbol{y} + \alpha \boldsymbol{s}^{(j)}\right) + \boldsymbol{c}^\top \left(\boldsymbol{y} + \alpha \boldsymbol{s}^{(j)}]\right) \\
=& \frac{1}{2}\boldsymbol{y}^\top G \boldsymbol{y} + \alpha \left(\boldsymbol{s}^{(i)}\right)^\top G\boldsymbol{y} + \frac{\alpha^2}{2}\left(\boldsymbol{s}^{(j)}\right)^\top G \boldsymbol{s}^{(j)} + \boldsymbol{c}^\top \boldsymbol{y} + \alpha \boldsymbol{c}^\top \boldsymbol{s}^{(j)} \\
=& f(\boldsymbol{y}) + \alpha \left(\boldsymbol{s}^{(j)}\right)^\top (G\boldsymbol{y} + \boldsymbol{c}) + \frac{\alpha^2}{2}\left(\boldsymbol{s}^{(j)}\right)^\top G \boldsymbol{s}^{(j)}.
\end{aligned} \tag{2.8.12}$$

At this point, note that $\boldsymbol{y} \in U_j$ and $\boldsymbol{x}^{(j)} \in U_j$ implies that

$$\boldsymbol{y} - \boldsymbol{x}^{(j)} \in \operatorname{span}\left\{\boldsymbol{s}^{(0)}, \ldots, \boldsymbol{s}^{(j-1)}\right\}.$$

The conjugacy of $\{\boldsymbol{s}^{(0)}, \ldots, \boldsymbol{s}^{(n-1)}\}$ then implies that

$$\left(\boldsymbol{s}^{(j)}\right)^\top G \left(\boldsymbol{y} - \boldsymbol{x}^{(j)}\right) = 0$$

which can be rearranged to

$$\left(\boldsymbol{s}^{(j)}\right)^\top G\boldsymbol{y} = \left(\boldsymbol{s}^{(j)}\right)^\top G\boldsymbol{x}^{(j)}. \tag{2.8.13}$$

Substituting (2.8.13) into (2.8.12), we have

$$f(\boldsymbol{x}) = f(\boldsymbol{y}) + \alpha \left(\boldsymbol{s}^{(j)}\right)^\top \left(G\boldsymbol{x}^{(j)} + \boldsymbol{c}\right) + \frac{\alpha^2}{2}\left(\boldsymbol{s}^{(j)}\right)^\top G \boldsymbol{s}^{(j)}. \tag{2.8.14}$$

Note that the right hand side of (2.8.14) is decoupled in that the first term depends on $\boldsymbol{y}$ only and the last two terms depend on α only. By our earlier assumption, since $\boldsymbol{x}^{(j)}$ yields the minimum of f on U_j, we must choose $\boldsymbol{y} = \boldsymbol{x}^{(j)}$ to minimize the first term. Minimizing the remaining two terms with respect to α requires

$$\begin{aligned}
&\frac{d}{d\alpha}\left(\alpha \left(\boldsymbol{s}^{(j)}\right)^\top \left(G\boldsymbol{x}^{(j)} + \boldsymbol{c}\right) + \frac{\alpha^2}{2}\left(\boldsymbol{s}^{(j)}\right)^\top G\boldsymbol{s}^{(j)}\right) \\
&\quad = \left(\boldsymbol{s}^{(j)}\right)^\top \left(G\boldsymbol{x}^{(j)} + \boldsymbol{c}\right) + \alpha \left(\boldsymbol{s}^{(j)}\right)^\top G\boldsymbol{s}^{(j)}
\end{aligned} \tag{2.8.15}$$

to equal zero, which, in turn, yields

$$\alpha = -\frac{\left(\boldsymbol{s}^{(j)}\right)^\top \left(G\boldsymbol{x}^{(j)} + \boldsymbol{c}\right)}{\left(\boldsymbol{s}^{(j)}\right)^\top G\boldsymbol{s}^{(j)}}.$$

Comparing with (2.8.11), we have $\alpha = \alpha_j^*$ which then shows that $\boldsymbol{x}^{(j+1)} = \boldsymbol{x}^{(j)} + \alpha_j^* \boldsymbol{s}^{(j)}$ minimizes f on U_{j+1}, as required. This proves the inductive step and the proof is complete.

This theorem is simple but important. All conjugate direction methods rely on this theorem. It shows that conjugacy plus exact line search implies quadratic termination. Also, the optimal step length in the ith exact line search depends directly on the corresponding conjugate search directions via the formula

$$\alpha_i^* = -\frac{\left(G\boldsymbol{x}^{(i)} + \boldsymbol{c}\right)^\top \boldsymbol{s}^{(i)}}{(\boldsymbol{s}^{(i)})^\top G\boldsymbol{s}^{(i)}}. \tag{2.8.16}$$

We now describe a version of the conjugate gradient method based on the one by Fletcher and Reeves [62]. The presentation given below follows that in [10]. At a point $\boldsymbol{x}^{(k)} \in \mathbb{R}^n$, instead of minimizing the function $f(\boldsymbol{x})$ in the direction of $-\boldsymbol{g}^{(k)}$, we will search in the direction of $\boldsymbol{s}^{(k)}$ generated by

$$\boldsymbol{s}^{(k)} = -\boldsymbol{g}^{(k)} + \beta_{k-1}\boldsymbol{s}^{(k-1)}, \quad k \geq 1, \tag{2.8.17}$$

$$\boldsymbol{s}^{(0)} = -\boldsymbol{g}^{(0)}, \tag{2.8.18}$$

where β_{k-1} is chosen such that $\boldsymbol{s}^{(k)}$ is G-conjugate to $\boldsymbol{s}^{(k-1)}$. More precisely, from (2.8.17) and (2.8.18), we have

$$\left(\boldsymbol{s}^{(k)}\right)^\top G\boldsymbol{s}^{(k-1)} = -\left(\boldsymbol{g}^{(k)}\right)^\top G\boldsymbol{s}^{(k-1)} + \beta_{k-1}\left(\boldsymbol{s}^{(k-1)}\right)^\top G\boldsymbol{s}^{(k-1)}. \tag{2.8.19}$$

Choosing

$$\beta_{k-1} = \frac{\left(\boldsymbol{g}^{(k)}\right)^\top G\boldsymbol{s}^{(k-1)}}{\left(\boldsymbol{s}^{(k-1)}\right)^\top G\boldsymbol{s}^{(k-1)}}, \tag{2.8.20}$$

it is clear from (2.8.19) that

$$\left(\boldsymbol{s}^{(k)}\right)^\top G\boldsymbol{s}^{(k-1)} = 0. \tag{2.8.21}$$

This shows that $\boldsymbol{s}^{(k)}$ is G-conjugate to $\boldsymbol{s}^{(k-1)}$.

Now, let $\boldsymbol{x}^{(k+1)}$ be generated from $\boldsymbol{x}^{(k)}$ through minimizing the function $f(\boldsymbol{x}^{(k)} + \alpha\boldsymbol{s}^{(k)})$ with respect to α, i.e.,

$$\min_{\alpha} f\left(\boldsymbol{x}^{(k)} + \alpha\boldsymbol{s}^{(k)}\right) = f\left(\boldsymbol{x}^{(k+1)}\right).$$

From (2.8.1), we have

$$f\left(\boldsymbol{x}^{(k)}+\alpha \boldsymbol{s}^{(k)}\right)=f\left(\boldsymbol{x}^{(k)}\right)+\alpha\left(\boldsymbol{g}^{(k)}\right)^{\top} \boldsymbol{s}^{(k)}+\frac{\alpha^2}{2}\left(\boldsymbol{s}^{(k)}\right)^{\top} G \boldsymbol{s}^{(k)}. \quad (2.8.22)$$

Minimizing this function with respect to α yields

$$\boldsymbol{x}^{(k+1)}=\boldsymbol{x}^{(k)}+\alpha_k \boldsymbol{s}^{(k)}, \quad (2.8.23)$$

where

$$\alpha_k=-\frac{\left(\boldsymbol{g}^{(k)}\right)^{\top} \boldsymbol{s}^{(k)}}{\left(\boldsymbol{s}^{(k)}\right)^{\top} G \boldsymbol{s}^{(k)}}. \quad (2.8.24)$$

Since $\boldsymbol{g}^{(k)}=G\boldsymbol{x}^{(k)}+\boldsymbol{c}$ and using (2.8.23), we obtain

$$\boldsymbol{g}^{(k+1)}=G\boldsymbol{x}^{(k+1)}+\boldsymbol{c}=G\left(\boldsymbol{x}^{(k)}+\alpha_k \boldsymbol{s}^{(k)}\right)+\boldsymbol{c}=\boldsymbol{g}^{(k)}+\alpha_k G \boldsymbol{s}^{(k)}. \quad (2.8.25)$$

By (2.8.17), it follows from the exact line search condition $\left(\boldsymbol{g}^{(k)}\right)^{\top} \boldsymbol{s}^{(k-1)}=0$ that

$$\left(\boldsymbol{g}^{(k)}\right)^{\top} \boldsymbol{s}^{(k)}=-\left(\boldsymbol{g}^{(k)}\right)^{\top} \boldsymbol{g}^{(k)}+\beta_{k-1}\left(\boldsymbol{g}^{(k)}\right)^{\top} \boldsymbol{s}^{(k-1)}=-\left(\boldsymbol{g}^{(k)}\right)^{\top} \boldsymbol{g}^{(k)}. \quad (2.8.26)$$

Since $\boldsymbol{s}^{(k-1)}$ is G-conjugate to $\boldsymbol{s}^{(k)}$, it is clear from (2.8.17) and (2.8.18) that

$$\begin{aligned}\left(\boldsymbol{s}^{(k)}\right)^{\top} G \boldsymbol{s}^{(k)} &=-\left(\boldsymbol{g}^{(k)}\right)^{\top} G \boldsymbol{s}^{(k)}+\beta_{k-1}\left(\boldsymbol{s}^{(k-1)}\right)^{\top} G \boldsymbol{s}^{(k)} \\ &=-\left(\boldsymbol{g}^{(k)}\right)^{\top} G \boldsymbol{s}^{(k)}.\end{aligned} \quad (2.8.27)$$

Substituting (2.8.26) and (2.8.27) into (2.8.24) yields

$$\alpha_k=-\frac{\left(\boldsymbol{g}^{(k)}\right)^{\top} \boldsymbol{g}^{(k)}}{\left(\boldsymbol{g}^{(k)}\right)^{\top} G \boldsymbol{s}^{(k)}}. \quad (2.8.28)$$

From (2.8.25), we have

$$\left(\boldsymbol{g}^{(k+1)}\right)^{\top} \boldsymbol{g}^{(k)}=\left(\boldsymbol{g}^{(k)}\right)^{\top} \boldsymbol{g}^{(k)}+\alpha_k\left(\boldsymbol{g}^{(k)}\right)^{\top} G \boldsymbol{s}^{(k)}. \quad (2.8.29)$$

Substituting (2.8.28) into (2.8.29) yields

$$\left(\boldsymbol{g}^{(k+1)}\right)^{\top} \boldsymbol{g}^{(k)}=\left(\boldsymbol{g}^{(k)}\right)^{\top} \boldsymbol{g}^{(k)}-\frac{\left(\boldsymbol{g}^{(k)}\right)^{\top} \boldsymbol{g}^{(k)}}{\left(\boldsymbol{g}^{(k)}\right)^{\top} G \boldsymbol{s}^{(k)}}\left(\boldsymbol{g}^{(k)}\right)^{\top} G \boldsymbol{s}^{(k)}=0. \quad (2.8.30)$$

This shows that $\boldsymbol{g}^{(k+1)}$ is orthogonal to $\boldsymbol{g}^{(k)}$.

Remark 2.8.1 *Let $\boldsymbol{x}^{(0)}$ be an initial point in $\mathbb{R}^n$. Then, $\boldsymbol{x}^{(1)}$ is calculated according to (2.8.23) with $k = 0$, where $\boldsymbol{s}^{(0)} = -\boldsymbol{g}^{(0)}$ and α_0 is obtained by minimizing $f\left(\boldsymbol{x}^{(0)} + \alpha \boldsymbol{s}^{(0)}\right)$ with respect to α. α_0 is given by (2.8.24) with $k = 0$. Let $\boldsymbol{x}^{(1)} = \boldsymbol{x}^{(0)} - \alpha_0 \boldsymbol{g}^{(0)}$. The gradient vector $\boldsymbol{g}^{(1)} = \left(\partial f(\boldsymbol{x})/\partial \boldsymbol{x}|_{\boldsymbol{x}=\boldsymbol{x}^{(1)}}\right)^\top$ of the function $f(\boldsymbol{x})$ is calculated at $\boldsymbol{x}^{(1)}$. Since α_0 is the minimum point of the function $f\left(\boldsymbol{x}^{(0)} - \alpha \boldsymbol{g}^{(0)}\right)$ with respect to α, we have*

$$0 = \left.\frac{df\left(\boldsymbol{x}^{(0)} - \alpha \boldsymbol{g}^{(0)}\right)}{d\alpha}\right|_{\alpha=\alpha_0} = -\left(\left.\frac{\partial f(\boldsymbol{x})}{\partial \boldsymbol{x}}\right|_{\boldsymbol{x}=\boldsymbol{x}^{(0)}-\alpha_0 \boldsymbol{g}^{(0)}}\right) \boldsymbol{g}^{(0)}$$
$$= -\left(\boldsymbol{g}^{(1)}\right)^\top \boldsymbol{g}^{(0)}. \tag{2.8.31}$$

This shows that $\boldsymbol{g}^{(1)}$ is orthogonal to $\boldsymbol{g}^{(0)}$ and hence to $\boldsymbol{s}^{(0)}$.

Theorem 2.8.3 *The sequences $\{\boldsymbol{s}^{(k)}\}$ and $\{\boldsymbol{g}^{(k)}\}$ generated according to (2.8.17), (2.8.18) and (2.8.25) are mutually G-conjugate and mutually orthogonal, respectively.*

Proof. We shall use mathematical induction. First, by (2.8.21) with $k = 1$, we note that $\boldsymbol{s}^{(0)}$ and $\boldsymbol{s}^{(1)}$ are G-conjugate. Next, by (2.8.31), $\boldsymbol{g}^{(0)}$ and $\boldsymbol{g}^{(1)}$ are orthogonal. Thus, the conclusion of the theorem for $k = 1$ is valid. Now, we suppose that $\boldsymbol{s}^{(0)}$, $\boldsymbol{s}^{(1)}$, $\ldots$, $\boldsymbol{s}^{(k-1)}$ are mutually G-conjugate and $\boldsymbol{g}^{(0)}, \boldsymbol{g}^{(1)}, \ldots, \boldsymbol{g}^{(k-1)}$ are mutually orthogonal for some $k \geq 2$. By (2.8.30), $\boldsymbol{g}^{(k)}$ is orthogonal to $\boldsymbol{g}^{(k-1)}$ and, by (2.8.21), $\boldsymbol{s}^{(k)}$ is G-conjugate to $\boldsymbol{s}^{(k-1)}$. By the induction hypotheses, we assume that $\boldsymbol{g}^{(k-1)}$ is orthogonal to all $\boldsymbol{g}^{(i)}$, $0 \leq i \leq k-2$ and $\boldsymbol{s}^{(k-1)}$ is G-conjugate to $\boldsymbol{s}^{(i)}$, $0 \leq i \leq k-2$. Then, we have, for $0 \leq i \leq k-2$,

$$\left(\boldsymbol{g}^{(k)}\right)^\top \boldsymbol{g}^{(i)} = \left(\boldsymbol{g}^{(k-1)} + \alpha_{k-1} G \boldsymbol{s}^{(k-1)}\right)^\top \boldsymbol{g}^{(i)}$$
$$= \left(\boldsymbol{g}^{(k-1)}\right)^\top \boldsymbol{g}^{(i)} + a_{k-1}\left(G \boldsymbol{s}^{(k-1)}\right)^\top \boldsymbol{g}^{(i)}$$
$$= a_{k-1}\left(G \boldsymbol{s}^{(k-1)}\right)^\top \boldsymbol{g}^{(i)}. \tag{2.8.32}$$

From (2.8.17) and (2.8.18), we can write $\boldsymbol{g}^{(i)}$ as:

$$\boldsymbol{g}^{(i)} = -\boldsymbol{s}^{(i)} + \beta_{i-1} \boldsymbol{s}^{(i-1)}, \tag{2.8.33}$$

where $\beta_{-1} = 0$, as $\boldsymbol{g}^{(0)} = -\boldsymbol{s}^{(0)}$, while for $i \geq 0$, β_i are determined by (2.8.20). Thus, from (2.8.32) and (2.8.33), it follows from the induction hypothesis that

$$\left(\boldsymbol{g}^{(k)}\right)^\top \boldsymbol{g}^{(i)} = \alpha_{k-1}\left(G \boldsymbol{s}^{(k-1)}\right)^\top \left(-\boldsymbol{s}^{(i)} + \beta_{i-1} \boldsymbol{s}^{(i-1)}\right)$$
$$= 0 \text{ for } i = 0, 1, \ldots, k-2. \tag{2.8.34}$$

Therefore, by (2.8.34) and (2.8.30), we conclude that $\boldsymbol{g}^{(0)}$, $\boldsymbol{g}^{(1)}, \ldots, \boldsymbol{g}^{(k)}$ are mutually orthogonal.

It remains to show that $\boldsymbol{s}^{(i)}$, $i = 0, 1, \ldots, k$, are mutually G-conjugate. We recall that $\boldsymbol{s}^{(k)}$ is G-conjugate to $\boldsymbol{s}^{(k-1)}$. By the induction hypothesis, $\boldsymbol{s}^{(k-1)}$ is G-conjugate to all $\boldsymbol{s}^{(i)}$, $0 \le i \le k-2$. Thus, for $0 \le i \le k-2$, it is clear from (2.8.17) and (2.8.18) that

$$\begin{aligned}\left(\boldsymbol{s}^{(k)}\right)^{\top} G\boldsymbol{s}^{(i)} &= \left(-\boldsymbol{g}^{(k)} + \beta_{k-1}\boldsymbol{s}^{(k-1)}\right)^{\top} G\boldsymbol{s}^{(i)} \\ &= -\left(\boldsymbol{g}^{(k)}\right)^{\top} G\boldsymbol{s}^{(i)} + \beta_{k-1}\left(\boldsymbol{s}^{(k-1)}\right)^{\top} G\boldsymbol{s}^{(i)} \\ &= -\left(\boldsymbol{g}^{(k)}\right)^{\top} G\boldsymbol{s}^{(i)}. \end{aligned} \tag{2.8.35}$$

From (2.8.25), we can write $G\boldsymbol{s}^{(i)}$ as:

$$G\boldsymbol{s}^{(i)} = \frac{\boldsymbol{g}^{(i+1)} - \boldsymbol{g}^{(i)}}{\alpha_i}. \tag{2.8.36}$$

From (2.8.24), we see that α_i is never 0 unless $\boldsymbol{g}^{(i)} = \mathbf{0}$. If $\boldsymbol{g}^{(i)} = \mathbf{0}$, then $\boldsymbol{x}^{(i)}$ is the minimizer of the function $f(\boldsymbol{x})$ and the method terminates.

Substituting (2.8.36) into (2.8.35), and then noting that $\boldsymbol{g}^{(k)}$ is orthogonal to all $\boldsymbol{g}^{(j)}$, $j = 0, 1, \ldots, k-1$, (and hence orthogonal to $\boldsymbol{g}^{(i)}$ and $\boldsymbol{g}^{(i+1)}$, for $i = 0, 1, \ldots, k-2$), we obtain

$$\left(\boldsymbol{s}^{(k)}\right)^{\top} G\boldsymbol{s}^{(i)} = -\left(\boldsymbol{g}^{(k)}\right)^{\top} G\boldsymbol{s}^{(i)} = -\frac{\left(\boldsymbol{g}^{(k)}\right)^{\top}\left(\boldsymbol{g}^{(i+1)} - \boldsymbol{g}^{(i)}\right)}{\alpha_i} = 0 \tag{2.8.37}$$

for $0 \le i \le k-2$. This concludes the induction step and shows that $\boldsymbol{s}^{(0)}$, $\boldsymbol{s}^{(1)}, \ldots, \boldsymbol{s}^{(k)}$ are mutually G-conjugate. This completes the proof.

Remark 2.8.2 *Combining* (2.8.20) *and* (2.8.36), *we obtain*

$$\begin{aligned}\beta_{k-1} &= \frac{\left(\boldsymbol{g}^{(k)}\right)^{\top}\left(\boldsymbol{g}^{(k)} - \boldsymbol{g}^{(k-1)}\right)}{(\boldsymbol{s}^{(k-1)})^{\top}\left(\boldsymbol{g}^{(k)} - \boldsymbol{g}^{(k-1)}\right)} \\ &= -\frac{\left(\boldsymbol{g}^{(k)}\right)^{\top}\boldsymbol{g}^{(k)}}{\left(\boldsymbol{s}^{(k-1)}\right)^{\top}\boldsymbol{g}^{(k-1)}}. \end{aligned} \tag{2.8.38}$$

The following *conjugate gradient method* is applicable to general minimization problems, not just those with a quadratic objective function:

Algorithm 2.8.1
Step 1. Choose a starting point $\boldsymbol{x}^{(0)}$.
Step 2. Set $\mathbf{s}^{(0)} = -\boldsymbol{g}^{(0)}$.
Step 3. Define $\boldsymbol{x}^{(1)} = \boldsymbol{x}^{(0)} + \alpha_0\boldsymbol{s}^{(0)}$, *where*

$$\alpha_0 = -\frac{\left(\boldsymbol{g}^{(0)}\right)^\top \boldsymbol{s}^{(0)}}{\left(\boldsymbol{s}^{(0)}\right)^\top G^{(0)} \boldsymbol{s}^{(0)}}$$

and $G^{(k)}$ denotes the Hessian of the function f evaluated at $\boldsymbol{x} = \boldsymbol{x}^{(k)}$.

Step 4. Compute

$$\boldsymbol{g}^{(1)} = \left(\left. \frac{\partial f(\boldsymbol{x})}{\partial \boldsymbol{x}} \right|_{\boldsymbol{x}=\boldsymbol{x}^{(1)}} \right)^\top.$$

Step 5. Set $\boldsymbol{s}^{(1)} = -\boldsymbol{g}^{(1)} + \beta_0 \boldsymbol{s}^{(0)}$, where

$$\beta_0 = \frac{\left(\boldsymbol{g}^{(1)}\right)^\top G^{(0)} \boldsymbol{s}^{(0)}}{\left(\boldsymbol{s}^{(0)}\right)^\top G^{(0)} \boldsymbol{s}^{(0)}}.$$

By Remark 2.8.2, β_0 can be expressed as:

$$\begin{aligned}\beta_0 &= \frac{\left(\boldsymbol{g}^{(1)}\right)^\top \left(\boldsymbol{g}^{(1)} - \boldsymbol{g}^{(0)}\right)}{\left(\boldsymbol{s}^{(0)}\right)^\top \left(\boldsymbol{g}^{(1)} - \boldsymbol{g}^{(0)}\right)} = -\frac{\left(\boldsymbol{g}^{(1)}\right)^\top \boldsymbol{g}^{(1)}}{\left(\boldsymbol{s}^{(0)}\right)^\top \boldsymbol{g}^{(0)}} \\ &= \frac{\left(\boldsymbol{g}^{(1)}\right)^\top \boldsymbol{g}^{(1)}}{(\boldsymbol{g}^{(0)})^\top \boldsymbol{g}^{(0)}}.\end{aligned}$$

This expression avoids the need to calculate $G\boldsymbol{s}^{(0)}$.

Step 6. General step. After reaching $\boldsymbol{x}^{(i)}$, we compute $\boldsymbol{g}^{(i)}$ and set

$$\boldsymbol{s}^{(i)} = -\boldsymbol{g}^{(i)} + \beta_{i-1} \boldsymbol{s}^{(i-1)},$$

where, by the orthogonality of $\{\boldsymbol{g}^{(i)}\}$,

$$\begin{aligned}\beta_{i-1} &= \frac{\left(\boldsymbol{g}^{(i)}\right)^\top \left(\boldsymbol{g}^{(i)} - \boldsymbol{g}^{(i-1)}\right)}{\left(\boldsymbol{s}^{(i-1)}\right)^\top \left(\boldsymbol{g}^{(i)} - \boldsymbol{g}^{(i-1)}\right)} = -\frac{\left(\boldsymbol{g}^{(i)}\right)^\top \boldsymbol{g}^{(i)}}{\left(\boldsymbol{s}^{(i-1)}\right)^\top \boldsymbol{g}^{(i-1)}} \\ &= \frac{\left(\boldsymbol{g}^{(i)}\right)^\top \boldsymbol{g}^{(i)}}{\left(\boldsymbol{g}^{(i-1)}\right)^\top \boldsymbol{g}^{(i-1)}}.\end{aligned}$$

Step 7. Set

$$\boldsymbol{x}^{(i+1)} = \boldsymbol{x}^{(i)} + \alpha_i \boldsymbol{s}^{(i)}, \tag{2.8.39}$$

where

$$\alpha_i = -\frac{\left(\boldsymbol{s}^{(i)}\right)^\top \boldsymbol{g}^{(i)}}{\left(\boldsymbol{s}^{(i)}\right)^\top G^{(i)} \boldsymbol{s}^{(i)}}.$$

Suppose that the Hessian matrices $G^{(i)}$, $i = 0, 1, 2, \ldots$, are not known. In this case, α_i can be found by a one-dimensional search instead of using the formula in Step 7. The quadratic interpolation method could be used to perform this one-dimensional search.

2.8.1 Convergence of the Conjugate Gradient Methods

Let us present the global convergence result of the Fletcher-Reeves method in the following theorem [232]:

Theorem 2.8.4 *Let $f : \mathbb{R}^n \to \mathbb{R}$ be a twice continuously differentiable function defined on a bounded level set*

$$L(\boldsymbol{x}^{(0)}) = \{\boldsymbol{x} \in \mathbb{R}^n : f(\boldsymbol{x}) \leq f(\boldsymbol{x}^{(0)})\}, \tag{2.8.40}$$

where $\boldsymbol{x}^{(0)}$ is a point in $\mathbb{R}^n$. Suppose that the sequence $\{\boldsymbol{x}^{(k)}\}$ is generated starting from $\boldsymbol{x}^{(0)}$ by the Fletcher-Reeves (FR) conjugate gradient method with inexact line search such that the conditions (2.7.1) *and* (2.7.2) *are satisfied. Then*

$$\liminf_{k\to\infty} \left\| \boldsymbol{g}^{(k)} \right\| = 0, \tag{2.8.41}$$

where $\|\cdot\|$ denotes the usual Euclidean norm.

Other conjugate gradient methods arise from different choices of β_k. For example, in the Polak-Ribiere-Polyak (PRP) method [204], β_k is chosen according to

$$\beta_k = \frac{\left(\boldsymbol{g}^{(k+1)}\right)^\top \left(\boldsymbol{g}^{(k+1)} - \boldsymbol{g}^{(k)}\right)}{\left(\boldsymbol{g}^{(k)}\right)^\top \boldsymbol{g}^{(k)}}. \tag{2.8.42}$$

In the quadratic case, the expression (2.8.42) for β_k is identical to that given by (2.8.38). This is because

$$\left(\boldsymbol{g}^{(k+1)}\right)^\top \boldsymbol{g}^{(k)} = 0.$$

For general functions, the two methods behave differently. The PRP formula given by (2.8.42) will yield $\beta_k \approx 0$ when $\boldsymbol{g}^{(k+1)} \approx \boldsymbol{g}^{(k)}$. Hence, $\boldsymbol{s}^{(k+1)} \approx -\boldsymbol{g}^{(k+1)}$. This means that the algorithm has a tendency of restarting automatically. Thus, it can overcome the deficiency of moving forward slowly. It is generally accepted that the PRP formula is more robust and efficient than the FR formula. Unfortunately, Theorem 2.8.5 is not valid for the PRP method (see [204]). However, we have the following two theorems. For their proofs, see [232].

Theorem 2.8.5 *Let the objective function $f : \mathbb{R}^n \to \mathbb{R}$ be three times continuously differentiable. Suppose that there exist constants $K_1 > K_2 > 0$ such that*

$$K_2 \|\boldsymbol{y}\|^2 \leq \boldsymbol{y}^\top \nabla^2 f(\boldsymbol{x}) \boldsymbol{y} \leq K_1 \|\boldsymbol{y}\|^2, \text{ for all } \boldsymbol{y} \in \mathbb{R}^n,\ \boldsymbol{x} \in L\left(\boldsymbol{x}^{(0)}\right), \tag{2.8.43}$$

where $L(\boldsymbol{x}^{(0)})$ is the bounded level set defined by (2.8.40). *Let the sequence $\{\boldsymbol{x}^{(k)}\}$ be generated by either the Polak-Ribiere-Polyak Conjugate Gradient* *or the Fletcher-Reeves Conjugate Gradient restart method with the exact line search. Then, there exists a constant $c > 0$, such that*

$$\limsup_{k_{(r)} \to \infty} \frac{\left\| \boldsymbol{x}^{(k_{(r)}+n)} - \boldsymbol{x}^* \right\|}{\left\| \boldsymbol{x}^{(k_{(r)})} - \boldsymbol{x}^* \right\|} \leq c < \infty, \tag{2.8.44}$$

where $k_{(r)}$ means that the methods restart after r iterations.

Theorem 2.8.6 *Let $f(\boldsymbol{x})$ be twice continuously differentiable and let $L\left(\boldsymbol{x}^{(0)}\right)$ be the level set defined by (2.8.40). Suppose that $L\left(\boldsymbol{x}^{(0)}\right)$ is bounded and that there is a constant $K > 0$ such that*

$$K \left\| \boldsymbol{y} \right\|^2 \leq \boldsymbol{y}^\top \nabla^2 f(\boldsymbol{x}) \boldsymbol{y}, \quad \forall \boldsymbol{y} \in \mathbb{R}^n \tag{2.8.45}$$

for all $\boldsymbol{x} \in L\left(\boldsymbol{x}^{(0)}\right)$. Then, the sequence $\left\{\boldsymbol{x}^{(k)}\right\}$ generated by the PRP method with the exact line search converges to a unique minimizer $\boldsymbol{x}^*$ of f.

2.9 Quasi-Newton Methods

Recall that $G^{(k)} = \nabla^2 f\left(\boldsymbol{x}^{(k)}\right)$ may not always be positive definite when $\boldsymbol{x}^{(k)}$ is far from a local minimum. Thus, Newton's method may not converge in such a situation. Quasi-Newton methods are based on the idea of approximating $\left(G^{(k)}\right)^{-1}$ at each iteration by a *symmetric positive definite matrix* $H^{(k+1)}$ which is updated at every iteration. The positive definiteness of the matrix $H^{(k+1)}$ ensures that the search direction generated by this method is a descent direction.

Algorithm 2.9.1
Step 1. Given $\boldsymbol{x}^{(0)}$, $H^{(0)}$. Set $k = 0$.
Step 2. Evaluate $f^{(k)} = f\left(\boldsymbol{x}^{(k)}\right)$, $\boldsymbol{g}^{(k)} = \left(\nabla f\left(\boldsymbol{x}^{(k)}\right)\right)^\top$, and $H^{(k)}$.
Step 3. Set $\boldsymbol{s}^{(k)} = -H^{(k)}\boldsymbol{g}^{(k)}$.
Step 4. Check for convergence. If $\left|\boldsymbol{s}^{(k)}\right| < \varepsilon$, stop.
Step 5. Set $\boldsymbol{x}^{(k+1)} = \boldsymbol{x}^{(k)} + \alpha^{(k)}\boldsymbol{s}^{(k)}$, where $\alpha^{(k)}$ is chosen by a line search.
Step 6. Update $H^{(k)}$ to $H^{(k+1)}$.
Step 7. Set $k := k + 1$, go to Step 2.

Usually, $H^{(0)} = I$. This implies that

$$\boldsymbol{s}^{(0)} = -\boldsymbol{g}^{(0)}.$$

This means that, we have the steepest descent direction at the first iteration.

Remark 2.9.1 *For Newton's method, it has fast local convergence properties. However, it requires information on second derivatives and $G^{(k)}$ may be indefinite. For quasi-Newton methods, their convergence properties are much* *better than that of the steepest descent method, but not as good as that of Newton's method. They require only information on first order partial derivatives. Also, for each* $k = 0, 1, \ldots$, *the matrix* $H^{(k)}$ *is always positive definite, and hence the corresponding* $\boldsymbol{s}^{(k)}$ *is a descent direction. Note that some quasi-Newton methods do not ensure that* $H^{(k)}$ *is positive definite. Those that do are also called* variable metric methods.

2.9.1 Approximation of the Inverse G^{-1}

The key idea in quasi-Newton methods is to approximate the inverse $(G^{(k)})^{-1}$ of the Hessian $G^{(k)}$ by $H^{(k+1)}$ at step k. This approximate matrix should be chosen to be positive definite so that the search direction generated is a descent direction. Let

$$\boldsymbol{\delta}^{(k)} = \boldsymbol{x}^{(k+1)} - \boldsymbol{x}^{(k)} \tag{2.9.1}$$

and

$$\boldsymbol{\gamma}^{(k)} = \boldsymbol{g}^{(k+1)} - \boldsymbol{g}^{(k)}. \tag{2.9.2}$$

Taylor's series expansion gives

$$\boldsymbol{g}^{(k+1)} = \boldsymbol{g}^{(k)} + G^{(k)}\left(\boldsymbol{x}^{(k+1)} - \boldsymbol{x}^{(k)}\right) + o\left(\|\boldsymbol{\delta}^{(k)}\|\right), \tag{2.9.3}$$

or, equivalently,

$$\boldsymbol{\gamma}^{(k)} = G^{(k)}\boldsymbol{\delta}^{(k)} + o\left(\|\boldsymbol{\delta}^{(k)}\|\right), \tag{2.9.4}$$

where $o(\|\boldsymbol{\xi}\|)$ is to be understood as a column vector such that

$$\lim_{\|\boldsymbol{\xi}\|\to 0} \frac{\|o\left(\|\boldsymbol{\xi}\|\right)\|}{\|\boldsymbol{\xi}\|} = 0.$$

This expansion is exact if $f(\boldsymbol{x})$ is a quadratic function. From (2.9.3), or equivalently (2.9.4), we see that $G^{(k)}$ depends on $\boldsymbol{\delta}^{(k)}$. For the case when $G^{(k)}$ is constant for all $k = 0, 1, 2, \ldots, n-1$, (i.e., $f(\boldsymbol{x})$ is a quadratic function), we have

$$\boldsymbol{\gamma}^{(k)} = G\boldsymbol{\delta}^{(k)}, \quad \forall k = 0, 1, \ldots, n-1.$$

We now return to (2.9.4) with higher order terms neglected, i.e.,

$$\boldsymbol{\gamma}^{(k)} = G^{(k)}\boldsymbol{\delta}^{(k)}, \quad \forall k = 0, 1, \ldots, n-1. \tag{2.9.5}$$

Note that $\boldsymbol{\delta}^{(k)}$ and hence $\boldsymbol{\gamma}^{(k)}$ are calculated after the line search and $H^{(k)}$ is used to calculate the direction of the search. Hence, it is usually expected that

$$H^{(k)}\boldsymbol{\gamma}^{(k)} \neq \boldsymbol{\delta}^{(k)}.$$

Thus, we choose $H^{(k+1)}$ such that

$$H^{(k+1)}\boldsymbol{\gamma}^{(k)} = \boldsymbol{\delta}^{(k)}. \tag{2.9.6}$$

Equation (2.9.6) is known as the *quasi-Newton condition.* This is the condition that the update formula should satisfy.

2.9.2 Rank Two Correction

Define

$$H^{(k+1)} = H^{(k)} + a\boldsymbol{u}\boldsymbol{u}^\top + b\boldsymbol{v}\boldsymbol{v}^\top. \tag{2.9.7}$$

Assume that the quasi-Newton condition (2.9.6) is satisfied. Then, by multiplying both sides of (2.9.7) by $\boldsymbol{\gamma}^{(k)}$, we obtain

$$\boldsymbol{\delta}^{(k)} = H^{(k)}\boldsymbol{\gamma}^{(k)} + a\boldsymbol{u}\boldsymbol{u}^\top\boldsymbol{\gamma}^{(k)} + b\boldsymbol{v}\boldsymbol{v}^\top\boldsymbol{\gamma}^{(k)}. \tag{2.9.8}$$

We choose a and b such that

$$a\boldsymbol{u}^\top\boldsymbol{\gamma}^{(k)} = 1 \text{ and } b\boldsymbol{v}^\top\boldsymbol{\gamma}^{(k)} = -1. \tag{2.9.9}$$

Let $\boldsymbol{u} = \boldsymbol{\delta}^{(k)}$ and $\boldsymbol{v} = H^{(k)}\boldsymbol{\gamma}^{(k)}$. Then, it follows from (2.9.9) that

$$a = \frac{1}{\left(\boldsymbol{\delta}^{(k)}\right)^\top\boldsymbol{\gamma}^{(k)}} \quad \text{and} \quad b = -\frac{1}{\left(\boldsymbol{\gamma}^{(k)}\right)^\top H^{(k)}\boldsymbol{\gamma}^{(k)}}. \tag{2.9.10}$$

Substituting (2.9.10) into (2.9.7) yields

$$H^{(k+1)} = H^{(k)} + \frac{\boldsymbol{\delta}^{(k)}\left(\boldsymbol{\delta}^{(k)}\right)^\top}{\left(\boldsymbol{\delta}^{(k)}\right)^\top\boldsymbol{\gamma}^{(k)}} - \frac{H^{(k)}\boldsymbol{\gamma}^{(k)}\left(\boldsymbol{\gamma}^{(k)}\right)^\top H^{(k)}}{\left(\boldsymbol{\gamma}^{(k)}\right)^\top H^{(k)}\boldsymbol{\gamma}^{(k)}}. \tag{2.9.11}$$

This is the *Davidon-Fletcher-Powell (DFP) formula.*

Remark 2.9.2 *Clearly, for any $\boldsymbol{a} \in \mathbb{R}^n$ and $\boldsymbol{b} \in \mathbb{R}^n$, it holds that*

$$(\boldsymbol{a}x - \boldsymbol{b})^\top(\boldsymbol{a}x - \boldsymbol{b}) \geq 0, \quad \textit{for all } x \in (-\infty, \infty).$$

This means that

$$\boldsymbol{a}^\top\boldsymbol{a}x^2 - 2\boldsymbol{a}^\top\boldsymbol{b}x + \boldsymbol{b}^\top\boldsymbol{b} \geq 0, \quad \textit{for all } x \in (-\infty, \infty). \tag{2.9.12}$$

The roots of the quadratic equation $\boldsymbol{a}^\top\boldsymbol{a}x^2 - 2\boldsymbol{a}^\top\boldsymbol{b}x + \boldsymbol{b}^\top\boldsymbol{b} = 0$ are given by

$$x = \frac{2\boldsymbol{a}^\top\boldsymbol{b} \pm \sqrt{(2\boldsymbol{a}^\top\boldsymbol{b})^2 - 4(\boldsymbol{a}^\top\boldsymbol{a})(\boldsymbol{b}^\top\boldsymbol{b})}}{2\boldsymbol{a}^\top\boldsymbol{a}}.$$

Clearly, (2.9.12) *is valid if and only if*

$$\left(\boldsymbol{a}^\top \boldsymbol{b}\right)^2 \le \left(\boldsymbol{a}^\top \boldsymbol{a}\right)\left(\boldsymbol{b}^\top \boldsymbol{b}\right).$$

Furthermore, (2.9.12) *holds as an equality only if* $\boldsymbol{a}$ *is parallel to* $\boldsymbol{b}$.

Theorem 2.9.1 *If* $H^{(k)}$ *is symmetric positive definite, then* $H^{(k+1)}$ *is also symmetric positive definite. (This ensures that the search direction at each iteration is downhill, and hence the function value is reduced along the search direction using a line search).*

Proof. For any $\boldsymbol{x} \in \mathbb{R}^n$, we have

$$\begin{aligned}\boldsymbol{x}^\top H^{(k+1)}\boldsymbol{x} &= \boldsymbol{x}^\top H^{(k)}\boldsymbol{x} + \boldsymbol{x}^\top \frac{\boldsymbol{\delta}^{(k)}\left(\boldsymbol{\delta}^{(k)}\right)^\top}{\left(\boldsymbol{\delta}^{(k)}\right)^\top \boldsymbol{\gamma}^{(k)}}\boldsymbol{x} - \boldsymbol{x}^\top \frac{H^{(k)}\boldsymbol{\gamma}^{(k)}\left(\boldsymbol{\gamma}^{(k)}\right)^\top H^{(k)}}{\left(\boldsymbol{\gamma}^{(k)}\right)^\top H^{(k)}\boldsymbol{\gamma}^{(k)}}\boldsymbol{x} \\ &= \boldsymbol{x}^\top H^{(k)}\boldsymbol{x} + \frac{\left(\boldsymbol{x}^\top \boldsymbol{\delta}^{(k)}\right)^2}{\left(\boldsymbol{\delta}^{(k)}\right)^\top \boldsymbol{\gamma}^{(k)}} - \frac{\left(\boldsymbol{x}^\top H^{(k)}\boldsymbol{\gamma}^{(k)}\right)^2}{(\boldsymbol{\gamma}^{(k)})^\top H^{(k)}\boldsymbol{\gamma}^{(k)}}.\end{aligned} \tag{2.9.13}$$

Let $\boldsymbol{a} = \left(H^{(k)}\right)^{\frac{1}{2}}\boldsymbol{x}$, and $\boldsymbol{b} = \left(H^{(k)}\right)^{\frac{1}{2}}\boldsymbol{\gamma}^{(k)}$, where $H^{1/2}$ is defined such that $H^{1/2}H^{1/2} = H$.

Then, we have

$$\begin{aligned}\boldsymbol{x}^\top H^{(k+1)}\boldsymbol{x} &= \boldsymbol{a}^\top \boldsymbol{a} - \frac{\left(\boldsymbol{a}^\top \boldsymbol{b}\right)^2}{\boldsymbol{b}^\top \boldsymbol{b}} + \frac{\left(\boldsymbol{x}^\top \boldsymbol{\delta}^{(k)}\right)^2}{\left(\boldsymbol{\delta}^{(k)}\right)^\top \boldsymbol{\gamma}^{(k)}} \\ &= \frac{\left(\boldsymbol{a}^\top \boldsymbol{a}\right)\left(\boldsymbol{b}^\top \boldsymbol{b}\right) - \left(\boldsymbol{a}^\top \boldsymbol{b}\right)^2}{\boldsymbol{b}^\top \boldsymbol{b}} + \frac{\left(\boldsymbol{x}^\top \boldsymbol{\delta}^{(k)}\right)^2}{\left(\boldsymbol{\delta}^{(k)}\right)^\top \boldsymbol{\gamma}^{(k)}}.\end{aligned} \tag{2.9.14}$$

Also, since $\boldsymbol{x}^{(k+1)}$ is the minimum point of $f(\boldsymbol{x})$ along the direction $\boldsymbol{\delta}^{(k)}$, it follows from (2.3.4) that

$$\left(\boldsymbol{\delta}^{(k)}\right)^\top \boldsymbol{g}^{(k+1)} = 0. \tag{2.9.15}$$

Thus,

$$\left(\boldsymbol{\delta}^{(k)}\right)^\top \boldsymbol{\gamma}^{(k)} = \left(\boldsymbol{\delta}^{(k)}\right)^\top \boldsymbol{g}^{(k+1)} - \left(\boldsymbol{\delta}^{(k)}\right)^\top \boldsymbol{g}^{(k)} = -\left(\boldsymbol{\delta}^{(k)}\right)^\top \boldsymbol{g}^{(k)}. \tag{2.9.16}$$

Now, by the definition of $\boldsymbol{\delta}^{(k)}$ given by (2.9.1) and noting that $\boldsymbol{x}^{(k+1)} = \boldsymbol{x}^{(k)} - \alpha_k H^{(k)}\boldsymbol{g}^{(k)}$, we obtain

$$\left(\boldsymbol{\delta}^{(k)}\right)^\top \boldsymbol{\gamma}^{(k)} = -\left(-\alpha_k H^{(k)}\boldsymbol{g}^{(k)}\right)^\top \boldsymbol{g}^{(k)} = \alpha_k \left(\boldsymbol{g}^{(k)}\right)^\top H^{(k)}\boldsymbol{g}^{(k)} > 0. \tag{2.9.17}$$

Substituting (2.9.17) into the denominator of the second term on the right hand side of (2.9.14), we obtain

$$x^\top H^{(k+1)} x = \frac{(a^\top a)(b^\top b) - (a^\top b)^2}{(b^\top b)} + \frac{(x^\top \delta^{(k)})^2}{\alpha_k (g^k)^\top H^{(k)} g^{(k)}}. \tag{2.9.18}$$

From Remark 2.9.2, we have

$$(a^\top a)(b^\top b) - (a^\top b)^2 \geq 0 \tag{2.9.19}$$

and (2.9.19) holds as an equality only if a is parallel to b. In this case, x is parallel to $\gamma^{(k)}$, i.e., there exists a constant λ such that

$$x = \lambda \gamma^{(k)} = \lambda G^{(k)} \delta^{(k)}.$$

This implies that

$$x^\top \delta^{(k)} = \lambda \left(\delta^{(k)}\right)^\top G \delta^{(k)} > 0. \tag{2.9.20}$$

Combining (2.9.18), (2.9.19) and (2.9.20), we have

$$x^\top H^{(k+1)} x > 0.$$

This completes the proof.

Remark 2.9.3 *Is* $\left(\delta^{(k)}\right)^\top \gamma^{(k)} > 0$*? The answer is positive as explained below.*

(i) Exact line search. From the exact line search condition (2.3.4) *(i.e.,* $\left(\delta^{(k)}\right)^\top g^{(k+1)} = 0$*),* $\delta^{(k)} = -\alpha H^{(k)} g^{(k)}$*,* $H^{(k)}$ *is positive definite (and hence* $\left(g^{(k)}\right)^\top H^{(k)} g^{(k)} > 0$*), and* $\alpha > 0$*, it follows that*

$$\left(\delta^{(k)}\right)^\top \gamma^{(k)} = \left(\delta^{(k)}\right)^\top g^{(k+1)} - \left(\delta^{(k)}\right)^\top g^{(k)} = \alpha \left(g^{(k)}\right)^\top H^{(k)} g^{(k)} > 0.$$

(ii) Inexact line search. Conditions on slope can easily be used to ensure $\left(\delta^{(k)}\right)^\top \gamma^{(k)} > 0$*. Recall that*

$$\left|(s^{(k)})^\top g^{(k+1)}\right| < -\sigma \left(s^{(k)}\right)^\top g^{(k)}, \quad 0 < \sigma < 1.$$

Since $H^{(k)}$ *is positive definite, it follows from Step 3 of Algorithm 2.9.1 that* $\left(s^{(k)}\right)^\top g^{(k)} = -\left(g^{(k)}\right)^\top H^{(k)} g^{(k)} < 0$*. Thus,*

$$\left(g^{(k+1)}\right)^\top s^{(k)} > \sigma \left(g^{(k)}\right)^\top s^{(k)}, \quad 0 < \sigma < 1.$$

Therefore,

$$\left(s^{(k)}\right)^\top \gamma^{(k)} = \alpha^{(k)} \left(s^{(k)}\right)^\top \left(g^{(k+1)} - g^{(k)}\right)$$

$$
\begin{aligned}
&= \alpha^{(k)} \left(\boldsymbol{g}^{(k+1)} - \boldsymbol{g}^{(k)} \right)^\top \boldsymbol{s}^{(k)} \\
&> \alpha^{(k)} \left(\sigma \boldsymbol{g}^{(k)} - \boldsymbol{g}^{(k)} \right)^\top \boldsymbol{s}^{(k)} \\
&= -(1-\sigma)\alpha^{(k)} \left(\boldsymbol{g}^{(k)} \right)^\top \boldsymbol{s}^{(k)} \\
&> 0.
\end{aligned}
$$

Theorem 2.9.2 *The DFP method is used to solve the following objective function:*

$$
f(\boldsymbol{x}) = \frac{1}{2}\boldsymbol{x}^\top G\boldsymbol{x} - \boldsymbol{c}^\top \boldsymbol{x}, \tag{2.9.21}
$$

where $G \in \mathbb{R}^{n\times n}$ is positive definite. Then, for $i = 0, 1, \ldots, m$, it holds that

$$
H^{(i+1)}\gamma^{(j)} = \boldsymbol{\delta}^{(j)}, \quad j = 0, 1, \ldots, i, \tag{2.9.22}
$$

$$
\left(\boldsymbol{\delta}^{(i)}\right)^\top G\boldsymbol{\delta}^{(j)} = 0, \quad j = 0, 1, \ldots, i-1. \tag{2.9.23}
$$

Furthermore, the method terminates at the step $m + 1 \leq n$. If $m = n - 1$, then

$$
H^{(n)} = G^{-1}. \tag{2.9.24}
$$

Proof. We shall show the validity of (2.9.22) and (2.9.23) by induction. For $i = 0$, it is clear from the quasi-Newton condition that

$$
H^{(1)}\gamma^{(0)} = \boldsymbol{\delta}^{(0)}. \tag{2.9.25}
$$

Thus, (2.9.22) with $i = 0$ is valid. Now we suppose that (2.9.22) and (2.9.23) are true for i. Then, it is required to show that they are also valid for $i + 1$. Note that

$$
\boldsymbol{g}^{(i+1)} \neq \mathbf{0}. \tag{2.9.26}
$$

Then, by the exact line search condition (2.3.4) and the induction hypothesis, it follows from (2.9.5) that, for $j \leq i$,

$$
\begin{aligned}
\left(\boldsymbol{g}^{(i+1)}\right)^\top \boldsymbol{\delta}^{(j)} &= \left(\boldsymbol{g}^{(j+1)}\right)^\top \boldsymbol{\delta}^{(j)} + \sum_{k=j+1}^{i} \left(\boldsymbol{g}^{(k+1)} - \boldsymbol{g}^{(k)}\right)^\top \boldsymbol{\delta}^{(j)} \\
&= \left(\boldsymbol{g}^{(j+1)}\right)^\top \boldsymbol{\delta}^{(j)} + \sum_{k=j+1}^{i} \left(\gamma^{(k)}\right)^\top \boldsymbol{\delta}^{(j)} \\
&= 0 + \sum_{k=j+1}^{i} \left(\boldsymbol{\delta}^{(k)}\right)^\top G\boldsymbol{\delta}^{(j)} \\
&= 0.
\end{aligned} \tag{2.9.27}
$$

Note that

$$\boldsymbol{\delta}^{(i+1)} = \boldsymbol{x}^{(i+2)} - \boldsymbol{x}^{(i+1)} \tag{2.9.28}$$

and

$$\boldsymbol{x}^{(i+2)} = \boldsymbol{x}^{(i+1)} - \alpha_{i+1} H^{(i+1)} \boldsymbol{g}^{(i+1)}. \tag{2.9.29}$$

Thus,

$$\boldsymbol{\delta}^{(i+1)} = -\alpha_{i+1} H^{(i+1)} \boldsymbol{g}^{(i+1)}. \tag{2.9.30}$$

Since (2.9.23) with i is valid by the induction hypothesis, it is clear from (2.9.30) and (2.9.27) that

$$\begin{aligned}\left(\boldsymbol{\delta}^{(i+1)}\right)^\top G\boldsymbol{\delta}^{(j)} &= -\alpha_{i+1}\left(\boldsymbol{g}^{(i+1)}\right)^\top H^{(i+1)}\boldsymbol{\gamma}^{(j)}\\ &= -\alpha_{i+1}\left(\boldsymbol{g}^{(i+1)}\right)^\top \boldsymbol{\delta}^{(j)}\\ &= 0. \end{aligned} \tag{2.9.31}$$

Thus, (2.9.23) with $i+1$ is valid. To show the validity of (2.9.22) with $i+1$, we need to show that

$$H^{(i+2)}\boldsymbol{\gamma}^{(j)} = \boldsymbol{\delta}^{(j)}, \quad j = 0, 1, \ldots, i+1.$$

From the quasi-Newton condition (2.9.6), we have

$$H^{(i+2)}\boldsymbol{\gamma}^{(i+1)} = \boldsymbol{\delta}^{(i+1)}. \tag{2.9.32}$$

For $j \le i$, it follows from (2.9.5) and (2.9.31) that

$$\left(\boldsymbol{\delta}^{(i+1)}\right)^\top \boldsymbol{\gamma}^{(j)} = \left(\boldsymbol{\delta}^{(i+1)}\right)^\top G\boldsymbol{\delta}^{(j)} = 0. \tag{2.9.33}$$

Now, by the induction hypothesis, (2.9.5) and (2.9.31), we have

$$\left(\boldsymbol{\gamma}^{(i+1)}\right)^\top H^{(i+1)}\boldsymbol{\gamma}^{(j)} = \left(\boldsymbol{\gamma}^{(i+1)}\right)^\top \boldsymbol{\delta}^{(j)} = \left(\boldsymbol{\delta}^{(i+1)}\right)^\top G\boldsymbol{\delta}^{(j)} = 0. \tag{2.9.34}$$

Thus, multiplying both sides of the DFP formula by $\gamma^{(j)}$, we obtain

$$\begin{aligned}H^{(i+2)}\boldsymbol{\gamma}^{(j)} =& H^{(i+1)}\boldsymbol{\gamma}^{(j)} + \frac{\boldsymbol{\delta}^{(i+1)}\left(\boldsymbol{\delta}^{(i+1)}\right)^\top \boldsymbol{\gamma}^{(j)}}{(\boldsymbol{\delta}^{(i+1)})^\top \boldsymbol{\gamma}^{(i+1)}}\\ &- \frac{H^{(i+1)}\boldsymbol{\gamma}^{(i+1)}\left(\boldsymbol{\gamma}^{(i+1)}\right)^\top H^{(i+1)}\boldsymbol{\gamma}^{(j)}}{\left(\boldsymbol{\gamma}^{(i+1)}\right)^\top H^{(i+1)}\boldsymbol{\gamma}^{(i+1)}}\\ =& H^{(i+1)}\boldsymbol{\gamma}^{(j)} + 0 - 0 = \boldsymbol{\delta}^{(j)}, \end{aligned} \tag{2.9.35}$$

for $j = 0, 1, \ldots, i+1$. Thus, (2.9.22) is valid.

Now, by (2.9.23), we note that $\boldsymbol{\delta}^{(i)}$, $i = 0.1, \ldots, m$, are G-conjugate. This means that the directions generated by the DFP method are mutually G-conjugate. Thus, by Theorem 2.8.2, the minimum of the function (2.9.1) is found in at most n iterations. This implies that there exists an $m \leq n-1$ such that the DFP method terminates after m iterations. For $m = n-1$, we have

$$H^{(n)}\boldsymbol{\gamma}^{(j)} = \boldsymbol{\delta}^{(j)}, \quad j = 0, 1, \ldots, n-1$$

and hence

$$H^{(n)}G\boldsymbol{\delta}^{(j)} = \boldsymbol{\delta}^{(j)}, \quad j = 0, 1, \ldots, n-1.$$

From Theorem 2.8.1, we see that $\boldsymbol{\delta}^{(i)}$, $i = 0, 1, \ldots, m$, are linearly independent. Thus,

$$H^{(n)} = G^{-1}.$$

This completes the proof.

The next theorem contains the results showing the convergence of the DFP method for a general objective function $f : \mathbb{R}^n \to \mathbb{R}$. Its proof can be found in [232].

Theorem 2.9.3 . *Consider the objective function $f : \mathbb{R}^n \to \mathbb{R}$. Suppose that the following conditions are satisfied:*

(a) f is twice continuously differentiable on an open convex set $D \subset \mathbb{R}^n$.
(b) There is a strict local minimizer $\boldsymbol{x}^* \in D$ such that $\nabla^2 f(\boldsymbol{x}^*)$ is symmetric and positive definite.
(c) There is a neighbourhood $\mathcal{N}_\varepsilon(\boldsymbol{x}^*)$ of $\boldsymbol{x}^*$ such that

$$\left\|\nabla^2 f\left(\widehat{\boldsymbol{x}}\right) - \nabla^2 f(\boldsymbol{x})\right\| \leq K \left\|\widehat{\boldsymbol{x}} - \boldsymbol{x}\right\|, \ \forall \boldsymbol{x}, \widehat{\boldsymbol{x}} \in \mathcal{N}_\varepsilon(\boldsymbol{x}^*),$$

where K is a positive constant.
Furthermore, suppose that the following condition is satisfied:

$$K \left\|\left(\nabla^2 f(\boldsymbol{x}^*)\right)^{-1}\right\| \varphi\left(\boldsymbol{x}^{(k)}, \boldsymbol{x}^{(k+1)}\right) \leq \frac{1}{3}$$

in $\mathcal{N}_\varepsilon(\boldsymbol{x}^*)$, where

$$\varphi\left(\boldsymbol{x}^{(k)}, \boldsymbol{x}^{(k+1)}\right) = \max\left\{\boldsymbol{x}^{(k)} - \boldsymbol{x}^*, \boldsymbol{x}^{(k+1)} - \boldsymbol{x}^*\right\}.$$

Then, the DFP method converges superlinearly.

2.9.3 BFGS Update Formula

We shall present another matrix update formula due to Broyden [32], Fletcher [60], Goldfarb [77] and Shanno [227]. This update formula is known as the *BFGS* formula. The update matrix $H^{(k+1)}$ is

$$\begin{aligned}H^{(k+1)} =&H^{(k)} - \frac{\boldsymbol{\delta}^{(k)} \left(\boldsymbol{\gamma}^{(k)}\right)^\top}{\left(\boldsymbol{\delta}^{(k)}\right)^\top \boldsymbol{\gamma}^{(k)}} H^{(k)} - H^{(k)} \frac{\boldsymbol{\gamma}^{(k)} \left(\boldsymbol{\delta}^{(k)}\right)^\top}{\left(\boldsymbol{\delta}^{(k)}\right)^\top \boldsymbol{\gamma}^{(k)}} \\ &+ \frac{\boldsymbol{\delta}^{(k)} \left(\boldsymbol{\gamma}^{(k)}\right)^\top H^{(k)} \boldsymbol{\gamma}^{(k)} \left(\boldsymbol{\delta}^{(k)}\right)^\top}{\left(\boldsymbol{\delta}^{(k)}\right)^\top \boldsymbol{\gamma}^{(k)} \left(\boldsymbol{\delta}^{(k)}\right)^\top \boldsymbol{\gamma}^{(k)}} + \frac{\boldsymbol{\delta}^{(k)} \left(\boldsymbol{\delta}^{(k)}\right)^\top}{\left(\boldsymbol{\delta}^{(k)}\right)^\top \boldsymbol{\gamma}^{(k)}}. \end{aligned} \tag{2.9.36}$$

This is obtained by approximating $G^{(k)}$ by $B^{(k+1)}$. Then, $H^{(k+1)}$ is chosen such that $H^{(k+1)} = \left(B^{(k+1)}\right)^{-1}$. To begin the derivation, we note that the quasi-Newton formula (2.9.6) is changed to

$$\boldsymbol{\gamma}^{(k)} = B^{(k+1)} \boldsymbol{\delta}^{(k)}. \tag{2.9.37}$$

We update $B^{(k)}$ by a rank two correction as in *DFP*. Thus,

$$B^{(k+1)} = B^{(k)} + \frac{\boldsymbol{\gamma}^{(k)} \left(\boldsymbol{\gamma}^{(k)}\right)^\top}{\left(\boldsymbol{\gamma}^{(k)}\right)^\top \boldsymbol{\delta}^{(k)}} - \frac{B^{(k)} \boldsymbol{\delta}^{(k)} \left(\boldsymbol{\delta}^{(k)}\right)^\top B^{(k)}}{\left(\boldsymbol{\delta}^{(k)}\right)^\top B^{(k)} \boldsymbol{\delta}^{(k)}}. \tag{2.9.38}$$

This is the *DFP* formula with B replacing H and $\boldsymbol{\delta}$ swapped with $\boldsymbol{\gamma}$.

The *BFGS* formula comes about by choosing $H^{(k+1)}$ such that

$$B^{(k+1)} H^{(k+1)} = I. \tag{2.9.39}$$

The *BFGS* method has all the properties that the *DFP* method has. Furthermore, the *BFGS* tends to do better than the *DFP* for low accuracy line searches. In addition, if the conditions

$$f\left(\boldsymbol{x}^{(k)}\right) - f\left(\boldsymbol{x}^{(k+1)}\right) \geq -\rho \left(\boldsymbol{g}^{(k)}\right)^\top \boldsymbol{\delta}^{(k)} \tag{2.9.40}$$

and

$$\left(\boldsymbol{g}^{(k+1)}\right)^\top \boldsymbol{s}^{(k)} \geq \sigma \left(\boldsymbol{g}^{(k)}\right)^\top \boldsymbol{s}^{(k)}, \text{ for } 0 < \rho \leq \sigma \tag{2.9.41}$$

hold in an inexact line search, then the *BFGS* method is globally convergent.

To derive the BFGS formula (2.9.36), we need some preliminary results. The first of these is the well-known result on the inversion of block matrices [109].

Lemma 2.9.1 *Let $A \in \mathbb{R}^{n\times n}$, $B \in \mathbb{R}^{m\times m}$, $C \in \mathbb{R}^{m\times n}$, $D \in \mathbb{R}^{n\times n}$ and $\triangle = B - CA^{-1}D$. Suppose that A^{-1} and $\triangle^{-1}$ exist. Then*

$$\begin{bmatrix} A & D \\ C & B \end{bmatrix}^{-1} = \begin{bmatrix} A^{-1} + E\triangle^{-1}F & -E\triangle^{-1} \\ -\triangle^{-1}F & \triangle^{-1} \end{bmatrix},$$

where $E = A^{-1}D$ and $F = CA^{-1}$.

The next result is from [275] but the origin of this result can be traced back to [15].

Lemma 2.9.2 *Let $A \in \mathbb{R}^{n\times n}$ be non-singular and let $\boldsymbol{u}, \boldsymbol{v} \in \mathbb{R}^n$ be such that $1 + \boldsymbol{v}^\top A^{-1}\boldsymbol{u} \neq 0$. Then, $A + \boldsymbol{u}\boldsymbol{v}^\top$ is non-singular. Furthermore,*

$$\left[A + \boldsymbol{u}\boldsymbol{v}^\top\right]^{-1} = A^{-1} - \frac{A^{-1}\boldsymbol{u}\boldsymbol{v}^\top A^{-1}}{1 + \boldsymbol{v}^\top A^{-1}\boldsymbol{u}}. \tag{2.9.42}$$

Proof. Let

$$\begin{bmatrix} 1 & \boldsymbol{v}^\top A^{-1} \\ \boldsymbol{0} & I \end{bmatrix} \begin{bmatrix} 1 & -\boldsymbol{v}^\top \\ \boldsymbol{u} & A \end{bmatrix} = \begin{bmatrix} 1 + \boldsymbol{v}^\top A^{-1}\boldsymbol{u} & \boldsymbol{0} \\ \boldsymbol{u} & A \end{bmatrix} = W_1 \tag{2.9.43}$$

and

$$\begin{bmatrix} 1 & \boldsymbol{0} \\ -\boldsymbol{u} & I \end{bmatrix} \begin{bmatrix} I & -\boldsymbol{v}^\top \\ \boldsymbol{u} & A \end{bmatrix} = \begin{bmatrix} 1 & -\boldsymbol{v}^\top \\ \boldsymbol{0} & A + \boldsymbol{u}\boldsymbol{v}^\top \end{bmatrix} = W_2. \tag{2.9.44}$$

By Lemma 2.9.1, we see that the inverses of W_1 and W_2 are, respectively, given by

$$W_1^{-1} = \begin{bmatrix} \frac{1}{1+\boldsymbol{v}^\top A^{-1}\boldsymbol{u}} & \boldsymbol{0} \\ \frac{-A^{-1}\boldsymbol{u}}{1+\boldsymbol{v}^\top A^{-1}\boldsymbol{u}} & A^{-1} \end{bmatrix} \tag{2.9.45}$$

and

$$W_2^{-1} = \begin{bmatrix} 1 & \boldsymbol{v}^\top (A + \boldsymbol{u}\boldsymbol{v}^\top)^{-1} \\ \boldsymbol{0} & (A + \boldsymbol{u}\boldsymbol{v}^\top)^{-1} \end{bmatrix}. \tag{2.9.46}$$

From (2.9.44) and (2.9.43), we obtain

$$W_2^{-1} = \begin{bmatrix} 1 & -\boldsymbol{v}^\top \\ \boldsymbol{u} & A \end{bmatrix}^{-1} \begin{bmatrix} I & 0 \\ \boldsymbol{u} & I \end{bmatrix} \tag{2.9.47}$$

and

$$W_1^{-1} = \begin{bmatrix} 1 & -\boldsymbol{v}^\top \\ \boldsymbol{u} & A \end{bmatrix}^{-1} \begin{bmatrix} 1 & \boldsymbol{v}^\top A^{-1} \\ \boldsymbol{0} & I \end{bmatrix}^{-1}, \tag{2.9.48}$$

respectively. Multiplying (2.9.48) by $\begin{bmatrix} 1 & \boldsymbol{v}^\top A^{-1} \\ \boldsymbol{0} & I \end{bmatrix}$ yields

$$\begin{bmatrix} 1 & -\boldsymbol{v}^\top \\ \boldsymbol{u} & A \end{bmatrix}^{-1} = W_1^{-1} \begin{bmatrix} 1 & \boldsymbol{v}^\top A^{-1} \\ \boldsymbol{0} & I \end{bmatrix}. \tag{2.9.49}$$

Substituting (2.9.49) into (2.9.47), and then using (2.9.45), we obtain

$$\begin{aligned} W_2^{-1} &= \begin{bmatrix} 1 & -\boldsymbol{v}^\top \\ \boldsymbol{u} & A \end{bmatrix}^{-1} \begin{bmatrix} I & 0 \\ \boldsymbol{u} & I \end{bmatrix} = W_1^{-1} \begin{bmatrix} 1 & \boldsymbol{v}^\top A^{-1} \\ \boldsymbol{0} & I \end{bmatrix} \begin{bmatrix} I & 0 \\ \boldsymbol{u} & I \end{bmatrix} \\ &= \begin{bmatrix} 1 & \frac{\boldsymbol{v}^\top A^{-1}}{1+\boldsymbol{v}^\top A^{-1}\boldsymbol{u}} \\ \boldsymbol{0} & A^{-1} - \frac{A^{-1}\boldsymbol{u}\boldsymbol{v}^\top A^{-1}}{1+\boldsymbol{v}^\top A^{-1}\boldsymbol{u}} \end{bmatrix}. \end{aligned} \tag{2.9.50}$$

Equating (2.9.46) and (2.9.50) gives

$$\left(A + \boldsymbol{u}\boldsymbol{v}^\top\right)^{-1} = A^{-1} - \frac{A^{-1}\boldsymbol{u}\boldsymbol{v}^\top A^{-1}}{1 + \boldsymbol{v}^\top A^{-1}\boldsymbol{u}}.$$

This completes the proof.

We now return to derive the inverse, $H_{BFGS}^{(k+1)}$, of $B_{BFGS}^{(k+1)}$. First, define

$$T = B^{(k)} + \frac{\boldsymbol{\gamma}^{(k)}\left(\boldsymbol{\gamma}^{(k)}\right)^\top}{\left(\boldsymbol{\gamma}^{(k)}\right)^\top \boldsymbol{\delta}^{(k)}}. \tag{2.9.51}$$

Then, by Lemma 2.9.2, we obtain

$$T^{-1} = H^{(k)} - \frac{H^{(k)}\boldsymbol{\gamma}^{(k)}(\boldsymbol{\gamma}^{(k)})^\top H^{(k)}}{\left(\boldsymbol{\delta}^{(k)}\right)^\top \boldsymbol{\gamma}^{(k)} + (\boldsymbol{\gamma}^{(k)})^\top H^{(k)}\boldsymbol{\gamma}^{(k)}} = H^{(k)} - \frac{H^{(k)}\boldsymbol{\gamma}^{(k)}(\boldsymbol{\gamma}^{(k)})^\top H^{(k)}}{\omega}, \tag{2.9.52}$$

where

$$H^{(k)} = \left[B^{(k)}\right]^{-1} \tag{2.9.53}$$

and

$$\omega = \left(\boldsymbol{\delta}^{(k)}\right)^\top \boldsymbol{\gamma}^{(k)} + (\boldsymbol{\gamma}^{(k)})^\top H^{(k)}\boldsymbol{\gamma}^{(k)}. \tag{2.9.54}$$

Thus, by (2.9.38), it is clear from Lemma 2.9.1 that

$$\begin{aligned} H_{BFGS}^{(k+1)} &= \left[B_{BFGS}^{(k+1)}\right]^{-1} = \left[T - \frac{B^{(k)}\boldsymbol{\delta}^{(k)}\left(\boldsymbol{\delta}^{(k)}\right)^\top B^{(k)}}{\left(\boldsymbol{\delta}^{(k)}\right)^\top B^{(k)}\boldsymbol{\delta}^{(k)}}\right]^{-1} \\ &= T^{-1} + \frac{T^{-1}B^{(k)}\boldsymbol{\delta}^{(k)}\left(\boldsymbol{\delta}^{(k)}\right)^\top B^{(k)}T^{-1}}{\left(\boldsymbol{\delta}^{(k)}\right)^\top B^{(k)}\boldsymbol{\delta}^{(k)} - \left(\boldsymbol{\delta}^{(k)}\right)^\top B^{(k)}T^{-1}B^{(k)}\boldsymbol{\delta}^{(k)}}. \end{aligned} \tag{2.9.55}$$

From (2.9.55), we have

$$\begin{aligned} &\left(\boldsymbol{\delta}^{(k)}\right)^\top B^{(k)}\boldsymbol{\delta}^{(k)} - \left(\boldsymbol{\delta}^{(k)}\right)^\top B^{(k)}T^{-1}B_k\boldsymbol{\delta}^{(k)} \\ &= \left(\boldsymbol{\delta}^{(k)}\right)^\top B^{(k)}\boldsymbol{\delta}^{(k)} - \left(\boldsymbol{\delta}^{(k)}\right)^\top B^{(k)}\left[H^{(k)} - \frac{H^{(k)}\boldsymbol{\gamma}^{(k)}(\boldsymbol{\gamma}^{(k)})^\top H^{(k)}}{\omega}\right] B^{(k)}\boldsymbol{\delta}^{(k)} \\ &= \frac{\left(\boldsymbol{\delta}^{(k)}\right)^\top \boldsymbol{\gamma}^{(k)}(\boldsymbol{\gamma}^{(k)})^\top\boldsymbol{\delta}^{(k)}}{\omega} = \frac{\left[\left(\boldsymbol{\delta}^{(k)}\right)^\top \boldsymbol{\gamma}^{(k)}\right]^2}{\omega}. \end{aligned} \tag{2.9.56}$$

Now, by (2.9.55) and (2.9.56), it follows from simple algebraic manipulations that

$$\frac{T^{-1}B^{(k)}\boldsymbol{\delta}^{(k)}\left(\boldsymbol{\delta}^{(k)}\right)^\top B^{(k)}T^{-1}}{\left(\boldsymbol{\delta}^{(k)}\right)^\top B^{(k)}\boldsymbol{\delta}^{(k)} - \left(\boldsymbol{\delta}^{(k)}\right)^\top B^{(k)}T^{-1}B^{(k)}\boldsymbol{\delta}^{(k)}}$$

$$
\begin{aligned}
=&\omega\frac{T^{-1}B^{(k)}\boldsymbol{\delta}^{(k)}\left(\boldsymbol{\delta}^{(k)}\right)^\top B^{(k)}T^{-1}}{\left[\left(\boldsymbol{\delta}^{(k)}\right)^\top\boldsymbol{\gamma}^{(k)}\right]^2}\\
=&\frac{\left(\omega I-H^{(k)}\boldsymbol{\gamma}^{(k)}(\boldsymbol{\gamma}^{(k)})^\top\right)\boldsymbol{\delta}^{(k)}\left(\boldsymbol{\delta}^{(k)}\right)^\top\left(\omega I-\boldsymbol{\gamma}^{(k)}\left(\boldsymbol{\gamma}^{(k)}\right)^\top H^{(k)}\right)}{\omega\left[\left(\boldsymbol{\delta}^{(k)}\right)^\top\boldsymbol{\gamma}^{(k)}\right]^2}\\
=&\frac{\omega^2\boldsymbol{\delta}^{(k)}\left(\boldsymbol{\delta}^{(k)}\right)^\top+H^{(k)}\boldsymbol{\gamma}^{(k)}\left(\boldsymbol{\gamma}^{(k)}\right)^\top\boldsymbol{\delta}^{(k)}\left(\boldsymbol{\delta}^{(k)}\right)^\top\boldsymbol{\gamma}^{(k)}\left(\boldsymbol{\gamma}^{(k)}\right)^\top H^{(k)}}{\omega\left[\left(\boldsymbol{\delta}^{(k)}\right)^\top\boldsymbol{\gamma}^{(k)}\right]^2}\\
&-\frac{\omega H^{(k)}\boldsymbol{\gamma}^{(k)}\left(\boldsymbol{\gamma}^{(k)}\right)^\top\boldsymbol{\delta}^{(k)}\left(\boldsymbol{\delta}^{(k)}\right)^\top+\omega\boldsymbol{\delta}^{(k)}\left(\boldsymbol{\delta}^{(k)}\right)^\top\boldsymbol{\gamma}^{(k)}\left(\boldsymbol{\gamma}^{(k)}\right)^\top H^{(k)}}{\omega\left[\left(\boldsymbol{\delta}^{(k)}\right)^\top\boldsymbol{\gamma}^{(k)}\right]^2}\\
=&\omega\frac{\boldsymbol{\delta}^{(k)}\left(\boldsymbol{\delta}^{(k)}\right)^\top}{\left[\left(\boldsymbol{\delta}^{(k)}\right)^\top\boldsymbol{\gamma}^{(k)}\right]^2}+\frac{H^{(k)}\boldsymbol{\gamma}^{(k)}\left(\boldsymbol{\gamma}^{(k)}\right)^\top H^{(k)}}{\omega}\\
&-\frac{H^{(k)}\boldsymbol{\gamma}^{(k)}\left(\boldsymbol{\delta}^{(k)}\right)^\top+\boldsymbol{\delta}^{(k)}\left(\boldsymbol{\gamma}^{(k)}\right)^\top H^{(k)}}{\left(\boldsymbol{\delta}^{(k)}\right)^\top\boldsymbol{\gamma}^{(k)}}.
\end{aligned}
\tag{2.9.57}
$$

Using (2.9.54), we obtain

$$
\omega\frac{\boldsymbol{\delta}^{(k)}\left(\boldsymbol{\delta}^{(k)}\right)^\top}{\left[\left(\boldsymbol{\delta}^{(k)}\right)^\top\boldsymbol{\gamma}^{(k)}\right]^2}=\frac{\left(\left(\boldsymbol{\delta}^{(k)}\right)^\top\boldsymbol{\gamma}^{(k)}+\left(\boldsymbol{\gamma}^{(k)}\right)^\top H^{(k)}\boldsymbol{\gamma}^{(k)}\right)\boldsymbol{\delta}^{(k)}\left(\boldsymbol{\delta}^{(k)}\right)^\top}{\left[\left(\boldsymbol{\delta}^{(k)}\right)^\top\boldsymbol{\gamma}^{(k)}\right]^2}.
\tag{2.9.58}
$$

Now, by (2.9.55), (2.9.52), (2.9.57) and (2.9.58), it follows that

$$
\begin{aligned}
H_{BFGS}^{(k+1)}=&H^{(k)}-\frac{H^{(k)}\boldsymbol{\gamma}^{(k)}\left(\boldsymbol{\gamma}^{(k)}\right)^\top H^{(k)}}{\omega}\\
&+\frac{\omega^2\boldsymbol{\delta}^{(k)}\left(\boldsymbol{\delta}^{(k)}\right)^\top+H^{(k)}\boldsymbol{\gamma}^{(k)}(\boldsymbol{\gamma}^{(k)})^\top\boldsymbol{\delta}^{(k)}\left(\boldsymbol{\delta}^{(k)}\right)^\top\boldsymbol{\gamma}^{(k)}\left(\boldsymbol{\gamma}^{(k)}\right)^\top H^{(k)}}{\omega\left[\left(\boldsymbol{\delta}^{(k)}\right)^\top\boldsymbol{\gamma}^{(k)}\right]^2}\\
&-\frac{\omega H^{(k)}\boldsymbol{\gamma}^{(k)}\left(\boldsymbol{\gamma}^{(k)}\right)^\top\boldsymbol{\delta}^{(k)}\left(\boldsymbol{\delta}^{(k)}\right)^\top+\omega\boldsymbol{\delta}^{(k)}\left(\boldsymbol{\delta}^{(k)}\right)^\top\boldsymbol{\gamma}^{(k)}\left(\boldsymbol{\gamma}^{(k)}\right)^\top H^{(k)}}{\omega\left[\left(\boldsymbol{\delta}^{(k)}\right)^\top\boldsymbol{\gamma}^{(k)}\right]^2}\\
=&H^{(k)}-\frac{H^{(k)}\boldsymbol{\gamma}^{(k)}\left(\boldsymbol{\gamma}^{(k)}\right)^\top H^{(k)}}{\omega}+\omega\frac{\boldsymbol{\delta}^{(k)}\left(\boldsymbol{\delta}^{(k)}\right)^\top}{\left[\left(\boldsymbol{\delta}^{(k)}\right)^\top\boldsymbol{\gamma}^{(k)}\right]^2}\\
&+\frac{H^{(k)}\boldsymbol{\gamma}^{(k)}\left(\boldsymbol{\gamma}^{(k)}\right)^\top H^{(k)}}{\omega}-\frac{H^{(k)}\boldsymbol{\gamma}^{(k)}\left(\boldsymbol{\delta}^{(k)}\right)^\top+\boldsymbol{\delta}^{(k)}\left(\boldsymbol{\gamma}^{(k)}\right)^\top H^{(k)}}{\left(\boldsymbol{\delta}^{(k)}\right)^\top\boldsymbol{\gamma}^{(k)}}
\end{aligned}
$$

$$
\begin{aligned}
&= H^{(k)} + \frac{\left[\left(\boldsymbol{\delta}^{(k)}\right)^\top \boldsymbol{\gamma}^{(k)} + \left(\boldsymbol{\gamma}^{(k)}\right)^\top H^{(k)} \boldsymbol{\gamma}^{(k)}\right] \boldsymbol{\delta}^{(k)} \left(\boldsymbol{\delta}^{(k)}\right)^\top}{\left[\left(\boldsymbol{\delta}^{(k)}\right)^\top \boldsymbol{\gamma}^{(k)}\right]^2} \\
&\quad - \frac{\boldsymbol{\delta}^{(k)} \left(\boldsymbol{\gamma}^{(k)}\right)^\top H^{(k)} + H^{(k)} \boldsymbol{\gamma}^{(k)} \left(\boldsymbol{\delta}^{(k)}\right)^\top}{\left(\boldsymbol{\delta}^{(k)}\right)^\top \boldsymbol{\gamma}^{(k)}} \\
&= \left[I - \frac{\boldsymbol{\delta}^{(k)} (\boldsymbol{\gamma}^{(k)})^\top}{\left(\boldsymbol{\delta}^{(k)}\right)^\top \boldsymbol{\gamma}^{(k)}}\right] H^{(k)} \left[I - \frac{\boldsymbol{\gamma}^{(k)} \left(\boldsymbol{\delta}^{(k)}\right)^\top}{\left(\boldsymbol{\delta}^{(k)}\right)^\top \boldsymbol{\gamma}^{(k)}}\right] \\
&\quad + \frac{\boldsymbol{\delta}^{(k)} \left(\boldsymbol{\delta}^{(k)}\right)^\top}{\left(\boldsymbol{\delta}^{(k)}\right)^\top \boldsymbol{\gamma}^{(k)}}.
\end{aligned}
\tag{2.9.59}
$$

This completes the derivation of the BFGS formula.

The next theorem contains the results on the convergence of the BFGS method. Its proof can be found in [232].

Theorem 2.9.4 *Consider the objective function $f : \mathbb{R}^n \to \mathbb{R}$. Suppose that the following conditions are satisfied:*

(a) f is twice continuously differentiable on an open convex set $D \subset \mathbb{R}^n$.
(b) There is a strong local minimizer $\boldsymbol{x}^* \in D$ such that $\nabla^2 f(\boldsymbol{x}^*)$ is symmetric and positive definite.
(c) There is a neighbourhood $\mathcal{N}_\varepsilon(\boldsymbol{x}^*)$ of $\boldsymbol{x}^*$ such that

$$\left\|\nabla^2 f(\widehat{\boldsymbol{x}}) - \nabla^2 f(\boldsymbol{x})\right\| \leq K_1 \left\|\widehat{\boldsymbol{x}} - \boldsymbol{x}\right\|, \quad \forall \boldsymbol{x},\, \widehat{\boldsymbol{x}} \in \mathcal{N}_\varepsilon(\boldsymbol{x}^*).$$

where K_1 is a positive constant.

Furthermore, for a positive constant K_2, suppose that the following condition is satisfied:

$$K_2 \left(\nabla^2 f(\boldsymbol{x}^*)\right)^{-1} \varphi\left(\boldsymbol{x}^{(k)}, \boldsymbol{x}^{(k+1)}\right) \leq \frac{1}{3}$$

in $\mathcal{N}_\varepsilon(\boldsymbol{x}^*)$, where

$$\varphi\left(\boldsymbol{x}^{(k)}, \boldsymbol{x}^{(k+1)}\right) = \max\left\{\boldsymbol{x}^{(k)} - \boldsymbol{x}^*, \boldsymbol{x}^{(k+1)} - \boldsymbol{x}^*\right\}.$$

Then, the sequence $\left\{\boldsymbol{x}^{(k)}\right\}$ generated by the BFGS method converges to $\boldsymbol{x}^*$ superlinearly.

In closing the chapter, we make some comments on the conjugate gradient methods and quasi-Newton methods in the following remark:

Remark 2.9.4 *If n is large, we may have problems storing the approximation to the inverse of Hessian in quasi-Newton methods. Thus, conjugate gradient methods are preferred. On the other hand, if n is not too large, then quasi-Newton methods tend to perform better.*

Chapter 3
Constrained Mathematical Programming

3.1 Introduction

The general constrained mathematical programming problem is to find an $\boldsymbol{x} \in \mathbb{R}^n$ to minimize the objective function

$$f(\boldsymbol{x}) \tag{3.1.1}$$

subject to the constraints

$$h_i(\boldsymbol{x}) = 0, \quad i = 1, \ldots, m, \tag{3.1.2}$$

$$h_i(\boldsymbol{x}) \leq 0, \quad i = m+1, \ldots, m+r, \tag{3.1.3}$$

where f and h_i, $i = 1, \ldots, m+r$, are continuously differentiable functions of the n-vector variable $\boldsymbol{x}$.

As mentioned in Chapter 2, there are already many excellent books on nonlinear optimization. For example, see [22], [61], [71], [172], [232] and [275]. In this chapter, we summarize some essential concepts and results in nonlinear constrained optimization methods. As for Chapter 2, this chapter is based on the lecture notes on optimization prepared and used by the authors. These lecture notes have also been used by their colleagues. In addition to these lecture notes, this chapter includes also some important results from those references mentioned above and in Section 2.1 of Chapter 2.

Definition 3.1.1 *A point $\boldsymbol{x}$ is said to be* feasible *if it satisfies the constraints* (3.1.2) *and* (3.1.3). *The set of all feasible points is called the* feasible region (or feasible set) *of the constraints. Let Ω denote the feasible region (or feasible set) throughout this chapter.*

Definition 3.1.2 *The j-th inequality constraint, $j \in \{m+1, m+2, \ldots, m+r\}$, is said to be* active *at $\boldsymbol{x}^*$ if $h_j(\boldsymbol{x}^*) = 0$.*

K. L. Teo et al., *Applied and Computational Optimal Control*, Springer Optimization and Its Applications 171,
https://doi.org/10.1007/978-3-030-69913-0_3

Definition 3.1.3 *Let $\boldsymbol{x}^*$ be a given point in $\mathbb{R}^n$. Then the* active set, $J(\boldsymbol{x}^*)$, *of inequality constraints at $\boldsymbol{x}^*$ is the set of indices corresponding to all those inequality constraints that are active, i.e.,*

$$J(\boldsymbol{x}^*) = \{j \in \{m+1, m+2, \ldots, m+r\} : h_j(\boldsymbol{x}^*) = 0\}. \tag{3.1.4}$$

Definition 3.1.4 *The point $\boldsymbol{x}^*$ is said to be a* regular point *of the constraints* (3.1.2)-(3.1.3) *if $\boldsymbol{x}^*$ satisfies all of the constraints and if the gradients of the equality and active inequality constraints*

$$\{\nabla_{\boldsymbol{x}} h_j(\boldsymbol{x}^*),\ j \in \{1, \ldots, m\} \cup J(\boldsymbol{x}^*)\} \tag{3.1.5}$$

are linearly independent, where

$$\nabla_{\boldsymbol{x}} h_j(\boldsymbol{x}^*) = \left.\frac{\partial h_j(\boldsymbol{x})}{\partial \boldsymbol{x}}\right|_{\boldsymbol{x}=\boldsymbol{x}^*}, \quad j = 1, 2, \ldots, m+r.$$

Remark 3.1.1 *Note that the condition stated in Definition 3.1.4 is known as* constraint qualification.

Definition 3.1.5 *The* Lagrangian *of the constrained optimization problem* (3.1.1)–(3.1.3) *is defined by*

$$L(\boldsymbol{x}, \boldsymbol{\lambda}) = f(\boldsymbol{x}) + \sum_{j=1}^{m} \lambda_j h_j(\boldsymbol{x}) + \sum_{j=m+1}^{m+r} \lambda_j h_j(\boldsymbol{x}), \tag{3.1.6}$$

where $\boldsymbol{\lambda} = [\lambda_1, \lambda_2, \ldots, \lambda_{m+r}]^\top$ is the vector of Lagrange multipliers.

Definition 3.1.6 *A point $\boldsymbol{x}^* \in \Omega$ is said to be a* relative minimum point *or a* local minimum point *of f over the feasible region Ω if there exists an $\varepsilon > 0$ such that $f(\boldsymbol{x}) \geq f(\boldsymbol{x}^*)$ for all $\boldsymbol{x} \in \mathcal{N}_\varepsilon(\boldsymbol{x}^*)$, where*

$$\mathcal{N}_\varepsilon(\boldsymbol{x}^*) = \{\boldsymbol{x} \in \Omega : |\boldsymbol{x} - \boldsymbol{x}^*| < \varepsilon\}.$$

If $f(\boldsymbol{x}) > f(\boldsymbol{x}^*)$ for all $\boldsymbol{x} \in \mathcal{N}_\varepsilon(\boldsymbol{x}^*)$ such that $\boldsymbol{x} \neq \boldsymbol{x}^*$, then $\boldsymbol{x}^*$ is said to be a strict local minimum point *of f over the feasible region Ω.*

We now state the well-known Karush-Kuhn-Tucker theorem without proof.

Theorem 3.1.1 (First Order Necessary Optimality Condition) *Let $\boldsymbol{x}^*$ be a local minimum point of Problem* (3.1.1)–(3.1.3). *If it is also a regular point of the constraints* (3.1.2)–(3.1.3), *then there exist λ_j^*, $j = 1, 2, \ldots, m+r$, not all equal to zero, such that*

$$\nabla_{\boldsymbol{x}} L(\boldsymbol{x}^*, \boldsymbol{\lambda}^*) = \left.\frac{\partial L(\boldsymbol{x}, \boldsymbol{\lambda}^*)}{\partial \boldsymbol{x}}\right|_{\boldsymbol{x}=\boldsymbol{x}^*} = \mathbf{0}^\top, \tag{3.1.7}$$

$$h_j(\boldsymbol{x}^*) = 0, \quad j = 1, 2, \ldots, m, \tag{3.1.8a}$$
$$h_j(\boldsymbol{x}^*) \leq 0, \quad j = m+1, m+2, \ldots, m+r, \tag{3.1.8b}$$

$$\lambda_j^* h_j(\boldsymbol{x}^*) = 0, \quad \lambda_j^* \geq 0, \quad j = m+1, m+2, \ldots, m+r. \tag{3.1.9}$$

In view of condition (3.1.9), we note that if the j-th inequality constraint is inactive, then $\lambda_j^* = 0$; and conversely, if $\lambda_j^* > 0$, then the *j-th* inequality constraint must be active.

In what follows, we assume that f and $h_j, j = 1, 2, \ldots, m+r$, are twice continuously differentiable.

Theorem 3.1.2 (Second Order Sufficient Optimality Condition) *The point* $\boldsymbol{x}^*$ *is a local minimum of Problem* (3.1.1)–(3.1.3) *if the conditions* (3.1.7)–(3.1.9) *are satisfied and, in addition, the Hessian,* $H^* = \nabla_{\boldsymbol{xx}} L(\boldsymbol{x}^*, \boldsymbol{\lambda}^*)$*, of the Lagrangian satisfies*

$$\boldsymbol{y}^\top H^* \boldsymbol{y} > 0, \quad \forall \boldsymbol{y} \in M^*, \quad \boldsymbol{y} \neq \boldsymbol{0}, \tag{3.1.10}$$

where $\nabla_{\boldsymbol{xx}} = (\nabla_{\boldsymbol{x}})^\top \nabla_{\boldsymbol{x}}$,

$$M^* = \{\boldsymbol{y} \in \mathbb{R}^n : \nabla_{\boldsymbol{x}} h_j(\boldsymbol{x}^*) \boldsymbol{y} = 0, \ j \in \{1, \ldots, m\} \cup J_+(\boldsymbol{x}^*)\} \tag{3.1.11}$$

and

$$J_+(\boldsymbol{x}^*) = \left\{ j \in J(\boldsymbol{x}^*) : \lambda_j^* > 0 \right\}. \tag{3.1.12}$$

If the Lagrange multipliers corresponding to all active inequality constraints are strictly positive (i.e., $J_+(\boldsymbol{x}^*) = J(\boldsymbol{x}^*)$), then M^* is a subspace, which is the tangent plane of the active constraints (including active inequality constraints and equality constraints). In this situation, we note, by virtue of condition (3.1.10), that the Hessian $H^*$of the Lagrangian must be positive definite on this subspace.

To solve large-scale constrained optimization problems, efficient numerical algorithms are required. There are many such algorithms, and each of them is efficient in its respective area of application. We refer the reader to several classics, such as [22], [54], [61], [71], [172], [187], [232] and [275] for detailed treatments of computational aspects of constrained optimization problems. We shall, however, elaborate somewhat on the method of sequential quadratic programming (see [61], [221], [222], [232] and [275]) in this chapter. This method has been recognized as one of the most efficient methods for solving small and medium size constrained optimization problems (see [222] and [232]). As the name suggests, the method of sequential quadratic programming computes the optimal solution as a sequence of quadratic programming problems. Each quadratic programming problem is solved, yielding the optimum search direction based on analytical gradient information for the objective and constraint functions. The quadratic programming problem to be solved at each step involves both linear equality and linear inequality constraints. In the next section, we discuss some basic techniques for solving a quadratic programming problem with only linear equality constraints. The solution of quadratic programming problems with both linear equality and linear inequality constraints via the active set strategy is outlined in Section 3.3. In Section 3.4, we summarize a constrained quasi-Newton method for solving a general linearly constrained optimization problem. In Section 3.5, we summarize the essential steps required in the sequential quadratic programming algorithm, making use of the materials outlined in Sections 3.2–3.4.

3.2 Quadratic Programming with Linear Equality Constraints

In this section, we consider a general class of quadratic optimization problems with linear equality constraints:

$$\text{minimize} \quad f(\boldsymbol{x}) = \frac{1}{2}\boldsymbol{x}^\top G\boldsymbol{x} + \boldsymbol{c}^\top \boldsymbol{x} \tag{3.2.1}$$

subject to

$$\boldsymbol{h}(\boldsymbol{x}) = A\boldsymbol{x} - \boldsymbol{b} = \boldsymbol{0}, \tag{3.2.2}$$

where $\boldsymbol{x} \in \mathbb{R}^n$ is the decision vector, $\boldsymbol{c} \in \mathbb{R}^n$, $G = G^\top \in \mathbb{R}^{n\times n}$, $A \in \mathbb{R}^{m\times n}$, $\boldsymbol{b} \in \mathbb{R}^m$, $m < n$ and $\text{rank}(A) = m$. This problem is referred to as Problem (QPE).

Note that Problem (QPE) without the linear constraint (3.2.2) is an unconstrained quadratic programming problem. It can be solved readily if the Hessian of f (which is equal to G) is positive definite. The first order necessary condition for optimality is that the gradient vanishes, i.e.,

$$\nabla_{\boldsymbol{x}} f(\boldsymbol{x}) = \frac{\partial f(\boldsymbol{x})}{\partial \boldsymbol{x}} = \boldsymbol{x}^\top G + \boldsymbol{c}^\top = \boldsymbol{0}^\top. \tag{3.2.3}$$

If the Hessian $\nabla_{\boldsymbol{x}\boldsymbol{x}} f = G$ is positive definite, then, by solving (3.2.3), we obtain the unique minimum solution:

$$\boldsymbol{x}^* = -G^{-1}\boldsymbol{c}. \tag{3.2.4}$$

There are many ways of solving Problem (QPE). We consider three of these: (1) direct elimination, (2) generalized elimination and (3) the method of Lagrange multipliers.

The direct elimination method seeks to remove the linear constraints by solving for some m independent variables in terms of the remaining $n - m$ variables. This subsequently reduces the problem to an unconstrained one that can be solved as shown in Chapter 2. Since A is of full rank, we can rearrange the order of the variables such that the first m columns of A are linearly independent, i.e.,

$$A = [A_1|A_2], \tag{3.2.5}$$

where $A_1 \in \mathbb{R}^{m\times m}$ is non-singular. Let

$$\boldsymbol{x} = \begin{bmatrix} \boldsymbol{x}_1 \\ \boldsymbol{x}_2 \end{bmatrix},$$

where $\boldsymbol{x}_1 \in \mathbb{R}^m$ and $\boldsymbol{x}_2 \in \mathbb{R}^{n-m}$. Then, by virtue of (3.2.2), we have

$$\boldsymbol{x}_1 = A_1^{-1}(\boldsymbol{b} - A_2\boldsymbol{x}_2). \tag{3.2.6}$$

Now we can partition the matrix G and the vector $\boldsymbol{c}$ accordingly:

$$G = \begin{bmatrix} G_{11} & G_{12} \\ G_{12}^\top & G_{22} \end{bmatrix} \quad \text{and} \quad \boldsymbol{c} = \begin{bmatrix} \boldsymbol{c}_1 \\ \boldsymbol{c}_2 \end{bmatrix}, \tag{3.2.7}$$

where $G_{11} \in \mathbb{R}^{m\times m}$, $G_{12} \in \mathbb{R}^{m\times(n-m)}$, $G_{22} \in \mathbb{R}^{(n-m)\times(n-m)}$, $\boldsymbol{c}_1 \in \mathbb{R}^m$ and $\boldsymbol{c}_2 \in \mathbb{R}^{n-m}$. Substitution of (3.2.6) and (3.2.7) into (3.2.1) yields

$$\begin{aligned} f(\boldsymbol{x}) =& \bar{f}(\boldsymbol{x}_2) \\ =& \frac{1}{2}\boldsymbol{x}_2^\top \Big[G_{22} + A_2^\top \left(A_1^{-1}\right)^\top G_{11}A_1^{-1}A_2 - A_2^\top \left(A_1^{-1}\right)^\top G_{12} \\ & - G_{12}^\top A_1^{-1}A_2\Big]\boldsymbol{x}_2 + \Big[\boldsymbol{b}^\top \left(A_1^{-1}\right)^\top G_{12} - \boldsymbol{b}^\top \left(A_1^{-1}\right)^\top G_{11}A_1^{-1}A_2 \\ & + \boldsymbol{c}_2^\top - \boldsymbol{c}_1^\top A_1^{-1}A_2\Big]\boldsymbol{x}_2 + \boldsymbol{c}_1^\top A_1^{-1}\boldsymbol{b} + \frac{1}{2}\boldsymbol{b}^\top \left(A_1^{-1}\right)^\top G_{11}A_1^{-1}\boldsymbol{b}. \end{aligned} \tag{3.2.8}$$

The first order necessary condition for stationarity of $\bar{f}(\boldsymbol{x}_2)$ shows that

$$\boldsymbol{x}_2^* = -\tilde{G}^{-1}\left[G_{12}^\top A_1^{-1}\boldsymbol{b} - A_2^\top \left(A_1^{-1}\right)^\top G_{11}A_1^{-1}\boldsymbol{b} + \boldsymbol{c}_2 - A_2^\top \left(A_1^{-1}\right)^\top \boldsymbol{c}_1\right], \tag{3.2.9}$$

where

$$\tilde{G} = \left[G_{22} + A_2^\top \left(A_1^{-1}\right)^\top G_{11}A_1^{-1}A_2 - A_2^\top \left(A_1^{-1}\right)^\top G_{12} - G_{12}^\top A_1^{-1}A_2\right]. \tag{3.2.10}$$

By the second order sufficient condition, $\boldsymbol{x}_2^*$ will be a minimizing (both locally and globally) solution of $\bar{f}$ if $\tilde{G}$ is positive definite. However, this assumption is not necessarily satisfied in general. The remaining solution is constructed from (3.2.6) by substitution of (3.2.9).

The direct elimination method is easy to understand, but it requires a significant amount of computation. The method of generalized elimination seeks to reduce this computational effort somewhat. Let E and F be real matrices of dimension $n \times m$ and $n \times (n-m)$, respectively, such that

$$[E|F] \quad \text{is non-singular}, \tag{3.2.11}$$

$$AE = I_m \tag{3.2.12}$$

and

$$AF = 0. \tag{3.2.13}$$

Note that E and F are not necessarily unique. In practice, E and F may be obtained by first selecting any matrix $Q \in \mathbb{R}^{n\times(n-m)}$ such that $\left[A^\top|Q\right]$ is non-singular. Defining

$$[E|F] = \left[\left[A^\top|Q\right]^{-1}\right]^\top, \tag{3.2.14}$$

it is then easy to verify that E and F satisfy (3.2.11), (3.2.12) and (3.2.13). Equation (3.2.13) implies that the columns of F are basis vectors for the null space of A, defined by $\{\boldsymbol{x} \in \mathbb{R}^n : A\boldsymbol{x} = 0\}$. The general solution of (3.2.2) can then be written as

$$\boldsymbol{x} = E\boldsymbol{b} + F\boldsymbol{y}, \tag{3.2.15}$$

where $\boldsymbol{y} \in \mathbb{R}^{n-m}$ is arbitrary. Substitution of (3.2.15) into (3.2.1) yields

$$f(\boldsymbol{x}) = \hat{f}(\boldsymbol{y}) = \frac{1}{2}\boldsymbol{y}^\top F^\top GF\boldsymbol{y} + (\boldsymbol{c} + GE\boldsymbol{b})^\top F\boldsymbol{y} + \left(\boldsymbol{c} + \frac{1}{2}GE\boldsymbol{b}\right)^\top E\boldsymbol{b}. \tag{3.2.16}$$

Clearly, if $F^\top GF$ is positive definite, then the unique minimizer $\boldsymbol{y}^*$ is given by

$$\boldsymbol{y}^* = -\left(F^\top GF\right)^{-1} F^\top (\boldsymbol{c} + GE\boldsymbol{b}). \tag{3.2.17}$$

The matrix $F^\top GF$ is referred to as the *reduced Hessian matrix*, while the vector $F^\top (\boldsymbol{c} + GE\boldsymbol{b})$ is referred to as the *reduced gradient*. $\boldsymbol{x}^*$ can then be easily obtained from (3.2.15).

The third method employs the idea of Lagrange multipliers. The Lagrangian function for the constrained problem (3.2.1) and (3.2.2) is

$$L(\boldsymbol{x}, \boldsymbol{\lambda}) = f(\boldsymbol{x}) + \boldsymbol{\lambda}^\top (A\boldsymbol{x} - \boldsymbol{b}). \tag{3.2.18}$$

Let $\boldsymbol{x}^*$ and $\boldsymbol{\lambda}^*$ denote the optimal values of $\boldsymbol{x}$ and $\boldsymbol{\lambda}$, respectively. Then the necessary conditions for optimality require that the gradients of L with respect to both $\boldsymbol{x}$ and $\boldsymbol{\lambda}$ vanish at $\boldsymbol{x} = \boldsymbol{x}^*$ and $\boldsymbol{\lambda} = \boldsymbol{\lambda}^*$, i.e.,

$$\nabla_{\boldsymbol{x}} L(\boldsymbol{x}^*, \boldsymbol{\lambda}^*) = (\boldsymbol{x}^*)^\top G + \boldsymbol{c}^\top + (\boldsymbol{\lambda}^*)^\top A = \boldsymbol{0}^\top, \tag{3.2.19}$$

$$\nabla_{\boldsymbol{\lambda}} L(\boldsymbol{x}^*, \boldsymbol{\lambda}^*) = (A\boldsymbol{x}^* - \boldsymbol{b})^\top = \boldsymbol{0}^\top. \tag{3.2.20}$$

Equations (3.2.19) and (3.2.20) constitute a set of linear simultaneous equations that can be expressed as

$$\begin{bmatrix} G & A^\top \\ A & 0 \end{bmatrix} \begin{bmatrix} \boldsymbol{x}^* \\ \boldsymbol{\lambda}^* \end{bmatrix} = \begin{bmatrix} -\boldsymbol{c} \\ \boldsymbol{b} \end{bmatrix}. \tag{3.2.21}$$

If G is positive definite and A has full rank, then it follows that

$$\begin{bmatrix} G & A^\top \\ A & 0 \end{bmatrix}$$

is non-singular and its inverse is given by

$$\begin{bmatrix} G & A^\top \\ A & 0 \end{bmatrix}^{-1} = \begin{bmatrix} H & P^\top \\ P & S \end{bmatrix}, \tag{3.2.22}$$

where

$$H = G^{-1} - G^{-1}A^\top \left(AG^{-1}A^\top\right)^{-1} AG^{-1}, \tag{3.2.23}$$

$$P = \left(AG^{-1}A^\top\right)^{-1} AG^{-1}, \tag{3.2.24}$$

$$S = -\left(AG^{-1}A^\top\right)^{-1}. \tag{3.2.25}$$

The solution of Problem (QPE) may then be expressed as

$$\boldsymbol{x}^* = -H\boldsymbol{c} + P^\top \boldsymbol{b}, \tag{3.2.26}$$

$$\boldsymbol{\lambda}^* = -P\boldsymbol{c} + S\boldsymbol{b}. \tag{3.2.27}$$

If we substitute (3.2.23) and (3.2.24) into (3.2.26), we have

$$\boldsymbol{x}^* = -G^{-1}\boldsymbol{c} + G^{-1}A^\top \left(AG^{-1}A^\top\right)^{-1} \left[AG^{-1}\boldsymbol{c} + \boldsymbol{b}\right], \tag{3.2.28}$$

which is a much more elegant form than that obtained by the direct elimination method in (3.2.6) and (3.2.9). The corresponding Lagrange multipliers, $\boldsymbol{\lambda}^*$, may be obtained similarly.

Finally, note that $\boldsymbol{x}^*$ and $\boldsymbol{\lambda}^*$ can also be generated by first finding any solution $\boldsymbol{x}$ that satisfies the constraints, i.e.,

$$A\boldsymbol{x} = \boldsymbol{b}. \tag{3.2.29}$$

Then

$$\boldsymbol{x}^* = \boldsymbol{x} - H\boldsymbol{p} \tag{3.2.30}$$

and

$$\boldsymbol{\lambda}^* = -P\boldsymbol{p}, \tag{3.2.31}$$

where

$$\boldsymbol{p} = \left(\nabla_{\boldsymbol{x}} f(\boldsymbol{x})\right)^\top = G\boldsymbol{x} + \boldsymbol{c}, \tag{3.2.32}$$

because

$$\begin{aligned} &\boldsymbol{x} - H\boldsymbol{p} \\ &\quad = \boldsymbol{x} - H(G\boldsymbol{x} + \boldsymbol{c}) \\ &\quad = \boldsymbol{x} - HG\boldsymbol{x} - H\boldsymbol{c} \\ &\quad = \boldsymbol{x} - \left(G^{-1} - G^{-1}A^\top \left(AG^{-1}A^\top\right)^{-1} AG^{-1}\right) G\boldsymbol{x} - H\boldsymbol{c} \ \text{(from (3.2.23))} \end{aligned}$$

$$= \boldsymbol{x} - G^{-1}G\boldsymbol{x} + G^{-1}A^\top \left(AG^{-1}A^\top\right)^{-1} AG^{-1}G\boldsymbol{x} - H\boldsymbol{c}$$
$$= G^{-1}A^\top \left(AG^{-1}A^\top\right)^{-1} \boldsymbol{b} - H\boldsymbol{c}$$
$$= -H\boldsymbol{c} + P^\top \boldsymbol{b} \text{ (from (3.2.24))}$$
$$= \boldsymbol{x}^* \text{ (from (3.2.26))}$$

and

$$\begin{aligned}
-P\boldsymbol{p} &= -P\left(G\boldsymbol{x} + \boldsymbol{c}\right) \\
&= -PG\boldsymbol{x} - P\boldsymbol{c} \\
&= -\left(AG^{-1}A^\top\right)^{-1} AG^{-1}G\boldsymbol{x} - P\boldsymbol{c} \text{ (from (3.2.24))} \\
&= -\left(AG^{-1}A^\top\right)^{-1} \boldsymbol{b} - P\boldsymbol{c} \text{ (from (3.2.29))} \\
&= -P\boldsymbol{c} + S\boldsymbol{b} \text{ (from (3.2.25))} \\
&= \boldsymbol{\lambda}^* \text{ (from (3.2.27))}.
\end{aligned}$$

3.3 Quadratic programming via Active Set Strategy

Problem (QPE) only involves linear equality constraints. In this section, let us consider the following general quadratic programming problem that also involves linear inequality constraints.

$$\text{minimize} \quad f\left(\boldsymbol{x}\right) = \frac{1}{2}\boldsymbol{x}^\top G\boldsymbol{x} + \boldsymbol{c}^\top \boldsymbol{x} \tag{3.3.1a}$$

subject to

$$h_i\left(\boldsymbol{x}\right) = \boldsymbol{a}_i^\top \boldsymbol{x} - b_i = 0, \quad i \in \mathcal{E}, \tag{3.3.1b}$$
$$h_i\left(\boldsymbol{x}\right) = \boldsymbol{a}_i^\top \boldsymbol{x} - b_i \leq 0, \quad i \in \mathcal{I}. \tag{3.3.1c}$$

This problem is referred to as Problem (QP), where the constraints have been partitioned into two sets. $\mathcal{E} = \{1, 2, \ldots, m\}$ is the set of indices corresponding to the equality constraints and $\mathcal{I} = \{m+1, 2, \ldots, m+r\}$ is the set of indices corresponding to the inequality constraints. Since the objective function is quadratic, the Hessian G is constant. The constraints are linear so that their gradients, $\nabla_{\boldsymbol{x}} h_i\left(\boldsymbol{x}\right) = \boldsymbol{a}_i^\top$, $i = 1, 2, \ldots, m+r$, are also constants. If G is positive definite, the problem is convex. In this case, the first order Karush-Kuhn-Tucker necessary conditions (3.1.7)–(3.1.9) are both necessary and sufficient. Thus, there exists a vector, $\boldsymbol{\lambda}^*$, of Lagrange multipliers, not all equal to zero, such that

$$G\boldsymbol{x}^* + \boldsymbol{c} = -\sum_{i \in \mathcal{E} \cup \mathcal{I}} \lambda_i^* \boldsymbol{a}_i, \tag{3.3.2}$$
$$h_i\left(\boldsymbol{x}^*\right) = 0, \quad i \in \mathcal{E}, \tag{3.3.3}$$

$$h_i(\boldsymbol{x}^*) \le 0, \quad i \in \mathcal{I}, \tag{3.3.4}$$

$$\lambda_i^* h_i(\boldsymbol{x}^*) = 0, \quad i \in \mathcal{I}, \tag{3.3.5}$$

$$\lambda_i^* \ge 0, \quad i \in \mathcal{I}. \tag{3.3.6}$$

Due to the existence of inequality constraints, Problem (QP) cannot be solved by any elimination method. Instead, it is solved via a series of Problems (QPE) via the *active set strategy.* To begin, we require some definitions.

A point $\boldsymbol{x}$ is said to be a *feasible solution* of Problem (QP) if it satisfies the constraints (3.3.1b) and (3.3.1c). Since all of these constraints are linear, it is easy to compute a feasible solution using the first phase of the simplex algorithm for linear programming.

The active set strategy is an iterative process. In the following definitions, the superscript $^{(k)}$ refers to a quantity at the k-th iteration of this process.

Definition 3.3.1 *The active set,* $\mathcal{A}^{(k)}$*, is defined by*

$$\mathcal{A}^{(k)} = \left\{ j \in \mathcal{E} \cup \mathcal{I} : h_j\left(\boldsymbol{x}^{(k)}\right) = \boldsymbol{a}_j^\top \boldsymbol{x}^{(k)} - b_j = 0 \right\}, \tag{3.3.7}$$

where $\boldsymbol{x}^{(k)}$ *is the* k*-th iterate solution.*

Definition 3.3.2 *The active constraint matrix,* $A^{(k)}$*, is defined by*

$$\begin{aligned} A^{(k)} &= \textit{Matrix formed by columns } \left[\nabla_{\boldsymbol{x}} h_i\left(\boldsymbol{x}^{(k)}\right)\right]^\top, \quad i \in \mathcal{A}^{(k)}, \\ &= \left[\boldsymbol{a}_i, \quad i \in \mathcal{A}^{(k)}\right]. \end{aligned} \tag{3.3.8}$$

Remark 3.3.1 *In view of* (3.3.1a)–(3.3.1c)*, we note that*

$$f\left(\boldsymbol{x}^{(k)} + \boldsymbol{d}\right) = f\left(\boldsymbol{x}^{(k)}\right) + \boldsymbol{d}^\top\left(G\boldsymbol{x}^{(k)} + \boldsymbol{c}\right) + \frac{1}{2}\boldsymbol{d}^\top G\boldsymbol{d}$$

and

$$h_i\left(\boldsymbol{x}^{(k)} + \boldsymbol{d}\right) = \boldsymbol{a}_i^\top\left(\boldsymbol{x}^{(k)} + \boldsymbol{d}\right) - b_i = \boldsymbol{a}_i^\top \boldsymbol{d} + h_i\left(\boldsymbol{x}^{(k)}\right).$$

The active set strategy can be summarized in the following algorithm.

Algorithm 3.3.1

Step 1. Set $k = 0$*. Select an initial feasible solution* $\boldsymbol{x}^{(0)}$ *and identify the corresponding active set* $\mathcal{A}^{(0)}$*.*

Step 2. Compute the search direction $\boldsymbol{d}^{(k)}$ *by solving the problem*

$$\textit{minimize } \frac{1}{2}\boldsymbol{d}^\top G\boldsymbol{d} + \boldsymbol{d}^\top\left(G\boldsymbol{x}^{(k)} + \boldsymbol{c}\right) \tag{3.3.9a}$$

$$\textit{subject to } \boldsymbol{a}_i^\top \boldsymbol{d} = 0, \quad i \in \mathcal{A}^{(k)}. \tag{3.3.9b}$$

If $\boldsymbol{d} = \boldsymbol{0}$ *solves problem* (3.3.9)*, go to Step 3. Otherwise, go to Step 4.*

Step 3. Since $\boldsymbol{d} = \mathbf{0}$ *solves problem* (3.3.9), $\boldsymbol{x}^{(k)}$ *solves the problem*

$$\text{minimize} \quad \frac{1}{2}\boldsymbol{x}^\top G\boldsymbol{x} + \boldsymbol{c}^\top \boldsymbol{x} \tag{3.3.10a}$$

$$\text{subject to} \quad \boldsymbol{a}_i^\top \boldsymbol{x} = b_i, \quad i \in \mathcal{A}^{(k)}. \tag{3.3.10b}$$

Compute the corresponding Lagrange multiplier vector (see (3.3.2)*):*

$$\boldsymbol{\lambda}^{(k)} = \left[\lambda_i^{(k)}, \quad i \in \mathcal{A}^{(k)}\right]^\top \tag{3.3.11}$$

by solving

$$-A^{(k)}\boldsymbol{\lambda}^{(k)} = G\boldsymbol{x}^{(k)} + \boldsymbol{c}, \tag{3.3.12}$$

where $A^{(k)}$ *is the active constraint matrix defined by* (3.3.8). *Select* j *such that*

$$\lambda_j^{(k)} = \min_{i \in \mathcal{A}^{(k)} \cap \mathcal{I}} \lambda_i^{(k)}. \tag{3.3.13}$$

If $\lambda_j^{(k)} \geq 0$, *terminate with* $\boldsymbol{x}^* = \boldsymbol{x}^{(k)}$. *Otherwise, set*

$$\mathcal{A}^{(k)} := \mathcal{A}^{(k)} \setminus \{j\} \tag{3.3.14}$$

and go to Step 2.

Step 4. Let $\mathbf{d}^{(k)} = \boldsymbol{d} \neq \mathbf{0}$ *be the solution to problem* (3.3.9). *Then, compute* $\alpha^{(k)}$ *according to* $\alpha^{(k)} = \min\{1, \bar{\alpha}^{(k)}\}$, *where*

$$\bar{\alpha}^{(k)} = \min_{\substack{i \in \mathcal{I} \setminus \mathcal{A}^{(k)} \\ \boldsymbol{a}_i^\top \boldsymbol{d}^{(k)} > 0}} \frac{b_i - \boldsymbol{a}_i^\top \boldsymbol{x}^{(k)}}{\boldsymbol{a}_i^\top \boldsymbol{d}^{(k)}}, \tag{3.3.15}$$

and set

$$\boldsymbol{x}^{(k+1)} = \boldsymbol{x}^{(k)} + \alpha^{(k)}\boldsymbol{d}^{(k)}. \tag{3.3.16}$$

If $\alpha^{(k)} < 1$, *set*

$$\mathcal{A}^{(k+1)} = \mathcal{A}^{(k)} \cup \{p\}, \tag{3.3.17}$$

where $p \in \mathcal{I} \setminus \mathcal{A}^{(k)}$ *is the index that achieves the minimum in* (3.3.15). *Otherwise, if* $\alpha^{(k)} = 1$, *set* $\mathcal{A}^{(k+1)} = \mathcal{A}^{(k)}$.

Step 5. Set $k := k + 1$ *and return to Step 2.*

It should be noted that if G is not positive definite, the stationary points may not be local minima. Special techniques are required for solving this class of indefinite quadratic programming problems.

The rationale underlying the steps in the active set strategy is as follows. At each iteration, the method seeks to solve a problem with equality constraints only, corresponding to the active set. At the k-th iteration, if all of the Lagrange multipliers associated with the active set are non-negative, the Karush-Kuhn-Tucker necessary condition is satisfied and a local optimum is reached. Otherwise, if some or all of the multipliers are strictly negative, the constraint corresponding to the most negative multiplier is removed from the active set. This procedure is carried out in Step 3 of the algorithm. Steps 2 and 4 describe the details of solving the equality only constrained problem associated with the active set. The iterate $\boldsymbol{x}^{(k)}$ may or may not be an optimal solution to this problem. We may shift the origin of the coordinate system to $\boldsymbol{x}^{(k)}$ and check if any non-zero local perturbation $\boldsymbol{d}$ solves the corresponding shifted problem (3.3.9). If the optimal solution is $\boldsymbol{d} = \boldsymbol{0}$, then proceed to Step 3 to check for the Lagrange multipliers. If $\boldsymbol{d} = \boldsymbol{d}^{(k)} \neq \boldsymbol{0}$, then we can reduce the cost function by updating $\boldsymbol{x}^{(k)}$ to $\boldsymbol{x}^{(k)} + \boldsymbol{d}^{(k)}$. This step may, however, cause some of the constraints to be violated. To prevent constraint violation, we change the update to $\boldsymbol{x}^{(k)} + \alpha^{(k)}\boldsymbol{d}^{(k)}$, where $\alpha^{(k)}$ is chosen such that the first non-active constraint in $\mathcal{I} \setminus \mathcal{A}^{(k)}$ becomes active. This is done only for constraints that increase in the $\boldsymbol{d}^{(k)}$ direction, i.e., for $\boldsymbol{a}_i^\top \boldsymbol{d}^{(k)} > 0$. The new active constraint is then included in the active set. The whole procedure is then repeated by returning to Step 2 after updating $\boldsymbol{x}^{(k+1)} = \boldsymbol{x}^{(k)} + \alpha^{(k)}\boldsymbol{d}^{(k)}$ and $k := k + 1$ in Step 5.

Note that in the computation of Lagrange multipliers associated with the active set in (3.3.12), the set of linear algebraic equations need not be solved independently at each step since only one constraint is removed or added each time. A pivoting strategy similar to that used in the simplex algorithm for linear programming can be used to significantly reduce the amount of computational effort required.

Example 3.3.1 Find an $\boldsymbol{x} \in \mathbb{R}^2$ such that the objective function

$$f(\boldsymbol{x}) = x_1^2 + x_2^2 - 4x_1 - 5x_2 + 2$$

is minimized subject to

$$\begin{aligned} h_1(\boldsymbol{x}) &= 2x_1 + x_2 - 2 \leq 0, \\ h_2(\boldsymbol{x}) &= -x_1 \leq 0, \\ h_3(\boldsymbol{x}) &= -x_2 \leq 0. \end{aligned}$$

Suppose we start at the feasible point $\boldsymbol{x}^{(0)} = 0$. The relevant gradients are

$$\begin{aligned} &(\nabla_{\boldsymbol{x}} f)^\top = \boldsymbol{g}(\boldsymbol{x}) = \begin{bmatrix} 2x_1 - 4 \\ 2x_2 - 5 \end{bmatrix}, \; \nabla_{\boldsymbol{x}} h_1 = [2, 1], \\ &\nabla_{\boldsymbol{x}} h_2 = [-1, 0], \; \nabla_{\boldsymbol{x}} h_3 = [0, -1], \end{aligned}$$

and the Hessian is

$$G = \begin{bmatrix} 2 & 0 \\ 0 & 2 \end{bmatrix}.$$

The active set is $\mathcal{A}^{(0)} = \{2, 3\}$, so that

$$A^{(0)} = \begin{bmatrix} -1 & 0 \\ 0 & -1 \end{bmatrix}.$$

Since two linearly independent constraints are active, $\boldsymbol{d} = \boldsymbol{0}$ is the solution to problem (3.3.9) of Step 2. Thus we move to Step 3. The Lagrange multipliers are found from the equation $-A^{(0)}\boldsymbol{\lambda}^{(0)} = \boldsymbol{g}^{(0)}$, i.e.,

$$-\begin{bmatrix} -1 & 0 \\ 0 & -1 \end{bmatrix} \begin{bmatrix} \lambda_2^{(0)} \\ \lambda_3^{(0)} \end{bmatrix} = \begin{bmatrix} -4 \\ -5 \end{bmatrix}.$$

The solution is $\lambda_2^{(0)} = -4$ and $\lambda_3^{(0)} = -5$, of which the most negative one is $\lambda_3^{(0)}$. Hence the third constraint is dropped. Now, we set $\mathcal{A}^{(0)} = \{2\}$ and go to Step 2.

Problem (3.3.9) is to minimize $\frac{1}{2}\boldsymbol{d}^\top G\boldsymbol{d} + \boldsymbol{d}^\top \boldsymbol{g}^{(0)}$ subject to $\nabla_{\boldsymbol{x}} h_2 \left(\boldsymbol{x}^{(0)}\right) \boldsymbol{d} = 0$. It has the solution $\boldsymbol{d}^\top = [0, 5/2]$. We now move to Step 4 to determine the step length along this direction of search. We have

$$\alpha^{(0)} = \min \left\{1, -h_1 \left(\boldsymbol{x}^{(0)}\right) / \nabla_{\boldsymbol{x}} h_1 \left(\boldsymbol{x}^{(0)}\right) \boldsymbol{d}\right\},$$

as the third constraint does not satisfy the criterion $\nabla_{\boldsymbol{x}} h_3 \left(\boldsymbol{x}^{(0)}\right) \boldsymbol{d} > 0$. This gives $\alpha^{(0)} = 4/5$, so

$$\boldsymbol{x}^{(1)} = \boldsymbol{x}^{(0)} + \alpha^{(0)} \boldsymbol{d}^{(0)} = \begin{bmatrix} 0 \\ 2 \end{bmatrix},$$

$$\mathcal{A}^{(1)} = \{1, 2\}.$$

Moving through Step 5, then Steps 2 and 3 again (two linearly independent constraints are active), the new Lagrange multipliers are found from

$$-\begin{bmatrix} 2 & -1 \\ 1 & 0 \end{bmatrix} \begin{bmatrix} \lambda_1^{(1)} \\ \lambda_2^{(1)} \end{bmatrix} = \begin{bmatrix} -4 \\ -1 \end{bmatrix},$$

to give $\lambda_1^{(1)} = 1$ and $\lambda_2^{(1)} = -2$. The second constraint is dropped, giving $\mathcal{A}^{(1)} = \{1\}$.

We now return to Step 2 where Problem (3.3.9) is solved again to give $\boldsymbol{d}^{(1)} = [1/5, -2/5]^\top$. Step 4 gives a step length of $\alpha^{(1)} = 1$, which, in turn, gives $\boldsymbol{x}^{(2)} = [1/5, 8/5]^\top$. Moving back to Step 2 with $\mathcal{A}^{(2)} = \{1\}$, we find that $\boldsymbol{d} = \boldsymbol{0}$ and then Step 3 gives $\lambda_1^{(2)} = 9/5$, which is greater than zero, so $\boldsymbol{x}^{(2)}$ is the optimal point. The corresponding optimal Lagrange multiplier is $\boldsymbol{\lambda}^* = [9/5, 0, 0]^\top$.

3.4 Constrained Quasi-Newton Method

In this section, we briefly describe a constrained quasi-Newton method for solving a general linearly constrained optimization problem. This method was initially developed by Fletcher [61].

Consider the linearly constrained optimization problem, where the objective function

$$f(\boldsymbol{x}) \tag{3.4.1a}$$

is to be minimized subject to

$$h_i(\boldsymbol{x}) = \boldsymbol{a}_i^{\top}\boldsymbol{x} - b_i = 0, \quad i \in \mathcal{E}, \tag{3.4.1b}$$

$$h_i(\boldsymbol{x}) = \boldsymbol{a}_i^{\top}\boldsymbol{x} - b_i \leq 0, \quad i \in \mathcal{I}, \tag{3.4.1c}$$

where $\boldsymbol{x} \in \mathbb{R}^n$ is the decision vector, f is a general nonlinear function in $\mathbb{R}^n$, $\mathcal{E} = \{1, 2, \ldots, m\}$, $\mathcal{I} = \{m+1, m+2, \ldots, m+r\}$, $\boldsymbol{a}_i$, $i \in \mathcal{E} \cup \mathcal{I}$, are n-vectors and b_i, $i \in \mathcal{E} \cup \mathcal{I}$, are real numbers. Since all the constraints are linear, $\nabla_{\boldsymbol{x}} h_i = \boldsymbol{a}_i^{\top}$, $i \in \mathcal{E} \cup \mathcal{I}$, are constant vectors.

A point $\boldsymbol{x}$ is said to be a *feasible point* of Problem (3.4.1) if it satisfies the constraints (3.4.1b)–(3.4.1c). Let Ξ be the set of all such feasible points.

Using the first phase of the simplex method for linear programming, we can compute an initial point $\boldsymbol{x}^{(0)}$ in Ξ. A new iterate $\boldsymbol{x}^{(k+1)} = \boldsymbol{x}^{(k)} + \boldsymbol{d}$ can be computed from the current iterate $\boldsymbol{x}^{(k)}$ by solving a quadratic programming subproblem. A quadratic model $q_k(\boldsymbol{d})$ is obtained by using the Taylor series expansion of the objective function about the point $\boldsymbol{x}^{(k)}$, truncated after the second order term, i.e.,

$$f\left(\boldsymbol{x}^{(k)} + \boldsymbol{d}\right) \approx q_k(\boldsymbol{d}) = f\left(\boldsymbol{x}^{(k)}\right) + \boldsymbol{d}^{\top}\boldsymbol{g}^{(k)} + \frac{1}{2}\boldsymbol{d}^{\top}H^{(k)}\boldsymbol{d},$$

where

$$\boldsymbol{g}^{(k)} = \left[\nabla_{\boldsymbol{x}} f\left(\boldsymbol{x}^{(k)}\right)\right]^{\top}$$

and

$$H^{(k)} = \nabla_{\boldsymbol{x}\boldsymbol{x}} f\left(\boldsymbol{x}^{(k)}\right).$$

Maintaining feasibility of the new iterate requires

$$h_i\left(\boldsymbol{x}^{(k)} + \boldsymbol{d}\right) = \boldsymbol{a}_i^{\top}\left(\boldsymbol{x}^{(k)} + \boldsymbol{d}\right) - b_i = \boldsymbol{a}_i^{\top}\boldsymbol{d} + h_i\left(\boldsymbol{x}^{(k)}\right) = 0, \ i \in \mathcal{E},$$

$$h_i\left(\boldsymbol{x}^{(k)} + \boldsymbol{d}\right) = \boldsymbol{a}_i^{\top}\left(\boldsymbol{x}^{(k)} + \boldsymbol{d}\right) - b_i = \boldsymbol{a}_i^{\top}\boldsymbol{d} + h_i\left(\boldsymbol{x}^{(k)}\right) \leq 0, \ i \in \mathcal{I}.$$

Thus, $\boldsymbol{x}^{(k+1)} = \boldsymbol{x}^{(k)} + \boldsymbol{d}$ is generated from $\boldsymbol{x}^{(k)}$ by solving the following quadratic programming subproblem, to be denoted as Problem $(\mathrm{QP})_k$: Find a $\boldsymbol{d} \in \mathbb{R}^n$ such that the objective function

$$\frac{1}{2}\boldsymbol{d}^{\top}B^{(k)}\boldsymbol{d}+\boldsymbol{d}^{\top}\boldsymbol{g}^{(k)} \tag{3.4.2a}$$

is minimized subject to

$$\boldsymbol{a}_i^{\top}\boldsymbol{d}+h_i\left(\boldsymbol{x}^{(k)}\right)=0, \quad i\in\mathcal{E}, \tag{3.4.2b}$$

$$\boldsymbol{a}_i^{\top}\boldsymbol{d}+h_i\left(\boldsymbol{x}^{(k)}\right)\leq 0, \quad i\in\mathcal{I}, \tag{3.4.2c}$$

where $B^{(k)}$ is a positive definite symmetric matrix that approximates the Hessian $H^{(k)}$ of the objective function at the point $x^{(k)}$. It is constructed according to the Broyden-Fletcher-Goldfarb-Shanno (BFGS) rank 2 updating formula (see Section 2.9.3):

$$B^{(k+1)}=B^{(k)}+\frac{\boldsymbol{\gamma}^{(k)}\left(\boldsymbol{\gamma}^{(k)}\right)^{\top}}{\left(\boldsymbol{\delta}^{(k)}\right)^{\top}\boldsymbol{\gamma}^{(k)}}-\frac{B^{(k)}\boldsymbol{\delta}^{(k)}\left(\boldsymbol{\delta}^{(k)}\right)^{\top}B^{(k)}}{\left(\boldsymbol{\delta}^{(k)}\right)^{\top}B^{(k)}\boldsymbol{\delta}^{(k)}}, \tag{3.4.3}$$

where

$$\boldsymbol{\delta}^{(k)}=\boldsymbol{x}^{(k+1)}-\boldsymbol{x}^{(k)}$$

and

$$\boldsymbol{\gamma}^{(k)}=\boldsymbol{g}^{(k+1)}-\boldsymbol{g}^{(k)}.$$

This formula ensures that if $B^{(0)}$ is symmetric positive definite, then so are all successive updates of the approximate Hessian matrices $B^{(k)}$ provided that

$$\left(\boldsymbol{\delta}^{(k)}\right)^{\top}\boldsymbol{\gamma}^{(k)}>0.$$

Note that if $\boldsymbol{x}^{(k)}$ is feasible, then $h_i\left(\boldsymbol{x}^{(k)}\right)=0$, $i\in\mathcal{A}^{(k)}\supseteq\mathcal{E}$, where $\mathcal{A}^{(k)}$ is as defined in Definition 3.3.1.

For each k, Problem $(\mathrm{QP})_k$ is solved by the active set method described in Section 3.3. The algorithm for solving Problem (3.4.1) can now be stated as follows.

Algorithm 3.4.1

Step 1. *Choose a point $\boldsymbol{x}^{(0)}\in\Xi$. (This can be achieved by the first phase of the simplex algorithm for linear programming.) Approximate $H^{(0)}$ by a symmetric positive definite matrix $B^{(0)}$. Choose an $\varepsilon>0$ and set $k=0$.*

Step 2. *Solve the quadratic programming subproblem, Problem $(QP)_k$.*

Step 3. *Consider the problem of minimizing $f\left(\boldsymbol{x}^{(k)}+\alpha\boldsymbol{d}^{(k)}\right)$ with respect to α, where $\boldsymbol{d}^{(k)}$ is the solution of Problem $(QP)_k$. Choose an approximate minimizer $\alpha^{(k)}\leq\min\left\{1,\bar{\alpha}^{(k)}\right\}$ for this problem (see Remark 3.4.1 below), where $\bar{\alpha}^{(k)}$ is defined by* (3.3.15).

Step 4. *If* $\left\|\boldsymbol{d}^{(k)}\right\| < \varepsilon$, *set* $\boldsymbol{x}^* = \boldsymbol{x}^{(k)}$ *and stop. Otherwise, set* $\boldsymbol{x}^{(k+1)} = \boldsymbol{x}^{(k)} + \alpha_k \boldsymbol{d}^{(k)}$.

Step 5. *Update* $B^{(k)}$ *according to the BFGS formula* (3.4.3).

Step 6. *Set* $k := k+1$ *and go to Step 2.*

Remark 3.4.1 *In Step 3 of Algorithm 3.4.1, we need to find an approximate minimizer* $\alpha^{(k)} \le \bar{\alpha}^{(k)}$ *of* $f\left(\boldsymbol{x}^{(k)} + \alpha \boldsymbol{d}^{(k)}\right)$. *It must be chosen such that the following two conditions are satisfied:*

(i) There is a sufficient function decrease (known as the Goldstein condition [see [78]])

$$f\left(\boldsymbol{x}^{(k)} + \alpha^{(k)}\boldsymbol{d}^{(k)}\right) \le f\left(\boldsymbol{x}^{(k)}\right) + \rho\alpha^{(k)}\left[\boldsymbol{d}^{(k)}\right]^\top \boldsymbol{g}\left(\boldsymbol{x}^{(k)}\right).$$

(ii) There is a sufficient slope improvement

$$\left|\left[\boldsymbol{g}\left(\boldsymbol{x}^{(k)} + \alpha^{(k)}\boldsymbol{d}^{(k)}\right)\right]^\top \boldsymbol{d}^{(k)}\right| \le -\eta\left[\boldsymbol{d}^{(k)}\right]^\top \boldsymbol{g}\left(\boldsymbol{x}^{(k)}\right).$$

Here, ρ *and* η *are constants satisfying* $0 < \rho < \eta < 1$. *If* $\eta = 0$, *then* $\alpha^{(k)}$ *is a stationary point of* $f\left(\boldsymbol{x}^{(k)} + \alpha \boldsymbol{d}^{(k)}\right)$ *with respect to* α. *Typical values of* η *are* $\eta = 0.9$ *(weak, i.e., not very accurate line search) and* $\eta = 0.1$ *(strong, i.e., fairly accurate line search). The parameter* ρ *is typically taken to be quite small, for example,* $\rho = 0.01$.

Remark 3.4.2 *The new point* $\boldsymbol{x}^{(k)} + \alpha^{(k)}\boldsymbol{d}^{(k)}$ *must be feasible. This will be the case provided that*

$$\alpha^{(k)} \le \bar{\alpha}^{(k)} = \min_{\substack{i \in \mathcal{I}\setminus\mathcal{A}^{(k)} \\ \boldsymbol{a}_i^\top \boldsymbol{d}^{(k)} > 0}} \frac{-h_i\left(\boldsymbol{x}^{(k)}\right)}{\boldsymbol{a}_i^\top \boldsymbol{d}^{(k)}}.$$

The condition $\alpha^{(k)} \le \min\left\{1, \bar{\alpha}^{(k)}\right\}$ *will ensure feasibility of* $\boldsymbol{x}^{(k)} + \alpha^{(k)}\boldsymbol{d}^{(k)}$. *As* $\boldsymbol{x}^{(0)}$ *is chosen to be a feasible point, Step 3 ensures that the algorithm generates a sequence of feasible points. The sufficient slope improvement condition ensures that* $\left[\boldsymbol{\delta}^{(k)}\right]^\top \boldsymbol{\gamma}^{(\kappa)} > 0$. *However, the constraint* $\alpha^{(k)} \le \min\{1, \bar{\alpha}^{(k)}\}$ *can destroy this property.*

3.5 Sequential Quadratic Programming Algorithm

The sequential quadratic programming algorithm is recognized as one of the most efficient algorithms for small and medium size nonlinearly constrained optimization problems. The theory was initiated by Wilson in [278] and was further developed by Han in [86] and [87], Powell in [207] and Schittkowski in [221] and [222]. In this section, we shall discuss some of the essential concepts of the algorithm without going into detail. The main references of this section are [222], [232] and [275]. For readers interested in details, see [232].

Consider the equality constrained optimization problem

$$\text{minimize} \quad f(\boldsymbol{x}) \tag{3.5.1}$$

subject to

$$h_i(\boldsymbol{x}) = 0, \quad i = 1, 2, \ldots, m. \tag{3.5.2}$$

The Lagrangian function is

$$L(\boldsymbol{x}, \boldsymbol{\lambda}) = f(\boldsymbol{x}) + \boldsymbol{\lambda}^\top \boldsymbol{h}(\boldsymbol{x}), \tag{3.5.3}$$

where $\boldsymbol{h} = [h_1, \ldots, h_m]^\top$ and $\boldsymbol{\lambda} = [\lambda_1, \ldots, \lambda_m]^\top \in \mathbb{R}^m$ is a Lagrange multiplier vector. A point $\boldsymbol{x} = [x_1, \ldots, x_n]^\top \in \mathbb{R}^n$ is a KKT point of (3.5.1)–(3.5.2) if and only if there exists a non-zero $\boldsymbol{\lambda} \in \mathbb{R}^m$ such that

$$\nabla_{\boldsymbol{x}} L(\boldsymbol{x}, \boldsymbol{\lambda}) = \nabla_{\boldsymbol{x}} f(\boldsymbol{x}) + \sum_{i=1}^{m} \lambda_i \nabla_{\boldsymbol{x}} h_i(\boldsymbol{x}) = \mathbf{0}^\top, \tag{3.5.4}$$

and

$$\nabla_{\boldsymbol{\lambda}} L(\boldsymbol{x}, \boldsymbol{\lambda}) = (\boldsymbol{h}(\boldsymbol{x}))^\top = \mathbf{0}^\top. \tag{3.5.5}$$

(3.5.4) and (3.5.5) may be rewritten as the following system of nonlinear equations:

$$W(\boldsymbol{x}, \boldsymbol{\lambda}) = \begin{bmatrix} (\nabla_{\boldsymbol{x}} L(\boldsymbol{x}, \boldsymbol{\lambda}))^\top \\ \boldsymbol{h}(\boldsymbol{x}) \end{bmatrix} = \begin{bmatrix} (\nabla_{\boldsymbol{x}} f(\boldsymbol{x}))^\top + \sum_{i=1}^{m} \lambda_i (\nabla_{\boldsymbol{x}} h_i(\boldsymbol{x}))^\top \\ \boldsymbol{h}(\boldsymbol{x}) \end{bmatrix} = \begin{bmatrix} \mathbf{0} \\ \mathbf{0} \end{bmatrix}. \tag{3.5.6}$$

We shall use Newton's method to find the solution of the nonlinear system (3.5.6). For a given iterate $\boldsymbol{x}^{(k)} \in \mathbb{R}^n$ and the corresponding Lagrange multiplier $\boldsymbol{\lambda}^{(k)} \in \mathbb{R}^m$, the next iterate $(\boldsymbol{x}^{(k+1)}, \boldsymbol{\lambda}^{(k+1)})$ is obtained by solving the following system of linear equations:

$$W\left(\boldsymbol{x}^{(k)}, \boldsymbol{\lambda}^{(k)}\right) + \begin{bmatrix} \nabla_{\boldsymbol{x}\boldsymbol{x}} L\left(\boldsymbol{x}^{(k)}, \boldsymbol{\lambda}^{(k)}\right) & \left(\nabla_{\boldsymbol{x}} \boldsymbol{h}\left(\boldsymbol{x}^{(k)}\right)\right)^\top \\ \nabla_{\boldsymbol{x}} \boldsymbol{h}\left(\boldsymbol{x}^{(k)}\right) & 0 \end{bmatrix} \begin{bmatrix} \boldsymbol{x}^{(k+1)} - \boldsymbol{x}^{(k)} \\ \boldsymbol{\lambda}^{(k+1)} - \boldsymbol{\lambda}^{(k)} \end{bmatrix} = \mathbf{0}. \tag{3.5.7}$$

From (3.5.6)–(3.5.7), it can be shown that

$$\begin{aligned} &\nabla_{\boldsymbol{x}\boldsymbol{x}} L\left(\boldsymbol{x}^{(k)}, \boldsymbol{\lambda}^{(k)}\right)\left(\boldsymbol{x}^{(k+1)} - \boldsymbol{x}^{(k)}\right) + \left(\nabla_{\boldsymbol{x}} \boldsymbol{h}\left(\boldsymbol{x}^{(k)}\right)\right)^\top \boldsymbol{\lambda}^{(k+1)} \\ &= -\left(\nabla_{\boldsymbol{x}} f\left(\boldsymbol{x}^{(k)}\right)\right)^\top \end{aligned} \tag{3.5.8}$$

and

$$\nabla_{\boldsymbol{x}} \boldsymbol{h}\left(\boldsymbol{x}^{(k)}\right)\left(\boldsymbol{x}^{(k+1)}-\boldsymbol{x}^{(k)}\right)=-\boldsymbol{h}\left(\boldsymbol{x}^{(k)}\right). \tag{3.5.9}$$

Let $\delta \boldsymbol{x}^{(k)}=\boldsymbol{x}-\boldsymbol{x}^{(k)}$. Then, (3.5.8) and (3.5.9) become

$$\nabla_{\boldsymbol{x}\boldsymbol{x}} L\left(\boldsymbol{x}^{(k)}, \boldsymbol{\lambda}^{(k)}\right) \delta \boldsymbol{x}^{(k)}+\left(\nabla_{\boldsymbol{x}} \boldsymbol{h}\left(\boldsymbol{x}^{(k)}\right)\right)^{\top} \boldsymbol{\lambda}^{(k+1)}=-\left(\nabla_{\boldsymbol{x}} f\left(\boldsymbol{x}^{(k)}\right)\right)^{\top} \tag{3.5.10}$$

and

$$\nabla_{\boldsymbol{x}} \boldsymbol{h}\left(\boldsymbol{x}^{(k)}\right) \delta \boldsymbol{x}^{(k)}=-\boldsymbol{h}\left(\boldsymbol{x}^{(k)}\right). \tag{3.5.11}$$

Solving the system of linear equations (3.5.10) and (3.5.11) gives ($\delta \boldsymbol{x}^{(k)}$, $\boldsymbol{\lambda}^{(k+1)}$). Then, we obtain the next iterate as

$$\boldsymbol{x}^{(k+1)}=\boldsymbol{x}^{(k)}+\delta \boldsymbol{x}^{(k)}. \tag{3.5.12}$$

This method is known as the Lagrange-Newton method for solving equality constrained optimization problems. If the constraints satisfy the constraint qualification conditions at the KKT point, then the Lagrange-Newton method is locally convergent and the convergence rate is of second order.

The increment $\delta \boldsymbol{x}^{(k)}$ resulting in the new iterate $\boldsymbol{x}^{(k+1)}$ given by (3.5.12) may also be considered as the solution of the following quadratic equality constrained optimization problem:

$$\text{minimize } f\left(\boldsymbol{x}^{(k)}\right)+\nabla_{\boldsymbol{x}} f\left(\boldsymbol{x}^{(k)}\right) \delta \boldsymbol{x}^{(k)}+\frac{1}{2}\left(\delta \boldsymbol{x}^{(k)}\right)^{\top} \nabla_{\boldsymbol{x}\boldsymbol{x}} L\left(\boldsymbol{x}^{(k)}, \boldsymbol{\lambda}^{(k)}\right) \delta \boldsymbol{x}^{(k)} \tag{3.5.13}$$

subject to

$$h_i\left(\boldsymbol{x}^{(k)}\right)+\nabla_{\boldsymbol{x}} h_i\left(\boldsymbol{x}^{(k)}\right) \delta \boldsymbol{x}^{(k)}=0, \quad i=1,2, \ldots, m. \tag{3.5.14}$$

To be more specific, we consider the Lagrangian for the problem (3.5.13)–(3.5.14), given by

$$\begin{aligned} &\mathcal{L}\left(\boldsymbol{x}^{(k)}, \boldsymbol{\lambda}^{(k)}, \delta \boldsymbol{x}^{(k)}, \boldsymbol{\lambda}\right) \\ &\quad=f\left(\boldsymbol{x}^{(k)}\right)+\nabla_{\boldsymbol{x}} f\left(\boldsymbol{x}^{(k)}\right) \delta \boldsymbol{x}^{(k)}+\frac{1}{2}\left(\delta \boldsymbol{x}^{(k)}\right)^{\top} \nabla_{\boldsymbol{x}\boldsymbol{x}} L\left(\boldsymbol{x}^{(k)}, \boldsymbol{\lambda}^{(k)}\right) \delta \boldsymbol{x}^{(k)} \\ &\qquad+\sum_{i=1}^{m} \lambda_i\left(h_i\left(\boldsymbol{x}^{(k)}\right)+\nabla_{\boldsymbol{x}} h_i\left(\boldsymbol{x}^{(k)}\right) \delta \boldsymbol{x}^{(k)}\right), \end{aligned} \tag{3.5.15}$$

where $\boldsymbol{\lambda}=[\lambda_1, \ldots, \lambda_m]^{\top}$. Then, by taking the gradients of the Lagrangian function $\mathcal{L}\left(\boldsymbol{x}^{(k)}, \boldsymbol{\lambda}^{(k)}, \delta \boldsymbol{x}^{(k)}, \boldsymbol{\lambda}\right)$ with respect to $\delta \boldsymbol{x}^{(k)}$ and $\boldsymbol{\lambda}$ and noting the equivalence of $\boldsymbol{\lambda}$ and $\boldsymbol{\lambda}^{(k+1)}$, we obtain (3.5.10)–(3.5.11).

We now move on to consider the following optimization problem with both equality and inequality constraints.

$$\text{minimize} \quad f(\boldsymbol{x}) \tag{3.5.16}$$

subject to

$$h_i(\boldsymbol{x}) = 0, \quad i = 1, 2, \ldots, m, \tag{3.5.17}$$
$$h_i(\boldsymbol{x}) \leq 0, \quad i = m+1, m+2, \ldots, m+r. \tag{3.5.18}$$

The function f is assumed to be twice continuously differentiable, and h_i, $i = 1, 2, \ldots, m+r$, are also assumed to be twice continuously differentiable. We introduce the Lagrangian function L for the constrained optimization problem (3.5.16)–(3.5.18) defined by

$$L(\boldsymbol{x}, \boldsymbol{\lambda}) = f(\boldsymbol{x}) + \sum_{i=1}^{m+r} \lambda_i h_i(\boldsymbol{x}). \tag{3.5.19}$$

Let $\boldsymbol{x}^{(k)}$ be the current iterate, and let $\boldsymbol{\lambda}^{(k)} \in \mathbb{R}^{m+r}$ be the corresponding Lagrange multiplier vector. To find the next iterate, we first construct the following quadratic programming subproblem:

$$\text{minimize} \quad f\left(\boldsymbol{x}^{(k)}\right) + \nabla_{\boldsymbol{x}} f\left(\boldsymbol{x}^{(k)}\right) \delta\boldsymbol{x}^{(k)} + \frac{1}{2}(\delta\boldsymbol{x}^{(k)})^{\top} \nabla_{\boldsymbol{x}\boldsymbol{x}} L\left(\boldsymbol{x}^{(k)}, \boldsymbol{\lambda}^{(k)}\right) \delta\boldsymbol{x}^{(k)} \tag{3.5.20}$$

subject to

$$h_i\left(\boldsymbol{x}^{(k)}\right) + \nabla_{\boldsymbol{x}} h_i\left(\boldsymbol{x}^{(k)}\right) \delta\boldsymbol{x}^{(k)} = 0, \quad i = 1, 2, \ldots, m, \tag{3.5.21}$$

and

$$h_i\left(\boldsymbol{x}^{(k)}\right) + \nabla_{\boldsymbol{x}} h_i\left(\boldsymbol{x}^{(k)}\right) \delta\boldsymbol{x}^{(k)} \leq 0, \quad i = m+1, 2, \ldots, m+r. \tag{3.5.22}$$

The next iterate is then given by

$$\left(\boldsymbol{x}^{(k+1)}, \boldsymbol{\lambda}^{(k+1)}\right) = \left(\boldsymbol{x}^{(k)} + \delta\boldsymbol{x}^{(k)}, \overline{\boldsymbol{\lambda}}\right), \tag{3.5.23}$$

where $\delta\boldsymbol{x}^{(k)}$ is the solution of the subproblem (3.5.20)–(3.5.22) and $\overline{\boldsymbol{\lambda}}$ is the optimal Lagrange multiplier vector for the same subproblem. Suppose that the optimal solution to the subproblem (3.5.20)–(3.5.18) is such that $\delta\boldsymbol{x}^{(k)} = \mathbf{0}$. Then, $\boldsymbol{x}^{(k)}$ must be a KKT point for the problem (3.5.16)–(3.5.18). For the subproblem (3.5.20)–(3.5.22), it is necessary to calculate the Hessian of the Lagrange function L defined by (3.5.19). If $\nabla_{\boldsymbol{x}\boldsymbol{x}} L\left(\boldsymbol{x}^{(k)}, \boldsymbol{\lambda}^{(k)}\right)$ fails to be positive definite, the solution method will fail. Thus, we will use an approximate matrix $B^{(k)}$ to replace $\nabla_{\boldsymbol{x}\boldsymbol{x}} L(\boldsymbol{x}^{(k)}, \boldsymbol{\lambda}^{(k)})$, where $B^{(k)}$ is to be updated in such a way that if $B^{(k)}$ is positive definite, then $B^{(k+1)}$ will also be positive definite. To generate a new iterate, we need conditions for the step length selection. For this, we will introduce a merit function. The SQP method for solving the nonlinear optimization problem with equality and inequality constraints may now be stated as follows.

Algorithm 3.5.1

Step 1. Let $k = 0$. *Choose a starting point* $\boldsymbol{x}^{(0)} \in \mathbb{R}^n$ *and choose a positive definite matrix* $B^{(0)}$.

Step 2. Solve the following subproblem to obtain $\left(\delta\boldsymbol{x}^{(k)}, \boldsymbol{\lambda}^{(k+1)}\right)$:

$$\text{minimize} \quad f\left(\boldsymbol{x}^{(k)}\right) + \nabla_{\boldsymbol{x}} f\left(\boldsymbol{x}^{(k)}\right)\delta\boldsymbol{x}^{(k)} + \frac{1}{2}\left(\delta\boldsymbol{x}^{(k)}\right)^{\top} B^{(k)}\delta\boldsymbol{x}^{(k)} \tag{3.5.24}$$

subject to

$$h_i\left(\boldsymbol{x}^{(k)}\right) + \nabla_{\boldsymbol{x}} h_i\left(\boldsymbol{x}^{(k)}\right)\delta\boldsymbol{x}^{(k)} = 0, \quad i = 1, 2, \ldots, m \tag{3.5.25}$$

and

$$h_i\left(\boldsymbol{x}^{(k)}\right) + \nabla_{\boldsymbol{x}} h_i\left(\boldsymbol{x}^{(k)}\right)\delta\boldsymbol{x}^{(k)} \leq 0, \quad i = m+1, 2, \ldots, m+r. \tag{3.5.26}$$

Step 3. If $\delta\boldsymbol{x}^{(k)} = \mathbf{0}$, *then the algorithm terminates and* $\boldsymbol{x}^{(k)}$ *is the KKT point for problem* (3.5.16)–(3.5.18). *Otherwise, set* $\boldsymbol{x}^{(k+1)} = \boldsymbol{x}^{(k)} + \alpha_k\delta\boldsymbol{x}^{(k)}$, *where* α_k *is determined by some step length rules, see below.*

Step 4. Update $B^{(k)}$ *to* $B^{(k+1)}$ *such that* $B^{(k+1)}$ *is positive definite. Return to Step 2.*

The next task is to derive the update formula for the matrix $B^{(k)}$. It is natural to consider using the BFGS update formula, i.e.,

$$B^{(k+1)} = B^{(k)} + \frac{\boldsymbol{\gamma}^{(k)}\left(\boldsymbol{\gamma}^{(k)}\right)^{\top}}{\left(\boldsymbol{\gamma}^{(k)}\right)^{\top}\boldsymbol{\delta}^{(k)}} - \frac{B^{(k)}\boldsymbol{\delta}^{(k)}\left(\boldsymbol{\delta}^{(k)}\right)^{\top}B^{(k)}}{(\boldsymbol{\delta}^{(k)})^{\top}B^{(k)}\boldsymbol{\delta}^{(k)}}, \tag{3.5.27}$$

where $\boldsymbol{\delta}^{(k)} = \boldsymbol{x}^{(k+1)} - \boldsymbol{x}^{(k)}$ and

$$\boldsymbol{\gamma}^{(k)} = \left(\nabla_{\boldsymbol{x}} L\left(\boldsymbol{x}^{(k+1)}, \boldsymbol{\lambda}^{(k+1)}\right)\right)^{\top} - \left(\nabla_{\boldsymbol{x}} L\left(\boldsymbol{x}^{(k)}, \boldsymbol{\lambda}^{(k)}\right)\right)^{\top}.$$

However, if this formula is used, then the new matrix $B^{(k+1)}$ is positive definite only if the condition $\left(\boldsymbol{\gamma}^{(k)}\right)^{\top}\boldsymbol{\delta}^{(k)} > 0$ is satisfied. This condition is always satisfied for the unconstrained case if the step length is chosen such that the following conditions are satisfied:

$$f(\boldsymbol{x} + \bar{\alpha}\boldsymbol{\delta}) \leq f(\boldsymbol{x}) + \rho\bar{\alpha}\boldsymbol{\delta}^{\top}(\nabla_{\boldsymbol{x}} f(\boldsymbol{x}))^{\top} \tag{3.5.28}$$

and

$$|\nabla_{\boldsymbol{x}} f(\boldsymbol{x} + \bar{\alpha}\boldsymbol{\delta})\boldsymbol{\delta}| \leq -\beta\boldsymbol{\delta}^{\top}(\nabla_{\boldsymbol{x}} f(\boldsymbol{x}))^{\top}, \tag{3.5.29}$$

where ρ and β are constants satisfying $0 < \rho < \beta < 1$. However, this is not always true for the constrained case. Therefore, the BFGS update should not be applied directly. It needs to be modified as suggested in [207]. More specifically, we introduce $\boldsymbol{\eta}^{(k)}$ to replace $\boldsymbol{\gamma}^{(k)}$, where $\boldsymbol{\eta}^{(k)}$ is given by

$$\boldsymbol{\eta}^{(k)} = \begin{cases} \boldsymbol{\gamma}^{(k)}, & \left(\boldsymbol{\gamma}^{(k)}\right)^\top \boldsymbol{\delta}^{(k)} \geq 0.2 \left(\boldsymbol{\delta}^{(k)}\right)^\top B^{(k)} \boldsymbol{\delta}^{(k)}, \\ \theta_k \boldsymbol{\gamma}^{(k)} + (1-\theta_k) B^{(k)} \boldsymbol{\delta}^{(k)}, & \text{otherwise}, \end{cases} \tag{3.5.30}$$

where

$$\theta_k = \frac{0.8 \left(\boldsymbol{\delta}^{(k)}\right)^\top B^{(k)} \boldsymbol{\delta}^{(k)}}{\left(\boldsymbol{\delta}^{(k)}\right)^\top B^{(k)} \boldsymbol{\delta}^{(k)} - (\boldsymbol{\gamma}^{(k)})^\top \boldsymbol{\delta}^{(k)}}. \tag{3.5.31}$$

Clearly,

$$\left(\boldsymbol{\eta}^{(k)}\right)^\top \boldsymbol{\delta}^{(k)} \geq 0.2 \left(\boldsymbol{\delta}^{(k)}\right)^\top B^{(k)} \boldsymbol{\delta}^{(k)} \tag{3.5.32}$$

and hence

$$\left(\boldsymbol{\delta}^{(k)}\right)^\top \boldsymbol{\eta}^{(k)} > 0. \tag{3.5.33}$$

Let $\boldsymbol{\eta}^{(k)}$ be chosen as defined by (3.5.30). Then, the modified BFGS formula is given ([207]) by

$$B^{(k+1)} = B^{(k)} + \frac{\boldsymbol{\eta}^{(k)} \left(\boldsymbol{\eta}^{(k)}\right)^\top}{\left(\boldsymbol{\eta}^{(k)}\right)^\top \boldsymbol{\delta}^{(k)}} - \frac{B^{(k)} \boldsymbol{\delta}^{(k)} \left(\boldsymbol{\delta}^{(k)}\right)^\top B^{(k)}}{\left(\boldsymbol{\delta}^{(k)}\right)^\top B^{(k)} \boldsymbol{\delta}^{(k)}}. \tag{3.5.34}$$

Now, by (3.5.33), it follows that if $B^{(k)}$ is positive definite, then $B^{(k+1)}$ is also positive definite.

We obtain the following subproblem:

$$\text{minimize} \quad f\left(\boldsymbol{x}^{(k)}\right) + \nabla_{\boldsymbol{x}} f\left(\boldsymbol{x}^{(k)}\right) \delta \boldsymbol{x}^{(k)} + \frac{1}{2} \left(\delta \boldsymbol{x}^{(k)}\right)^\top B^{(k)} \delta \boldsymbol{x}^{(k)} \tag{3.5.35}$$

subject to

$$h_i\left(\boldsymbol{x}^{(k)}\right) + \nabla_{\boldsymbol{x}} h_i\left(\boldsymbol{x}^{(k)}\right) \delta \boldsymbol{x}^{(k)} = 0, \quad i = 1, 2, \ldots, m, \tag{3.5.36}$$

and

$$h_i\left(\boldsymbol{x}^{(k)}\right) + \nabla_{\boldsymbol{x}} h_i\left(\boldsymbol{x}^{(k)}\right) \delta \boldsymbol{x}^{(k)} \leq 0, \quad i = m+1, 2, \ldots, m+r. \tag{3.5.37}$$

This is a quadratic programming problem and can hence be solved by the active set method described in Section 3.3. Let $\delta \boldsymbol{x}^{(k)}$ be the solution of this quadratic programming problem, and let $\overline{\boldsymbol{\lambda}}^{(k)} = \left[\overline{\lambda}_1^{(k)}, \ldots, \overline{\lambda}_{m+r}^{(k)}\right]^\top$ be the corresponding optimal Lagrange multiplier vector for this problem, i.e., it satisfies

$$\nabla_{\boldsymbol{x}} f\left(\boldsymbol{x}^{(k)}\right) \delta \boldsymbol{x}^{(k)}+\left(\delta \boldsymbol{x}^{(k)}\right)^{\top} B^{(k)}+\sum_{i=1}^{m+r} \overline{\lambda}_i^{(k)}\left(\nabla_{\boldsymbol{x}} h_i\left(\boldsymbol{x}^{(k)}\right)\right)^{\top}=0, \tag{3.5.38}$$

$$\overline{\lambda}_i^{(k)}\left[h_i\left(\boldsymbol{x}^{(k)}\right)+\nabla_{\boldsymbol{x}} h_i\left(\boldsymbol{x}^{(k)}\right) \delta \boldsymbol{x}^{(k)}\right]=0, \quad i=m+1,2, \ldots, m+r, \tag{3.5.39}$$

and

$$\overline{\lambda}_i^{(k)} \geq 0, \quad i=m+1,2, \ldots, m+r. \tag{3.5.40}$$

Then, the new estimates $\boldsymbol{x}^{(k+1)}, \boldsymbol{\lambda}^{(k+1)}$ and $B^{(k+1)}$ may be determined by

$$\boldsymbol{x}^{(k+1)} =\boldsymbol{x}^{(k)}+\alpha_k \boldsymbol{d}^{(k)}, \tag{3.5.41}$$

$$\boldsymbol{\lambda}^{(k+1)} =\boldsymbol{\lambda}^{(k)}+\alpha_k\left(\overline{\boldsymbol{\lambda}}^{(k)}-\boldsymbol{\lambda}^{(k)}\right), \tag{3.5.42}$$

$$B^{(k+1)} =B^{(k)}+\frac{\boldsymbol{\gamma}^{(k)}\left(\boldsymbol{\gamma}^{(k)}\right)^{\top}}{\left(\boldsymbol{\delta}^{(k)}\right)^{\top} \boldsymbol{\gamma}^{(k)}}-\frac{B^{(k)} \boldsymbol{\delta}^{(k)}\left(\boldsymbol{\delta}^{(k)}\right)^{\top} B^{(k)}}{\left(\boldsymbol{\delta}^{(k)}\right)^{\top} B^{(k)} \boldsymbol{\delta}^{(k)}}, \tag{3.5.43}$$

where

$$\boldsymbol{\delta}^{(k)}=\boldsymbol{x}^{(k+1)}-\boldsymbol{x}^{(k)} \tag{3.5.44}$$

and

$$\boldsymbol{\gamma}^{(k)}=\left[\nabla_{\boldsymbol{x}} L\left(\boldsymbol{x}^{(k+1)}, \boldsymbol{\lambda}^{(k+1)}\right)\right]^{\top}-\left[\nabla_{\boldsymbol{x}} L\left(\boldsymbol{x}^{(k)}, \boldsymbol{\lambda}^{(k)}\right)\right]^{\top}. \tag{3.5.45}$$

We introduce a merit function to determine the step length in such a way that the value of the cost function is reduced while the feasibility is maintained. The merit function given below is suggested in [87], which is an L_1-penalty function:

$$P(\boldsymbol{x}, \boldsymbol{\sigma})=f(\boldsymbol{x})+\sum_{i=1}^m \sigma_i\left|h_i(\boldsymbol{x})\right|+\sum_{i=m+1}^{m+r} \sigma_i \max \left\{0,-h_i(\boldsymbol{x})\right\}, \tag{3.5.46}$$

where $\boldsymbol{\sigma}=\left[\sigma_1, \ldots, \sigma_{m+r}\right]^{\top}$. The parameters σ_i, $i=1,2, \ldots, m+r$, should be chosen (see [207]) such that $P(\boldsymbol{x}, \boldsymbol{\sigma})$ is locally decreasing along the direction $\delta \boldsymbol{x}$. More specifically, for each $i=1,2, \ldots, m+r$, define

$$\sigma_i^{(k)}= \begin{cases}\left|\overline{\lambda}_i^{(0)}\right|, & \text { if } k=0, \\ \max \left\{\left|\overline{\lambda}_i^{(k)}\right|, \frac{1}{2}\left(\sigma_i^{(k-1)}+\left|\overline{\lambda}_i^{(k)}\right|\right)\right\}, & \text { if } k \geq 1,\end{cases} \tag{3.5.47}$$

where $\overline{\boldsymbol{\lambda}}^{(k)}$ is the optimal Lagrange multiplier of the subproblem (3.5.35)–(3.5.37), i.e., it satisfies (3.5.38)–(3.5.40). Clearly, $\left|\overline{\lambda}_i^{(k)}\right| \leq \sigma_i^{(k)}$.

To determine the convergence and convergence rate of the SQP method, we need the following two lemmas from [275].

Lemma 3.5.1 *Suppose that* $h_i(\boldsymbol{x})$, $i \in I$, *are given continuously differentiable functions, where* $I = \{1, \ldots, r\}$. *Let* $\Phi(\boldsymbol{x}) = \max_{i \in I}\{h_i(\boldsymbol{x})\}$. *Then, the directional derivative,* $\Phi'(\boldsymbol{x}; \delta\boldsymbol{x})$, *of the function* $\Phi(\boldsymbol{x})$ *along any direction* $\delta\boldsymbol{x}$ *exists and*

$$\Phi'(\boldsymbol{x}; \delta\boldsymbol{x}) = \max_{i \in I(x)} \{\nabla_{\boldsymbol{x}} h_i(\boldsymbol{x}) \delta\boldsymbol{x}\}, \tag{3.5.48}$$

where

$$I(\boldsymbol{x}) = \{i \in I : h_i(\boldsymbol{x}) = \Phi(\boldsymbol{x})\}. \tag{3.5.49}$$

Lemma 3.5.2 *Consider Problem* (3.5.16)-(3.5.18), *where* $f(\boldsymbol{x})$ *and* $h_i(\boldsymbol{x})$, $i = 1, 2, \ldots, m + r$, *are continuously differentiable functions. Suppose that* $B^{(k)}$ *is positive definite and that* $\left(\delta\boldsymbol{x}^{(k)}, \overline{\boldsymbol{\lambda}}^{(k)}\right)$ *is a KKT point of the subproblem* (3.5.35)-(3.5.37) *for which* $\delta\boldsymbol{x}^{(k)} \neq \mathbf{0}$ *and* $\left|\overline{\lambda}_i^{(k)}\right| \leq \sigma_i^{(k)}$, $i = 1, 2, \ldots, m + r$. *Then*

$$P'\left(\boldsymbol{x}^{(k)}, \boldsymbol{\sigma}^{(k)}; \delta\boldsymbol{x}^{(k)}\right) < 0. \tag{3.5.50}$$

With the help of Lemmas 3.5.1 and 3.5.2, the step size α_k can be chosen (see [87]) such that

$$P\left(\boldsymbol{x}^{(k)} + \alpha_k \delta\boldsymbol{x}^{(k)}, \boldsymbol{\sigma}^{(k)}\right) < \max_{1 \leq \alpha \leq \beta} P\left(\boldsymbol{x}^{(k)} + \alpha \delta\boldsymbol{x}^{(k)}, \boldsymbol{\sigma}^{(k)}\right) + \varepsilon_k, \tag{3.5.51}$$

where β is a given positive number and

$$\sum_{k=1}^{\infty} \varepsilon_k < \infty. \tag{3.5.52}$$

The following theorem is established in [87].

Theorem 3.5.1 *Let* $f(\boldsymbol{x})$ *and* $h_i(\boldsymbol{x})$, $i = 1, 2, \ldots, m + r$, *be continuously differentiable functions. Suppose that there exist constants* $K_1 > 0$ *and* $K_2 > 0$ *such that*

$$K_1 \|\boldsymbol{x}\|^2 \leq \boldsymbol{x}^\top B^{(k)} \boldsymbol{x} \leq K_2 \|\boldsymbol{x}\|^2, \ \forall k \geq 1 \tag{3.5.53}$$

and for all $\boldsymbol{x} \in \mathbb{R}^n$. *Furthermore, it is assumed that there exists a vector* $\boldsymbol{\sigma} > \mathbf{0}$ *satisfying* $\left|\overline{\lambda}_i\right| \leq \sigma_i$, $= 1, 2, \ldots, m + r$, $\forall k \geq 1$. *Let the step size be chosen such that conditions* (3.5.51)–(3.5.52) *are satisfied. Then the sequence of the points* $\left\{\boldsymbol{x}^{(k)}\right\}$ *generated by Algorithm 3.5.1 is such that it either terminates at a KKT point or its accumulation point is a KKT point.*

To proceed further, let the following conditions be satisfied.

Assumption 3.5.1 *Let* $f(\boldsymbol{x})$ *and* $h_i(\boldsymbol{x})$, $i = 1, 2, \ldots, m + r$, *be continuously differentiable functions.*

Assumption 3.5.2 *The sequence* $\{\boldsymbol{x}^{(k)}\}$ *converges to* $\boldsymbol{x}^*$.

Assumption 3.5.3 *$\boldsymbol{x}^*$ is a KKT point and*

$$\nabla_{\boldsymbol{x}} h_i(\boldsymbol{x}^*), \quad i \in \mathcal{A}(\boldsymbol{x}^*), \tag{3.5.54}$$

are linearly independent, where $\mathcal{A}(\boldsymbol{x}^*) = \{i \in \mathcal{E} \cup \mathcal{I} : h_i(\boldsymbol{x}^*) = 0\}$, $\mathcal{E} = \{1, 2, \ldots, m\}$ and $\mathcal{I} = \{m+1, 2, \ldots, m+r\}$. Let $A(\boldsymbol{x}^*)$ be the $n \times |\mathcal{A}(\boldsymbol{x}^*)|$ matrix consisting of the vectors given by (3.5.54). For any non-zero vector $\delta\boldsymbol{x}$, if it satisfies

$$(A(\boldsymbol{x}^*))^\top \delta\boldsymbol{x} = \mathbf{0}, \tag{3.5.55}$$

then it holds that

$$(\delta\boldsymbol{x})^\top W(\boldsymbol{x}^*, \boldsymbol{\lambda}^*)\delta\boldsymbol{x} \neq 0, \tag{3.5.56}$$

where $W(\boldsymbol{x}, \boldsymbol{\lambda})$ is defined by (3.5.6) and $\boldsymbol{\lambda}^*$ is the Lagrange multiplier at $\boldsymbol{x}^*$.

Assumption 3.5.4 *If k is sufficiently large, then $\delta\boldsymbol{x}^{(k)}$ is a solution of*

$$\min_{\delta x \in \mathbb{R}^n} \nabla_{\boldsymbol{x}} f\left(\boldsymbol{x}^{(k)}\right) \delta\boldsymbol{x} + \frac{1}{2}\left(\delta\boldsymbol{x}^{(k)}\right)^\top B^{(k)} \delta\boldsymbol{x}^{(k)} \tag{3.5.57}$$

subject to

$$h_i\left(\boldsymbol{x}^{(k)}\right) + \left(\delta\boldsymbol{x}^{(k)}\right)^\top \left(\nabla_{\boldsymbol{x}} h_i\left(\boldsymbol{x}^{(k)}\right)\right)^\top = 0, \quad i \in \mathcal{A}\left(\boldsymbol{x}^{(k)}\right). \tag{3.5.58}$$

Now, suppose that there exists a k_0 such that Assumption 3.5.4 is satisfied for $k > k_0$. Then, there exists a $\overline{\boldsymbol{\lambda}}^{(k)} \in \mathbb{R}^{|\mathcal{A}(\boldsymbol{x}^{(k)})|}$ such that

$$\nabla_{\boldsymbol{x}} f\left(\boldsymbol{x}^{(k)}\right) + B^{(k)} \delta\boldsymbol{x}^{(k)} = A\left(\boldsymbol{x}^{(k)}\right) \overline{\boldsymbol{\lambda}}^{(k)} \tag{3.5.59}$$

and

$$A\left(\boldsymbol{x}^{(k)}\right)^\top \delta\boldsymbol{x}^{(k)} = -\widehat{h}\left(x^{(k)}\right), \tag{3.5.60}$$

for all $k > k_0$, where $\widehat{h}(\boldsymbol{x})$ is the vector whose elements are $h_i(\boldsymbol{x})$, $i \in \mathcal{A}\left(\boldsymbol{x}^{(k)}\right)$.

Theorem 3.5.2 *Suppose that Assumptions 3.5.1–3.5.4 are satisfied. Then, $\delta\boldsymbol{x}^{(k)}$ is a superlinearly convergent step, i.e.,*

$$\lim_{k \to \infty} \frac{\left\|\boldsymbol{x}^{(k)} + \delta\boldsymbol{x}^{(k)} - \boldsymbol{x}^*\right\|}{\left\|\boldsymbol{x}^{(k)} - \boldsymbol{x}^*\right\|} = 0, \tag{3.5.61}$$

if and only if

$$\lim_{k \to \infty} \frac{\left\|P_k\left(B_k - W(\boldsymbol{x}^*, \boldsymbol{\lambda}^*)\delta\boldsymbol{x}^{(k)}\right)\right\|}{\left\|\delta\boldsymbol{x}^{(k)}\right\|} = 0, \tag{3.5.62}$$

where P_k is a projection from $\mathbb{R}^n$ onto the null space of $\left(A\left(\boldsymbol{x}^{(k)}\right)\right)^\top$, i.e.,

$$P_k = I - A\left(\boldsymbol{x}^{(k)}\right)\left(A\left(\boldsymbol{x}^{(k)}\right)\right)^\top \left(A(\boldsymbol{x}^{(k)})\right)^{-1} \left(A\left(\boldsymbol{x}^{(k)}\right)\right)^\top. \tag{3.5.63}$$

The proof of this result may be found in [275].

As a concluding remark for this chapter, we note that any constrained optimal control problem can be reduced to a mathematical programming problem by using the control parametrization technique. Throughout this book, the sequential quadratic programming algorithm will be the basis on which all these mathematical programming problems are solved. The main reason is that the sequential quadratic programming algorithm is recognized as one of the most efficient algorithms for small and medium size nonlinearly constrained optimization problems.

However, we wish to note that there are other methods in the literature for solving general nonlinearly constrained mathematical programming problems. For example, see [61] and [71].

Chapter 4
Optimization Problems Subject to Continuous Inequality Constraints

4.1 Introduction

In this chapter, we shall present two computational approaches to solve a general class of optimization problems subject to continuous inequality constraints. The first approach is known as the constraint transcription method, while the other approach is referred to as an exact penalty function method. To begin, we first consider the problem of finding a feasible solution to a system of nonlinear inequality constraints. Using the constrained transcription method with a local smoothing technique, the problem is approximated by an unconstrained optimization problem. This approach is extended to find a feasible solution of a system of continuous inequality constraints. We then move on to consider the optimization problems subject to continuous inequality constraints. The constraint transcription method is used in conjunction with a local smoothing method to develop two computational methods to solve this general optimization problem with continuous inequality constraints. We then move on to introduce the second approach (i.e., the exact penalty function approach) to solving the same class of continuous inequality constrained optimization problems.

The main references for this chapter are [76, 103, 259, 300, 301].

4.2 Constraint Transcription Technique

In this section, we consider the following system of nonlinear inequality constraints:

$$h_j(\boldsymbol{x}) \leq 0, \quad j = 1, 2, \ldots, m,$$

K. L. Teo et al., *Applied and Computational Optimal Control*, Springer Optimization and Its Applications 171,
https://doi.org/10.1007/978-3-030-69913-0_4

where $\boldsymbol{x} = [x_1, \ldots, x_n]^\top \in \mathbb{R}^n$ and, for each $j = 1, 2, \ldots, m$, $h_j : \mathbb{R}^n \to \mathbb{R}$ is a continuously differentiable function.

The technique to be presented is referred to as *the constraint transcription technique.* Using this technique, we can find a feasible solution of a system of nonlinear inequality constraints in a finite number of iterations by solving an associated unconstrained optimization problem. The same technique is extended to find a feasible solution to a system of continuous inequality constraints. The resulting algorithms are presented in Sections 4.2.1 and 4.2.2, respectively. In Sections 4.3 and 4.4, we show that by using the constraint transcription technique in conjunction with the local smoothing technique, an optimization problem subject to continuous inequality constraints can be solved as a conventional constrained optimization problem or as an unconstrained optimization problem.

4.2.1 Inequality Constraints

Consider the following problem: Find an $\boldsymbol{x} \in \mathbb{R}^n$ such that

$$h_j(\boldsymbol{x}) \leq 0, \quad j = 1, 2, \ldots, m, \tag{4.2.1a}$$

and

$$\boldsymbol{a} \leq \boldsymbol{x} \leq \boldsymbol{b}, \tag{4.2.1b}$$

where $\boldsymbol{x} = [x_1, x_2, \ldots, x_n]^\top$, $\boldsymbol{a} = [a_1, a_2, \ldots, a_n]^\top$ and $\boldsymbol{b} = [b_1, b_2, \ldots, b_n]^\top$. Here, a_i, $i = 1, 2, \ldots, n$, and b_i, $i = 1, 2, \ldots, n$, are given constants, and we assume that for each $j = 1, 2, \ldots, m$, h_j is continuously differentiable with respect to $\boldsymbol{x}$.

Given Problem (4.2.1), we can construct a corresponding unconstrained optimization problem as follows. Let $\varepsilon > 0$ and consider

$$\text{minimize} \left\{ J_\varepsilon(\boldsymbol{x}) = \sum_{j=1}^{m} \Phi_\varepsilon(h_j(\boldsymbol{x})) \right\} \tag{4.2.2a}$$

subject to

$$\boldsymbol{a} \leq \boldsymbol{x} \leq \boldsymbol{b}, \tag{4.2.2b}$$

where $\Phi_\varepsilon(\cdot)$ is defined by

$$\Phi_\varepsilon(h) = \begin{cases} h, & \text{if } h \geq \varepsilon, \\ \dfrac{(h+\varepsilon)^2}{4\varepsilon}, & \text{if } -\varepsilon \leq h \leq \varepsilon, \\ 0, & \text{if } h \leq -\varepsilon, \end{cases} \tag{4.2.3}$$

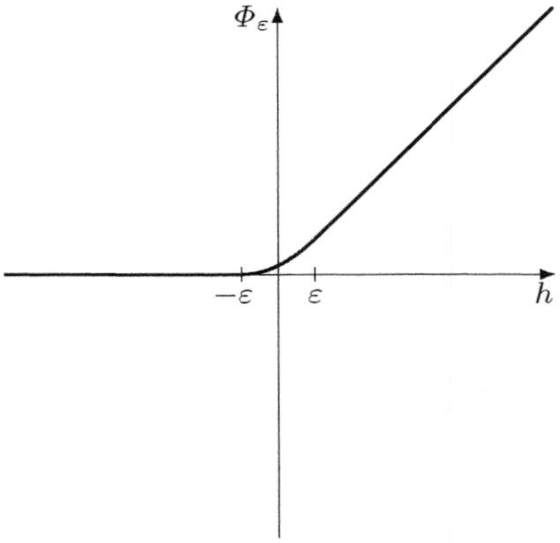

Fig. 4.2.1: Smoothing constraint transcription

see Figure 4.2.1.

Note that $\Phi_\varepsilon(\cdot)$ is differentiable. This unconstrained optimization problem is referred to as Problem (4.2.2).

Note that the boundedness constraints (4.2.2b) in Problem (4.2.2) can also be incorporated in the cost function using the constraint transcription defined by (4.2.3). We choose to leave them as such because most optimization algorithms are able to handle these boundedness constraints efficiently.

Theorem 4.2.1 (Necessary Condition) *Let* $\boldsymbol{x}^0 \in \mathbb{R}^n$ *be a solution of Problem* (4.2.1). *Then,*

$$J_\varepsilon\left(\boldsymbol{x}^0\right) \leq \frac{m\varepsilon}{4}. \tag{4.2.4}$$

Proof. In view of the constraint transcription defined in (4.2.3), it is clear from (4.2.1a) that

$$\Phi_\varepsilon\left(h_j\left(\boldsymbol{x}^0\right)\right) \leq \frac{\varepsilon}{4}.$$

Thus, the conclusion follows readily from (4.2.2a).

Theorem 4.2.2 (Sufficient Condition) *Let* $\boldsymbol{x}^0$, $\boldsymbol{a} \leq \boldsymbol{x}^0 \leq \boldsymbol{b}$, *be such that*

$$J_\varepsilon\left(\boldsymbol{x}^0\right) \leq \frac{\varepsilon}{4}. \tag{4.2.5}$$

Then, $\boldsymbol{x}^0$ *is a feasible solution of Problem* (4.2.1).

Proof. Suppose $\boldsymbol{x}^0$ is not feasible. Then there exists some j such that

$$h_j\left(\boldsymbol{x}^0\right) > 0.$$

This, in turn, implies that

$$\Phi_\varepsilon\left(h_j\left(\boldsymbol{x}^0\right)\right) > \frac{\varepsilon}{4},$$

and hence

$$J_\varepsilon\left(\boldsymbol{x}^0\right) > \frac{\varepsilon}{4}.$$

This is a contradiction, and hence the proof is complete.

Based on these two simple theorems, we can devise a computational algorithm for finding a feasible solution of Problem (4.2.1) in a finite number of iterations. There are many algorithms available in the literature for solving unconstrained optimization problems (such as the quasi-Newton method, see Chapter 2), and most of these can readily incorporate the boundedness constraints (4.2.1b). Suppose we employ such an algorithm to minimize (4.2.2a) subject to (4.2.2b), and also suppose that it is capable of finding a global optimal solution. Then, at the k-th step of that algorithm with iterate $\boldsymbol{x}^k$, we insert the following procedure:

Algorithm 4.2.1

Step 1. *If* $J(\boldsymbol{x}^{(k)}) > m\varepsilon/4$, *go to Step 3. Otherwise, go to Step 2.*

Step 2. *If the constraints* (4.2.1a) *are satisfied, go to Step 4. Otherwise, go to Step 3.*

Step 3. *Obtain the next iterate* $\boldsymbol{x}^{(k+1)}$. *Set* $k := k+1$, *and go to Step 1.*

Step 4. *Stop.* $\boldsymbol{x}^{(k)}$ *is a feasible solution.*

We can show that Algorithm 4.2.1 terminates in a finite number of iterations under a mild assumption.

Theorem 4.2.3 *Let* $\{\boldsymbol{x}^{(k)}\}$ *be a sequence of admissible points generated by Algorithm 4.2.1. If*

$$\lim_{k\to\infty} J_\varepsilon\left(\boldsymbol{x}^{(k)}\right) = 0, \tag{4.2.6}$$

then there exists a positive integer k_0 *such that* $\boldsymbol{x}^{(k)}$ *is a feasible solution of Problem* (4.2.1) *for all* $k > k_0$.

Proof. Since $\lim_{k\to\infty} J_\varepsilon\left(\boldsymbol{x}^{(k)}\right) = 0$, there exists a positive integer N such that

$$J_\varepsilon\left(\boldsymbol{x}^{(k)}\right) \le \varepsilon/4$$

for all $k > N$. Thus, by Theorem 4.2.2, $\boldsymbol{x}^{(N+1)}$ is a feasible solution of Problem (4.2.1).

In view of Theorem 4.2.3, we see that Algorithm 4.2.1 will find a feasible solution of Problem (4.2.1) in a finite number of iterations if (4.2.6) holds. In fact, if the optimization algorithm simply produces a point $\boldsymbol{x}^*$ such that the sufficient condition (4.2.5) is satisfied, then $\boldsymbol{x}^*$ is a feasible solution of the inequality constraints (4.2.1), so (4.2.6) is not strictly necessary for finite convergence.

Remark 4.2.1 *Note that Algorithm 4.2.1 can produce a non-feasible solution, which corresponds to a local minimum point for the corresponding unconstrained optimization problem. In such a case, the solution should obviously be rejected. The algorithm may be restarted from another initial point in an attempt to reach a global minimum.*

Remark 4.2.2 *ε is required to be positive, but, in practice, it does not need to be very small. Note that if we choose $\varepsilon = 0$, then Φ_ε becomes nonsmooth. In such a case, the necessary and sufficient condition for the solvability of Problem (4.2.1) is $J_\varepsilon = 0$. This, however, may require an infinite number of iterations and is thus not recommended. In general, an appropriate choice of ε will enhance the convergence characteristic of the algorithm. If the feasible space is sufficiently large, ε can be larger (and hence speed up convergence). Otherwise, ε should be made small.*

4.2.2 Continuous Inequality Constraints

Here, we shall extend the technique presented in Section 4.2.1 to find a feasible solution of the following continuous inequality constraints:

$$h_j(\boldsymbol{x}, t) \leq 0, \quad \forall\ t \in [0, T], \quad j = 1, 2, \ldots, m, \tag{4.2.7a}$$

and

$$\boldsymbol{a} \leq \boldsymbol{x} \leq \boldsymbol{b}, \tag{4.2.7b}$$

where $0 < T < \infty$.

For convenience, let this continuous inequality constrained problem be referred to as Problem (4.2.7).

For any $\varepsilon > 0$, we consider the following optimization problem:

$$\min \left\{ J_\varepsilon(\boldsymbol{x}) = \int_0^T \sum_{j=1}^m \Phi_\varepsilon(h_j(\boldsymbol{x}, t))dt \right\} \tag{4.2.8a}$$

subject to

$$\boldsymbol{a} \leq \boldsymbol{x} \leq \boldsymbol{b}, \tag{4.2.8b}$$

where $\Phi_\varepsilon(\cdot)$ is defined by (4.2.3) and, for each $j = 1, 2, \ldots, m$, h_j is assumed to satisfy the following conditions.

Assumption 4.2.1 *$\int_0^T \Phi_\varepsilon(h_j(\boldsymbol{x}, t))dt$ exists for all $\boldsymbol{x}$.*

Assumption 4.2.2 *$h_j(\boldsymbol{x}, t)$ is continuous in $t \in [0, T]$, and $\partial h_j(\boldsymbol{x}, t)/\partial t$ is piecewise continuous in $t \in [0, T]$ for each $\boldsymbol{x}$.*

Assumption 4.2.3 *$h_j(\boldsymbol{x}, t)$ is continuously differentiable with respect to $\boldsymbol{x}$ for all $t \in [0, T]$, except, possibly, at a finite number of points in $[0, T]$.*

Theorem 4.2.4 (Necessary Condition) *Let $\boldsymbol{x}^0 \in \mathbb{R}^n$ be a solution of Problem* (4.2.2)*. Then,*

$$J_\varepsilon\left(\boldsymbol{x}^0\right) \leq \frac{m\varepsilon T}{4}. \tag{4.2.9}$$

Proof. In view of the constraint transcription defined in (4.2.3), it is clear from (4.2.7a) that

$$\Phi_\varepsilon\left(h_j\left(\boldsymbol{x}^0, t\right)\right) \leq \frac{\varepsilon}{4}, \qquad \forall\, t \in [0, T]\,,\ j = 1, 2, \ldots, m. \tag{4.2.10}$$

Thus, the conclusion follows readily from (4.2.8a).

Our next task is to derive a sufficient condition for Problem (4.2.7). We first require the following lemma.

Lemma 4.2.1 *Let f be a non-negative valued function defined on $[0, T]$. If f is continuous on $[0, T]$ and df/dt is piecewise continuous on $[0, T]$, then*

$$\int_0^T f(t)dt \geq \frac{\widehat{f}}{2} \min\left\{\frac{\widehat{f}}{M}, T\right\}, \tag{4.2.11a}$$

where

$$M = \max_{t \in [0,T]} \left|\frac{df(t)}{dt}\right| \tag{4.2.11b}$$

and

$$\widehat{f} = \max_{t \in [0,T]} f(t). \tag{4.2.11c}$$

Proof. Let $t_0 \in [0, T]$ be such that $f(t_0) = \widehat{f}$. There are three cases to be considered:

$$t_0 + \frac{\widehat{f}}{M} \leq T, \tag{4.2.12a}$$

$$t_0 - \frac{\widehat{f}}{M} \geq 0, \tag{4.2.12b}$$

$$t_0 - \frac{\widehat{f}}{M} < 0 \quad \textit{and} \quad t_0 + \frac{\widehat{f}}{M} > T. \tag{4.2.12c}$$

Case (4.2.12a)**:** Define

$$h(t) = f(t_0) - M(t - t_0). \tag{4.2.13}$$

Then, for $t \in \left[t_0, t_0 + \widehat{f}/M\right]$,

$$f(t) - h(t) = f(t_0) + \int_{t_0}^t \frac{df(s)}{ds}\, ds - f(t_0) + M \int_{t_0}^t ds$$

$$= \int_{t_0}^{t} \left[\frac{df(s)}{ds} + M \right] ds \geq 0.$$

Hence,

$$\int_0^T f(t)\,dt \geq \int_{t_0}^{t_0+(\widehat{f}/M)} f(t)\,dt \geq \int_{t_0}^{t_0+(\widehat{f}/M)} h(t)\,dt = \frac{1}{2}\frac{\left(\widehat{f}\right)^2}{M}. \quad (4.2.14)$$

Case (4.2.12b): Define

$$h(t) = f(t_0) + M(t - t_0). \quad (4.2.15)$$

Then, for $t \in \left[kt_0 - \widehat{f}/M, t_0\right]$,

$$\begin{aligned} f(t) - h(t) &= f(t_0) + \int_{t_0}^{t} \frac{df(s)}{ds}\,ds - f(t_0) - M\int_{t_0}^{t} ds \\ &= \int_{t_0}^{t} \left[\frac{df(s)}{ds} - M\right] ds \\ &= \int_{t}^{t_0} \left[M - \frac{df(s)}{ds}\right] ds \geq 0. \end{aligned}$$

Hence,

$$\int_0^T f(t)\,dt \geq \int_{t_0-(\widehat{f}/M)}^{t_0} f(t)\,dt \geq \int_{t_0-(\widehat{f}/M)}^{t_0} h(t)\,dt = \frac{1}{2}\frac{\left(\widehat{f}\right)^2}{M}. \quad (4.2.16)$$

Case (4.2.12c): Define

$$h(t) = \begin{cases} f(t_0) - M(t - t_0),\, t \geq t_0, \\ f(t_0) + M(t - t_0),\, t < t_0. \end{cases} \quad (4.2.17)$$

Then,

$$f(t) - h(t) = \left\{ \begin{array}{l} \displaystyle\int_{t_0}^{t} \left[\frac{df(s)}{ds} + M\right] ds, \quad t \geq t_0 \\ \displaystyle\int_{t}^{t_0} \left[M - \frac{df(s)}{ds}\right] ds,\, t < t_0 \end{array} \right\} \geq 0, \ \forall t \in [0, T],$$

and hence,

$$\int_0^T f(t)\,dt \geq \int_0^T h(t)\,dt$$

$$
\begin{aligned}
&= \int_{t_0}^{T} h(t)\,dt + \int_0^{t_0} h(t)\,dt \\
&= (T - t_0)\left[\widehat{f} - \frac{M}{2}(T - t_0)\right] + t_0\left[\widehat{f} - \frac{M\,t_0}{2}\right] \\
&= T\,\widehat{f} - \frac{M}{2}(T - t_0)^2 - \frac{M}{2}(t_0)^2.
\end{aligned}
\tag{4.2.18}
$$

Note that

$$
t_0 + \frac{\widehat{f}}{M} > T \quad \Rightarrow \quad M(T - t_0) < \widehat{f},
$$

$$
t_0 - \frac{\widehat{f}}{M} < 0 \quad \Rightarrow \quad M\,t_0 < \widehat{f}.
$$

We have

$$
\begin{aligned}
T\,\widehat{f} - \frac{M}{2}(T - t_0)^2 - \frac{M}{2}(t_0)^2 &\geq T\,\widehat{f} - \frac{1}{2}\left[(T - t_0)\,\widehat{f} + t_0\,\widehat{f}\right] \\
&= T\,\widehat{f} - \frac{1}{2}T\,\widehat{f} = \frac{1}{2}T\,\widehat{f}.
\end{aligned}
$$

Thus, it follows from (4.2.18) that

$$
\int_0^T f(t)\,dt \geq \frac{1}{2}T\,\widehat{f}. \tag{4.2.19}
$$

Combining (4.2.14), (4.2.16) and (4.2.19), we obtain

$$
\int_0^T f(t)\,dt \geq \min\left\{\frac{\left(\widehat{f}\right)^2}{2M}, \frac{1}{2}T\,\widehat{f}\right\} = \frac{\widehat{f}}{2}\min\left\{\frac{\widehat{f}}{M}, T\right\}.
$$

Thus, the proof is complete.

With the help of Lemma 4.2.1, we are in a position to present the sufficient condition for Problem (4.2.7).

Theorem 4.2.5 *Let $\boldsymbol{x}^0$ be an admissible vector such that*

$$
J_\varepsilon(\boldsymbol{x}^0) \leq \frac{\varepsilon}{8}\min\left\{\frac{\varepsilon}{4M}, T\right\}, \tag{4.2.20}
$$

where

$$
M = \max\left\{\left|\frac{\partial g_j(\boldsymbol{x}^0, t)}{\partial t}\right| : t \in [0, T],\ \ j = 1, 2, \ldots, m\right\}. \tag{4.2.21}
$$

Then $\boldsymbol{x}^0$ is a solution of Problem (4.2.7).

Proof. Suppose $\boldsymbol{x}^0$ is not feasible. Then, by virtue of Assumption 4.2.2, there exist some $j \in \{1, 2, \ldots, m\}$ and an open set $\theta_j \subseteq [0, T]$ with positive measure such that

$$g_j(\boldsymbol{x}^0, t) > 0, \quad \forall\, t \in \theta_j. \tag{4.2.22}$$

By the second part of Assumption 4.2.2, there exists a positive constant M satisfying (4.2.21). Now, we define

$$\bar{M} = \max\left\{\left|\frac{\partial \Phi_\varepsilon(h_j(\boldsymbol{x}^0, t))}{\partial t}\right| : t \in [0, T],\ \ j = 1, 2, \ldots, m\right\}. \tag{4.2.23}$$

Clearly,

$$\left|\frac{\partial \Phi_\varepsilon(h_j(\boldsymbol{x}^0, t))}{\partial t}\right| \le \left|\frac{\partial \Phi_\varepsilon(h_j(\boldsymbol{x}^0, t))}{\partial h_j}\right| \left|\frac{\partial (h_j(\boldsymbol{x}^0, t))}{\partial t}\right|. \tag{4.2.24}$$

Since

$$\left|\frac{\partial \Phi_\varepsilon(h_j(\boldsymbol{x}^0, t))}{\partial h_j}\right| \le 1, \quad \forall\, t \in [0, T],$$

it is clear that

$$\bar{M} \le M. \tag{4.2.25}$$

Thus, by Lemma 4.2.1, we have

$$\int_0^T \Phi_\varepsilon(h_j(\boldsymbol{x}, t))\, dt \ge \frac{\Phi_\varepsilon^*}{2} \min\left\{\frac{\Phi_\varepsilon^*}{\bar{M}}, T\right\}, \tag{4.2.26}$$

where

$$\Phi_\varepsilon^* = \max_{t \in [0, T]} \Phi_\varepsilon(h_j(\boldsymbol{x}, t)). \tag{4.2.27}$$

By (4.2.22) and (4.2.3), we have

$$\Phi_\varepsilon^* > \frac{\varepsilon}{4}. \tag{4.2.28}$$

Thus, it follows from (4.2.28) and (4.2.25) that

$$\frac{\Phi_\varepsilon^*}{2} \min\left\{\frac{\Phi_\varepsilon^*}{\bar{M}}, T\right\} > \frac{\varepsilon}{8} \min\left\{\frac{\varepsilon}{4\bar{M}}, T\right\} \ge \frac{\varepsilon}{8} \min\left\{\frac{\varepsilon}{4M}, T\right\}. \tag{4.2.29}$$

This is, however, a contradiction to (4.2.20). Thus, the proof is complete.

Note that the gradient of the cost function (4.2.8a) with respect to each $\boldsymbol{x} \in [\boldsymbol{a}, \boldsymbol{b}]$ is given by

$$\nabla J_\varepsilon(\boldsymbol{x}) = \int_0^T \sum_{j=1}^m \frac{\partial \Phi_\varepsilon(h_j(\boldsymbol{x}, t))}{\partial \boldsymbol{x}} dt, \tag{4.2.30}$$

where

$$\frac{\partial \Phi_\varepsilon(h_j(\boldsymbol{x},t))}{\partial \boldsymbol{x}} = \begin{cases} \dfrac{\partial h_j(\boldsymbol{x},t)}{\partial \boldsymbol{x}}, & \text{if } h_j(\boldsymbol{x},t) \geq \varepsilon, \\ \left(\dfrac{h_j(\boldsymbol{x},t)+\varepsilon}{2\varepsilon}\right)\dfrac{\partial h_j(\boldsymbol{x},t)}{\partial \boldsymbol{x}}, & \text{if } -\varepsilon \leq h_j(\boldsymbol{x},t) \leq \varepsilon, \\ 0, & \text{if } -\varepsilon \leq h_j(\boldsymbol{x},t). \end{cases} \tag{4.2.31}$$

Thus, Problem (4.2.8) can be viewed as a standard unconstrained optimization problem and hence is solvable by any efficient unconstrained optimization technique such as the quasi-Newton method (see Chapter 2). We may summarize these findings in the following algorithm.

Algorithm 4.2.2 *At the k-th iteration of the unconstrained optimization algorithm with iterate $\boldsymbol{x}^{(k)}$, we insert the following steps:*

Step 1. If $J_\varepsilon(\boldsymbol{x}^{(k)}) > \frac{m\varepsilon T}{4}$, go to Step 3; otherwise, go to Step 2.
Step 2. Check if the constraints (4.2.7a) *are satisfied. If so, go to Step 4; otherwise, go to Step 3.*
Step 3. Continue the next iteration of the unconstrained optimization method to obtain $\boldsymbol{x}^{(k+1)}$. Set $k := k+1$, and go to Step 1.
Step 4. Stop. $\boldsymbol{x}^{(k)}$ is a feasible solution.

The following theorem shows that Algorithm 4.2.2 terminates in a finite number of iterations.

Theorem 4.2.6 *Let $\{\boldsymbol{x}^{(k)}\}$ be a sequence of admissible points generated by Algorithm 4.2.2. If*

$$\lim_{k\to\infty} J_\varepsilon\left(\boldsymbol{x}^{(k)}\right) = 0, \tag{4.2.32}$$

then there exists a positive integer k_0 such that $\boldsymbol{x}^{(k)}$ is a feasible solution of Problem (4.2.7) *for all $k > k_0$.*

Proof. Since

$$\lim_{k\to\infty} J_\varepsilon\left(\boldsymbol{x}^{(k)}\right) = 0,$$

there exists a positive integer N such that

$$J_\varepsilon\left(\boldsymbol{x}^{(k)}\right) \leq \frac{\varepsilon}{8}\min\left\{\frac{\varepsilon}{4M}, T\right\} \tag{4.2.33}$$

for all $k > N$. Thus, by Theorem 4.2.5, $\boldsymbol{x}^{(N+1)}$ is a feasible solution of Problem (4.2.7). This completes the proof.

As in the previous case, it is clear from Theorem 4.2.6 that Algorithm 4.2.2 will find a feasible solution of Problem (4.2.7) in a finite number of iterations if (4.2.32) holds. Again, it is possible that a solution can still be obtained in a finite number of steps even when (4.2.32) is not satisfied. Also, the algorithm may converge to a non-global minimum, in which case one may repeat it with a different starting point in the hope of reaching the global optimum.

4.3 Continuous Inequality Constraint Transcription Approach

We consider a class of optimization problems subject to continuous inequality constraints, where the cost function

$$f(\boldsymbol{x}) \tag{4.3.1a}$$

is minimized subject to inequality constraints

$$h_j(\boldsymbol{x}) \leq 0, \quad j = 1, 2, \ldots, p \tag{4.3.1b}$$

and continuous inequality constraints

$$\phi_j(\boldsymbol{x}, t) \leq 0, \quad \forall\, t \in [0, T], \; j = 1, 2, \ldots, m, \tag{4.3.1c}$$

where $\boldsymbol{x} \in \mathbb{R}^n$ is the parameter vector to be found and T is such that $0 < T < \infty$. We assume throughout this section that the following conditions are satisfied.

Assumption 4.3.1 h_j, $j = 1, \ldots, p$, *are continuously differentiable in* $\mathbb{R}^n$.

Assumption 4.3.2 $\phi_j : \mathbb{R}^n \times \mathbb{R} \to \mathbb{R}$, $j = 1, \ldots, m$, *satisfy Assumptions* (4.2.2) *and* (4.2.3).

For convenience, let this constrained optimization problem be referred to as Problem (4.3.1).

We shall present two computational methods for solving Problem (4.3.1). For the first method, the constraint transcription introduced in Section 4.2 is used to transform the continuous inequality constraints into equality constraints. However, these equality constraints are nonsmooth. Thus a local smoothing technique, also introduced in Section 4.2, is used to approximate these equality constraint functions by smooth inequality constraints. Then we construct a sequence of approximate optimization problems subject to conventional inequality constraints. Each of these approximate problems can be viewed as a conventional optimization problem and hence can be solved by existing optimization techniques, such as those reported in Chapter 3. With this ground work, we can construct an effective algorithm for solving Problem (4.3.1). The proposed algorithm will depend on two parameters, one controlling the smoothing and the other controlling the position of solutions (with respect to feasibility–infeasibility) of a sequence of conventional optimization problems. It will be shown that the algorithm produces a sequence of suboptimal parameter vectors approaching to an optimal parameter vector from inside the feasible region of Problem (4.3.1). For the second method, we first apply the same constraint transcription and smoothing technique that were suggested in Section 4.2 to transform the continuous inequality constraints. Then, a special penalty function will be used to incorporate these transformed constraints into the cost function, thus forming a new augmented cost function. We show that this is different from usual penalty function methods in that the penalty weighting factor does not need to approach infinity to yield a feasible solution. As for the first method, the smoothing technique gives rise to a sequence of approximate problems, each of which can be regarded as a conventional unconstrained optimization problem. Thus, each of the approximate problems can be solved by existing optimization techniques such as the quasi-Newton method. The algorithm for solving (4.3.1) then involves constructing and solving a number of approximate problems. It depends on two parameters, one controlling the smoothing and the other controlling the position of solutions with respect to the feasible region. It is further shown that, as for the first method, the proposed algorithm produces a sequence of suboptimal parameter vectors which approaches an optimal parameter vector of (4.3.1) from inside the feasible region of (4.3.1). The advantage of the second method is that it effectively turns Problem (4.3.1) into an unconstrained problem that is easier to solve. Thus it can handle problems of much larger dimensions.

4.3.1 The First Method

The method presented in this subsection first appeared in [103].

For each $j = 1, 2, \ldots, m$, define

$$G_j(\boldsymbol{x}) = \int_0^T \max\{\phi_j(\boldsymbol{x}, t), 0\}\, dt.$$

Since ϕ_j is continuously differentiable in $\boldsymbol{x}$ and t, $\max\{\phi_j(\boldsymbol{x}, t), 0\}$ is a continuous function of t for each $\boldsymbol{x} \in \mathbb{R}^n$. Thus, for each $j = 1, 2, \ldots, m$, the corresponding continuous constraint (4.3.1c) is equivalent to

$$G_j(\boldsymbol{x}) = 0. \tag{4.3.2}$$

For convenience, let us rewrite Problem (4.3.1) with (4.3.1c) replaced by (4.3.2) as follows:
Minimize the cost function

$$f(\boldsymbol{x}) \tag{4.3.3a}$$

subject to

$$h_j(\boldsymbol{x}) \leq 0, \quad j = 1, 2, \ldots, p \tag{4.3.3b}$$

and

$$G_j(\boldsymbol{x}) = 0, \quad j = 1, 2, \ldots, m. \tag{4.3.3c}$$

This problem is referred to as Problem (4.3.3). Clearly, Problem (4.3.1) is equivalent to Problem (4.3.3).

Note that for each $j = 1, 2, \ldots, m$, $G_j(\boldsymbol{x})$ is nonsmooth in $\boldsymbol{x}$. Hence, standard optimization routines can have difficulty with these equality constraints.

Let

$$\Theta = \{\boldsymbol{x} \in \mathbb{R}^n : h_j(\boldsymbol{x}) \leq 0,\ j = 1, 2, \ldots, p\}, \tag{4.3.4}$$

and let $\mathcal{F}$ be the feasible region of Problem (4.3.1) defined by

$$\begin{aligned} \mathcal{F} &= \{\boldsymbol{x} \in \Theta : \phi_j(\boldsymbol{x}, t) \leq 0, \quad \forall t \in [0, T], \quad j = 1, 2, \ldots, m\} \\ &= \{\boldsymbol{x} \in \Theta : G_j(\boldsymbol{x}) = 0,\ j = 1, 2, \ldots, m\}. \end{aligned} \tag{4.3.5}$$

Furthermore, let $\mathring{\Theta}$ (respectively, $\mathring{\mathcal{F}}$) be the interior of the set Θ (respectively, $\mathcal{F}$) in the sense that

$$\mathring{\Theta} = \{\boldsymbol{x} \in \mathbb{R}^n : h_j(\boldsymbol{x}) < 0,\ j = 1, 2, \ldots, p\} \tag{4.3.6}$$

and

$$\mathring{\mathcal{F}} = \left\{\boldsymbol{x} \in \mathring{\Theta} : \phi_j(\boldsymbol{x}, t) < 0, \quad \forall\, t \in [0, T], \quad j = 1, 2, \ldots, m\right\}. \tag{4.3.7}$$

We assume that the following condition is satisfied.

Assumption 4.3.3 $\mathring{\mathcal{F}} \neq \emptyset$.

The smoothing technique is to replace $\max\{\phi_j(\boldsymbol{x}, t), 0\}$ by $\Phi_\varepsilon(\phi_j(\boldsymbol{x}, t))$, where $\Phi_\varepsilon(\cdot)$ is defined by (4.2.3). Note that for each $j = 1, 2, \ldots, m$, $\Phi_\varepsilon(\phi_j(\boldsymbol{x}, t))$ is continuously differentiable in $\boldsymbol{x}$. However, its second derivative is discontinuous at those points where $\phi_j(\boldsymbol{x}, t) = \pm\varepsilon$. For each $j = 1, 2, \ldots, m$, define

$$G_{j,\varepsilon}(\boldsymbol{x}) = \int_0^T \Phi_\varepsilon(\phi_j(\boldsymbol{x}, t))\, dt. \tag{4.3.8}$$

We assume that the following condition is satisfied.

Assumption 4.3.4 *$\int_0^T \Phi_\varepsilon(\phi_j(\boldsymbol{x}, t))\, dt$ exists for all $\boldsymbol{x}$.*

Clearly, for each $j = 1, \ldots, m$, $G_{j,\varepsilon}$ is continuously differentiable in $\boldsymbol{x}$. We now define two related approximate problems. The first approximate problem is as follows: Minimize

$$f(\boldsymbol{x}) \tag{4.3.9a}$$

subject to

$$\boldsymbol{x} \in \Theta \tag{4.3.9b}$$

and

$$G_{j,\varepsilon}(\boldsymbol{x}) = 0, \quad j = 1, 2, \ldots, m. \tag{4.3.9c}$$

This problem is referred to as Problem (4.3.9). Note that the equality constraints specified in (4.3.9c) do not satisfy the regular point condition, which is also known as the constraint qualification (see Chapter 3). Thus, it is not advisable to solve it numerically as such. For this reason, we consider our second approximate problem as follows: Minimize

$$f(\boldsymbol{x}) \tag{4.3.10a}$$

subject to

$$\boldsymbol{x} \in \Theta \tag{4.3.10b}$$

and

$$G_{j,\varepsilon}(\boldsymbol{x}) \leq \tau, \quad j = 1, 2, \ldots, m. \tag{4.3.10c}$$

This problem is referred to as Problem (4.3.10). Let $\mathcal{F}_\varepsilon$ be the feasible region of Problem (4.3.9) defined by

$$\mathcal{F}_\varepsilon = \{\boldsymbol{x} \in \Theta : G_{j,\varepsilon}(\boldsymbol{x}) = 0, \quad j = 1, 2, \ldots, m\}. \tag{4.3.11}$$

Then, for each $\varepsilon > 0$, $\mathcal{F}_\varepsilon \subset \mathcal{F}$. We assume that the following condition is satisfied.

Assumption 4.3.5 *Let $\boldsymbol{x}^*$ be an optimal parameter vector of Problem* (4.3.1). *Then, there exists a parameter vector $\boldsymbol{x} \in \overset{\circ}{\mathcal{F}}$ such that*

$$\alpha\boldsymbol{x} + (1-\alpha)\boldsymbol{x}^* \in \overset{\circ}{\mathcal{F}}$$

for all $\alpha \in (0, 1]$.

For the subsequent convergence result, we need the following assumption.

Assumption 4.3.6 *The set Θ defined by* (4.3.4) *is compact.*

Remark 4.3.1 *As an alternative to Assumption* (4.3.6), *we can impose the requirement that*

$$f(\boldsymbol{x}) \to \infty \text{ as } \|\boldsymbol{x}\| \to \infty.$$

We now relate the solutions of Problem (4.3.1) and Problem (4.3.9) as $\varepsilon \to 0$ in the following theorem.

Theorem 4.3.1 *Let $\boldsymbol{x}^*$ be an optimal solution to Problem* (4.3.1), *and let $\boldsymbol{x}_\varepsilon^*$ be an optimal solution to Problem* (4.3.9). *Then,*

$$\lim_{\varepsilon \to 0} f(\boldsymbol{x}_\varepsilon^*) = f(\boldsymbol{x}^*).$$

Proof. By Assumption 4.3.5, there exists an $\bar{\boldsymbol{x}} \in \overset{\circ}{\mathcal{F}}$ such that

$$\boldsymbol{x}_\alpha = \alpha\bar{\boldsymbol{x}} + (1-\alpha)\boldsymbol{x}^* = \boldsymbol{x}^* + \alpha(\bar{\boldsymbol{x}} - \boldsymbol{x}^*) \in \overset{\circ}{\mathcal{F}}, \quad \forall \alpha \in (0, 1]. \tag{4.3.12}$$

For any $\delta_1 > 0$, there exists an $\alpha_1 \in (0,1]$ such that

$$f(\boldsymbol{x}^*) \leq f(\boldsymbol{x}_\alpha) \leq f(\boldsymbol{x}^*) + \delta_1, \quad \forall \alpha \in (0, \alpha_1). \tag{4.3.13}$$

Choose $\alpha_2 = \alpha_1/2$. Then it is clear that $\boldsymbol{x}_{\alpha_2} \in \overset{\circ}{\mathcal{F}}$. Thus, there exists a $\delta_2 > 0$ such that

$$\phi_j(\boldsymbol{x}_{\alpha_2}, t) < -\delta_2, \quad \forall\, t \in [0, T] \text{ and } j = 1, 2, \ldots, m.$$

If we choose $\varepsilon = \delta_2$, then $\boldsymbol{x}_{\alpha_2} \in \mathcal{F}_\varepsilon$. Thus it follows that

$$f(\boldsymbol{x}_\varepsilon^*) \leq f(\boldsymbol{x}_{\alpha_2}). \tag{4.3.14}$$

From (4.3.13) and (4.3.14), we have

$$f(\boldsymbol{x}^*) \leq f(\boldsymbol{x}_\varepsilon^*) \leq f(\boldsymbol{x}^*) + \delta_1.$$

Letting $\varepsilon \to 0$ and noting that $\delta_1 > 0$ is arbitrary, the conclusion of the theorem follows.

Theorem 4.3.2 *Let $\boldsymbol{x}^*$ and $\boldsymbol{x}_\varepsilon^*$ be as in Theorem 4.3.1. Then, the sequence $\{\boldsymbol{x}_\varepsilon^*\}$ has an accumulation point. Furthermore, any accumulation point of the sequence $\{\boldsymbol{x}_\varepsilon^*\}$ is an optimal parameter vector of Problem* (4.3.1).

Proof. By Assumption 4.3.6, we note that $\{\boldsymbol{x}_\varepsilon^*\}$ is in a compact set. Thus, there exists a subsequence, denoted again by the original sequence, such that

$$\lim_{\varepsilon \to 0} \|\boldsymbol{x}_\varepsilon^* - \hat{\boldsymbol{x}}\| = 0. \tag{4.3.15}$$

We shall show that $\hat{\boldsymbol{x}} \in \mathcal{F}$. Suppose not. Then, there exist a j and a non-zero interval $I \subset [0, T]$ such that

$$\max\{\phi_j(\hat{\boldsymbol{x}}, t), 0\} = \phi_j(\hat{\boldsymbol{x}}, t) > 0, \ \forall\, t \in I.$$

However, $\boldsymbol{x}_\varepsilon^* \in \mathcal{F}_\varepsilon \subset \mathcal{F}$ so that $\phi_j(\boldsymbol{x}_\varepsilon^*, t) \leq -\varepsilon$. Since, for each t, $\phi_j(\cdot, t)$ is continuous in Θ, it follows from (4.3.15) that

$$|\phi_j(\boldsymbol{x}_\varepsilon^*, t) - \phi_j(\hat{\boldsymbol{x}}, t)| \to 0, \text{ as } \varepsilon \to 0.$$

Furthermore, $|\phi_j(\boldsymbol{x}_\varepsilon^*, t) - \phi_j(\hat{\boldsymbol{x}}, t)|$ is uniformly bounded in $\Theta \times [0, T]$. Hence, by the Lebesgue Dominated Convergence Theorem (Theorem A.1.10), we have

$$\lim_{\varepsilon \to 0} \int_I |\phi_j(\boldsymbol{x}_\varepsilon^*, t) - \phi_j(\hat{\boldsymbol{x}}, t)|\, dt = 0.$$

Therefore,

$$0 < \int_I \phi_j(\hat{\boldsymbol{x}}, t)\, dt \leq \lim_{\varepsilon \to 0} \int_I |\phi_j(\boldsymbol{x}_\varepsilon^*, t) - \phi_j(\hat{\boldsymbol{x}}, t)|\, dt = 0.$$

This contradiction shows that $\hat{\boldsymbol{x}}$ is feasible.

Recall that f is continuous on $\mathbb{R}^n$. Thus, by (4.3.15), we have

$$\lim_{\varepsilon \to 0} f\left(\boldsymbol{x}_{\varepsilon}^{*}\right) = f\left(\hat{\boldsymbol{x}}\right).$$

Combining this result and Theorem 4.3.1, the conclusion follows.

Theorem 4.3.3 *For any $\varepsilon > 0$, there exists a $\tau(\varepsilon) > 0$ such that for all τ, $0 < \tau < \tau(\varepsilon)$, any feasible solution of Problem* (4.3.10) *is also a feasible point of Problem* (4.3.1).

Proof. We first recall that for each $j = 1, 2, \ldots, m$, ϕ_j is continuously differentiable in $\Theta \times [0, T]$. Since $\Theta \times [0, T]$ is compact, there exists a positive constant m_j such that

$$\left| \frac{\partial \phi_j\left(\boldsymbol{x}, t\right)}{\partial t} \right| \leq m_j, \quad \forall\ (\boldsymbol{x}, t) \in \Theta \times [0, T]. \tag{4.3.16}$$

Next, for any $\varepsilon > 0$, define

$$k_{j,\varepsilon} = \frac{\varepsilon}{16} \min\left\{ \frac{T}{2}, \frac{\varepsilon}{2m_j} \right\}. \tag{4.3.17}$$

It suffices to show that

$$\mathcal{F}_{j,\varepsilon,\tau} \subset \mathcal{F}_j \tag{4.3.18}$$

for any τ such that

$$0 < \tau < k_{j,\varepsilon}, \tag{4.3.19}$$

where

$$\mathcal{F}_{j,\varepsilon,\tau} = \left\{ \boldsymbol{x} \in \Theta : G_{j,\varepsilon}\left(\boldsymbol{x}\right) \leq \tau \right\} \tag{4.3.20}$$

and

$$\mathcal{F}_j = \left\{ \boldsymbol{x} \in \Theta : G_j\left(\boldsymbol{x}\right) = 0 \right\}. \tag{4.3.21}$$

Assume the contrary. Then there exists an $\boldsymbol{x} \in \Theta$ such that

$$G_{j,\varepsilon}\left(\boldsymbol{x}\right) \leq \tau < k_{j,\varepsilon}, \tag{4.3.22}$$

but

$$G_j\left(\boldsymbol{x}\right) > 0. \tag{4.3.23}$$

Since ϕ_j is a continuously differentiable function in $[0, T]$, (4.3.23) implies that there exists a $\bar{t} \in [0, T]$ such that

$$\phi_j\left(\boldsymbol{x}, \bar{t}\right) > 0. \tag{4.3.24}$$

Again by continuity, there exists an interval $I_j \subset [0, T]$ containing $\bar{t}$ such that

$$\phi_j\left(\boldsymbol{x}, \bar{t}\right) > -\frac{\varepsilon}{2}. \tag{4.3.25}$$

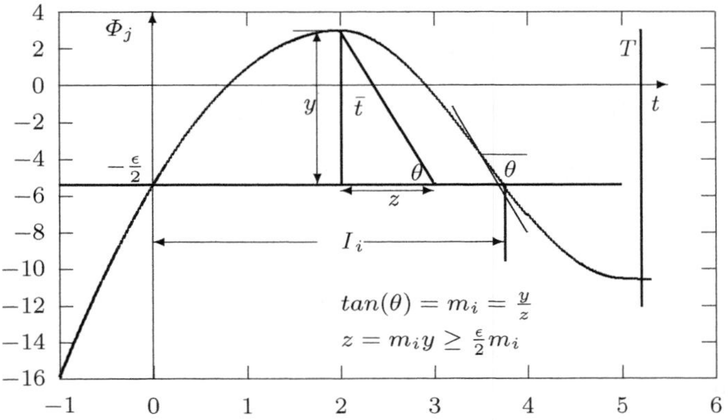

Fig. 4.3.1: Geometrical interpretation of Theorem 4.3.3

See Figure 4.3.1. Using (4.3.16), it is clear from (4.3.25) that the length $|I_j|$ of the interval I_j must satisfy

$$|I_j| \geq \min\left\{\frac{T}{2}, \frac{\varepsilon}{2m_j}\right\}. \tag{4.3.26}$$

From the definition of $G_{j,\varepsilon}(\boldsymbol{x})$, note that

$$\begin{aligned} G_{j,\varepsilon}(\boldsymbol{x}) &= \int_0^T g_{j,\varepsilon}(\boldsymbol{x},t)\, dt \geq \int_{I_j} g_{j,\varepsilon}(\boldsymbol{x},t)\, dt \geq \int_{I_j} \min_{t\in I_j} g_{j,\varepsilon}(\boldsymbol{x},t)\, dt \\ &\geq \min_{t\in I_j}\left\{(\phi_j(\boldsymbol{x},t)+\varepsilon)^2/4\varepsilon\right\}|I_j| \geq \frac{\varepsilon}{16}\min\left\{\frac{T}{2}, \frac{\varepsilon}{2m_j}\right\} \\ &= k_{j,\varepsilon}. \end{aligned}$$

This is a contradiction to (4.3.22). Thus, the proof is complete.

Algorithm 4.3.1 *(Note that the integral in* (4.3.8) *is replaced by any suitable quadrature with positive weights in any computational scheme.)*

Step 1. Choose $\varepsilon = 10^{-1}$, $\tau = \varepsilon T/4$ *(say).*

Step 2. Solve Problem (4.3.10) *to an accuracy of* $\max\{10^{-8}, \varepsilon 10^{-3}\}$ *to give* $\boldsymbol{x}^*_{\varepsilon,\tau}$.

Step 3. Check the feasibility of $\phi_j(\boldsymbol{x}^*_{\varepsilon,\tau}) \leq 0$ *for all* $j = 1, 2, \ldots, m$ *at the quadrature points. If* $\boldsymbol{x}^*_{\varepsilon,\tau}$ *is feasible, go to Step 4. Otherwise, set* $\tau = \tau/2$. *If* $\tau < 10^{-10}$, *stop, abnormal exit. Else, go to Step 2.*

Step 4. Set $\varepsilon = \varepsilon/10$, and $\tau = \tau/10$. If $\varepsilon > 10^{-7}$, go to Step 2. Otherwise, stop, successful exit.

Remark 4.3.2 *Note that the number of quadrature points may need to be increased as $\varepsilon \to 0$ and that for large values of ε, a small number of quadrature points are sufficient.*

Remark 4.3.3 *From Theorem 4.3.2, we see that the halving process of τ in Step 2 of the algorithm only needs to be carried out a finite number of times. Thus, the algorithm produces a sequence of suboptimal parameter vectors to Problem* (4.3.1)*, where each of them is in the feasible region of* (4.3.1).

Remark 4.3.4 *From the proof of Theorem 4.3.3, we see that ε and τ are closely related. At the solution of a particular problem, if a constraint is active over a large fraction of $[0,T]$, then it appears that $\tau = O(\varepsilon)$. On the other hand, if the constraint is active at only one point in $[0,T]$, then $\tau = O(\varepsilon^2)$.*

Theorem 4.3.4 *Let $\{\boldsymbol{x}^*_{\varepsilon,\tau}\}$ be a sequence of the suboptimal parameter vectors produced by the above algorithm. Then,*

$$f\left(\boldsymbol{x}^*_{\varepsilon,\tau}\right) \to f\left(\boldsymbol{x}^*\right),$$

and any accumulation point of $\{\boldsymbol{x}^*_{\varepsilon,\tau}\}$ is a solution of Problem (4.3.1).

Proof. Clearly,

$$f\left(\boldsymbol{x}^*_{\varepsilon}\right) \le f\left(\boldsymbol{x}^*_{\varepsilon,\tau}\right) \le f\left(\boldsymbol{x}^*\right).$$

Since $f(\boldsymbol{x}^*_\varepsilon) \to f(\boldsymbol{x}^*)$, it follows that $f\left(\boldsymbol{x}^*_{\varepsilon,\tau}\right) \to f(\boldsymbol{x}^*)$.

To prove the second part of the theorem, we note from (4.3.A6) that the sequence $\{\boldsymbol{x}^*_{\varepsilon,\tau}\}$ is in a compact set. Thus, the existence of an accumulation point is assured. On this basis, the conclusion follows easily from the continuity of the function f.

Example 4.3.1 We choose an example that was used in [81] and [241]. Minimize

$$f(\boldsymbol{x}) = \frac{x_2\left(122 + 17x_1 + 6x_3 - 5x_2 + x_1x_3\right) + 180x_3 - 36x_1 + 1224}{x_2\left(408 + 56x_1 - 50x_2 + 60x_3 + 10x_1x_3 - 2x_1^2\right)} \tag{4.3.27a}$$

subject to

$$\phi(\boldsymbol{x},\omega) \le 0, \ \forall\, \omega \in \Omega, \tag{4.3.27b}$$

where

$$\phi(\boldsymbol{x},\omega) = \mathcal{J}(T(\boldsymbol{x},\omega)) - 3.33\left[\mathcal{R}(T(\boldsymbol{x},\omega))\right]^2 + 1.0,$$

and where $\Omega = \left[10^{-6}, 30\right]$, $i^2 = -1$, $T(\boldsymbol{x},\omega) = 1 + H(\boldsymbol{x}, i\omega)\, G[i\omega]$,

$$H(\boldsymbol{x}, x) = x_1 + x_2/s + x_3 s,$$

$$G(s) = \frac{1}{(s+3)(s_2+2s+2)},$$

and $\mathcal{R}(\cdot)$ and $\mathcal{J}(\cdot)$ denote the real and imaginary parts of their arguments, respectively. Finally, there are simple bounds on the variables:

$$0 \leq x_1 \leq 100, \quad 0.1 \leq x_2 \leq 100, \quad 0 \leq x_3 \leq 100. \tag{4.3.27c}$$

We apply Algorithm 4.3.1 to the problem and report the results in Table 4.3.1. In the table, we report a "failure" parameter F. $F = 0$ indicates normal termination, and a minimum has been found. $F = 2$ indicates that the convergence criteria are not satisfied, but no downhill search could be determined. $F = 3$ means the maximum number of iterations, 50, has been exceeded. The number of function evaluations is given by n_f with a '$*$' indicating when the maximum number of iterations was reached. ε, τ and f and x_1, x_2 and x_3 are self-explanatory. $\bar{\omega}$ gives the approximate value of ω where ϕ attains its maximum over Ω at the computed $\boldsymbol{x}$ value. The value of $g(\boldsymbol{x})$ reported gives an idea of feasibility/infeasibility over 2001 equally spaced points in Ω, which is a much finer partition than that used for the quadrature formula. A '$*$' in these columns indicates that the solution found is well within the feasible region. Finally, λ is the Lagrange multiplier for the constraint. Note, in particular, the good failure record of $F = 0$ for all iterations, leaving no doubt that a solution has been found.

Table 4.3.1: Problem (4.3.27)

ε	τ	$\bar{\omega}$	g	f	$x1$	$x2$	$x3$	n_f	F	λ
10^{-1}	10^{-2}	4.095	-2.9×10^{-2}	0.2223351	28.23895	22.40785	19.51560	25	0	0.0
10^{-2}	10^{-3}	5.655	1.2×10^{-3}	0.1745032	17.03023	45.46166	34.69278	48	0	-0.440
	5×10^{-4}	5.655	-1.4×10^{-3}	0.1747744	16.97336	45.46475	34.59242	5	0	-0.706
10^{-3}	5×10^{-5}	5.655	5.3×10^{-6}	0.1746270	17.00433	45.46298	34.64676	11	0	-0.975
10^{-4}	5×10^{-6}	5.670	1.8×10^{-4}	0.1746131	16.75351	45.45822	34.80058	19	0	-0.991
10^{-5}	5×10^{-7}	5.670	7.9×10^{-5}	0.1746124	16.75367	45.45822	34.80085	8	0	-0.991
10^{-6}	5×10^{-8}	5.670	2.7×10^{-5}	0.1746123	16.75368	45.45823	34.80088	6	0	-0.990
10^{-7}	5×10^{-9}	5.670	8.8×10^{-6}	0.1746123	16.75368	45.45823	34.80088	6	0	-0.990

Table 4.3.2 shows the effect of increasing the quadrature accuracy as ε is decreased for our algorithm. The number of quadrature points at each iteration is indicated by nq. While less function evaluations are required for the earlier iterations as expected, more are required for the later iterations.

Table 4.3.2: Problem (4.3.27) with variable quadrature points

ϵ	τ	nq	$\bar{\omega}$	g	f	$x1$	$x2$	$x3$	nf	F	λ
10^{-1}	10^{-2}	8	3.720	-8.5×10^{-2}	0.2417111	26.28261	20.49562	16.42453	13	0	0.0
10^{-2}	10^{-3}	16	*	*	0.1788802	15.94624	31.52512	34.15840	30	0	0.0
10^{-3}	10^{-4}	32	5.610	-5.3×10^{-4}	0.1747647	17.80857	44.56686	34.13024	34	0	−0.325
10^{-4}	10^{-5}	64	5.670	1.6×10^{-4}	0.1746221	16.77723	44.63652	34.76034	28	0	−0.490
10^{-5}	10^{-6}	128	5.670	7.5×10^{-5}	0.1746210	16.77749	44.63660	34.76076	8	0	−0.693
	5×10^{-7}	128	5.670	7.4×10^{-5}	0.1746214	16.77740	44.63670	34.76061	6	0	−0.980
10^{-6}	5×10^{-8}	256	5.655	2.4×10^{-5}	0.1746128	16.82795	45.48089	34.75696	18	0	−1.394
	2.5×10^{-8}	256	5.655	2.4×10^{-5}	0.1746129	16.82788	45.48090	34.75699	5	0	−1.969
10^{-7}	2.5×10^{-9}	512	5.655	6.3×10^{-6}	0.1746168	16.96054	45.48194	34.67752	29	0	−1.901

4.3.2 The Second Method

In this section, we shall introduce a second computational procedure for solving Problem (4.3.1). It was originally proposed in [259]. We first apply the same constraint transcription introduced in Section 4.2 to the continuous inequality constraints. For each $j = 1, 2, \ldots, m$, define

$$G_j(\boldsymbol{x}) = \int_0^T \max\{\phi_j(\boldsymbol{x}, t), 0\}dt. \tag{4.3.28}$$

We assume once more that Assumptions 4.3.1, 4.3.2 and 4.3.4 are satisfied. The continuous inequality constraints (4.3.1c) are equivalent to

$$G_j(\boldsymbol{x}) = 0, \quad j = 1, 2, \ldots, m. \tag{4.3.29}$$

Let Θ, $\mathring{\Theta}$, $\mathcal{F}$ and $\mathring{\mathcal{F}}$ be as defined in (4.3.4), (4.3.6), (4.3.5) and (4.3.7), respectively. We further assume that Assumptions 4.3.3, 4.3.5, and 4.3.6 are satisfied.

Note that, for each $j = 1, 2, \ldots, m$, $G_j(\boldsymbol{x})$ is, in general, nonsmooth in $\boldsymbol{x}$. Consequently, standard optimization routines would have difficulties with this type of equality constraints. The smoothing technique is to replace $\max\{\phi_j(\boldsymbol{x}, t), 0\}$ with $\phi_{j,\varepsilon}(\boldsymbol{x}, t)$, where

$$\phi_{j,\varepsilon}(\boldsymbol{x}, t) = \begin{cases} 0, & \text{if } \phi_j(\boldsymbol{x}, t) < -\varepsilon, \\ (\phi_j(\boldsymbol{x}, t) + \varepsilon)^2/4\varepsilon, & \text{if } -\varepsilon \le \phi_j(\boldsymbol{x}, t) \le \varepsilon, \\ \phi_j(\boldsymbol{x}, t), & \text{if } \phi_j(\boldsymbol{x}, t) > \varepsilon. \end{cases} \tag{4.3.30}$$

For each $j = 1, 2, \ldots, m$, define

$$G_{j,\varepsilon}(\boldsymbol{x}) = \int_0^T \phi_{j,\varepsilon}(\boldsymbol{x}, t)dt. \tag{4.3.31}$$

Then, since for each $j = 1, 2, \ldots, m$, $\phi_{j,\varepsilon}(\boldsymbol{x}, t)$ is continuously differentiable in $\boldsymbol{x}$, so is $G_{j,\varepsilon}(\boldsymbol{x})$. Let

$$\begin{aligned}\mathcal{F}_\varepsilon &= \{\boldsymbol{x} \in \Theta : G_{j,\varepsilon}(\boldsymbol{x}) = 0, \quad j = 1, 2, \ldots, m\} \\ &= \{\boldsymbol{x} \in \Theta : \phi_j(\boldsymbol{x}, t) \leq -\varepsilon, \quad \forall\, t \in [0, T], \quad j = 1, 2, \ldots, m\}. \end{aligned} \tag{4.3.32}$$

Then, clearly, $\mathcal{F}_\varepsilon \subset \mathcal{F}$ for each $\varepsilon > 0$.

We now define an approximate problem, where the smoothed continuous constraints are treated as penalty functions.

For $\gamma > 0$, determine an $\boldsymbol{x} \in \Theta$ that minimizes the cost function

$$f(\boldsymbol{x}) + \gamma \sum_{j=1}^{m} G_{j,\varepsilon}(\boldsymbol{x}). \tag{4.3.33}$$

This problem is referred to as Problem (4.3.33). The following result ensures the feasibility of a solution to Problem (4.3.33) with respect to Problem (4.3.1).

Theorem 4.3.5 *There exists a* $\gamma(\varepsilon) > 0$ *such that for all* $\gamma > \gamma(\varepsilon)$*, any solution to Problem* (4.3.33) *is also a feasible point of Problem* (4.3.1).

Proof. Let $\boldsymbol{x}^*_{\varepsilon,\gamma}$ denote an optimal solution to Problem (4.3.33). Then,

$$f\left(\boldsymbol{x}^*_{\varepsilon,\gamma}\right) + \gamma \sum_{j=1}^{m} G_{j,\varepsilon}\left(\boldsymbol{x}^*_{\varepsilon,\gamma}\right) \leq f(\boldsymbol{x}) + \gamma \sum_{j=1}^{m} G_{j,\varepsilon}(\boldsymbol{x}), \tag{4.3.34}$$

for all $\boldsymbol{x} \in \Theta$. Let $\boldsymbol{x}_\varepsilon \in \mathcal{F}_\varepsilon$ be fixed. Then, by the definition of $G_{j,\varepsilon}$,

$$G_{j,\varepsilon}(\boldsymbol{x}_\varepsilon) = 0$$

for $j = 1, 2, \ldots, m$. Now, since Θ is compact and f is continuous, there exists an $\bar{\boldsymbol{x}} \in \Theta$ such that $f(\bar{\boldsymbol{x}}) \leq f(\boldsymbol{x})$ for all $\boldsymbol{x} \in \Theta$. Clearly, $f(\bar{\boldsymbol{x}}) \leq f\left(\boldsymbol{x}^*_{\varepsilon,\gamma}\right)$. Adding the penalty term to each side, using the definition of $\boldsymbol{x}_\varepsilon$, and (4.3.34), we have

$$f(\bar{\boldsymbol{x}}) + \gamma \sum_{j=1}^{m} G_{j,\varepsilon}\left(\boldsymbol{x}^*_{\varepsilon,\gamma}\right) \leq f\left(\boldsymbol{x}^*_{\varepsilon,\gamma}\right) + \gamma \sum_{j=1}^{m} G_{j,\varepsilon}\left(\boldsymbol{x}^*_{\varepsilon,\gamma}\right) \leq f(\boldsymbol{x}_\varepsilon). \tag{4.3.35}$$

Rearranging (4.3.35) gives

$$\gamma \sum_{j=1}^{m} G_{j,\varepsilon}\left(\boldsymbol{x}^*_{\varepsilon,\gamma}\right) \leq f(\boldsymbol{x}_\varepsilon) - f(\bar{\boldsymbol{x}}). \tag{4.3.36}$$

Letting $z = f(\boldsymbol{x}_\varepsilon) - f(\bar{\boldsymbol{x}})$ in (4.3.36), we get

$$\sum_{j=1}^{m} G_{j,\varepsilon}\left(\boldsymbol{x}_{\varepsilon,\gamma}^{*}\right) \leq \frac{z}{\gamma}.$$

By Theorem 4.3.3, we recall that for any $\varepsilon > 0$, there exists a $\tau(\varepsilon)$ such that for all $0 < \tau < \tau(\varepsilon)$, if $G_{j,\varepsilon}(\boldsymbol{x}) < \tau$, then $\boldsymbol{x} \in \mathcal{F}$. Thus, by choosing $\gamma(\varepsilon) \geq z/\tau(\varepsilon)$, it follows that for all $\gamma > \gamma(\varepsilon)$, $\sum_{j=1}^{m} G_{j,\varepsilon}\left(\boldsymbol{x}_{\varepsilon,\gamma}^{*}\right) < \tau$. Consequently, $G_{j,\varepsilon}\left(\boldsymbol{x}_{\varepsilon,\gamma}^{*}\right) < \tau$, $j = 1, 2, \ldots, m$, and hence $\boldsymbol{x}_{\varepsilon,\gamma}^{*} \in \mathcal{F}$. This completes the proof.

Theorem 4.3.6 *Let $\boldsymbol{x}^*$ be an optimal solution to Problem* (4.3.1), *and let $\boldsymbol{x}_{\varepsilon,\gamma}^{*}$ be an optimal solution to Problem* (4.3.33), *in which γ is chosen appropriately to ensure that $\boldsymbol{x}_{\varepsilon,\gamma}^{*} \in \mathcal{F}$. Then,*

$$\lim_{\varepsilon \to 0} f\left(\boldsymbol{x}_{\varepsilon,\gamma}^{*}\right) = f(\boldsymbol{x}^*).$$

Proof. By Assumption (4.3.5), there exists an $\bar{\boldsymbol{x}} \in \mathring{\mathcal{F}}$ such that

$$\boldsymbol{x}_\alpha = \alpha\bar{\boldsymbol{x}} + (1-\alpha)\boldsymbol{x}^* = \boldsymbol{x}^* + \alpha(\bar{\boldsymbol{x}} - \boldsymbol{x}^*) \in \mathring{\mathcal{F}}$$

for all $\alpha \in (0, 1]$. Now, for any $\delta_1 > 0$, there exists an $\alpha_1 \in (0, 1]$ such that

$$f(\boldsymbol{x}^*) \leq f(\boldsymbol{x}_\alpha) \leq f(\boldsymbol{x}^*) + \delta_1, \tag{4.3.37}$$

for all $\alpha \in (0, \alpha_1)$. Choose $\alpha_2 = \alpha_1/2$. Then it is clear that $\boldsymbol{x}_{\alpha_2} \in \mathring{\mathcal{F}}$. Thus, there exists a $\delta_2 > 0$ such that

$$\max_{t\in[0,T]} \phi_j(\boldsymbol{x}_{\alpha_2}, t) < -\delta_2, \quad \text{for} \quad j = 1, 2, \ldots, m.$$

If we choose $\varepsilon = \delta_2$, then $\boldsymbol{x}_{\alpha_2}$ satisfies $G_{j,\varepsilon}(\boldsymbol{x}_{\alpha_2}) = 0$, $j = 1, 2, \ldots, m$. Using this and from the definition of $\boldsymbol{x}_{\varepsilon,\gamma}^{*}$, we have

$$f\left(\boldsymbol{x}_{\varepsilon,\gamma}^{*}\right) + \gamma \sum_{j=1}^{m} G_{j,\varepsilon}\left(\boldsymbol{x}_{\varepsilon,\gamma}^{*}\right) \leq f(\boldsymbol{x}_{\alpha_2}) + \gamma \sum_{j=1}^{m} G_{j,\varepsilon}(\boldsymbol{x}_{\alpha_2}) = f(\boldsymbol{x}_{\alpha_2}).$$

Noting that the penalty term is non-negative, we get

$$f\left(\boldsymbol{x}_{\varepsilon,\gamma}^{*}\right) \leq f(\boldsymbol{x}_{\alpha_2}). \tag{4.3.38}$$

Combining (4.3.38) with (4.3.37) and remembering that $\boldsymbol{x}_{\varepsilon,\gamma}^{*}$ is feasible for Problem (4.3.1), we obtain

$$f(\boldsymbol{x}^*) \leq f\left(\boldsymbol{x}_{\varepsilon,\gamma}^{*}\right) \leq f(\boldsymbol{x}^*) + \delta_1.$$

Letting $\varepsilon \to 0$ and noting that $\delta_1 > 0$ is arbitrary, the result follows.

4.3.2.1 Solution Method

Based on the results presented in Theorems 4.3.5 and 4.3.6, we now propose the following algorithm for solving Problem (4.3.1). Essentially, the idea of algorithm is to reduce ε from an initial value to a suitably small value in a number of stages. The value of γ is adjusted at each stage to ensure that the solution obtained remains feasible. We only use a simple updating scheme here.

Algorithm 4.3.2 *(Note that the integral in* (4.3.31) *is replaced by any suitable quadrature with positive weights.)*

Step 1. Choose $\varepsilon = 10^{-1}, \gamma = 1$ *and a starting point* $\boldsymbol{x} \in \Theta$.

Step 2. Solve Problem (4.3.33) *to an accuracy of* $\max\{10^{-8}, \varepsilon 10^{-3}\}$ *to give* $\boldsymbol{x}^*_{\varepsilon,\gamma}$.

Step 3. Check the feasibility of $\phi_j\left(\boldsymbol{x}^*_{\varepsilon,\gamma}, t\right) \leq 0$ *for all* $j = 1, 2, \ldots, m$ *at the quadrature points. If* $\boldsymbol{x}^*_{\varepsilon,\gamma}$ *is feasible, go to Step 4. Else, set* $\gamma = 2\gamma$. *If* $\gamma > 10^5$, *we have an abnormal exit. Else, go to Step 2, using* $\boldsymbol{x}^*_{\varepsilon,\gamma}$ *as the next starting point.*

Step 4. Set $\varepsilon = \varepsilon/10$. *If* $\varepsilon > 10^{-7}$, *go to Step 2, using* $\boldsymbol{x}^*_{\varepsilon,\gamma}$ *as the next starting point. Else, we have a successful exit.*

Remark 4.3.5 *Note that it is advisable to use a fixed quadrature scheme. Also, a small number of quadrature points are sufficient for a large* ε *and the number of the quadrature points may need to be increased as* $\varepsilon \to 0$.

Remark 4.3.6 *Using Theorem 4.3.5, we see that the doubling of* γ *in Step 3 needs only to be carried out a finite number of times. Consequently, the algorithm produces a sequence of suboptimal parameter vectors to Problem* (4.3.1), *and each of these is in the feasible region of Problem* (4.3.1). *Convergence of the cost is assured by Theorem 4.3.6. Since* Θ *is a compact set, we know that the above sequence has an accumulation point. Furthermore, each accumulation point is a solution of Problem* (4.3.1).

Remark 4.3.7 *Note also that for each* $\varepsilon > 0$ *and* $\gamma > 0$, *Problem* (4.3.33) *is constructed using the concept of penalty functions. However, due to the special structure of the penalty function used, the penalty weighting factor* γ *does not need to go to infinity. In [259], the method was thus referred to as an exact penalty function method. However, if the solution obtained for Problem* (4.3.33) *is a local solution, it is not known whether or not it is a local solution of the original problem* (4.3.1). *Thus, to avoid confusion, the phrase* exact penalty function method *will be used to refer to the method introduced in Section 4.4.*

Example 4.3.2 We solve Example 4.3.1 once more with the proposed penalty method. The integral in (4.3.31) is calculated via the trapezoidal rule with 256 quadrature points. The optimization routine used is based on the quasi-Newton method (see Chapter 2). The required accuracy asked of the optimization routine is $\max\{10^{-8}, 10^{-3}\varepsilon\}$.

The results are given in Table 4.3.3. The number of function evaluations made at each iteration of the algorithm is given in the column n_f. The columns marked ε, γ, f, x_1, x_2 and x_3 are self-explanatory. Finally we check the maximum value of $g(\boldsymbol{x})$ for 2001 equally spaced points of the interval Ω. This is given in the column marked g, while $\bar{\omega}$ indicates the value of ω at which this maximum occurred.

It should be noted that, at each iteration, the optimization routine returned with a 'failure' parameter of '0'. This indicates that the optimum has been found for each iteration. The algorithm only ensures that $\phi(\boldsymbol{x}, \omega) \leq 0$ for the quadrature points, and this is reflected in the value of g as ε becomes small. However, noting that the interval between quadrature points is of the order of 10^{-1}, it is quite reasonable to expect a constraint violation between quadrature points of the order of 10^{-4}. An encouraging point to note about the algorithm is that it tends to keep the solution of the approximate problems more to the inside of the feasible region.

Table 4.3.3: Iterations for Algorithm 4.3.2

ε	γ	$\bar{\omega}$	g	f	$x1$	$x2$	$x3$	nf
10^{-1}	1.0	5.48	-6.7×10^{-2}	0.1822723	16.18185	48.86607	32.07144	21
10^{-2}	1.0	5.48	-6.7×10^{-2}	0.1822723	16.18185	48.86607	32.07144	1
10^{-3}	1.0	5.63	6.2×10^{-4}	0.1747277	17.57267	48.63336	34.44601	11
10^{-3}	2.0	5.63	-7.3×10^{-5}	0.1748006	17.55490	48.63305	34.42106	4
10^{-4}	2.0	5.63	-8.7×10^{-6}	0.1747937	17.55597	48.63278	34.42370	3
10^{-5}	2.0	5.69	3.7×10^{-4}	0.1746830	16.53537	48.11178	35.01657	17
10^{-6}	2.0	5.69	3.7×10^{-4}	0.1746830	16.53538	48.11178	35.01658	4
10^{-7}	2.0	5.69	3.7×10^{-4}	0.1746830	16.53541	48.11173	35.01657	6

4.4 Exact Penalty Function Method

This section basically comes from [300, 301]. Consider the following optimization problem with continuous inequality constraints:

$$\text{minimize} \quad f(\boldsymbol{x}) \tag{4.4.1}$$

subject to

$$h_j(\boldsymbol{x}, t) \leq 0, \quad \forall t \in [0, T], \quad j = 1, 2, \ldots, m. \tag{4.4.2}$$

Let this problem be referred to as Problem (P). An exact penalty function method is introduced for this continuous inequality constrained optimization problem. The summation of the integrals of the exact penalty functions is appended to the cost function forming a new cost function. It is shown that a minimizer of the cost function can be obtained without requiring the penalty parameter to go to $+\infty$. Furthermore, any local minimizer of the unconstrained optimization problem when the penalty parameter is sufficiently large is a local minimizer of the original problem. This property is not shared by the approaches reported in Section 4.3. Clearly, this is a major advancement.

For Problem (P), it is clear that for each $\boldsymbol{x} \in \mathbb{R}^n$, $\max\{h_j(\boldsymbol{x},t),0\}$ is a continuous function of t, since h_j is continuously differentiable. Define

$$S_\varepsilon = \{(\boldsymbol{x},\varepsilon) \in \mathbb{R}^n \times \mathbb{R}_+ : h_j(\boldsymbol{x},t) \leq \varepsilon^\gamma W_j, \ \forall\, t \in [0,T], \ j = 1,2,\ldots,m\}, \tag{4.4.3}$$

where $\mathbb{R}_+ = \{\alpha \in \mathbb{R} \ : \ \alpha \geq 0\}$, $W_j \in (0,1)$, $j = 1,2,\ldots,m$, are fixed constants and γ is a positive real number. Clearly, Problem (P) is equivalent to the following problem, which is denoted as Problem $(\widetilde{P})$,

$$\text{minimize} \quad f(\boldsymbol{x}) \tag{4.4.4a}$$

subject to

$$(\boldsymbol{x},\varepsilon) \in S_0, \tag{4.4.4b}$$

where S_0 is simply S_ε with $\varepsilon = 0$.

We assume that the following conditions are satisfied.

Assumption 4.4.1 *There exists a global minimizer of Problem (P), implying that $f(\boldsymbol{x})$ is bounded from below on S_0.*

Assumption 4.4.2 *The number of distinct local minimum values of the objective function of Problem (P) is finite.*

Assumption 4.4.3 *Let $\mathbb{A}$ denote the set of all local minimizers of Problem (P). If $\boldsymbol{x}^* \in \mathbb{A}$, then $\mathbb{A}(\boldsymbol{x}^*) = \{\boldsymbol{x} \in \mathbb{A} : f(\boldsymbol{x}) = f(\boldsymbol{x}^*)\}$ is a compact set.*

We introduce a new exact penalty function $f_\sigma(\boldsymbol{x},\varepsilon)$ as follows:

$$f_\sigma(\boldsymbol{x},\varepsilon) = \begin{cases} f(\boldsymbol{x}), & \text{if } \varepsilon = 0, h_j(\boldsymbol{x},t) \leq 0, t \in [0,T], \\ f(\boldsymbol{x}) + \varepsilon^{-\alpha}\Delta(\boldsymbol{x},\varepsilon) + \sigma\varepsilon^\beta, & \text{if } \varepsilon > 0, \\ +\infty & \text{otherwise}, \end{cases} \tag{4.4.5}$$

where $\Delta(\boldsymbol{x},\varepsilon)$, which is referred to as the *constraint violation*, is defined by

$$\Delta(\boldsymbol{x},\varepsilon) = \sum_{j=1}^{m} \int_0^T [\max\{0, h_j(\boldsymbol{x},t) - \varepsilon^\gamma W_j\}]^2\, dt, \tag{4.4.6}$$

α and γ are positive real numbers, $\beta > 2$ and $\sigma > 0$ is a penalty parameter. We now introduce a surrogate optimization problem as follows:

$$\min \quad f_\sigma(\boldsymbol{x}, \varepsilon) \tag{4.4.7a}$$

subject to

$$(\boldsymbol{x}, \varepsilon) \in \mathbb{R}^n \times [0, +\infty). \tag{4.4.7b}$$

Let this problem be referred to as Problem (P_σ). Intuitively, during the process of minimizing $f_\sigma(\boldsymbol{x}, \varepsilon)$, if σ is increased, ε^β should be reduced, meaning that ε should be reduced as β is fixed. Thus $\varepsilon^{-\alpha}$ will be increased, and hence the constraint violation will also be reduced. This means that the value of $[\max\{0, h_j(\boldsymbol{x}, t) - \varepsilon^\gamma W_j\}]^2$ must go down, eventually leading to the satisfaction of the continuous inequality constraints, i.e.,

$$h_j(\boldsymbol{x}, t) \leq 0, \ \forall\, t \in [0, T], \ j = 1, 2, \ldots, m.$$

Let $\{\sigma_k\}_{k \in \mathbb{N}}$ be a sequence such that $\sigma_k \to \infty$ as $k \to \infty$. We will prove in the next section that, under some mild assumptions, if the parameter σ_k is sufficiently large and $\left(\boldsymbol{x}^{(k),*}, \varepsilon^{(k),*}\right)$ is a local minimizer of Problem (P_{σ_k}), then $\varepsilon^{(k),*} \to \varepsilon^* = 0$, and $\boldsymbol{x}^{(k),*} \to \boldsymbol{x}^*$ with $\boldsymbol{x}^*$ being a local minimizer of Problem (P). The importance of this result is quite obvious.

4.4.1 Convergence Analysis

Taking the gradients of $f_\sigma(\boldsymbol{x}, \varepsilon)$ with respect to $\boldsymbol{x}$ and ε gives

$$\frac{\partial f_\sigma(\boldsymbol{x}, \varepsilon)}{\partial \boldsymbol{x}} = \frac{\partial f(\boldsymbol{x})}{\partial \boldsymbol{x}} + 2\varepsilon^{-\alpha} \sum_{j=1}^{m} \int_0^T \max\{0, h_j(\boldsymbol{x}, t) - \varepsilon^\gamma W_j\} \frac{\partial h_j(\boldsymbol{x}, t)}{\partial \boldsymbol{x}} dt \tag{4.4.8}$$

and

$$\begin{aligned}\frac{\partial f_\sigma(\boldsymbol{x}, \varepsilon)}{\partial \varepsilon} =& - \alpha\varepsilon^{-\alpha-1} \sum_{j=1}^{m} \int_0^T [\max\{0, h_j(\boldsymbol{x}, t) - \varepsilon^\gamma W_j\}]^2 \, dt \\ & - 2\gamma\varepsilon^{\gamma-\alpha-1} \sum_{j=1}^{m} \int_0^T \max\{0, h_j(\boldsymbol{x}, t) - \varepsilon^\gamma W_j\} \, W_j dt + \sigma\beta\varepsilon^{\beta-1} \\ =& \varepsilon^{-\alpha-1} \left\{ -\alpha \sum_{j=1}^{m} \int_0^T [\max\{0, h_j(\boldsymbol{x}, t) - \varepsilon^\gamma W_j\}]^2 \, dt \right.\end{aligned}$$

$$+2\gamma\sum_{j=1}^{m}\int_0^T \max\{0, h_j(\boldsymbol{x},t)-\varepsilon^\gamma W_j\}(-\varepsilon^\gamma W_j)dt\Bigg\}+\sigma\beta\varepsilon^{\beta-1}. \tag{4.4.9}$$

For every positive integer k, let $\left(\boldsymbol{x}^{(k),*},\varepsilon^{(k),*}\right)$ be a local minimizer of Problem (P_{σ_k}). To obtain the main result, we need the following lemma.

Lemma 4.4.1 *Let $\left(\boldsymbol{x}^{(k),*},\varepsilon^{(k),*}\right)$ be a local minimizer of Problem (P_{σ_k}). Suppose that $f_{\sigma_k}\left(\boldsymbol{x}^{(k),*},\varepsilon^{(k),*}\right)$ is finite and that $\varepsilon^{(k),*}>0$. Then,*

$$\left(\boldsymbol{x}^{(k),*},\varepsilon^{(k),*}\right)\notin S_\varepsilon,$$

where S_ε is defined by (4.4.3).

Proof. Since $\left(\boldsymbol{x}^{(k),*},\varepsilon^{(k),*}\right)$ is a local minimizer of Problem (P_{σ_k}) and $\varepsilon^{(k),*}>0$, we have

$$\frac{\partial f_{\sigma_k}\left(\boldsymbol{x}^{(k),*},\varepsilon^{(k),*}\right)}{\partial\varepsilon}=0. \tag{4.4.10}$$

Let us assume that the conclusion of the lemma is false. Then, we have

$$h_j\left(\boldsymbol{x}^{(k),*},\varepsilon^{(k),*}\right)\le\left(\varepsilon^{(k),*}\right)^\gamma W_j,\quad \forall t\in[0,T],\ \ j=1,2,\ldots,m. \tag{4.4.11}$$

Thus, by (4.4.9)–(4.4.11), we obtain

$$0=\frac{\partial f_{\sigma_k}\left(\boldsymbol{x}^{(k),*},\varepsilon^{(k),*}\right)}{\partial\varepsilon}=\sigma_k\beta\varepsilon^{\beta-1}.$$

This is a contradiction, hence completing the proof.

To continue, we introduce the following definition.

Definition 4.4.1 *It is said that the constraint qualification is satisfied for the continuous inequality constraints* (4.4.2) *at $\boldsymbol{x}=\overline{\boldsymbol{x}}$, if the following implication is valid. Suppose that*

$$\int_0^T\sum_{j=1}^{m}\varphi_j(t)\frac{\partial h_j(\overline{\boldsymbol{x}},\boldsymbol{t})}{\partial\boldsymbol{x}}dt=0.$$

Then, $\varphi_j(t)=0$, $\forall t\in[0,T]$, $j=1,2,\ldots,m$.

Let the conditions of Lemma 4.4.1 be satisfied. Then, we have the following theorem.

Theorem 4.4.1 *Suppose that $\left(\boldsymbol{x}^{(k),*},\varepsilon^{(k),*}\right)$ is a local minimizer of Problem (P_{σ_k}) such that $f_{\sigma_k}\left(\boldsymbol{x}^{(k),*},\varepsilon^{(k),*}\right)$ is finite and $\varepsilon^{(k),*}>0$. If $\left(\boldsymbol{x}^{(k),*},\varepsilon^{(k),*}\right)\to(\boldsymbol{x}^*,\varepsilon^*)$ as $k\to+\infty$, and the constraint qualification is satisfied for the continuous inequality constraints* (4.4.2) *at $\boldsymbol{x}=\boldsymbol{x}^*$, then $\varepsilon^*=0$ and $\boldsymbol{x}^*\in S_0$.*

Proof. From Lemma 4.4.1, it follows that $\left(\boldsymbol{x}^{(k),*}, \varepsilon^{(k),*}\right) \notin S_{\varepsilon^{(k),*}}$. Thus, by (4.4.8), we have

$$\begin{aligned}
&\frac{\partial f_{\sigma_k}\left(\boldsymbol{x}^{(k),*}, \varepsilon^{(k),*}\right)}{\partial \boldsymbol{x}} \\
&= \frac{\partial f\left(\boldsymbol{x}^{(k),*}\right)}{\partial \boldsymbol{x}} + 2\left(\varepsilon^{(k),*}\right)^{-\alpha} \cdot \\
&\sum_{j=1}^{m} \int_0^T \max\left\{0, h_j\left(\boldsymbol{x}^{(k),*}, t\right) - \left(\varepsilon^{(k),*}\right)^{\gamma} W_j\right\} \frac{\partial h_j\left(\boldsymbol{x}^{(k),*}, t\right)}{\partial \boldsymbol{x}} dt \\
&= 0. \qquad (4.4.12)
\end{aligned}$$

Similarly, by (4.4.9), we have

$$\begin{aligned}
&\frac{\partial f_{\sigma_k}\left(\boldsymbol{x}^{(k),*}, \varepsilon^{(k),*}\right)}{\partial \varepsilon} \\
&= -\alpha\left(\varepsilon^{(k),*}\right)^{-\alpha-1} \sum_{j=1}^{m} \int_0^T \left[\max\left\{0, h_j\left(\boldsymbol{x}^{(k),*}, t\right) - \left(\varepsilon^{(k),*}\right)^{\gamma} W_j\right\}\right]^2 dt \\
&\quad - 2\gamma\left(\varepsilon^{(k),*}\right)^{\gamma-\alpha-1} \sum_{j=1}^{m} \int_0^T \max\left\{0, h_j\left(\boldsymbol{x}^{(k),*}, t\right) - \left(\varepsilon^{(k),*}\right)^{\gamma} W_j\right\} W_j dt \\
&\quad + \sigma_k \beta\left(\varepsilon^{(k),*}\right)^{\beta-1} \\
&= \left(\varepsilon^{(k),*}\right)^{-\alpha-1} \left\{-\alpha \sum_{j=1}^{m} \int_0^T \left[\max\left\{0, h_j\left(\boldsymbol{x}^{(k),*}, t\right) - \left(\varepsilon^{(k),*}\right)^{\gamma} W_j\right\}\right]^2 dt \right. \\
&\quad \left. + 2\gamma \sum_{j=1}^{m} \int_0^T \max\left\{0, h_j\left(\boldsymbol{x}^{(k),*}, t\right) - \left(\varepsilon^{(k),*}\right)^{\gamma} W_j\right\} \left(-\left(\varepsilon^{(k),*}\right)^{\gamma} W_j\right) dt \right\} \\
&\quad + \sigma_k \beta\left(\varepsilon^{(k),*}\right)^{\beta-1} \\
&= 0. \qquad (4.4.13)
\end{aligned}$$

Suppose that $\varepsilon^{(k),*} \to \varepsilon^* \neq 0$. Then, by (4.4.13), we observe that its first term tends to a finite value, while the last term tends to infinity as $\sigma_k \to +\infty$, when $k \to +\infty$. This is impossible for the validity of (4.4.13). Thus, $\varepsilon^* = 0$.

Now, by (4.4.12), we obtain

$$\begin{aligned}
&\left(\varepsilon^{(k),*}\right)^{\alpha} \frac{\partial f\left(\boldsymbol{x}^{(k),*}\right)}{\partial \boldsymbol{x}} \\
&\quad + 2\sum_{j=1}^{m} \int_0^T \max\left\{0, h_j\left(\boldsymbol{x}^{(k),*}, t\right) - \left(\varepsilon^{(k),*}\right)^{\gamma} W_j\right\} \frac{\partial h_j\left(\boldsymbol{x}^{(k),*}, t\right)}{\partial \boldsymbol{x}} dt \\
&= 0. \qquad (4.4.14)
\end{aligned}$$

Thus,

$$\begin{aligned}
&\lim_{k\to+\infty}\left\{\left(\varepsilon^{(k),*}\right)^{\alpha}\frac{\partial f\left(\boldsymbol{x}^{(k),*}\right)}{\partial \boldsymbol{x}}\right.\\
&\left.+2\sum_{j=1}^{m}\int_0^T \max\left\{0, h_j\left(\boldsymbol{x}^{(k),*},t\right)-\left(\varepsilon^{(k),*}\right)^{\gamma}W_j\right\}\frac{\partial h_j\left(\boldsymbol{x}^{(k),*},t\right)}{\partial \boldsymbol{x}}dt\right\}\\
&=2\sum_{j=1}^{m}\int_0^T \max\left\{0, h_j(\boldsymbol{x}^*,t)\right\}\frac{\partial h_j\left(\boldsymbol{x}^*,t\right)}{\partial \boldsymbol{x}}dt\\
&=0.
\end{aligned} \tag{4.4.15}$$

Since the constraint qualification is satisfied for the continuous inequality constraints (4.4.2) at $\boldsymbol{x}=\boldsymbol{x}^*$, it follows that, for each $j=1,2,\ldots,m$,

$$\max\left\{0, h_j(\boldsymbol{x}^*,t)\right\}=0$$

for each $t\in[0,T]$. This, in turn, implies that, for each $j=1,2,\ldots,m$, $h_j(\boldsymbol{x}^*,t)\le 0$, $\forall t\in[0,T]$. The proof is complete.

Corollary 4.4.1 *If $\boldsymbol{x}^{(k),*}\to\boldsymbol{x}^*\in S_0$ and $\varepsilon^{(k),*}\to\varepsilon^*=0$, then*

$$\Delta(\boldsymbol{x}^{(k),*},\varepsilon^{(k),*})\to\Delta(\boldsymbol{x}^*,\varepsilon^*)=0.$$

Proof. The conclusion follows readily from the definition of $\Delta(\boldsymbol{x},\varepsilon)$ and the continuity of $h_j(\boldsymbol{x},t)$.

For the exact penalty function constructed in (4.4.5), we have the following results.

Theorem 4.4.2 *Assume that $h_j\left(\boldsymbol{x}^{(k),*},\omega\right)=o\left(\left(\varepsilon^{(k),*}\right)^{\delta}\right)$, $\delta>0$, $j=1,2,\ldots,m$. Suppose that $\gamma>\alpha$, $\delta>\alpha$, $-\alpha-1+2\delta>0$ and $2\gamma-\alpha-1>0$. Then, as $\varepsilon^{(k),*}\to\varepsilon^*=0$ and $\boldsymbol{x}^{(k),*}\to\boldsymbol{x}^*\in S_0$, it holds that*

$$f_{\sigma_k}\left(\boldsymbol{x}^{(k),*},\varepsilon^{(k),*}\right)\to f(\boldsymbol{x}^*) \tag{4.4.16}$$

and

$$\nabla_{(\boldsymbol{x},\varepsilon)}f_{\sigma_k}\left(\boldsymbol{x}^{(k),*},\varepsilon^{(k),*}\right)\to(\nabla f(\boldsymbol{x}^*),0). \tag{4.4.17}$$

Proof. By the conditions of the theorem, it follows that, for $\varepsilon\neq 0$,

$$\begin{aligned}
&\lim_{\substack{\varepsilon^{(k),*}\to\varepsilon^*=0\\ \boldsymbol{x}^{(k),*}\to\boldsymbol{x}^*\in S_0}} f_{\sigma_k}\left(\boldsymbol{x}^{(k),*},\varepsilon^{(k),*}\right)\\
&=\lim_{\substack{\varepsilon^{(k),*}\to\varepsilon^*=0\\ \boldsymbol{x}^{(k),*}\to\boldsymbol{x}^*\in S_0}}\left\{f\left(\boldsymbol{x}^{(k),*}\right)+\left(\varepsilon^{(k),*}\right)^{-\alpha}\cdot\right.
\end{aligned}$$

$$\sum_{j=1}^{m}\int_{0}^{T}\left[\max\left\{0,h_j\left(\boldsymbol{x}^{(k),*},t\right)-\left(\varepsilon^{(k),*}\right)^{\gamma}W_j\right\}\right]^2dt+\sigma_k\left(\varepsilon^{(k),*}\right)^{\beta}\Bigg\}$$

$$=f(\boldsymbol{x}^*)+\lim_{\substack{\boldsymbol{\varepsilon}^{(k),*}\to\varepsilon^*=0\\ \boldsymbol{x}^{(k),*}\to\boldsymbol{x}^*\in S_0}}\frac{\sum_{j=1}^{m}\int_0^T\left[\max\left\{0,h_j\left(\boldsymbol{x}^{(k),*},t\right)-\left(\varepsilon^{(k),*}\right)^{\gamma}W_j\right\}\right]^2dt}{(\varepsilon^{(k),*})^{\alpha}}. \tag{4.4.18}$$

For the second term of the right hand side of (4.4.18), it is clear from Lemma 4.4.1 that

$$\lim_{\substack{\boldsymbol{\varepsilon}^{(k),*}\to\varepsilon^*=0\\ \boldsymbol{x}^{(k),*}\to\boldsymbol{x}^*\in S_0}}\frac{\sum_{j=1}^{m}\int_0^T\left[\max\left\{0,h_j\left(\boldsymbol{x}^{(k),*},t\right)-\left(\varepsilon^{(k),*}\right)^{\gamma}W_j\right\}\right]^2dt}{\left(\varepsilon^{(k),*}\right)^{\alpha}}$$

$$=\lim_{\substack{\boldsymbol{\varepsilon}^{(k),*}\to\varepsilon^*=0\\ \boldsymbol{x}^{(k),*}\to\boldsymbol{x}^*\in S_0}}\sum_{j\in J'}\int_0^T\left[\left(\varepsilon^{(k),*}\right)^{-\frac{\alpha}{2}}h_j\left(\boldsymbol{x}^{(k),*},t\right)-\left(\varepsilon^{(k),*}\right)^{\gamma-\frac{\alpha}{2}}W_j\right]^2dt, \tag{4.4.19}$$

where

$$J'=\left\{j\in[1,2,\ldots,m]:h_j\left(\boldsymbol{x}^{(k),*},t\right)-\left(\varepsilon^{(k),*}\right)^{\gamma}W_j\geq 0\right\}.$$

Since $\gamma>\alpha$, $h_j\left(\boldsymbol{x}^{(k),*},t\right)=o\left((\varepsilon^{(k),*})^{\delta}\right)$ and $\delta>\alpha$, we have

$$\lim_{\substack{\boldsymbol{\varepsilon}^{(k),*}\to\varepsilon^*=0\\ \boldsymbol{x}^{(k),*}\to\boldsymbol{x}^*\in S_0}}\sum_{j\in J'}\int_0^T\left[\left(\varepsilon^{(k),*}\right)^{-\frac{\alpha}{2}}h_j\left(\boldsymbol{x}^{(k),*},t\right)-\left(\varepsilon^{(k),*}\right)^{\gamma-\frac{\alpha}{2}}W_j\right]^2dt=0. \tag{4.4.20}$$

Combining (4.4.18)–(4.4.20) gives

$$\lim_{\substack{\boldsymbol{\varepsilon}^{(k),*}\to\varepsilon^*=0\\ \boldsymbol{x}^{(k),*}\to\boldsymbol{x}^*\in S_0}}f_{\sigma_k}\left(\boldsymbol{x}^{(k),*},\varepsilon^{(k),*}\right)=f(\boldsymbol{x}^*). \tag{4.4.21}$$

Similarly, we have

$$\lim_{\substack{\boldsymbol{\varepsilon}^{(k),*}\to\varepsilon^*=0\\ \boldsymbol{x}^{(k),*}\to\boldsymbol{x}^*\in S_0}}\nabla_{(\boldsymbol{x},\varepsilon)}f_{\sigma_k}\left(\boldsymbol{x}^{(k),*},\varepsilon^{(k),*}\right)$$

$$=\lim_{\substack{\boldsymbol{\varepsilon}^{(k),*}\to\varepsilon^*=0\\ \boldsymbol{x}^{(k),*}\to\boldsymbol{x}^*\in S_0}}\left[\nabla_{\boldsymbol{x}}f_{\sigma_k}\left(\boldsymbol{x}^{(k),*},\varepsilon^{(k),*}\right),\nabla_{\varepsilon}f_{\sigma_k}\left(\boldsymbol{x}^{(k),*},\varepsilon^{(k),*}\right)\right], \tag{4.4.22}$$

where

$$
\begin{aligned}
&\lim_{\substack{\varepsilon^{(k),*}\to\varepsilon^*=0\\ \boldsymbol{x}^{(k),*}\to\boldsymbol{x}^*\in S_0}} \nabla_{\boldsymbol{x}} f_{\sigma_k}\left(\boldsymbol{x}^{(k),*},\varepsilon^{(k),*}\right)\\
&= \lim_{\substack{\varepsilon^{(k),*}\to\varepsilon^*=0\\ \boldsymbol{x}^{(k),*}\to\boldsymbol{x}^*\in S_0}} \left\{\frac{\partial f\left(\boldsymbol{x}^{(k),*}\right)}{\partial \boldsymbol{x}} \quad +2\left(\varepsilon^{(k),*}\right)^{-\alpha}\cdot\right.\\
&\left.\sum_{j=1}^{m}\int_0^T\left[\max\left\{0,h_j\left(\boldsymbol{x}^{(k),*},t\right)-\left(\varepsilon^{(k),*}\right)^{\gamma}W_j\right\}\right]\frac{\partial h_j\left(\boldsymbol{x}^{(k),*},t\right)}{\partial \boldsymbol{x}}dt\right\}\\
&= \nabla_{\boldsymbol{x}} f(\boldsymbol{x}^*) \quad + \lim_{\substack{\varepsilon^{(k),*}\to\varepsilon^*=0\\ \boldsymbol{x}^{(k),*}\to\boldsymbol{x}^*\in S_0}} 2\sum_{j\in J'}\int_0^T\left[\left(\varepsilon^{(k),*}\right)^{-\alpha}h_j\left(\boldsymbol{x}^{(k),*},t\right)\right.\\
&\quad\left.-\left(\varepsilon^{(k),*}\right)^{\gamma-\alpha}W_j\right]\frac{\partial h_j\left(\boldsymbol{x}^{(k),*},t\right)}{\partial \boldsymbol{x}}dt\\
&= \nabla_{\boldsymbol{x}} f(\boldsymbol{x}^*),
\end{aligned}
\tag{4.4.23}
$$

while

$$
\begin{aligned}
&\lim_{\substack{\varepsilon^{(k),*}\to\varepsilon^*=0\\ \boldsymbol{x}^{(k),*}\to\boldsymbol{x}^*\in S_0}} \nabla_{\varepsilon} f_{\sigma_k}\left(\boldsymbol{x}^{(k),*},\varepsilon^{(k),*}\right)\\
&= \lim_{\substack{\varepsilon^{(k),*}\to\varepsilon^*=0\\ \boldsymbol{x}^{(k),*}\to\boldsymbol{x}^*\in S_0}} \left\{\left(\varepsilon^{(k),*}\right)^{-\alpha-1}\cdot\right.\\
&\left\{-\alpha\sum_{j=1}^{m}\int_0^T\left[\max\left\{0,h_j\left(\boldsymbol{x}^{(k),*},t\right)-\left(\varepsilon^{(k),*}\right)^{\gamma}W_j\right\}\right]^2dt\right.\\
&\quad\left.+2\gamma\sum_{j=1}^{m}\int_0^T\max\left\{0,h_j\left(\boldsymbol{x}^{(k),*},t\right)-\left(\varepsilon^{(k),*}\right)^{\gamma}W_j\right\}\left(-\left(\varepsilon^{(k),*}\right)^{\gamma}W_j\right)dt\right\}\\
&\quad\left.+\sigma_k\beta\left(\varepsilon^{(k),*}\right)^{\beta-1}\right\}\\
&= \lim_{\substack{\varepsilon^{(k),*}\to\varepsilon^*=0\\ \boldsymbol{x}^{(k),*}\to\boldsymbol{x}^*\in S_0}} \left\{-\alpha\sum_{j\in J'}\int_0^T\left[h_j\left(\boldsymbol{x}^{(k),*},t\right)\left(\varepsilon^{(k),*}\right)^{-\frac{\alpha+1}{2}}\right.\right.\\
&\quad\left.-\left(\varepsilon^{(k),*}\right)^{\gamma-\frac{\alpha+1}{2}}W_j\right]^2dt+2\gamma\sum_{j\in J'}\int_0^T\left[h_j\left(\boldsymbol{x}^{(k),*},t\right)-\left(\varepsilon^{(k),*}\right)^{\gamma}W_j\right]\cdot\\
&\left.\left(-\left(\varepsilon^{(k),*}\right)^{\gamma}W_j\right)\left(\varepsilon^{(k),*}\right)^{-\alpha-1}dt\right\}\\
&= 0.
\end{aligned}
\tag{4.4.24}
$$

Thus, the proof is complete.

The exactness of the penalty function is shown in the following theorem.

Theorem 4.4.3 *There exists a $k_0 > 0$, such that for any $k \geq k_0$, every local minimizer $\left(\boldsymbol{x}^{(k),*}, \varepsilon^{(k),*}\right)$ of the penalty problem with finite $f_{\sigma_k}(\boldsymbol{x}^*, \varepsilon^*)$ has the form $(\boldsymbol{x}^*, 0)$, where $\boldsymbol{x}^*$ is a local minimizer of Problem (P).*

Proof. Let us assume that the conclusion is false. Then, there exists a subsequence of $\{\left(\boldsymbol{x}^{(k),*}, \varepsilon^{(k),*}\right)\}$, which is denoted by the original sequence, such that for any $k_0 > 0$, there exists a $k' > k_0$ satisfying $\varepsilon^{(k'),*} \neq 0$. By Theorem 4.4.1, we have

$$\varepsilon^{(k),*} \to \varepsilon^* = 0, \ \boldsymbol{x}^{(k),*} \to \boldsymbol{x}^* \in S_0, \text{ as } k \to +\infty.$$

Since $\varepsilon^{(k),*} \neq 0$ for all k, it follows from dividing (4.4.13) by $\left(\varepsilon^{(k),*}\right)^{\beta-1}$ that

$$\begin{aligned}
&\left(\varepsilon^{(k),*}\right)^{-\alpha-\beta}\left\{-\alpha\sum_{j=1}^{m}\int_0^T\left[\max\left\{0, h_j\left(\boldsymbol{x}^{(k),*}, t\right) - \left(\varepsilon^{(k),*}\right)^{\gamma}W_j\right\}\right]^2 dt\right.\\
&\left.+2\gamma\sum_{j=1}^{m}\int_0^T \max\left\{0, h_j\left(\boldsymbol{x}^{(k),*}, t\right) - \left(\varepsilon^{(k),*}\right)^{\gamma}W_j\right\}\left(\left(-\varepsilon^{(k),*}\right)^{\gamma}W_j\right) dt\right\}\\
&+\sigma_k\beta\\
&= 0.
\end{aligned} \tag{4.4.25}$$

This is equivalent to

$$\begin{aligned}
&\left(\varepsilon^{(k),*}\right)^{-\alpha-\beta}\left\{-\alpha\sum_{j=1}^{m}\int_0^T\left[\max\left\{0, h_j\left(\boldsymbol{x}^{(k),*}, t\right) - \left(\varepsilon^{(k),*}\right)^{\gamma}W_j\right\}\right]^2 dt\right.\\
&+2\gamma\sum_{j=1}^{m}\int_0^T\left[\max\left\{0, h_j\left(\boldsymbol{x}^{(k),*}, t\right) - \left(\varepsilon^{(k),*}\right)^{\gamma}W_j\right\}\left(\left(-\varepsilon^{(k),*}\right)^{\gamma}W_j\right)\right.\\
&+\max\left\{0, h_j\left(\boldsymbol{x}^{(k),*}, t\right) - \left(\varepsilon^{(k),*}\right)^{\gamma}W_j\right\}h_j\left(\boldsymbol{x}^{(k),*}, t\right)\\
&\left.\left.-\max\left\{0, h_j\left(\boldsymbol{x}^{(k),*}, t\right) - \left(\varepsilon^{(k),*}\right)^{\gamma}W_j\right\}h_j\left(\boldsymbol{x}^{(k),*}, t\right)\right] dt\right\}+\sigma_k\beta\\
&= 0.
\end{aligned} \tag{4.4.26}$$

Note that

$$\begin{aligned}
&\max\left\{0, h_j\left(\boldsymbol{x}^{(k),*}, t\right) - \left(\varepsilon^{(k),*}\right)^{\gamma}W_j\right\}\left(\left(-\varepsilon^{(k),*}\right)^{\gamma}W_j\right)\\
&\quad+\max\left\{0, h_j\left(\boldsymbol{x}^{(k),*}, t\right) - \left(\varepsilon^{(k),*}\right)^{\gamma}W_j\right\}h_j\left(\boldsymbol{x}^{(k),*}, t\right)\\
&\quad=\max\left\{0, h_j\left(\boldsymbol{x}^{(k),*}, t\right) - \left(\varepsilon^{(k),*}\right)^{\gamma}W_j\right\}\left[h_j\left(\boldsymbol{x}^{(k),*}, t\right) - \left(\varepsilon^{(k),*}\right)^{\gamma}W_j\right]
\end{aligned}$$

$$= \left[\max\left\{0, h_j\left(\boldsymbol{x}^{(k),*}, t\right) - \left(\varepsilon^{(k),*}\right)^{\gamma} W_j\right\}\right]^2. \tag{4.4.27}$$

Combining (4.4.26) and (4.4.27) yields

$$\left(\varepsilon^{(k),*}\right)^{-\alpha-\beta}(2\gamma-\alpha)\left\{\sum_{j=1}^{m}\int_0^T\left[\max\left\{0, h_j\left(\boldsymbol{x}^{(k),*}, t\right) - \left(\varepsilon^{(k),*}\right)^{\gamma} W_j\right\}\right]^2 dt\right\}$$
$$+\sigma_k\beta$$
$$= 2\gamma\left(\varepsilon^{(k),*}\right)^{-\alpha-\beta}\sum_{j=1}^{m}\int_0^T \max\left\{0, h_j\left(\boldsymbol{x}^{(k),*}, t\right) - \left(\varepsilon^{(k),*}\right)^{\gamma} W_j\right\}\cdot$$
$$h_j\left(\boldsymbol{x}^{(k),*}, t\right) dt. \tag{4.4.28}$$

Define

$$y^k = \left(\varepsilon^{(k),*}\right)^{-\alpha-\beta}\sum_{j=1}^{m}\int_0^T \max\left\{0, h_j\left(\boldsymbol{x}^{(k),*}, t\right) - \left(\varepsilon^{(k),*}\right)^{\gamma} W_j\right\} dt \tag{4.4.29}$$

and

$$z^k = y^k / \left|y^k\right|. \tag{4.4.30}$$

Clearly,

$$\lim_{k\to+\infty}\left|z^k\right| = |z^*| = 1, \tag{4.4.31}$$

where $|\cdot|$ denotes the modulus. The usual Euclidean norm is denoted as $\|\cdot\|$.

Dividing (4.4.14) by $\left|y^k\right|$ and $\left(\varepsilon^{(k),*}\right)^{\alpha}$ yields

$$\frac{\frac{\partial f\left(\boldsymbol{x}^{(k),*}\right)}{\partial \boldsymbol{x}}}{\left|y^k\right|} + \frac{2\left(\varepsilon^{(k),*}\right)^{-\alpha}}{\left|y^k\right|}\sum_{j=1}^{m}\int_0^T \max\left\{0, h_j\left(\boldsymbol{x}^{(k),*}, t\right) - \left(\varepsilon^{(k),*}\right)^{\gamma} W_j\right\}\cdot$$
$$\frac{\partial h_j\left(\boldsymbol{x}^{(k),*}, t\right)}{\partial \boldsymbol{x}} dt$$
$$= 0. \tag{4.4.32}$$

Note that $\boldsymbol{x}^{(k),*} \to \boldsymbol{x}^*$ as $k \to +\infty$ and that $\frac{\partial f(\boldsymbol{x})}{\partial \boldsymbol{x}}$ and, for each $j = 1, 2, \ldots, m$, h_j and $\frac{\partial h_j(\cdot\ ,t)}{\partial \boldsymbol{x}}$ are continuous in $\mathbb{R}^n$ for each $t \in [0, T]$. Clearly, $[0, T]$ is a compact set. Thus, it can be shown that there are constants $\hat{K}$ and $\overline{K}$, independent of k, such that

$$\left\|\frac{\partial f\left(\boldsymbol{x}^{(k),*}\right)}{\partial \boldsymbol{x}}\right\| \leq \hat{K} \tag{4.4.33}$$

and

$$\left\|\frac{\partial h_j\left(\boldsymbol{x}^{(k),*},t\right)}{\partial \boldsymbol{x}}\right\| \leq \overline{K},\ \forall t \in [0,T],\ j=1,2,\ldots,m. \tag{4.4.34}$$

Note that

$$\begin{aligned}
&\frac{1}{|y^k|\left(\varepsilon^{(k),*}\right)^{\beta}}\\
=&\frac{1}{|\left(\varepsilon^{(k),*}\right)^{-\alpha-\beta}\sum_{j=1}^{m}\int_0^T \max\left\{0,h_j\left(\boldsymbol{x}^{(k),*},t\right)-\left(\varepsilon^{(k),*}\right)^{\gamma}W_j\right\}dt|\left(\varepsilon^{(k),*}\right)^{\beta}}\\
=&\frac{1}{|\sum_{j=1}^{m}\int_0^T \max\left\{0,h_j\left(\boldsymbol{x}^{(k),*},t\right)-\left(\varepsilon^{(k),*}\right)^{\gamma}W_j\right\}dt|\left(\varepsilon^{(k),*}\right)^{-\alpha}}.
\end{aligned} \tag{4.4.35}$$

Recalling the assumption stated in Theorem 4.4.2, we have $h_j\left(\boldsymbol{x}^{(k),*},t\right) = o\left(\left(\varepsilon^{(k),*}\right)^{\delta}\right)$, $\gamma > \alpha$ and $\delta > \alpha$. Thus,

$$\begin{aligned}
&\lim_{k\to+\infty}\left|\sum_{j=1}^{m}\int_0^T \max\left\{0,h_j\left(\boldsymbol{x}^{(k),*},t\right)-\left(\varepsilon^{(k),*}\right)^{\gamma}W_j\right\}dt\right|\left(\varepsilon^{(k),*}\right)^{-\alpha}\\
&=\lim_{k\to+\infty}\left|\sum_{j=1}^{m}\int_0^T \max\left\{0,o\left(\left(\varepsilon^{(k),*}\right)^{\delta}\right)-\left(\varepsilon^{(k),*}\right)^{\gamma}W_j\right\}dt\right|\left(\varepsilon^{(k),*}\right)^{-\alpha}\\
&=\lim_{k\to+\infty}\left|\sum_{j=1}^{m}\int_0^T \max\left\{0,o\left(\left(\varepsilon^{(k),*}\right)^{\delta}\right)\left(\varepsilon^{(k),*}\right)^{-\alpha}-\left(\varepsilon^{(k),*}\right)^{\gamma-\delta}W_j\right\}dt\right|\\
&=\lim_{k\to+\infty}\left|\sum_{j=1}^{m}\int_0^T \max\left\{0,\frac{o\left(\left(\varepsilon^{(k),*}\right)^{\delta}\right)}{\left(\varepsilon^{(k),*}\right)^{\delta}}\left(\varepsilon^{(k),*}\right)^{\delta-\alpha}-\left(\varepsilon^{(k),*}\right)^{\gamma-\delta}W_j\right\}dt\right|\\
&=0.
\end{aligned} \tag{4.4.36}$$

Therefore,

$$\frac{1}{|y^k|\left(\varepsilon^{(k),*}\right)^{\beta}} \to +\infty, \quad k \to +\infty. \tag{4.4.37}$$

From (4.4.33) and (4.4.37), it is clear that

$$\frac{\frac{\partial f\left(\boldsymbol{x}^{(k),*}\right)}{\partial \boldsymbol{x}}}{|y^k|\left(\varepsilon^{(k),*}\right)^{\beta}} \to +\infty, \quad k \to +\infty. \tag{4.4.38}$$

On the other hand, by (4.4.29) and (4.4.30), we have

$$\left| \frac{2\left(\varepsilon^{(k),*}\right)^{-\alpha-\beta}}{|y^k|} \sum_{j=1}^{m} \int_0^T \max\left\{0, h_j\left(\boldsymbol{x}^{(k),*}, t\right) - \left(\varepsilon^{(k),*}\right)^{\gamma} W_j\right\} \cdot \frac{\partial h_j\left(\boldsymbol{x}^{(k),*}, t\right)}{\partial \boldsymbol{x}} dt \right|$$

$$\leq \frac{2\left(\varepsilon^{(k),*}\right)^{-\alpha-\beta}}{|y^k|} \sum_{j=1}^{m} \int_0^T \left| \max\left\{0, h_j\left(\boldsymbol{x}^{(k),*}, t\right) - \left(\varepsilon^{(k),*}\right)^{\gamma} W_j\right\} \cdot \frac{\partial h_j\left(\boldsymbol{x}^{(k),*}, t\right)}{\partial \boldsymbol{x}} \right| dt$$

$$= \frac{2\left(\varepsilon^{(k),*}\right)^{-\alpha-\beta}}{|y^k|} \sum_{j=1}^{m} \int_0^T \max\left\{0, h_j\left(\boldsymbol{x}^{(k),*}, t\right) - \left(\varepsilon^{(k),*}\right)^{\gamma} W_j\right\} \cdot \left| \frac{\partial h_j\left(\boldsymbol{x}^{(k),*}, t\right)}{\partial \boldsymbol{x}} \right| dt$$

$$\leq \frac{2\left(\varepsilon^{(k),*}\right)^{-\alpha-\beta}}{|y^k|} \sum_{j=1}^{m} \int_0^T \max\left\{0, h_j\left(\boldsymbol{x}^{(k),*}, t\right) - \left(\varepsilon^{(k),*}\right)^{\gamma} W_j\right\} \overline{K} dt$$

$$= 2\overline{K} z^k, \tag{4.4.39}$$

where z^k is defined by (4.4.30). Clearly, $\left|z^k\right| = 1$. Thus, it follows that $2\overline{K}z^k$ is bounded uniformly with respect to k. This together with (4.4.38) is a contradiction to (4.4.32). Thus, the proof is complete.

We may now conclude that, under some mild assumptions and the constraint qualification condition, when the parameter σ is sufficiently large, a local minimizer of Problem (P_σ) is a local minimizer of Problem (P).

4.4.2 Algorithm and Numerical Results

Unconstrained optimization techniques such as quasi-Newton methods or conjugate gradient methods can be used to solve Problem (P_δ). Here, the one implemented is the optimization tool *fmincon* within the MATLAB environment. The integral appearing in $f_\sigma(\boldsymbol{x}, \varepsilon)$ is approximated by using Simpson's Rule. The global error of Simpson's Rule is of the order of h^4, where h is the discretization step size. Thus, a reasonable accuracy can easily be achieved for the integration if the discretization step size is sufficiently small. In the following, we define various terms used in the algorithm.

σ is the penalty parameter which is to be increased in every iteration.
$\bar{t}$ is the point at which $\max_{1\le j\le m} h_j\left(\boldsymbol{x}^{(k),*}, \bar{t}\right) = \max_{1\le j\le m}\max_{t\in[0,T]} h_j\left(\boldsymbol{x}^{(k),*}, t\right)$.
$\bar{h}$ is the value of $\max_{1\le j\le m}\max_{t\in[0,t]} h_j\left(\boldsymbol{x}^{(k),*}, t\right)$.
f is the objective function value.
ε is a new variable that is introduced in the construction of the exact penalty function.
$\underline{\varepsilon}$ is a lower bound of $\varepsilon^{(k),*}$, which is introduced to avoid $\varepsilon^{(k),*} \to 0$.

With the new exact penalty function, we can construct an efficient algorithm, which is given below.

Algorithm 4.4.1

Step 1. *Set the iteration index $k = 0$. Set $\sigma^{(1)} = 10$, $\varepsilon^{(1)} = 0.1$, $\underline{\varepsilon} = 10^{-9}$ and $\beta > 2$, and choose an initial point $(\boldsymbol{x}_0, \varepsilon_0)$. The values of γ and α are chosen depending on the specific structure of Problem (P).*

Step 2. *Solve Problem (P_{σ_k}), and let $\left(\boldsymbol{x}^{(k),*}, \varepsilon^{(k),*}\right)$ be the minimizer obtained.*

Step 3. *If $\varepsilon^{(k),*} > \underline{\varepsilon}$ and $\sigma^{(k)} < 10^8$, set $\sigma^{(k+1)} = 10 \times \sigma^{(k)}$ and $k := k + 1$. Go to Step 2 with $\left(\boldsymbol{x}^{(k),*}, \varepsilon^{(k),*}\right)$ as the initial point in the next optimization process. Else set $\varepsilon^{(k),*} = \underline{\varepsilon}$, and then go to Step 4.*

Step 4. *Check the feasibility of $\boldsymbol{x}^{(k),*}$, i.e., whether or not*

$$\max_{1\le j\le m}\max_{t\in[0,T]} h_j\left(\boldsymbol{x}^{(k),*}, t\right) \le 0.$$

If $\boldsymbol{x}^{(k),*}$ is feasible, then it is a local minimizer of Problem (P). Exit. Else go to Step 5.

Step 5. *Adjust the parameters α, β and γ such that conditions of Theorem 4.4.2 are satisfied. Set $\sigma^{(k+1)} = 10\sigma^{(k)}$, $\varepsilon^{(k+1)} = 0.1\varepsilon^{(k)}$ and $k := k + 1$. Go to Step 2.*

Remark 4.4.1 *In Step 3, if $\varepsilon^{(k),*} > \underline{\varepsilon}$, we obtain from Lemma 4.4.1 that $\boldsymbol{x}^{(k),*}$ cannot be a feasible point, meaning that the penalty parameter σ may not be large enough. Thus we need to increase σ. If $\sigma_k > 10^8$, but $\varepsilon^{(k),*} > \varepsilon^*$ still, then we should adjust the value of α, β and γ such that conditions assumed in the Theorem 4.4.2 are satisfied and go to Step 2.*

Remark 4.4.2 *Clearly, we cannot check the feasibility of $h_j(\boldsymbol{x}, t) \le 0$, $j = 1, 2, \ldots, m$, for every $t \in [0, T]$. In practice, we choose a set, $\widehat{T}$, which contains a dense enough set of points in $[0, T]$. We then check the feasibility of $h_j(\boldsymbol{x}, t) \le 0$ over $\widehat{T}$ for each $j = 1, 2, \ldots, m$.*

Remark 4.4.3 *Although we have proven that a local minimizer of the exact penalty function optimization problem (P_{σ_k}) will converge to a local minimizer of the original problem (P), in actual computation we need to set a lower bound $\underline{\varepsilon} = 10^{-9}$ for $\varepsilon^{(k),*}$ so as to avoid division by zero.*

Example 4.4.1 The following example is taken from [81], and it was also used for testing the numerical algorithms in [259, 261, 296] and Section 4.3. In this problem, the objective function

$$f(\boldsymbol{x}) = \frac{x_2(122 + 17x_1 + 6x_3 - 5x_2 + x_1x_3) + 180x_3 - 36x_1 + 1224}{x_2(408 + 56x_1 - 50x_2 + 60x_3 + 10x_1x_3 - 2x_1^2)} \tag{4.4.40}$$

is to be minimized subject to

$$h(\boldsymbol{x}, t) \le 0 \ , \ \forall \ t \in [T_1, T_2], \tag{4.4.41}$$

$$0 \le x_1, x_3 \le 100, \ 0.1 \le x_2 \le 100, \tag{4.4.42}$$

where $[T_1, T_2] = [10^{-6}, 30]$,

$$h(\boldsymbol{x}, t) = \Im\varphi(\boldsymbol{x}, t) - 3.33[\Re\varphi(\boldsymbol{x}, t)]^2 + 1.0,$$

$$\varphi(\boldsymbol{x}, t) = 1 + H(\boldsymbol{x}, it)G(it), \quad (i = \sqrt{-1})$$

$$H(\boldsymbol{x}, s) = x_1 + x_2/s + x_3 s,$$

and

$$G(s) = \frac{1}{(s+3)(s^2 + 2s + 2)}.$$

Here, $\Im\varphi(\boldsymbol{x}, t)$ and $\Re\varphi(\boldsymbol{x}, t)$ are, respectively, the imaginary and real parts of $\varphi(\boldsymbol{x}, t)$. The initial point is $\left[x_1^0, x_2^0, x_3^0\right]^\top = [50, 50, 50]^\top$. Actually, we can start from any point within the boundedness constraints (4.4.42).

For the continuous inequality constraint (4.4.41), the corresponding exact penalty function $f_\sigma(x, \varepsilon)$ is defined by (4.4.5) with the constraint violation $\Delta(x, \varepsilon)$ given by

$$\Delta(\boldsymbol{x}, \varepsilon) = \int_{T_1}^{T_2} \left[\max\left\{0, \Im\varphi(\boldsymbol{x}, t) - 3.33[\Re\varphi(\boldsymbol{x}, t)]^2 + 1.0 - \varepsilon^\gamma W_j\right\}\right]^2 dt.$$

Simpson's Rule with $[T_1, T_2] = [10^{-6}, 30]$ being divided into 3000 equal subintervals is used to evaluate the integral. The value obtained is very accurate. Also, these discretized points define a dense subset $\widehat{T}$ of $[T_1, T_2]$. We check the feasibility of the continuous inequality constraint by evaluating the values of the function h over $\widehat{T}$. Results obtained are given in Table 4.4.1.

As we can see that, as the penalty parameter, σ, is increased, the minimizer approaches to the boundary of the feasible region. When σ is sufficiently large, we obtain a feasible point. It has the same objective function value as that obtained in Section 4.3. However, for the minimizer obtained in Section 4.3, there are some minor violations of the continuous inequality constraints (4.4.41).

Table 4.4.1: Results for Example 4.4.1

σ	$\bar{t}$	$\bar{h}$	f^*	x_1^*	x_2^*	x_3^*	ε
10	5.66	1.1012×10^{-4}	0.17469205	16.961442	45.496567	34.677990	0.001976
10^2	5.65	1.3205×10^{-5}	0.17469506	16.959419	45.496640	34.674199	0.000261
10^3	5.66	1.31695×10^{-6}	0.17469547	16.967833	45.495363	34.668668	0.000054
10^4	5.66	5.839365×10^{-7}	0.17469569	16.987820	45.498793	34.657147	0.000019
10^5	5.66	5.070583×10^{-8}	0.17469635	16.981243	45.497607	34.660896	0.000011
10^6	5.66	-1.82251×10^{-7}	0.17469633	16.980628	45.497520	34.661238	0.000003

Example 4.4.2 Consider the problem:

$$\min \; x_1^2 + (x_2 - 3)^2$$
$$\text{subject to } \; x_2 - 2 + x_1 \sin\left(\frac{t}{x_2 - \omega}\right) \leq 0, \; \forall t \in [0, \pi],$$
$$-1 \leq x_1 \leq 1, \; 0 \leq x_2 \leq 2,$$

where ω is a parameter that controls the frequency of the constraint. As in [261], ω is chosen as 2.032. In this case, the corresponding exact penalty function $f_\sigma(x, \varepsilon)$ is defined by (4.4.5) with the constraint violation given by

$$\Delta(x, \varepsilon) = \int_0^{\pi} \left[\max\left\{0, x_2 - 2 + x_1 \sin\left(\frac{t}{x_2 - \omega}\right) - \varepsilon^{\gamma} W_j \right\}\right]^2 dt.$$

Simpson's Rule with the interval $[0, \pi]$ being divided into 1000 equal subintervals is used to evaluate the integral. These discretized points also form a dense subset $\widehat{T}$ of the interval $[0, \pi]$. The feasibility check is carried out over $\widehat{T}$. We use Algorithm 4.4.1 with the initial point taken as $[x_1^0, x_2^0]^\top = [0.5, 0.5]^\top$. The solution of this problem is $[x_1^*, x_2^*]^\top = [0, 2]^\top$ with the objective function value $f^* = 1$. The results of the algorithm are presented in Table 4.4.2.

Table 4.4.2: Results for Example 4.4.2

σ	$\bar{t}$	$\bar{h}$	f^*	x_1^*	x_2^*	ε
10	1.41	3.9583799×10^{-7}	1.000002326	4×10^{-7}	1.9999992	1.62×10^{-3}
10^2	1.51	-2.309265×10^{-8}	1.000000582	1.769×10^{-7}	1.9999998	3.0310×10^{-4}
10^3	1.51	-1.629325×10^{-8}	1.00000047	1.438×10^{-7}	1.9999998	9.6×10^{-5}

It is observed that for sufficiently large σ, the minimizer obtained is such that the continuous inequality constraints are satisfied for all $t \in [0, \pi]$.

Example 4.4.3 Consider the problem:

$$\min\ (x_1 + x_2 - 2)^2 + (x_1 - x_2)^2 + 30[\min\{0, x_1 - x_2\}]^2$$
$$\text{subject to}\ \ x_1 \cos t + x_2 \sin t - 1 \le 0,\ \ \forall\, t \in [0, \pi].$$

Again, Simpson's Rule with the interval $[0, \pi]$ being partitioned into 1000 equal subintervals is used to evaluate the corresponding constraint violation in the exact penalty function. These discretized points also define a dense subset $\widehat{T}$ of the interval $[0, \pi]$, which is to be used for checking the feasibility of the continuous inequality constraint. Now, by using Algorithm 4.4.1 with the initial point taken as $[x_1^0, x_2^0]^\top = [0.5, 0.5]$, the results obtained are reported in Table 4.4.3.

Table 4.4.3: Results for Example 4.4.3

σ	$\bar{t}$	$\bar{h}$	f^*	x_1^*	x_2^*	ε
10	0.784	0.1312984779	0.2778915304	0.7999488587	0.7999487935	0.2512652637
10^2	0.786	0.0155757133	0.3352776626	0.7181206321	0.7181203152	0.0274357241
10^3	0.79	0.0016566706	0.3423203957	0.7082782234	0.7082782249	0.0002872258
10^4	0.79	−0.0000022403	0.343138432	0.7071227384	0.7072248258	0.0000389857

By comparing our results with those obtained in [81, 103, 259, 261], it is observed that the objective values are almost the same. However, for our minimizer, it is a feasible point, while those obtained in [81, 103, 259, 261] are not.

4.5 Exercises

4.1. Consider an optimization problem for which the cost function

$$F_\alpha(\boldsymbol{x}) = \sum_{i=1}^{N} f_{i,\alpha}(\boldsymbol{x}) \tag{4.5.1}$$

is to be minimized with respect to $\boldsymbol{x} \in \mathbb{R}^n$, where

$$f_{i,\alpha}(\boldsymbol{x}) = \begin{cases} (1-\alpha)|f_i(\boldsymbol{x})| & f_i(\boldsymbol{x}) \le 0, \\ \alpha f_i(\boldsymbol{x}) & f_i(\boldsymbol{x}) > 0. \end{cases} \tag{4.5.2}$$

f_i, $i = 1, \ldots, N$, are real-valued continuously differentiable functions in $\boldsymbol{x} \in \mathbb{R}^n$, and $\alpha \in (0, 1)$ is given. For each $i = 1, \ldots, N$, $f_{i,\alpha}$ is called a 'lop-sided' function. Each of these functions is non-differentiable at those $\boldsymbol{x}$ where

$f_i(\boldsymbol{x}) = 0$. The following smoothing method was introduced in [105].

$$f_{i,\alpha,\delta}(\boldsymbol{x}) = \begin{cases} \alpha f_i(\boldsymbol{x}) & \text{if } \alpha f_i(\boldsymbol{x}) > \delta \\ \left(\delta^2 + (\alpha f_i(\boldsymbol{x}))^2\right)/2\delta & \text{if } 0 \le \alpha f_i(\boldsymbol{x}) \le \delta \\ \left(\delta^2 + ((1-\alpha) f_i(\boldsymbol{x}))^2\right)/2\delta & \text{if } -\delta \le (1-\alpha) f_i(\boldsymbol{x}) \le \delta \\ (\alpha - 1) f_i(\boldsymbol{x}) & \text{if } (\alpha-1) f_i(\boldsymbol{x}) < -\delta, \end{cases} \tag{4.5.3}$$

where $\delta > 0$ is a smoothing parameter.

(i) Show that for each $i = 1, \ldots, N$, $f_{i,\alpha,\delta}$ is continuously differentiable in $\boldsymbol{x} \in \mathbb{R}^n$.
(ii) Show that for each $i = 1, \ldots, N$, $f_{i,\alpha,\delta}$ and $f_{i,\alpha}$ have their minima at the same value of $\boldsymbol{x}$.
(iii) Show that

$$0 \le f_{i,\alpha,\delta}(\boldsymbol{x}) - f_{i,\alpha}(\boldsymbol{x}) \le \delta/2. \tag{4.5.4}$$

Replacing $f_{i,\alpha}(\boldsymbol{x})$ with $f_{i,\alpha,\delta}(\boldsymbol{x})$, the cost function (4.5.1) is then approximated by

$$F_{\alpha,\delta}(\boldsymbol{x}) = \sum_{i=1}^{N} f_{i,\alpha,\delta}(\boldsymbol{x}). \tag{4.5.5}$$

Clearly,

$$F_{\alpha,\delta}(\boldsymbol{x}) - F_{\alpha}(\boldsymbol{x}) \le \frac{N\delta}{2}. \tag{4.5.6}$$

(iv) Let $\boldsymbol{x}^{\delta,*}$ and $\boldsymbol{x}^*$ minimize (4.5.1) and (4.5.5), respectively. Show that

$$0 \le F_{\alpha,\delta}(\boldsymbol{x}^{\delta}) - F_{\alpha}(\boldsymbol{x}^*) \le \frac{N\delta}{2}. \tag{4.5.7}$$

(v) Let

$$\boldsymbol{x}^* = \arg\min F_{\alpha}(\boldsymbol{x}) \tag{4.5.8}$$

and

$$\boldsymbol{x}^{\delta,*} = \arg\min F_{\alpha,\delta}(\boldsymbol{x}). \tag{4.5.9}$$

If there exists a unique minimizer of $F_{\alpha}(\boldsymbol{x})$, show that

$$\lim_{\delta \to 0} \boldsymbol{x}^{\delta,*} = \boldsymbol{x}^*. \tag{4.5.10}$$

4.2. Consider the following optimization problem (see [242]). The cost function

$$J(\boldsymbol{x}) = \int_a^b |F(\boldsymbol{x}, t)|\, dt \tag{4.5.11a}$$

is to be minimized with respect to $\boldsymbol{x} = [x_i, \ldots, x_n]^\top \in \mathbb{R}^n$ and subject to the following constraints:

$$h_j(\boldsymbol{x}) = 0, \qquad j = 1, \ldots, N_e \tag{4.5.11b}$$

$$g_j(\boldsymbol{x}) \leq 0, \qquad j = N_e, \ldots, N \tag{4.5.11c}$$

and

$$\alpha_i \leq x_i \leq \beta_i, \quad i = 1, \ldots, n, \tag{4.5.11d}$$

where a and b are real constants,

$$\alpha = [\alpha_1, \ldots, \alpha_n]^\top, \qquad \beta = [\beta_1, \ldots, \beta_n]^\top$$

and α_i, $i = 1, \ldots, n$, and β_i, $i = 1, \ldots, n$, are real constants. Let this problem be referred to as Problem (P). Clearly, Problem (P) is a nonlinearly constrained nonsmooth optimization problem, where the nonsmooth function appears in the cost function (4.5.11a). Note that the nonsmoothness is due to $|F(\boldsymbol{x}, t)|$, which, for each $t \in [a, b]$, is non-negative. Recall the smoothing approximation introduced in [242] as follows:

$$F_\varepsilon(\boldsymbol{x}, t) = \begin{cases} |F(\boldsymbol{x}, t)|, & \text{if} \quad |F(\boldsymbol{x}, t)| \geq \frac{\varepsilon}{2} \\ \left[(F(\mathbf{x}, t))^2 + \frac{\varepsilon^2}{4}\right] / \varepsilon, & \text{if} \quad |F(\boldsymbol{x}, t)| < \frac{\varepsilon}{2}. \end{cases} \tag{4.5.12}$$

Show that

(i) For each $t \in [a, b]$, $F_\varepsilon(\boldsymbol{x}, t)$ is continuously differentiable with respect to $\boldsymbol{x}$.
(ii) $F_\varepsilon(\boldsymbol{x}, t) \geq |F(\boldsymbol{x}, t)|$ for each $(\boldsymbol{x}, t) \in \mathbb{R}^n \times [a, b]$.
(iii) For each $(\boldsymbol{x}, t) \in \mathbb{R}^n \times [a, b]$, $|F_\varepsilon(\boldsymbol{x}, t) - |F(\boldsymbol{x}, t)|| \leq \frac{\varepsilon}{4}$.
(iv) For each $t \in [a, b]$, $\boldsymbol{x}$ minimizes $|F(\boldsymbol{x}, t)|$ if and only if it minimizes $F_\varepsilon(\boldsymbol{x}, t)$. With $|F(\boldsymbol{x}, t)|$ approximated by $F_\varepsilon(\boldsymbol{x}, t)$, the cost function (4.5.11a) becomes

$$J_\varepsilon(\boldsymbol{x}) = \int_a^b F_\varepsilon(\boldsymbol{x}, t) dt. \tag{4.5.13}$$

This approximate problem, referred to as Problem (P_ε), is to minimize the cost function (4.5.13) subject to the constraints (4.5.11b)–(4.5.11d).
For each $\varepsilon > 0$, let $\boldsymbol{x}_\varepsilon^*$ be an optimal solution to Problem (P_ε). Furthermore, let $\boldsymbol{x}^*$ be the optimal solution of Problem (P). Show that

$$0 \leq J_\varepsilon(\boldsymbol{x}_\varepsilon^*) - J(\boldsymbol{x}^*) \leq \frac{\varepsilon(b - a)}{4}.$$

(v) Assume that $J(\boldsymbol{x}) \to \infty$ as $|\boldsymbol{x}| \to \infty$, where $|\cdot|$ denotes the Euclidean norm. Show that there exists an accumulation point of the sequence $\{\boldsymbol{x}_\varepsilon^*\}$ when $\varepsilon \to 0$. Furthermore, show that any such accumulation point is an optimal solution of Problem (P).

4.3. Consider the following optimization problem:

$$\min f(x) \tag{4.5.14a}$$

subject to

$$h_i(x) = 0, i = 1, 2, \dots, m \tag{4.5.14b}$$

$$h_i(x) \leq 0, i = m + 1, 2, \dots, N, \tag{4.5.14c}$$

where f and $h_i, i = 1, 2, \dots, N,$, are continuously differentiable functions.

(i) Define the constraint violation of the form of (4.4.6) and then the exact penalty function of the form of (4.4.5).
(ii) Write down the surrogate optimization problem of the form of (4.4.7).
(iii) Develop the exact penalty function algorithm as of Algorithm 4.4.1.
(iv) Show that a local minimum of the surrogate optimization problem is a local minimum of the problem (4.5.12)

Chapter 5
Discrete Time Optimal Control Problems

5.1 Introduction

Discrete time optimal control problems arise naturally in many multi-stage control and inventory problems where time enters discretely in a natural fashion. In this chapter, we first use the dynamic programming approach to study a class of discrete time optimal control problems. We then move on to consider a general class of constrained discrete time optimal control problems in canonical form. An efficient algorithm supported by a rigorous convergence analysis will be developed for solving this constrained discrete time optimal control problem. We also consider a class of discrete time optimal control problems subject to terminal and all-time-step inequality constraints involving state and control variables. It is then shown that this optimal control problem can be solved as a constrained discrete time optimal control problem in canonical form via the constraint transcription introduced in Chapter 4. Three examples are computed using the algorithm developed in this chapter so as to demonstrate the efficiency and effectiveness of the algorithm. The main references of the chapter are [58], Chapter 11 of [253] and [258, 265, 266, 282].

5.2 Dynamic Programming Approach

Consider the system of difference equations

$$\boldsymbol{x}(i+1) = \boldsymbol{f}(i, \boldsymbol{x}(i), \boldsymbol{u}(i)), \quad i = 0, 1, \ldots, N-1, \tag{5.2.1}$$

$$\boldsymbol{x}(0) = \boldsymbol{x}^{(0)}, \tag{5.2.2}$$

K. L. Teo et al., *Applied and Computational Optimal Control*, Springer Optimization and Its Applications 171,
https://doi.org/10.1007/978-3-030-69913-0_5

where $\boldsymbol{x} \in \mathbb{R}^n$ and $\boldsymbol{u} \in \mathbb{R}^r$ are the state and control vectors, respectively, and $\boldsymbol{x}^{(0)} \in \mathbb{R}^n$ is a given initial state. Furthermore, $\boldsymbol{f}(i,\cdot,\cdot) : \mathbb{R}^n \times \mathbb{R}^r$ is a given function. For each $i = 0, 1, \ldots, N-1$,

$$\boldsymbol{u}(i) \in U_i, \tag{5.2.3}$$

where U_i is a given compact and convex subset of $\mathbb{R}^r$. For notational convenience, the set of feasible control sequences starting at time $k \in \{0, 1, \ldots, N-1\}$ is defined by

$$\mathcal{U}_k = \{\{\boldsymbol{u}(k), \boldsymbol{u}(k+1), \ldots, \boldsymbol{u}(N-1)\} : \boldsymbol{u}(i) \in U_i, i = k, k+1, \ldots, N-1\}. \tag{5.2.4}$$

We then consider the following discrete time optimal control problem:

$$\min_{\{\boldsymbol{u}(0),\ldots,\boldsymbol{u}(N-1)\}\in\mathcal{U}_0} \left\{ \Phi_0(\boldsymbol{x}(N)) + \sum_{i=0}^{N-1} \mathcal{L}_0(i, \boldsymbol{x}(i), \boldsymbol{u}(i)) \right\} \tag{5.2.5}$$

subject to (5.2.1)–(5.2.2), where $\mathcal{L}_0(i,\cdot,\cdot) : \mathbb{R}^n \times \mathbb{R}^r \to \mathbb{R}$ and $\Phi_0 : \mathbb{R}^r \to \mathbb{R}$ are given functions. Let this problem be referred to Problem (P).

Dynamic programming involves a sequence of problems embedded in the original Problem (P). Each of these embedded problems shares the dynamics, cost functional, and control constraints of the original but has a different initial state and a different starting time. More specifically, for each $\boldsymbol{\xi} = [\xi_1, \ldots, \xi_n]^\top \in \mathbb{R}^n$, and $k \in \{0, \ldots, N-1\}$, we consider the following problem:

$$\min_{\{\boldsymbol{u}(k),\ldots,\boldsymbol{u}(N-1)\}\in\mathcal{U}_k} \left\{ \Phi_0(\boldsymbol{x}(N)) + \sum_{i=k}^{N-1} \mathcal{L}_0(i, \boldsymbol{x}(i), \boldsymbol{u}(i)) \right\} \tag{5.2.6}$$

subject to

$$\boldsymbol{x}(i+1) = \boldsymbol{f}(i, \boldsymbol{x}(i), \boldsymbol{u}(i)), \quad i = k, k+1, \ldots, N-1, \tag{5.2.7}$$

$$\boldsymbol{x}(k) = \boldsymbol{\xi}. \tag{5.2.8}$$

Let problem (5.2.6)–(5.2.8) be referred to as Problem $(P_{k,\boldsymbol{\xi}})$. As k and the initial state $\boldsymbol{\xi}$ can vary, we are dealing with a family of optimal control problems.

The result presented in the following theorem is known as the Principle of Optimality:

Theorem 5.2.1 *Suppose that $\{\boldsymbol{u}^*(k), \ldots, \boldsymbol{u}^*(N-1)\}$ is an optimal control for Problem $(P_{k,\boldsymbol{\xi}})$ and that $\{\boldsymbol{x}^*(k) = \boldsymbol{\xi}, \boldsymbol{x}^*(k+1), \ldots, \boldsymbol{x}^*(N)\}$ is the corresponding optimal trajectory. Then, for any j such that $k \le j \le N-1$, $\{\boldsymbol{u}^*(j), \boldsymbol{u}^*(j+1) \ldots, \boldsymbol{u}^*(N-1)\}$ is an optimal control for Problem $(P_{j,\boldsymbol{x}^*(j)})$.*

Proof. Suppose that the conclusion is false. Then, there exists a control

$$\{\widehat{\boldsymbol{u}}(j), \widehat{\boldsymbol{u}}(j+1), \ldots, \widehat{\boldsymbol{u}}(N-1)\}$$

with the corresponding trajectory

$$\{\widehat{\boldsymbol{x}}(j) = \boldsymbol{x}^*(j), \widehat{\boldsymbol{x}}(j+1), \ldots, \widehat{\boldsymbol{x}}(N)\}$$

such that

$$\Phi_0(\widehat{\boldsymbol{x}}(N)) + \sum_{i=j}^{N-1} \mathcal{L}_0(i, \widehat{\boldsymbol{x}}(i), \widehat{\boldsymbol{u}}(i)) < \Phi_0(\boldsymbol{x}^*(N)) + \sum_{i=j}^{N-1} \mathcal{L}_0(i, \boldsymbol{x}^*(i), \boldsymbol{u}^*(i)). \tag{5.2.9}$$

Now construct the control $\{\widetilde{\boldsymbol{u}}(k), \widetilde{\boldsymbol{u}}(k+1), \ldots, \widetilde{\boldsymbol{u}}(N-1)\}$ as follows:

$$\widetilde{\boldsymbol{u}}(i) = \begin{cases} \boldsymbol{u}^*(i), \; i = k, \ldots, j-1, \\ \widehat{\boldsymbol{u}}(i), \quad i = j, \ldots, N. \end{cases} \tag{5.2.10}$$

The corresponding trajectory, which starts from state $\boldsymbol{\xi}$ at time k, is $\{\widetilde{\boldsymbol{x}}(k), \widetilde{\boldsymbol{x}}(k+1), \ldots, \widetilde{\boldsymbol{x}}(N)\}$, where

$$\widetilde{\boldsymbol{x}}(i) = \begin{cases} \boldsymbol{x}^*(i), \; i = k, \ldots, j, \\ \widehat{\boldsymbol{x}}(i), \quad i = j+1, \ldots, N. \end{cases} \tag{5.2.11}$$

From (5.2.9)–(5.2.11), it follows that

$$\begin{aligned} &\Phi_0(\widetilde{\boldsymbol{x}}(N)) + \sum_{i=k}^{N-1} \mathcal{L}_0(i, \widetilde{\boldsymbol{x}}(i), \widetilde{\boldsymbol{u}}(i)) \\ =&\Phi_0(\widehat{\boldsymbol{x}}(N)) + \sum_{i=k}^{j-1} \mathcal{L}_0(i, \boldsymbol{x}^*(i), \boldsymbol{u}^*(i)) + \sum_{i=j}^{N-1} \mathcal{L}_0(i, \widehat{\boldsymbol{x}}(i), \widehat{\boldsymbol{u}}(i)) \\ <&\Phi_0(\boldsymbol{x}^*(N)) + \sum_{i=k}^{N-1} \mathcal{L}_0(i, \boldsymbol{x}^*(i), \boldsymbol{u}^*(i)) \end{aligned} \tag{5.2.12}$$

Equation (5.2.12) shows that $\{\boldsymbol{u}^*(k), \ldots, \boldsymbol{u}^*(N-1)\}$ is not an optimal control for Problem $(P_{k,\boldsymbol{\xi}})$ which is a contradiction to the hypothesis of the theorem. Hence the proof is complete.

To continue, we assume that an optimal solution to $(P_{k,\boldsymbol{\xi}})$ exists for each k, $0 \leq k \leq N-1$, and for each $\boldsymbol{\xi} \in \mathbb{R}^n$. Let $V(k, \boldsymbol{\xi})$ be the minimum value of the cost functional (5.2.6) for Problem $(P_{k,\boldsymbol{\xi}})$. It is called the *value function.*

Theorem 5.2.2 *The value function V satisfies the following backward recursive relation. For any k, $0 \leq k \leq N-1$,*

$$V(k, \boldsymbol{\xi}) = \min_{\boldsymbol{u} \in U_k} \left\{ \mathcal{L}_0(k, \boldsymbol{\xi}, \boldsymbol{u}) + V(k+1, \boldsymbol{f}(k, \boldsymbol{\xi}, \boldsymbol{u})) \right\}, \tag{5.2.13}$$

with the final condition

$$V(N, \boldsymbol{\xi}) = \Phi_0(\boldsymbol{\xi}). \tag{5.2.14}$$

Proof. Equation (5.2.14) is obvious. Let $\boldsymbol{\xi} \in \mathbb{R}^n$. Then, by Theorem 5.2.1,

$$\begin{aligned}
V(k, \boldsymbol{\xi}) &= \min_{\{\boldsymbol{u}(k),\ldots,\boldsymbol{u}(N-1)\}\in\mathcal{U}_k} \left\{ \Phi_0(\boldsymbol{x}(N)) + \sum_{i=k}^{N-1} \mathcal{L}_0(i, \boldsymbol{x}(i), \boldsymbol{u}(i)) \right\} \\
&= \min_{\boldsymbol{u}(k)\in U_k} \left\{ \mathcal{L}_0(k, \boldsymbol{\xi}, \boldsymbol{u}(k)) \right. \\
&\quad \left. + \min_{\{\boldsymbol{u}(k+1),\ldots,\boldsymbol{u}(N-1)\}\in\mathcal{U}_{k+1}} \left[\Phi_0(\boldsymbol{x}(N)) + \sum_{i=k+1}^{N-1} \mathcal{L}_0(i, \boldsymbol{x}(i), \boldsymbol{u}(i)) \right] \right\} \\
&= \min_{\boldsymbol{u}(k)\in U_k} \left\{ \mathcal{L}_0(k, \boldsymbol{\xi}, \boldsymbol{u}(k)) + V(k+1, \boldsymbol{x}(k+1)) \right\} \\
&= \min_{\boldsymbol{u}(k)\in U_k} \left\{ \mathcal{L}_0(k, \boldsymbol{\xi}, \boldsymbol{u}(k)) + V(k+1, \boldsymbol{f}(k, \boldsymbol{\xi}, \boldsymbol{u}(k))) \right\},
\end{aligned}$$

which proves (5.2.13).

Remark 5.2.1 *During the process of finding the minimum in the recursive equation* (5.2.13)–(5.2.14), *we obtain the optimal control* $\boldsymbol{v}$ *in feedback form. This optimal feedback control can be used for all initial conditions. However, Dynamic Programming requires us to compute and store the values of* V *and* $\boldsymbol{v}$ *for all* k *and* $\boldsymbol{x}$. *Therefore, unless we can find a closed-form analytic solution to* (5.2.13)–(5.2.14), *the Dynamic Programming formulation will inevitably lead to an enormous amount of storage and computation. This phenomena is known as the curse of dimensionality. It seriously hinders the applicability of Dynamic Programming to problems where we cannot obtain a closed-form analytic solution to* (5.2.13)-(5.2.14).

For illustration, let us look at a simple example, where all the functions involved are scalar.

Example 5.2.1 Minimize

$$(x(N))^2 + \sum_{i=1}^{N-1} \left\{ (x(i))^2 + (u(i))^2 \right\} \tag{5.2.15}$$

subject to

$$x(i+1) = x(i) + u(i), \quad i = 0, 1, \ldots, N-1, \tag{5.2.16}$$

$$x(0) = x^{(0)}. \tag{5.2.17}$$

This problem is clearly in the form of Problem (P) with $\Phi_0(x) = x^2$, $\mathcal{L}_0(x, u) = x^2 + u^2$, $f(x, u) = x + u$, and $u(i) \in \mathbb{R}$, $i = 0, 1, \ldots, N-1$.

Solution Let $V(k,\xi)$ be the value function for the corresponding Problem $(P_{k,\xi})$ starting from the arbitrary state ξ at step k. By Theorem 5.2.2, we have

$$\begin{aligned} V(k,\xi) &= \min_{u\in\mathbb{R}} \{\mathcal{L}_0(\xi,u) + V(k+1, f(\xi,u))\} \\ &= \min_{u\in\mathbb{R}} \{\xi^2 + u^2 + V(k+1, \xi+u)\} \end{aligned} \tag{5.2.18}$$

and

$$V(N,\xi) = \xi^2, \tag{5.2.19}$$

where $V(N,\xi)$ denotes the value function for the minimization problem starting from the arbitrary state ξ at step $k = N$.

For $k = N-1$, we start from the arbitrary state ξ at step $N-1$. If we denote $u(N-1) = u$,

$$\begin{aligned} V(N-1,\xi) &= \min_{u\in\mathbb{R}} \{\xi^2 + u^2 + V(N, \xi+u)\} \\ &= \min_{u\in\mathbb{R}} \{\xi^2 + u^2 + (\xi+u)^2\}. \end{aligned} \tag{5.2.20}$$

Minimizing the expression in the brackets of (5.2.20) with respect to u gives

$$\frac{\partial\{\xi^2 + u^2 + (\xi+u)^2\}}{\partial u} = 0 \Rightarrow 2u + 2(\xi+u) = 0 \Rightarrow u = -\frac{1}{2}\xi.$$

This implies that the optimal control for step $N-1$ starting at state ξ is

$$u^*(N-1) = -\frac{1}{2}\xi. \tag{5.2.21}$$

The corresponding value function is

$$V(N-1,\xi) = \xi^2 + \left(-\frac{1}{2}\xi\right)^2 + \left(\xi - \frac{1}{2}\xi\right)^2 = \frac{5}{4}\xi^2 + \frac{1}{4}\xi^2 = \frac{3}{2}\xi^2. \tag{5.2.22}$$

Continuing this process, at the next step, we have

$$\begin{aligned} V(N-2,\xi) &= \min_{u\in\mathbb{R}} \{\xi^2 + u^2 + V(N-1, \xi+u)\} \\ &= \min_{u\in\mathbb{R}} \left\{\xi^2 + u^2 + \frac{3}{2}(\xi+u)^2\right\}. \end{aligned}$$

Minimizing the right hand side with respect to u gives

$$\frac{\partial\{\xi^2 + u^2 + \frac{3}{2}(\xi+u)^2\}}{\partial u} = 0 \Rightarrow 2u + 3(\xi+u) = 0 \Rightarrow u = -\frac{3}{5}\xi.$$

This implies that the optimal control for step $N-2$ starting at state ξ is

$$u^*(N-2) = -\frac{3}{5}\xi. \tag{5.2.23}$$

The corresponding value function is

$$V(N-2,\xi) = \xi^2 + \left(-\frac{3}{5}\xi\right)^2 + \frac{3}{2}\left(\xi - \frac{3}{5}\xi\right)^2 = \frac{34}{25}\xi^2 + \frac{6}{25}\xi^2 = \frac{8}{5}\xi^2. \tag{5.2.24}$$

Based on the above results, it is reasonable to guess that

$$V(k,\xi) = c_k\xi^2, \tag{5.2.25}$$

for some $c_k > 0$. Let us show that this guess is correct. We begin by assuming the form of (5.2.25). Then, by (5.2.18), we have

$$\begin{aligned} V(k,\xi) &= \min_{u\in\mathbb{R}}\{\xi^2 + u^2 + V(k+1,\xi+u)\} \\ &= \min_{u\in\mathbb{R}}\{\xi^2 + u^2 + c_{k+1}(\xi+u)^2\}. \end{aligned} \tag{5.2.26}$$

Minimizing with respect to u gives

$$\frac{\partial}{\partial u}\{\xi^2 + u^2 + c_{k+1}(\xi+u)^2\} = 0 \;\Rightarrow\; 2u + 2c_{k+1}(\xi+u) = 0 \;\Rightarrow\; u = -\frac{c_{k+1}}{1+c_{k+1}}\xi. \tag{5.2.27}$$

This implies that the optimal control for step k starting at state ξ is

$$u^*(k) = -\frac{c_{k+1}}{1+c_{k+1}}\xi. \tag{5.2.28}$$

Substituting (5.2.28) into (5.2.26) gives

$$\begin{aligned} V(k,\xi) =& \xi^2 + \left(\frac{c_{k+1}}{1+c_{k+1}}\right)^2\xi^2 + c_{k+1}\left(x - \frac{c_{k+1}}{1+c_{k+1}}\xi\right)^2 \\ =& \left[1 + \frac{(c_{k+1})^2}{(1+c_{k+1})^2} + \frac{c_k+1}{(1+c_{k+1})^2}\right]\xi^2 \\ =& \left[1 + \frac{c_{k+1}(c_{k+1}+1)}{(1+c_{k+1})^2}\right]\xi^2 = \left(1 + \frac{c_{k+1}}{1+c_{k+1}}\right)\xi^2 \\ =& \frac{1+2c_{k+1}}{1+c_{k+1}}\xi^2, \end{aligned}$$

which is of the assumed form of (5.2.25). Thus, we conclude that

$$V(k,\xi) = c_k\xi^2, \tag{5.2.29}$$

with

$$c_k = \frac{1+2c_{k+1}}{1+c_{k+1}}, \quad \text{and} \quad c_N = 1. \tag{5.2.30}$$

(5.2.30) allows us to determine the coefficients of the value function recursively, i.e.,

$$c_{N-1} = \frac{3}{2},\; c_{N-2} = \frac{8}{5},\; \ldots .$$

Therefore, we can easily find the solution for the problem for any number of steps. The optimal control law resulting from the above dynamic programming solution is

$$u^*(k) = -\frac{c_{k+1}}{1+c_{k+1}}x^*(k). \tag{5.2.31}$$

Note that the control law is expressed in terms of the current state of the system, i.e., the control is expressed in closed-loop or feedback form. This is a great advantage of dynamic programming over Pontryagin type theory: it makes the dynamic programming solution more robust with respect to modeling errors and noise.

Notice how we use the dynamic programming solution in practice. We compute c_k, $k = 0, 1, \ldots, N$, from (5.2.30), starting with $c_N = 1$. We then run the system using the corresponding controls in the reverse order.

Consider the case $N = 3$. From (5.2.30),

$$\begin{aligned} c_3 &= 1, \\ c_2 &= \frac{1+2c_3}{1+c_3} = \frac{1+2(1)}{1+1} = \frac{3}{2}, \\ c_1 &= \frac{1+2c_2}{1+c_2} = \frac{1+2(\frac{3}{2})}{1+\frac{3}{2}} = \frac{8}{5}, \\ c_0 &= \frac{1+2c_1}{1+c_1} = \frac{1+2(\frac{8}{5})}{1+\frac{8}{5}} = \frac{21}{13}. \end{aligned}$$

Now, starting with the given initial condition $x(0) = x^{(0)}$, we obtain

$$u^*(0) = -\frac{c_1}{1+c_1}x^*(0) = -\frac{\frac{8}{5}}{1+\frac{8}{5}}x^{(0)} = -\frac{8}{13}x^{(0)}.$$

Then

$$x^*(1) = x^*(0) + u^*(0) = x^{(0)} - \frac{8}{13}x^{(0)} = \frac{5}{13}x^{(0)},$$

and

$$\begin{aligned} u^*(1) &= -\frac{c_2}{1+c_2}x^*(1) = -\frac{\frac{3}{2}}{1+\frac{3}{3}}x^*(1) \\ &= -\frac{3}{5}x^*(1) = -\frac{3}{5}\times\frac{5}{13}x^{(0)} = -\frac{3}{13}x^{(0)}. \end{aligned}$$

Next,

$$x^*(2) = x^*(1) + u^*(1) = \frac{5}{13}x^{(0)} - \frac{3}{13}x^{(0)} = \frac{2}{13}x^{(0)},$$

and

$$u^*(2) = -\frac{c_3}{1+c_3}x^*(2) = -\frac{1}{1+1}x^*(2) = -\frac{1}{2}x^*(2) = -\frac{1}{2}\times\frac{2}{13}x^{(0)} = -\frac{1}{13}x^{(0)}.$$

Finally,

$$x^*(3) = x^*(2) + u^*(2) = \frac{2}{13}x^{(0)} - \frac{1}{13}x^{(0)} = \frac{1}{13}x^{(0)}.$$

Using (5.2.29), the optimal cost can be determined in terms of the initial condition, i.e.,

$$V(0, x^{(0)}) = c_0\left(x^{(0)}\right)^2 = \frac{21}{13}\left(x^{(0)}\right)^2.$$

Remark 5.2.2 *The following general discrete time linear quadratic regulator (LQR) problem can be solved in a similar way.*

$$\begin{aligned} &\textit{minimize} \quad \sum_{i=1}^{N-1} \left\{(\boldsymbol{x}(k))^\top Q\boldsymbol{x}(k) + (\boldsymbol{u}(k))^\top R\boldsymbol{u}(k)\right\} \\ &\textit{subject to} \quad \boldsymbol{x}(k+1) = A\boldsymbol{x}(k) + B\boldsymbol{u}(k), \quad k = 0, 1, \ldots, M \\ &\qquad\qquad\quad \boldsymbol{x}(0) = \boldsymbol{x}^{(0)}, \end{aligned}$$

where $\boldsymbol{x}(\cdot) \in \mathbb{R}^n$, $\boldsymbol{u}(\cdot) \in \mathbb{R}^r$ and A, B, Q, R are constant matrices with appropriate dimensions. Furthermore, we need to assume that $Q \geq 0$ (i.e., Q is positive semi-definite) and $R > 0$ (i.e., R is positive definite).

We now consider a modified version of Problem (P) which includes a terminal state constraint. The dynamics are given by

$$\boldsymbol{x}(i+1) = \boldsymbol{f}(\boldsymbol{x}(i), \boldsymbol{u}(i)), \qquad i = 0, 1, \ldots, N-1, \tag{5.2.32}$$

with the initial condition

$$\boldsymbol{x}(0) = \boldsymbol{x}^{(0)} \tag{5.2.33}$$

and the terminal state constraint

$$\boldsymbol{x}(N) = \hat{\boldsymbol{x}}. \tag{5.2.34}$$

Here, $\boldsymbol{x} = [x_1, \ldots, x_n]^\top \in \mathbb{R}^n$ and $\boldsymbol{u} = [u_1, \ldots, u_r]^\top \in \mathbb{R}^r$ are, respectively, the state and control vectors, $\boldsymbol{f} : \mathbb{R}^n \times \mathbb{R}^n \mapsto \mathbb{R}^n$ is a given function, and $\boldsymbol{x}^{(0)} = [x_1^{(0)}, \ldots, x_n^{(0)}]^\top \in \mathbb{R}^n$ and $\hat{\boldsymbol{x}} = [\hat{x}_1, \ldots, \hat{x}_n]^\top \in \mathbb{R}^n$ are given vectors. Once again, we assume that for each $i = 0, 1, \ldots, N-1$,

$$\boldsymbol{u}(i) \in U_i, \tag{5.2.35}$$

where U_i is a given compact and convex subset of $\mathbb{R}^r$, and $\mathcal{U}_k$, which is defined by 5.2.4, is the set of feasible control sequences starting at time $k \in 0, 1, ..., N-1$. The problem is then described by

$$\min_{\{\boldsymbol{u}(0),\ldots,\boldsymbol{u}(N-1)\}\in\mathcal{U}_0} \sum_{i=0}^{N-1} \mathcal{L}(\boldsymbol{x}(i),\boldsymbol{u}(i)) \tag{5.2.36}$$

subject to (5.2.32)–(5.2.34), where $\mathcal{L}_0 : \mathbb{R}^n \times \mathbb{R}^r \to \mathbb{R}$ is a given function. We denote this problem as Problem $(\widehat{P})$. The main difference to Problem (P) is that the system must finish at $\hat{\boldsymbol{x}}$ when $i = N$.

As we did for Problem (P), we consider a family of problems which are embedded in Problem $(\widehat{P})$. These are identical to Problem $(\widehat{P})$ except for the initial time and state. For each $\boldsymbol{\xi} = [\xi,\ldots,\xi]^\top \in \mathbb{R}^n$, and $k \in \{0,\ldots,N-1\}$, we consider the following problem:

$$\min_{\{\boldsymbol{u}(k),\ldots,\boldsymbol{u}(N-1)\}\in\mathcal{U}_k} \sum_{i=k}^{N-1} \mathcal{L}(\boldsymbol{x}(i),\boldsymbol{u}(i)) \tag{5.2.37}$$

subject to

$$\boldsymbol{x}(i+1) = \boldsymbol{f}(\boldsymbol{x}(i),\boldsymbol{u}(i)), \qquad i = 0,1,\ldots,N-1, \tag{5.2.38}$$

with

$$\boldsymbol{x}(k) = \boldsymbol{\xi} \tag{5.2.39}$$

and

$$\boldsymbol{x}(N) = \hat{\boldsymbol{x}}. \tag{5.2.40}$$

We denote this as Problem $(\widehat{P}_{k,\boldsymbol{\xi}})$. The value function for Problem $(\widehat{P}_{k,\boldsymbol{\xi}})$ is the minimum value of the cost functional (5.2.37) and denoted again by $V(k,\boldsymbol{\xi})$. The following two results are the equivalent of Theorems 5.2.1 and 5.2.2. The proofs are almost identical and hence omitted.

Theorem 5.2.3 *Suppose that $\{\boldsymbol{u}^*(k),\ldots,\boldsymbol{u}^*(N-1)\}$ is an optimal control for Problem $(\widehat{P}_{k,\boldsymbol{\xi}})$, and that $\{\boldsymbol{x}^*(k) = \boldsymbol{\xi},\ \boldsymbol{x}^*(k+1),\ldots,\boldsymbol{x}^*(N) = \widehat{\boldsymbol{x}}\}$ is the corresponding optimal trajectory. Then, for any j, $k \le j \le N-1$, $\{\boldsymbol{u}^*(j),\ldots,\boldsymbol{u}^*(N-1)\}$ is an optimal control for Problem $(\widehat{P}_{j,\boldsymbol{x}^*(j)})$.*

Theorem 5.2.4 *The value function V satisfies the following backward recursive relation. For k, $0 \le k \le N-2$,*

$$V(k,\boldsymbol{\xi}) = \min_{\boldsymbol{u}\in U_k} \left\{\mathcal{L}_0(k,\boldsymbol{\xi},\boldsymbol{u}) + V\left(k+1,\boldsymbol{f}(k,\boldsymbol{\xi},\boldsymbol{u})\right)\right\}, \tag{5.2.41}$$

with the terminal condition

$$V(N-1,\boldsymbol{\xi}) = \min_{\boldsymbol{u}\in U_{N-1}} \{\mathcal{L}_0(\boldsymbol{\xi},\boldsymbol{u})\} \tag{5.2.42}$$

subject to

$$\boldsymbol{x}(N) = \boldsymbol{f}(\boldsymbol{\xi},\boldsymbol{u}) = \widehat{\boldsymbol{x}}. \tag{5.2.43}$$

Let us consider a simple example.

Example 5.2.2 Minimize

$$\sum_{i=0}^{N-1} \left\{ (x(i))^2 + (u(i))^2 \right\}$$

subject to

$$\begin{aligned} x(i+1) &= x(i) + u(i), \quad i = 0, 1, \ldots, N-1, \\ x(0) &= x^{(0)}, \\ x(N) &= 0. \end{aligned}$$

Solution In the notation of Problem $\left(\widehat{P}\right)$, $\mathcal{L}_0(x, u) = x^2 + u^2$, $f(x, u) = x + u$, $\hat{x} = 0$, and $u(i) \in \mathbb{R}$, $i = 0, 1, \ldots, N-1$. For $k = 0, 1, \ldots, N$, (5.2.41)–(5.2.43) become

$$V(k, \xi) = \min_{u \in \mathbb{R}} \left\{ \xi^2 + u^2 + V(k+1, \xi + u) \right\}, \ k = 0, 1, \ldots, N-2. \quad (5.2.44)$$

$$V(N-1, \xi) = \min_{u \in \mathbb{R}} \left\{ \xi^2 + u^2 \right\} \quad (5.2.45)$$

subject to

$$x(N) = f(\xi, u(N-1)) = \xi + u(N-1) = 0. \quad (5.2.46)$$

This is the basic dynamic programming backward recursive relation. For illustration, let us first consider the case $N = 3$. For $i = N - 1 = 2$, we have

$$V(2, \xi) = \min_{u \in \mathbb{R}} \left\{ \xi^2 + u^2 \right\}$$

subject to

$$x(3) = \xi + u = 0.$$

This can only be satisfied with $u^*(2) = -\xi$ which, in turn, leads to

$$V(2, \xi) = 2\xi^2.$$

For $i = 1$, (5.2.44) gives

$$V(1, \xi) = \min_{u \in \mathbb{R}} \left\{ \xi^2 + u^2 + V(2, \xi + u) \right\} = \min_{u \in \mathbb{R}} \left\{ \xi^2 + u^2 + 2(\xi + u)^2 \right\}.$$

Aiming to minimize the right hand side, we have

$$\frac{\partial \left\{ \xi^2 + u^2 + 2(\xi + u)^2 \right\}}{\partial u} = 0 \ \Rightarrow \ 2u + 4(\xi + u) = 0 \ \Rightarrow \ u^*(1) = -\frac{2}{3}\xi$$

which then leads to

$$V(1, \xi) = \frac{5}{3}\xi^2.$$

For $i = 0$, (5.2.44) gives

$$V(0,\xi) = \min_{u\in\mathbb{R}} \left\{\xi^2 + u^2 + V(1, \xi + u)\right\} = \min_{u\in\mathbb{R}} \left\{\xi^2 + u^2 + \frac{5}{3}(\xi + u)^2\right\}.$$

Again minimizing the right hand side, we have

$$\frac{\partial\left\{\xi^2 + u^2 + 2(\xi + u)^2\right\}}{\partial u} = 0 \;\Rightarrow\; 2u + \frac{10}{3}(\xi + u) = 0 \;\Rightarrow\; u^*(0) = -\frac{5}{8}\xi.$$

with

$$V(0,\xi) = \frac{5}{3}\xi^2.$$

To obtain the solution for the case with arbitrary N, we conjecture that

$$V(k,\xi) = c_k\xi^2,$$

for some $c_k > 0$. To prove this conjecture, we substitute $V(k+1,\xi) = c_{k+1}\xi^2$ into the right hand side of (5.2.44). Thus, for $k = 0, 1, \ldots, N-2$, we have

$$V(k,\xi) = \min_{u\in\mathbb{R}} \left\{\xi^2 + u^2 + c_{k+1}(\xi + u)^2\right\}.$$

Minimizing the contents of the brackets, we have

$$\frac{\partial\left\{\xi^2+u^2 + c_{k+1}(\xi+u)^2\right\}}{\partial u} = 0 \;\Rightarrow\; 2u+2c_{k+1}(\xi+u)=0 \;\Rightarrow\; u^*(k) = -\frac{c_{k+1}}{1+c_{k+1}}\xi,$$

which, in turn, yields

$$V(k,\xi) = \frac{1+2c_{k+1}}{1+c_{k+1}}\xi^2.$$

Thus,

$$c_k = \frac{1+2c_{k+1}}{1+c_{k+1}}. \tag{5.2.47}$$

For $k = N-1$, it follows from (5.2.45) and (5.2.46) that

$$c_{N-1}\xi^2 = \min_{u\in\mathbb{R}} \left\{\xi^2 + u^2\right\}$$

subject to $x(N) = \xi + u = 0$, i.e., $u = -\xi$. Substituting into the above equation yields $c_{N-1}\xi^2 = 2\xi^2$ which gives $c_{N-1} = 2$. By (5.2.47), we have $c_{N-2} = \frac{5}{3}$ and $c_{N-3} = \frac{13}{8}$.

Therefore, for $N = 3$, $c_0 = \frac{13}{8}$, $c_1 = \frac{5}{3}$, $c_2 = 2$, and the final answer is

$$V(0,\xi) = c_0\xi^2 = \frac{13}{8}\xi^2,$$

or, remembering that $x(0) = x^{(0)}$, $V\left(0, x^{(0)}\right) = \frac{13}{8}\left(x^{(0)}\right)^2$. The corresponding optimal controls and states can then be calculated as follows:

$$\begin{aligned}
u^*(0) &= -\frac{c_1}{1+c_1}x^{(0)} = -\frac{5}{8}\left(x^{(0)}\right) \Rightarrow x^*(1) = x(0) + u^*(0) = \frac{3}{8}x^{(0)},\\
u^*(1) &= -\frac{c_2}{1+c_2}x^*(1) = -\frac{2}{3}x^*(1) \Rightarrow x^*(2) = x^*(1) + u^*(1) = \frac{1}{3}x^*(1) = \frac{1}{8}x^{(0)},\\
u^*(2) &= -x^*(2) \Rightarrow x^*(3) = x^*(2) + u^*(2) = 0,
\end{aligned}$$

where the latter equations result from the first step of our analysis. Note that, once again, it is possible to express the optimal control in a feedback form.

5.2.1 Application to Portfolio Optimization

In this section, we present an application of the dynamic programming method to a multi-period portfolio optimization problem subject to probabilistic risks. The main reference for this section is [235].

5.2.1.1 Problem Formulation

We consider a multi-period portfolio optimization problem, where an investor is going to invest in N possible risky assets S_j, $j = 1, \ldots, N$, with a positive initial wealth of M_0. The investment will be made at the beginning of the first period of a T-period portfolio planning horizon. Then, the wealth will be reallocated to these N risky assets at the beginning of the following $T-1$ consecutive time periods. The investor will claim the final wealth at the end of the Tth period.

Let x_{tj} be the fraction of wealth at the end of period $t-1$ invested in asset S_j at the beginning of period t. Denote $\boldsymbol{x}_t = [x_{t1}, \ldots, x_{tN}]^\top$. Here we assume that the whole investment process is a self-financing process. Thus, the investor will not increase the investment nor put aside funds in any period in the portfolio planning horizon. In other words, the total funds in the portfolio at the end of period $t-1$ will be allocated to those risky assets at the beginning of period t. Thus,

$$\sum_{j=1}^{N} x_{tj} = 1, \quad t = 1, \ldots, T. \tag{5.2.48}$$

Moreover, it is assumed that short selling of the risky assets is not allowed at any time. Hence, we have

$$x_{tj} \geq 0, \quad t = 1, \ldots, T, \; j = 1, \ldots, N. \tag{5.2.49}$$

Let R_{tj} denote the rate of return of asset S_j for period t. Define $\boldsymbol{R}_t = [R_{t1}, \ldots, R_{tN}]^\top$. Here, R_{tj} is assumed to follow a normal distribution with mean r_{tj} and standard deviation σ_{tj}. We further assume that vectors $\boldsymbol{R}_t$, $t = 1, \ldots, T$, are statistically independent, and the mean $E(\boldsymbol{R}_t) = \boldsymbol{r}_t = [r_{t1}, \ldots, r_{tN}]^\top$ is calculated by averaging the returns over a fixed window of time τ.

Let

$$r_{tj} = \frac{1}{\tau} \sum_{i=t-\tau}^{t-1} R_{ji}, \quad t = 1, \ldots, T, \; j = 1, \ldots, N. \tag{5.2.50}$$

We assume that in any time period, there are no two distinct assets in the portfolio that have the same level of expected return as well as standard deviation, i.e., for any $1 \leq t \leq T$, there exist no i and j such that $i \neq j$, but $r_{ti} = r_{tj}$, and $\sigma_{ti} = \sigma_{tj}$.

Let V_t denote the total wealth of the investor at the end of period t. Clearly, we have

$$V_t = V_{t-1} \left(1 + \boldsymbol{R}_t^\top \boldsymbol{x}_t \right), \quad t = 1, \ldots, T, \tag{5.2.51}$$

with $V_0 = M_0$.

First recall the definition of probabilistic risk measure, which was introduced in [233] for the single-period probabilistic risk measure.

$$w_p(\boldsymbol{x}) = \min_{1 \leq j \leq N} \Pr\{ |R_j x_j - r_j x_j| \leq \theta \varepsilon \}, \tag{5.2.52}$$

where θ is a constant to adjust the risk level, and ε denotes the average risk of the entire portfolio, which is calibrated by the function below.

$$\varepsilon = \frac{1}{N} \sum_{j=1}^{N} \sigma_j. \tag{5.2.53}$$

The whole idea of this risk measure in (5.2.52) is to locate the single asset with greatest deviation in the portfolio. With this 'biggest risk' mitigated, the risk of the whole portfolio can be substantially reduced as well. For multi-period portfolio optimization, the single 'biggest risk' should be selected over all risky assets and over the entire planning horizon. Thus, we define

$$w_p(\boldsymbol{x}) = \min_{1 \leq t \leq T} \min_{1 \leq j \leq N} \Pr\{ |R_{tj} x_{tj} - r_{tj} x_{tj}| \leq \theta \varepsilon \}, \tag{5.2.54}$$

where $\boldsymbol{x} = [\boldsymbol{x}_1^\top, \ldots, \boldsymbol{x}_T^\top]^\top$, θ is the same as defined in (5.2.52) to be a constant adjusting the risk level, and ε denotes the average risk of the entire portfolio over T periods, which is calibrated by

$$\varepsilon = \frac{1}{T \times N} \sum_{t=1}^{T} \sum_{j=1}^{N} \sigma_{tj}. \tag{5.2.55}$$

Assume that the investor is rational and risk-averse, who wants to maximize the terminal wealth as well as minimize the risk in the investment. Thus, the portfolio selection problem can be formulated as a bi-criteria optimization problem stated as follows:

$$\max \Big(\min_{1\le t\le T} \min_{1\le j\le N} f(x_{tj}),\ E(V_T) \Big), \tag{5.2.56a}$$

$$\text{s.t. } V_t = V_{t-1}\left(1 + \boldsymbol{R}_t^\top \boldsymbol{x}_t\right), \quad t = 1,\dots,T, \tag{5.2.56b}$$

$$\sum_{j=1}^{N} x_{tj} = 1, \quad t = 1,\dots,T, \tag{5.2.56c}$$

$$x_{tj} \ge 0, \quad t = 1,\dots,T,\ j = 1,\dots,N, \tag{5.2.56d}$$

where $f(x_{tj}) = \Pr\{\,|R_{tj}x_{tj} - r_{tj}x_{tj}| \le \theta\varepsilon\,\}$.

To reach the optimality for the above bi-criteria optimization problem, we recall the following definition which is based on Theorem 3.1 in [268].

Definition 5.2.1 *A solution $\boldsymbol{x}^* = \{\boldsymbol{x}_1^*,\dots,\boldsymbol{x}_T^*\}$ satisfying (5.2.56c) and (5.2.56d) is said to be a* Pareto-minimal *solution for the bi-criteria optimization problem (5.2.56) if there does not exist a solution $\tilde{\boldsymbol{x}} = \{\tilde{\boldsymbol{x}}_1,\dots,\tilde{\boldsymbol{x}}_N\}$ satisfying (5.2.56c) and (5.2.56d) such that*

$$\min_{1\le t\le T} \min_{1\le j\le N} f(x_{tj}^*) \le \min_{1\le t\le T} \min_{1\le j\le N} f(\tilde{x}_{tj}),\ \textit{and}\ E(V_T^*|\boldsymbol{x}^*) \le E(V_T^*|\tilde{\boldsymbol{x}}),$$

for which at least one of the inequalities holds strictly.

We can transform problem (5.2.56) into an equivalent bi-criteria optimization problem by adding another decision variable y, and $T \times N$ constraints.

$$\max \Big(y,\ E(V_T) \Big), \tag{5.2.57a}$$

$$\text{s.t. } V_t = V_{t-1}\left(1 + \boldsymbol{R}_t^\top \boldsymbol{x}_t\right), \quad t = 1,\dots,T, \tag{5.2.57b}$$

$$y \le f(x_{tj}), \quad t = 1,\dots,T,\ j = 1,\dots,N, \tag{5.2.57c}$$

$$\sum_{j=1}^{N} x_{tj} = 1, \quad t = 1,\dots,T, \tag{5.2.57d}$$

$$x_{tj} \ge 0, \quad t = 1,\dots,T,\ j = 1,\dots,N, \tag{5.2.57e}$$

where $y \le f(x_{tj})$ is the $(N \times (t-1) + j)$th probabilistic constraint. The optimization process trying to maximize y will eventually push the value of y to be equal to $\min_{1\le t\le T} \min_{1\le j\le N} f(x_{tj})$. Thus, the optimization problem (5.2.57) is equivalent to the optimization problem (5.2.56).

5.2.1.2 Analytical Solution

Recall that in Section 5.2.1.1, the return, R_{tj}, of asset S_j in period t, follows the normal distribution with mean r_{tj} and standard deviation σ_{tj}. It is clear that $R_{tj} - r_{tj}$ is a linear mixture of normal distributions with mean 0 and standard deviation σ_{tj}.

Let $q_{tj} = R_{tj} - r_{tj}$. Then it follows that

$$f(x_{tj}) = \Pr\{ |R_{tj}x_{tj} - r_{tj}x_{tj}| \leq \theta\varepsilon \} = \Pr\left\{ |R_{tj} - r_{tj}| \leq \frac{\theta\varepsilon}{x_{tj}} \right\}$$
$$= 2\int_0^{\frac{\theta\varepsilon}{x_{tj}}} \frac{1}{\sqrt{2\pi}\sigma_{tj}} \exp\left\{ -\frac{q_{tj}^2}{2\sigma_{tj}^2} \right\} dq_{tj}. \tag{5.2.58}$$

By the property of cumulative distribution function (5.2.58), $f(x_{tj})$ is clearly a monotonically strictly decreasing function with respect to x_{tj}.

As mentioned in the optimization problem (5.2.57), in the process of maximizing the objective function, the value of y is to be pushed to reach $\min_{1\leq t\leq T}\min_{1\leq j\leq N} f(x_{tj})$. As a result, if we choose y to be an arbitrary but fixed real number between 0 and 1 (because y is a value of probability), using the monotonic property of $f(x_{tj})$, we can find an upper bound for x_{tj}. Let this upper bound be denoted as U_{tj}, which is given by

$$U_{tj} = \min\{ 1, \hat{x}_{tj} \}, \tag{5.2.59}$$

where $\hat{x}_{tj} = f^{-1}(y)$.

Consequently, for a fixed value of y, the optimization problem (5.2.57) is equivalent to the following discrete time optimal control problem:

$$\max E(V_T), \tag{5.2.60a}$$
$$\text{s.t. } V_t = V_{t-1}\left(1 + \boldsymbol{R}_t^\top \boldsymbol{x}_t\right), \quad t = 1, \ldots, T, \tag{5.2.60b}$$
$$\sum_{j=1}^{N} x_{tj} = 1, \quad t = 1, \ldots, T, \tag{5.2.60c}$$
$$0 \leq x_{tj} \leq U_{tj}, \quad t = 1, \ldots, T,\ j = 1, \ldots, N. \tag{5.2.60d}$$

We use dynamic programming to solve problem (5.2.60).

For time periods $k = 1, \ldots, T$, we define a series of optimal control problems with the same dynamics, cost functional and constraints but different initial states and initial times. These problems are stated as follows:

$$\max E(V_T), \tag{5.2.61a}$$
$$\text{s.t. } V_t = V_{t-1}\left(1 + \boldsymbol{R}_t^\top \boldsymbol{x}_t\right), \quad t = k, \ldots, T, \tag{5.2.61b}$$
$$V_{k-1} = \xi, \tag{5.2.61c}$$

$$\sum_{j=1}^{N} x_{tj} = 1, \quad t = k, \ldots, T, \tag{5.2.61d}$$

$$0 \leq x_{tj} \leq U_{tj}, \quad t = k, \ldots, T, \ j = 1, \ldots, N, \tag{5.2.61e}$$

where ξ is a variable. At the time $k-1$, the number of steps to go is $T-k+1$. This is a family of optimal control problems determined by the initial time $k-1$ and initial state ξ. We use Problem ($\mathrm{P}_{k-1,\xi}$) to denote each different problem. Suppose that $\{\boldsymbol{x}_k^*, \ldots, \boldsymbol{x}_T^*\}$ is an optimal control for Problem ($\mathrm{P}_{k-1,\xi}$), and that $\{V_{k-1}^* = \xi, V_k^*, \ldots, V_T^*\}$ is the corresponding optimal trajectory. Then it follows from Theorem 5.2.4 that, for any $k \leq k^{'} \leq T$, $\{\boldsymbol{x}_{k'}^*, \ldots, \boldsymbol{x}_T^*\}$ is an optimal control for Problem $(P_{k'-1, V_{k'-1}^*})$. To continue, we utilize an auxiliary *value function* denoted by $F(k-1,\xi)$ to be the maximum value of the objective function for Problem ($\mathrm{P}_{k-1,\xi}$). The following value function $F(k-1,\xi)$ is defined as

$$F(k-1,\xi) = \max\left\{ F\left(k, \xi\left(1 + \boldsymbol{R}_k^\top \boldsymbol{x}_k\right)\right) : \boldsymbol{x}_k \in \mathcal{X}_k \right\}, \quad 1 \leq k \leq T, \tag{5.2.62}$$

with terminal condition

$$F(T,\xi) = E(V_T), \tag{5.2.63}$$

where $\mathcal{X}_k = \left\{ \boldsymbol{x}_k : \sum_{j=1}^{N} x_{kj} = 1 \text{ and } 0 \leq x_{kj} \leq U_{kj}, j = 1, \ldots, N \right\}$.

Now, we shall solve the discrete time optimal control problem (5.2.60) backwards using dynamic programming method.

For $k = T$, we start from the state $V_{T-1} = \xi$ at time $T-1$. Then, from (5.2.62) and (5.2.63), the value function becomes

$$\begin{aligned} F(T-1,\xi) &= \max\left\{ F\left(T, \xi\left(1 + \boldsymbol{R}_T^\top \boldsymbol{x}_T\right)\right) : \boldsymbol{x}_T \in \mathcal{X}_T \right\} \\ &= \max\left\{ E\left(\xi\left(1 + \boldsymbol{R}_T^\top \boldsymbol{x}_T\right)\right) : \boldsymbol{x}_T \in \mathcal{X}_T \right\} \\ &= \max\left\{ \xi\left(1 + \boldsymbol{r}_T^\top \boldsymbol{x}_T\right) : \boldsymbol{x}_T \in \mathcal{X}_T \right\}. \end{aligned} \tag{5.2.64}$$

Consequently, Problem ($\mathrm{P}_{T-1,\xi}$) becomes a linear programming problem stated as below.

$$\max \ \xi\left(1 + \boldsymbol{r}_T^\top \boldsymbol{x}_T\right), \tag{5.2.65a}$$

$$\text{s.t.} \ \sum_{j=1}^{N} x_{Tj} = 1, \tag{5.2.65b}$$

$$0 \leq x_{Tj} \leq U_{Tj}, \quad j = 1, \ldots, N. \tag{5.2.65c}$$

The optimal solution to this linear programming problem has been obtained in [233]. The result is quoted in the following as a lemma:

Lemma 5.2.1 *Let the assets be sort in such an order that* $r_{T1} \geq r_{T2} \geq \cdots \geq r_{TN}$. *Then, there exists an integer* $n \leq N$ *such that*

$$\sum_{j=1}^{n-1} U_{Tj} < 1, \text{ and } \sum_{j=1}^{n} U_{Tj} \geq 1,$$

and

$$x_{Tj}^* = \begin{cases} U_{Tj}, & j=1,\ldots,n-1, \\ 1 - \sum_{j=1}^{n-1} U_{Tj}, & j=n, \\ 0, & j>n, \end{cases} \tag{5.2.66}$$

is an optimal solution to problem (5.2.65).

Thus, it follows from (5.2.64) and (5.2.66) that the corresponding value function is given by

$$F(T-1,\xi) = \xi\left(1 + \boldsymbol{r}_T^\top \boldsymbol{x}_T^*\right).$$

Note that $\boldsymbol{x}_T^*$ in (5.2.66) is independent from ξ as the solution is a function of U_{Tj}, which, from (5.2.58) and (5.2.59), depends only on the mean and standard deviation of R_{Tj}.

Continuing with this process, we obtain the value function at $k = T-1$.

$$\begin{aligned} F(T-2,\xi) &= \max\left\{ F\left(T-1, \xi\left(1 + \boldsymbol{R}_{T-1}^\top \boldsymbol{x}_{T-1}\right)\right) : \boldsymbol{x}_{T-1} \in \mathcal{X}_{T-1} \right\} \\ &= \max\left\{ E\left(\xi\left(1 + \boldsymbol{R}_{T-1}^\top \boldsymbol{x}_{T-1}\right)\left(1 + \boldsymbol{r}_T^\top \boldsymbol{x}_T^*\right)\right) : \boldsymbol{x}_{T-1} \in \mathcal{X}_{T-1} \right\} \\ &= \max\left\{ \xi\left(1 + \boldsymbol{r}_{T-1}^\top \boldsymbol{x}_{T-1}\right)\left(1 + \boldsymbol{r}_T^\top \boldsymbol{x}_T^*\right) : \boldsymbol{x}_{T-1} \in \mathcal{X}_{T-1} \right\}. \end{aligned} \tag{5.2.67}$$

Again, it is easy to show that this maximizing problem is equivalent to the following linear programming problem:

$$\max \quad \xi\left(1 + \boldsymbol{r}_{T-1}^\top \boldsymbol{x}_{T-1}\right)\left(1 + \boldsymbol{r}_T^\top \boldsymbol{x}_T^*\right), \tag{5.2.68a}$$

$$s.t. \quad \sum_{j=1}^{N} x_{(T-1)j} = 1, \tag{5.2.68b}$$

$$0 \leq x_{(T-1)j} \leq U_{(T-1)j}, \quad j = 1, \ldots, N. \tag{5.2.68c}$$

Since $\boldsymbol{x}_T^*$ is solved in (5.2.65) and known, the above problem is in the same form as (5.2.65). Thus, it can be solved in a similar manner as that for (5.2.66). Details are given below as a lemma.

Lemma 5.2.2 *Let the assets be sort in such an order that* $r_{(T-1)1} \geq r_{(T-1)2} \geq \cdots \geq r_{(T-1)N}$. *Then, there exists an integer* $n \leq N$ *such that*

$$\sum_{j=1}^{n-1} U_{(T-1)j} < 1, \text{ and } \sum_{j=1}^{n} U_{(T-1)j} \geq 1,$$

and

$$x^*_{(T-1)j} = \begin{cases} U_{(T-1)j}, & j=1,\ldots,n-1, \\ 1 - \sum_{j=1}^{n-1} U_{(T-1)j}, & j=n, \\ 0, & j>n, \end{cases} \tag{5.2.69}$$

is an optimal solution to problem (5.2.68).

From Lemma 5.2.2, the corresponding value function is

$$F(T-2,\xi) = \xi \left(1 + \boldsymbol{r}_{T-1}^\top \boldsymbol{x}_{T-1}^*\right)\left(1 + \boldsymbol{r}_T^\top \boldsymbol{x}_T^*\right).$$

Similarly, the optimal solution $\boldsymbol{x}_{T-1}^*$ has no dependency on the state ξ. It depends only on fixed values of $U_{(T-1)j}$, $j = 1, \ldots, N$.

From Lemmas 5.2.1 and 5.2.2, it is reasonable to postulate that

$$\begin{aligned} F(k-1,\xi) = \xi \left(1 + \boldsymbol{r}_{k-1}^\top \boldsymbol{x}_{k-1}^*\right)\left(1 + \boldsymbol{r}_k^\top \boldsymbol{x}_k^*\right) \cdots \quad 7\left(1 + \boldsymbol{r}_T^\top \boldsymbol{x}_T^*\right), \\ k = T, \ldots, 1. \end{aligned} \tag{5.2.70}$$

Moreover, the optimal solution for every period k, $k = 1, \ldots, T$, can be written in a unified form as given below.

Theorem 5.2.5 *For any $k = 1, \ldots, T$, let the assets be sort in such an order that $r_{k1} \geq r_{k2} \geq \cdots \geq r_{kN}$. Then, there exists an integer $n \leq N$ such that*

$$\sum_{j=1}^{n-1} U_{kj} < 1, \text{ and } \sum_{j=1}^{n} U_{kj} \geq 1,$$

and

$$x^*_{kj} = \begin{cases} U_{kj}, & j=1,\ldots,n-1, \\ 1 - \sum_{j=1}^{n-1} U_{kj}, & j=n, \\ 0, & j>n, \end{cases} \tag{5.2.71}$$

where U_{kj} is defined as in (5.2.58) and (5.2.59). Thus, with $\boldsymbol{x}_k^* = [x_{k1}^*, \ldots, x_{kN}^*]^\top$, $\{\boldsymbol{x}_1^*, \ldots, \boldsymbol{x}_T^*\}$ is an optimal solution to problem (5.2.60).

5.2.1.3 Numerical Simulations

We use daily return data of stocks on the ASX100 dated from 01/01/2007 to 30/11/2011. The portfolio takes effect from 01/01/2009 and is open for trading for 3 years. 3 years is chosen since professionally managed portfolios (e.g., Aberdeen Asset Management, JBWere) usually list the average holding period as 3–5 years. At the beginning of each month, the funds in the portfolio are allocated based on the updated asset return data. We use 2 years historical data to decide the portfolio allocation each month. For the first portfolio allocation on 01/01/2009, the return data from 01/01/2007 to 31/12/2008 is used to evaluate the expected return and standard deviation of each stock. In the month following, the return data from 01/02/2007 to 31/01/2009 is used to evaluate the updated expected return and standard deviation, and this goes on.

The formulation of the corresponding portfolio optimization problem is as defined in Section 5.2.1.1. Assume the investor has an initial wealth of $M_0 =$ 1,000,000 dollars. There are $N = 100$ stocks to choose from for a portfolio of investment holding period of $T = 36$ months. The average risk of the portfolio, ε, over T periods is calculated as in (5.2.55). Table 5.2.1 shows the portfolio returns for various combination of θ (risk adjusting parameter), and y (lower bound of the probabilistic constraint). By changing the value of θ and/or y, the investor is able to alter the portfolio composition to cater for different risk tolerance levels. The lower the value of θ, the more diversified the portfolio can be, while the lower the value of y, the less stringent the probabilistic constraint.

From Table 5.2.1, it can be seen that the expected portfolio return increases when θ increases. Similarly, the expected portfolio return increases when y decreases. This makes sense since when θ increases, the portfolio selection consists of a much smaller number of selected ‘better-performing’ stocks. When y is lower, the risk is higher and hence the return is generally expected to be higher.

The historical price index value of ASX100, composed of 100 large-cap and mid-cap stocks, was 3067.90 and 3329.40 at the end of December 2008 and at the end of December 2011, respectively. This translates to a return of 8.52% for a portfolio which comprises the entire stock selection of ASX100.

From Table 5.2.1, it can be seen that when $\theta > 0.5$, the expected returns following our portfolio selection criteria outperforms the market index return. The multi-period model outperforms the passive single-period model with a period of 3 years. Table 5.2.2 shows the expected returns using the single-period model with a period of 3 years.

When $\theta = 0.1$ and $y = 0.95$, solving the problem with Theorem 5.2.5 suggests a total wealth in portfolio of 1,043,223.89 dollars at the end of the 3 year investment, which is a return of 4.32%. Comparing this with the result of the single-period investment strategy, the multi-period solution outperforms it by more than one fold (the single-period portfolio has a total return of about 2%).

Table 5.2.1: Multi-period—expected portfolio returns for selected θ and y

y \ θ	0.01	0.1	0.2	0.5	1
0.95	0.59%	4.32%	6.35%	10.78%	15.38%
0.90	0.71%	4.77%	7.01%	11.80%	16.85%
0.80	0.96%	5.48%	8.09%	13.40%	19.25%

Table 5.2.2: Single-period—expected portfolio returns for selected θ and y

y \ θ	0.01	0.1	0.2	0.5	1
0.95	−2.01%	2.00%	3.31%	3.30%	4.72%
0.90	−1.89%	2.23%	3.66%	3.83%	5.14%
0.80	−1.69%	2.65%	3.53%	4.38%	5.89%

5.3 Discrete Time Optimal Control Problems with Canonical Constraints

Consider a process described by the following system of difference equations:

$$\boldsymbol{x}(k+1) = \boldsymbol{f}(k, \boldsymbol{x}(k), \boldsymbol{u}(k)), \qquad k = 0, 1, \ldots, M-1, \tag{5.3.1a}$$

where

$$\boldsymbol{x} = [x_1, \ldots, x_n]^\top \in \mathbb{R}^n, \quad \boldsymbol{u} = [u_1, \ldots u_r]^\top \in \mathbb{R}^r,$$

are, respectively, state and control vectors; and $\boldsymbol{f} = [f_1, \ldots, f_n]^\top \in \mathbb{R}^n$ is a given functional. The initial condition for the system of difference equations (5.3.1a) is

$$\boldsymbol{x}(0) = \boldsymbol{x}^0, \tag{5.3.1b}$$

where $\boldsymbol{x}^0 = \left[x_1^0, \ldots, x_n^0\right]^\top \in \mathbb{R}^n$ is a given vector.

Define

$$U = \{\boldsymbol{\nu} = [v_1, \ldots v_r]^\top \in \mathbb{R}^r : \alpha_i \leq v_i \leq \beta_i, \ \ i = 1, \ldots, r\}, \tag{5.3.2}$$

where α_i, $i = 1, \ldots, r$ and β_i, $i = 1, \ldots, r$, are given real numbers. Note that U is a compact and convex subset of $\mathbb{R}^r$. Let $\boldsymbol{u}$ denote a control sequence $\{\boldsymbol{u}(k) : k = 0, 1, \ldots, M-1\}$ in U. Then, $\boldsymbol{u}$ is called an admissible control. Let $\mathcal{U}$ denote the class of all such admissible controls.

For each $\boldsymbol{u} \in \mathcal{U}$, let $\boldsymbol{x}(k \mid u)$, $k = 0, 1, \ldots, M$, be a sequence in $\mathbb{R}^n$ such that the system of difference equations (5.3.1a) with the initial condition (5.3.1b) is satisfied. This discrete time function is called the solution of the system (5.3.1) corresponding to the control $\boldsymbol{u} \in \mathcal{U}$.

We now consider the following class of discrete time optimal control problems in canonical formulation:

$$g_0(\boldsymbol{u}) = \Phi_0(\boldsymbol{x}(M \mid \boldsymbol{u})) + \sum_{k=0}^{M-1} \mathcal{L}_0(k, \boldsymbol{x}(k \mid \boldsymbol{u}), \boldsymbol{u}(k)) \tag{5.3.3}$$

is minimized subject to $\boldsymbol{u} \in \mathcal{U}$ and the following constraints (in canonical form):

$$g_i(\boldsymbol{u}) = 0, \qquad i = 1, 2, \ldots, N_e \tag{5.3.4a}$$

$$g_i(\boldsymbol{u}) \leq 0, \qquad i = N_e + 1, \ldots, N, \tag{5.3.4b}$$

where

$$g_i(\boldsymbol{u}) = \Phi_i(\boldsymbol{x}(M \mid \boldsymbol{u})) + \sum_{k=0}^{M-1} \mathcal{L}_i(k, \boldsymbol{x}(k \mid \boldsymbol{u}), \boldsymbol{u}(k)). \tag{5.3.5}$$

Let this discrete time optimal control problem be referred to as Problem (P). The following conditions are assumed throughout this section:

Assumption 5.3.1 *For each $k = 0, 1, \ldots, M-1$, $\boldsymbol{f}(k, \cdot, \cdot)$ is continuously differentiable on $\mathbb{R}^n \times \mathbb{R}^r$.*

Assumption 5.3.2 *For each $i = 0, 1, \ldots, N$, Φ_i is continuously differentiable on $\mathbb{R}^n$.*

Assumption 5.3.3 *For each $i = 0, 1, \ldots, N$, and for each $k = 0, 1, \ldots, M-1$, $\mathcal{L}_i(k, \cdot, \cdot)$ is continuously differentiable on $\mathbb{R}^n \times \mathbb{R}^r$.*

Note that the canonical constraint functionals have the same form as the cost functional. This feature is to allow gradient formulae of the cost and constraint functionals to be computed in a unified way.

5.3.1 Gradient Formulae

In this section, our aim is to derive gradient formulae for the cost as well as the constraint functionals. Define

$$\boldsymbol{u} = \left[(\boldsymbol{u}(0))^\top, (\boldsymbol{u}(1))^\top, \ldots, (\boldsymbol{u}(M-1))^\top\right]^\top. \tag{5.3.6}$$

Let the control vector $\boldsymbol{u}$ be perturbed by $\varepsilon\hat{\boldsymbol{u}}$, where $\varepsilon > 0$ is a small real number and $\hat{\boldsymbol{u}}$ is an arbitrary but fixed perturbation of $\boldsymbol{u}$ given by

$$\hat{\boldsymbol{u}} = \left[(\hat{\boldsymbol{u}}(0))^\top, (\hat{\boldsymbol{u}}(1))^\top, \ldots, (\hat{\boldsymbol{u}}(M-1))^\top\right]^\top. \tag{5.3.7}$$

Then, we have

$$\boldsymbol{u}_\varepsilon = \boldsymbol{u} + \varepsilon\hat{\boldsymbol{u}} = \left[(\boldsymbol{u}(0,\varepsilon))^\top, (\boldsymbol{u}(1,\varepsilon))^\top, \ldots, (\boldsymbol{u}(M-1,\varepsilon))^\top\right]^\top, \tag{5.3.8}$$

where

$$\boldsymbol{u}(k,\varepsilon) = \boldsymbol{u}(k) + \varepsilon\hat{\boldsymbol{u}}(k), \qquad k = 0, 1, \ldots, M-1. \tag{5.3.9}$$

Consequently, the state of the system will be perturbed, and so are the cost and constraint functionals.

Define

$$\boldsymbol{x}(k,\varepsilon) = \boldsymbol{x}(k \mid \boldsymbol{u}_\varepsilon), \qquad k = 1, 2, \ldots, M. \tag{5.3.10}$$

Then,

$$\boldsymbol{x}(k+1,\varepsilon) = \boldsymbol{f}(k, \boldsymbol{x}(k,\varepsilon), \boldsymbol{u}(k,\varepsilon)). \tag{5.3.11}$$

The variation of the state for $k = 0, 1, \ldots, M-1$ is

$$\begin{aligned}\triangle\boldsymbol{x}(k+1) &= \left.\frac{d\boldsymbol{x}(k+1,\varepsilon)}{d\varepsilon}\right|_{\varepsilon=0} \\ &= \frac{\partial \boldsymbol{f}(k,\boldsymbol{x}(k),\boldsymbol{u}(k))}{\partial \boldsymbol{x}(k)}\triangle\boldsymbol{x}(k) + \frac{\partial \boldsymbol{f}(k,\boldsymbol{x}(k),\boldsymbol{u}(k))}{\partial \boldsymbol{u}(k)}\hat{\boldsymbol{u}}(k) \end{aligned} \tag{5.3.12a}$$

with

$$\triangle\boldsymbol{x}(0) = 0. \tag{5.3.12b}$$

For the i–th functional ($i = 0$ denotes the cost functional), we have

$$\begin{aligned}\frac{\partial g_i(\boldsymbol{u})}{\partial \boldsymbol{u}}\hat{\boldsymbol{u}} &= \lim_{\varepsilon\to 0}\frac{g_i(\boldsymbol{u}_\varepsilon) - g_i(\boldsymbol{u})}{\varepsilon} \equiv \left.\frac{dg_i(\boldsymbol{u}_\varepsilon)}{d\varepsilon}\right|_{\varepsilon=0} \\ &= \frac{\partial \Phi_i(\boldsymbol{x}(M))}{\partial \boldsymbol{x}(M)}\triangle\boldsymbol{x}(M) + \sum_{k=0}^{M-1}\left[\frac{\partial \mathcal{L}_i(k,\boldsymbol{x}(k),\boldsymbol{u}(k))}{\partial \boldsymbol{x}(k)}\triangle\boldsymbol{x}(k) \right. \\ &\quad \left. + \frac{\partial \mathcal{L}_i(k,\boldsymbol{x}(k),\boldsymbol{u}(k))}{\partial \boldsymbol{u}(k)}\hat{\boldsymbol{u}}(k)\right]. \end{aligned} \tag{5.3.13}$$

For each $i = 0, 1, \ldots, N$, define the Hamiltonian

$$H_i(k,\boldsymbol{x}(k),\boldsymbol{u}(k),\boldsymbol{\lambda}^i(k+1)) = \mathcal{L}_i(k,\boldsymbol{x}(k),\boldsymbol{u}(k)) + (\boldsymbol{\lambda}^i(k+1))^\top \boldsymbol{f}(k,\boldsymbol{x}(k),\boldsymbol{u}(k)), \tag{5.3.14}$$

where $\boldsymbol{\lambda}^i(k) \in \mathbb{R}^n$, $k = M, M-1, \ldots, 1$, is referred to as the costate sequence for the i–th canonical constraint. Then, it follows from (5.3.13) that

$$\begin{aligned}\frac{\partial g_i(\boldsymbol{u})}{\partial \boldsymbol{u}}\hat{\boldsymbol{u}} =& \frac{\partial \Phi_i(\boldsymbol{x}(M))}{\partial \boldsymbol{x}(M)}\triangle\boldsymbol{x}(M) \\ &+ \sum_{k=0}^{M-1}\left\{\frac{\partial H_i(k,\boldsymbol{x}(k),\boldsymbol{u}(k),\boldsymbol{\lambda}^i(k+1))}{\partial \boldsymbol{x}(k)}\Delta\boldsymbol{x}(k)\right.\end{aligned}$$

$$
\begin{aligned}
&- (\boldsymbol{\lambda}^i(k+1))^\top \frac{\partial \boldsymbol{f}(k, \boldsymbol{x}(k), \boldsymbol{u}(k))}{\partial \boldsymbol{x}(k)} \triangle \boldsymbol{x}(k) \\
&+ \frac{\partial H_i(k, \boldsymbol{x}(k), \boldsymbol{u}(k), \boldsymbol{\lambda}^i(k+1))}{\partial \boldsymbol{u}(k)} \hat{\boldsymbol{u}}(k) \\
&- (\boldsymbol{\lambda}^i(k+1))^\top \frac{\partial \boldsymbol{f}(k, \boldsymbol{x}(k), \boldsymbol{u}(k))}{\partial \boldsymbol{u}(k)} \hat{\boldsymbol{u}}(k) \Bigg\}. \qquad (5.3.15)
\end{aligned}
$$

By (5.3.12), we have

$$
\Delta \boldsymbol{x}(k+1) = \frac{\partial \boldsymbol{f}(k, \boldsymbol{x}(k), \boldsymbol{u}(k))}{\partial \boldsymbol{x}(k)} \triangle \boldsymbol{x}(k) + \frac{\partial \boldsymbol{f}(k, \boldsymbol{x}(k), \boldsymbol{u}(k))}{\partial \boldsymbol{u}(k)} \hat{\boldsymbol{u}}(k). \qquad (5.3.16)
$$

Let the costate $\boldsymbol{\lambda}^i(k)$ be determined by the following system of difference equations:

$$
(\boldsymbol{\lambda}^i(k))^\top = \frac{\partial H_i(k, \boldsymbol{x}(k), \boldsymbol{u}(k), \boldsymbol{\lambda}^i(k+1))}{\partial \boldsymbol{x}(k)}, \quad k = M-1, M-2, \ldots, 1, \qquad (5.3.17a)
$$

and

$$
(\boldsymbol{\lambda}^i(M))^\top = \frac{\partial \Phi_i(\boldsymbol{x}(M))}{\partial \boldsymbol{x}(M)}. \qquad (5.3.17b)
$$

Hence, from (5.3.15)–(5.3.17), it follows from (5.3.12) that

$$
\begin{aligned}
&\frac{\partial g_i(\boldsymbol{u})}{\partial \boldsymbol{u}} \hat{\boldsymbol{u}} \\
&= \left[\frac{\partial H_i(0, \boldsymbol{x}(0), \boldsymbol{u}(0), \boldsymbol{\lambda}^i(1))}{\partial \boldsymbol{u}(0)}, \ldots, \frac{\partial H_i(M-1, \boldsymbol{x}(M-1), \boldsymbol{u}(M-1), \boldsymbol{\lambda}^i(M))}{\partial \boldsymbol{u}(M-1)} \right] \hat{\boldsymbol{u}}.
\end{aligned}
$$

Since $\hat{\boldsymbol{u}}$ is arbitrary, we have the following gradient formula:

$$
\begin{aligned}
&\frac{\partial g_i(\boldsymbol{u})}{\partial \boldsymbol{u}} \\
&= \left[\frac{\partial H_i(0, \boldsymbol{x}(0), \boldsymbol{u}(0), \boldsymbol{\lambda}^i(1))}{\partial \boldsymbol{u}(0)}, \ldots, \frac{\partial H_i(M-1, \boldsymbol{x}(M-1), \boldsymbol{u}(M-1), \boldsymbol{\lambda}^i(M))}{\partial \boldsymbol{u}(M-1)} \right]. \qquad (5.3.18)
\end{aligned}
$$

We summarize this result in the following theorem.

Theorem 5.3.1 *Consider Problem (P). For each $i = 0, 1, \ldots, N$, the gradient of $g_i(\boldsymbol{u})$ with respect to $\boldsymbol{u}$, where*

$$
\boldsymbol{u} = \left[(\boldsymbol{u}(0))^\top, (\boldsymbol{u}(1))^\top, \ldots, (\boldsymbol{u}(M-1))^\top \right]^\top,
$$

is given by (5.3.18).

5.3.2 A Unified Computational Approach

Problem (P) is essentially nonlinear mathematical programming problem where the variables are the control parameters in the vector $\boldsymbol{u}$. It can be solved by using any suitable numerical optimization technique, such as sequential quadratic programming (see Section 3.5). In order to apply any such method, for each $\boldsymbol{u} \in \mathcal{U}$, we need to be able to compute the corresponding values of the cost functional $g_0(\boldsymbol{u})$ and the constraint functionals $g_i(\boldsymbol{u})$, $i = 1, \ldots, N$, where $g_i(\boldsymbol{u})$ is defined by (5.3.3) if $i = 0$, and by (5.3.5) if $i = 1, \ldots, N$, as well as their respective gradients.

To calculate the values of cost functional and the constraint functionals corresponding to each $\boldsymbol{u} \in \mathcal{U}$, the first task is to calculate the solution of the system (5.3.1) corresponding to each $\boldsymbol{u} \in \mathcal{U}$. This is presented as an algorithm for future reference.

Algorithm 5.3.1 *For each given $\boldsymbol{u} \in \mathcal{U}$, compute the solution $\boldsymbol{x}(k \mid \boldsymbol{u})$, $k = 0, 1, \ldots, M$ of system* (5.3.1) *by solving the difference equations* (5.3.1a) *forward in time from $k = 0$ to $k = M$ with initial condition* (5.3.1b).

With the information obtained in Algorithm 5.3.1, the values of g_i corresponding to each $\boldsymbol{u} \in \mathcal{U}$ can be easily calculated by the following simple algorithm:

Algorithm 5.3.2 *For a given $\boldsymbol{u} \in \mathcal{U}$,*

Step 1. Use Algorithm 5.3.1 to solve for $\boldsymbol{x}(k \mid \boldsymbol{u})$, $k = 0, 1, \ldots, M$. Thus, $\boldsymbol{x}(k \mid \boldsymbol{u})$ is known for each $k = 0, 1, \ldots, M$. This implies that

(a) $\Phi_i(\boldsymbol{x}(M \mid \boldsymbol{u}))$, $i = 0, 1, \ldots, N$, are known and
(b) $\mathcal{L}_i(k, \boldsymbol{x}(k \mid \boldsymbol{u}), \boldsymbol{u}(k))$, $i = 0, 1, \ldots, N$, are known for each $k = 0, 1, \ldots, M - 1$. Hence, the summations

$$\sum_{k=0}^{M-1} \mathcal{L}_i(k, \boldsymbol{x}(k \mid \boldsymbol{u}), \boldsymbol{u}(k)), \quad i = 0, 1, \ldots, N,$$

can be readily determined.

Step 2. Calculate the values of the cost functional (with $i = 0$) and the constraint functionals (with $i = 1, \ldots, N$) as follows:

$$g_i(\boldsymbol{u}) = \Phi_i(\boldsymbol{x}(M \mid \boldsymbol{u})) + \sum_{k=0}^{M-1} \mathcal{L}_i(k, \boldsymbol{x}(k \mid \boldsymbol{u}), \boldsymbol{u}(k)), \quad i = 0, 1, \ldots, N.$$

By using Theorem 5.3.1, we can calculate the gradients of the cost functional and the canonical constraint functionals as stated in the following algorithm:

Algorithm 5.3.3 *For each* $i = 0, 1, \ldots, N$, *and for a given* $\boldsymbol{u} \in \mathcal{U}$,

Step 1. *Use Algorithm 5.3.1 to solve system* (5.3.1) *to obtain* $\boldsymbol{x}(k \mid \boldsymbol{u})$, $k = 0, 1, \ldots, M$.

Step 2. *Solve the system of the costate difference equations* (5.3.17) *backward in time from* $k = M, M-1, \ldots, 1$. *Let* $\boldsymbol{\lambda}^i(k \mid \boldsymbol{u})$ *be the solution obtained.*

Step 3. *Compute the gradient of* g_i *according to* (5.3.18).

5.4 Problems with Terminal and All-Time-Step Inequality Constraints

Consider the system (5.3.1), and let $\mathcal{U}$ be the class of admissible controls defined in Section 5.3. Two sets of nonlinear terminal state constraints are specified as follows:

$$\Psi_i(\boldsymbol{x}(M \mid \boldsymbol{u})) = 0, \qquad i = 1, \ldots, N_0 \tag{5.4.1a}$$

$$\Psi_i(\boldsymbol{x}(M \mid \boldsymbol{u})) \le 0, \qquad i = N_0 + 1, \ldots, N_1, \tag{5.4.1b}$$

where Ψ_i, $i = 1, \ldots, N_1$, are given real-valued functions defined in $\mathbb{R}^n$. Furthermore, we also consider the following set of all-time-step inequality constraints on the state and control variables.

$$h_i(k, \boldsymbol{x}(k \mid u), \boldsymbol{u}(k)) \le 0, \qquad \forall\, k \in \mathcal{M}, \qquad i = 1, \ldots, N_2, \tag{5.4.2}$$

where

$$\mathcal{M} = \{0, 1, \ldots, M-1\},$$

and h_i, $i = 1, \ldots, N_2$, are given real-valued functions defined in $\mathcal{M} \times \mathbb{R}^n \times \mathbb{R}^r$.

If $\boldsymbol{u} \in \mathcal{U}$ satisfies the constraints (5.4.1) and (5.4.2), then it is called a *feasible control.* Let $\mathcal{F}$ be the class of all feasible controls.

Consider the following problem, to be referred to as Problem (Q): Given the system (5.3.1), find a control $\boldsymbol{u} \in \mathcal{F}$ such that the cost functional

$$g_0(\boldsymbol{u}) = \Phi_0(\boldsymbol{x}(M \mid \boldsymbol{u})) + \sum_{k=0}^{M-1} \mathcal{L}_0(k, \boldsymbol{x}(k \mid \boldsymbol{u}), \boldsymbol{u}(k)) \tag{5.4.3}$$

is minimized over $\mathcal{F}$.

The following conditions are assumed throughout this section:

Assumption 5.4.1 *For each* $k = 0, 1, \ldots, M-1$, $\boldsymbol{f}(k, \cdot, \cdot)$ *satisfies Assumption* (5.3.1).

Assumption 5.4.2 *For each* $i = 1, \ldots, N_1$, Ψ_i *is continuously differentiable on* $\mathbb{R}^n$.

Assumption 5.4.3 *For each* $i = 1, \ldots, N_2$, *and for each* $k = 0, 1, \ldots, M - 1$, $h_i(k, \cdot, \cdot)$ *is continuously differentiable on* $\mathbb{R}^n \times \mathbb{R}^r$.

Assumption 5.4.4 Φ_0 *satisfies Assumption 5.3.2.*

Assumption 5.4.5 *For each* $k = 0, 1, \ldots, M-1$, $\mathcal{L}_0(k, \cdot, \cdot)$ *satisfies Assumption 5.3.3.*

Define

$$\begin{aligned}\Theta = \big\{\boldsymbol{u} \in \mathcal{U} : \;& \Psi_i(\boldsymbol{x}(M \mid \boldsymbol{u})) = 0, \; i = 1, \ldots, N_0; \\ & \Psi_i(\boldsymbol{x}(M \mid \boldsymbol{u})) \leq 0, \; i = N_e + 1, \ldots, N_1\big\}\end{aligned} \tag{5.4.4}$$

and

$$\mathcal{F} = \{\boldsymbol{u} \in \Theta : h_i(k, \boldsymbol{x}(k \mid \boldsymbol{u}), \boldsymbol{u}(k)) \leq 0, \; \forall k \in \mathcal{M}, \; i = 1, \ldots N_2\}. \tag{5.4.5}$$

Note that the terminal constraints (5.4.1) are already in canonical form (5.3.5). Although the all-time-step inequality constraints (5.4.2) are not in canonical form, they can be approximated by a sequence of canonical constraints via the constraint transcription introduced in Section 4.3. Details are given in the next section.

5.4.1 Constraint Approximation

Consider Problem (Q) of Section 5.4. Note that for each $i = 1, \ldots, N_2$, the corresponding all-time-step inequality constraint in (5.4.2) is equivalent to

$$g_i(\boldsymbol{u}) = \sum_{k=0}^{M-1} \max\{h_i(k, \boldsymbol{x}(k \mid \boldsymbol{u}), \boldsymbol{u}(k)), 0\} = 0. \tag{5.4.6}$$

For convenience, let Problem (Q) with (5.4.2) replaced by (5.4.6) again be denoted by Problem (Q). Recall the set Θ defined by (5.4.4). Then, it is clear that the set $\mathcal{F}$ of feasible controls can also be written as

$$\mathcal{F} = \{\boldsymbol{u} \in \Theta : g_i(\boldsymbol{u}) = 0, \; i = 1, \ldots, N_2\}. \tag{5.4.7}$$

As in Section 4.2, we approximate the nonsmooth function

$$\max\{h_i(k, \boldsymbol{x}(k \mid \boldsymbol{u}), \boldsymbol{u}(k)), 0\}$$

by

$$\mathcal{L}_{i,\varepsilon}(k, \boldsymbol{x}(k \,|\, \boldsymbol{u}), \boldsymbol{u}(k))$$

$$= \begin{cases} h_i(k, \boldsymbol{x}(k \mid \boldsymbol{u}), \boldsymbol{u}(k)), & \text{if } h_i(k, \boldsymbol{x}(k \mid \boldsymbol{u}), \boldsymbol{u}(k)) \geq 0, \\ \dfrac{(h_i(k, \boldsymbol{x}(k \mid \boldsymbol{u}), \boldsymbol{u}(k)) + \epsilon)^2}{4\epsilon}, & \text{if } -\epsilon < h_i(k, \boldsymbol{x}(k \mid \boldsymbol{u}), \boldsymbol{u}(k)) < \epsilon, \\ 0, & \text{if } h_i(k, \boldsymbol{x}(k \mid \boldsymbol{u}), \boldsymbol{u}(k)) \leq -\epsilon. \end{cases} \tag{5.4.8}$$

Then, for each $i = 1, \ldots, N_2$, let

$$g_{i,\varepsilon}(\boldsymbol{u}) = \sum_{k=0}^{M-1} \mathcal{L}_{i,\varepsilon}(k, \boldsymbol{x}(k \mid \boldsymbol{u}), \boldsymbol{u}(k)). \tag{5.4.9}$$

We may now define the following approximate problem:

Problem (Q_ε): Problem (Q) with (5.4.6) replaced by

$$-\frac{\varepsilon}{4} + g_{i,\varepsilon}(\boldsymbol{u}) \leq 0, \qquad i = 1, \ldots, N_2. \tag{5.4.10}$$

Note that (5.4.10) is slightly more restrictive in the sense that it is a sufficient, but not a necessary condition, for (5.4.6) to hold.

Let

$$\begin{aligned} \mathcal{D} = \big\{ \boldsymbol{u} \in \mathcal{U} : \ & \Psi_i(\boldsymbol{x}(M \mid \boldsymbol{u})) = 0, \ i = 1, \ldots, N_e; \\ & \Psi_i(\boldsymbol{x}(M \mid \boldsymbol{u})) \leq 0, \ i = N_e + 1, \ldots, N_1 \big\}, \end{aligned} \tag{5.4.11}$$

$$\mathcal{F}_\varepsilon = \{ \boldsymbol{u} \in \mathcal{D} : -(\varepsilon/4) + g_{i,\varepsilon}(\boldsymbol{u}) \leq 0, \quad i = 1, \ldots, N_2 \}, \tag{5.4.12}$$

and

$$\mathcal{F}^0 = \{ \boldsymbol{u} \in \mathcal{D} : h_i(k, \boldsymbol{x}(k \mid \boldsymbol{u}), \boldsymbol{u}(k)) < 0, \ k \in \mathcal{M}, \ i = 1, \ldots, N_2 \}. \tag{5.4.13}$$

Clearly, Problem (Q_ε) is also equivalent to: Find a control $\boldsymbol{u} \in \mathcal{F}_\varepsilon$ such that the cost functional (5.4.3) is minimized over $\mathcal{F}_\varepsilon$.

We assume that the following condition is satisfied.

Assumption 5.4.6 *For any $\boldsymbol{u}$ in $\mathcal{F}$ and any $\delta > 0$, there exists a $\bar{\boldsymbol{u}} \in \mathcal{F}^0$ such that*

$$\max_{0 \leq k \leq M-1} |\boldsymbol{u}(k) - \bar{\boldsymbol{u}}(k)| \leq \delta.$$

Lemma 5.4.1 *If $\boldsymbol{u}_\varepsilon$ is a feasible control vector of Problem (Q_ε), then it is also a feasible control vector of Problem (Q).*

Proof. Suppose $\boldsymbol{u}_\varepsilon$ is not a feasible control vector of Problem (Q). Then, there exist some $i = 1, \ldots, N_2$, and $k \in \mathcal{M}$ such that

$$h_i(k, \boldsymbol{x}(k \mid \boldsymbol{u}_\varepsilon), \boldsymbol{u}_\varepsilon(k)) > 0.$$

This, in turn, implies that

$$\mathcal{L}_{i,\varepsilon}(k, \boldsymbol{x}(k \mid \boldsymbol{u}_\varepsilon), \boldsymbol{u}_\varepsilon(k)) > \varepsilon/4$$

and hence,

$$g_{i,\varepsilon}(\boldsymbol{u}_\varepsilon) > \frac{M\,\varepsilon}{4} > \frac{\varepsilon}{4}.$$

That is,

$$-\frac{\varepsilon}{4} + g_{i,\varepsilon}(\boldsymbol{u}_\varepsilon) > 0.$$

This is a contradiction to the constraints specified in (5.4.10), and thus the proof is complete.

For each $\varepsilon > 0$, Problem (Q_ε) can be regarded as a nonlinear mathematical programming problem. Since the constraints appearing in (5.4.11) and (5.4.12) are in canonical form, their gradient formulae as well as that of $g_0(\boldsymbol{u})$ can be readily computed as explained in Section 5.3.2. Hence, Problem (Q_ε) can be solved by any efficient optimization technique, such as the SQP technique presented in Chapter 3. In view of Lemma 5.4.1, we see that any feasible control vector of Problem (Q_ε) is in $\mathcal{F}$, and hence is a suboptimal control vector for Problem (Q). Thus, by adjusting $\varepsilon > 0$ in such a way that $\varepsilon \to 0$, we obtain a sequence of approximate problems (Q_ε), each being solved as a nonlinear mathematical programming problem. We shall now investigate certain convergence properties of this approximation scheme.

5.4.2 Convergence Analysis

The aim of this section is to provide a convergence analysis for the approximation scheme proposed in the last subsection.

Theorem 5.4.1 *Let $\boldsymbol{u}^*$ and $\boldsymbol{u}_\varepsilon^*$ be optimal controls of Problems (Q) and (Q_ε), respectively. Then, there exists a subsequence of $\{\boldsymbol{u}_\varepsilon^*\}$, which is again denoted by the original sequence, and a control $\bar{\boldsymbol{u}} \in \mathcal{F}$ such that*

$$\lim_{\varepsilon\to 0} |\boldsymbol{u}_\varepsilon^*(k) - \bar{\boldsymbol{u}}(k)| = 0, \qquad k = 0, 1, .., M-1. \tag{5.4.14}$$

Furthermore, $\bar{\boldsymbol{u}}$ is an optimal control of Problem (Q).

Proof. Since U is a compact subset of $\mathbb{R}^r$, and since $\{\boldsymbol{u}_\varepsilon^*\}$ as a sequence in ε is in U, it is clear that there exists a subsequence, which is again denoted by the original sequence, and a control parameter vector $\bar{\boldsymbol{u}} \in \mathcal{U}$ such that

$$\lim_{\varepsilon\to 0} |\boldsymbol{u}_\varepsilon^*(k) - \bar{\boldsymbol{u}}(k)| = 0, \qquad k = 0, 1, \ldots, M-1. \tag{5.4.15}$$

By induction, it follows from Assumption 5.4.1 that, for each $k = 0, 1, \ldots, M$,

$$\lim_{\varepsilon\to 0} |\boldsymbol{x}(k \mid \boldsymbol{u}_\varepsilon^*) - \boldsymbol{x}(k \mid \bar{\boldsymbol{u}})| = 0. \tag{5.4.16}$$

Thus, by Assumptions 5.4.2 and 5.4.3, we have

$$\lim_{\varepsilon \to 0} \Psi_i(\boldsymbol{x}(M \mid \boldsymbol{u}_\varepsilon^*)) = \Psi_i(\boldsymbol{x}(M \mid \bar{\boldsymbol{u}})), \quad i = 1, \ldots, N_1 \tag{5.4.17}$$

and, for each $k \in \mathcal{M}$,

$$\lim_{\varepsilon \to 0} h_i(k, \boldsymbol{x}(k \mid \boldsymbol{u}_\varepsilon^*), \boldsymbol{u}_\varepsilon^*(k)) = h_i(k, \boldsymbol{x}(k \mid \bar{\boldsymbol{u}}), \bar{\boldsymbol{u}}(k)), \quad i = 1, \ldots, N_2. \tag{5.4.18}$$

By Lemma 5.4.1, $\boldsymbol{u}_\varepsilon^* \in \mathcal{F}$ for all $\varepsilon > 0$. Thus, it follows from (5.4.17) and (5.4.18) that $\bar{\boldsymbol{u}} \in \mathcal{F}$.

Next, by Assumptions 5.4.1 and 5.4.2, we deduce from (5.4.15) and (5.4.16) that

$$\lim_{\varepsilon \to 0} g_0(\boldsymbol{u}_\varepsilon^*) = g_0(\bar{\boldsymbol{u}}). \tag{5.4.19}$$

For any $\delta_1 > 0$, Assumption 5.4.6 asserts that there exists a $\hat{\boldsymbol{u}} \in \mathcal{F}^0$ such that

$$|\boldsymbol{u}^*(k) - \hat{\boldsymbol{u}}(k)| \le \delta_1, \qquad \forall\, k = 0, 1, \ldots, M-1. \tag{5.4.20}$$

By Assumption 5.4.1 and induction, we can show that, for any $\rho_1 > 0$, there exists a $\delta_1 > 0$ such that for all $k = 0, 1, \ldots, M$

$$|\boldsymbol{x}(k \mid \boldsymbol{u}^*) - \boldsymbol{x}(k \mid \hat{\boldsymbol{u}})| \le \rho_1, \tag{5.4.21}$$

whenever (5.4.20) is satisfied.

Using (5.4.20), (5.4.21) and Assumptions 5.4.1 and 5.4.2, it follows that, for any $\rho_2 > 0$, there exists a $\hat{\boldsymbol{u}} \in \mathcal{F}^0$ such that

$$g_0(\boldsymbol{u}^*) \le g(\hat{\boldsymbol{u}}) \le g_0(\boldsymbol{u}^*) + \rho_2. \tag{5.4.22}$$

Since $\hat{\boldsymbol{u}} \in \mathcal{F}^0$, we have

$$h_i(k, \boldsymbol{x}(k \mid \hat{\boldsymbol{u}}), \hat{\boldsymbol{u}}(k)) < 0, \qquad k \in \mathcal{M}, \qquad i = 1, \ldots, N_2,$$

and hence there exists a $\delta_2 > 0$ such that

$$h_i(k, \boldsymbol{x}(k \mid \hat{\boldsymbol{u}}), \hat{\boldsymbol{u}}(k)) \le -\delta_2, \qquad k \in \mathcal{M}, \qquad i = 1, \ldots, N_2. \tag{5.4.23}$$

Thus, in view of (5.4.12), we see that

$$\hat{\boldsymbol{u}} \in \mathcal{F}_\varepsilon$$

for all ε, $0 \le \varepsilon \le \delta_2$. Therefore,

$$g_0(\boldsymbol{u}_\varepsilon^*) \le g_0(\hat{\boldsymbol{u}}^p). \tag{5.4.24}$$

Using (5.4.22) and (5.4.23), and noting that $\boldsymbol{u}_\varepsilon^* \in \mathcal{F}$, we obtain

$$g_0(\boldsymbol{u}^*) \le g_0(\boldsymbol{u}_\varepsilon^*) \le g_0(\boldsymbol{u}^*) + \rho_2. \tag{5.4.25}$$

Since $\rho_2 > 0$ is arbitrary, it follows that

$$\lim_{\varepsilon \to 0} g_0(\boldsymbol{u}_\varepsilon^*) = g_0(\boldsymbol{u}^*). \tag{5.4.26}$$

Combining (5.4.19) and (5.4.26), we conclude that $\bar{\boldsymbol{u}}$ is an optimal control of Problem (Q), and the proof is complete.

5.4.3 Illustrative Examples

In this section, three numerical examples are solved to illustrate the proposed computational method.

Example 5.4.1 This problem concerns the vertical ascent of a rocket. The original formulation (continuous time version) and numerical solution to the problem are taken from [54]. A discrete time version of the control process is obtained by the Euler scheme to discretize the system equations, where the time step is taken as 1 s as in [54].

$$x_1(k+1) = x_1(k) - \boldsymbol{u}(k) \tag{5.4.27a}$$

$$x_2(k+1) = x_2(k) + x_3(k) \tag{5.4.27b}$$

$$x_3(k+1) = x_3(k) + \frac{[V\boldsymbol{u}(k) - Q(\boldsymbol{x}(k))]}{x_1(k)} - g, \tag{5.4.27c}$$

where $x_1(k)$ is the mass of the rocket; $x_2(k)$ is the altitude (km) above the earth's surface; $x_3(k)$ is the rocket velocity (km/s); $\boldsymbol{u}(k)$ is the mass flow rate; $V = 2$ is the constant gas nozzle velocity; $g = 0.01$ km/s^2 is the acceleration due to gravity (assumed constant); $Q(x_2(k), x_3(k))$ is the aerodynamic drag defined by the formula:

$$Q(x_2(k), x_3(k)) = 0.05 \exp(0.01 x_2(k))(x_3(k))^2.$$

The initial state of the rocket is $x_1(0) = 1$, $x_2(0) = 0$, $x_3(0) = 0$; at the terminal time $M = 100$, the final value of the mass is constrained to be 20% of the initial mass, i.e., $x_1(M) = 0.2$. The bounds on the control are

$$0 \le \boldsymbol{u}(k) \le 0.04$$

for $k = 0, 1, \ldots, M-1$.

The control objective is to maximize the rocket's peak altitude by suitable choice of the mass flow rate. In other words, we want to minimize

$$g_0 = -k_1 x_2(M) \tag{5.4.28}$$

subject to the terminal state constraint

$$\Psi_1 = k_2(x_1(M) - 0.2) = 0, \tag{5.4.29}$$

where $k_1 = 0.01$ and $k_2 = 10$ are the appropriate weighting factors.

The problem is solved by the discrete time optimal control software, DMISER [58]. Table 5.4.1 summarizes the computed results for this example. The original continuous time problem is solved by MISER [104], and the solution is found to be consistent with the present result.

Table 5.4.1: Numerical results for Example 5.4.1

M	g_0	Ψ_1	$x_1(M)$	$x_2(M)$
100	−0.36454188	-0.167×10^{-14}	0.2	36.45

Example 5.4.2 The original problem is taken from [189] and the same problem was also considered in [54]. The control process (discretized by the Euler scheme using the time step $h = 0.02$) is described by the difference equations:

$$x_1(k+1) = x_1(k) + 0.02x_2(k) \tag{5.4.30a}$$

$$x_2(k+1) = 0.98x_2(k) + 0.02\boldsymbol{u}(k), \tag{5.4.30b}$$

where $k = 0, 1, \ldots, M-1$, with initial state

$$x_1(0) = 0, \qquad x_2(0) = -1. \tag{5.4.30c}$$

The problem is to minimize

$$g_0 = (x_1(M))^2 + (x_2(M))^2 + \sum_{k=0}^{M-1} \left[(x_1(k))^2 + (x_2(k))^2 + 0.005(\boldsymbol{u}(k))^2\right]$$

subject to the all-time-step constraints

$$h(k, \boldsymbol{x}(k)) = -8(0.02k - 0.05)^2 + x^2(k) + 0.5 \leq 0 \tag{5.4.31}$$

for $k = 1, \ldots, M$. For computational purpose, we let $M = 50$.

By using the constraint transcription method described in Section 4.2, we obtain a sequence of approximate optimal control problems (Q_ε). For each $\varepsilon > 0$, it is a problem of minimizing a function of 50 variables with one constraint. Each of these problems is solvable by DMISER. Table 5.4.2 summarizes the computed results.

Note that all the constraints in (5.4.31) are satisfied.

Example 5.4.3 Consider the following first order system:

$$x(k+1) = 0.5\boldsymbol{x}(k) + \boldsymbol{u}(k)$$

$$x(0) = 1,$$

Table 5.4.2: Numerical results for example 5.4.2

M	g_0	$-\frac{\varepsilon}{4}+g_{i,\epsilon}$
50	9.1415507	0.429×10^{-9}

where $k = 0, 1, \ldots, M-1$, with $M = 50$. The cost functional

$$g_0 = \sum_{k=1}^{M-1} \left[(\boldsymbol{x}(k))^2 + (\boldsymbol{u}(k))^2\right]$$

is to be minimized with the control bounded by

$$-1 \leq \boldsymbol{u}(k) \leq 1,$$

for $k = 0, 1, \ldots, M-1$. The computed result is $g_0 = 1.1327822$.

5.5 Discrete Time Time-Delayed Optimal Control Problem

The main reference of this section is [131]. Consider a process described by the following system of difference equations with time-delay:

$$\boldsymbol{x}(k+1) = \boldsymbol{f}(k, \boldsymbol{x}(k), \boldsymbol{x}(k-h), \boldsymbol{u}(k), \boldsymbol{u}(k-h)),\ k = 0, 1, \ldots, M-1, \quad (5.5.1a)$$

where

$$\boldsymbol{x} = [x_1, \ldots, x_n]^\top \in \mathbb{R}^n, \boldsymbol{u} = [u_1, \ldots u_r]^\top \in \mathbb{R}^r,$$

are, respectively, the state and control vectors, while $\boldsymbol{f} = [f_1, \ldots, f_n]^\top \in \mathbb{R}^n$ is a given $0 < h < M$. Here, we consider the case where there is only one time delay. The extension to the case involving many time-delays is straightforward but is more involved in terms of notation.

The initial functions for the state and control functions are

$$\boldsymbol{x}(k) = \boldsymbol{\phi}(k),\ k = -h, -h+1, \ldots, -1,\ x(0) = x^0, \quad (5.5.1b)$$

$$\boldsymbol{u}(k) = \boldsymbol{\gamma}(k),\ k = -h, -h+1, \ldots, -1, \quad (5.5.1c)$$

where

$$\boldsymbol{\phi}(k) = [\phi_1, \ldots, \phi_n]^\top,\ \boldsymbol{\gamma}(k) = [\gamma_1, \ldots, \gamma_r]^\top,$$

are given functions from $k = -h, -h+1, \ldots -1$ into $\mathbb{R}^n$ and $\mathbb{R}^r$, respectively, and x^0 is a given vector in $\mathbb{R}^n$. Define

$$U = \{\boldsymbol{\nu} = [v_1, \ldots v_r]^\top \in \mathbb{R}^r : \alpha_i \le v_i \le \beta_i,\ i = 1, \ldots, r\}, \tag{5.5.2}$$

where α_i, $i = 1, \ldots, r$, and β_i, $i = 1, \ldots, r$, are given real numbers. Note that U is a compact and convex subset of $\mathbb{R}^r$.

Consider the all-time-step inequality constraints on the state and control variables given below:

$$h_i\left(k, \boldsymbol{x}(k), \boldsymbol{u}(k)\right) \le 0,\ k = 0, 1, \ldots M-1;\ i = 1, \ldots, N_2, \tag{5.5.3}$$

where h_i, $i = 1, \ldots, N_2$, are given real-valued functions.

A control sequence $\boldsymbol{u} = \{u(0), \ldots, u(M-1)\}$ is said to be an admissible control if $\boldsymbol{u}(k) \in U$, $k = 0, \ldots, M-1$, where U is defined by (5.5.2). Let $\mathcal{U}$ be the class of all such admissible controls. If a $\boldsymbol{u} \in \mathcal{U}$ is such that the all-time-step inequality constraints (5.5.3) are satisfied, then it is called a feasible control. Let $\mathcal{F}$ be the class of all such feasible controls.

We now state our problem formally as follows:

Problem (Q) Given system (5.5.1a), (5.5.1b), (5.5.1c), find a control $\boldsymbol{u} \in \mathcal{F}$ such that the cost functional

$$g_0(\boldsymbol{u}) = \Phi_0(\boldsymbol{x}(M)) + \sum_{k=0}^{M-1} \mathcal{L}_0(k, \boldsymbol{x}(k), \boldsymbol{x}(k-h), \boldsymbol{u}(k), \boldsymbol{u}(k-h)) \tag{5.5.4}$$

is minimized over $\mathcal{F}$, where Φ_0 and $\mathcal{L}_0$ are given real-valued functions.

5.5.1 Approximation

To begin, we first note that the all-time-step inequality constraints are equivalent to the following equality constraints:

$$g_i(\boldsymbol{u}) = \sum_{k=0}^{M-1} \max\{h_i\left(k, \boldsymbol{x}(k), \boldsymbol{u}(k)\right), 0\} = 0,\ i = 1, \ldots, N_2. \tag{5.5.5}$$

Thus, the set $\mathcal{F}$ of feasible controls can be written as

$$\mathcal{F} = \left\{\boldsymbol{u}(k) \in U, k = 0, \ldots, M-1 : g_i(\boldsymbol{u}) = 0,\ i = 1, \ldots, N_2, \right\}, \tag{5.5.6}$$

where U is defined by (5.5.2). However, the functions appearing in (5.5.5) are nonsmooth. Thus, for each $i = 1, \ldots, N_2$, we shall approximate the nonsmooth function $\max\{h_i\left(k, \boldsymbol{x}(k), \boldsymbol{u}(k)\right), 0\}$ by a smooth function $\mathcal{L}_{i,\varepsilon}\left(k, \boldsymbol{x}(k), \boldsymbol{u}(k)\right)$ given by

$$\mathcal{L}_{i,\varepsilon} = \begin{cases} 0, & \text{if } h_i < -\varepsilon, \\ (h_i + \varepsilon)^2 / 4\varepsilon, & \text{if } -\varepsilon \leq h_i \leq \varepsilon, \\ h_i, & \text{if } h_i > \varepsilon, \end{cases} \tag{5.5.7}$$

where $\varepsilon > 0$ is an adjustable constant with small value. Then, the all-time-step inequality constraints (5.5.3) are approximated by the inequality constraints in canonical form defined by

$$-\frac{\varepsilon}{4} + g_{i,\varepsilon}(\boldsymbol{u}) \leq 0, \ i = 1, \ldots, N_2, \tag{5.5.8}$$

where

$$g_{i,\varepsilon}(\boldsymbol{u}) = \sum_{i=1}^{N_2} \sum_{k=0}^{M-1} \mathcal{L}_{i,\varepsilon} k, \boldsymbol{x}(k), \boldsymbol{u}(k).$$

Define

$$\mathcal{F}_\varepsilon = \left\{ \boldsymbol{u}(k) \in U, \ k = 0, \ldots, M-1 : -\frac{\varepsilon}{4} + g_{i,\varepsilon}(\boldsymbol{u}) \leq 0, \ i = 1, \ldots, N_2 \right\}. \tag{5.5.9}$$

Now, we can define a sequence of approximate problems (Q_ε), where $\varepsilon > 0$, below.

Problem $(\mathbf{Q}_\varepsilon)$ Problem (Q) with (5.5.5) replaced by

$$G_\varepsilon(\boldsymbol{u}) = -\frac{\varepsilon}{4} + g_\varepsilon(\boldsymbol{u}) \leq 0, \ i = 1, \ldots, N_2. \tag{5.5.10}$$

In Problem (Q_ε), our aim is to find a control $\boldsymbol{u}$ in $\mathcal{F}_\varepsilon$ such that the cost functional (5.5.4) is minimized over $\mathcal{F}_\varepsilon$. For each $\varepsilon > 0$, Problem (Q_ε) is a special case of a general class of discrete time optimal control problems with time-delay and subject to canonical constraints defined below.

Problem (P) Given system (5.5.1a)–(5.5.1c), find an admissible control $\boldsymbol{u} \in \mathcal{U}$ such that the cost functional

$$g_0(\boldsymbol{u}) = \Phi_0(\boldsymbol{x}(M)) + \sum_{k=0}^{M-1} \mathcal{L}_0(k, \boldsymbol{x}(k), \boldsymbol{x}(k-h), \boldsymbol{u}(k), \boldsymbol{u}(k-h)) \tag{5.5.11}$$

is minimized over $\mathcal{U}$ subject to the following constraints in canonical form:

$$g_i(\boldsymbol{u}) = 0, \ i = 1, 2, \ldots, N_e, \tag{5.5.12a}$$

$$g_i(\boldsymbol{u}) \leq 0, \ i = N_e + 1, \ldots, N, \tag{5.5.12b}$$

where

$$g_i(\boldsymbol{u}) = \Phi_i(\boldsymbol{x}(M)) + \sum_{k=0}^{M-1} \mathcal{L}_i(k, \boldsymbol{x}(k), \boldsymbol{x}(k-h), \boldsymbol{u}(k), \boldsymbol{u}(k-h)). \tag{5.5.13}$$

We shall develop an efficient computational method for solving Problem (P) in the next section. In the rest of this section, our aim is to establish the required convergence properties of Problems (Q_ε) to Problem (Q). We assume that the following conditions are satisfied. These conditions are now quite standard in optimal control algorithms.

Assumption 5.5.1 *For each* $k = 0, 1, \ldots, M-1$, $f(k, \cdot, \cdot, \cdot, \cdot)$ *is continuously differentiable on* $\mathbb{R}^n \times \mathbb{R}^n \times \mathbb{R}^r \times \mathbb{R}^r$.

Assumption 5.5.2 *For each* $i = 1, \ldots, N_2$, *and for each* $k = 0, 1, \ldots, M - 1$, $h_i(k, \cdot, \cdot)$ *is continuously differentiable on* $\mathbb{R}^n \times \mathbb{R}^r$.

Assumption 5.5.3 Φ_0 *is continuously differentiable on* $\mathbb{R}^n$.

Assumption 5.5.4 *For each* $k = 0, 1, \ldots, M - 1$, $\mathcal{L}_0(k, \cdot, \cdot, \cdot, \cdot)$ *is continuously differentiable on* $\mathbb{R}^n \times \mathbb{R}^n \times \mathbb{R}^r \times \mathbb{R}^r$.

Assumption 5.5.5 *For any control* $\boldsymbol{u}$ *in* $\mathcal{F}$, *there exists a control* $\bar{\boldsymbol{u}} \in \mathcal{F}^0$ *such that* $\alpha\bar{\boldsymbol{u}} + (1-\alpha)\boldsymbol{u} \in \mathcal{F}^0$ *for all* $\alpha \in (0, 1]$.

Remark 5.5.1 *Under Assumption 5.5.5, it can be shown that for any* $\boldsymbol{u}$ *in* $\mathcal{F}$ *and* $\delta > 0$, *there exists a* $\bar{\boldsymbol{u}} \in \mathcal{F}^0$ *such that*

$$\max_{0 \le k \le M-1} |\boldsymbol{u}(k) - \bar{\boldsymbol{u}}(k)| \le \delta,$$

where $\mathcal{F}^0$ *is the interior of* $\mathcal{F}$, *meaning that if* $u \in \mathcal{F}^0$, *then*

$$h_i(k, \boldsymbol{x}(k), \boldsymbol{u}(k)) < 0, \; i = 1, \ldots, N_2,$$

for all $k = 0, 1, \ldots M - 1$.

In what follows, we shall present an algorithm for solving Problem (Q) as a sequence of Problems (Q_ε).

Algorithm 5.5.1

Step 1. Set $\varepsilon = \varepsilon_0$.

Step 2. Solve Problem (Q_ε) *as a nonlinear programming problem, obtaining an optimal solution.*

Step 3. Set $\varepsilon = \varepsilon/10$, *and go to Step 2.*

Remark 5.5.2 ε_0 *is usually set as* 1.0×10^{-2}; *and the algorithm is terminated as 'successful exit' when* $\varepsilon < 10^{-7}$.

The convergence properties of Problems (Q_ε) to Problem (Q) are given by the following theorems. Their proofs are left as exercises.

Theorem 5.5.1 *Let $\boldsymbol{u}^*$ be an optimal control of Problem (Q) and let $\boldsymbol{u}_\varepsilon^*$ be an optimal control of Problem (Q_ε). Then,*

$$\lim_{\varepsilon\to 0} g_0(\boldsymbol{u}_\varepsilon^*) = g_0(\boldsymbol{u}^*).$$

Proof. By Assumption 5.5.5, there exists a $\bar{\boldsymbol{u}} \in \mathcal{F}^0$ such that

$$\boldsymbol{u}_\alpha \equiv \alpha\bar{\boldsymbol{u}} + (1+\alpha)\boldsymbol{u}^* \in \mathcal{F}^0,\ \forall \alpha \in (0,1].$$

Thus, for any $\delta_1 > 0, \exists$ an $\alpha_1 \in (0,1]$ such that

$$g_0(\boldsymbol{u}^*) \leq g_0(\boldsymbol{u}_\alpha) \leq g_0(\boldsymbol{u}^*) + \delta_1,\ \forall \alpha \in (0,1]. \tag{5.5.14}$$

Choose $\alpha_2 = \alpha_1/2$. Then, it is clear that $\alpha_2 \in \mathcal{F}^0$. Thus, there exists a $\delta_2 > 0$ such that

$$h_i(k, \boldsymbol{x}(k\,|\boldsymbol{u}_{\alpha_2}), \boldsymbol{u}_{\alpha_2}) < -\delta_2,\ i = 1, \ldots, N_2,$$

for all $k,\ 0 \leq k \leq M-1$. Let $\varepsilon = \delta_2$. Then, it follows from the definition of $\mathcal{L}_{i,\varepsilon}$ given by (5.5.7) that $\mathcal{L}_{i,\varepsilon} = 0$. Thus, (5.5.10) is satisfied and hence $\boldsymbol{u}_{\alpha_2} \in \mathcal{F}_\varepsilon$. Let $\boldsymbol{u}_\varepsilon^*$ be an optimal control of Problem (Q(ε)). Clearly, $\boldsymbol{u}_\varepsilon^* \in \mathcal{F}_\varepsilon$ and

$$g_0(\boldsymbol{u}_\varepsilon^*) \leq g_0(\boldsymbol{u}_{\alpha_2}). \tag{5.5.15}$$

However,

$$g_0(\boldsymbol{u}^*) \leq g_0(\boldsymbol{u}_\varepsilon^*). \tag{5.5.16}$$

Thus, if follows from (5.5.14), (5.5.15) and (5.5.16) that

$$g_0(\boldsymbol{u}^*) \leq g_0(\boldsymbol{u}^*) \leq g_0(\boldsymbol{u}_{\alpha_2}) \leq g_0(\boldsymbol{u}^*) + \delta_1.$$

Letting $\varepsilon \to 0$ and noting that $\delta_1 > 0$ is arbitrary, the conclusion of the theorem follows readily. This completes the proof.

Theorem 5.5.2 *Let $\boldsymbol{u}_\varepsilon^*$ and $\boldsymbol{u}^*$ be optimal controls of Problems (Q_ε) and (Q), respectively. Then, there exists a subsequence of $\{\boldsymbol{u}_\varepsilon^*\}$, which is again denoted by the original sequence, and a control $\bar{\boldsymbol{u}} \in \mathcal{F}$ such that, for each $k = 0, 1, .., M-1$,*

$$\lim_{\varepsilon\to 0} |\boldsymbol{u}_\varepsilon^*(k) - \bar{\boldsymbol{u}}(k)| = 0. \tag{5.5.17}$$

Furthermore, $\bar{\boldsymbol{u}}$ is an optimal control of Problem (Q).

Proof. Since U is a compact subset of $\mathbb{R}^r$, and $\{\boldsymbol{u}_\varepsilon^*\}$, as a sequence in ε, is such that $\boldsymbol{u}_\varepsilon^*(k) \in U$, for $k = 0, 1, \ldots, M-1$, it is clear that there exists a subsequence, which is again denoted by the original sequence, and a control parameter vector $\bar{\boldsymbol{u}} \in \mathcal{U}$ such that, for each $k = 0, 1, \ldots, M-1$,

$$\lim_{\varepsilon\to 0} |\boldsymbol{u}_\varepsilon^*(k) - \bar{\boldsymbol{u}}(k)| = 0. \tag{5.5.18}$$

By induction, we can show, by using Assumption 5.5.1 and (5.5.18), that, for each $k = 0, 1, \ldots, M$,

$$\lim_{\varepsilon \to 0} |\boldsymbol{x}(k \,|\boldsymbol{u}_\varepsilon^*) - \boldsymbol{x}(k \,|\bar{\boldsymbol{u}})| = 0. \tag{5.5.19}$$

Thus, by Assumption 5.5.2, we have, for each $k = 0, 1, \ldots, M$,

$$\lim_{\varepsilon \to 0} h_i(k, \boldsymbol{x}(k \,|\boldsymbol{u}_\varepsilon^*), \boldsymbol{u}_\varepsilon^*(k)) = h_i(k, \boldsymbol{x}(k \,|\bar{\boldsymbol{u}}), \bar{\boldsymbol{u}}(k)), \quad i = 1, \ldots, N_2. \tag{5.5.20}$$

By Lemma 1, $\boldsymbol{u}_\varepsilon^* \in \mathcal{F}$ for all $\varepsilon > 0$. Thus, it follows from (5.5.20) that $\bar{\boldsymbol{u}} \in \mathcal{F}$. Next, by Assumption 5.5.1, we deduce from (5.5.18) and (5.5.19) that

$$\lim_{\varepsilon \to 0} g_0(\boldsymbol{u}_\varepsilon^*) = g_0(\bar{\boldsymbol{u}}). \tag{5.5.21}$$

For any $\delta_1 > 0$, it follows from Remark 3.1 that there exists a $\hat{\boldsymbol{u}} \in \mathcal{F}^0$ such that, for each $k = 0, 1, \ldots, M-1$,

$$|\boldsymbol{u}^*(k) - \hat{\boldsymbol{u}}(k)| \leq \delta_1. \tag{5.5.22}$$

By Assumption 5.5.1 and induction, we can show that, for any $\rho_1 > 0$, there exists a $\delta_1 > 0$ such that for each $k = 0, 1, \ldots, M$,

$$|\boldsymbol{x}(k \,|\boldsymbol{u}^*) - \boldsymbol{x}(k \,|\hat{\boldsymbol{u}})| \leq \rho_1, \tag{5.5.23}$$

whenever (5.5.22) is satisfied. Using (5.5.22), (5.5.23) and Assumption 5.5.4, it follows that, for any $\rho_2 > 0$, there exists a $\hat{\boldsymbol{u}} \in \mathcal{F}^0$ such that

$$g_0(\boldsymbol{u}^*) \leq g_0(\hat{\boldsymbol{u}}) \leq g_0(\boldsymbol{u}^*) + \rho_2. \tag{5.5.24}$$

Since $\hat{\boldsymbol{u}} \in \mathcal{F}^0$, we have, for each $k = 0, 1, \ldots, M$,

$$h_i(k, \boldsymbol{x}(k \,|\hat{\boldsymbol{u}}), \hat{\boldsymbol{u}}(k)) < 0, \ \ i = 1, \ldots, N_2,$$

and hence there exists a $\delta > 0$ such that, for each $k = 0, 1, \ldots, M$,

$$h_i(k, \boldsymbol{x}(k \,|\hat{\boldsymbol{u}}), \hat{\boldsymbol{u}}(k)) \leq -\delta, \ \ i = 1, \ldots, N_2. \tag{5.5.25}$$

Thus, in view of (5.5.9), we see that

$$\hat{\boldsymbol{u}} \in \mathcal{F}_\varepsilon,$$

for all ε, $0 \leq \varepsilon \leq \delta$. Therefore,

$$g_0(\boldsymbol{u}_\varepsilon^*) \leq g_0(\hat{\boldsymbol{u}}). \tag{5.5.26}$$

Using (5.5.24) and (5.5.26), and noting that $u_\varepsilon^* \in \mathcal{F}$, we obtain

$$g_0(\boldsymbol{u}^*) \leq g_0(\boldsymbol{u}_\varepsilon^*) \leq g_0(\boldsymbol{u}^*) + \rho_2. \tag{5.5.27}$$

Since $\rho_2 > 0$ is arbitrary, it follows that

$$\lim_{\varepsilon \to 0} g_0(\boldsymbol{u}_\varepsilon^*) = g_0(\boldsymbol{u}^*). \tag{5.5.28}$$

Combining (5.5.21) and (5.5.28), we conclude that $\bar{\boldsymbol{u}}$ is an optimal control of Problem (Q). This completes the proof.

5.5.2 Gradients

To calculate the gradients of the cost and constraint functionals, we will derive the required gradient formulas corresponding to each control sequence $\boldsymbol{u} = \{u(0), \ldots, u(M-1)\}$ as follows:

For each $i = 0, 1, \ldots, N$, let

$$H_i\left(k, \boldsymbol{x}(k), \boldsymbol{y}(k), \boldsymbol{z}(k), \boldsymbol{u}(k), \boldsymbol{v}(k), \boldsymbol{w}(k), \boldsymbol{\lambda}^i(k+1), \bar{\boldsymbol{\lambda}}^i(k)\right)$$

be the corresponding Hamiltonian sequence defined by

$$\begin{aligned} &H_i\left(k, \boldsymbol{x}(k), \boldsymbol{y}(k), \boldsymbol{z}(k), \boldsymbol{u}(k), \boldsymbol{v}(k), \boldsymbol{w}(k), \boldsymbol{\lambda}^i(k+1), \bar{\boldsymbol{\lambda}}^i(k)\right) \\ =&\mathcal{L}_i(k, \boldsymbol{x}(k), \boldsymbol{y}(k), \boldsymbol{u}(k), \boldsymbol{v}(k)) \\ &+ \mathcal{L}_i(k+h, \boldsymbol{z}(k), \boldsymbol{x}(k), \boldsymbol{w}(k), \boldsymbol{u}(k))e(M-k-h) \\ &+ \left(\boldsymbol{\lambda}^i(k+1)\right)^\top \boldsymbol{f}(k, \boldsymbol{x}(k), \boldsymbol{y}(k), \boldsymbol{u}(k), \boldsymbol{v}(k)) \\ &+ \left(\bar{\boldsymbol{\lambda}}^i(k)\right)^\top \boldsymbol{f}(k+h, \boldsymbol{z}(k), \boldsymbol{x}(k), \boldsymbol{w}(k), \boldsymbol{u}(k))e(M-k-h), \end{aligned} \tag{5.5.29}$$

where $e(\cdot)$ denotes the unit step function defined by

$$e(k) = \begin{cases} 1, \ k \geq 0 \\ 0, \ k < 0, \end{cases} \tag{5.5.30}$$

and

$$\boldsymbol{y}(k) = \boldsymbol{x}(k-h), \tag{5.5.31a}$$
$$\boldsymbol{z}(k) = \boldsymbol{x}(k+h), \tag{5.5.31b}$$
$$\boldsymbol{v}(k) = \boldsymbol{u}(k-h), \tag{5.5.31c}$$
$$\boldsymbol{w}(k) = \boldsymbol{u}(k+h), \tag{5.5.31d}$$
$$\bar{\boldsymbol{\lambda}}^i(k) = \boldsymbol{\lambda}^i(k+h+1). \tag{5.5.31e}$$

For each control u, $\boldsymbol{\lambda}^i$ is the solution of the following costate system:

$$\left(\boldsymbol{\lambda}^i(k)\right)^\top = \frac{\partial H_i(k)}{\partial \boldsymbol{x}(k)}, \ k = M-1, M-2, \ldots, 0, \tag{5.5.32}$$

with boundary conditions

$$\left(\boldsymbol{\lambda}^i(M)\right)^\top = \frac{\partial \Phi_i(\boldsymbol{x}(M))}{\partial \boldsymbol{x}(M)}, \tag{5.5.33a}$$

$$\boldsymbol{\lambda}^i(k) = 0, \ k > M. \tag{5.5.33b}$$

We set

$$\boldsymbol{z}(k) = 0, \ \forall k = M-h+1, M-h+2, \ldots, M, \tag{5.5.34}$$

and

$$\boldsymbol{w}(k) = 0, \ \forall k = M-h, M-h+1, \ldots, M. \tag{5.5.35}$$

Then, the gradient formulas for the cost functionals (for $i = 0$) and constraint functionals (for $i = 1, \ldots, N$) are given in the following theorem.

Theorem 5.5.3 *Let $g_i(\boldsymbol{u})$, $i = 0, 1, \ldots, N$, be defined by (5.5.11) (the cost functional for $i = 0$) and (5.5.13) (the constraint functionals for $i = 1, \ldots, N$). Then, for each $i = 0, 1, \ldots, N$, the gradient of the function $g_i(\boldsymbol{u})$ is given by*

$$\frac{\partial g_i(\boldsymbol{u})}{\partial u} = \left[\frac{\partial H_i(0)}{\partial \boldsymbol{u}(0)}, \frac{\partial H_i(1)}{\partial \boldsymbol{u}(1)}, \ldots, \frac{\partial H_i(M-1)}{\partial \boldsymbol{u}(M-1)}\right], \tag{5.5.36}$$

where

$$H_i(k) = H_i\left(k, \boldsymbol{x}(k), \boldsymbol{y}(k), \boldsymbol{z}(k), \boldsymbol{u}(k), \boldsymbol{v}(k), \boldsymbol{w}(k), \boldsymbol{\lambda}^i(k+1), \bar{\boldsymbol{\lambda}}^i(k)\right),$$
$$k = 0, 1, \ldots, M-1.$$

Proof. Define

$$\boldsymbol{u} = \left[(u(0))^\top, (u(1))^\top, \ldots, (u(M-1))^\top\right]^\top. \tag{5.5.37}$$

Let the control $\boldsymbol{u}$ be perturbed by $\varepsilon\hat{\boldsymbol{u}}$, where $\varepsilon > 0$ is a small real number and $\hat{\boldsymbol{u}}$ is an arbitrary but fixed perturbation of $\boldsymbol{u}$ given by

$$\hat{\boldsymbol{u}} = \left[(\hat{u}(0))^\top, (\hat{u}(0))^\top, \ldots, (\hat{u}(M-1))^\top\right]^\top. \tag{5.5.38}$$

Then, we have

$$\boldsymbol{u}_\varepsilon = \boldsymbol{u} + \varepsilon\hat{\boldsymbol{u}} = \left[(u(0,\varepsilon))^\top, (u(1,\varepsilon))^\top, \ldots, (u(M-1,\varepsilon))^\top\right]^\top, \tag{5.5.39}$$

where

$$\boldsymbol{u}(k,\varepsilon) = \boldsymbol{u}(k) + \varepsilon\hat{\boldsymbol{u}}(k), \ k = 0, 1, \ldots, M-1. \tag{5.5.40}$$

Let the perturbed solution be denoted by

$$\boldsymbol{x}(k,\varepsilon) = \boldsymbol{x}(k \mid \boldsymbol{u}_\varepsilon),\ k = 1, 2, \dots, M. \tag{5.5.41}$$

Then,

$$\boldsymbol{x}(k+1,\varepsilon) = \boldsymbol{f}(k, \boldsymbol{x}(k,\varepsilon), \boldsymbol{y}(k,\varepsilon), \boldsymbol{u}(k,\varepsilon), \boldsymbol{v}(k,\varepsilon)). \tag{5.5.42}$$

The variation of the state for $k = 0, 1, \dots, M-1$ is

$$\begin{aligned}\triangle\boldsymbol{x}(k+1) &= \left.\frac{d\boldsymbol{x}(k+1,\varepsilon)}{d\varepsilon}\right|_{\varepsilon=0} \\ &= \frac{\partial \boldsymbol{f}(k, \boldsymbol{x}(k), \boldsymbol{y}(k), \boldsymbol{u}(k), \boldsymbol{v}(k))}{\partial \boldsymbol{x}(k)} \triangle\boldsymbol{x}(k) \\ &+ \frac{\partial \boldsymbol{f}(k, \boldsymbol{x}(k), \boldsymbol{y}(k), \boldsymbol{u}(k), \boldsymbol{v}(k))}{\partial \boldsymbol{y}(k)} \triangle\boldsymbol{y}(k) \\ &+ \frac{\partial \boldsymbol{f}(k, \boldsymbol{x}(k), \boldsymbol{y}(k), \boldsymbol{u}(k), \boldsymbol{v}(k))}{\partial \boldsymbol{u}(k)} \hat{\boldsymbol{u}}(k) \\ &+ \frac{\partial \boldsymbol{f}(k, \boldsymbol{x}(k), \boldsymbol{y}(k), \boldsymbol{u}(k), \boldsymbol{v}(k))}{\partial \boldsymbol{v}(k)} \triangle\boldsymbol{v}(k),\end{aligned} \tag{5.5.43a}$$

where

$$\triangle\boldsymbol{x}(k) = 0,\ k \le 0, \tag{5.5.43b}$$

$$\triangle\boldsymbol{u}(k) = 0,\ k < 0. \tag{5.5.43c}$$

From (5.5.43b) and (5.5.43c), we obtain

$$\triangle\boldsymbol{y}(k) = 0,\ k = 0, 1, \dots, h, \tag{5.5.44a}$$

and

$$\triangle\boldsymbol{v}(k) = 0,\ k = 0, 1, \dots, h-1. \tag{5.5.44b}$$

Define

$$\bar{\mathcal{L}}_i = \mathcal{L}_i(k, \boldsymbol{x}(k), \boldsymbol{y}(k), \boldsymbol{u}(k), \boldsymbol{v}(k)), \tag{5.5.45a}$$

$$\hat{\mathcal{L}}_i = \mathcal{L}_i(k+h, \boldsymbol{z}(k), \boldsymbol{x}(k), \boldsymbol{w}(k), \boldsymbol{u}(k)), \tag{5.5.45b}$$

$$\bar{\boldsymbol{f}} = \boldsymbol{f}(k, \boldsymbol{x}(k), \boldsymbol{y}(k), \boldsymbol{u}(k), \boldsymbol{v}(k)), \tag{5.5.45c}$$

$$\hat{\boldsymbol{f}} = \boldsymbol{f}(k+h, \boldsymbol{z}(k), \boldsymbol{x}(k), \boldsymbol{w}(k), \boldsymbol{u}(k)), \tag{5.5.45d}$$

$$\bar{H}_i = H_i(k). \tag{5.5.45e}$$

By chain rule and (5.5.45a), it follows that

$$\frac{\partial g_i(\boldsymbol{u})}{\partial \boldsymbol{u}} \hat{\boldsymbol{u}} = \lim_{\varepsilon \to 0} \frac{g_i(\boldsymbol{u}_\varepsilon) - g_i(\boldsymbol{u})}{\varepsilon} \equiv \left.\frac{dg_i(\boldsymbol{u}_\varepsilon)}{d\varepsilon}\right|_{\varepsilon=0}$$

$$= \frac{\partial \Phi_i(x(M))}{\partial \boldsymbol{x}(M)} \triangle \boldsymbol{x}(M) + \sum_{k=0}^{M-1} \left[\frac{\partial \bar{\mathcal{L}}_i}{\partial \boldsymbol{x}(k)} \triangle \boldsymbol{x}(k) \right.$$

$$\left. + \frac{\partial \bar{\mathcal{L}}_i}{\partial \boldsymbol{y}(k)} \triangle \boldsymbol{y}(k) + \frac{\partial \bar{\mathcal{L}}_i}{\partial \boldsymbol{u}(k)} \hat{\boldsymbol{u}}(k) \frac{\partial \bar{\mathcal{L}}_i}{\partial \boldsymbol{v}(k)} \triangle \boldsymbol{v}(k) \right]. \quad (5.5.46)$$

From (5.5.31a), (5.5.31c) and (5.5.45b), we have

$$\sum_{k=0}^{M-1} \left\{ \left(\frac{\partial \bar{\mathcal{L}}_i}{\partial \boldsymbol{y}(k)} \right) \triangle \boldsymbol{y}(k) + \left(\frac{\partial \bar{\mathcal{L}}_i}{\partial \boldsymbol{v}(k)} \right) \triangle v(k) \right\}$$

$$= \sum_{k=0}^{M-1} e(M-k-h) \left[\left(\frac{\partial \hat{\mathcal{L}}_i}{\partial \boldsymbol{x}(k)} \right) \triangle \boldsymbol{x}(k) + \left(\frac{\partial \hat{\mathcal{L}}_i}{\partial \boldsymbol{u}(k)} \right) \hat{\boldsymbol{u}}(k) \right]. \quad (5.5.47)$$

Substituting (5.5.47) into (5.5.46), and then using (5.5.32) and (5.5.45a)–(5.5.45e), we obtain

$$\frac{\partial g_i(\boldsymbol{u})}{\partial \boldsymbol{u}} \hat{\boldsymbol{u}} = \left(\frac{\partial \Phi_i(\boldsymbol{x}(M))}{\partial \boldsymbol{x}(k)} \right) \triangle \boldsymbol{x}(M) + \sum_{k=0}^{M-1} \left[\left(\frac{\partial \bar{H}_i}{\partial \boldsymbol{x}(k)} \right) \triangle \boldsymbol{x}(k) \right.$$

$$+ \left(\frac{\partial \bar{H}_i}{\partial \boldsymbol{u}(k)} \right) \hat{\boldsymbol{u}}(k) - \left(\boldsymbol{\lambda}^i(k+1) \right)^\top \frac{\partial \bar{\boldsymbol{f}}}{\partial \boldsymbol{x}} \triangle \boldsymbol{x}(k)$$

$$- \left(\bar{\boldsymbol{\lambda}}^i(k) \right)^\top \frac{\partial \hat{\boldsymbol{f}}}{\partial \boldsymbol{x}(k)} \triangle \boldsymbol{x}(k) e(M-k-h)$$

$$- \left(\boldsymbol{\lambda}^i(k+1) \right)^\top \frac{\partial \bar{\boldsymbol{f}}}{\partial \boldsymbol{u}(k)} \hat{\boldsymbol{u}}(k)$$

$$\left. - \left(\bar{\boldsymbol{\lambda}}^i(k) \right)^\top \frac{\partial \hat{\boldsymbol{f}}}{\partial \boldsymbol{u}(k)} \triangle \boldsymbol{u}(k) e(M-k-h) \right]. \quad (5.5.48)$$

Using (5.5.33b) and the definition of $e(\cdot)$, it follows that

$$\sum_{k=0}^{M-1} \left(\bar{\boldsymbol{\lambda}}^i(k) \right)^T \left[\frac{\partial \hat{\boldsymbol{f}}}{\partial \boldsymbol{x}(k)} \triangle \boldsymbol{x}(k) + \frac{\partial \hat{\boldsymbol{f}}}{\partial \boldsymbol{u}(k)} \triangle \boldsymbol{u}(k) \right] e(M-k-h)$$

$$= \sum_{k=0}^{M-h-1} \left(\bar{\boldsymbol{\lambda}}^i(k) \right)^T \left[\frac{\partial \hat{\boldsymbol{f}}}{\partial \boldsymbol{x}(k)} \triangle \boldsymbol{x}(k) + \frac{\partial \hat{\boldsymbol{f}}}{\partial \boldsymbol{u}(k)} \triangle \boldsymbol{u}(k) \right] e(M-k-h)$$

$$= \sum_{k=h}^{M-1} \left(\boldsymbol{\lambda}^i(k+1) \right)^\top \left[\frac{\partial \bar{\boldsymbol{f}}}{\partial \boldsymbol{y}(k)} \triangle \boldsymbol{y}(k) + \frac{\partial \bar{\boldsymbol{f}}}{\partial \boldsymbol{v}(k)} \triangle \boldsymbol{v}(k) \right]. \quad (5.5.49)$$

As $\triangle \boldsymbol{y}(k) = 0$, for $0 \le k \le h$, and $\triangle \boldsymbol{v}(k) = 0$, $0 < k \le h$, we have

$$\sum_{k=h}^{M-1}(\boldsymbol{\lambda}^i(k+1))^\top\left[\frac{\partial\bar{\boldsymbol{f}}}{\partial\boldsymbol{y}(k)}\triangle\boldsymbol{y}(k)+\frac{\partial\bar{\boldsymbol{f}}}{\partial\boldsymbol{v}(k)}\triangle\boldsymbol{v}(k)\right]$$
$$=\sum_{k=0}^{M-1}(\boldsymbol{\lambda}^i(k+1))^\top\left[\frac{\partial\bar{\boldsymbol{f}}}{\partial\boldsymbol{y}(k)}\triangle\boldsymbol{y}(k)+\frac{\partial\bar{\boldsymbol{f}}}{\partial\boldsymbol{v}(k)}\triangle\boldsymbol{v}(k)\right]. \tag{5.5.50}$$

Combining (5.5.49) and (5.5.50), we obtain

$$\sum_{k=0}^{M-1}(\bar{\boldsymbol{\lambda}}^i(k))^\top\left[\frac{\partial\hat{\boldsymbol{f}}}{\partial\boldsymbol{x}(k)}\triangle\boldsymbol{x}(k)+\frac{\partial\hat{\boldsymbol{f}}}{\partial\boldsymbol{u}(k)}\triangle\boldsymbol{u}(k)\right]e(M-k-h)$$
$$=\sum_{k=0}^{M-1}(\boldsymbol{\lambda}^i(k+1))^\top\left[\frac{\partial\bar{\boldsymbol{f}}}{\partial\boldsymbol{y}(k)}\triangle\boldsymbol{y}(k)+\frac{\partial\bar{\boldsymbol{f}}}{\partial\boldsymbol{v}(k)}\triangle\boldsymbol{v}(k)\right]. \tag{5.5.51}$$

From (5.5.43a) and (5.5.51), it follows from (5.5.48) that

$$\frac{\partial g_i(\boldsymbol{u})}{\partial\boldsymbol{u}}\hat{\boldsymbol{u}}=\left(\frac{\partial\Phi_i(x(M))}{\partial\boldsymbol{x}(k)}\right)\triangle\boldsymbol{x}(M)+\sum_{k=0}^{M-1}\left\{\left(\frac{\partial\bar{H}_i}{\partial\boldsymbol{x}(k)}\right)\triangle\boldsymbol{x}(k)\right.$$
$$\left.+\left(\frac{\partial\bar{H}_i}{\partial\boldsymbol{u}(k)}\right)\hat{\boldsymbol{u}}(k)-(\boldsymbol{\lambda}^i(k+1))^\top\triangle\boldsymbol{x}(k+1)\right\}. \tag{5.5.52}$$

Thus, by (5.5.32), (5.5.45c) and (5.5.52), we obtain

$$\frac{\partial g_i(\boldsymbol{u})}{\partial\boldsymbol{u}}\hat{\boldsymbol{u}}=\left(\frac{\partial\Phi_i(\boldsymbol{x}(M))}{\partial\boldsymbol{x}(k)}\right)\triangle\boldsymbol{x}(M)+\sum_{k=0}^{M-2}\left[\left(\frac{\partial\bar{H}_i(k)}{\partial\boldsymbol{x}(k)}\right)\triangle\boldsymbol{x}(k)\right.$$
$$\left.-\left(\frac{\partial\bar{H}_i(k+1)}{\partial\boldsymbol{x}(k)}\right)\triangle\boldsymbol{x}(k+1)\right]+\frac{\partial\bar{H}_i(M-1)}{\partial\boldsymbol{x}(k)}\triangle\boldsymbol{x}(M-1)$$
$$-(\boldsymbol{\lambda}^i(M))^\top\triangle\boldsymbol{x}(M)+\sum_{k=0}^{M-1}\left[\left(\frac{\partial\bar{H}_i}{\partial\boldsymbol{u}(k)}\right)\hat{\boldsymbol{u}}(k)\right]. \tag{5.5.53}$$

Therefore, by substituting (5.5.33a) and (5.5.43b) into (5.5.47), it follows that

$$\frac{\partial g_i(\boldsymbol{u})}{\partial\boldsymbol{u}}\hat{\boldsymbol{u}}=\left[\frac{\partial H_i(0)}{\partial\boldsymbol{u}(0)},\frac{\partial H_i(1)}{\partial\boldsymbol{u}(1)},\ldots,\frac{\partial H_i(M-1)}{\partial\boldsymbol{u}(M-1)}\right]\hat{\boldsymbol{u}}.$$

Since $\hat{\boldsymbol{u}}$ is arbitrary, we obtain

$$\frac{\partial g_i(\boldsymbol{u})}{\partial\boldsymbol{u}}=\left[\frac{\partial H_i(0)}{\partial\boldsymbol{u}(0)},\frac{\partial H_i(1)}{\partial\boldsymbol{u}(1)},\ldots,\frac{\partial H_i(M-1)}{\partial\boldsymbol{u}(M-1)}\right].$$

This completes the proof.

5.5.3 A Tactical Logistic Decision Analysis Problem

In this section, we consider a tactical logistic decision analysis problem studied [12] . It is a problem of decision making for the distribution of resources within a network of support, where the network seeks to mimic how logistic support might be delivered in a military area of operations. The optimal control model of tactical logistic decision analysis problem formulated by Baker and Shi [12] are as follows:

$$\boldsymbol{x}(t+1) = A\boldsymbol{x}(t) + B_0\boldsymbol{u}(t) + B_1\boldsymbol{u}(t-1), \tag{5.5.54a}$$
$$\boldsymbol{x}(0) = \boldsymbol{x}_0, \boldsymbol{u}(-1) = 0, \tag{5.5.54b}$$

$$\boldsymbol{x}_{\min} \le \boldsymbol{x}(t) \le \boldsymbol{x}_{\max}, \tag{5.5.55a}$$
$$\boldsymbol{u}_{\min} \le \boldsymbol{u}(t) \le \boldsymbol{u}_{\max}, \tag{5.5.55b}$$

where

$$A = \begin{bmatrix} 0.95 & 0 & 0 & 0 & 0 \\ 0 & 0.9 & 0 & 0 & 0 \\ 0 & 0 & 0.75 & 0 & 0 \\ 0 & 0 & 0 & 0.75 & 0 \\ 0 & 0 & 0 & 0 & 0.85 \end{bmatrix}, B_0 = \begin{bmatrix} 0 & -1 & -1 & 0 & 0 & 0 & 0 & 0 \\ 0 & 0 & 0 & -1 & -1 & -1 & 0 & 0 \\ 0 & 0 & 0 & 0 & 0 & 0 & -1 & 0 \\ 0 & 0 & 0 & 0 & 0 & 0 & 0 & -1 \\ 0 & 0 & 0 & 0 & 0 & 0 & 0 & 0 \end{bmatrix},$$

$$B_1 = \begin{bmatrix} 0.95 & 0 & 0 & 0 & 0 & 0 & 0 & 0 \\ 0 & 0.87 & 0 & 0 & 0 & 0 & 0 & 0 \\ 0 & 0 & 0 & 0 & 0.75 & 0 & 0 & 0.7 \\ 0 & 0 & 0.8 & 0 & 0 & 0.8 & 0.7 & 0 \\ 0 & 0 & 0 & 0.85 & 0 & 0 & 0 & 0 \end{bmatrix}, \boldsymbol{x}_0 = \begin{bmatrix} 3500 \\ 800 \\ 400 \\ 400 \\ 200 \end{bmatrix}.$$

The cost functional is

$$\begin{aligned} G = & \frac{1}{2}(\boldsymbol{x}(T))^\top Q\boldsymbol{x}(T) \\ & + \sum_{t=0}^{T-1} \frac{1}{2}\left\{(\boldsymbol{x}(T))^\top Q\boldsymbol{x}(t) + (\boldsymbol{u}(t))^\top R\boldsymbol{u}(t)\right\}, \end{aligned} \tag{5.5.56}$$

where

$$Q = \begin{bmatrix} 1 & 0 & 0 & 0 & 0 \\ 0 & 2 & 0 & 0 & 0 \\ 0 & 0 & 3 & 0 & 0 \\ 0 & 0 & 0 & 1.5 & 0 \\ 0 & 0 & 0 & 0 & 2.5 \end{bmatrix}, \quad R = \begin{bmatrix} 1 & 0 & 0 & 0 & 0 & 0 & 0 & 0 \\ 0 & 5 & 0 & 0 & 0 & 0 & 0 & 0 \\ 0 & 0 & 5 & 0 & 0 & 0 & 0 & 0 \\ 0 & 0 & 0 & 2.5 & 0 & 0 & 0 & 0 \\ 0 & 0 & 0 & 0 & 3 & 0 & 0 & 0 \\ 0 & 0 & 0 & 0 & 0 & 4 & 0 & 0 \\ 0 & 0 & 0 & 0 & 0 & 0 & 2 & 0 \\ 0 & 0 & 0 & 0 & 0 & 0 & 0 & 2 \end{bmatrix}.$$

The logistic network for the example is as shown in Figure 5.5.1.

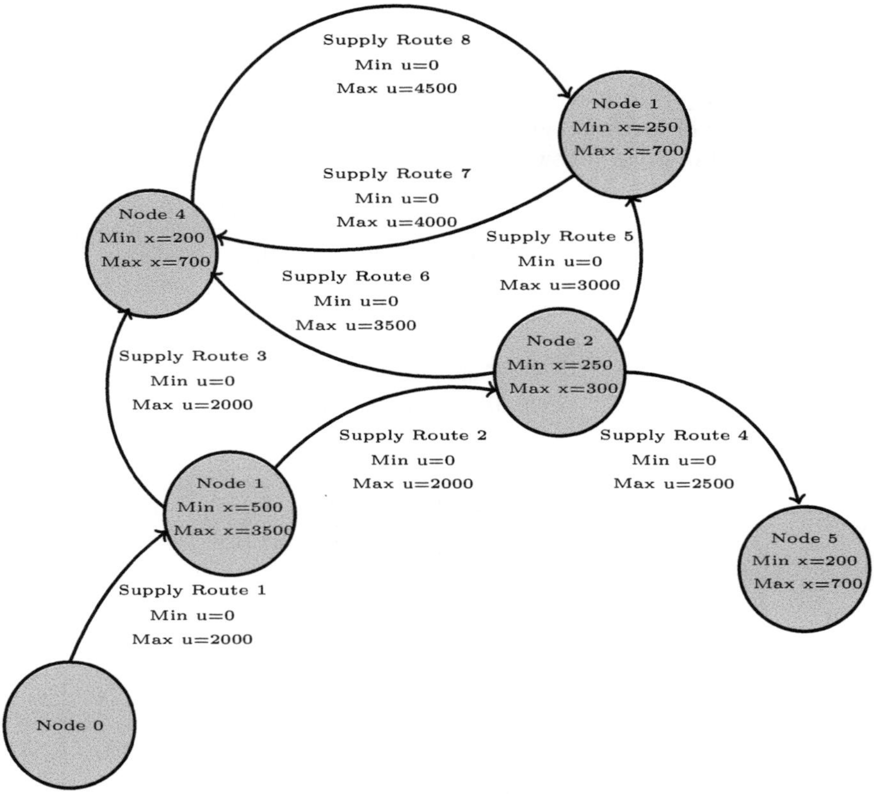

Fig. 5.5.1: The logistic network

The constraints (5.5.55a) and (5.5.55b) can be rewritten as

$$g_i(\boldsymbol{u}) = x_{i,\min} - x_i(k) \leq 0, \;\; k = 0, 1, \ldots M-1,$$
$$i = 1, 2, \ldots, 5, \tag{5.5.57}$$
$$g_i(\boldsymbol{u}) = x_i(k) - x_{i,\max} \leq 0, \;\; k = 0, 1, \ldots M-1,$$
$$i = 6, 7, \ldots, 10. \tag{5.5.58}$$

The optimal cost functional value obtained by using Algorithm 5.5.1 is 1.68×10^7, which is much less than that obtained in [12], which is 3.5×10^7. The optimal control and the corresponding optimal state obtained using our method are depicted in Figures 5.5.2, 5.5.3, 5.5.4, 5.5.5, 5.5.6 and 5.5.7. By careful examination of these figures, we see that the constraints on the control

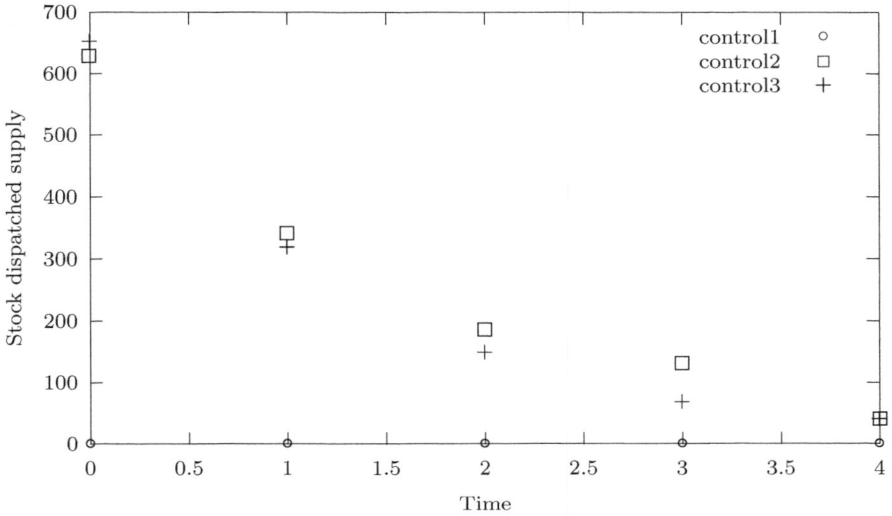

Fig. 5.5.2: Stock dispatched supply 1–3

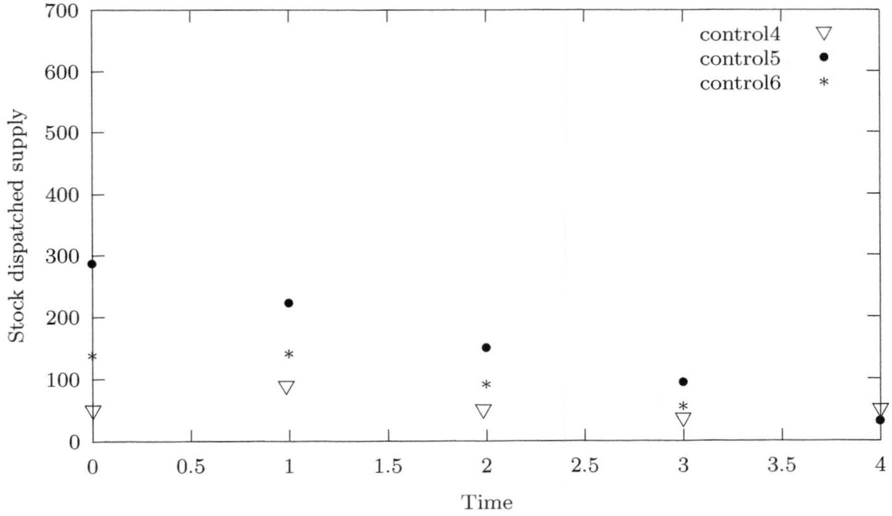

Fig. 5.5.3: Stock dispatched supply 4–6

and the all-time-step constraints are satisfied at each time point. However, all-time-step constraints in [12] are not always satisfied at each time point.

From Figure 5.5.2, 5.5.3 and 5.5.4, we see that $u_1(k) = 0$ for $k = 0, 1, \ldots, 4$, indicating no stock being dispatched along the supply route 1 to Node 1. This is because $u_1(k)$ could only contribute extra stock to Node 1 through the supply route 1 from Node 0, and the initial stock in Node 1 is large, twice as

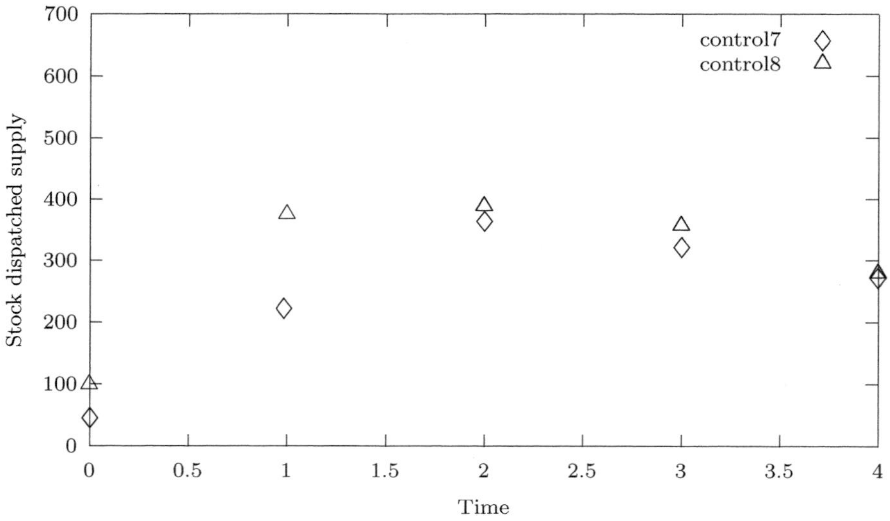

Fig. 5.5.4: Stock dispatched supply 7–8

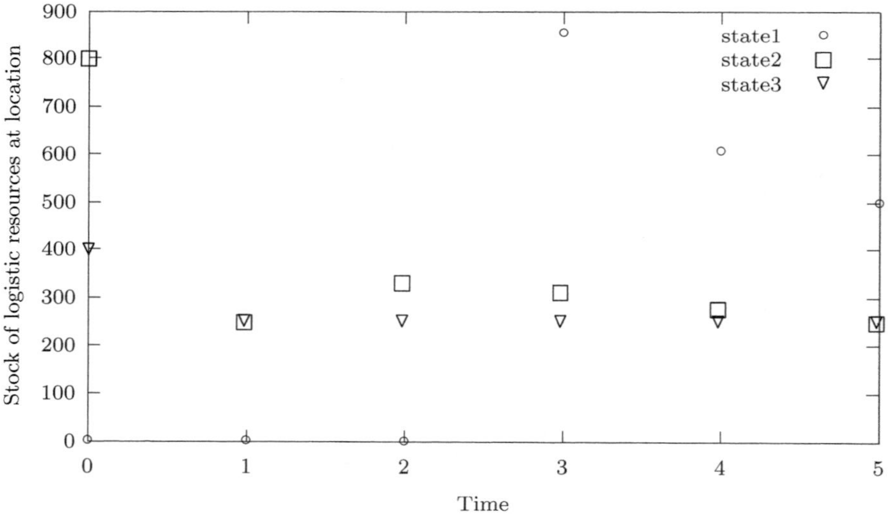

Fig. 5.5.5: Stock at location 1–3

large as those in the other nodes. Thus, it is clear that stock should be moved out of Node 1 to other nodes quickly through the supply routes 2 and 3 so as to decrease the cost of holding the stock in Node 1. Also from Figure 5.5.2, 5.5.3 and 5.5.4, we see that $u_2(k)$ and $u_3(k)$ are very large at $k = 0$, meaning that a large amount of stock is dispatched from Node 1 to the other nodes

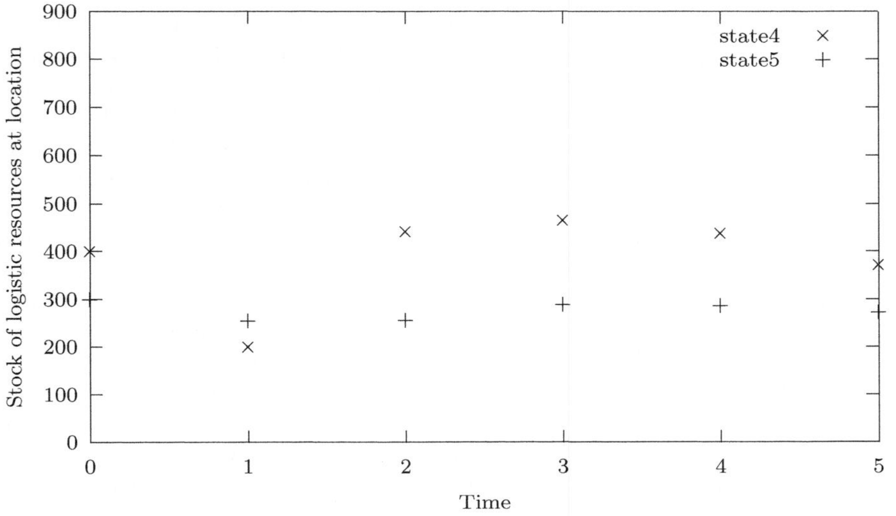

Fig. 5.5.6: Stock at location 4–5

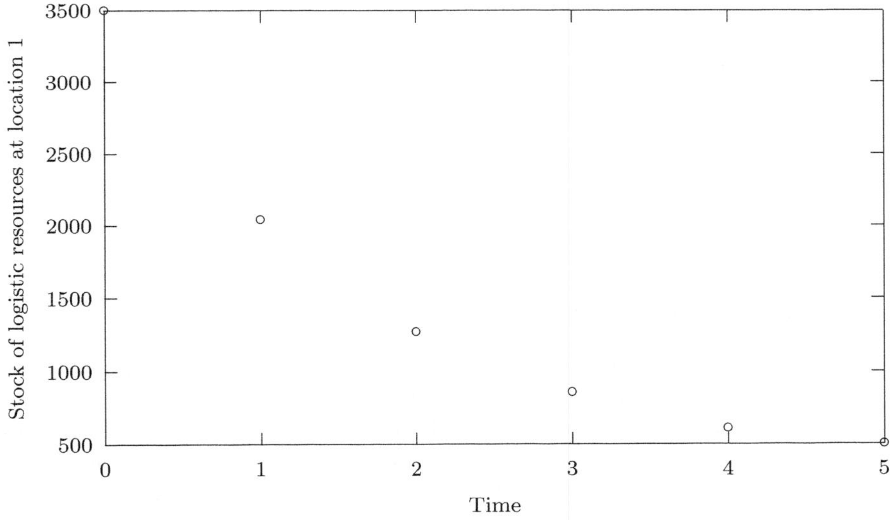

Fig. 5.5.7: Stock at location 1

of the network at $k = 0$. From the structure of the network, it is clear that there is only one supply route to Node 5 with no supply route coming out of it. This means that Node 5 is a pure receiver of stock from other nodes. In view of the limits imposed on the maximum stock in various nodes, we see from Figures 5.5.5 and 5.5.6 that the amount of stock that is moved along the supply route 4 to Node 5 is low for $k = 0, 1, \ldots, 4$. The structure of the network depicted in Figure 5.5.1 clearly reveals that there are 4 supply routes (i.e., supply routes 3, 6 and 7) to Node 4 with only one supply route (i.e., supply route 8) coming out of it. For Node 3, there are 2 supply routes (i.e., supply routes 2 and 4) in and only 1 (i.e., supply route 7) out. By virtue of these observations, the amounts of stock along the supply routes for which the stocks are moved out should be large. This is confirmed in Figure 5.5.2, 5.5.3 and 5.5.4 that $u_k(k)$ and $u_8(k)$, which denote, respectively, the amounts of stock being moved out from Node 3 along the supply route 7 and Node 4 along the supply route 8 are large for $k = 1, \ldots, 4$. Their values are quite low at $k = 0$. This is due to the appearance of time-delay along the supply routes, which indicates that nodes cannot receive stock instantaneously. The stocks arrive with delay. For Node 2 , there are 3 supply routes (i.e., supply routes 4, 5 and 6) out but only 1 supply route (i.e., supply route 2) in. As shown in Figures 5.5.2, 5.5.3 and 5.5.4, we see that the amounts of the stock being moved out along the supply routes 4, 5 and 6 are relatively small.

5.6 Exercises

5.6.1 *Give a proof of Theorem 5.2.3.*

5.6.2 *Give a proof of Theorem 5.2.4.*

5.6.3 *Give a proof of Lemma 5.2.1.*

5.6.4 *Give a proof of Lemma 5.2.2.*

5.6.5 *Give a proof of Theorem 5.2.5.*

5.6.6 *Give a proof of Theorem 5.5.3.*

5.6.7 *A woman owns an initial sum of money a, and knows that she will live exactly N years. At the beginning of year i she owns the sum \$$x(i)$. She selects a fraction $v(i)$, $0 \leq v(i) \leq 1$, to invest, and consumes the remainder, $c(i) = (1 - v(i))x(i)$. She derives a benefit $B(c(i))$ from this, where B is a given monotonically increasing smooth function. The invested sum accrues interest at the rate r, so that $x(i+1) = (1+r)x(i)v(i)$. The woman wishes to maximize the total benefit, i.e., the sum of all the $B(c(i))$ over the remaining N years of her life, and has no particular desire to have any money left to bequeath after death. Formulate the problem so that it can be attacked by the dynamic programming, and write down the dynamic programming equation for the process.* Do not solve the dynamic programming equation.

5.6.8 *The problem of investing money over a period of N years can be represented as a multistage decision process where each year \$$x$ is subdivided into \$$u$, where $x \geq u \geq 0$, which are invested, and \$$(x - u)$ which are spent.*

The process is assumed to have the dynamic equation, $x(i+1) = au(i)$ where the integer i counts the years and the growth rate "a" ($a > 0$) is a constant. The satisfaction from money spent in any year is described by a function $H(i) = H(x(i) - u(i))$ and the objective is to maximize the total satisfaction $I = \sum_{i=1}^{N} H(i)$ over N years. Define an optimal return function and write the dynamic programming functional recurrence equation for the process. Take $H = (x - u)^{1/2}$ and find the optimal spending policy

(i) over 2 years
(ii) over N years.

5.6.9 *Same as Exercise 5.6.8, but with H taken as*

$$H = \log(x - u).$$

5.6.10 *For the discrete time process*

$$x(k+1) = 2\boldsymbol{x}(k) + \boldsymbol{u}(k).$$

It is desired to minimize the cost functional,

$$I = 50(x(2))^2 + \sum_{k=0}^{1} (\boldsymbol{u}(k))^2$$

for a two-step process starting from $x(0) = 1$ with the finishing state $x(2)$ free.

(i) Use the dynamic programming to find the optimal controls $u(i)$, which are to be real numbers satisfying $|u(i)| \leq 2$.
(ii) Solve the problem in part (i) when the initial state is given by $x(0) = 10$.

5.6.11

(a) A farm grows wheat as a yearly crop and has available unlimited acreage and free labour. Of the total grain $x(i)$ tones available at the start of each year i, $u(i)$ tones ($0 \leq u(i) \leq x(i)$) is planted and the remainder $x(i) - u(i)$ is sold during the year. The planted wheat produces a new crop, $x(i+1) = au(i)$, ($a =$ constant > 1). It is assumed that A tones of grain (i.e., $x(1) = A$) is provided to start the venture which is to run for 4 years. The desire is to maximize

$$I = \sum_{i=1}^{4} \{x(i) - u(i)\}$$

the total amount of grain sold over the period. Use the dynamic programming to find the optimal planting and selling policy.

(b) Use the dynamic programming to find the optimal planting and selling policy for the problem in part (a) with the following modifications:

(i) The project is to finish with A tones left at the end of the fourth year.

(ii) As in (b)(i), but the amount of land available for cultivation is limited so that no more than aA tones can be sown in any year, i.e., $0 \leq u(i) \leq aA$.

5.6.12 *Consider the unconstrained nonlinear discrete time optimal control problem with dynamics governed by*

$$\boldsymbol{x}(k+1) = \boldsymbol{f}(k, \boldsymbol{x}(k), \boldsymbol{u}(k)), \qquad k = 0, 1, \ldots, M-1 \tag{5.6.1a}$$

$$\boldsymbol{x}(0) = \boldsymbol{x}^0, \tag{5.6.1b}$$

where $\boldsymbol{x}(k) \in \mathbb{R}^n$, and $\boldsymbol{u}(k) \in \mathbb{R}^r$. The sequence $\boldsymbol{u}(k)$, $k = 1, \ldots, M-1$, is to be determined such that the cost functional

$$g_0(\boldsymbol{u}) = \Phi_0(\boldsymbol{x}(M)) + \sum_{k=1}^{M-1} \mathcal{L}_0(k, \boldsymbol{x}(k), \boldsymbol{u}(k)) \tag{5.6.2}$$

is minimized. Let $\boldsymbol{u}^*$ be an optimal control, and let $\boldsymbol{x}^*$ be the corresponding solution of the system (5.6.1). *Use the technique of Section 4.1 to show that*

$$\frac{\partial H(k, \boldsymbol{x}^*(k), \boldsymbol{u}^*(k), \boldsymbol{\lambda}^*(k+1))}{\partial \boldsymbol{u}(k)} = 0, \quad k = 0, 1, \ldots, M-1, \tag{5.6.3}$$

where

$$\begin{aligned} &H(k, \boldsymbol{x}^*(k), \boldsymbol{u}^*(k), \boldsymbol{\lambda}^*(k+1)) \\ =&\mathcal{L}_0(k, \boldsymbol{x}^*(k), \boldsymbol{u}^*(k)) + (\boldsymbol{\lambda}^*(k+1))^\top \boldsymbol{f}(k, \boldsymbol{x}^*(k), \boldsymbol{u}^*(k)), \end{aligned} \tag{5.6.4}$$

and

$$(\boldsymbol{\lambda}^*(k))^\top = \frac{\partial H(k, \boldsymbol{x}^*(k), \boldsymbol{u}^*(k), \boldsymbol{\lambda}^*(k+1))}{\partial \boldsymbol{x}(k)} \tag{5.6.5a}$$

$$\boldsymbol{\lambda}^*(M) = \frac{\partial \Phi_0(\boldsymbol{x}^*(M))}{\partial \boldsymbol{x}(M)}. \tag{5.6.5b}$$

5.6.13 *Consider the problem of Exercise 5.6.12. Let $\boldsymbol{u}(k) \in U$ for all $k = 0, 1, \ldots, M-1$, where U is a compact and convex subset of $\mathbb{R}^r$. Show that the first order necessary condition for optimality is the same as that given in Exercise 5.6.12, except with* (5.6.3) *replaced by*

$$\sum_{k=0}^{M-1} \frac{\partial H(k, \boldsymbol{x}(k), \boldsymbol{u}(k), \boldsymbol{\lambda}(k+1))}{\partial \boldsymbol{u}(k)} (\boldsymbol{u}(k) - \boldsymbol{u}^*(k)) \geq 0$$

for all $\boldsymbol{u}(k) \in U$, $k = 0, 1, \ldots, M-1$.

5.6.14 *Consider the discrete time process governed by the following system of difference equations:*

$$\boldsymbol{x}(k+1) = A(k)\boldsymbol{x}(k) + B(k)\boldsymbol{u}(k), \quad k = 0, 1, \ldots, M-1,$$

where $\boldsymbol{x}(k) \in \mathbb{R}^n$, $\boldsymbol{u}(k) \in \mathbb{R}^r$, $A(k) \in \mathbb{R}^{n\times n}$, *and* $B(k) \in \mathbb{R}^{n\times r}$. *The control sequence* $\boldsymbol{u}(k), k = 0, 1, \ldots, M-1$, *is to be determined such that the quadratic cost functional*

$$\begin{aligned} g_0(\boldsymbol{u}) = &\frac{1}{2}(\boldsymbol{x}(M))^\top Q(M)\boldsymbol{x}(M) \\ &+ \sum_{k=0}^{M-1} \left\{ \frac{1}{2}(\boldsymbol{x}(k))^\top Q(k)\boldsymbol{x}(k) + \frac{1}{2}(\boldsymbol{u}(k))^\top R(k)\boldsymbol{u}(k) \right\} \end{aligned}$$

is minimized, where $Q(k) = (Q(k))^\top \in \mathbb{R}^{n\times n}$ *is positive semi-definite, and* $R(k) = (R(k))^\top \in \mathbb{R}^{r\times r}$ *is positive definite.*

(a) Show that the optimal control is

$$\boldsymbol{u}(k) = -(R(k))^{-1}(B(k))^\top \boldsymbol{\lambda}(k+1),$$

where $\boldsymbol{\lambda}(k)$ *is the costate vector governed by the following system of difference equations:*

$$\begin{aligned} \boldsymbol{\lambda}(k+1) &= (A(k))^\top \boldsymbol{\lambda}(k+1) + Q(k)\boldsymbol{x}(k) \\ \boldsymbol{\lambda}(M) &= Q(M)\boldsymbol{x}(M). \end{aligned}$$

(b) By assuming that the costate and state vectors are related through the relationship:

$$\boldsymbol{\lambda}(k) = S(k)\boldsymbol{x}(k),$$

where $S(k) = (S(k))^\top$. *Show that the symmetric matrix* $S(k)$ *is governed by the discrete matrix Riccati equation:*

$$\begin{aligned} &S(k) = Q(k) + (A(k))^\top \left((S(k+1))^{-1} + B(k)(R(k))^{-1}(B(k))^\top\right)^{-1} A(k) \\ &S(M) = Q(M). \end{aligned}$$

(c) Show that the system of optimal state equations is given by

$$\boldsymbol{x}(k+1) = \left(I + B(k)(R(k))^{-1}(B(k))^\top S(k+1)\right)^{-1} A(k)\boldsymbol{x}(k).$$

5.6.15 *A combined discrete time optimal control and optimal parameter selection problem is defined in the canonical form as follows:*

$$\text{minimize} \quad g_0(\boldsymbol{u}, \boldsymbol{\zeta}),$$

where

$$g_0(\boldsymbol{u}, \boldsymbol{\zeta}) = \Phi_0(\boldsymbol{x}(M \mid \boldsymbol{u}, \boldsymbol{\zeta}), \boldsymbol{\zeta}) + \sum_{k=0}^{M-1} \mathcal{L}_0(k, \boldsymbol{x}(k \mid \boldsymbol{u}, \boldsymbol{\zeta}), \boldsymbol{u}(k), \boldsymbol{\zeta}),$$

subject to

$$g_i(\boldsymbol{u}, \boldsymbol{\zeta}) = 0, \qquad i = 1, \ldots, N_e$$
$$g_i(\boldsymbol{u}, \boldsymbol{\zeta}) \geq 0, \qquad i = 1, \ldots, N,$$

where

$$g_i(\boldsymbol{u}, \boldsymbol{\zeta}) = \Phi_i(\boldsymbol{x}(M \mid \boldsymbol{u}, \boldsymbol{\zeta}), \boldsymbol{\zeta}) + \sum_{k=0}^{M-1} \mathcal{L}_i(k, \boldsymbol{x}(k \mid \boldsymbol{u}, \boldsymbol{\zeta}), \boldsymbol{u}(k), \boldsymbol{\zeta}),$$

and $\boldsymbol{\zeta} \in \mathbb{R}^s$ *is the system parameter vector to be optimized together with* $\boldsymbol{u}$. *Derive the corresponding gradient formulae for the cost functional and the constraint functionals with respect to* $\boldsymbol{\zeta}$ *and* $\boldsymbol{u}$.

5.6.16 *Consider the first order system*

$$x(k+1) = \frac{1}{2}\boldsymbol{x}(k) + \boldsymbol{u}(k)$$
$$x(0) = 1,$$

where $k = 0, 1, \ldots, M$. *The optimal control problem is find* $\boldsymbol{u}(k)$, $k = 0, 1, \ldots, M-1$, *such that*

$$g_0(\boldsymbol{u}) = \sum_{k=0}^{M} \{(\boldsymbol{x}(k))^2 + (\boldsymbol{u}(k))^2\}$$

is minimized. Solve the problem for $M = 4$, $M = 6$ *and* $M = 10$. *Comment on your results.*

5.6.17 *Problem* (Q_ε) *constructed in Section 5.4.1 can be solved as a nonlinear optimization problem. Give details of the computational procedure.*

Chapter 6
Elements of Optimal Control Theory

6.1 Introduction

There are already many excellent books devoted solely to the detailed exposition of optimal control theory. We refer the interested reader to [3, 4, 8, 11, 18–21, 29, 33, 40, 59, 64, 74, 83, 90, 121, 130, 198, 201, 206, 226, 276], just to name a few. Texts dealing with the optimal control of partial differential equations include [5, 37, 149, 250]. The aim of this chapter is to give a brief account of some fundamental optimal control theory results for systems described by ordinary differential equations.

In the next section, we present the basic formulation of an unconstrained optimal control problem and derive the first order necessary condition known as the Euler-Lagrange equations. In Section 6.3, we consider a class of linear quadratic optimal control problems. This class is important because the problem can be solved analytically and the optimal control so obtained is in closed loop form. In Section 6.4, the well-known Pontryagin minimum principle is briefly discussed. The Maximum Principle is then used to introduce singular control and time optimal control in Sections 6.5 and 6.6, respectively. In Section 6.7, a version of the optimality conditions for optimal control problems subject to constraints is presented and an illustrative example is solved using these conditions. To conclude this chapter, Bellman's dynamic programming principle is included in Section 6.8. For results on the existence of optimal controls, we refer the interested reader to [40].

The main references of this chapter are the lecture notes prepared and used by the authors, [88] and Chapter 4 of [253].

K. L. Teo et al., *Applied and Computational Optimal Control*, Springer Optimization and Its Applications 171,
https://doi.org/10.1007/978-3-030-69913-0_6

6.2 First Order Necessary Condition: Euler-Lagrange Equations

We shall begin with the simplest optimal control formulation from which we derive the first order necessary conditions for optimality, also known as the Euler-Lagrange equations. More complex classes of optimal control problems will be discussed in later sections. Consider a dynamical system described by the following system of differential equations:

$$\frac{d\boldsymbol{x}(t)}{dt} = \boldsymbol{f}(t, \boldsymbol{x}(t), \boldsymbol{u}(t)), \quad t \in [0, T], \tag{6.2.1a}$$

with initial condition:

$$\boldsymbol{x}(0) = \boldsymbol{x}^0, \tag{6.2.1b}$$

where

$$\boldsymbol{x}(t) = [x_1(t), x_2(t), \ldots, x_n(t)]^\top, \quad \boldsymbol{u}(t) = [u_1(t), u_2(t), \ldots, u_m(t)]^\top$$

are, respectively, referred to as the state and control. $\boldsymbol{f} = [f_1, \ldots, f_n]^\top$ is assumed to be continuously differentiable with respect to all its arguments and $\boldsymbol{x}^0$ is a given vector in $\mathbb{R}^n$. In many applications, the independent variable t denotes time, but there are some, such as problems in shape optimization, where t has a different meaning. We have assumed that the given process starts at $t = 0$ and ends at the fixed terminal time $T > 0$. A process that starts from $t_0 \neq 0$ may be readily transformed to satisfy this assumption by a suitable shifting of the time scale.

For now, any piecewise continuous function $\boldsymbol{u}$ from $[0, T]$ into $\mathbb{R}^m$ may be taken as an admissible control. Let $\mathcal{U}$ be the class of all such admissible controls.

We now consider an optimal control problem where a control $\boldsymbol{u} \in \mathcal{U}$ is to be chosen such that the cost functional

$$g_0(\boldsymbol{u}) = \Phi_0(\boldsymbol{x}(T)) + \int_0^T \mathcal{L}_0(t, \boldsymbol{x}(t), \boldsymbol{u}(t))dt \tag{6.2.2}$$

is minimized, where Φ_0 and $\mathcal{L}_0$ are continuously differentiable with respect to their respective arguments. Note that the cost functional can be regarded as depending on $\boldsymbol{u}$ only, as $\boldsymbol{x}$ is implicitly determined by $\boldsymbol{u}$ from (6.2.1). The dynamical constraint (6.2.1) can be appended to the cost functional by introducing the appropriate Lagrange multiplier $\boldsymbol{\lambda} \in \mathbb{R}^n$ as follows.

$$\bar{g}_0(\boldsymbol{u}) = \Phi_0(\boldsymbol{x}(T)) + \int_0^T \left\{ \mathcal{L}_0(t, \boldsymbol{x}(t), \boldsymbol{u}(t)) + (\boldsymbol{\lambda}(t))^\top \left[\boldsymbol{f}(t, \boldsymbol{x}(t), \boldsymbol{u}(t)) - \frac{d\boldsymbol{x}(t)}{dt} \right] \right\} dt. \tag{6.2.3}$$

For convenience, we define the Hamiltonian function as

$$H(t, \boldsymbol{x}, \boldsymbol{u}, \boldsymbol{\lambda}) = \mathcal{L}_0(t, \boldsymbol{x}, \boldsymbol{u}) + \boldsymbol{\lambda}^\top \boldsymbol{f}(t, \boldsymbol{x}, \boldsymbol{u}). \tag{6.2.4}$$

Note that the appended cost functional $\bar{g}_0$ is identical to the original g_0 if the dynamical constraint is satisfied. The time-dependent Lagrange multiplier is referred to as the *costate vector*. It is also known as the *adjoint vector*. Substituting (6.2.4) into (6.2.3) and integrating the last term by parts, we have

$$\begin{aligned}\bar{g}_0(\boldsymbol{u}) =& \Phi_0(\boldsymbol{x}(T)) - (\boldsymbol{\lambda}(T))^\top \boldsymbol{x}(T) + (\boldsymbol{\lambda}(0))^\top \boldsymbol{x}(0) \\ &+ \int_0^T \left\{ H(t, \boldsymbol{x}(t), \boldsymbol{u}(t), \boldsymbol{\lambda}(t)) + \frac{d(\boldsymbol{\lambda}(t))^\top}{dt} \boldsymbol{x}(t) \right\} dt. \end{aligned} \tag{6.2.5}$$

For a small variation $\delta \boldsymbol{u}$ in $\boldsymbol{u}$, the corresponding first order variations in $\boldsymbol{x}$ and $\bar{g}_0$ are $\delta \boldsymbol{x}$ and $\delta \bar{g}_0$, respectively, where $\delta \bar{g}_0$ is obtained by the chain rule:

$$\begin{aligned}\delta \bar{g}_0 &= \lim_{\epsilon \to 0} \frac{\bar{g}_0(\boldsymbol{u} + \epsilon \delta \boldsymbol{u}) - \bar{g}_0(\boldsymbol{u})}{\epsilon} \\ &= \left[\frac{\partial \Phi_0(\boldsymbol{x}(T))}{\partial \boldsymbol{x}} - (\boldsymbol{\lambda}(T))^\top \right] \delta \boldsymbol{x}(T) + (\boldsymbol{\lambda}(0))^\top \delta \boldsymbol{x}(0) \\ &\quad + \int_0^T \left\{ \left[\frac{\partial H(t, \boldsymbol{x}(t), \boldsymbol{u}(t), \boldsymbol{\lambda}(t))}{\partial \boldsymbol{x}} + \frac{d(\boldsymbol{\lambda}(t))^\top}{dt} \right] \delta \boldsymbol{x}(t) \right. \\ &\quad \left. + \frac{\partial H(t, \boldsymbol{x}(t), \boldsymbol{u}(t), \boldsymbol{\lambda}(t))}{\partial \boldsymbol{u}} \delta \boldsymbol{u}(t) \right\} dt. \end{aligned} \tag{6.2.6}$$

Since $\boldsymbol{\lambda}(t)$ is arbitrary so far, we may choose

$$\frac{d\boldsymbol{\lambda}(t)}{dt} = - \left[\frac{\partial H(t, \boldsymbol{x}(t), \boldsymbol{u}(t), \boldsymbol{\lambda}(t))}{\partial \boldsymbol{x}} \right]^\top \tag{6.2.7a}$$

with boundary condition

$$\boldsymbol{\lambda}(T) = \left[\frac{\partial \Phi_0(\boldsymbol{x}(T))}{\partial \boldsymbol{x}} \right]^\top. \tag{6.2.7b}$$

As the initial condition $\boldsymbol{x}(0)$ is fixed, $\delta \boldsymbol{x}(0)$ vanishes and (6.2.6) reduces to

$$\delta \bar{g}_0 = \int_0^T \left\{ \frac{\partial H(t, \boldsymbol{x}(t), \boldsymbol{u}(t), (t))}{\partial \boldsymbol{u}} \delta \boldsymbol{u}(t) \right\} dt. \tag{6.2.8}$$

For a local minimum, $\delta \bar{g}_0$ is required to vanish for any arbitrary $\delta \boldsymbol{u}$. Therefore, it is necessary that

$$\left[\frac{\partial H(t, \boldsymbol{x}(t), \boldsymbol{u}(t), (t))}{\partial \boldsymbol{u}} \right]^\top = \boldsymbol{0} \tag{6.2.9}$$

for all $t \in [0, T]$, except possibly on a finite set (i.e., on a set consisting of a finite number of points). Note that this condition holds only if $\boldsymbol{u}$ is unconstrained. In the case of control bounds, the Pontryagin Maximum Principle to be discussed later is required.

Equations (6.2.1), (6.2.7) and (6.2.9) are the well-known Euler-Lagrange equations. Note that (6.2.7) is a set of ordinary differential equations in $\boldsymbol{\lambda}$ with boundary conditions specified at the terminal time T. We shall summarize these results in a theorem.

Theorem 6.2.1 *Let $\boldsymbol{u}^*(t)$ be a local optimal control for the cost functional (6.2.2), and let $\boldsymbol{x}^*(t)$ and $\boldsymbol{\lambda}^*(t)$ be, respectively, the corresponding optimal state and costate. Then, it is necessary that*

$$\begin{aligned} \frac{d\boldsymbol{x}^*(t)}{dt} &= \left[\frac{\partial H(t, \boldsymbol{x}^*(t), \boldsymbol{u}^*(t), \boldsymbol{\lambda}^*(t))}{\partial \boldsymbol{\lambda}}\right]^\top \\ &= \boldsymbol{f}(t, \boldsymbol{x}^*(t), \boldsymbol{u}^*(t)), \end{aligned} \tag{6.2.10a}$$

$$\boldsymbol{x}^*(0) = \boldsymbol{x}^0, \tag{6.2.10b}$$

$$\frac{d\boldsymbol{\lambda}^*(t)}{dt} = -\left[\frac{\partial H(t, \boldsymbol{x}^*(t), \boldsymbol{u}^*(t), \boldsymbol{\lambda}^*(t))}{\partial \boldsymbol{x}}\right]^\top, \tag{6.2.10c}$$

$$\boldsymbol{\lambda}^*(T) = \left[\frac{\partial \Phi_0(\boldsymbol{x}^*(T))}{\partial \boldsymbol{x}}\right]^\top, \tag{6.2.10d}$$

and, for all $t \in [0, T]$, except possibly on a finite subset of $[0, T]$,

$$\left[\frac{\partial H(t, \boldsymbol{x}^*(t), \boldsymbol{u}^*(t), \boldsymbol{\lambda}^*(t))}{\partial \boldsymbol{u}}\right]^\top = \mathbf{0}, \tag{6.2.10e}$$

where a finite set denotes a set that contains only a finite number of points.

Note that (6.2.10a)–(6.2.10d) constitute $2n$ differential equations with n boundary conditions for $\boldsymbol{x}^*$ specified at $t = 0$ and n boundary conditions for $\boldsymbol{\lambda}^*$ specified at $t = T$. This is referred to as a *two-point boundary-value problem* (TPBVP). In principle, the dependence on $\boldsymbol{u}^*$ can be removed by solving for $\boldsymbol{u}^*$ as a function of $\boldsymbol{x}^*$ and $\boldsymbol{\lambda}^*$ from the m algebraic equations in (6.2.10e) via the implicit function theorem, provided that the Hessian, $H_{\boldsymbol{uu}} = (\nabla_{\boldsymbol{u}})^\top(\nabla_{\boldsymbol{u}} H)$, is non-singular at the optimal point. In practice, however, an analytic solution of (6.2.10e) is often not possible. Even if it could be determined, the resulting TPBVP is likely to be difficult to solve.

Example 6.2.1 Consider the following simple problem.

$$\min \left\{ g_0(u) = \int_0^1 \{(x(t))^2 + (u(t))^2\}\, dt \right\} \tag{6.2.11}$$

subject to

$$\frac{dx(t)}{dt} = u(t) \tag{6.2.12a}$$

$$x(0) = 1. \tag{6.2.12b}$$

The corresponding Hamiltonian function is

$$H = x^2 + u^2 + \lambda u. \tag{6.2.13}$$

By Theorem 6.2.1, we have

$$\frac{d\lambda(t)}{dt} = -2x(t) \tag{6.2.14a}$$

$$\lambda(1) = 0 \tag{6.2.14b}$$

$$\frac{\partial H}{\partial u} = 2u(t) + \lambda(t) = 0. \tag{6.2.15}$$

Substituting (6.2.15) into (6.2.12a) gives

$$\frac{dx(t)}{dt} = -\frac{\lambda(t)}{2}. \tag{6.2.16}$$

Differentiating (6.2.16) with respect to t and then using (6.2.14a), we have

$$\frac{d^2x(t)}{dt^2} - x(t) = 0. \tag{6.2.17}$$

Clearly, the solution of (6.2.17) is given by

$$x(t) = Ae^t + Be^{-t}, \tag{6.2.18}$$

where A and B are constants to be determined by the boundary conditions. Some elementary algebra then leads to the exact solution of the resulting TPBVP:

$$u^*(t) = \frac{-2e}{1+e^2}\sinh(1-t) \tag{6.2.19}$$

$$x^*(t) = \frac{2e}{1+e^2}\cosh(1-t), \tag{6.2.20}$$

and

$$g_0^* = \frac{e^4-1}{(1+e^2)^2}. \tag{6.2.21}$$

While the above example could be solved analytically, it is far less complex than those one can expect to encounter in practice. In fact, most practical optimal control problems, especially those with hard constraints, cannot be solved analytically.

6.3 The Linear Quadratic Theory

An important family of unconstrained optimal control problems is known as linear quadratic problems. There are basically two types: regulator problems and tracking problems. Both are intimately related. We shall first consider the general regulator problem, where the system dynamics are governed by *linear* differential equations, while the performance index is *quadratic* in the state and control:

$$\min\Bigg\{g_0(\boldsymbol{u}) = \frac{1}{2}(\boldsymbol{x}(T))^\top S^f \boldsymbol{x}(T) + \frac{1}{2}\int_0^T \{(\boldsymbol{x}(t))^\top Q(t)\boldsymbol{x}(t) + (\boldsymbol{u}(t))^\top R(t)\boldsymbol{u}(t)\}\, dt\Bigg\} \tag{6.3.1}$$

subject to

$$\frac{d\boldsymbol{x}(t)}{dt} = A(t)\boldsymbol{x}(t) + B(t)\boldsymbol{u}(t) \tag{6.3.2a}$$

$$\boldsymbol{x}(0) = \boldsymbol{x}^0, \tag{6.3.2b}$$

where $\boldsymbol{x} \in \mathbb{R}^n$, $\boldsymbol{u} \in \mathbb{R}^r$, S^f, Q, $A \in \mathbb{R}^{n\times n}$, $R \in \mathbb{R}^{r\times r}$ and $B \in \mathbb{R}^{n\times r}$. Here, T is fixed, S^f and $Q(t)$ are symmetric positive semi-definite matrices, while $R(t)$ is symmetric positive definite for all $t \in [0, T]$. The regulator problem seeks to regulate the state to be as close as possible to zero while minimizing the control effort.

The corresponding Hamiltonian function is

$$H = \frac{1}{2}\{\boldsymbol{x}^\top Q\boldsymbol{x} + \boldsymbol{u}^\top R\boldsymbol{u}\} + \boldsymbol{\lambda}^\top\{A\boldsymbol{x} + B\boldsymbol{u}\}. \tag{6.3.3}$$

Applying the first order necessary conditions (6.2.10), we have

$$\frac{d\boldsymbol{x}(t)}{dt} = A(t)\boldsymbol{x}(t) + B(t)\boldsymbol{u}(t) \tag{6.3.4a}$$

$$\boldsymbol{x}(0) = \boldsymbol{x}^0 \tag{6.3.4b}$$

$$\frac{d\boldsymbol{\lambda}(t)}{dt} = -Q(t)\boldsymbol{x}(t) - (A(t))^\top \boldsymbol{\lambda}(t) \tag{6.3.4c}$$

$$\boldsymbol{\lambda}(T) = S^f \boldsymbol{x}(T) \tag{6.3.4d}$$

$$R(t)\boldsymbol{u}(t) + (B(t))^\top \boldsymbol{\lambda}(t) = 0, \tag{6.3.4e}$$

where we have suppressed the superscript * on the optimal control, state and costate for clarity. Since $R(t)$ is positive definite, the last equation (6.3.4e) immediately yields

$$\boldsymbol{u}(t) = -(R(t))^{-1}(B(t))^\top \boldsymbol{\lambda}(t). \tag{6.3.5}$$

(6.3.4a) and (6.3.4c) with $\boldsymbol{u}(t)$ given by (6.3.5) can be written in the form of the following linear homogeneous system of differential equations in $\boldsymbol{x}$ and $\boldsymbol{\lambda}$:

$$\frac{d}{dt}\begin{bmatrix} \boldsymbol{x}(t) \\ \boldsymbol{\lambda}(t) \end{bmatrix} = \begin{bmatrix} A(t) & -B(t)(R(t))^{-1}(B(t))^\top \\ -Q(t) & -(A(t))^\top \end{bmatrix} \begin{bmatrix} \boldsymbol{x}(t) \\ \boldsymbol{\lambda}(t) \end{bmatrix}. \tag{6.3.6}$$

Since the boundary conditions for $\boldsymbol{x}$ and $\boldsymbol{\lambda}$ are prescribed at two different end points, (6.3.6) presents a linear homogeneous TPBVP. Unlike the general nonlinear TPBVP problem, this one can be solved analytically in two ways: the *transition matrix method* or the *backward sweep method* [33].

The transition matrix method assumes the existence of two transition matrices $X(t)$ and $\Lambda(t) \in \mathbb{R}^{n\times n}$ such that

$$\boldsymbol{x}(t) = X(t)\boldsymbol{x}(T) \tag{6.3.7}$$

and

$$\boldsymbol{\lambda}(t) = \Lambda(t)\boldsymbol{x}(T). \tag{6.3.8}$$

It is easy to see that the transition matrices must also satisfy (6.3.6), i.e.,

$$\frac{d}{dt}\begin{bmatrix} X(t) \\ \Lambda(t) \end{bmatrix} = \begin{bmatrix} A(t) & -B(t)(R(t))^{-1}(B(t))^\top \\ -Q(t) & -(A(t)))^\top \end{bmatrix} \begin{bmatrix} X(t) \\ \Lambda(t) \end{bmatrix} \tag{6.3.9}$$

along with the boundary conditions

$$X(T) = I \tag{6.3.10a}$$

and

$$\Lambda(T) = S^f. \tag{6.3.10b}$$

(6.3.9) with the boundary conditions (6.3.10) gives rise to a final value problem, as the system can be integrated backwards in time, starting at the terminal time T. If $X(t)$ is non-singular for all $t \in [0,T]$, then, by (6.3.5), (6.3.7) and (6.3.8), it follows that

$$\boldsymbol{u}(t) = -(R(t))^{-1}(B(t))^\top \Lambda(t)(X(t))^{-1}\boldsymbol{x}(t). \tag{6.3.11}$$

(6.3.11) relates the optimal control at time t to the state at time t and hence it is called a *closed loop* (or *feedback*) control law. It can be written as

$$\boldsymbol{u}(t) = -G(t)\boldsymbol{x}(t) \tag{6.3.12a}$$

with the feedback gain matrix given by

$$G(t) = (R(t))^{-1}(B(t))^\top \Lambda(t)(X(t))^{-1}. \tag{6.3.12b}$$

The feedback gain matrix can be evaluated once the system (6.3.9) is solved with the boundary conditions (6.3.10).

Note that the control law may also be expressed in terms of the initial state, i.e.,

$$\boldsymbol{u}(t) = -\hat{G}(t)\boldsymbol{x}(0), \tag{6.3.13}$$

where

$$\hat{G}(t) = (R(t))^{-1}(B(t))^{\top}\Lambda(t)(X(0))^{-1}. \tag{6.3.14}$$

The proof is left as an exercise.

The transition matrix method is conceptually easy, though difficulty often arises during actual computation of the inverse of $X(t)$. The reader is referred to [33] for further elaboration of the numerical difficulties involved.

The backward sweep method is more popular by virtue of its computational efficiency. It assumes a linear relationship between the state and the costate of the form:

$$\boldsymbol{\lambda}(t) = S(t)\boldsymbol{x}(t). \tag{6.3.15}$$

Direct differentiation of (6.3.15) yields

$$\begin{aligned}\frac{d\boldsymbol{\lambda}(t)}{dt} &= -Q(t)\boldsymbol{x}(t) - (A(t))^{\top}\boldsymbol{\lambda}(t)\\ &= \frac{dS(t)}{dt}\boldsymbol{x}(t) + S(t)\frac{d\boldsymbol{x}(t)}{dt}\\ &= \frac{dS(t)}{dt}\boldsymbol{x}(t) + S(t)A(t)\boldsymbol{x}(t) + S(t)B(t)\boldsymbol{u}(t).\end{aligned} \tag{6.3.16}$$

Substituting (6.3.5) and (6.3.15) into (6.3.16) yields

$$\left[\frac{dS(t)}{dt} + S(t)A(t) + (A(t))^{\top}S(t) + Q(t)\right. \tag{6.3.17}$$

$$\left. - S(t)B(t)(R(t))^{-1}(B(t))^{\top}S(t)\right]\boldsymbol{x}(t) = 0. \tag{6.3.18}$$

Since (6.3.18) must hold for arbitrary $\boldsymbol{x}$, it is necessary that, for all $t \in [0, T]$,

$$\frac{dS(t)}{dt} = -S(t)A(t) - (A(t))^{\top}S(t) - Q(t) + S(t)B(t)(R(t))^{-1}(B(t))^{\top}S(t). \tag{6.3.19a}$$

The boundary condition is obtained from (6.3.15) and (6.3.4d):

$$S(T) = S^f. \tag{6.3.19b}$$

(6.3.19) is a nonlinear nonhomogeneous differential equation known as the *matrix Riccati equation*. Since the boundary condition is specified only at the end point, it can be solved directly by integrating backwards from T to 0. The reader may like to check that $S(t)$ is also symmetric, and hence there are really only $\frac{n}{2}(n+1)$ differential equations to be solved. Once $S(t)$ is determined, a closed loop control law may again be established from (6.3.15) and (6.3.5):

$$\boldsymbol{u}(t) = -G(t)\boldsymbol{x}(t), \tag{6.3.20a}$$

where

$$G(t) = (R(t))^{-1}(B(t))^\top S(t). \tag{6.3.20b}$$

Once $S(t)$ is determined, the initial condition for the costate is given by (6.3.15):

$$\boldsymbol{\lambda}(0) = S(0)\boldsymbol{x}^0. \tag{6.3.21}$$

Hence, the optimal state and costate may be obtained by direct integration of (6.3.6) forward in time starting at $t = 0$. By comparing (6.3.20) with (6.3.11), it is obvious that

$$S(t) = \Lambda(t)(X(t))^{-1}. \tag{6.3.22}$$

This implies that the matrix Riccati equation may be solved alternatively by the procedure given in the transition matrix method, i.e., by solving the $2n^2$ linear differential equations in (6.3.9), computing the inverse of X and then computing $S(t) = \Lambda(t)(X(t))^{-1}$. In principle, it may be easier to solve the linear differential equation. However, there are efficient numerical algorithms [44] for solving the nonlinear Riccati equation directly. The need to solve about four times as many differential equations and to invert an $n \times n$ matrix at each time point t for the transition matrix method is often not warranted.

As an extension to the linear quadratic regulator problem, the *tracking problem* seeks to track a desired reference trajectory $\boldsymbol{r}(t)$ with a linear combination of the states over the interval $[0, T]$. Again, subject to the previous linear system (6.3.2), we have the following cost functional:

$$\begin{aligned} \min \boldsymbol{u}\Big\{ g(\boldsymbol{u}) &= \frac{1}{2}[\boldsymbol{y}(T) - \boldsymbol{r}(T)]^\top S^f [\boldsymbol{y}(T) - \boldsymbol{r}(T)] \\ &+ \frac{1}{2}\int_0^T \{[\boldsymbol{y}(t) - \boldsymbol{r}(t)]^\top Q(t)[\boldsymbol{y}(t) - \boldsymbol{r}(t)] + (\boldsymbol{u}(t))^\top R(t)\boldsymbol{u}(t)\}\, dt \Big\}, \end{aligned} \tag{6.3.23}$$

where

$$\boldsymbol{y}(t) = C(t)\boldsymbol{x}(t) \tag{6.3.24}$$

and

$$S \geq 0, \qquad Q(t) \geq 0, \qquad R(t) > 0 \tag{6.3.25}$$

(i.e., S^f is symmetric and positive semi-definite, $Q(t)$ is symmetric and positive semi-definite for each $t \in [0, T]$, and $R(t)$ is symmetric and positive definite for each $t \in [0, T]$.)

If we assume a linear relationship for the state and costate, i.e.,

$$\boldsymbol{\lambda}(t) = S(t)\boldsymbol{x}(t) + \boldsymbol{\mu}(t) \tag{6.3.26}$$

and following an argument similar to the previous analysis, it can be shown that the optimal closed loop control law is given by

$$\boldsymbol{u}(t) = -\,G(t)\boldsymbol{x}(t) - (R(t))^{-1}(B(t))^\top \boldsymbol{\mu}(t) \tag{6.3.27a}$$

$$G(t) = (R(t))^{-1}(B(t))^\top S(t) \tag{6.3.27b}$$

$$\frac{dS(t)}{dt} = -\,(A(t))^\top S(t) - S(t)A(t) + S(t)B(t)(R(t))^{-1}(B(t))^\top S(t) - (C(t))^\top Q(t)C(t) \tag{6.3.27c}$$

$$S(T) = (C(T))^\top S^f C(T) \tag{6.3.27d}$$

$$\frac{d\boldsymbol{\mu}(t)}{dt} = (-(A(t))^\top + (G(t))^\top (B(t))^\top)\boldsymbol{\mu}(t) + (C(t))^\top Q(t)\boldsymbol{r}(t) \tag{6.3.27e}$$

$$\boldsymbol{\mu}(T) = -\,(C(T))^\top S^f \boldsymbol{r}(T). \tag{6.3.27f}$$

6.4 Pontryagin Maximum Principle

There are already many books written solely on the Pontryagin minimum principle and its applications. In this section, we merely point out some fundamental results and briefly investigate some applications.

The Euler-Lagrange equations for the unconstrained optimal control problem of Section 6.2 require that the Hamiltonian function must be stationary with respect to the control, i.e., $\dfrac{\partial H}{\partial \boldsymbol{u}} = \mathbf{0}$ at optimality.

Consider the case when the control is constrained to lie in a subset U of $\mathbb{R}^r$, where U, known as the *control restraint set*, is generally a compact subset of $\mathbb{R}^r$. In this situation, the optimality conditions obtained in Section 6.2 do not make sense if the optimal control happens to lie on the boundary of U for any positive subinterval of the planning horizon $[0, T]$. To cater for this and more general situations, some fundamental results due to Pontryagin and his co-workers [206] will be stated without proof in the next two theorems.

Let U be a compact subset of $\mathbb{R}^r$. Any piecewise continuous function from $[0, T]$ into U is said to be an admissible control. Let $\mathcal{U}$ be the class of all such admissible controls.

Now we consider the problem where the cost functional (6.2.2) is to be minimized over $\mathcal{U}$ subject to the dynamical system (6.2.1). We refer to this as Problem $(P1)$.

Theorem 6.4.1 *Consider Problem $(P1)$. If $\boldsymbol{u}^* \in \mathcal{U}$ is an optimal control, and $\boldsymbol{x}^*(t)$ and $\boldsymbol{\lambda}^*(t)$ are the corresponding optimal state and costate, then it is necessary that*

$$\frac{d\boldsymbol{x}^*(t)}{dt} = \left[\frac{\partial H(t, \boldsymbol{x}^*(t), \boldsymbol{u}^*(t), \boldsymbol{\lambda}^*(t))}{\partial \boldsymbol{\lambda}}\right]^\top = \boldsymbol{f}(t, \boldsymbol{x}^*(t), \boldsymbol{u}^*(t)), \tag{6.4.1a}$$

$$\boldsymbol{x}^*(0) = \boldsymbol{x}^0, \tag{6.4.1b}$$

$$\frac{d\boldsymbol{\lambda}^*(t)}{dt} = -\left[\frac{\partial H(t, \boldsymbol{x}^*(t), \boldsymbol{u}^*(t), \boldsymbol{\lambda}^*(t))}{\partial \boldsymbol{x}}\right]^\top, \tag{6.4.1c}$$

$$\boldsymbol{\lambda}^*(T) = \left[\frac{\partial \Phi_0(\boldsymbol{x}^*(T))}{\partial \boldsymbol{x}}\right]^\top \tag{6.4.1d}$$

and

$$\min_{\boldsymbol{v} \in U} H(t, \boldsymbol{x}^*(t), \boldsymbol{v}, \boldsymbol{\lambda}^*(t)) = H(t, \boldsymbol{x}^*(t), \boldsymbol{u}^*(t), \boldsymbol{\lambda}^*(t)) \tag{6.4.1e}$$

for all $t \in [0, T]$, *except possibly on a finite subset of* $[0, T]$.

Remark 6.4.1 *Note that the condition* (6.4.1e) *in the above theorem may also be written as*

$$H(t, \boldsymbol{x}^*(t), \boldsymbol{u}^*(t), \boldsymbol{\lambda}^*(t)) \leq H(t, \boldsymbol{x}^*(t), \boldsymbol{v}, \boldsymbol{\lambda}^*(t)) \tag{6.4.2}$$

for all $\boldsymbol{v} \in U$, *and for all* $t \in [0, T]$, *except possibly on a finite subset of* $[0, T]$. *Note furthermore that the necessary condition* (6.4.1e) *(and hence* (6.4.2)*) reduces to the stationary condition* (6.2.10e) *if the Hamiltonian function* H *is continuously differentiable and if* $U = \mathbb{R}^r$.

To motivate the second theorem, we add an additional terminal constraint to the dynamical system (6.2.1) as follows:

$$\boldsymbol{x}(T) = \boldsymbol{x}^f, \tag{6.4.3}$$

where $\boldsymbol{x}^f$ is a given vector in $\mathbb{R}^n$. Our second problem may now be stated as: Subject to the dynamical system (6.2.1) together with the terminal condition (6.4.3), find a control $\boldsymbol{u} \in \mathcal{U}$ such that the cost functional (6.2.2) is minimized over $\mathcal{U}$.
For convenience, let this second optimal control problem be referred to as Problem ($P2$).

Theorem 6.4.2 *Consider Problem* ($P2$). *If* $\boldsymbol{u}^* \in \mathcal{U}$ *is an optimal control, and* $\boldsymbol{x}^*(t)$ *and* $\boldsymbol{\lambda}(t)$ *are the corresponding optimal state and costate, then it is necessary that*

$$\frac{d\boldsymbol{x}^*(t)}{dt} = \left[\frac{\partial H(t, \boldsymbol{x}^*(t), \boldsymbol{u}^*(t), \boldsymbol{\lambda}^*(t))}{\partial \boldsymbol{\lambda}}\right]^\top = \boldsymbol{f}(t, \boldsymbol{x}^*(t), \boldsymbol{u}^*(t)), \tag{6.4.4a}$$

$$\boldsymbol{x}^*(0) = \boldsymbol{x}^0, \tag{6.4.4b}$$

$$\boldsymbol{x}^*(T) = \boldsymbol{x}^f, \tag{6.4.4c}$$

$$\frac{d\boldsymbol{\lambda}^*(t)}{dt} = -\left[\frac{\partial H(t, \boldsymbol{x}^*(t), \boldsymbol{u}^*(t), \boldsymbol{\lambda}^*(t))}{\partial \boldsymbol{x}}\right]^\top \tag{6.4.4d}$$

and

$$\min_{\boldsymbol{v}\in U} H(t,\boldsymbol{x}^*(t),\boldsymbol{v},\boldsymbol{\lambda}^*(t)) = H(t,\boldsymbol{x}^*(t),\boldsymbol{u}^*(t),\boldsymbol{\lambda}^*(t)) \tag{6.4.4e}$$

for all $t \in [0,T]$, *except possibly on a finite subset of* $[0,T]$.

Remark 6.4.2 *As noted in Remark 6.4.1, the condition* (6.4.4e) *in Theorem 6.4.2 is equivalent to*

$$H(t,\boldsymbol{x}^*(t),\boldsymbol{u}^*(t),\boldsymbol{\lambda}^*(t)) \le H(t,\boldsymbol{x}^*(t),\boldsymbol{v},\boldsymbol{\lambda}^*(t)) \tag{6.4.5}$$

for all $\boldsymbol{v} \in U$, *and for all* $t \in [0,T]$, *except possibly on a finite subset of* $[0,T]$.

To illustrate the applicability of the Pontryagin Maximum Principle, we consider some examples.

Example 6.4.1 The original problem that motivates this example is due to Thompson [262]. Suppose the quality state of a machine can be measured by $x(t)$. The deterioration of the state is governed by the first order differential equation:

$$\frac{dx(t)}{dt} = -bx(t) + u(t), \; t \in [0,T] \tag{6.4.6a}$$

$$x(0) = x_0, \tag{6.4.6b}$$

where b is the natural rate of deterioration and $u(t)$ is the non-dimensional maintenance effort that serves to retard the deterioration and is subjected to the maintenance budget constraint

$$0 \le u(t) \le \bar{u}, \text{ for all } t \in [0,T]. \tag{6.4.7}$$

The optimal maintenance policy thus seeks to maximize the net discounted payoff (i.e., productivity benefit minus maintenance cost) plus the salvage value, assuming that the productivity is proportional to the machine quality, i.e.,

$$\max \left\{ g_0(u) = e^{-rT} S\, x(T) + \int_0^T e^{-rt}[px(t) - u(t)]dt \right\}, \tag{6.4.8}$$

where r, S, p, and T are, respectively, the interest rate, salvage value per unit terminal quality, the productivity per unit quality and the sale date of the machine.

Before applying the Pontryagin Maximum Principle presented in Theorem 6.4.1 to this problem, we first note that

$$\max \{g_0(u)\} = -\min \{-g_0(u)\}.$$

Thus, it is clear that the objective functional (6.4.8) is equivalent to

$$\min \left\{ -g_0(u) = -e^{-rT} S\ x(T) - \int_0^T e^{-rt}[px(t) - u(t)]dt \right\}. \tag{6.4.9}$$

We now write down the corresponding Hamiltonian function for the optimal control problem with (6.4.9) as the objective functional:

$$H(t, x, u, \lambda) = e^{-rt}(u - px) + \lambda(-bx + u). \tag{6.4.10}$$

The corresponding costate equation is

$$\frac{d\lambda^*(t)}{dt} = -\frac{\partial H}{\partial x} = b\lambda^*(t) + pe^{-rt}, \tag{6.4.11a}$$

with the boundary condition

$$\lambda^*(T) = \frac{\partial \left\{ -e^{-rT} S\ x(T) \right\}}{\partial x(T)} = -Se^{-rT}. \tag{6.4.11b}$$

The Pontryagin Maximum Principle asserts that

$$\begin{aligned} &H(t, x^*(t), u^*(t), \lambda^*(t)) \\ &= \min_{0 \le v \le \bar{u}} \left\{ v(\lambda^*(t) + e^{-rt}) - px^*(t)e^{-rt} - b\lambda^*(t)x^*(t) \right\}. \end{aligned} \tag{6.4.12}$$

The solution of the costate equation (6.4.11) is

$$\lambda^*(t) = \left(-S + \frac{p}{r+b} \right) e^{-(b+r)T+bt} - \frac{p}{r+b} e^{-rt}. \tag{6.4.13}$$

Assume that $\frac{p}{r+b} > 1 > S$. Then, $\lambda^*(t)$ grows from a negative value in a monotonically increasing manner. In (6.4.12), the minimization with respect to v involves only the first term in the curly brackets of (6.4.12). If $\lambda^*(t)+e^{-rt}$ is negative, v should be as large as possible, and if $\lambda^*(t) + e^{-rt}$ is positive, v should be as small as possible. The resulting optimal maintenance policy is therefore of the bang-bang type, i.e.,

$$u^*(t) = \begin{cases} \bar{u}, & 0 \le t \le t^* \\ 0, & t^* < t \le T, \end{cases} \tag{6.4.14}$$

where t^* is the time when $\lambda^*(t) + e^{-rt}$ switches from being negative to being positive. Since both λ^* and e^{-rt} are monotonic, there can only be one such switching point. By solving for the zero of $\{\lambda^*(t) + e^{-rt}\}$, we obtain

$$t^* = T + \frac{1}{b+r} \ln \left[\frac{\frac{p}{r+b} - 1}{\frac{p}{r+b} - S} \right]. \tag{6.4.15}$$

Example 6.4.2 Recall the student problem discussed as Example 1.2.1 in Chapter 1. In the notation of this chapter, we can state the problem in the following standard form:

$$\text{Minimize} \quad g(u) = \int_0^T u(t)\,dt$$

subject to

$$\begin{aligned} \frac{dx(t)}{dt} =& bu(t) - cx(t), \\ x(0) =& k_0, \\ x(T) =& k_T, \end{aligned}$$

and $0 \leq u(t) \leq \bar{w}$ for all $0 \leq t < T$. Here, $u(t)$ denotes the rate of work done at time t, $x(t)$ is the knowledge level at time t and T is the total time in weeks. Furthermore, $c > 0$, $b > 0$, $\bar{w}$, k_0, and k_T are constants as described in Example 1.2.1. We assume that $k_0 < k_T$ (i.e., the student's initial knowledge level is insufficient to pass the examination) and also that $\bar{w}$ is sufficiently large so that the final knowledge level can be reached (otherwise the problem would be infeasible). The Hamiltonian is given by

$$H = u + \lambda(bu - cx) = u(1 + \lambda b) - c\lambda x.$$

The optimal costate must satisfy

$$\frac{d\lambda^*(t)}{dt} = -\frac{\partial H}{\partial x} = c\lambda^*(t).$$

This yields $\lambda^*(t) = Ke^{ct}$ for some constant K, which is a strictly monotone function (note that $K = 0$ would lead to $u^*(t) = 0$ for all t, which would lead to a loss of knowledge and the student unable to reach the final knowledge level). Since Pontryagin's Maximum Principle requires the minimization of H with respect to u, we must have

$$u^*(t) = \begin{cases} \bar{w}, \text{ if } 1 + \lambda^*(t)b < 0, \\ 0, \ \text{ if } 1 + \lambda^*(t)b > 0. \end{cases}$$

Now, $K > 0$ would result in $\lambda^*(t) > 0$ for all t, which, in turn, leads to $1 + \lambda^*(t)b > 0$ for all t. As this forces $u^*(t) = 0$ for all t, the student could again not reach the required final knowledge level. Hence we must have $K < 0$, which means that $\lambda^*(t) < 0$ for all t and it is monotonically decreasing. If $1 + \lambda^*(0)b < 0$, it follows that $1 + \lambda^*(t)b < 0$ for all t, which means that $u^*(t) = \bar{w}$ for all t. The more likely scenario is that $1 + \lambda^*(0)b > 0$ and $1 + \lambda^*(t)b$ then decreases, becoming negative after a time t^*. This means the optimal control is of bang-bang type with

$$u^*(t) = \begin{cases} 0, & \text{if } 0 \le t < t^*, \\ \bar{w}, & \text{if } t^* \le t \le T. \end{cases}$$

We can now derive the complete solution. For $0 \le t < t^*$, $\dfrac{dx(t)}{dt} = -cx(t)$. Together with the initial condition, this results in

$$x^*(t) = k_0 e^{-ct}, \quad 0 \le t < t^*.$$

For $t^* \le t \le T$, we have $\dfrac{dx(t)}{dt} = b\bar{w} - cx(t)$. Together with the terminal state constraint, this results in

$$x^*(t) = \frac{b\bar{w}}{c} + \left(k_T - \frac{b\bar{w}}{c}\right) e^{c(T-t)}, \quad t^* \le t \le T.$$

By equating the two forms of the optimal state trajectory at t^*, we can derive

$$t^* = \frac{\ln\left[\frac{ck_0}{b\bar{w}} - \frac{ck_T}{b\bar{w}} e^{cT} + e^{cT}\right]}{c}.$$

To conclude this section, we wish to note that there exist several versions of the proof for the Pontryagin Maximum Principle. What appears in Pontryagin's original book [206] is somewhat complex. For the proof of a simplified version, we refer the reader to [3].

6.5 Singular Control

In this section, we wish to consider a situation where the Pontryagin minimum principle fails to determine a unique value of the optimal control. This leads to what is known as a *singular control.* Let us illustrate this situation by means of the following simple example. For more details, see [40, 69, 72, 73] and the references cited therein.

Assume that there are two existing highways. We wish to construct a new highway to link point A on the first highway and point B on the second highway as shown in Figure 6.5.1. Let L be the horizontal distance between point A and point B, and let $y_e(\ell)$, $0 < \ell < L$, be the existing terrain on which a new highway is to be built. Let the profile of this highway be denoted by $y(\ell)$, $0 < \ell < L$. The vertical height from A to $y_e(0)$ is denoted by ξ, and the vertical height from point B to $y_e(L)$ is denoted by η. Clearly, the new highway will be of no practical value if it is too steep for cars to drive on. Thus, we must make sure that the slope of the new highway be within the allowable limits, as given by

$$-S_1 \le \frac{dy(\ell)}{d\ell} \le S_2,\ 0 \le \ell \le L, \tag{6.5.1}$$

where S_1 and S_2 are the given positive constants.

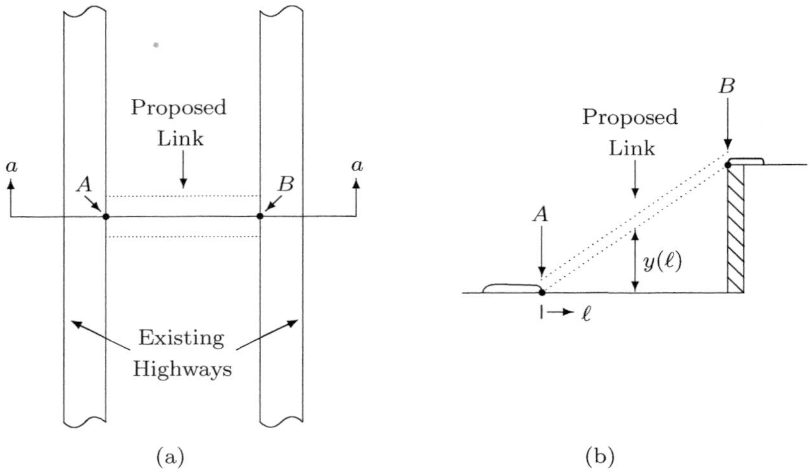

Fig. 6.5.1: A hypothetical case of highway construction. (**a**) Plan view. (**b**) Section $a - a$

Our aim is to accomplish this job with minimum cost, where the total cost is the sum of the costs of cutting and filling of earth. If the costs of cutting and filling are the same, then the minimum cost is, in fact, equivalent to the minimum filling and cutting of earth.

We are now in a position to write down the mathematical model of this optimal control problem. To simplify the presentation, we shall assume that

$$L = 2,\ \xi = 0,\ \eta = 1,\ S_1 = S_2 = 1,\ y_e(\ell) = 0 \text{ for } 0 \le \ell \le 2. \tag{6.5.2}$$

With these specifications, the corresponding version of the model is

$$\min \left\{ g_0(u) = \int_0^2 (y(\ell))^2 d\ell \right\} \tag{6.5.3a}$$

subject to

$$\frac{dy(\ell)}{d\ell} = u(\ell),\ 0 \le \ell \le 2, \tag{6.5.3b}$$

$$y(0) = 0 \tag{6.5.3c}$$

$$y(2) = 1 \tag{6.5.3d}$$

and

$$-1 \leq u(\ell) \leq 1, \ 0 \leq l \leq 2. \tag{6.5.3e}$$

According to Theorem 6.4.2, the corresponding Hamiltonian function is

$$H = y^2 + u\lambda \tag{6.5.4}$$

and the costate system is

$$\frac{d\lambda(\ell)}{d\ell} = -\frac{\partial H}{\partial y} = -2y(\ell). \tag{6.5.5}$$

Minimizing H with respect to u, the optimal control takes the form

$$u^*(\ell) = \begin{cases} 1, & \text{if } \lambda(\ell) < 0, \\ -1, & \text{if } \lambda(\ell) > 0, \\ \text{undetermined}, & \text{if } \lambda(\ell) = 0. \end{cases} \tag{6.5.6}$$

Let us assume that we can ignore the third case of (6.5.6) for the time being. Two possibilities remain, namely $u^*(\ell) = 1$ for $\lambda(\ell) < 0$ and $u^*(\ell) = -1$ for $\lambda(\ell) > 0$.

Let us explore each of these cases. We integrate the differential equations (6.5.3b) and (6.5.5) in turn and we let K_1 , K_2 , K_3 and K_4 be the constants of integration that arise in this process.

(i) $\lambda(\ell) < 0$:
In this case, $u^*(\ell) = 1$ and $\dfrac{dy(\ell)}{d\ell} = 1$. Hence,

$$y(\ell) = \ell + K_1. \tag{6.5.7}$$

By (6.5.5), we obtain

$$\lambda(\ell) = -(\ell)^2 - 2K_1\ell + K_2 < 0. \tag{6.5.8}$$

(ii) $\lambda(\ell) > 0$:
In this case, $u^*(\ell) = -1$ and $\dfrac{dy(\ell)}{d\ell} = -1$. Hence,

$$y(\ell) = -\ell + K_3. \tag{6.5.9}$$

By (6.5.5), we have

$$\lambda(\ell) = (\ell)^2 - 2K_3\ell + K_4 > 0. \tag{6.5.10}$$

Let us plot the relationship between y and λ for these cases by first eliminating the variable ℓ. Consider Case (i). Then, from (6.5.7), we have

$$\ell = y - K_1.$$

Hence, for $\lambda < 0$,

$$\lambda = -(y - K_1)^2 - 2K_1(y - K_1) + K_2 = -y^2 + (K_2 + (K_1)^2) < 0. \qquad (6.5.11)$$

Now consider Case (ii). Then, by a similar argument, it follows from (6.5.9) that

$$\ell = K_3 - y.$$

Hence, for $\lambda > 0$,

$$\lambda = (K_3 - y)^2 - 2K_3(K_3 - y) + K_4 = y^2 + (K_4 - (K_3)^2) > 0. \qquad (6.5.12)$$

These are the desired relationships between y and λ. They are plotted on the phase diagram of Figure 6.5.2. In the top half-plane of the figure, λ is positive,

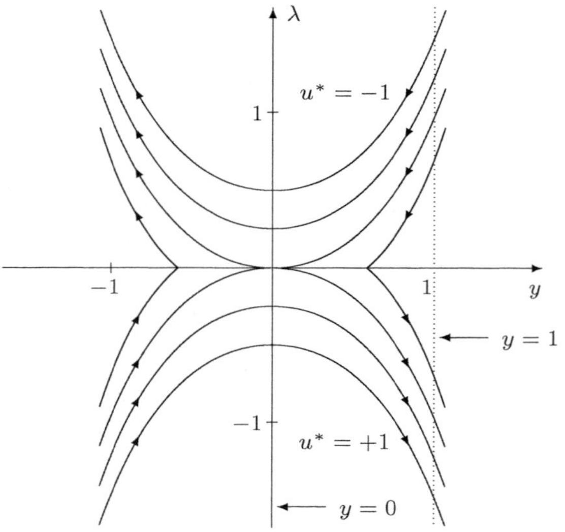

Fig. 6.5.2: Phase orbits in the y-λ plane

$u^*(\ell) = -1$, and (6.5.12) applies. In the bottom half-plane, λ is negative, $u^*(\ell) = 1$, and (6.5.11) applies. The curves are parabolas, facing upwards for (6.5.12) and downwards for (6.5.11). Any upwards-facing parabola that reaches the y-axis is interrupted at this point and replaced by a suitable segment of a downwards-facing parabola below the axis. Directions of travel along the phase orbits are indicated with arrows. At any point $(y(\ell), \lambda(\ell))$ in the phase plane, the travel direction is obtained from (6.5.5). If $y(\ell)$ is positive, $\lambda(\ell)$ decreases with ℓ. Otherwise, $\lambda(\ell)$ increases with ℓ.

The phase orbits shown on the figure are the only possible candidates for optimal solutions. However, not all these orbits lead to feasible solutions, since they do not all meet the *boundary conditions*, i.e., (6.5.3c), $y(0) = 0$, and (6.5.3d), $y(2) = 1$. In terms of the phase diagram, a feasible solution must start on the λ-axis ($y(0) = 0$ when $\ell = 0$), and it must end on the vertical line $y = 1$ when $\ell = 2$.

Inspecting the phase diagram carefully, we see that points on the λ-axis with $\lambda > 0$ are no use. We can never go to the desired final value of $y = 1$ from such initial points. Thus, we must start from or below the λ-axis. This, in turn, implies that we must use the control $u^*(\ell) = 1$ if it is to be an optimal control. Thus, (6.5.11) applies throughout. But, from (6.5.3c) and (6.5.7), we have

$$y(0) = 0 + K_1 = 0.$$

Thus, $y(\ell) = \ell$ and hence $y(2) = 2$. This clearly does not satisfy the final condition (6.5.3d) (i.e., $y(2) = 1$).

Where does this leave us? If we re-examine (6.5.6), there are, in fact, three possibilities. The second is not feasible in any case, and the exclusive use of the first has been shown to also not give a feasible answer. We are therefore required to consider the third possibility, $\lambda = 0$ and u^* undetermined, (6.5.6). This situation leads to a *singular control*, because *the Maximum Principle* (6.4.4e) fails to directly determine a unique value of the optimal control.

Let us investigate this singular case further. Suppose that $\lambda = 0$ not merely at a single point in $[0, 2]$, but in some finite open interval, say (ℓ_1, ℓ_2). Then, the derivative $d\lambda(\ell)/d\ell$ must also vanish and, by (6.5.5), we obtain $y(\ell) = 0$ in that same interval. But then the derivative $dy(\ell)/d\ell = 0$ as well and, by (6.5.3b), we get $u^*(\ell) = 0$. The unique singular control solution is therefore given by

$$y(\ell) = u^*(\ell) = \lambda(\ell) = 0. \tag{6.5.13}$$

On the phase diagram presented in Figure 6.5.2, this solution corresponds to the *origin*, i.e., a single point in the plane rather than a curve. There are other points in the plane with $y = 0$, namely all points on the λ-axis. But none of these others can be maintained so that the differential equations (6.5.3b) and (6.5.5) are satisfied over some finite interval, since (6.5.5) implies that $\lambda(\ell)$ cannot remain constant, in particular equal to zero, unless $2y$ vanishes.

The desired final optimal solution now emerges in two parts. From $\ell = 0$ until some distance $\widehat{\ell}$, the new link should follow the existing terrain. We start to fill the existing terrain from $\widehat{\ell}$ onwards in such a way that the slope at each point ℓ is always at its maximum allowable limit. In this way, the desired height at $\ell = 2$, $y(2) = 1$, will be reached exactly. The remaining question is to determine the value of $\widehat{\ell}$. First, we recall that

$$y(2) = 1 \tag{6.5.14}$$

$$y(\ell) = 0, \quad 0 \leq \ell \leq \widehat{\ell}, \tag{6.5.15}$$

and

$$y(\ell) = \ell + K_1, \quad \widehat{\ell} \le \ell \le 2. \tag{6.5.16}$$

Now, from (6.5.14) and (6.5.16), we obtain

$$K_1 = -1. \tag{6.5.17}$$

Thus, by (6.5.16), we have

$$y(\ell) = \ell - 1, \ \widehat{\ell} \le \ell \le 2. \tag{6.5.18}$$

Since $y(\widehat{\ell}) = 0$, by (6.5.15), it follows from (6.5.18) that $0 = \widehat{\ell} - 1$ and therefore $\widehat{\ell} = 1$. Finally, we can conclude that if the new highway is to be built with minimum cost, its slope must be

$$u^*(\ell) = \begin{cases} 0, \ 0 \le \ell \le 1 \\ 1, \ 1 \le \ell \le 2. \end{cases} \tag{6.5.19}$$

Hence, the corresponding profile of the new highway is

$$y^*(\ell) = \begin{cases} 0, \quad 0 \le \ell \le 1 \\ \ell - 1, \ 1 \le \ell \le 2. \end{cases} \tag{6.5.20}$$

This clearly agrees with our intuition, although we can quite easily imagine how uncomfortable it would be to drive on such a highway—the slope of the highway is at its maximum allowable limit from the point $\ell = 1$ onwards. This is due to the fact that we did not take driving comfort into account in the problem formulation.

Note that the singular control problem considered in this section is only a very simple one. For more information regarding singular control, we refer the interested reader to [40, 69, 72, 73]. Note also that computational algorithms to be introduced in subsequent chapters work equally well regardless of whether the optimal solution to be obtained is a singular one or otherwise.

6.6 Time Optimal Control

This section is devoted to a class of time optimal control problems. To begin, we assume that any piecewise continuous function from $[0, \infty)$ into U may be taken as an admissible control, where U is the control restraint set defined in Section 6.4. Let $\mathcal{U}$ be the class of all such admissible controls.

We may now state the class of time optimal control problems formally as: Subject to the system (6.2.1) together with the final condition:

$$\boldsymbol{x}(T) = \boldsymbol{x}^f, \tag{6.6.1}$$

find a control $\boldsymbol{u} \in \mathcal{U}$ such that T is minimized, where $\boldsymbol{x}^f$ is a given vector in $\mathbb{R}^n$.

For convenience, let this time optimal control problem be referred to as Problem (TP). Note that the cost functional of Problem (TP) can be written as

$$\min \int_0^T dt. \tag{6.6.2}$$

With this cost functional, the corresponding Hamiltonian function is given by

$$H(t, \boldsymbol{x}, \boldsymbol{u}, \boldsymbol{\lambda}) = 1 + (\boldsymbol{\lambda})^\top \boldsymbol{f}(t, \boldsymbol{x}, \boldsymbol{u}). \tag{6.6.3}$$

The corresponding version of the Pontryagin Maximum Principle is now stated without proof in the following theorem.

Theorem 6.6.1 *Consider Problem* (TP). *Let* $\boldsymbol{u}^* \in \mathcal{U}$ *be an optimal control, let* $\boldsymbol{x}^*$ *be the corresponding optimal state, and let* T^* *be the minimum time such that the final condition* (6.6.1) *is satisfied. Then, there exists a function*

$$\boldsymbol{\lambda}^* = [\lambda_1^*, \ldots, \lambda_n^*]^\top : [0, T^*] \to \mathbb{R}^n,$$

which is not identically zero such that

$$\frac{d\boldsymbol{x}^*(t)}{dt} = \left[\frac{\partial H(t, \boldsymbol{x}^*(t), \boldsymbol{u}^*(t), \boldsymbol{\lambda}^*(t))}{\partial \boldsymbol{\lambda}}\right]^\top = \boldsymbol{f}(t, \boldsymbol{x}^*(t), \boldsymbol{u}^*(t)) \tag{6.6.4a}$$

$$\boldsymbol{x}^*(0) = \boldsymbol{x}^0 \tag{6.6.4b}$$

$$\boldsymbol{x}^*(T^*) = \boldsymbol{x}^f \tag{6.6.4c}$$

$$\frac{d\boldsymbol{\lambda}^*(t)}{dt} = -\left[\frac{\partial H(t, \boldsymbol{x}^*(t), \boldsymbol{u}^*(t), \boldsymbol{\lambda}^*(t))}{\partial \boldsymbol{x}}\right]^\top \tag{6.6.4d}$$

and

$$\min_{\boldsymbol{v} \in U} H(t, \boldsymbol{x}^*(t), \boldsymbol{v}, \boldsymbol{\lambda}^*(t)) = H(t, \boldsymbol{x}^*(t), \boldsymbol{u}^*(t), \boldsymbol{\lambda}^*(t)) \tag{6.6.4e}$$

for all $t \in [0, T^*]$, *except possibly on a finite subset of* $[0, T^*]$. *Also,*

$$H(T^*, \boldsymbol{x}^*(T^*), \boldsymbol{u}^*(T^*), \boldsymbol{\lambda}^*(T^*)) = 0. \tag{6.6.4f}$$

Furthermore, if the system (6.2.1a) *is autonomous (i.e.,* $\boldsymbol{f}$ *does not depend on* t *explicitly), then*

$$H(t, \boldsymbol{x}^*(t), \boldsymbol{u}^*(t), \boldsymbol{\lambda}^*(t)) = 0, \ \textit{for all } t \in [0, T^*]. \tag{6.6.4g}$$

The proof of this theorem is rather involved. Since the emphasis of this text is on computational methods for optimal control, we omit the proof and refer the interested reader to [3]. We illustrate the application of the theorem with the following example.

Example 6.6.1 Consider the motion of a particle of unit mass governed by the following differential equation:

$$\frac{d^2x(t)}{dt^2} = u(t)$$

with the following boundary conditions

$$x(0) = x_1^0, \quad \frac{dx(0)}{dt} = x_2^0$$

$$x(T) = 0, \quad \frac{dx(T)}{dt} = 0.$$

Our aim is to find a control u with

$$|u(t)| \leq 1, \text{ for all } t \geq 0$$

such that T is minimized.

Define $x_1 = x$ and $x_2 = \dot{x}$. Then, the problem may be re-stated as

$$\min \int_0^T dt$$

subject to

$$\begin{aligned}\frac{dx_1(t)}{dt} =& x_2(t)\\ \frac{dx_2(t)}{dt} =& u(t)\end{aligned}$$

with the initial conditions

$$x_1(0) = x_1^0, \quad x_2(0) = x_2^0,$$

the terminal state constraints

$$x_1(T) = 0, \quad x_2(T) = 0,$$

and the control constraints

$$| \ u(t)| \leq 1, \text{ for all } t \geq 0.$$

The Hamiltonian function is $H(\boldsymbol{x}, u, \boldsymbol{\lambda}) = 1 + x_2\lambda_1 + u\lambda_2$ and the costate system is

$$\frac{d\lambda_1(t)}{dt} = -\frac{\partial H}{\partial x_1} = 0$$

$$\frac{d\lambda_2(t)}{dt} = -\frac{\partial H}{\partial x_2} = -\lambda_1(t).$$

Minimizing the Hamiltonian function with respect to u, we obtain

$$u^*(t) = \begin{cases} -1, & \text{if } \lambda_2(t) > 0, \\ 1, & \text{if } \lambda_2(t) < 0, \\ \text{undetermined}, & \text{if } \lambda_2(t) = 0. \end{cases}$$

The costate system can be easily solved to yield

$$\lambda_1(t) = C$$
$$\lambda_2(t) = -Ct + D,$$

where C and D are arbitrary constants. From the linearity of $\lambda_2(t)$, it follows that $\lambda_2(t)$ either changes sign at most once or is identically zero over the time horizon. However, we can rule out the latter case, since this would require $C = D = 0$, leading to $\lambda_1(t) = 0$, for all $t \geq 0$, and hence

$$H(\boldsymbol{x}(t), u^*(t), \boldsymbol{\lambda}(t)) = 1,$$

which contradicts the condition $H(\boldsymbol{x}(t), u^*(t), \boldsymbol{\lambda}(t)) = 0$. We conclude therefore that $\lambda_2(t)$ can change sign at most once. This, in turn, implies that $u^*(t)$ changes sign at most once. In other words, $u^*(t)$ is either $+1$ or -1 with a possible switch at some time $\hat{t}$. This type of control is called a *bang-bang control.*

If $u^*(t) = +1$, $\frac{dx_2(t)}{dt} = 1 \Rightarrow x_2(t) = t + A$, where A is an arbitrary constant. Clearly,

$$\frac{1}{2}(x_2)^2 = \frac{1}{2}t^2 + At + \frac{1}{2}A^2$$

and

$$\frac{dx_1(t)}{dt} = x_2(t) = t + A \Rightarrow x_1(t) = \frac{1}{2}t^2 + At + B,$$

where B is also arbitrary. Eliminating t from the last two equations, we obtain

$$x_1 = \frac{1}{2}(x_2)^2 + \text{constant}.$$

This describes a family of parabolas shown as solid curves in Figure 6.6.1. Possible movements along these phase trajectories are indicated by arrows and can be deduced from $\frac{dx_2(t)}{dt} = 1$.

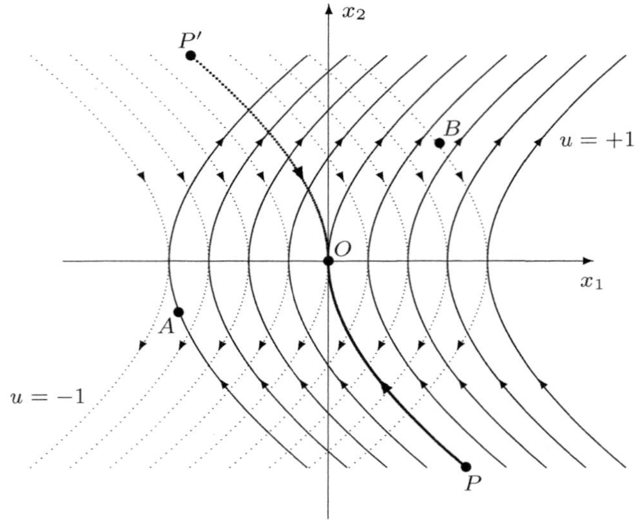

Fig. 6.6.1: Bang-bang optimal control

If $u^*(t) = -1$, $\frac{dx_2(t)}{dt} = -1 \Rightarrow x_2(t) = -t + A'$, where A' is once more an arbitrary constant. Clearly,

$$\frac{1}{2}(x_2)^2 = \frac{1}{2}t^2 - A't + \frac{1}{2}(A')^2$$

and

$$\frac{dx_1(t)}{dt} = x_2(t) = -t + A' \Longrightarrow x_1(t) = -\frac{1}{2}t^2 + A't + B',$$

where B' is also arbitrary. Eliminating t from these two equations, we obtain

$$x_1 = -\frac{1}{2}(x_2)^2 + \text{constant}.$$

This is another family of parabolas indicated by the broken lines in Figure 6.6.1. Again, possible movements along these phase trajectories follow from $\frac{dx_2(t)}{dt} = -1$ and are indicated by arrows.

Any potential optimal trajectory must lie on one of the parabolas shown in Figure 6.6.1. It is clear that there are only two curves that will take us to the origin. The combination of these two curves is the curve marked by POP', and it plays a special role in the solution of the problem. Consider two distinct starting points, A and B, as indicated.

Case 1: Suppose we are starting at A. In this situation, an optimal trajectory must follow the solid parabola up until it reaches the curve POP' and it then follows this curve to the origin. Noting the control values that give rise to the two sections of this optimal trajectory, the time optimal control is clearly given by

$$u^*(t) = \begin{cases} 1, & \text{for } t \in [0, \hat{t}), \\ -1, & \text{for } t \in [\hat{t}, T^*], \end{cases}$$

where $\hat{t}$ is the time at which POP' is reached, and T^* is the time at which the trajectory reaches the origin.

Case 2: Suppose we are starting at B. In this situation, an optimal trajectory must follow the broken parabola down until it reaches POP' and it then follows this curve to the origin. In this case, the time optimal control is clearly given by

$$u^*(t) = \begin{cases} -1, & \text{for } t \in [0, \hat{t}), \\ 1, & \text{for } t \in [\hat{t}, T^*], \end{cases}$$

where $\hat{t}$ and T^* have the same meaning as in Case 1.

POP' is known as the *switching curve*, since the optimal control changes sign once the corresponding optimal trajectory reaches this curve. Note that it does not matter where we start initially. For any starting point in the phase plane, we must travel directly to the switching curve and then stay on it to the origin.

Remark 6.6.1 *Note that the control law we end up with is in closed loop form, since we are able to specify $u^*(t)$ entirely in terms of the current state of the system. The law can be stated simply as follows: take the path that leads onto the switching curve, and then ride the switching curve home.*

6.7 Continuous State Constraints

In this section, we extend the Maximum Principle to a more general class of problems also involving continuous inequality constraints on a function of the state. The main reference for this section is [88]. Consider the constraints

$$\boldsymbol{h}(t, \boldsymbol{x}(t)) \leq \boldsymbol{0}, \text{ for all } t \in [0, T], \tag{6.7.1}$$

$$\boldsymbol{a}(T, \boldsymbol{x}(T)) \leq \boldsymbol{0}, \tag{6.7.2}$$

and

$$\boldsymbol{b}(T, \boldsymbol{x}(T)) = 0, \tag{6.7.3}$$

where $\boldsymbol{h} = [h_1, \ldots, h_q]^\top$, $\boldsymbol{a} = [a_1, \ldots, a_r]^\top$, and $\boldsymbol{b} = [b_1, \ldots, b_s]^\top$ are assumed to be continuously differentiable with respect to all their arguments. In addition, higher order differentiability of $\boldsymbol{h}$ may be required as detailed later.

Instead of the class of admissible controls defined in Section 6.4, we assume here that the control is governed by the more general constraint

$$\boldsymbol{g}(t, \boldsymbol{u}(t)) \leq \mathbf{0}, \text{ for all } t \in [0, T], \tag{6.7.4}$$

where $\boldsymbol{g} = [g_1, \ldots, g_c]^\top$ is continuously differentiable in all its arguments. We then consider the problem of finding a measurable control $\boldsymbol{u}(\cdot)$ that minimizes the cost functional (6.2.2) subject to the dynamical system (6.2.1) and subject to the constraints (6.7.1)–(6.7.4). We refer to this state constrained optimal control problem as Problem (SP).

Throughout this section, we make the following assumptions.

Assumption 6.7.1 *For all possible values of T and $\boldsymbol{x}(T)$, the terminal state constraints* (6.7.2)–(6.7.3) *are such that the following condition is satisfied.*

$$\text{rank} \begin{bmatrix} \partial \boldsymbol{a}/\partial \boldsymbol{x} & \text{diag}(\boldsymbol{a}) \\ \partial \boldsymbol{b}/\partial \boldsymbol{x} & 0 \end{bmatrix} = r + s, \tag{6.7.5}$$

where $\text{diag}(\boldsymbol{a})$ *denotes the $r \times r$ diagonal matrix with $a_1, \ldots, a_r$ along the main diagonal.* (6.7.5) *is known as the constraint qualification for the constraints* (6.7.2)–(6.7.3). *It ensures that the gradients of the equality and active inequality constraints with respect to $\boldsymbol{x}$ are linearly independent.*

Assumption 6.7.2 *For all possible $\boldsymbol{u}(t)$ and t,*

$$\text{rank}\, [\partial \boldsymbol{g}/\partial \boldsymbol{u} \;\; \text{diag}(\boldsymbol{g})] = c, \tag{6.7.6}$$

where $\text{diag}(\boldsymbol{g})$ *denotes the $c \times c$ diagonal matrix with $g_1, \ldots, g_c$ along the main diagonal. The constraint qualification* (6.7.6) *means that the gradients with respect to $\boldsymbol{u}$ of all active components of* (6.7.4) *must be linearly independent.*

Before we can state a similar constraint qualification assumption for (6.7.1), we need to introduce some additional concepts. For each h_i, $i = 1, \ldots, q$, define

$$\begin{aligned}
h_i^0(t, \boldsymbol{x}, \boldsymbol{u}) &= h_i(t, \boldsymbol{x}), \\
h_i^1(t, \boldsymbol{x}, \boldsymbol{u}) &= \frac{dh_i^0}{dt} = \frac{\partial h_i}{\partial \boldsymbol{x}} \boldsymbol{f}(t, \boldsymbol{x}, \boldsymbol{u}) + \frac{\partial h_i}{\partial t} \\
h_i^2(t, \boldsymbol{x}, \boldsymbol{u}) &= \frac{dh_i^1}{dt} = \frac{\partial h_i^1}{\partial \boldsymbol{x}} \boldsymbol{f}(t, \boldsymbol{x}, \boldsymbol{u}) + \frac{\partial h_i^1}{\partial t} \\
&\vdots \\
h_i^p(t, \boldsymbol{x}, \boldsymbol{u}) &= \frac{dh_i^{p-1}}{dt} = \frac{\partial h_i^{p-1}}{\partial \boldsymbol{x}} \boldsymbol{f}(t, \boldsymbol{x}, \boldsymbol{u}) + \frac{\partial h_i^{p-1}}{\partial t}
\end{aligned} \tag{6.7.7}$$

for any integer $p > 0$. For each $i = 1, \ldots, q$, let p_i be such that

$$\frac{\partial h_i^j}{\partial \boldsymbol{u}} = \mathbf{0} \text{ for all } 0 \leq j \leq p_i - 1 \text{ and } \frac{\partial h_i^{p_i}}{\partial \boldsymbol{u}} \neq \mathbf{0}. \tag{6.7.8}$$

Then we say that the state constraint $h_i \leq 0$ *is of order* p_i. Note that many practical problems have state constraints of order 1 only.

Consider the state constraint $h_i \leq 0$ for a particular $i \in [1, \ldots, q]$. We say that a subinterval $(t_1, t_2) \subset [0, T]$ with $t_1 < t_2$ is an *interior interval* of the trajectory $\boldsymbol{x}(\cdot)$ if $h_i(t, \boldsymbol{x}(t)) < 0$ for all $t \in (t_1, t_2)$. An interval $[\tau_1, \tau_2] \subset [0, T]$ with $\tau_1 < \tau_2$ is called a *boundary interval* if $h_i(t, \boldsymbol{x}(t)) = 0$ for all $t \in [\tau_1, \tau_2]$. An instant $t_{en} \in [0, T]$ is called an *entry time* if there is an interior interval that ends at t_{en} and a corresponding boundary interval that starts at t_{en}. Similarly, an instant $t_{ex} \in [0, T]$ is called an *exit time* if a boundary interval finishes and an interior interval commences at t_{ex}. If $h_i(\tau^*, \boldsymbol{x}(\tau^*)) = 0$ and if $h_i(t, \boldsymbol{x}(t)) < 0$ just before and just after τ^*, then τ^* is referred to as a *contact time.* We collectively refer to entry, exit and contact times as *junction times.* Finally, let us assume for notational convenience that boundary intervals do not intersect. If this is not the case, a more elaborate statement of the next assumption would be required.

Assumption 6.7.3 *On any boundary interval* $[\tau_1, \tau_2]$,

$$\operatorname{rank} \begin{bmatrix} \frac{\partial h_1^{p_1}}{\partial \boldsymbol{u}} \\ \vdots \\ \frac{\partial h_{q'}^{p_{q'}}}{\partial \boldsymbol{u}} \end{bmatrix} = q', \tag{6.7.9}$$

where we assume that the constraints have been ordered so that

$$h_i(t, \boldsymbol{x}(t)) = 0, \ i = 1, \ldots, q' \leq q \ \textit{and} \ h_i(t, \boldsymbol{x}(t)) < 0, \ i = q' + 1, \ldots, q,$$

for $t \in [\tau_1, \tau_2]$ *and* p_i *is the order of the constraint* $h_i \leq 0$.

Since we have replaced the class of admissible control in previous sections with (6.7.4) here, further assumptions may be required to guarantee the existence of an optimal solution of Problem (SP). We refer the interested reader to [88] for a more detailed analysis of this issue.

To state the first order optimality conditions for Problem (SP), we follow the so-called *direct adjoining approach* where the state constraints (6.7.1) and control constraints (6.7.4) are directly adjoined to the Hamiltonian to form the Lagrangian. Thus, we define

$$H(t, \boldsymbol{x}, \boldsymbol{u}, \lambda_0, \boldsymbol{\lambda}) = \lambda_0 \mathcal{L}_0(t, \boldsymbol{x}, \boldsymbol{u}) + \boldsymbol{\lambda}^\top \boldsymbol{f}(t, \boldsymbol{x}, \boldsymbol{u}) \tag{6.7.10}$$

and

$$L(t, \boldsymbol{x}, \boldsymbol{u}, \lambda_0, \boldsymbol{\lambda}, \boldsymbol{\mu}, \boldsymbol{\nu}) = H(t, \boldsymbol{x}, \boldsymbol{u}, \lambda_0, \boldsymbol{\lambda}) + \boldsymbol{\mu}^\top \boldsymbol{g}(t, \boldsymbol{u}) + \boldsymbol{\nu}^\top \boldsymbol{h}(t, \boldsymbol{x}), \tag{6.7.11}$$

where

$$\lambda_0 \geq 0 \tag{6.7.12}$$

is a constant, $\boldsymbol{\lambda}(\cdot) \in \mathbb{R}^n$ is the costate vector, and $\boldsymbol{\mu}(\cdot) \in \mathbb{R}^c$ and $\boldsymbol{\nu}(\cdot) \in \mathbb{R}^q$ are Lagrange multiplier functions. At any time $t \in [0, T]$, we also define the feasible control region

$$\Omega(t) = \{\boldsymbol{u} \in \mathbb{R}^m \,|\, \boldsymbol{g}(t, \boldsymbol{u}) \leq 0\}. \tag{6.7.13}$$

Theorem 6.7.1 *Consider Problem (SP). Suppose that $\boldsymbol{u}^*(\cdot)$, where $\boldsymbol{u}^*(t) \in \Omega(t)$, $t \in [0, T)$, is an optimal control that is right continuous with left hand limits and also assume that the constraint qualification* (6.7.6) *holds for every pair $\{\boldsymbol{u}, t\}$, $t \in [0, T]$ with $\boldsymbol{u} \in \Omega(t)$. Denote the corresponding optimal state as $\boldsymbol{x}^*(\cdot)$ and assume that it has finitely many junction times. Then there exist a constant $\lambda_0 \geq 0$, a piecewise continuous costate trajectory $\boldsymbol{\lambda}^*(\cdot)$ whose continuous segments are absolutely continuous, piecewise continuous multiplier functions $\boldsymbol{\mu}^*(\cdot)$ and $\boldsymbol{\nu}^*(\cdot)$, a vector $\boldsymbol{\eta}(\tau_i) \in \mathbb{R}^q$ for each discontinuity point τ_i of $\boldsymbol{\lambda}^*(\cdot)$, and $\boldsymbol{\alpha} \in \mathbb{R}^r$, $\boldsymbol{\beta} \in \mathbb{R}^s$, $\boldsymbol{\gamma} \in \mathbb{R}^q$ such that*

$$(\lambda_0, \boldsymbol{\lambda}(t), \boldsymbol{\mu}(t), \boldsymbol{\nu}(t), \boldsymbol{\alpha}, \boldsymbol{\beta}, \boldsymbol{\gamma}, \boldsymbol{\eta}(\tau_1), \boldsymbol{\eta}(\tau_2), \ldots) \neq \boldsymbol{0}$$

for every $t \in [0, T]$ and such that the following conditions hold almost everywhere.

$$\frac{d\boldsymbol{x}^*(t)}{dt} = \left[\frac{\partial H(t, \boldsymbol{x}^*(t), \boldsymbol{u}^*(t), \lambda_0, \boldsymbol{\lambda}^*(t))}{\partial \boldsymbol{\lambda}}\right]^\top = \boldsymbol{f}(t, \boldsymbol{x}^*(t), \boldsymbol{u}^*(t)), \tag{6.7.14a}$$

$$\boldsymbol{x}^*(0) = \boldsymbol{x}^0, \tag{6.7.14b}$$

$$\frac{d\boldsymbol{\lambda}^*(t)}{dt} = -\left[\frac{\partial L(t, \boldsymbol{x}^*(t), \boldsymbol{u}^*(t), \lambda_0, \boldsymbol{\lambda}^*(t), \boldsymbol{\mu}^*(t), \boldsymbol{\nu}^*(t))}{\partial \boldsymbol{x}}\right]^\top, \tag{6.7.14c}$$

$$\begin{aligned}\boldsymbol{\lambda}^*(T^-) = {} & \lambda_0 \left[\frac{\partial \Phi_0(\boldsymbol{x}^*(T))}{\partial \boldsymbol{x}}\right]^\top + \boldsymbol{\alpha}^\top \left[\frac{\partial \boldsymbol{a}(T, \boldsymbol{x}^*(T))}{\partial \boldsymbol{x}}\right]^\top \\ & + \boldsymbol{\beta}^\top \left[\frac{\partial \boldsymbol{b}(T, \boldsymbol{x}^*(T))}{\partial \boldsymbol{x}}\right]^\top + \boldsymbol{\gamma}^\top \left[\frac{\partial \boldsymbol{h}(T, \boldsymbol{x}^*(T))}{\partial \boldsymbol{x}}\right]^\top,\end{aligned} \tag{6.7.14d}$$

$$\boldsymbol{\alpha} \geq \boldsymbol{0},\ \boldsymbol{\gamma} \geq \boldsymbol{0},\ \boldsymbol{\alpha}^\top \boldsymbol{a}(T, \boldsymbol{x}^*(T)) = \boldsymbol{\gamma}^\top \boldsymbol{h}(T, \boldsymbol{x}^*(T)) = 0, \tag{6.7.14e}$$

where T^- denotes the limit from the left,

$$\begin{aligned}&\frac{\partial L(t, \boldsymbol{x}^*(t), \boldsymbol{u}^*(t), \lambda_0, \boldsymbol{\lambda}^*(t), \boldsymbol{\mu}^*(t), \boldsymbol{\nu}^*(t))}{\partial \boldsymbol{u}} \\ &\quad = \frac{\partial H(t, \boldsymbol{x}^*(t), \boldsymbol{u}^*(t), \lambda_0, \boldsymbol{\lambda}^*(t))}{\partial \boldsymbol{u}} + (\boldsymbol{\mu}^*(t))^\top \frac{\partial \boldsymbol{g}(t, \boldsymbol{u}^*(t))}{\partial \boldsymbol{u}} \\ &\quad = \boldsymbol{0}^\top\end{aligned} \tag{6.7.14f}$$

$$\boldsymbol{\mu}^*(t) \geq \mathbf{0}, \quad (\boldsymbol{\mu}^*(t))^\top \boldsymbol{g}(t, \boldsymbol{u}^*(t)) = 0, \quad \boldsymbol{g}(t, \boldsymbol{u}^*(t)) \leq 0, \tag{6.7.14g}$$

$$\boldsymbol{\nu}^*(t) \geq \mathbf{0}, \quad (\boldsymbol{\nu}^*(t))^\top \boldsymbol{h}(t, \boldsymbol{x}^*(t)) = 0, \quad \boldsymbol{h}(t, \boldsymbol{x}^*(t)) \leq 0, \tag{6.7.14h}$$

$$\begin{aligned} &\frac{dH(t, \boldsymbol{x}^*(t), \boldsymbol{u}^*(t), \lambda_0, \boldsymbol{\lambda}^*(t))}{dt} \\ =& \frac{dL(t, \boldsymbol{x}^*(t), \boldsymbol{u}^*(t), \lambda_0, \boldsymbol{\lambda}^*(t), \boldsymbol{\mu}^*(t), \boldsymbol{\nu}^*(t))}{dt} \\ \triangleq& \frac{\partial L(t, \boldsymbol{x}^*(t), \boldsymbol{u}^*(t), \lambda_0, \boldsymbol{\lambda}^*(t), \boldsymbol{\mu}^*(t), \boldsymbol{\nu}^*(t))}{\partial t}, \end{aligned} \tag{6.7.14i}$$

and

$$\min_{\boldsymbol{v} \in \Omega(t)} H(t, \boldsymbol{x}^*(t), \boldsymbol{v}, \lambda_0, \boldsymbol{\lambda}^*(t)) = H(t, \boldsymbol{x}^*(t), \boldsymbol{u}^*(t), \lambda_0, \boldsymbol{\lambda}^*(t)). \tag{6.7.14j}$$

Furthermore, for any time τ in a boundary interval and for any contact time τ, the costate $\boldsymbol{\lambda}^*(\cdot)$ may have a discontinuity given by the following jump conditions.

$$\boldsymbol{\lambda}^*(\tau^-) = \boldsymbol{\lambda}^*(\tau^+) + (\boldsymbol{\eta}(\tau))^\top \boldsymbol{h}(\tau, \boldsymbol{x}^*(\tau)), \tag{6.7.14k}$$

$$\begin{aligned} &H(\tau^-, \boldsymbol{x}^*(\tau^-), \boldsymbol{u}^*(\tau^-), \lambda_0, \boldsymbol{\lambda}^*(\tau^-)) \\ &= H(\tau^+, \boldsymbol{x}^*(\tau^+), \boldsymbol{u}^*(\tau^+), \lambda_0, \boldsymbol{\lambda}^*(\tau^+)) - (\boldsymbol{\eta}(\tau))^\top \boldsymbol{h}(\tau, \boldsymbol{x}^*(\tau)), \end{aligned} \tag{6.7.14l}$$

$$\boldsymbol{\eta}(\tau) \geq \mathbf{0}, \ (\boldsymbol{\eta}(\tau))^\top \boldsymbol{h}(\tau, \boldsymbol{x}^*(\tau)) = 0, \tag{6.7.14m}$$

where τ^+ and τ^- denote the left hand side and right hand side limits, respectively. Finally, suppose that τ is a junction time corresponding to a first order constraint $h_i(t, \boldsymbol{x}) \leq 0$ for some $i \in \{1, \ldots, q\}$. Recalling the definition of h_i^1 in (6.7.7), *if*

$$h_i^1(\tau^-, \boldsymbol{x}^*(\tau^-), \boldsymbol{u}^*(\tau^-)) < 0, \tag{6.7.15}$$

or

$$h_i^1(\tau^+, \boldsymbol{x}^*(\tau^+), \boldsymbol{u}^*(\tau^+)) > 0, \tag{6.7.16}$$

i.e., if the entry or exit of the state into or out of a boundary interval is non-tangential, then the costate $\boldsymbol{\lambda}$ is continuous at τ.

Remark 6.7.1 *The condition*

$$(\lambda_0, \boldsymbol{\lambda}(t), \boldsymbol{\mu}(t), \boldsymbol{\nu}(t), \boldsymbol{\alpha}, \boldsymbol{\beta}, \boldsymbol{\gamma}, \boldsymbol{\eta}(\tau_1), \boldsymbol{\eta}(\tau_2), \ldots) \neq \mathbf{0}$$

for every t can help us to distinguish a normal case ($\lambda_0 = 1$) *from an abnormal case* ($\lambda_0 = 0$).

Remark 6.7.2 *Conditions* (6.7.14i) *and* (6.7.14l) *are equivalent to the requirement that $H(t, \boldsymbol{x}^*(t), \boldsymbol{u}^*(t), \lambda_0, \boldsymbol{\lambda}^*(t))$ is constant in the autonomous case, i.e., when $\boldsymbol{f}$, $\mathcal{L}$, $\boldsymbol{g}$, and $\boldsymbol{h}$ do not depend on t explicitly.*

Remark 6.7.3 *In most practical cases, $\boldsymbol{\lambda}^*$ and H will only jump at junction times. However, a discontinuity may also occur in the interior of boundary interval.*

Remark 6.7.4 *Parts of Theorem 6.7.1 have been proven by a range of different authors, see [88] and the references cited therein. In particular, a complete proof that establishes the existence of various multipliers may be found in [184]. A more general version of Theorem 6.7.1, where $\boldsymbol{g}$ also depends on $\boldsymbol{x}$, is stated as an informal theorem in [88]. However, no formal proof has as yet been established for this case.*

Example 6.7.1 Consider the problem of minimizing

$$g_0(u) = \int_0^3 x(t)\,dt \tag{6.7.17}$$

subject to

$$\frac{dx(t)}{dt} = u(t) \tag{6.7.18}$$

$$x(0) = 1, \tag{6.7.19}$$

$$-1 \le u(t) \le 1, \text{ for all } t \in [0,3], \tag{6.7.20}$$

$$x(t) \ge 0, \text{ for all } t \in [0,3], \tag{6.7.21}$$

and the terminal state constraint

$$x(3) = 1. \tag{6.7.22}$$

In the notation of Problem (SP), we have $\mathcal{L}_0 = x$, $\Phi_0 = 0$, $f = u$, $h = -x \le 0$, $g_1 = -u - 1 \le 0$, $g_2 = u - 1 \le 0$ and $b = x - 1 = 0$. Applying Theorem 6.7.1, we have

$$\begin{aligned} H =& x + \lambda u, \\ L =& x + \lambda u + \mu_1(-1-u) + \mu_2(u-1) - \nu x, \end{aligned}$$

where

$$\begin{aligned} &\frac{d\lambda}{dt} = -\frac{\partial L}{\partial x} = -1 + \nu, \\ &\lambda(3) = \beta - \gamma, \\ &\mu_1 \ge 0, \quad \mu_1(-1-u) = 0, \text{ for all } t \in [0,3], \\ &\mu_2 \ge 0, \quad \mu_2(u-1) = 0, \text{ for all } t \in [0,3], \\ &\nu \ge 0, \quad \nu(-x) = 0, \text{ for all } t \in [0,3], \end{aligned}$$

and $\gamma \ge 0$ along with $\gamma h(x(3)) = 0$. Note that the last condition gives $\gamma x(3) = \gamma = 0$. Furthermore,

$$\frac{\partial L}{\partial u} = \lambda - \mu_1 + \mu_2 = 0. \tag{6.7.23}$$

While it is possible to derive the optimal solution from the above conditions, it is obvious from the problem statement that

$$x^*(t) = \begin{cases} 1-t, & 0 \le t < 1, \\ 0, & 1 \le t < 2, \\ t-2, & 2 \le t \le 3, \end{cases}$$

and

$$u^*(t) = \begin{cases} -1, & 0 \le t < 1, \\ 0, & 1 \le t < 2, \\ 1, & 2 \le t < 3. \end{cases}$$

We can then derive the corresponding costate and multipliers by satisfying the requirements of Theorem 6.7.1.

For $t \in [1, 2]$, $u^*(t) = 0$ requires that $\mu_1^*(t) = \mu_2^*(t) = 0$. (6.7.23) then yields $\lambda^*(t) = 0$. It follows that $d\lambda/dt = -1 + \nu^*(t) = 0$, which results in $\nu^*(t) = 1$.

For $t \in [0, 1)$, $x^*(t) > 0$ implies that $\nu^*(t) = 0$ and hence $d\lambda^*/dt = -1$ so that $\lambda^*(t) = A - t$, for some constant A. Since the state enters the boundary interval non-tangentially, Theorem 6.7.1 requires λ^* to be continuous at $t = 1$. $\lambda^*(1^-) = A - 1 = \lambda^*(1) = 0$ yields $A = 1$ so that $\lambda^*(t) = 1 - t$ on $[0, 1)$. $u^*(t) = -1$ on $[0, 1)$ means that $\mu_2^*(t) = 0$ and, by virtue of (6.7.23), $\mu_1^*(t) = \lambda^*(t) = 1 - t$ on the same interval.

For $t \in (2, 3]$, $x^*(t) > 0$ again implies that $\nu^*(t) = 0$ and hence $d\lambda^*/dt = -1$ so that $\lambda^*(t) = B - t$, for some constant B. Since the state exits the boundary interval non-tangentially, continuity of λ^* is required at $t = 2$, so $\lambda^*(2^+) = B - 2 = \lambda^*(2) = 0$, i.e., $B = 2$ and $\lambda^*(t) = 2 - t$ in this interval. $u^*(t) = 1$ means that $\mu_1^*(t) = 0$, so, using (6.7.23), we have $\mu_2^*(t) = -\lambda^*(t) = t - 2$.

Clearly, the costate and multipliers derived above satisfy all the requirements of Theorem 6.7.1.

6.8 The Bellman Dynamic Programming

In contrast to the Pontryagin Maximum Principle, Bellman's principle of optimality for dynamic programming is surprisingly clear and intuitive and the proof is almost trivial. Nevertheless, it has made a tremendous contribution in a wide range of applications ranging from decision science to various engineering disciplines. Its application in optimal control theory has led to the derivation of the famous Hamilton-Jacobi-Bellman (HJB) partial differential equation that can be used for constructing nonlinear optimal feedback control laws.

We introduce *Bellman's principle of optimality* via a simple diagram as shown in Figure. 6.8.1.

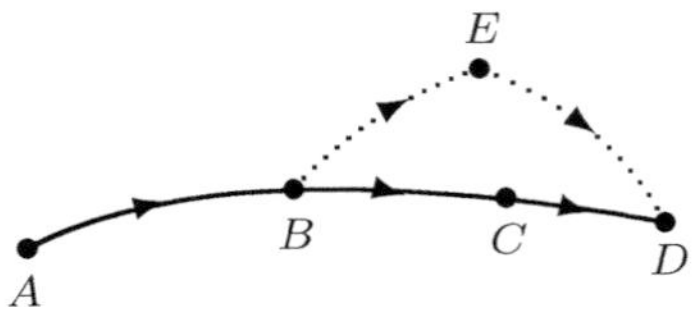

Fig. 6.8.1: Proof of Bellman's principle of optimality

Suppose we wish to travel from the point A to the point D by the shortest possible path and suppose this optimal path is given by $ABCD$, where the intermediate points B and C represent some intermediate stages of the journey. Now let us suppose that we start from point B and wish to travel to point D by the shortest possible path. Bellman's principle of optimality then asserts that the optimal path will be given by BCD. Though this answer appears trivial, we shall prove it nevertheless. Suppose there exists another optimal path from B to D, indicated by the broken line BED. Then, since distance is additive, this implies that $ABED$ is a shorter path than $ABCD$. This contradicts the original assumption and the proof is complete.

In the context of optimal control, we are dealing with a dynamical process that evolves with time. Once the initial state and initial time are fixed, a unique optimal control to minimize the cost functional can be determined. Hence the optimal control is a function of the initial state and time. Each optimal state is associated with an optimal path called an *extremal*. If the process starts at a different initial state and time, a different optimal extremal will result. For all possible initial points $(\boldsymbol{\xi}, t)$ in the state–time space, a family of extremals may be computed, which is referred to as the *field of extremals* in classical calculus of variation terminology. Each extremal yields a corresponding cost functional that can be regarded as a function of the initial state and time, i.e.,

$$V(\boldsymbol{\xi}, t) = \min \{g_0(\boldsymbol{\xi}, t \mid \boldsymbol{u}) : \boldsymbol{u} \in \mathcal{U}\}, \tag{6.8.1}$$

where

$$g_0(\boldsymbol{\xi}, t \mid \boldsymbol{u}) = \Phi_0(\boldsymbol{x}(T \mid \boldsymbol{u})) + \int_t^T \mathcal{L}_0(\tau, \boldsymbol{x}(\tau \mid \boldsymbol{u}), \boldsymbol{u}(t))d\tau, \tag{6.8.2}$$

$$\mathcal{U} = \{\boldsymbol{u}\colon \boldsymbol{u} \text{ is measurable on } [0, T] \text{ with } \boldsymbol{u}(t) \in U, \text{ for all } t \in [0, T]\},$$

and, for each $\boldsymbol{u} \in \mathcal{U}$, $\boldsymbol{x}(\cdot \mid \boldsymbol{u})$ is determined from $(\boldsymbol{\xi}, t)$ by

$$\frac{d\boldsymbol{x}(\tau)}{d\tau} = \boldsymbol{f}(\tau, \boldsymbol{x}(\tau), \boldsymbol{u}(\tau)), \tag{6.8.3a}$$

$$\boldsymbol{x}(t) = \boldsymbol{\xi}. \tag{6.8.3b}$$

Here, the function V is called the *value function*, and $g_0(\boldsymbol{\xi}, t \mid \boldsymbol{u})$ denotes the cost functional corresponding to the control $\boldsymbol{u}$ over the interval $[t, T]$ starting from $\boldsymbol{x}(t) = \boldsymbol{\xi}$. Moreover, the control restraint set U is, in general, taken as a compact subset of $\mathbb{R}^r$.

For convenience, let Problem $(P(\boldsymbol{\xi}, t))$ denote the optimal control problem in which the cost functional (6.8.2) is to be minimized over $\mathcal{U}$ subject to the dynamical system (6.8.3).

We may now apply Bellman's principle of optimality to derive a partial differential equation for the value function $V(\boldsymbol{\xi}, t)$. Suppose t is the current time and $t + \Delta t$ is a future time infinitesimally close to t. Then, by virtue of the system (6.8.3), it follows that corresponding to each $\boldsymbol{u} \in \mathcal{U}$, the state at $t + \Delta t$ is

$$\boldsymbol{x}(t + \Delta t \mid \boldsymbol{u}) = \boldsymbol{\xi} + \Delta\boldsymbol{\xi}, \tag{6.8.4}$$

where

$$\Delta\boldsymbol{\xi} = \boldsymbol{f}(t, \boldsymbol{\xi}, \boldsymbol{u}(t))\Delta t + O\left(\Delta t^2\right). \tag{6.8.5}$$

Now, let Problem $(P(\boldsymbol{\xi} + \Delta\boldsymbol{\xi}, t + \Delta t))$ be Problem $(P(\boldsymbol{\xi}, t))$ with the initial condition (6.8.3b) and the cost functional (6.8.2) replaced by (6.8.4) and

$$g_0(\boldsymbol{\xi}+\Delta\boldsymbol{\xi}, t+\Delta t \mid \boldsymbol{u}) = \Phi_0(\boldsymbol{x}(T \mid \boldsymbol{u}))+\int_{t+\Delta t}^{T} \mathcal{L}_0(\tau, \boldsymbol{x}(\tau \mid \boldsymbol{u}), \boldsymbol{u}(\tau))d\tau, \tag{6.8.6}$$

respectively.

Clearly, the cost functional (6.8.2) can be expressed as

$$g_0(\boldsymbol{\xi}, t \mid \boldsymbol{u}) = \Phi_0(\boldsymbol{x}(T \mid \boldsymbol{u})) + \int_{t+\Delta t}^{T} \mathcal{L}_0(\tau, \boldsymbol{x}(\tau \mid \boldsymbol{u}), \boldsymbol{u}(\tau))d\tau \tag{6.8.7}$$

$$+ \int_{t}^{t+\Delta t} \mathcal{L}_0(\tau, \boldsymbol{x}(\tau \mid \boldsymbol{u}), \boldsymbol{u}(\tau))d\tau. \tag{6.8.8}$$

If we employ a control $\tilde{\boldsymbol{u}} \in \mathcal{U}$ given by

$$\tilde{\boldsymbol{u}}(\tau) = \begin{cases} \boldsymbol{u}^*(\tau), \text{ for } \tau \in (t + \Delta t, T] \\ \boldsymbol{u}(\tau), \text{ for } \tau \in [t, t + \Delta t), \end{cases} \tag{6.8.9}$$

where $\boldsymbol{u}^* \in \mathcal{U}$ is an optimal control for Problem $(P(t + \Delta t, \boldsymbol{\xi} + \Delta\boldsymbol{\xi}))$, we may rewrite (6.8.8) as

$$g_0(\boldsymbol{\xi}, t \mid \tilde{\boldsymbol{u}}) = V(\boldsymbol{\xi} + \Delta\boldsymbol{\xi}, t + \Delta t) + \int_{t}^{t+\Delta t} \mathcal{L}_0(\tau, \boldsymbol{x}(\tau \mid \boldsymbol{u}), \boldsymbol{u}(\tau))d\tau, \tag{6.8.10}$$

where

$$V(\boldsymbol{\xi} + \Delta\boldsymbol{\xi}, t + \Delta t) = \min\{g_0(\boldsymbol{\xi} + \Delta\boldsymbol{\xi}, t + \Delta t \mid \boldsymbol{u}) : \boldsymbol{u} \in \mathcal{U}\}. \tag{6.8.11}$$

Obviously,

$$g_0(\boldsymbol{\xi}, t \mid \tilde{\boldsymbol{u}}) \geq V(\boldsymbol{\xi}, t), \tag{6.8.12}$$

which, in turn, becomes an equality if optimal control is used in the interval $[t, t + \Delta t]$. Thus,

$$V(\boldsymbol{\xi}, t) = \min \Big\{ V(\boldsymbol{\xi} + \Delta\boldsymbol{\xi}, t + \Delta t) + \int_t^{t+\Delta t} \mathcal{L}_0(\tau, \boldsymbol{x}(\tau \mid \boldsymbol{u}), \boldsymbol{u}(\tau)) d\tau$$
$$: \boldsymbol{u} \text{ measurable on } [t, t + \Delta t] \text{ with } \boldsymbol{u}(\tau) \in U \Big\}. \tag{6.8.13}$$

Using a Taylor series expansion of the value function $V(\boldsymbol{\xi} + \Delta\boldsymbol{\xi}, t + \Delta t)$, and then invoking (6.8.5), we have

$$\begin{aligned} &V(\boldsymbol{\xi} + \Delta\boldsymbol{\xi}, t + \Delta t) \\ &\quad = V(\boldsymbol{\xi}, t) + \frac{\partial V(\boldsymbol{\xi}, t)}{\partial \boldsymbol{\xi}} \Delta\boldsymbol{\xi} + \frac{\partial V(\boldsymbol{\xi}, t)}{\partial t} \Delta t + O\left(\Delta t^2\right) \\ &\quad = V(\boldsymbol{\xi}, t) + \frac{\partial V(\boldsymbol{\xi}, t)}{\partial \boldsymbol{\xi}} \boldsymbol{f}(t, \boldsymbol{\xi}, \boldsymbol{u}(t)) \Delta t + \frac{\partial V(\boldsymbol{\xi}, t)}{\partial t} \Delta t + O\left(\Delta t^2\right). \end{aligned} \tag{6.8.14}$$

Next, since Δt is infinitesimally small, it is clear from using the initial condition (6.8.3b) that

$$\int_t^{t+\Delta t} \mathcal{L}_0(\tau, \boldsymbol{x}(\tau \mid \boldsymbol{u}), \boldsymbol{u}(\tau)) d\tau = \mathcal{L}_0(t, \boldsymbol{\xi}, \boldsymbol{u}(t)) \Delta t + O\left(\Delta t^2\right). \tag{6.8.15}$$

Substituting (6.8.14) and (6.8.15) into (6.8.13), and then noting that $V(\boldsymbol{\xi}, t)$ is independent of $\boldsymbol{u}(t)$, we obtain

$$\begin{aligned} &-\frac{\partial V(\boldsymbol{\xi}, t)}{\partial t} \Delta t \\ &\quad = \min \left\{ \frac{\partial V(\boldsymbol{\xi}, t)}{\partial \boldsymbol{\xi}} \boldsymbol{f}(t, \boldsymbol{\xi}, \boldsymbol{u}(t)) \Delta t + \mathcal{L}_0(t, \boldsymbol{\xi}, \boldsymbol{u}(t)) \Delta t + O\left(\Delta t^2\right) : \boldsymbol{u}(t) \in U \right\}. \end{aligned} \tag{6.8.16}$$

Thus, dividing (6.8.16) by Δt, and then letting $\Delta t \to 0$, it follows that

$$-\frac{\partial V(\boldsymbol{\xi}, t)}{\partial t} = \min \left\{ \frac{\partial V(\boldsymbol{\xi}, t)}{\partial \boldsymbol{\xi}} \boldsymbol{f}(t, \boldsymbol{\xi}, \boldsymbol{v}) + \mathcal{L}_0(t, \boldsymbol{\xi}, \boldsymbol{v}) : \boldsymbol{v} \in U \right\}. \tag{6.8.17}$$

(6.8.17) is the well-known Hamilton-Jacobi-Bellman (HJB) equation that is a partial differential equation in $\boldsymbol{\xi}$ and t. It is to be solved subject to the boundary condition:

$$V(\boldsymbol{\xi}, T) = \Phi_0(\boldsymbol{\xi}). \tag{6.8.18}$$

The validity of (6.8.18) is obvious from the definition of the value function.

There are many numerical methods available in the literature for solving HJB partial differential equations. See, for example, [98, 99, 130, 188, 208, 273, 274, 305–307, 309]. However, except for linear quadratic optimal control problems, these HJB equations are rather difficult to solve when the dimension of ξ is larger than 2. On the other hand, if it can be solved, the optimal control obtained from the minimization process of (6.8.17) is dependent on the state. It thus furnishes a natural nonlinear optimal feedback control law.

Note that if the Hamiltonian function is defined as

$$H(t, \boldsymbol{x}, \boldsymbol{u}, \boldsymbol{\lambda}) = \mathcal{L}_0(t, \boldsymbol{x}, \boldsymbol{u}) + \boldsymbol{\lambda}^\top \boldsymbol{f}(t, \boldsymbol{x}, \boldsymbol{u}), \tag{6.8.19}$$

then (6.8.17) may be rewritten as

$$-\frac{\partial V(\boldsymbol{\xi}, t)}{\partial t} = \min \left\{ H\left(t, \boldsymbol{\xi}, \boldsymbol{v}, \left(\frac{\partial V(\boldsymbol{\xi}, t)}{\partial \boldsymbol{\xi}}\right)^\top\right) : \boldsymbol{v} \in U \right\}. \tag{6.8.20}$$

In fact, it can even be shown that the optimal costate is

$$\boldsymbol{\lambda}^*(t) = \left[\frac{\partial V(\boldsymbol{\xi}, t)}{\partial \boldsymbol{\xi}}\right]^\top. \tag{6.8.21}$$

We shall conclude this section by re-deriving the optimal feedback control law of a linear quadratic regulator via the HJB equation (compare Section 6.3). Recall that the linear quadratic regulator problem's formulation is

$$\begin{aligned} \min \ \Big\{ g_0(\boldsymbol{u}) = \frac{1}{2}(\boldsymbol{x}(T))^\top S^f \boldsymbol{x}(T) \\ + \frac{1}{2}\int_0^T \left\{ (\boldsymbol{x}(t))^\top Q(t)\boldsymbol{x}(t) + (\boldsymbol{u}(t))^\top R(t)\boldsymbol{u}(t) \right\} dt \Big\} \end{aligned} \tag{6.8.22}$$

subject to

$$\frac{d\boldsymbol{x}(t)}{dt} = A(t)\boldsymbol{x}(t) + B(t)\boldsymbol{u}(t). \tag{6.8.23}$$

Let $V(\boldsymbol{\xi}, t)$ be the corresponding value function if the process is assumed to start at $t < T$ from some state $\boldsymbol{\xi}$. The associated HJB equation is

$$-\frac{\partial V(\boldsymbol{\xi},t)}{\partial t} = \min\left\{\frac{\partial V(\boldsymbol{\xi},t)}{\partial \boldsymbol{\xi}}(A(t)\boldsymbol{\xi}+B(t)\boldsymbol{v})+\frac{1}{2}\boldsymbol{\xi}^\top Q(t)\boldsymbol{\xi}+\frac{1}{2}\boldsymbol{v}^\top R(t)\boldsymbol{v} : \boldsymbol{v}\in\mathbb{R}^r\right\} \tag{6.8.24a}$$

with boundary condition

$$V(\boldsymbol{\xi},T) = \frac{1}{2}\boldsymbol{\xi}^\top S^f \boldsymbol{\xi}. \tag{6.8.24b}$$

Given the form of (6.8.24), we may speculate that $V(\boldsymbol{\xi},t)$ takes the form:

$$V(\boldsymbol{\xi},t) = \frac{1}{2}\boldsymbol{\xi}^\top S(t)\boldsymbol{\xi}, \tag{6.8.25}$$

where $S(t)$ is a symmetric matrix. Substitution of (6.8.25) into (6.8.24a) yields

$$-\frac{1}{2}\boldsymbol{\xi}^\top\left\{\frac{dS(t)}{dt}\right\}\boldsymbol{\xi} = \min\left\{\boldsymbol{\xi}^\top S(t)(A(t)\boldsymbol{\xi}+B(t)\boldsymbol{v})+\frac{1}{2}\boldsymbol{\xi}^\top Q(t)\boldsymbol{\xi}+\frac{1}{2}\boldsymbol{v}^\top R(t)\boldsymbol{v} : \boldsymbol{v}\in\mathbb{R}^r\right\}. \tag{6.8.26}$$

Let $\boldsymbol{u}^*(t)$ denote the $\boldsymbol{v}$ that solves the minimization problem. Clearly, $\boldsymbol{u}^*(t)$ is given by

$$\boldsymbol{u}^*(t) = -(R(t))^{-1}(B(t))^\top S(t)\boldsymbol{\xi}, \tag{6.8.27}$$

which simply confirms the optimal state feedback control law of (6.3.20).

Substituting the minimizing control $\boldsymbol{u}^*(t)$ of (6.8.27) into (6.8.26), we have

$$-\frac{1}{2}\boldsymbol{\xi}^\top\dot{S}(t)\boldsymbol{\xi} = \boldsymbol{\xi}^\top S(t)(A(t)\boldsymbol{\xi}-B(t)(R(t))^{-1}(B(t))^\top S(t)\boldsymbol{\xi})+\frac{1}{2}\boldsymbol{\xi}^\top Q(t)\boldsymbol{\xi} + \frac{1}{2}\boldsymbol{\xi}^\top S(t)B(t)(R(t))^{-1}(B(t))^\top S(t)\boldsymbol{\xi}. \tag{6.8.28}$$

This can be simplified to

$$\boldsymbol{\xi}^\top\left[\frac{dS(t)}{dt}+2S(t)A(t)-S(t)B(t)(R(t))^{-1}(B(t))^\top S(t)+Q(t)\right]\boldsymbol{\xi}=0. \tag{6.8.29}$$

Since $S(t)$ is symmetric, it follows that

$$\boldsymbol{\xi}^\top(2S(t)A(t))\boldsymbol{\xi} = \boldsymbol{\xi}^\top\left(S(t)A(t)+(A(t))^\top S(t)\right)\boldsymbol{\xi}+\boldsymbol{\xi}^\top\left(S(t)A(t)-(A(t))^\top S(t)\right)\boldsymbol{\xi}. \tag{6.8.30}$$

Also, since $S(t)A(t) - (A(t))^\top S(t)$ is antisymmetric, the second term on the right hand side of (6.8.30) vanishes and (6.8.29) therefore reduces to

$$\boldsymbol{\xi}^\top \left[\frac{dS(t)}{dt} + S(t)A(t) + (A(t))^\top S(t) \right.$$
$$\left. - S(t)B(t)(R(t))^{-1}(B(t))^\top S(t) + Q(t) \right] \boldsymbol{\xi} = 0. \tag{6.8.31}$$

Since (6.8.31) must hold true for an arbitrary $\boldsymbol{\xi}$, we conclude that

$$\frac{dS(t)}{dt} = - S(t)A(t) - (A(t))^\top S(t)$$
$$+ S(t)B(t)(R(t))^{-1}(B(t))^\top S(t) - Q(t), \tag{6.8.32}$$

which is exactly the same matrix Riccati differential equation derived previously in equation (6.3.19).

The boundary condition follows from (6.8.25) and (6.8.24b):

$$S(T) = S^f. \tag{6.8.33}$$

After solving the matrix Riccati equation (6.8.32) with the boundary condition (6.8.33), the feedback control law is obtained readily from (6.8.27).

To close this chapter, we wish to emphasize that the class of linear quadratic optimal control problems is one of the very few cases that can be solved analytically via the dynamic programming approach.

6.9 Exercises

6.9.1 *Show the validity of* (6.2.9) *from* (6.2.8).

6.9.2 *Show that Equation* (6.3.9) *is valid.*

6.9.3 *Show that Equation* (6.3.11) *is valid.*

6.9.4 *Show that Equation* (6.3.27) *is valid.*

6.9.5 *A continuous time controlled process is described by*

$$\frac{dx(t)}{dt} = u(t)$$

with the initial condition

$$x(0) = 1,$$

and the terminal condition

$$x(1) = \textit{free}.$$

The objective is to find a real-valued control u defined on $[0,1]$, which minimizes the cost functional

$$g_0(u) = \int_0^1 \left\{ (x(t))^2 + (u(t))^2 \right\} dt.$$

(i) Write down the Hamiltonian for the problem.

(ii) Show that the optimal control can be expressed as a function of the adjoint variable λ by the relation

$$u^*(t) = -\frac{\lambda(t)}{2}.$$

(iii) Write down the corresponding two-point boundary-value problem .

(iv) Show that the solution for λ is

$$\lambda(t) = \frac{2\left[-e^{-(1-t)} + e^{(1-t)}\right]}{e + e^{-1}}.$$

(v) Determine the optimal control and the optimal trajectory.

6.9.6 *Use an appropriate version of the Pontryagin Maximum Principle to solve the problem of minimizing*

$$g_0(u) = T$$

subject to

$$\begin{aligned} \frac{dx(t)}{dt} =& - x(t) + u(t), \ \ t \in [0,T), \\ x(0) =& 1, \\ x(T) =& 0, \end{aligned}$$

and

$$|u(t)| \le 1, \ \ \textit{for all } t \in [0,T].$$

6.9.7 *Use the Pontryagin Maximum Principle to solve the problem of minimizing*

$$g_0(u) = \int_0^1 (x(t))^2 \ dt$$

subject to

$$\frac{dx(t)}{dt} = -x(t) + u(t), \ t \in [0,1),$$
$$x(0) = 1,$$
$$x(1) = 0,$$

and

$$|u(t)| \leq 1, \ \textit{for all } t \in [0,1].$$

6.9.8 *Consider the following problem.*

$$\textit{minimize}\left\{ g_0(u) = \int_0^1 (u(t))^2 dt \right\}$$

subject to

$$\frac{dx(t)}{dt} = u(t)$$
$$x(0) = a$$
$$x(1) = 0.$$

Use Pontryagin Maximum Principle to show that

$$x^*(t) = a(1-t).$$

6.9.9 *Use Pontryagin Maximum Principle to solve the following problem.*

$$\textit{minimize}\left\{ g_0(u) = \int_0^1 (x(t))^2 dt \right\}$$

subject to

$$\frac{dx(t)}{dt} = u(t), \quad 0 < t \leq 1$$
$$x(0) = 1,$$
$$-1 \leq u(t) \leq 1.$$

6.9.10 *Use Pontryagin Maximum Principle to solve the following problem. Given the system*

$$\frac{dx(t)}{dt} = x(t) + u(t)$$

with the boundary conditions

$$x(0) = a, \quad x(T) = 0,$$

find a real-valued control u, defined in $[0,T]$, such that the cost functional

$$g_0 = \int_0^T (u(t))^2 dt$$

is minimized. Here, T is fixed. What is the optimal trajectory?

6.9.11 *A continuous time controlled process has differential equation*

$$\frac{dx(t)}{dt} = u(t)$$

and we wish to minimize the cost functional

$$g_0 = \frac{1}{2} \int_0^T \left[(x(t))^2 + (u(t))^2\right] dt.$$

Write down the Riccati equation for optimal control and hence find the optimal control law and the optimal trajectory.

6.9.12 *Consider the necessary condition* (6.4.1e). *If the Hamiltonian function H is continuously differentiable and $U = \mathbb{R}^r$, show that the stationary condition* (6.2.10e) *must be satisfied.*

6.9.13 *Consider the highway construction problem described by* (6.5.3). *If the comfort of driving is taken into consideration, then the cost functional* (6.5.3c) *should be replaced by*

$$g_0(u) = \int_0^2 (u(l))^2 dl.$$

Solve the modified problem by using the Pontryagin Maximum Principle.

6.9.14 *Consider a firm that produces a single product and sells it in a market that can absorb no more than M dollars of the product per unit time. It is assumed that if the firm does no advertising, its rate of sales at any point in time will decrease at a rate proportional to the rate of sales at that time. If the firm advertises, the rate of change of the rate of sales will have an additional term that increases proportional to the rate of advertising, but this increase affects only the share of the market that is not already purchasing the product. Define $S(t)$ and $A(t)$ to be the rate of sales and the rate of advertising at time t, respectively. Let $x(t) = \frac{S(t)}{M}$, let $\gamma > 0$ be a constant, and define $v(t) = \frac{A(t)}{M}$. Then, $1 - x(t)$ represents the share of the market affected by advertising. Under these assumptions and with a suitable scale for the time variable, we have*

$$\frac{dx(t)}{dt} = -x(t) + \gamma v(t)[1 - x(t)]. \tag{6.9.1}$$

It is assumed that there is an upper bound on the advertising rate, i.e.,

$$0 \leq A(t) \leq \bar{A}. \tag{6.9.2}$$

This implies that

$$0 \le v(t) \le a, \tag{6.9.3}$$

where $a = \frac{\bar{A}}{M}$. *The problem is to find an admissible* $v(t)$ *that will maximize the profit functional*

$$J(v) = \int_0^T \{x(t) - v(t)\}dt \tag{6.9.4}$$

subject to the system equation (6.9.1) *and the constraint* (6.9.3), *along with the initial condition*

$$x(0) = x_0, \tag{6.9.5}$$

where T *is a fixed time.*

(a) *Write down the problem in the standard form as a minimization problem.*

(b) *Write down the Hamiltonian for this problem.*

(c) *Show that the optimal control is of bang-bang type, with*

$$v^*(t) = \begin{cases} a, & \text{if } \gamma(x(t)-1)\lambda(t) > 1, \\ 0, & \text{if } \gamma(x(t)-1)\lambda(t) < 1, \\ \text{undetermined}, & \text{if } \gamma(x(t)-1)\lambda(t) = 1. \end{cases}$$

(d) *Write down the corresponding two-point boundary-value problem.*

(e) *Show that the optimal advertising policy involves not advertising near the final time* $t = T$.

(f) *Assume that the optimal control function has the form*

$$v^*(t) = \begin{cases} a, & \text{for } 0 < t < \tau, \\ 0, & \text{for } \tau < t < T, \end{cases}$$

where τ *is a constant yet to be determined. Under this assumption, give an explicit solution for* $\lambda(t)$ *in the region* $\tau < t < T$.

(g) *Under the same assumption, give an explicit solution for* $x(t)$ *in the region* $0 < t < \tau$.

(h) *Hence, or otherwise, show that* τ *is a solution of the equation*

$$\gamma\left(1 - e^{-(T-\tau)}\right)\left(1 - \left(x_0 - \frac{\gamma a}{1+\gamma a}\right)e^{-(1+\gamma a)\tau} - \frac{\gamma a}{1+\gamma a}\right) = 1.$$

6.9.15 *(i) Given the following dynamical system*

$$\frac{dx(t)}{dt} = u(t)$$
$$x(0) = 0.$$

Let the state x be such that the following constraint is satisfied

$$x(t) \geq 0, \forall t \in [0, 2].$$

Find a control u that satisfies the constraints:

$$-1 \leq u(t) \leq 1, \forall t \in [0, 2]$$

such that the cost function

$$J = \int_0^2 x(t)dt$$

is minimized.

(ii) Solve the following optimal control problem

$$\min\ J = \int_0^3 e^{-\rho t} u(t)dt$$

subject to

$$\frac{dx_1(t)}{dt} = u_1(t), \quad x_1(0) = 4$$
$$\frac{dx_2(t)}{dt} = u_1(t) - u_2(t), \quad x_2(0) = 4$$
$$0 \leq u_i(t) \leq 1, \forall t \in [0, 3], i = 1, 2$$
$$x_i(t) \geq 0, \forall t \in [0, 3], i = 1, 2.$$

6.9.16 *(i) Solve the following optimal control problem*

$$\min\ J = \int_0^{10} \{(10 - t)u_1(t) + tu_2(t)\}\, dt$$

subject to

$$\frac{dx_1(t)}{dt} = u_1(t), x_1(0) = 4$$
$$\frac{dx_2(t)}{dt} = u_1(t) - u_2(t), x_2(0) = 4$$
$$0 \leq u_i(t) \leq 1, \forall t \in [0, 3], i = 1, 2$$
$$x_i(t) \geq 0, \forall t \in [0, 3], i = 1, 2.$$

(ii) Solve the following optimal control problem

$$\min\ J = \int_0^5 u(t)dt$$

subject to

$$\begin{aligned} \frac{dx(t)}{dt} &= u(t) - x(t) \\ x(0) &= 1 \\ 0 \leq &u(t) \leq 1,\ \forall t \in [0,5] \\ x(t) &\geq 0.7 - 0.2t,\ \forall t \in [0,5]. \end{aligned}$$

6.9.17 *For the problem:*

$$\textit{minimize} \left\{ g_0(v) = \int_0^1 \left\{ (x(t)^2 + (v(t))^2 \right\} dt \right\}$$

subject to

$$\begin{aligned} \frac{dx(t)}{dt} &= v(t) \\ x(0) &= 1, \end{aligned}$$

and no constraint on $v(t)$, show that the dynamic programming equation yields the optimal control policy

$$v(t) = -\frac{e^{2(1-t)} - 1}{e^{2(1-t)} + 1} x(t).$$

6.9.18 *A system follows the equation:*

$$\frac{dx(t)}{dt} = u(t).$$

The process of ultimate interest starts at time $t = 0$ and finishes at $t = T$ with the cost functional:

$$g_0 = (x(T))^2 + \int_0^T (u(t))^2 dt,$$

where T is fixed.

(i) Use Bellman's dynamic programming to find the optimal control u^*.

(ii) Substitute for the optimal control in the dynamical equation to find the optimal trajectory $x^*(t)$ when $x(0) = x_0$. What is the value of $x^*(T)$?

6.9.19 *Consider the following optimal control problem:*

$$\textit{minimize } g_0(u) = (x(T))^2 + \int_0^T (x(t))^2 + (u(t))^2 dt$$

subject to

$$\frac{dx(t)}{dt} = -x(t) + u(t), \;\; t \in [0, T),$$

where T is fixed.

(i) *Use Bellman's dynamic programming to find the optimal control $u^*(t)$.*

(ii) *Substitute the optimal control into the dynamical equation to find the optimal trajectory $x^*(t)$ when $x(0) = x_0$. What is the value of $g_0(u^*)$?*

Chapter 7
Gradient Formulae for Optimal Parameter Selection Problems

7.1 Introduction

The main theme of this chapter is to derive gradient formulae for the cost and constraint functionals of several types of optimal parameter selection problems with respect to various types of parameters. An optimal parameter selection problem can be regarded as a special type of optimal control problems in which the controls are restricted to be constant functions of time. Many optimization problems involving dynamical systems can be formulated as optimal parameter selection problems. Examples include parameter identification problems and controller parameter design problems. In fact, optimal control problems with their control being parametrized are essentially reduced to optimal parameter selection problems. Thus, when developing numerical techniques for solving complex optimal control problems, it is essential to ensure the solvability of the resulting optimal parameter selection problems.

In Section 7.2, we consider a class of optimal parameter selection problems where the system dynamics are described by ordinary differential equations without time-delay arguments. Gradient formulae for the cost functional as well as for the constraint functionals are then derived. With these gradient formulae, the optimal parameter selection problems can be readily solved using gradient-based mathematical programming algorithms, such as sequential quadratic programming (see Section 3.5).

In Sections 7.3–7.4, we consider a class of optimal control problems in which the control takes a special structure. To be more precise, for each $k = 1, \ldots, r$, the $k-$th component of the control function is a piecewise constant function over the planning horizon $[0, T]$ with jumps at $t_1^k, \ldots, t_{M_k}^k$. In Section 7.3, only the heights of the piecewise constant control functions are regarded as decision parameters and gradient formulae of the cost and constraint functionals with respect to these heights are derived on the

K. L. Teo et al., *Applied and Computational Optimal Control*, Springer Optimization and Its Applications 171,
https://doi.org/10.1007/978-3-030-69913-0_7

basis of the formulae in Section 7.2. In Section 7.4, the switching times of the piecewise constant control functions are regarded as decision parameters. Gradient formulae of the cost functional as well as the constraint functionals with respect to these switching times are then derived. However, these gradient formulae may not exist when two or more switching times collapse into one during the optimization process [148, 151, 168]. Thus, optimization techniques based on these gradient formulae are often ineffective. To overcome this difficulty, the concept of a time scaling transformation, which was originally called the control parametrization enhancing transform (CPET), is introduced. More specifically, by applying the time scaling transformation, the switching times are mapped into fixed knots. The equivalent transformed problem is a standard optimal parameter selection problem of the form considered in Section 7.2. Later in Section 7.4, we consider a class of combined optimal parameter selection and optimal control problems in which the control takes the same structure as that of Section 7.3 with variable heights and variable switching times. After applying the time scaling transformation, gradient formulae of the cost and constraint functionals in the transformed problem with respect to these heights, switching times and system parameters are summarized. Furthermore, we consider the applications of these formulae to the special cases of discrete valued optimal control problems as well as the optimal control of switched systems in Section 7.4.

In Section 7.5, we extend the results of Section 7.2 to the case involving time-delayed arguments. In Section 7.6, we consider a class of optimal parameter selection problems with multiple characteristic time points in the cost and constraint functionals. The main references for this chapter are [89, 125, 142, 148, 168, 215] and Chapter 5 of [253].

7.2 Optimal Parameter Selection Problems

Consider a process described by the following system of differential equations on the fixed time interval $(0, T]$.

$$\frac{d\boldsymbol{x}(t)}{dt} = \boldsymbol{f}(t, \boldsymbol{x}(t), \boldsymbol{\zeta}), \tag{7.2.1a}$$

where $\boldsymbol{x} = [x_1, \ldots, x_n]^\top \in \mathbb{R}^n$ and $\boldsymbol{\zeta} = [\zeta_1, \ldots, \zeta_s]^\top \in \mathbb{R}^s$ are, respectively, the state and system parameter vectors, and $\boldsymbol{f} = [f_1, \ldots, f_n]^\top : [0, T] \times \mathbb{R}^n \times \mathbb{R}^s \mapsto \mathbb{R}^n$. The initial condition for the system of differential equations (7.2.1a) is

$$\boldsymbol{x}(0) = \boldsymbol{x}^0(\boldsymbol{\zeta}), \tag{7.2.1b}$$

where $\boldsymbol{x}^0 = \left[x_1^0, \ldots, x_n^0\right]^\top \in \mathbb{R}^n$ is a given vector valued function of the system parameter vector $\boldsymbol{\zeta}$.

Define

$$\mathcal{Z} = \{\boldsymbol{\zeta} = [\zeta_1, \ldots, \zeta_s]^\top \in \mathbb{R}^s : \ a_i \leq \zeta_i \leq b_i, \ \ i = 1, \ldots, s\}, \tag{7.2.2}$$

where a_i and b_i, $i = 1, \ldots, s$, are real numbers. Clearly, $\mathcal{Z}$ is a compact and convex subset of $\mathbb{R}^s$. For each $\boldsymbol{\zeta} \in \mathbb{R}^s$, let $\boldsymbol{x}(\cdot \mid \boldsymbol{\zeta})$ be the corresponding solution of the system (7.2.1). We may now define an optimal parameter selection problem as follows.

Problem $(P1)$**:** Given the system (7.2.1), find a system parameter vector $\boldsymbol{\zeta} \in \mathcal{Z}$ such that the cost functional

$$g_0(\boldsymbol{\zeta}) = \Phi_0(\boldsymbol{x}(T \mid \boldsymbol{\zeta}), \boldsymbol{\zeta}) + \int_0^T \mathcal{L}_0(t, \boldsymbol{x}(t \mid \boldsymbol{\zeta}), \boldsymbol{\zeta})\, dt \tag{7.2.3}$$

is minimized subject to the equality constraints (in canonical form)

$$g_i(\boldsymbol{\zeta}) = \Phi_i(\boldsymbol{x}(\tau_i \mid \boldsymbol{\zeta}), \boldsymbol{\zeta}) + \int_0^{\tau_i} \mathcal{L}_i(t, \boldsymbol{x}(t \mid \boldsymbol{\zeta}), \boldsymbol{\zeta})\, dt = 0, \qquad i = 1, \ldots, N_e, \tag{7.2.4a}$$

and subject to the inequality constraints (in canonical form)

$$g_i(\boldsymbol{\zeta}) = \Phi_i(\boldsymbol{x}(\tau_i \mid \boldsymbol{\zeta}), \boldsymbol{\zeta}) + \int_0^{\tau_i} \mathcal{L}_i(t, \boldsymbol{x}(t \mid \boldsymbol{\zeta}), \boldsymbol{\zeta})\, dt \leq 0, \qquad i = N_e + 1, \ldots, N. \tag{7.2.4b}$$

Here, $\Phi_i : \mathbb{R}^n \times \mathbb{R}^s \mapsto \mathbb{R}$ and $\mathcal{L}_i : [0, T] \times \mathbb{R}^n \times \mathbb{R}^s \mapsto \mathbb{R}$, $i = 0, 1, \ldots, N$, are given real-valued functions and, for each $i = 0, 1, \ldots, N$, τ_i is referred to as the *characteristic time* for the i–th constraint with $0 < \tau_i \leq T$. For subsequent notational convenience, we also define $\tau_0 = T$.
We assume throughout that the following conditions are satisfied.

Assumption 7.2.1 *For each $i = 0, \ldots, N$, and for each compact subset V of $\mathbb{R}^s$, there exists a positive constant K such that, for all $(t, \boldsymbol{x}, \boldsymbol{\zeta}) \in [0, T] \times \mathbb{R}^n \times V$,*

$$|\boldsymbol{f}(t, \boldsymbol{x}, \boldsymbol{\zeta})| \ \leq K(1 + |\boldsymbol{x}|)$$

and

$$|\mathcal{L}_i(t, \boldsymbol{x}, \boldsymbol{\zeta})| \ \leq K(1 + |\boldsymbol{x}|).$$

Assumption 7.2.2 *$\boldsymbol{f}$ and $\mathcal{L}_i$, $i = 0, 1, \ldots, N$, together with their partial derivatives with respect to each of the components of $\boldsymbol{x}$ and $\boldsymbol{\zeta}$ are piecewise continuous on $[0, T]$ for each $(\boldsymbol{x}, \boldsymbol{\zeta}) \in \mathbb{R}^n \times \mathbb{R}^s$, and continuous on $\mathbb{R}^n \times \mathbb{R}^s$ for each $t \in [0, T]$.*

Assumption 7.2.3 *Φ_i, $i = 0, 1, \ldots, N$, are continuously differentiable with respect to $\boldsymbol{x}$ and $\boldsymbol{\zeta}$. Furthermore, $\boldsymbol{x}^0$ is continuously differentiable with respect to $\boldsymbol{\zeta}$.*

Remark 7.2.1 *From the theory of differential equations, we note that the system* (7.2.1) *admits a unique solution,* $\boldsymbol{x}(\cdot \mid \boldsymbol{\zeta})$*, corresponding to each* $\boldsymbol{\zeta} \in \mathcal{Z}$.

Remark 7.2.2 *The constraints given by* (7.2.4a) *and* (7.2.4b) *are said to be in canonical form.*

Remark 7.2.3 *For each* $i = 1, \ldots, N$*, the presence of a characteristic time* τ_i *for the* $i-th$ *constraint is to cater for the possibility of an interior point constraint.*

7.2.1 Gradient Formulae

To solve Problem $(P1)$ as a mathematical programming problem, we require the gradient of the functional g_i for each $i = 0, 1, \ldots, N$. Since the functionals depend implicitly on the system parameter vector via the dynamical system, the derivation of the gradients requires some care. The two common approaches are known as the *variational (or sensitivity) method* (see, for example, [111, 142, 144, 151, 158, 159, 169, 268]) and the *costate method* (see, for example, [75, 89, 148, 164, 168, 169, 180, 181, 238, 245, 246, 253]). We illustrate both methods below.

In the variational method, we first define an auxiliary dynamic system for each component of $\boldsymbol{\zeta}$. For $k = 1, \ldots, s$, consider

$$\frac{d\boldsymbol{\psi}^k(t)}{dt} = \frac{\partial \boldsymbol{f}(t, \boldsymbol{x}(t \mid \boldsymbol{\zeta}), \boldsymbol{\zeta})}{\partial \boldsymbol{x}} \boldsymbol{\psi}^k(t) + \frac{\partial \boldsymbol{f}(t, \boldsymbol{x}(t \mid \boldsymbol{\zeta}), \boldsymbol{\zeta})}{\partial \zeta_k}, \quad t \in [0, T], \tag{7.2.5}$$

along with the initial condition

$$\boldsymbol{\psi}^k(0) = \frac{\partial \boldsymbol{x}^0(\boldsymbol{\zeta})}{\partial \zeta_k}. \tag{7.2.6}$$

For each $\boldsymbol{\zeta} \in \mathcal{Z}$, let $\boldsymbol{\psi}^k(\cdot \mid \boldsymbol{\psi})$ denote the corresponding solution of (7.2.5)–(7.2.6).

Theorem 7.2.1 *For each* $\boldsymbol{\zeta} \in \mathcal{Z}$,

$$\frac{\partial \boldsymbol{x}(t \mid \boldsymbol{\zeta})}{\partial \zeta_k} = \boldsymbol{\psi}^k(t), \quad t \in [0, T]. \tag{7.2.7}$$

Proof. For any $t \in [0, T]$, the solution of (7.2.1) corresponding to $\boldsymbol{\zeta} \in \mathcal{Z}$ may be written as

$$\boldsymbol{x}(t \mid \boldsymbol{\zeta}) = \boldsymbol{x}^0(\boldsymbol{\zeta}) + \int_0^t \boldsymbol{f}(s, \boldsymbol{x}(s \mid \boldsymbol{\zeta}), \boldsymbol{\zeta})\, ds. \tag{7.2.8}$$

Differentiating both sides of (7.2.8) with respect to ζ_k, we obtain

$$\frac{\partial \boldsymbol{x}(t|\boldsymbol{\zeta})}{\partial \zeta_k} = \frac{\partial \boldsymbol{x}^0(\boldsymbol{\zeta})}{\partial \zeta_k} + \int_0^t \left[\frac{\partial \boldsymbol{f}(s, \boldsymbol{x}(s|\boldsymbol{\zeta}), \boldsymbol{\zeta})}{\partial \boldsymbol{x}} \frac{\partial \boldsymbol{x}(s|\boldsymbol{\zeta})}{\partial \zeta_k} + \frac{\partial \boldsymbol{f}(s, \boldsymbol{x}(s|\boldsymbol{\zeta}), \boldsymbol{\zeta})}{\partial \zeta_k} \right] ds. \tag{7.2.9}$$

Substituting $t = 0$ into (7.2.9) gives

$$\frac{\partial \boldsymbol{x}(0 \mid \boldsymbol{\zeta})}{\partial \zeta_k} = \frac{\partial \boldsymbol{x}^0(\boldsymbol{\zeta})}{\partial \zeta_k}, \tag{7.2.10}$$

while differentiating both sides of (7.2.9) with respect to t yields

$$\frac{d}{dt}\left(\frac{\partial \boldsymbol{x}(t \mid \boldsymbol{\zeta})}{\partial \zeta_k} \right) = \frac{\partial \boldsymbol{f}(t, \boldsymbol{x}(t \mid \boldsymbol{\zeta}), \boldsymbol{\zeta})}{\partial \boldsymbol{x}} \frac{\partial \boldsymbol{x}(t \mid \boldsymbol{\zeta})}{\partial \zeta_k} + \frac{\partial \boldsymbol{f}(t, \boldsymbol{x}(t \mid \boldsymbol{\zeta}), \boldsymbol{\zeta})}{\partial \zeta_k}. \tag{7.2.11}$$

Noting the equivalence of (7.2.11) and (7.2.5) as well as that of(7.2.10) and (7.2.6), the result follows.

Using (7.2.7), the gradients of the cost and constraint functionals can then be calculated on the basis of the chain rule of differentiation as follows. For each $i = 0, 1, \ldots, N$,

$$\begin{aligned}
\frac{\partial g_i(\boldsymbol{\zeta})}{\partial \zeta_k} =& \frac{\partial \Phi_i(\boldsymbol{x}(\tau_i \mid \boldsymbol{\zeta}), \boldsymbol{\zeta})}{\partial \zeta_k} + \frac{\partial \Phi_i(\boldsymbol{x}(\tau_i \mid \boldsymbol{\zeta}), \boldsymbol{\zeta})}{\partial \boldsymbol{x}} \frac{\partial \boldsymbol{x}(\tau_i \mid \boldsymbol{\zeta})}{\partial \zeta_k} \\
&+ \int_0^{\tau_i} \left[\frac{\partial \mathcal{L}_i(t, \boldsymbol{x}(t \mid \boldsymbol{\zeta}), \boldsymbol{\zeta})}{\partial \boldsymbol{x}} \frac{\partial \boldsymbol{x}(t \mid \boldsymbol{\zeta})}{\partial \zeta_k} + \frac{\partial \mathcal{L}_i(t, \boldsymbol{x}(t \mid \boldsymbol{\zeta}), \boldsymbol{\zeta})}{\partial \zeta_k} \right] dt \\
=& \frac{\partial \Phi_i(\boldsymbol{x}(\tau_i \mid \boldsymbol{\zeta}), \boldsymbol{\zeta})}{\partial \zeta_k} + \frac{\partial \Phi_i(\boldsymbol{x}(\tau_i \mid \boldsymbol{\zeta}), \boldsymbol{\zeta})}{\partial \boldsymbol{x}} \boldsymbol{\psi}^k(\tau_i \mid \boldsymbol{\zeta}) \\
&+ \int_0^{\tau_i} \left[\frac{\partial \mathcal{L}_i(t, \boldsymbol{x}(t \mid \boldsymbol{\zeta}), \boldsymbol{\zeta})}{\partial \boldsymbol{x}} \boldsymbol{\psi}^k(t \mid \boldsymbol{\zeta}) + \frac{\partial \mathcal{L}_i(t, \boldsymbol{x}(t \mid \boldsymbol{\zeta}), \boldsymbol{\zeta})}{\partial \zeta_k} \right] dt.
\end{aligned} \tag{7.2.12}$$

The gradient derivation via the costate method is carried out as follows. For each $i = 0, 1, \ldots, N$, let the corresponding Hamiltonian H_i be defined by

$$H_i(t, \boldsymbol{x}, \boldsymbol{\zeta}, \boldsymbol{\lambda}) = \mathcal{L}_i(t, \boldsymbol{x}, \boldsymbol{\zeta}) + \left(\boldsymbol{\lambda}^i(t)\right)^\top \boldsymbol{f}(t, \boldsymbol{x}, \boldsymbol{\zeta}), \tag{7.2.13}$$

where, for each $\boldsymbol{\zeta} \in \mathbb{R}^s$, $\boldsymbol{\lambda}^i(t)$ is governed by the system

$$\frac{d\boldsymbol{\lambda}^i(t)}{dt} = -\left[\frac{\partial H_i\left(t, \boldsymbol{x}(t \mid \boldsymbol{\zeta}), \boldsymbol{\zeta}, \boldsymbol{\lambda}^i(t)\right)}{\partial \boldsymbol{x}} \right]^\top, \quad t \in [0, \tau_i) \tag{7.2.14a}$$

with the condition

$$\boldsymbol{\lambda}^i(\tau_i) = \left[\frac{\partial \Phi_i(\boldsymbol{x}(\tau_i \mid \boldsymbol{\zeta}), \boldsymbol{\zeta})}{\partial \boldsymbol{x}} \right]^\top. \tag{7.2.14b}$$

Here, $\boldsymbol{x}(\cdot \mid \boldsymbol{\zeta})$ denotes the solution of the system (7.2.1) corresponding to $\boldsymbol{\zeta} \in \mathbb{R}^s$, the right hand side of (7.2.14a) denotes the transpose of the gradient of H_i with respect to $\boldsymbol{x}$ evaluated at $\boldsymbol{x}(t \mid \boldsymbol{\zeta})$, and the right hand side of (7.2.14b) is to be understood similarly. The system (7.2.14) is known as the *costate system*. Let $\boldsymbol{\lambda}^i(\cdot \mid \boldsymbol{\zeta})$ be the solution of this costate system corresponding to $\boldsymbol{\zeta} \in \mathbb{R}^s$.

Theorem 7.2.2 *Consider Problem (P1). For each $i = 0, 1, \ldots, N$, the gradient of the functional g_i is given by*

$$\frac{\partial g_i(\boldsymbol{\zeta})}{\partial \boldsymbol{\zeta}} = \frac{\partial \Phi_i(\boldsymbol{x}(\tau_i \mid \boldsymbol{\zeta}), \boldsymbol{\zeta})}{\partial \boldsymbol{\zeta}} + \left(\boldsymbol{\lambda}^i(0 \mid \boldsymbol{\zeta})\right)^\top \frac{\partial \boldsymbol{x}^0(\boldsymbol{\zeta})}{\partial \boldsymbol{\zeta}} + \int_0^{\tau_i} \frac{\partial H_i\left(t, \boldsymbol{x}(t \mid \boldsymbol{\zeta}), \boldsymbol{\zeta}, \boldsymbol{\lambda}^i(t \mid \boldsymbol{\zeta})\right)}{\partial \boldsymbol{\zeta}} \, dt. \tag{7.2.15}$$

Proof. Let $\boldsymbol{\zeta} \in \mathbb{R}^s$ be given and let $\boldsymbol{\rho} \in \mathbb{R}^s$ be arbitrary but fixed. Define

$$\boldsymbol{\zeta}(\varepsilon) = \boldsymbol{\zeta} + \varepsilon \boldsymbol{\rho}, \tag{7.2.16}$$

where $\varepsilon > 0$ is an arbitrarily small real number. For brevity, let $\boldsymbol{x}(\cdot)$ and $\boldsymbol{x}(\cdot; \varepsilon)$ denote, respectively, the solution of the system (7.2.1) corresponding to $\boldsymbol{\zeta}$ and $\boldsymbol{\zeta}(\varepsilon)$. Clearly, from (7.2.1), we have

$$\boldsymbol{x}(t) = \boldsymbol{x}^0(\boldsymbol{\zeta}) + \int_0^t \boldsymbol{f}(s, \boldsymbol{x}(s), \boldsymbol{\zeta}) \, ds \tag{7.2.17}$$

and

$$\boldsymbol{x}(t; \varepsilon) = \boldsymbol{x}^0(\boldsymbol{\zeta}(\varepsilon)) + \int_0^t \boldsymbol{f}(s, \boldsymbol{x}(s; \varepsilon), \boldsymbol{\zeta}(\varepsilon)) \, ds. \tag{7.2.18}$$

Thus,

$$\begin{aligned} \triangle \boldsymbol{x}(t) &= \left.\frac{d\boldsymbol{x}(t; \varepsilon)}{d\varepsilon}\right|_{\varepsilon=0} \\ &= \frac{\partial \boldsymbol{x}^0(\boldsymbol{\zeta})}{\partial \boldsymbol{\zeta}} \boldsymbol{\rho} + \int_0^t \left\{ \frac{\partial \boldsymbol{f}(s, \boldsymbol{x}(s), \boldsymbol{\zeta})}{\partial \boldsymbol{x}} \triangle \boldsymbol{x}(s) + \frac{\partial \boldsymbol{f}(s, \boldsymbol{x}(s), \boldsymbol{\zeta})}{\partial \boldsymbol{\zeta}} \boldsymbol{\rho} \right\} ds. \end{aligned} \tag{7.2.19}$$

Clearly,

$$\frac{d(\triangle \boldsymbol{x}(t))}{dt} = \frac{\partial \boldsymbol{f}(t, \boldsymbol{x}(t), \boldsymbol{\zeta})}{\partial \boldsymbol{x}} \triangle \boldsymbol{x}(t) + \frac{\partial \boldsymbol{f}(t, \boldsymbol{x}(t), \boldsymbol{\zeta})}{\partial \boldsymbol{\zeta}} \boldsymbol{\rho} \tag{7.2.20a}$$

$$\triangle \boldsymbol{x}(0) = \frac{\partial \boldsymbol{x}^0(\boldsymbol{\zeta})}{\partial \boldsymbol{\zeta}} \boldsymbol{\rho}. \tag{7.2.20b}$$

Now $g_i(\boldsymbol{\zeta}(\varepsilon))$ can be expressed as

$$\begin{aligned} g_i(\boldsymbol{\zeta}(\varepsilon)) =& \Phi_i(\boldsymbol{x}(\tau_i;\varepsilon),\boldsymbol{\zeta}(\varepsilon)) + \int_0^{\tau_i} \Big\{ H_i\left(t,\boldsymbol{x}(t;\varepsilon),\boldsymbol{\zeta}(\varepsilon),\boldsymbol{\lambda}^i(t)\right) \\ & - \left(\boldsymbol{\lambda}^i(t)\right)^\top \boldsymbol{f}(t,\boldsymbol{x}(t;\varepsilon),\boldsymbol{\zeta}(\varepsilon)) \Big\}\, dt, \end{aligned} \tag{7.2.21}$$

where $\boldsymbol{\lambda}^i$ is as yet arbitrary. Thus

$$\begin{aligned} \triangle g_i(\boldsymbol{\zeta}) =& \left.\frac{dg_i(\boldsymbol{\zeta}(\varepsilon))}{d\varepsilon}\right|_{\varepsilon=0} \\ =& \frac{\partial g_i(\boldsymbol{\zeta})}{\partial \boldsymbol{\zeta}}\boldsymbol{\rho} \\ =& \triangle\Phi_i(\boldsymbol{x}(\tau_i),\boldsymbol{\zeta}) + \int_0^{\tau_i} \Big\{ \triangle H_i\left(t,\boldsymbol{x}(t),\boldsymbol{\zeta},\boldsymbol{\lambda}^i(t)\right) \\ & - \left(\boldsymbol{\lambda}^i(t)\right)^\top \triangle\boldsymbol{f}(t,\boldsymbol{x}(t),\boldsymbol{\zeta}) \Big\} dt, \end{aligned} \tag{7.2.22}$$

where

$$\triangle\Phi_i(\boldsymbol{x}(\tau_i),\boldsymbol{\zeta}) = \frac{\partial\Phi_i(\boldsymbol{x}(\tau_i),\boldsymbol{\zeta})}{\partial \boldsymbol{x}}\triangle\boldsymbol{x}(\tau_i) + \frac{\partial\Phi_i(\boldsymbol{x}(\tau_i),\boldsymbol{\zeta})}{\partial \boldsymbol{\zeta}}\boldsymbol{\rho}, \tag{7.2.23}$$

$$\triangle\boldsymbol{f}(t,\boldsymbol{x}(t),\boldsymbol{\zeta}) = \frac{d(\triangle\boldsymbol{x}(t))}{dt}, \tag{7.2.24}$$

and

$$\begin{aligned} & \triangle H_i\left(t,\boldsymbol{x}(t),\boldsymbol{\zeta},\boldsymbol{\lambda}^i(t)\right) \\ =& \frac{\partial H_i\left(t,\boldsymbol{x}(t),\boldsymbol{\zeta},\boldsymbol{\lambda}^i(t)\right)}{\partial \boldsymbol{x}}\triangle\boldsymbol{x}(t) + \frac{\partial H_i\left(t,\boldsymbol{x}(t),\boldsymbol{\zeta},\boldsymbol{\lambda}^i(t)\right)}{\partial \boldsymbol{\zeta}}\boldsymbol{\rho}. \end{aligned} \tag{7.2.25}$$

Choose $\boldsymbol{\lambda}^i$ to be the solution of the costate system (7.2.14) corresponding to $\boldsymbol{\zeta}$. Then, by substituting (7.2.14a) into (7.2.25), we obtain

$$\begin{aligned} & \triangle H_i\left(t,\boldsymbol{x}(t),\boldsymbol{\zeta},\boldsymbol{\lambda}^i(t)\right) \\ =& -\frac{d\left(\boldsymbol{\lambda}^i(t)\right)^\top}{dt}\triangle\boldsymbol{x}(t) + \frac{\partial H_i\left(t,\boldsymbol{x}(t),\boldsymbol{\zeta},\boldsymbol{\lambda}^i(t)\right)}{\partial \boldsymbol{\zeta}}\boldsymbol{\rho}. \end{aligned} \tag{7.2.26}$$

Hence, (7.2.22) yields

$$\begin{aligned} \frac{\partial g_i(\boldsymbol{\zeta})}{\partial \boldsymbol{\zeta}}\boldsymbol{\rho} =& \frac{\partial\Phi_i(\boldsymbol{x}(\tau_i),\boldsymbol{\zeta})}{\partial \boldsymbol{x}}\triangle\boldsymbol{x}(\tau_i) + \frac{\partial\Phi_i(\boldsymbol{x}(\tau_i),\boldsymbol{\zeta})}{\partial \boldsymbol{\zeta}}\boldsymbol{\rho} \\ & + \int_0^{\tau_i} \left\{ -\frac{d}{dt}\left[\left(\boldsymbol{\lambda}^i(t)\right)^\top \triangle\boldsymbol{x}(t)\right] + \frac{\partial H_i\left(t,\boldsymbol{x}(t),\boldsymbol{\zeta},\boldsymbol{\lambda}^i(t)\right)}{\partial \boldsymbol{\zeta}}\boldsymbol{\rho} \right\} dt \end{aligned}$$

$$
\begin{aligned}
&= \frac{\partial \Phi_i(\boldsymbol{x}(\tau_i), \boldsymbol{\zeta})}{\partial \boldsymbol{x}} \triangle \boldsymbol{x}(\tau_i) + \frac{\partial \Phi_i(\boldsymbol{x}(\tau_i), \boldsymbol{\zeta})}{\partial \boldsymbol{\zeta}} \boldsymbol{\rho} - \left(\boldsymbol{\lambda}^i(\tau_i)\right)^\top \triangle \boldsymbol{x}(\tau_i) \\
&+ \left(\boldsymbol{\lambda}^i(0)\right)^\top \triangle \boldsymbol{x}(0) + \int_0^{\tau_i} \left\{ \frac{\partial H_i\left(t, \boldsymbol{x}(t), \boldsymbol{\zeta}, \boldsymbol{\lambda}^i(t)\right)}{\partial \boldsymbol{\zeta}} \boldsymbol{\rho} \right\} dt. \quad (7.2.27)
\end{aligned}
$$

Substituting (7.2.14b) and (7.2.20b) into (7.2.27), we have

$$
\begin{aligned}
\frac{\partial g_i(\boldsymbol{\zeta})}{\partial \boldsymbol{\zeta}} \boldsymbol{\rho} =& \frac{\partial \Phi_i(\boldsymbol{x}(\tau_i), \boldsymbol{\zeta})}{\partial \boldsymbol{\zeta}} \boldsymbol{\rho} + (\boldsymbol{\lambda}^i(0))^\top \frac{\partial \boldsymbol{x}^0(\boldsymbol{\zeta})}{\partial \boldsymbol{\zeta}} \boldsymbol{\rho} \\
&+ \int_0^{\tau_i} \left\{ \frac{\partial H_i\left(t, \boldsymbol{x}(t), \boldsymbol{\zeta}, \boldsymbol{\lambda}^i(t)\right)}{\partial \boldsymbol{\zeta}} \boldsymbol{\rho} \right\} dt. \qquad (7.2.28)
\end{aligned}
$$

Since $\boldsymbol{\rho}$ is arbitrary, (7.2.15) follows readily from (7.2.28) and the proof is complete.

Remark 7.2.4 *The choice between gradient formulae* (7.2.12) *and* (7.2.15) *is problem dependent. If the number of system parameters, s, is large, the use of* (7.2.12) *requires the solution of a large number of auxiliary systems compared to the use of* (7.2.15)*, which requires the solution of just one costate system. On the other hand, the costate system associated with* (7.2.15) *needs to be solved backwards along the time horizon and requires the solution of the state dynamics to be stored beforehand, whereas* (7.2.12) *can be solved forward in time alongside the state dynamics. Nevertheless, the use of* (7.2.15) *is generally more efficient in a computational sense and hence is described in more detail in the next subsection.*

7.2.2 A Unified Computational Approach

The optimal parameter selection problem, Problem $(P1)$, is essentially a nonlinear mathematical programming problem which is typically solved by a gradient-based numerical algorithm. At each iteration of a gradient-based numerical optimization algorithm, it is necessary to compute the value of the cost functional and the values of all the constraint functionals as well as their respective gradients. We shall do this in a unified manner. To be more precise, the cost functional and all the constraint functionals are treated the same way in as far as the computations of their values and their respective gradients are concerned. To compute these values and the respective gradients, the first task is to calculate the solution of the system (7.2.1) corresponding to each $\boldsymbol{\zeta} \in \mathcal{Z}$. This is presented as an algorithm for future reference.

Algorithm 7.2.1 *For each given* $\boldsymbol{\zeta} \in \mathcal{Z}$*, compute the solution* $\boldsymbol{x}(\cdot \mid \boldsymbol{\zeta})$ *of the system* (7.2.1) *by solving the differential equations* (7.2.1a) *forward in time from* $t = 0$ *to* $t = T$ *with the initial condition* (7.2.1b).

With the information obtained in Algorithm (7.2.1), the values of $g_i(\boldsymbol{\zeta})$, $i = 0, 1, \ldots, N$, corresponding to each $\boldsymbol{\zeta} \in \mathcal{Z}$ can be easily calculated by the following simple algorithm.

Algorithm 7.2.2 *For each given $\boldsymbol{\zeta} \in \mathcal{Z}$, compute the corresponding value of $g_i(\boldsymbol{\zeta})$ from (7.2.3) (respectively, (7.2.4)) if $i = 0$ (respectively, $i = 1, \ldots, N$).*

In view of Theorem 7.2.2, we see that the derivations of the gradient formulae for the cost functional and the canonical constraint functionals are the same. For each $i = 0, 1, \ldots, N$, the gradient of the corresponding $g_i(\boldsymbol{\zeta})$ may be computed using the following algorithm.

Algorithm 7.2.3 *For a given $\boldsymbol{\zeta} \in \mathcal{Z}$.*

Step 1 *Solve the costate differential equation (7.2.14) backward in time from $t = \tau_i$ to $t = 0$, where $\boldsymbol{x}(\cdot \mid \boldsymbol{\zeta})$ is the solution of the system (7.2.1) corresponding to $\boldsymbol{\zeta} \in \mathcal{Z}$. Let $\boldsymbol{\lambda}^i(\cdot \mid \boldsymbol{\zeta})$ be the solution of the costate system (7.2.14).*

Step 2 *The gradient of $g_i(\boldsymbol{\zeta})$ with respect to $\boldsymbol{\zeta} \in \mathcal{Z}$ is computed from (7.2.15).*

Remark 7.2.5 *In the above algorithm, the solution $\boldsymbol{x}(\cdot \mid \boldsymbol{\zeta})$ of the system (7.2.1) corresponding to each $\boldsymbol{\zeta} \in \mathcal{Z}$ is computed by Algorithm 7.2.1.*

7.3 Control Parametrization

The most basic form of control parametrization assumes a piecewise constant form for each control function. In this section, we described a class of optimal control problems with piecewise constant controls and derive the gradients of the cost and constraint functionals with respect to the heights of the controls. Consider the process described by the following system of nonlinear differential equations on the fixed time interval $(0, T]$:

$$\frac{d\boldsymbol{x}(t)}{dt} = \boldsymbol{f}(t, \boldsymbol{x}(t), \boldsymbol{u}(t)), \tag{7.3.1a}$$

where

$$\boldsymbol{x} = [x_1, \ldots, x_n]^\top \in \mathbb{R}^n, \quad \boldsymbol{u} = [u_1, \ldots, u_r]^\top \in \mathbb{R}^r$$

are, respectively, the state and control vectors. The initial condition for the differential equation (7.3.1a) is

$$\boldsymbol{x}(0) = \boldsymbol{x}^0, \tag{7.3.1b}$$

where $\boldsymbol{x}^0 \in \mathbb{R}^n$ is a given vector.

For each $k = 1, \ldots, r$, the k−th component, u_k, of the control $\boldsymbol{u}$ is a piecewise constant function over the interval $[0, T]$ with jumps at $t_1, \ldots, t_M$. In other words, it takes a constant value until the next switching time is reached, at which point the value changes instantaneously to another constant, which is held until the next switching time. Mathematically, u_k may be expressed as:

$$u_k(t) = \sum_{j=0}^{M} h_k^j \chi_{[t_j, t_{j+1})}(t), \tag{7.3.2}$$

where

$$\chi_I(t) = \begin{cases} 1, & \text{if } t \in I, \\ 0, & \text{otherwise.} \end{cases} \tag{7.3.3}$$

h_k^j, $j = 0, 1, \ldots, M$, are decision parameters satisfying

$$a_k \le h_k^j \le b_k, \quad k = 1, \ldots, r \quad \text{and} \quad j = 0, 1, \ldots, M, \tag{7.3.4}$$

where a_k and b_k, $k = 1, \ldots, r$, are given constants. Furthermore, $t_0 = 0$, $t_{M+1} = T$, and the given switching times t_j, $j = 1, \ldots, M$, satisfy

$$0 \le t_1 \le \cdots \le t_M \le T. \tag{7.3.5}$$

Let

$$\boldsymbol{h} = \left[(\boldsymbol{h}^0)^\top, (\boldsymbol{h}^1)^\top, \ldots, (\boldsymbol{h}^M)^\top\right]^\top \in \mathbb{R}^{(M+1)r}, \tag{7.3.6}$$

where

$$\boldsymbol{h}^j = \left[h_1^j, \ldots, h_r^j\right]^\top \in \mathbb{R}^r, \quad j = 0, 1, \ldots, M. \tag{7.3.7}$$

Let Λ be the set of all those decision parameter vectors $\boldsymbol{h}$ which satisfy (7.3.4). For convenience, for any $\boldsymbol{h} \in \Lambda$, the corresponding control is written as $\boldsymbol{u}(\cdot \mid \boldsymbol{h})$. Let $\mathcal{U}$ denote the set of all such controls. Clearly, each control in $\mathcal{U}$ is determined uniquely by a decision parameter vector $\boldsymbol{h}$ in Λ and vice versa.

Let $\boldsymbol{x}(\cdot \mid \boldsymbol{h})$ denote the solution of the system (7.3.1) corresponding to $\boldsymbol{u}(\cdot \mid \boldsymbol{h}) \in \mathcal{U}$ (and hence corresponding to $\boldsymbol{h} \in \Lambda$). We may now state the optimal control problem as follows.

Problem $(P2)$**:** Given the system (7.3.1), find a decision parameter vector $\boldsymbol{h} \in \Lambda$ such that the cost functional

$$g_0(\boldsymbol{h}) = \Phi_0(\boldsymbol{x}(T \mid \boldsymbol{h})) + \int_0^T \mathcal{L}_0(t, \boldsymbol{x}(t \mid \boldsymbol{h}), \boldsymbol{u}(t \mid \boldsymbol{h}))\, dt \tag{7.3.8}$$

is minimized subject to the equality constraints (in canonical form)

$$g_i(\boldsymbol{h}) = \Phi_i(\boldsymbol{x}(\tau_i \mid \boldsymbol{h})) + \int_0^{\tau_i} \mathcal{L}_i(t, \boldsymbol{x}(t \mid \boldsymbol{h}), \boldsymbol{u}(t \mid \boldsymbol{h}))\, dt = 0,$$
$$i = 1, \ldots, N_e, \tag{7.3.9a}$$

and the inequality constraints (in canonical form)

$$g_i(\boldsymbol{h}) = \Phi_i(\boldsymbol{x}(\tau_i \mid \boldsymbol{h})) + \int_0^{\tau_i} \mathcal{L}_i(t, \boldsymbol{x}(t \mid \boldsymbol{h}), \boldsymbol{u}(t \mid \boldsymbol{h}))\, dt \le 0,$$
$$i = N_e + 1, \ldots, N, \tag{7.3.9b}$$

where $\Phi_i : \mathbb{R}^n \mapsto \mathbb{R}$ and $\mathcal{L}_i : [0, T] \times \mathbb{R}^n \times \mathbb{R}^r \mapsto \mathbb{R}$, $i = 0, 1, \ldots, N$, are given real-valued functions.

We assume that the corresponding versions of Assumptions 7.2.1, 7.2.3 and 7.4.1 are satisfied throughout this section.

Problem $(P2)$ is, in essence, a special case of Problem $(P1)$ where the system parameter vector is now the vector $\boldsymbol{h}$ of control heights. In view of (7.3.2), we may write the dynamics (7.3.1a) in the following more detailed form.

$$\frac{d\boldsymbol{x}(t)}{dt} = \boldsymbol{f}\left(t, \boldsymbol{x}(t), \boldsymbol{h}^j\right), \quad t \in [t_j, t_{j+1}),\ j = 0, \ldots, M. \tag{7.3.10}$$

Similarly, for each $i = 0, 1, \ldots, N$, the corresponding functional defined by (7.3.8)–(7.3.9) may be written as

$$g_i(\boldsymbol{h}) = \Phi_i(\boldsymbol{x}(\tau_i \mid \boldsymbol{h})) + \sum_{j=0}^{M_i - 1} \int_{t_j}^{t_{j+1}} \mathcal{L}_i(t, \boldsymbol{x}(t \mid \boldsymbol{h}), \boldsymbol{h}^j)\, dt + \int_{t_{M_i}}^{\tau_i} \mathcal{L}_i\left(t, \boldsymbol{x}(t \mid \boldsymbol{h}), \boldsymbol{h}^{M_i}\right) dt, \tag{7.3.11}$$

where $\tau_0 = T$ and $M_i \in \{0, 1, \ldots, M\}$ is such that $\tau_i \in [t_{M_i}, t_{M_i+1})$.

To derive the gradients of each functional with respect to the control heights, we can once again follow either the variational or the costate approach as we did in Section 7.2. All the formulae we derived in Section 7.2 are still valid if we replace $\boldsymbol{\zeta}$ with $\boldsymbol{h}$. However, due to the special forms of (7.3.10) and (7.3.11), we can state the gradient formulae with respect to $\boldsymbol{h}$ in a manner that allows for a more efficient computational implementation.

For each $j = 0, \ldots, M$ and each $k = 1, \ldots, r$, consider the variational system

$$\frac{d\boldsymbol{\psi}_k^j(t)}{dt} = \frac{\partial \boldsymbol{f}(t, \boldsymbol{x}(t \mid \boldsymbol{h}), \boldsymbol{u}(t \mid \boldsymbol{h}))}{\partial \boldsymbol{x}} \boldsymbol{\psi}_k^j(t) + \frac{\partial \boldsymbol{f}(t, \boldsymbol{x}(t \mid \boldsymbol{h}), \boldsymbol{u}(t \mid \boldsymbol{h}))}{\partial h_k^j}, \quad t \in [0, T], \tag{7.3.12}$$

along with the initial condition

$$\boldsymbol{\psi}_k^j(0) = \mathbf{0}. \tag{7.3.13}$$

In view of (7.3.10), this variational system may also be written as

$$\begin{aligned} \frac{d\boldsymbol{\psi}_k^j(t)}{dt} &= \frac{\partial \boldsymbol{f}\left(t, \boldsymbol{x}(t \mid \boldsymbol{h}), \boldsymbol{h}^l\right)}{\partial \boldsymbol{x}} \boldsymbol{\psi}_k^j(t) + \frac{\partial \boldsymbol{f}\left(t, \boldsymbol{x}(t \mid \boldsymbol{h}), \boldsymbol{h}^l\right)}{\partial h_k^l}, \\ &= \frac{\partial \boldsymbol{f}\left(t, \boldsymbol{x}(t \mid \boldsymbol{h}), \boldsymbol{h}^l\right)}{\partial \boldsymbol{x}} \boldsymbol{\psi}_k^j(t) + \delta_k^l \frac{\partial \boldsymbol{f}\left(t, \boldsymbol{x}(t \mid \boldsymbol{h}), \boldsymbol{u}(t \mid \boldsymbol{h})\right)}{\partial u_k}, \end{aligned}$$

$$t \in [t_l, t_{l+1}), \ l = j, \ldots, M, \tag{7.3.14}$$

with

$$\psi_k^j(t) = \mathbf{0}, \quad t \in [0, t_j), \tag{7.3.15}$$

where

$$\delta_k^l = \begin{cases} 1, \text{ if } k = l, \\ 0, \text{ otherwise.} \end{cases}$$

In other words, the integration to determine $\psi_k^j(\cdot)$ only needs to commence at $t = t_j$. Let $\psi_k^j(t \mid \boldsymbol{h})$ denote the solution of (7.3.14)–(7.3.15) corresponding to $\boldsymbol{h} \in \Lambda$. We may then state the following theorem.

Theorem 7.3.1 *For each $i \in \{0, 1 \ldots, N\}$, the gradient of the functional g_i with respect to h_k^j, $k = 1, \ldots, r$, $j = 0, \ldots, M$, is given by*

$$\begin{aligned}
\frac{\partial g_i(\boldsymbol{h})}{\partial h_k^j} &= \frac{\partial \Phi_i(\boldsymbol{x}(\tau_i \mid \boldsymbol{h}))}{\partial \boldsymbol{x}} \psi_k^j(\tau_i) \\
&\quad + \int_0^{\tau_i} \left[\frac{\partial \mathcal{L}_i(t, \boldsymbol{x}(t \mid \boldsymbol{h}), \boldsymbol{u}(t \mid \boldsymbol{h}))}{\partial \boldsymbol{x}} \psi_k^j(t) + \frac{\partial \mathcal{L}_i(t, \boldsymbol{x}(t \mid \boldsymbol{h}), \boldsymbol{u}(t \mid \boldsymbol{h}))}{\partial h_k^j} \right] dt \\
&= \frac{\partial \Phi_i(\boldsymbol{x}(\tau_i \mid \boldsymbol{h}))}{\partial \boldsymbol{x}} \psi_k^j(\tau_i) \\
&\quad + \sum_{l=0}^{M_i - 1} \int_{t_l}^{t_{l+1}} \left[\frac{\partial \mathcal{L}_i\left(t, \boldsymbol{x}(t \mid \boldsymbol{h}), \boldsymbol{h}^l\right)}{\partial \boldsymbol{x}} \psi_k^j(t) + \frac{\partial \mathcal{L}_i\left(t, \boldsymbol{x}(t \mid \boldsymbol{h}), \boldsymbol{h}^l\right)}{\partial h_k^j} \right] dt \\
&\quad + \int_{t_{M_i}}^{\tau_i} \left[\frac{\partial \mathcal{L}_i\left(t, \boldsymbol{x}(t \mid \boldsymbol{h}), \boldsymbol{h}^{M_i}\right)}{\partial \boldsymbol{x}} \psi_k^j(t) + \frac{\partial \mathcal{L}_i\left(t, \boldsymbol{x}\left(t \mid \boldsymbol{h}\right), \boldsymbol{h}^{M_i}\right)}{\partial h_k^j} \right] dt \\
&= \sum_{l=0}^{M_i - 1} \int_{t_l}^{t_{l+1}} \frac{\partial \mathcal{L}_i\left(t, \boldsymbol{x}(t|\boldsymbol{h}), \boldsymbol{h}^l\right)}{\partial \boldsymbol{x}} \psi_k^j(t) dt + \int_{t_{M_i}}^{\tau_i} \frac{\partial \mathcal{L}_i\left(t, \boldsymbol{x}(t|\boldsymbol{h}), \boldsymbol{h}^{M_i}\right)}{\partial \boldsymbol{x}} \psi_k^j(t) dt \\
&\quad + \begin{cases} \int_{t_j}^{t_{j+1}} \frac{\partial \mathcal{L}_i\left(t, \boldsymbol{x}(t|\boldsymbol{h}), \boldsymbol{h}^j\right)}{\partial h_k^j} dt, \text{ if } [t_j, t_{j+1}] \subset [0, t_{M_i}], \\ \int_{t_{M_i}}^{\tau_i} \frac{\partial \mathcal{L}_i\left(t, \boldsymbol{x}(t|\boldsymbol{h}), \boldsymbol{h}^j\right)}{\partial h_k^j} dt, \ \text{ if } [t_j, t_{j+1}] = [t_{M_i}, t_{M_i+1}], \\ 0, \qquad\qquad\qquad\qquad \text{ if } [t_j, t_{j+1}] \subset [\tau_i, T], \end{cases}
\end{aligned} \tag{7.3.16}$$

where the subsequent equalities follow due to (7.3.11) *and τ_0 and M_i are the same as defined in that equation.*

The costate approach to determine the gradients also follows from the corresponding results in Section 7.2. Replacing $\boldsymbol{\zeta}$ with $\boldsymbol{h}$ in Equations (7.2.13)–(7.2.15), we obtain the following results. For each $i = 0, 1, \ldots, N$, let the Hamiltonian H_i be defined by

$$\begin{aligned}
H_i(t, \boldsymbol{x}(t \mid \boldsymbol{h}), \boldsymbol{u}(t \mid \boldsymbol{h}), \boldsymbol{\lambda}^i(t)) &= \mathcal{L}_i(t, \boldsymbol{x}(t \mid \boldsymbol{h}), \boldsymbol{u}(t \mid \boldsymbol{h})) \\
&\quad + (\boldsymbol{\lambda}^i(t))^\top \boldsymbol{f}(t, \boldsymbol{x}(t \mid \boldsymbol{h}), \boldsymbol{u}(t \mid \boldsymbol{h})),
\end{aligned} \tag{7.3.17}$$

where, for each $\boldsymbol{h} \in \Lambda$, $\boldsymbol{\lambda}^i(\cdot)$ is governed by the system

$$\frac{d\boldsymbol{\lambda}^i(t)}{dt} = -\left[\frac{\partial H_i(t, \boldsymbol{x}(t \mid \boldsymbol{h}), \boldsymbol{u}(t \mid \boldsymbol{h}), \boldsymbol{\lambda}^i(t))}{\partial \boldsymbol{x}}\right]^\top, \quad t \in [0, \tau_i) \tag{7.3.18a}$$

with the condition

$$\boldsymbol{\lambda}^i(\tau_i) = \left[\frac{\partial \Phi_i(\boldsymbol{x}(\tau_i \mid \boldsymbol{h}))}{\partial \boldsymbol{x}}\right]^\top. \tag{7.3.18b}$$

Let $\boldsymbol{\lambda}^i(\cdot \mid \boldsymbol{h})$ denote the solution of this costate system corresponding to $\boldsymbol{h} \in \Lambda$. Note that, by virtue of (7.3.2), for $t \in [t_j, t_{j+1})$,

$$H_i(t, \boldsymbol{x}(t \mid \boldsymbol{h}), \boldsymbol{u}(t \mid \boldsymbol{h}), \boldsymbol{\lambda}^i(t \mid \boldsymbol{h})) = H_i(t, \boldsymbol{x}(t \mid \boldsymbol{h}), \boldsymbol{h}^j, \boldsymbol{\lambda}^i(t \mid \boldsymbol{h})).$$

We may then state the gradient formula in the following theorem.

Theorem 7.3.2 *For each $i = 0, 1, \ldots, N$ and $j = 0, 1, \ldots, M$, the gradient of the functional g_i with respect to $\boldsymbol{h}^j$ is given by*

$$\begin{aligned}\frac{\partial g_i(\boldsymbol{h})}{\partial \boldsymbol{h}^j} &= \int_0^{\tau_i} \frac{\partial H_i(t, \boldsymbol{x}(t \mid \boldsymbol{h}), \boldsymbol{u}(t \mid \boldsymbol{h}), \boldsymbol{\lambda}^i(t \mid \boldsymbol{h}))}{\partial \boldsymbol{h}^j} dt \\ &= \begin{cases} \int_{t_j}^{t_{j+1}} \frac{\partial H_i(t, \boldsymbol{x}(t|\boldsymbol{h}), \boldsymbol{h}^j, \boldsymbol{\lambda}^i(t|\boldsymbol{h}))}{\partial \boldsymbol{h}^j} dt, & \text{if } [t_j, t_{j+1}] \subset [0, t_{M_i}], \\ \int_{t_{M_i}}^{\tau_i} \frac{\partial H_i(t, \boldsymbol{x}(t|\boldsymbol{h}), \boldsymbol{h}^j, \boldsymbol{\lambda}^i(t|\boldsymbol{h}))}{\partial \boldsymbol{h}^j} dt, & \text{if } [t_j, t_{j+1}] = [t_{M_i}, t_{M_{i+1}}], \\ 0, & \text{if } [t_j, t_{j+1}] \subset [\tau_i, T], \end{cases}\end{aligned} \tag{7.3.19}$$

where, once again, the latter equality in (7.3.19) *follows from* (7.3.11) *and τ_0 and M_i are the same as defined in that equation.*

For implementation purposes, it is also convenient to express these gradients in a component-wise form as follows.

$$\frac{\partial g_i(\boldsymbol{h})}{\partial h_k^j} = \begin{cases} \int_{t_j}^{t_{j+1}} \frac{\partial H_i(t, \boldsymbol{x}(t|\boldsymbol{h}), \boldsymbol{u}(t|\boldsymbol{h}), \boldsymbol{\lambda}^i(t|\boldsymbol{h}))}{\partial u_k} dt, & \text{if } [t_j, t_{j+1}] \subset [0, t_{M_i}], \\ \int_{t_{M_i}}^{\tau_i} \frac{\partial H_i(t, \boldsymbol{x}(t|\boldsymbol{h}), \boldsymbol{u}(t|\boldsymbol{h}), \boldsymbol{\lambda}^i(t|\boldsymbol{h}))}{\partial u_k} dt, & \text{if } [t_j, t_{j+1}] = [t_{M_i}, t_{M_{i+1}}], \\ 0, & \text{if } [t_j, t_{j+1}] \subset [\tau_i, T]. \end{cases} \tag{7.3.20}$$

Remark 7.3.1 *As M is typically much larger than N, the gradients based on the variational approach require a significantly larger amount of computation and are hence not used as much in practice.*

Remark 7.3.2 *We have assumed here that each component of the control $\boldsymbol{u}$ has the same set of switching times. This is purely for notational convenience. In practice, there is no reason for this restriction.*

7.4 Switching Times as Decision Parameters

Consider once more the dynamics described by (7.3.1) and assume that the control takes the form (7.3.2). In contrast to Section 7.3, we now assume that the control heights are given and that the switching times of the control are decision parameters. In other words, we consider h_k^j, $j = 0, 1, \ldots, M_k$, $k = 1, \ldots, r$, to be given and regard the switching times t_j, $j = 1, \ldots, M$, as decision parameters.

$\boldsymbol{\vartheta} = [t_1, \ldots, t_M]^\top$ is called the *switching vector*. Let Ξ be the set which consists of all those vectors $\boldsymbol{\vartheta} \in \mathbb{R}^M$ such that the constraints (7.3.5) are satisfied. Furthermore, let $\mathcal{U}$ denote the set of all the corresponding control functions. For convenience, for any $\boldsymbol{\vartheta} \in \Xi$, the corresponding control in $\mathcal{U}$ is written as $\boldsymbol{u}(\cdot \mid \boldsymbol{\vartheta})$. Clearly, each control in $\mathcal{U}$ is determined uniquely by a switching vector $\boldsymbol{\vartheta}$ in Ξ and vice versa.

Let $\boldsymbol{x}(\cdot \mid \boldsymbol{\vartheta})$ denote the solution of the system (7.3.1) corresponding to the control $\boldsymbol{u}(\cdot \mid \boldsymbol{\vartheta}) \in \mathcal{U}$ (and hence to the switching vector $\boldsymbol{\vartheta} \in \Xi$). We may then state the canonical optimal control problem as follows.

Problem $(P3)$**:** Given the system (7.3.1), find a switching vector $\boldsymbol{\vartheta} \in \Xi$ such that the cost functional

$$g_0(\boldsymbol{\vartheta}) = \Phi_0(\boldsymbol{x}(T \mid \boldsymbol{\vartheta})) + \int_0^T \mathcal{L}_0(t, \boldsymbol{x}(t \mid \boldsymbol{\vartheta}), \boldsymbol{u}(t \mid \boldsymbol{\vartheta}))\, dt, \tag{7.4.1}$$

is minimized subject to the equality constraints

$$g_i(\boldsymbol{\vartheta}) = \Phi_i(\boldsymbol{x}(T \mid \boldsymbol{\vartheta})) + \int_0^T \mathcal{L}_i(t, \boldsymbol{x}(t \mid \boldsymbol{\vartheta}), \boldsymbol{u}(t \mid \boldsymbol{\vartheta}))\, dt = 0, i = 1, \ldots, N_e, \tag{7.4.2a}$$

and the inequality constraints

$$g_i(\boldsymbol{\vartheta}) = \Phi_i(\boldsymbol{x}(T \mid \boldsymbol{\vartheta})) + \int_0^T \mathcal{L}_i(t, \boldsymbol{x}(t \mid \boldsymbol{\vartheta}), \boldsymbol{u}(t \mid \boldsymbol{\vartheta}))\, dt \leq 0, i = N_e + 1, \ldots, N, \tag{7.4.2b}$$

where $\Phi_i : \mathbb{R}^n \mapsto \mathbb{R}$ and $\mathcal{L}_i : [0, T] \times \mathbb{R}^n \times \mathbb{R}^r \mapsto \mathbb{R}$, $i = 0, 1, \ldots, N$, are given real-valued functions.

Let Ω be the subset of Ξ which consists of all those switching vectors $\boldsymbol{\vartheta}$ such that the constraints (7.4.2) are satisfied. Furthermore, let $\mathcal{F}$ be the corresponding subset of $\mathcal{U}$.

Remark 7.4.1 *Comparing the constraints* (7.2.4) *with the constraints* (7.4.2), *we observe that all the characteristic times in the constraints* (7.4.2) *are taken as* T. *This choice is for the sake of simplicity. In fact, the results to follow are also valid for the general case, albeit with more cumbersome notation.*

We assume that the corresponding versions of Assumptions 7.2.1 and 7.2.3 are satisfied. However, Assumption 7.2.2 needs to be replaced by a stronger version given below.

Assumption 7.4.1 *$\boldsymbol{f}$ and $\mathcal{L}_i$, $i = 0, 1, \ldots, N$, together with their partial derivatives with respect to each of the components of $\boldsymbol{x}$ and $\boldsymbol{u}$ are continuous on $[0, T] \times \mathbb{R}^n \times \mathbb{R}^r$.*

7.4.1 Gradient Computation

Note that the appearance of the switching times in Problem $(P3)$ is of a different nature than the appearance of $\boldsymbol{\zeta}$ in Problem $(P1)$, so we cannot simply adapt the gradient formulae derived in Section 7.2. Once again, we derive the new gradient formulae using both the variational and the costate approach. Throughout this section, recall that, by virtue of (7.3.2), we have

$$\boldsymbol{u}(t \mid \boldsymbol{\vartheta}) = \boldsymbol{h}^l, \quad t \in [t_l, t_{l+1}),$$

for each $l = 0, 1, \ldots, M$.

For each $j = 1, \ldots, M$, consider the following variational system.

$$\frac{d\boldsymbol{\phi}^j(t)}{dt} = \frac{\partial \boldsymbol{f}(t, \boldsymbol{x}(t \mid \boldsymbol{\vartheta}), \boldsymbol{u}(t \mid \boldsymbol{\vartheta}))}{\partial \boldsymbol{x}} \boldsymbol{\phi}^j(t), \quad t \in (t_j, T), \tag{7.4.3}$$

with jump condition

$$\boldsymbol{\phi}^j\left(t_j^+\right) = \boldsymbol{f}\left(t_j, \boldsymbol{x}(t_j \mid \boldsymbol{\vartheta}), \boldsymbol{h}^{j-1}\right) - \boldsymbol{f}\left(t_j, \boldsymbol{x}(t_j \mid \boldsymbol{\vartheta}), \boldsymbol{h}^j\right) \tag{7.4.4}$$

and initial condition

$$\boldsymbol{\phi}^j(t) = 0, \quad t \in [0, t_j). \tag{7.4.5}$$

In (7.4.4), we set $t_j^+ = T$ if $t_j = T$. Let $\boldsymbol{\phi}^j(\cdot \mid \boldsymbol{\vartheta})$ denote the unique right-continuous solution of (7.4.3)–(7.4.5).

Theorem 7.4.1 *Consider $j \in \{1, \ldots, M\}$ and assume that $t_{j-1} < t_j < t_{j+1}$. Then for all time points $t \neq t_j$,*

$$\frac{\partial \boldsymbol{x}(t \mid \boldsymbol{\vartheta})}{\partial t_j} = \lim_{\varepsilon \to 0} \frac{\boldsymbol{x}(t \mid \boldsymbol{\vartheta} + \varepsilon \boldsymbol{e}^j) - \boldsymbol{x}(t \mid \boldsymbol{\vartheta})}{\varepsilon} = \boldsymbol{\phi}^j(t \mid \boldsymbol{\vartheta}). \tag{7.4.6}$$

Proof. For any $t \in [0, T]$, we have

$$\boldsymbol{x}(t \mid \boldsymbol{\vartheta}) = \boldsymbol{x}^0 + \int_0^t \boldsymbol{f}(s, \boldsymbol{x}(s \mid \boldsymbol{\vartheta}), \boldsymbol{u}(s \mid \boldsymbol{\vartheta}))dt. \tag{7.4.7}$$

There are two cases to consider. If $t < t_j$, then $\boldsymbol{x}(t \mid \boldsymbol{\vartheta})$ clearly does not depend on t_j and thus

$$\frac{\partial \boldsymbol{x}(t \mid \boldsymbol{\vartheta})}{\partial t_j} = 0 \text{ for all } t < t_j. \tag{7.4.8}$$

Assume now that $t > t_j$. Then we may rewrite (7.4.7) as follows.

$$\boldsymbol{x}(t|\boldsymbol{\vartheta}) = \boldsymbol{x}^0 + \int_0^{t_{j-1}} \boldsymbol{f}(s, \boldsymbol{x}(s|\boldsymbol{\vartheta}), \boldsymbol{u}(s|\boldsymbol{\vartheta}))dt + \int_{t_{j-1}}^{t_j} \boldsymbol{f}\left(s, \boldsymbol{x}(s|\boldsymbol{\vartheta}), \boldsymbol{h}^{j-1}\right) dt$$
$$+ \begin{cases} \displaystyle\int_{t_j}^{t} \boldsymbol{f}\left(s, \boldsymbol{x}(s|\boldsymbol{\vartheta}), \boldsymbol{h}^j\right) dt, & \text{if } t \le t_{j+1}, \\ \displaystyle\int_{t_j}^{t_{j+1}} \boldsymbol{f}\left(s, \boldsymbol{x}(s|\boldsymbol{\vartheta}), \boldsymbol{h}^j\right) dt \\ \displaystyle\quad + \int_{t_{j+1}}^{t} \boldsymbol{f}(s, \boldsymbol{x}(s|\boldsymbol{\vartheta}), \boldsymbol{u}(s|\boldsymbol{\vartheta}))dt, \text{ otherwise}. \end{cases} \tag{7.4.9}$$

Differentiating both sides of (7.4.9) with respect to t_j, we have

$$\begin{aligned}\frac{\partial \boldsymbol{x}(t \mid \boldsymbol{\vartheta})}{\partial t_j} =& \boldsymbol{f}\left(t_j, \boldsymbol{x}(t_j|\boldsymbol{\vartheta}), \boldsymbol{h}^{j-1}\right) - \boldsymbol{f}\left(t_j, \boldsymbol{x}(t_j \mid \boldsymbol{\vartheta}), \boldsymbol{h}^j\right) \\ &+ \int_{t_j}^{t} \frac{\partial \boldsymbol{f}(s, \boldsymbol{x}(s \mid \boldsymbol{\vartheta}), \boldsymbol{u}(s \mid \boldsymbol{\vartheta}))}{\partial \boldsymbol{x}} \frac{\partial \boldsymbol{x}(s \mid \boldsymbol{\vartheta})}{\partial t_j} dt. \end{aligned} \tag{7.4.10}$$

Taking the limit as t approaches t_j from above of both sides of (7.4.10), we obtain

$$\lim_{t \to t_j^+} \frac{\partial \boldsymbol{x}(t \mid \boldsymbol{\vartheta})}{\partial t_j} = \frac{\partial \boldsymbol{x}(t_j^+ \mid \boldsymbol{\vartheta})}{\partial t_j} = \boldsymbol{f}\left(t_j, \boldsymbol{x}(t_j|\boldsymbol{\vartheta}), \boldsymbol{h}^{j-1}\right) - \boldsymbol{f}\left(t_j, \boldsymbol{x}(t_j|\boldsymbol{\vartheta}), \boldsymbol{h}^j\right). \tag{7.4.11}$$

Differentiating both sides of (7.4.10) with respect to t, we have

$$\frac{d}{dt}\left(\frac{\partial \boldsymbol{x}(t \mid \boldsymbol{\vartheta})}{\partial t_j}\right) = \frac{\partial \boldsymbol{f}(t, \boldsymbol{x}(t|\boldsymbol{\vartheta}), \boldsymbol{u}(s|\boldsymbol{\vartheta}))}{\partial \boldsymbol{x}} \frac{\partial \boldsymbol{x}(t \mid \boldsymbol{\vartheta})}{\partial t_j}, \quad t > t_j. \tag{7.4.12}$$

Noting the equivalence of (7.4.3)–(7.4.5) and (7.4.8)–(7.4.12), the conclusion follows.

Remark 7.4.2 *Using (7.4.8), it is easy to see that the limit of the state variation as t approaches t_j from below is*

$$\lim_{t \to t_j^-} \frac{\partial \boldsymbol{x}(t \mid \boldsymbol{\vartheta})}{\partial t_j} = \frac{\partial \boldsymbol{x}(t_j^- \mid \boldsymbol{\vartheta})}{\partial t_j} = 0.$$

Comparing this with (7.4.11), *we can conclude that the state variation with respect to* t_j *does not exist at* $t = t_j$ *but has a jump condition there instead. As we show below, however, this does not prevent us from calculating the required gradients of the cost and constraint functionals.*

Note that the technical caveats in Theorem 7.4.1 ($t \neq t_j$ and $t_{j-1} < t_j < t_{j+1}$) are not needed for the state variation with respect to the control heights. Thus, optimizing the switching times is much more difficult than optimizing the control heights. This is one reason for the popularity of the time scaling transformation to be introduced in the next subsection which allows one to circumvent the difficulties caused by variable switching times.

Using Theorem 7.4.1, we can derive the partial derivatives of the cost and constraint functionals with respect to the switching times (assuming that all switching times are distinct). First, recall that the cost and constraint functionals are defined by

$$g_i(\boldsymbol{\vartheta}) = \Phi_i(\boldsymbol{x}(T \mid \boldsymbol{\vartheta})) + \int_0^T \mathcal{L}_i(t, \boldsymbol{x}(t \mid \boldsymbol{\vartheta}), \boldsymbol{u}(t \mid \boldsymbol{\vartheta}))dt, \quad i = 0, 1, \ldots, N.$$

Using the chain rule of differentiation, we obtain

$$\frac{\partial}{\partial t_j}\{\Phi_i(\boldsymbol{x}(T \mid \boldsymbol{\vartheta}))\} = \frac{\partial \Phi_i(\boldsymbol{x}(T \mid \boldsymbol{\vartheta}))}{\partial \boldsymbol{x}} \frac{\partial \boldsymbol{x}(T \mid \boldsymbol{\vartheta})}{\partial t_j}. \tag{7.4.13}$$

Furthermore, using the Leibniz rule for differentiating integrals, we obtain

$$\begin{aligned}
&\frac{\partial}{\partial t_j}\left\{\int_0^T \mathcal{L}_i(t, \boldsymbol{x}(t \mid \boldsymbol{\vartheta}), \boldsymbol{u}(t \mid \boldsymbol{\vartheta}))dt\right\} \\
=&\frac{\partial}{\partial t_j}\left\{\sum_{l=0}^{M}\int_{t_l}^{t_{l+1}} \mathcal{L}_i\left(t, \boldsymbol{x}(t \mid \boldsymbol{\vartheta}), \boldsymbol{h}^l\right)dt\right\} \\
=&\mathcal{L}_i\left(t_j, \boldsymbol{x}(t_j \mid \boldsymbol{\vartheta}), \boldsymbol{h}^{j-1}\right) - \mathcal{L}_i\left(t_j, \boldsymbol{x}(t_j \mid \boldsymbol{\vartheta}), \boldsymbol{h}^{j}\right) \\
&+\int_0^T \frac{\partial \mathcal{L}_i(t, \boldsymbol{x}(t \mid \boldsymbol{\vartheta}), \boldsymbol{u}(t \mid \boldsymbol{\vartheta}))}{\partial \boldsymbol{x}} \frac{\partial \boldsymbol{x}(t \mid \boldsymbol{\vartheta})}{\partial t_j} dt.
\end{aligned} \tag{7.4.14}$$

Applying the Leibniz rule to interchange the order of differentiation and integration is valid here because the partial derivative of $\boldsymbol{x}(\cdot \mid \boldsymbol{\vartheta})$ with respect to t_j exists at all points in the interior of the interval $[t_l, t_{l+1}]$. Recall that the Leibniz rule does not require differentiability at the end points. Combining (7.4.13) and (7.4.14) with Theorem 7.4.1 yields the following gradient formulae.

Theorem 7.4.2 *For each* $i = 0, 1, \ldots, N$ *and* $j = 1, \ldots, M$,

$$\frac{\partial g_i(\boldsymbol{\vartheta})}{\partial t_j} = \frac{\partial \Phi_i(\boldsymbol{x}(T \mid \vartheta))}{\partial \boldsymbol{x}} \boldsymbol{\phi}^j(T \mid \vartheta)$$

$$+ \mathcal{L}_i\left(t_j, \boldsymbol{x}(t_j \mid \boldsymbol{\vartheta}), \boldsymbol{h}^{j-1}\right) - \mathcal{L}_i\left(t_j, \boldsymbol{x}(t_j \mid \boldsymbol{\vartheta}), \boldsymbol{h}^j\right) + \int_0^T \frac{\partial \mathcal{L}_i(t, \boldsymbol{x}(t \mid \boldsymbol{\vartheta}), \boldsymbol{u}(t \mid \boldsymbol{\vartheta}))}{\partial \boldsymbol{x}} \boldsymbol{\phi}^j(t \mid \boldsymbol{\vartheta}) dt. \tag{7.4.15}$$

These gradient formulae are based on the variational system. The gradient formulae derived based on the costate system are given below.

For each $i = 0, 1, \ldots, N$, and for each switching vector $\boldsymbol{\vartheta} \in \Xi$, we consider the following system of differential equations.

$$\frac{d\boldsymbol{\lambda}^i(t)}{dt} = -\left[\frac{\partial H_i\left(t, \boldsymbol{x}(t \mid \boldsymbol{\vartheta}), \boldsymbol{u}(t \mid \boldsymbol{\vartheta}), \boldsymbol{\lambda}^i(t)\right)}{\partial \boldsymbol{x}}\right]^\top, \quad t \in [0, T) \tag{7.4.16a}$$

with

$$\boldsymbol{\lambda}^i(T) = \left[\frac{\partial \Phi_i(\boldsymbol{x}(T \mid \boldsymbol{\vartheta}))}{\partial \boldsymbol{x}}\right]^\top, \tag{7.4.16b}$$

where

$$H_i(t, \boldsymbol{x}, \boldsymbol{u}, \boldsymbol{\lambda}) = \mathcal{L}_i(t, \boldsymbol{x}, \boldsymbol{u}) + \boldsymbol{\lambda}^\top \boldsymbol{f}(t, \boldsymbol{x}, \boldsymbol{u}). \tag{7.4.17}$$

The system (7.4.16) is called the costate system for the cost functional if $i = 0$ and for the i–th constraint functional if $i \neq 0$. Let $\boldsymbol{\lambda}^i(\cdot \mid \boldsymbol{\vartheta})$ denote the solution of the costate system (7.4.16) corresponding to $\boldsymbol{\vartheta} \in \Xi$. The gradient of each functional g_i, $i = 0, \ldots, N$, with respect to the switching time t_j, $j = 1, \ldots, M$, is given in the following theorem.

Theorem 7.4.3 *Consider $j \in \{1, \ldots, M\}$ and assume that $t_{j-1} < t_j < t_{j+1}$. For each $i = 0, 1, \ldots, N$, the gradient of the functional g_i with respect to t_j is given by*

$$\frac{\partial g_i(\boldsymbol{\vartheta})}{\partial t_j} = H_i\left(t_j, \boldsymbol{x}(t_j \,|\, \boldsymbol{\vartheta}), \boldsymbol{h}^{j-1}, \boldsymbol{\lambda}^i(t_j \,|\, \boldsymbol{\vartheta})\right) - H_i(t_j, \boldsymbol{x}(t_j \,|\, \boldsymbol{\vartheta}), \boldsymbol{h}^j, \boldsymbol{\lambda}^i(t_j \,|\, \boldsymbol{\vartheta})). \tag{7.4.18}$$

Proof. Although this result has been proven in [253], a somewhat more direct proof is given here. Let $\boldsymbol{v} : [0; T] \to \mathbb{R}^n$ be any absolutely continuous function. Then g_i may be written as

$$\begin{aligned} g_i(\boldsymbol{\vartheta}) =& \Phi_i(\boldsymbol{x}(T \,|\, \boldsymbol{\vartheta})) + \sum_{l=0}^{M} \int_{t_l}^{t_{l+1}} \mathcal{L}_i\left(t, \boldsymbol{x}(t \,|\, \boldsymbol{\vartheta}), \boldsymbol{h}^l\right) dt \\ =& \Phi_i(\boldsymbol{x}(T \,|\, \boldsymbol{\vartheta})) + \sum_{l=0}^{M} \int_{t_l}^{t_{l+1}} H_i\left(t, \boldsymbol{x}(t \,|\, \boldsymbol{\vartheta}), \boldsymbol{h}^l, \boldsymbol{v}(t)\right) dt \\ & - \sum_{l=0}^{M} \int_{t_l}^{t_{l+1}} \boldsymbol{v}(t) \frac{d\boldsymbol{x}(t \,|\, \boldsymbol{\vartheta})}{dt} dt. \end{aligned}$$

Using integration by parts on the last term, we have

$$
\begin{aligned}
g_i(\boldsymbol{\vartheta}) =& \Phi_j(\boldsymbol{x}(T\,|\,\boldsymbol{\vartheta})) + \sum_{l=0}^{M} \int_{t_l}^{t_{l+1}} H_i\left(t, \boldsymbol{x}(t\,|\,\boldsymbol{\vartheta}), \boldsymbol{h}^l, \boldsymbol{v}(t)\right) dt \\
& - \sum_{l=0}^{M} \Bigg\{ (\boldsymbol{v}(t_{l+1}))^{\top} \boldsymbol{x}(t_{l+1}\,|\,\boldsymbol{\vartheta}) \\
& - (\boldsymbol{v}(t_l))^{\top} \boldsymbol{x}(t_l\,|\,\boldsymbol{\vartheta}) - \int_{t_l}^{t_{l+1}} \left(\frac{d\boldsymbol{v}(t)}{dt}\right)^{\top} \boldsymbol{x}(t\,|\,\boldsymbol{\vartheta}) dt \Bigg\} \\
=& \Phi_i(\boldsymbol{x}(T\,|\,\boldsymbol{\vartheta})) - (\boldsymbol{v}(t_{M+1}))^{\top} \boldsymbol{x}(t_{M+1}\,|\,\boldsymbol{\vartheta}) + (v(t_0))^{\top} \boldsymbol{x}(t_0\,|\,\boldsymbol{\vartheta}) \\
& + \sum_{l=0}^{M} \int_{t_l}^{t_{l+1}} \left\{ H_i\left(t, \boldsymbol{x}(t\,|\,\boldsymbol{\vartheta}), \boldsymbol{h}^l, \boldsymbol{v}(t)\right) + \left(\frac{d\boldsymbol{v}(t)}{dt}\right)^{\top} \boldsymbol{x}(t\,|\,\boldsymbol{\vartheta}) \right\} dt \\
=& \Phi_i(\boldsymbol{x}(T\,|\,\boldsymbol{\vartheta})) - (\boldsymbol{v}(T))^{\top} \boldsymbol{x}(T\,|\,\boldsymbol{\vartheta}) + (v(0))^{\top} \boldsymbol{x}(0\,|\,\boldsymbol{\vartheta}) \\
& + \sum_{l=0}^{j-2} \int_{t_l}^{t_{l+1}} \left\{ H_i\left(t, \boldsymbol{x}(t\,|\,\boldsymbol{\vartheta}), \boldsymbol{h}^l, \boldsymbol{v}(t)\right) + \left(\frac{d\boldsymbol{v}(t)}{dt}\right)^{\top} \boldsymbol{x}(t\,|\,\boldsymbol{\vartheta}) \right\} dt \\
& + \int_{t_{j-1}}^{t_j} \left\{ H_i\left(t, \boldsymbol{x}(t\,|\,\boldsymbol{\vartheta}), \boldsymbol{h}^{j-1}, \boldsymbol{v}(t)\right) + \left(\frac{d\boldsymbol{v}(t)}{dt}\right)^{\top} \boldsymbol{x}(t\,|\,\boldsymbol{\vartheta}) \right\} dt \\
& + \int_{t_j}^{t_{j+1}} \left\{ H_i\left(t, \boldsymbol{x}(t\,|\,\boldsymbol{\vartheta}), \boldsymbol{h}^{j}, \boldsymbol{v}(t)\right) + \left(\frac{d\boldsymbol{v}(t)}{dt}\right)^{\top} \boldsymbol{x}(t\,|\,\boldsymbol{\vartheta}) \right\} dt \\
& + \sum_{l=j+1}^{M} \int_{t_l}^{t_{l+1}} \left\{ H_i\left(t, \boldsymbol{x}(t\,|\,\boldsymbol{\vartheta}), \boldsymbol{h}^l, \boldsymbol{v}(t)\right) + \left(\frac{d\boldsymbol{v}(t)}{dt}\right)^{\top} \boldsymbol{x}(t\,|\,\boldsymbol{\vartheta}) \right\} dt.
\end{aligned}
$$

Using the Leibniz rule, this equation can be differentiated with respect to t_j to give

$$
\begin{aligned}
\frac{\partial g_i(\boldsymbol{\vartheta})}{\partial t_j} =& \frac{\partial \Phi_i(\boldsymbol{x}(T\,|\,\boldsymbol{\vartheta}))}{\partial \boldsymbol{x}} \frac{\partial \boldsymbol{x}(T\,|\,\boldsymbol{\vartheta})}{\partial t_j} - (\boldsymbol{v}(T))^{\top} \frac{d\boldsymbol{x}(T\,|\,\boldsymbol{\vartheta})}{\partial t_j} \\
& + H_i\left(t_j, \boldsymbol{x}(t_j\,|\,\boldsymbol{\vartheta}), \boldsymbol{h}^{j-1}, \boldsymbol{v}(t_j)\right) - H_i\left(t_j, \boldsymbol{x}(t_j\,|\,\boldsymbol{\vartheta}), \boldsymbol{h}^{j}, \boldsymbol{v}(t_j)\right) \\
& + \sum_{l=0}^{M} \int_{t_l}^{t_{l+1}} \Bigg\{ \frac{\partial H_i\left(t, \boldsymbol{x}(t\,|\,\boldsymbol{\vartheta}), \boldsymbol{h}^l, \boldsymbol{v}(t)\right)}{\partial \boldsymbol{x}} \frac{\partial \boldsymbol{x}(t\,|\,\boldsymbol{\vartheta})}{\partial t_j} \\
& - \left(\frac{d\boldsymbol{v}(t)}{dt}\right)^{\top} \frac{\partial \boldsymbol{x}(t\,|\,\boldsymbol{\vartheta})}{\partial t_j} \Bigg\} dt.
\end{aligned}
$$

Setting $\boldsymbol{v}(\cdot) = \boldsymbol{\lambda}^i(\cdot\,|\,\boldsymbol{\vartheta}))$ and then applying (7.4.16), we obtain

$$
\frac{\partial g_i(\boldsymbol{\vartheta})}{\partial t_j} = H_i\left(t_j, \boldsymbol{x}(t_j\,|\,\boldsymbol{\vartheta}), \boldsymbol{h}^{j-1}, \boldsymbol{v}(t_j)\right) - H_i\left(t_j, \boldsymbol{x}(t_j\,|\,\boldsymbol{\vartheta}), \boldsymbol{h}^{j}, \boldsymbol{v}(t_j)\right),
$$

as required.

Remark 7.4.3 *Equations* (7.4.15) *and* (7.4.18) *both give the partial derivatives of the canonical functionals* g_i, $j = 0, 1, \ldots, N$, *with respect to the switching times. Since the state trajectory and the solutions of the variational and costate systems depend continuously on* $\boldsymbol{\vartheta}$, *the derivative formulae in* (7.4.15) *and* (7.4.18) *also depend continuously on* $\boldsymbol{\vartheta}$. *In principle, either* (7.4.15) *or* (7.4.18) *can be used in conjunction with a gradient-based optimization method to optimize the switching times. However, there are several difficulties with this approach:*

(i) The variational systems for the switching times contain a jump condition.
(ii) The partial derivatives of the canonical functionals with respect to t_j *only exist when the switching times are distinct.*
(iii) It is cumbersome to integrate the state and variational or costate systems numerically when the switching times are variable, especially when two or more switching times are close together.

For the reasons given in Remark 7.4.3, it is less popular to use the gradient formulae given by (7.4.15) and (7.4.18) in practice. Instead, a time scaling transformation is typically used to transform a problem with variable switching times into an equivalent problem with fixed switching times. This is discussed in the next section.

7.4.2 Time Scaling Transformation

The purpose of this section is to demonstrate that an optimal control problem with piecewise constant controls and variable switching times can be readily transformed into an equivalent optimal control problem with piecewise constant controls and fixed switching times. Since gradients for the transformed problem with fixed switching times can be readily determined using the formulae in Section 7.3, we can thus avoid the difficulties mentioned in Remark 7.4.3. In early publications see, for example, [125, 126], the transformation described in this section was referred to as the *Control Parametrization Enhancing Technique* (CPET), while later works simply refer to it as a time scaling transformation.

Consider Problem $(P3)$ once more where the switching vector $\boldsymbol{\vartheta} = [t_1, \ldots, t_M]^\top$ is the decision variable. For notational convenience, the equivalent problem with fixed switching times to be introduced below is more easily expressed in terms of the durations between individual switching times. These durations are given by

$$\gamma_j = t_{j+1} - t_j, \quad j = 0, 1, \ldots, M, \tag{7.4.19}$$

where we recall that $t_0 = 0$ and $t_M = T$. Let $\boldsymbol{\gamma} = [\gamma_0, \gamma_1, \ldots, \gamma_M]^\top$. Note that constraints (7.3.5) satisfied by $\boldsymbol{\vartheta}$ are equivalent to the constraints

$$\gamma_j \geq 0, \quad j = 0, 1, \ldots, M, \tag{7.4.20}$$

on $\boldsymbol{\gamma}$. Furthermore, $\boldsymbol{\gamma}$ must also satisfy the additional constraint

$$\gamma_0 + \gamma_1 + \cdots + \gamma_M = T. \tag{7.4.21}$$

The basic idea of the time scaling transformation is to replace the original time horizon $[0, T]$ containing the variable switching times t_j, $j = 1, \ldots, M$, with a new time horizon $[0, M+1]$ with fixed switching times at $1, 2, \ldots, M$. We use s to denote 'time' in the new time horizon. The relationship between $t \in [0, T]$ and $s \in [0, M+1]$ can be defined by the differential equation

$$\frac{dt(s)}{ds} = v(s) \tag{7.4.22a}$$

with the initial condition

$$t(0) = 0 \tag{7.4.22b}$$

and the terminal condition

$$t(M+1) = T, \tag{7.4.22c}$$

where the scalar valued function $v(s)$ is called the *time scaling control*. It is defined by

$$v(s) = \sum_{j=0}^{M} \gamma_j \chi_{[j,j+1)}(s), \tag{7.4.23}$$

where $\chi_I(\cdot)$ is the indicator function defined by (7.3.3) and γ_j, $j = 0, 1, \ldots, M$, are the durations defined above. Note that

$$0 \leq \gamma_i \leq T, \; i = 0, 1, \ldots, M, \tag{7.4.24}$$

where the upper bound follows from (7.4.21). For easy reference, the time scaling control $v(s)$ is written as $v(s \mid \boldsymbol{\gamma})$. Let Γ be the set containing all those vectors $\boldsymbol{\gamma} = [\gamma_0, \gamma_1, \ldots, \gamma_M]^\top \in \mathbb{R}^{M+1}$ satisfying (7.4.24). Clearly, $v(s \mid \boldsymbol{\gamma})$ is uniquely determined by $\boldsymbol{\gamma} \in \Gamma$ and vice versa. By virtue of (7.4.24), the solution of (7.4.22) is monotonically non-decreasing. For each $s \in [0, M+1]$, we have

$$\begin{aligned} t(s) &= \int_0^s v(\tau \mid \boldsymbol{\gamma}) d\tau \\ &= \begin{cases} \gamma_0 s, & \text{if } s \in [0,1], \\ \displaystyle\sum_{j=0}^{l-1} \gamma_j + \gamma_l (s - l), & \text{if } s \in [l, l+1]. \end{cases} \end{aligned} \tag{7.4.25}$$

Note in particular that (7.4.25) results in

$$t(j) = t_j, \ j = 0, 1, \ldots, M+1,$$

so that the desired mapping between the variable switching times in $[0, T]$ and the fixed switching times in $[0, M+1]$ has been achieved and that (7.4.22c) is equivalent to (7.4.21). Furthermore, the piecewise constant control

$$\boldsymbol{u}(t) = \sum_{j=0}^{M} \boldsymbol{h}^j \chi_{[t_j, t_{j+1})}(t)$$

for $t \in [0, T)$ is equivalent to

$$\tilde{\boldsymbol{u}}(s) = \boldsymbol{u}(t(s)) = \sum_{j=0}^{M} \boldsymbol{h}^j \chi_{[j, j+1)}(s) \tag{7.4.26}$$

for $s \in [0, M+1)$. We also adopt the notation

$$\tilde{\boldsymbol{x}}(s) = \boldsymbol{x}(t(s)). \tag{7.4.27}$$

Using (7.4.22a) and the chain rule, we then obtain the transformed dynamics

$$\frac{d\tilde{\boldsymbol{x}}(s)}{ds} = \frac{d\boldsymbol{x}(t(s))}{dt}\frac{dt(s)}{ds} = v(s \mid \boldsymbol{\gamma})\boldsymbol{f}(t(s), \tilde{\boldsymbol{x}}(s), \tilde{\boldsymbol{u}}(s)) \tag{7.4.28}$$

with the initial condition

$$\tilde{\boldsymbol{x}}(0) = \boldsymbol{x}^0. \tag{7.4.29}$$

For each $\boldsymbol{\gamma} \in \Gamma$, let $t(\cdot \mid \boldsymbol{\gamma})$ denote the solution of (7.4.22a)–(7.4.22b) and let $\tilde{\boldsymbol{x}}(\cdot \mid \boldsymbol{\gamma})$ denote the solution of (7.4.28)–(7.4.29).

Finally, note that we may write (7.4.22a) in differential form as $dt = v(s)ds$. Hence, for any function $\mathcal{L}(t)$ defined on $[0, T]$, we have

$$\int_0^T \mathcal{L}(t)\, dt = \int_0^{M+1} \mathcal{L}(t(s))\, v(s)\, ds.$$

The equivalent transformed optimal control problem may then be stated as follows.
Problem $(TP3)$: Given the combined system (7.4.22a) and (7.4.28) with the initial conditions (7.4.22b) and (7.4.29), find a vector $\boldsymbol{\gamma} \in \Gamma$ such that the cost functional

$$\tilde{g}_0(\boldsymbol{\gamma}) = \Phi_0(\tilde{\boldsymbol{x}}(M+1 \mid \boldsymbol{\gamma})) + \int_0^{M+1} \mathcal{L}_0(t(s \mid \boldsymbol{\gamma}), \tilde{\boldsymbol{x}}(s \mid \boldsymbol{\gamma}), \tilde{\boldsymbol{u}}(s))\, v(s \mid \boldsymbol{\gamma})\, ds \tag{7.4.30}$$

is minimized subject to the equality constraints

$$\tilde{g}_i(\boldsymbol{\gamma}) = \Phi_i(\tilde{\boldsymbol{x}}(M+1\,|\,\boldsymbol{\gamma})) + \int_0^{M+1} \mathcal{L}_i(t(s\,|\,\boldsymbol{\gamma}), \tilde{\boldsymbol{x}}(s\,|\,\boldsymbol{\gamma}), \tilde{\boldsymbol{u}}(s))\, v(s \mid \boldsymbol{\gamma})\, ds = 0,$$
$$i = 1, \ldots, N_e, \tag{7.4.31a}$$

the inequality constraints

$$\tilde{g}_i(\boldsymbol{\gamma}) = \Phi_i(\tilde{\boldsymbol{x}}(M+1\,|\,\boldsymbol{\gamma})) + \int_0^{M+1} \mathcal{L}_i(t(s\,|\,\boldsymbol{\gamma}), \tilde{\boldsymbol{x}}(s\,|\,\boldsymbol{\gamma}), \tilde{\boldsymbol{u}}(s))\, v(s \mid \boldsymbol{\gamma})\, ds \leq 0,$$
$$i = N_e + 1, \ldots, N, \tag{7.4.31b}$$

and the additional equality constraint

$$\tilde{g}_{N+1}(\boldsymbol{\gamma}) = t(M+1\,|\,\boldsymbol{\gamma}) - T = 0. \tag{7.4.32}$$

Before continuing, let us compare and contrast the equivalent Problems $(P3)$ and $(TP3)$. Problem $(TP3)$ has an additional state variable, $t(\cdot)$, an additional associated differential equation, (7.4.22a), and an additional equality constraint, (7.4.32). Note that the latter is of the same canonical form as the other constraints, with $\Phi_{N+1} = t(M+1 \mid \boldsymbol{\gamma}) - T$ and $\mathcal{L}_{N+1} = 0$. Furthermore, Problem $(TP3)$ has an additional scalar control function $v(\cdot \mid \boldsymbol{\gamma})$ sharing the fixed switching times $j = 1, \ldots, M$ with $\tilde{\boldsymbol{u}}(\cdot)$ and with variable heights. While the transformation of Problem $(P3)$ to Problem $(TP3)$ has added some complexity, Problem $(TP3)$ is clearly of the same form as Problem $(P2)$ in Section 7.3 in that the switching times of the controls are fixed and the only variables are control heights (the components of $\boldsymbol{\gamma}$ defining $v(\cdot \mid \boldsymbol{\gamma})$). Hence, we can apply the gradient formulae of Section 7.3 to determine the gradients of the cost and constraint functionals in Problem $(TP3)$. For reference, these are detailed as follows.

Consider Problem $(TP3)$. For each $i = 0, 1, \ldots, N$, define the Hamiltonian

$$\begin{aligned}
&\tilde{H}_i\left(t(s\,|\,\boldsymbol{\gamma}), \tilde{\boldsymbol{x}}(s\,|\,\boldsymbol{\gamma}), v(s\,|\,\boldsymbol{\gamma}), \tilde{\boldsymbol{u}}(s), \tilde{\lambda}_v^i(s\,|\,\boldsymbol{\gamma}), \tilde{\boldsymbol{\lambda}}^i(s\,|\,\boldsymbol{\gamma})\right) \\
=&\mathcal{L}_i\left(t(s\,|\,\boldsymbol{\gamma}), \tilde{\boldsymbol{x}}(s\,|\,\boldsymbol{\gamma}), \tilde{\boldsymbol{u}}(s)\right) + \tilde{\lambda}_v^i(s\,|\,\boldsymbol{\gamma})\, v(s\,|\,\boldsymbol{\gamma}) \\
&+ \left(\tilde{\boldsymbol{\lambda}}^i(s\,|\,\boldsymbol{\gamma})\right)^\top v(s\,|\,\boldsymbol{\gamma}) \boldsymbol{f}(t(s\,|\,\boldsymbol{\gamma}), \tilde{\boldsymbol{x}}(s\,|\,\boldsymbol{\gamma}), \tilde{\boldsymbol{u}}(s)),
\end{aligned} \tag{7.4.33}$$

where, for each $\boldsymbol{\gamma} \in \Gamma$, $\tilde{\lambda}_v^i(s \mid \boldsymbol{\gamma})$ and $\tilde{\boldsymbol{\lambda}}^i(s \mid \boldsymbol{\gamma})$ denote the solution of the costate system

$$\frac{d}{ds}\begin{bmatrix} \tilde{\lambda}_v^i(s) \\ \tilde{\boldsymbol{\lambda}}^i(s) \end{bmatrix}$$

$$= \begin{bmatrix} -\dfrac{\partial \tilde{H}_i\left(t(s), \tilde{\boldsymbol{x}}(s\,|\,\boldsymbol{\gamma}), v(s\,|\,\boldsymbol{\gamma}), \tilde{\boldsymbol{u}}(s), \tilde{\lambda}_v^i(s), \tilde{\boldsymbol{\lambda}}^i(s)\right)}{\partial t} \\ -\left(\dfrac{\partial \tilde{H}_i\left(t(s), \tilde{\boldsymbol{x}}(s\,|\,\boldsymbol{\gamma}), v(s\,|\,\boldsymbol{\gamma}), \tilde{\boldsymbol{u}}(s), \tilde{\lambda}_v^i(s), \tilde{\boldsymbol{\lambda}}^i(s)\right)}{\partial \tilde{\boldsymbol{x}}}\right)^\top \end{bmatrix}, s \in [0, M+1), \tag{7.4.34}$$

with the terminal condition

$$\begin{bmatrix} \tilde{\lambda}_v^i(M+1) \\ \tilde{\boldsymbol{\lambda}}^i(M+1) \end{bmatrix} = \begin{bmatrix} 0 \\ \left(\dfrac{\partial \Phi_i(\tilde{\boldsymbol{x}}(M+1\,|\,\boldsymbol{\gamma}))}{\partial \tilde{\boldsymbol{x}}}\right)^\top \end{bmatrix}. \tag{7.4.35}$$

The gradients of the cost and constraint functionals in Problem (TP3) may now be given in the following Theorem.

Theorem 7.4.4 *The gradient of the functional $\tilde{g}_i$ with respect to γ_j, $j \in \{0, 1, \ldots, M\}$, is*

$$\frac{\partial \tilde{g}_i(\boldsymbol{\gamma})}{\partial \gamma_j} = \int_j^{j+1} \frac{\partial \tilde{H}_i\left(t(s\,|\,\boldsymbol{\gamma}), \tilde{\boldsymbol{x}}(s\,|\,\boldsymbol{\gamma}), \gamma_j, \boldsymbol{h}^j, \tilde{\lambda}_v^i(s\,|\,\boldsymbol{\gamma}), \tilde{\boldsymbol{\lambda}}^i(s\,|\,\boldsymbol{\gamma})\right)}{\partial \gamma_j} ds. \tag{7.4.36}$$

Remark 7.4.4 *Note that for computational purposes,* (7.4.36) *may also be written as*

$$\frac{\partial \tilde{g}_i(\boldsymbol{\gamma})}{\partial \gamma_j} = \int_j^{j+1} \frac{\partial \tilde{H}_i\left(t(s\,|\,\boldsymbol{\gamma}), \tilde{\boldsymbol{x}}(s\,|\,\boldsymbol{\gamma}), v(s\,|\,\boldsymbol{\gamma}), \tilde{\boldsymbol{u}}(s), \tilde{\lambda}_v^i(s\,|\,\boldsymbol{\gamma}), \tilde{\boldsymbol{\lambda}}^i(s\,|\,\boldsymbol{\gamma})\right)}{\partial v} ds. \tag{7.4.37}$$

Remark 7.4.5 *While we neglected to include the gradient of $g_{N+1}(\boldsymbol{\gamma})$ in Theorem 7.4.4, note that, due to the canonical nature of this functional, it is quite easy to determine this gradient using formulae similar to* (7.4.33)–(7.4.36)*. However, since $g_{N+1}(\boldsymbol{\gamma}) = \gamma_0 + \gamma_1 + \cdots + \gamma_M - T$, it is even easier to see that*

$$\frac{\partial g_{N+1}(\boldsymbol{\gamma})}{\partial \gamma_j} = 1, \quad j = 0, 1, \ldots, M. \tag{7.4.38}$$

Remark 7.4.6 *Finally, note that since $g_i(\boldsymbol{\vartheta}) = \tilde{g}_i(\boldsymbol{\gamma})$ for each $i \in \{0, 1, \ldots, N\}$ and $\gamma_j = t_{j+1} - t_j$, $j = 0, 1, \ldots, M$, it is possible to generate the gradients of $g_i(\boldsymbol{\vartheta})$ with respect to each t_j from the formulae above. For each $j \in \{1, \ldots, M\}$, we have*

$$\frac{\partial g_i(\boldsymbol{\vartheta})}{\partial t_j} = \frac{\partial \tilde{g}_i(\boldsymbol{\gamma})}{\partial \gamma_j}\frac{\partial \gamma_j}{\partial t_j} + \frac{\partial \tilde{g}_i(\boldsymbol{\gamma})}{\partial \gamma_{j-1}}\frac{\partial \gamma_j}{\partial t_j}$$

$$
\begin{aligned}
&= \frac{\partial \tilde{g}_i(\boldsymbol{\gamma})}{\partial \gamma_j} - \frac{\partial \tilde{g}_i(\boldsymbol{\gamma})}{\partial \gamma_{j-1}} \\
&= \int_j^{j+1} \frac{\partial \tilde{H}_i\Big(t(s\,|\,\boldsymbol{\gamma}), \tilde{\boldsymbol{x}}(s\,|\,\boldsymbol{\gamma}), \gamma_j, \boldsymbol{h}^j, \tilde{\lambda}_v^i(s\,|\,\boldsymbol{\gamma}), \tilde{\boldsymbol{\lambda}}^i(s\,|\,\boldsymbol{\gamma})\Big)}{\partial \gamma_j} ds \\
&\quad - \int_{j-1}^{j} \frac{\partial \tilde{H}_i\Big(t(s\,|\,\boldsymbol{\gamma}), \tilde{\boldsymbol{x}}(s\,|\,\boldsymbol{\gamma}), \gamma_{j-1}, \boldsymbol{h}^{j-1}, \tilde{\lambda}_v^i(s\,|\,\boldsymbol{\gamma}), \tilde{\boldsymbol{\lambda}}^i(s\,|\,\boldsymbol{\gamma})\Big)}{\partial \gamma_{j-1}} ds.
\end{aligned}
\tag{7.4.39}
$$

Remark 7.4.7 *Note that the terminal constraint* (7.4.32) *is equivalent to*

$$
\tilde{g}_{N+1}(\boldsymbol{\gamma}) = \int_0^{M+1} v(s)\, ds - T = 0 \tag{7.4.40}
$$

which is also in canonical form with $\Phi_{N+1} = -T$ *and* $\mathcal{L}_{N+1} = v(s)$.

Remark 7.4.8 *Suppose the functions* $\boldsymbol{f}$ *and* $\mathcal{L}_i$, $i = 0, 1, \ldots, N$, *in Problem* $(P3)$ *do not depend explicitly on time. Then, there is no need to include the differential equation* (7.4.22a) *and its initial condition* (7.4.22b) *in the corresponding equivalent transformed problem* $(TP3)$ *if we also replace the terminal constraint* (7.4.32) *by* (7.4.40). *In this case, the solution to Problem* $(TP3)$ *can be used to construct the solution of Problem* $(P3)$ *by using* (7.4.25) *in order to obtain* t *as a function of* s. *Indeed,* (7.4.25) *can be used instead of* (7.4.22a)–(7.4.22b) *to define the time scaling transformation in these cases [170, 171].*

Remark 7.4.9 *The components of* $\boldsymbol{\gamma}$ *can also be regarded as system parameters instead of arising from the construction of the time scaling control function* $v(\cdot)$ *[215].*

Remark 7.4.10 *Note that the transformed problem has only fixed time points where the state differential equations are discontinuous. All locations of the discontinuities of the state differential equations are thus known and fixed during the optimization process. Even when two or more of the switching times in the original time scale coalesce, the number of these locations remains unchanged in the transformed problem.*

Remark 7.4.11 *Although the allocation of* $s = 1, 2, \ldots, M$ *as fixed switching times in transformed time horizon* $[0, M+1]$ *is easy to follow and widely used, large values of* M *may result in numerical scaling issues for the transformed problem when* $[0, M+1]$ *becomes too long. Instead, the time scaling transformation can also be formulated so that* $[0, T]$ *is transformed to* $[0, 1]$ *with fixed switching times located at* ξ_j, $j = 1, \ldots, M$. *The choice of the* $\xi_j \in [0, 1]$ *is arbitrary as long as* $\xi_1 > 0$, $\xi_j < \xi_{j+1}$, $j = 1, \ldots, M-1$, *and* $\xi_M < 1$. *A natural choice typically used in practice is* $\xi_j = j/(M+1)$, $j = 1, \ldots, M$ *[215].*

In the next few subsections, we illustrate several important classes of optimal control problems where the time scaling transformation can be used.

7.4.3 Combined Piecewise Constant Control and Variable System Parameters

In this section, we consider a general class of optimal control problems with piecewise constant controls, variable switching times for the control and variable system parameters. This class encapsulates a large range of practical optimal control problems. To deal with the variable control switching times, we then apply the time scaling transformation and state the gradient formulae with respect to each set of decision variables.

Consider a process described by the following system of differential equations on the fixed time interval $(0, T]$.

$$\frac{d\boldsymbol{x}(t)}{dt} = \boldsymbol{f}(t, \boldsymbol{x}(t), \boldsymbol{u}(t), \boldsymbol{\zeta}), \tag{7.4.41a}$$

where $\boldsymbol{x} = [x_1, \ldots, x_n]^\top \in \mathbb{R}^n$, $\boldsymbol{u} = [u_1, \ldots, u_r]^\top \in \mathbb{R}^r$ and $\boldsymbol{\zeta} = [\zeta_1, \ldots, \zeta_s]^\top \in \mathbb{R}^s$ are, respectively, the state, control and system parameter vectors, $\boldsymbol{f} : [0, T] \times \mathbb{R}^n \times \mathbb{R}^s \mapsto \mathbb{R}^n$, and $\boldsymbol{f} = [f_1, \ldots, f_n]^\top \in \mathbb{R}^n$. The initial condition for the system of differential equations (7.4.41a) is

$$\boldsymbol{x}(0) = \boldsymbol{x}^0(\boldsymbol{\zeta}), \tag{7.4.41b}$$

where $\boldsymbol{x}^0 = [x_1^0, \ldots, x_n^0]^\top \in \mathbb{R}^n$ is a given vector valued function of the system parameter vector $\boldsymbol{\zeta}$.

As in Section 7.3, we assume that each component of $\boldsymbol{u}(\cdot)$ is a piecewise constant function of the form (7.3.2) with possible jumps at the variable switching times $t_1, \ldots, t_M$. Let $\boldsymbol{h}$ and Λ be as defined in Section 7.3 and let $\boldsymbol{\vartheta}$ and Ξ be as defined in Section 7.4. Then for each $\boldsymbol{h} \in \Lambda$ and $\boldsymbol{\vartheta} \in \Xi$, we denote the corresponding control as $\boldsymbol{u}(\cdot \mid \boldsymbol{h}, \boldsymbol{\vartheta})$. Furthermore, assume that $\boldsymbol{\zeta} \in \mathcal{Z}$, where $\mathcal{Z}$ is as defined in Section 7.2. For each $(\boldsymbol{h}, \boldsymbol{\vartheta}, \boldsymbol{\zeta}) \in \mathbb{R}^{(M+1)r} \times \mathbb{R}^M \times \mathbb{R}^s$, let $\boldsymbol{x}(\cdot \mid \boldsymbol{h}, \boldsymbol{\vartheta}, \boldsymbol{\zeta})$ be the corresponding solution of the system (7.2.1). We may now define an optimal parameter selection problem as follows.

Problem $(P4)$**:** Given the system (7.4.41), find $\boldsymbol{h} \in \Lambda$, $\boldsymbol{\vartheta} \in \Xi$ and $\boldsymbol{\zeta} \in \mathcal{Z}$ such that the cost functional

$$g_0(\boldsymbol{h}, \boldsymbol{\vartheta}, \boldsymbol{\zeta}) = \Phi_0(\boldsymbol{x}(T \mid \boldsymbol{h}, \boldsymbol{\vartheta}, \boldsymbol{\zeta}), \boldsymbol{\zeta}) + \int_0^T \mathcal{L}_0(t, \boldsymbol{x}(t \mid \boldsymbol{h}, \boldsymbol{\vartheta}, \boldsymbol{\zeta}), \boldsymbol{u}(t \mid \boldsymbol{h}, \boldsymbol{\vartheta}), \boldsymbol{\zeta})\, dt \tag{7.4.42}$$

is minimized subject to the equality constraints

$$g_i(\boldsymbol{h}, \boldsymbol{\vartheta}, \boldsymbol{\zeta}) = \Phi_i(\boldsymbol{x}(\tau_i \,|\, \boldsymbol{h}, \boldsymbol{\vartheta}, \boldsymbol{\zeta}), \boldsymbol{\zeta}) + \int_0^T \mathcal{L}_i(t, \boldsymbol{x}(t \,|\, \boldsymbol{h}, \boldsymbol{\vartheta}, \boldsymbol{\zeta}), \boldsymbol{u}(t \,|\, \boldsymbol{h}, \boldsymbol{\vartheta}), \boldsymbol{\zeta}) \, dt = 0,$$
$$i = 1, \ldots, N_e, \tag{7.4.43a}$$

and subject to the inequality constraints

$$g_i(\boldsymbol{h}, \boldsymbol{\vartheta}, \boldsymbol{\zeta}) = \Phi_i(\boldsymbol{x}(\tau_i \,|\, \boldsymbol{h}, \boldsymbol{\vartheta}, \boldsymbol{\zeta}), \boldsymbol{\zeta}) + \int_0^T \mathcal{L}_i(t, \boldsymbol{x}(t \,|\, \boldsymbol{h}, \boldsymbol{\vartheta}, \boldsymbol{\zeta}), \boldsymbol{u}(t \,|\, \boldsymbol{h}, \boldsymbol{\vartheta}), \boldsymbol{\zeta}) \, dt \leq 0,$$
$$i = N_e + 1, \ldots, N. \tag{7.4.43b}$$

Here, $\Phi_i : \mathbb{R}^n \times \mathbb{R}^s \mapsto \mathbb{R}$ and $\mathcal{L}_i : [0, T] \times \mathbb{R}^n \times \mathbb{R}^r \times \mathbb{R}^s \mapsto \mathbb{R}$, $i = 0, 1, \ldots, N$, are given real-valued functions. We assume throughout this section that the corresponding versions of Assumptions 7.2.1–7.4.1 are satisfied.

Consider the time scaling transformation (7.4.22a)–(7.4.22b) and let $\boldsymbol{\gamma}$, Γ and $v(\cdot \mid \boldsymbol{\gamma})$ be as defined in Section 7.4.2. Let

$$\tilde{\boldsymbol{u}}(s \mid \boldsymbol{h}) = \boldsymbol{u}(t(s) \mid \boldsymbol{h}, \boldsymbol{\vartheta}) = \sum_{j=0}^{M} \boldsymbol{h}^j \chi_{[j,j+1)}(s) \tag{7.4.44}$$

for $s \in [0, M+1)$. Similarly, we define $\tilde{\boldsymbol{x}}(s) = \boldsymbol{x}(t(s))$, for $s \in [0, M+1]$. Using (7.4.22a), the transformed dynamics are

$$\frac{d\tilde{\boldsymbol{x}}(s)}{ds} = v(s \mid \boldsymbol{\gamma}) \boldsymbol{f}\left(t(s), \tilde{\boldsymbol{x}}(s), \tilde{\boldsymbol{u}}(s \mid \boldsymbol{h}, \boldsymbol{\zeta})\right) \tag{7.4.45}$$

with the initial condition

$$\tilde{\boldsymbol{x}}(0) = \boldsymbol{x}^0(\boldsymbol{\zeta}). \tag{7.4.46}$$

For each $\boldsymbol{\gamma} \in \Gamma$, let $t(\cdot \mid \boldsymbol{\gamma})$ denote the corresponding solution of (7.4.22a)–(7.4.22b). In order to simplify notation somewhat, let the triple $(\boldsymbol{h}, \boldsymbol{\gamma}, \boldsymbol{\zeta})$ be denoted by $\boldsymbol{\theta}$. Then for each $\boldsymbol{\theta} = (\boldsymbol{h}, \boldsymbol{\gamma}, \boldsymbol{\zeta}) \in \Lambda \times \Gamma \times \mathcal{Z}$, let $\tilde{\boldsymbol{x}}(\cdot \mid \boldsymbol{\theta})$ denote the corresponding solution of (7.4.45)–(7.4.46). The transformed version of Problem $(P4)$ is given as follows.

Problem $(TP4)$: Given the combined system (7.4.22a) and (7.4.45) with the initial conditions (7.4.22b) and (7.4.46), find a combined vector $\boldsymbol{\theta} = (\boldsymbol{h}, \boldsymbol{\gamma}, \boldsymbol{\zeta}) \in \Lambda \times \Gamma \times \mathcal{Z}$ such that the cost functional

$$\tilde{g}_0(\boldsymbol{\theta}) = \Phi_0(\boldsymbol{x}(M+1 \,|\, \boldsymbol{\theta}), \boldsymbol{\zeta}) + \int_0^{M+1} \mathcal{L}_0(t(s \,|\, \boldsymbol{\gamma}), \tilde{\boldsymbol{x}}(s \,|\, \boldsymbol{\theta}), \tilde{\boldsymbol{u}}(s \,|\, \boldsymbol{h}), \boldsymbol{\zeta}) v(s \,|\, \boldsymbol{\gamma}) ds \tag{7.4.47}$$

is minimized over $\Lambda \times \Gamma \times \mathcal{Z}$ subject to equality constraints

$$\tilde{g}_i(\boldsymbol{\theta}) = \Phi_i(\boldsymbol{x}(M+1\,|\,\boldsymbol{\theta}), \boldsymbol{\zeta}) + \int_0^{M+1} \mathcal{L}_i(t(s\,|\,\gamma), \tilde{\boldsymbol{x}}(s\,|\,\boldsymbol{\theta}), \tilde{\boldsymbol{u}}(s\,|\,\boldsymbol{h}), \boldsymbol{\zeta}) v(s\,|\,\gamma) ds = 0,$$
$$i = 1, \ldots, N_e, \tag{7.4.48}$$

the inequality constraints

$$\tilde{g}_i(\boldsymbol{\theta}) = \Phi_i(\boldsymbol{x}(M+1\,|\,\boldsymbol{\theta}), \boldsymbol{\zeta}) + \int_0^{M+1} \mathcal{L}_i(t(s\,|\,\gamma), \tilde{\boldsymbol{x}}(s\,|\,\boldsymbol{\theta}), \tilde{\boldsymbol{u}}(s\,|\,\boldsymbol{h}), \boldsymbol{\zeta}) v(s\,|\,\gamma) ds \leq 0,$$
$$i = N_e + 1, \ldots, N, \tag{7.4.49}$$

and the additional equality constraint

$$\tilde{g}_{N+1}(\gamma) = t(M+1\,|\,\gamma) - T = 0. \tag{7.4.50}$$

Note that all the controls in the transformed problem have fixed switching times and variable heights, just like those in Section 7.3. The gradients of the cost and constraint functionals in Problem $(TP4)$ with respect to each of $\boldsymbol{h}$, γ and $\boldsymbol{\zeta}$ are summarized in the following theorem, which follows from the corresponding results in Sections 7.2 and 7.3.

Consider Problem $(TP4)$. For each $i = 0, 1, \ldots, N$, define the Hamiltonian

$$\begin{aligned} &\tilde{H}_i\left(t(s\,|\,\gamma), \tilde{\boldsymbol{x}}(s\,|\,\boldsymbol{\theta}), v(s\,|\,\gamma), \tilde{\boldsymbol{u}}(s\,|\,\boldsymbol{h}), \boldsymbol{\zeta}, \tilde{\lambda}_v^i(s\,|\,\gamma), \tilde{\boldsymbol{\lambda}}^i(s\,|\,\boldsymbol{\theta})\right) \\ = {}& \mathcal{L}_i\left(t(s\,|\,\gamma), \tilde{\boldsymbol{x}}(s\,|\,\boldsymbol{\theta}), \tilde{\boldsymbol{u}}(s\,|\,\boldsymbol{h}), \boldsymbol{\zeta}\right) + \tilde{\lambda}_v^i(s\,|\,\gamma)\, v(s\,|\,\gamma) \\ &+ \left(\tilde{\boldsymbol{\lambda}}^i(s\,|\,\boldsymbol{\theta})\right)^{\top} v(s\,|\,\gamma) \boldsymbol{f}(t(s\,|\,\gamma), \tilde{\boldsymbol{x}}(s\,|\,\boldsymbol{\theta}), \tilde{\boldsymbol{u}}(s\,|\,\boldsymbol{h}), \boldsymbol{\zeta}) \end{aligned} \tag{7.4.51}$$

where, for each $\boldsymbol{\theta} = (\boldsymbol{h}, \gamma, \boldsymbol{\zeta}) \in \Lambda \times \Gamma \times \mathcal{Z}$, $\tilde{\lambda}_v^i(s\,|\,\gamma)$ and $\tilde{\boldsymbol{\lambda}}^i(s\,|\,\boldsymbol{\theta})$ denote the solution of the costate system

$$\frac{d}{ds}\begin{bmatrix} \tilde{\lambda}_v^i(s) \\ \tilde{\boldsymbol{\lambda}}^i(s) \end{bmatrix} = \begin{bmatrix} -\dfrac{\partial \tilde{H}_i\left(t(s\,|\,\gamma), \tilde{\boldsymbol{x}}(s\,|\,\boldsymbol{\theta}), v(s\,|\,\gamma), \tilde{\boldsymbol{u}}(s\,|\,\boldsymbol{h}), \boldsymbol{\zeta}, \tilde{\lambda}_v^i(s), \tilde{\boldsymbol{\lambda}}^i(s)\right)}{\partial t} \\ -\left(\dfrac{\partial \tilde{H}_i\left(t(s\,|\,\gamma), \tilde{\boldsymbol{x}}(s\,|\,\boldsymbol{\theta}), v(s\,|\,\gamma), \tilde{\boldsymbol{u}}(s\,|\,\boldsymbol{h}), \boldsymbol{\zeta}, \tilde{\lambda}_v^i(s), \tilde{\boldsymbol{\lambda}}^i(s)\right)}{\partial \tilde{\boldsymbol{x}}}\right)^{\top} \end{bmatrix},$$
$$s \in [0, M+1), \tag{7.4.52}$$

with the terminal condition

$$\begin{bmatrix} \tilde{\lambda}_v^i(M+1) \\ \tilde{\boldsymbol{\lambda}}^i(M+1) \end{bmatrix} = \begin{bmatrix} 0 \\ \left(\dfrac{\partial \Phi_i(\tilde{\boldsymbol{x}}(M+1\,|\,\boldsymbol{\theta}), \boldsymbol{\zeta})}{\partial \tilde{\boldsymbol{x}}}\right)^{\top} \end{bmatrix}. \tag{7.4.53}$$

Theorem 7.4.5 *The gradients of the functional $\tilde{g}_i$ with respect to $\boldsymbol{h}^j$ and γ_j, $j = 0, 1, \ldots, M$, as well as $\boldsymbol{\zeta}$ are given by*

$$\frac{\partial \tilde{g}_i(\boldsymbol{\theta})}{\partial \boldsymbol{h}^j} = \int_j^{j+1} \frac{\partial \tilde{H}_i\left(t(s\,|\,\gamma), \tilde{\boldsymbol{x}}(s\,|\,\boldsymbol{\theta}), \gamma_j, \boldsymbol{h}^j, \tilde{\lambda}_v^i(s\,|\,\gamma), \tilde{\boldsymbol{\lambda}}^i(s\,|\,\boldsymbol{\theta})\right)}{\partial \boldsymbol{h}^j} ds, \quad (7.4.54)$$

$$\frac{\partial \tilde{g}_i(\boldsymbol{\theta})}{\partial \gamma_j} = \int_j^{j+1} \frac{\partial \tilde{H}_i\left(t(s\,|\,\gamma), \tilde{\boldsymbol{x}}(s\,|\,\boldsymbol{\theta}), \gamma_j, \boldsymbol{h}^j, \boldsymbol{\zeta}, \tilde{\lambda}_v^i(s\,|\,\gamma), \tilde{\boldsymbol{\lambda}}^i(s\,|\,\boldsymbol{\theta})\right)}{\partial \gamma_j} ds, \quad (7.4.55)$$

and

$$\frac{\partial \tilde{g}_i(\boldsymbol{\theta})}{\partial \boldsymbol{\zeta}} = \frac{\partial \Phi_i(\tilde{\boldsymbol{x}}(M+1\,|\,\boldsymbol{\theta}), \boldsymbol{\zeta})}{\partial \boldsymbol{\zeta}} + \left(\tilde{\boldsymbol{\lambda}}^i(0\,|\,\boldsymbol{\theta})\right)^\top \frac{\partial \boldsymbol{x}^0(\boldsymbol{\zeta})}{\partial \boldsymbol{\zeta}} + \int_0^{M+1} \frac{\partial \tilde{H}_i\left(t(s\,|\,\gamma), \tilde{\boldsymbol{x}}(s\,|\,\boldsymbol{\theta}), v(s\,|\,\gamma), \tilde{\boldsymbol{u}}(s\,|\,\boldsymbol{h}), \boldsymbol{\zeta}, \tilde{\lambda}_v^i(s\,|\,\gamma), \tilde{\boldsymbol{\lambda}}^i(s\,|\,\boldsymbol{\theta})\right)}{\partial \boldsymbol{\zeta}} ds, \quad (7.4.56)$$

respectively.

Remark 7.4.12 *Again, for computational implementation,* (7.4.54) *and* (7.4.55) *may be written as:*

$$\frac{\partial \tilde{g}_i(\boldsymbol{\theta})}{\partial \boldsymbol{h}^j} = \int_j^{j+1} \frac{\partial \tilde{H}_i\left(t(s\,|\,\gamma), \tilde{\boldsymbol{x}}(s\,|\,\boldsymbol{\theta}), v(s\,|\,\gamma), \tilde{\boldsymbol{u}}(s\,|\,\boldsymbol{h}), \tilde{\lambda}_v^i(s\,|\,\gamma), \tilde{\boldsymbol{\lambda}}^i(s\,|\,\boldsymbol{\theta})\right)}{\partial \tilde{\boldsymbol{u}}} ds \quad (7.4.57)$$

and

$$\frac{\partial \tilde{g}_i(\boldsymbol{\theta})}{\partial \gamma_j} = \int_j^{j+1} \frac{\partial \tilde{H}_i\left(t(s\,|\,\gamma), \tilde{\boldsymbol{x}}(s\,|\,\boldsymbol{\theta}), v(s\,|\,\gamma), \tilde{\boldsymbol{u}}(s\,|\,\boldsymbol{h}), \boldsymbol{\zeta}, \tilde{\lambda}_v^i(s\,|\,\gamma), \tilde{\boldsymbol{\lambda}}^i(s\,|\,\boldsymbol{\theta})\right)}{\partial v} ds, \quad (7.4.58)$$

respectively. Also, the gradients of $g_{N+1}(\gamma)$ with respect to $\boldsymbol{h}$ and $\boldsymbol{\zeta}$ are clearly equal to zero, while those with respect to the components of $\boldsymbol{\gamma}$ are given once again by (7.4.38).

Remark 7.4.13 *Problem* ($P4$) *did not allow for individual characteristic times for the constraint functionals like Problems* ($P1$) *and* ($P2$) *did. This is because fixed time points like the $\tau_i \in [0, T]$, $i = 1, \ldots, N$, in Problems* ($P1$) *and* ($P2$) *are actually transformed to variable time points in $[0, M+1]$ with the time scaling transformation* (7.4.22a)–(7.4.22b). *While it is possible to allow for individual characteristic times by incorporating them with the switching times of the controls and introducing some additional constraints, the notation for this procedure becomes very cumbersome. We choose not to do it here.*

7.4.4 Discrete Valued Optimal Control Problems and Optimal Control of Switched Systems

In many practical optimal control problems, the values of the control components may only be chosen from a discrete set rather than from an interval defined by an upper and a lower bound, as we assumed in Section 7.3. The main references for this section are [122, 126, 127, 297].

Consider once more the dynamics (7.3.1) and assume that each component of the control takes the piecewise constant form (7.3.2). Let $\boldsymbol{h}$ and $\boldsymbol{h}^j$, $j = 0, 1, \ldots, M$, be as defined by (7.3.6) and (7.3.7), respectively. Instead of the individual control heights being bounded above and below by (7.3.4), though, we now assume that

$$\boldsymbol{h}^j \in \left\{\bar{\boldsymbol{h}}^1, \bar{\boldsymbol{h}}^2, \ldots, \bar{\boldsymbol{h}}^q\right\}, j = 0, 1, \ldots, M, \tag{7.4.59}$$

where each $\bar{\boldsymbol{h}}^l$, $l = 1, \ldots, q$, is a given fixed vector in $\mathbb{R}^r$. Let $\bar{\Lambda}$ be the set of all those decision parameter vectors $\boldsymbol{h}$ which satisfy (7.4.59). We also assume that the switching times of the controls are decision variables, so let $\boldsymbol{\vartheta}$ and Ξ be defined as for Problem ($P3$). Furthermore, let $\bar{\mathcal{U}}$ denote the set of all the corresponding control functions. For convenience, for any $(\boldsymbol{\vartheta}, \boldsymbol{h}) \in \Xi \times \bar{\Lambda}$, the corresponding control in $\bar{\mathcal{U}}$ is written as $\boldsymbol{u}(\cdot \mid \boldsymbol{\vartheta}, \boldsymbol{h})$. Let $\boldsymbol{x}(\cdot \mid \boldsymbol{\vartheta}, \boldsymbol{h})$ denote the solution of the system (7.3.1) corresponding to the control $\boldsymbol{u}(\cdot \mid \boldsymbol{\vartheta}, \boldsymbol{h}) \in \mathcal{U}$ (and hence to $(\boldsymbol{\vartheta}, \boldsymbol{h}) \in \Xi \times \bar{\Lambda}$). We may then state the canonical discrete valued optimal control problem as follows.

Problem ($P5$)**:** Given the system (7.3.1), find a combined vector $(\boldsymbol{\vartheta}, \boldsymbol{h}) \in \Xi \times \bar{\Lambda}$ such that the cost functional

$$g_0(\boldsymbol{\vartheta}, \boldsymbol{h}) = \Phi_0(\boldsymbol{x}(T \mid \boldsymbol{\vartheta}, \boldsymbol{h})) + \int_0^T \mathcal{L}_0(t, \boldsymbol{x}(t \mid \boldsymbol{\vartheta}, \boldsymbol{h}), \boldsymbol{u}(t \mid \boldsymbol{\vartheta}, \boldsymbol{h}))\, dt \tag{7.4.60}$$

is minimized subject to the equality constraints

$$g_i(\boldsymbol{\vartheta}, \boldsymbol{h}) = \Phi_i(\boldsymbol{x}(T \mid \boldsymbol{\vartheta}, \boldsymbol{h})) + \int_0^T \mathcal{L}_i(t, \boldsymbol{x}(t \mid \boldsymbol{\vartheta}, \boldsymbol{h}), \boldsymbol{u}(t \mid \boldsymbol{\vartheta}, \boldsymbol{h}))\, dt = 0, \quad i = 1, \ldots, N_e, \tag{7.4.61a}$$

and the inequality constraints

$$g_i(\boldsymbol{\vartheta}, \boldsymbol{h}) = \Phi_i(\boldsymbol{x}(T \mid \boldsymbol{\vartheta}, \boldsymbol{h})) + \int_0^T \mathcal{L}_i(t, \boldsymbol{x}(t \mid \boldsymbol{\vartheta}, \boldsymbol{h}), \boldsymbol{u}(t \mid \boldsymbol{\vartheta}, \boldsymbol{h}))\, dt \le 0, \quad i = N_e + 1, \ldots, N, \tag{7.4.61b}$$

where $\Phi_i : \mathbb{R}^n \mapsto \mathbb{R}$ and $\mathcal{L}_i : [0, T] \times \mathbb{R}^n \times \mathbb{R}^r \mapsto \mathbb{R}$, $i = 0, 1, \ldots, N$, are given real-valued functions.

Before going on to discuss the solution strategies for Problem $(P5)$, let us consider another common class of optimal control problems and show that it is equivalent to Problem $(P5)$. Suppose that instead of a single dynamical system, there is a finite set of distinct dynamical systems each of which can be invoked on any subinterval of the time horizon $[0, T]$. The state of the system is then determined as follows. Starting with the given initial condition at $t = 0$, the first dynamical system active of the first subinterval $[0, t_1]$ of $[0, T]$ is integrated up to t_1. $\boldsymbol{x}(t_1)$ then becomes the initial state for the next subinterval $[t_1, t_2]$. Starting with $\boldsymbol{x}(t_1)$ the second dynamical system active on $[t_1, t_2]$ is then integrated forward in time until t_2 to get $\boldsymbol{x}(t_2)$. The process continues in the same manner until we reach the terminal time. Mathematically, we can describe the overall dynamics as

$$\frac{d\boldsymbol{x}(t)}{dt} = \boldsymbol{f}^{v_i}(t, \boldsymbol{x}(t)), \quad t \in (t_i, t_{i+1}],\ i = 0, 1, \ldots, M, \tag{7.4.62}$$

$$\boldsymbol{x}(0) = \boldsymbol{x}^0, \tag{7.4.63}$$

where $v_i \in \{0, 1, \ldots, q\}$ and $\{\boldsymbol{f}^1, \boldsymbol{f}^2, \ldots, \boldsymbol{f}^q\}$ is a given set of candidate dynamical systems. The *switching sequence* $\boldsymbol{v} = [v_0, v_1, \ldots, v_M]^\top \in \mathbb{R}^{M+1}$ is a decision variable which determines the sequence in which the dynamical systems are to be invoked. Let

$$\mathcal{V} = \left\{[v_0, v_1, \ldots, v_M]^\top \in \mathbb{R}^{M+1} \,|\, v_i \in \{1, \ldots, q\},\ i = 0, 1, \ldots, M\right\} \tag{7.4.64}$$

be the set of feasible switching sequences. Suppose that the switching times t_i, $i = 1, \ldots, M$, are also decision variables and let $\boldsymbol{\vartheta}$ and Ξ be defined as for Problem $(P5)$. For each $(\boldsymbol{\vartheta}, \boldsymbol{v}) \in \Xi \times \mathcal{V}$, let $\boldsymbol{x}(\cdot \mid \boldsymbol{\vartheta}, \boldsymbol{v})$ denote the corresponding solution of (7.4.62)–(7.4.63). Then we can define the following canonical *optimal switching control problem.*

Problem $(P6)$**:** Given the system (7.4.62)–(7.4.63), find a combined vector $(\boldsymbol{\vartheta}, \boldsymbol{v}) \in \Xi \times \mathcal{V}$ such that the cost functional

$$g_0(\boldsymbol{\vartheta}, \boldsymbol{v}) = \Phi_0(\boldsymbol{x}(T \mid \boldsymbol{\vartheta}, \boldsymbol{v})) + \int_0^T \mathcal{L}_0(t, \boldsymbol{x}(t \mid \boldsymbol{\vartheta}, \boldsymbol{v}))\, dt, \tag{7.4.65}$$

is minimized subject to the equality constraints

$$g_i(\boldsymbol{\vartheta}, \boldsymbol{v}) = \Phi_i(\boldsymbol{x}(T \mid \boldsymbol{\vartheta}, \boldsymbol{v})) + \int_0^T \mathcal{L}_i(t, \boldsymbol{x}(t \mid \boldsymbol{\vartheta}, \boldsymbol{v}))\, dt = 0,\ i = 1, \ldots, N_e, \tag{7.4.66a}$$

and the inequality constraints

$$g_i(\boldsymbol{\vartheta}, \boldsymbol{v}) = \Phi_i(\boldsymbol{x}(T \mid \boldsymbol{\vartheta}, \boldsymbol{v})) + \int_0^T \mathcal{L}_i(t, \boldsymbol{x}(t \mid \boldsymbol{\vartheta}, \boldsymbol{v}))\, dt \leq 0,\ i = N_e + 1, \ldots, N, \tag{7.4.66b}$$

where $\Phi_i : \mathbb{R}^n \mapsto \mathbb{R}$ and $\mathcal{L}_i : [0,T] \times \mathbb{R}^n \mapsto \mathbb{R}$, $i = 0, 1, \ldots, N$, are given real-valued functions.

Problem $(P6)$ is effectively a special case of Problem $(P5)$. This can be easily seen if we adopt the notation $\boldsymbol{f}^i(t, \boldsymbol{x}(t)) = \tilde{\boldsymbol{f}}(t, \boldsymbol{x}(t), i)$, $i = 1, \ldots, q$, i.e., the index i denoting the choice of system at time t can also be thought of as a scalar discrete valued control function. Similarly, Problem $(P5)$ may be seen as a special case of Problem $(P6)$ if we adopt the notation $\boldsymbol{f}(t, \boldsymbol{x}(t), \bar{\boldsymbol{h}}^j) = \bar{\boldsymbol{f}}^j(t, \boldsymbol{x}(t))$, $j = 1, \ldots, q$, at time t and if we choose to neglect the dependence of the objective and constraint functional integrands on the control.

Indeed, since both problems clearly involve continuous (switching times) and discrete (control values or system indices) decision variables, they belong to the class of mixed discrete optimization problems. Most of the existing solution methods, which we detail further below, treat the continuous and discrete variables separately. Typically, a bilevel formulation is constructed where the switching times are the variables of the lower level problem and the discrete variables appear at the upper level. Thus, the lower level problems generally take the form of Problem $(P3)$, where a piecewise constant control or system sequence is fixed and only the switching times are variable. The time scaling transformation leading to Problem $(TP3)$ can then be invoked to effectively solve the lower level problem for the optimal switching times. More generally, discrete valued optimal control and optimal switching control problems also involve additional continuous valued control functions and system parameters. In these cases the lower level problems most often take the form of Problem $(P4)$ and the same time scaling transformation leading to Problem $(TP4)$ can then be invoked to solve these. The upper level problems typically involve only discrete variables and a range of discrete optimization algorithms can be employed to solve these. Examples include branch-and-bound [66], simulated annealing [127], and filled function methods (see Appendix A3 and [283]).

A simple and effective solution strategy which avoids the need for a bilevel formulation and discrete variables works as follows. Let $\bar{M}$ be the expected number of switchings in an optimal solution of Problem $(P5)$. Consider the fixed control height vector

$$\bar{\boldsymbol{h}} = \Big[(\bar{\boldsymbol{h}}^1)^\top, (\bar{\boldsymbol{h}}^2)^\top, \ldots, (\bar{\boldsymbol{h}}^q)^\top, (\bar{\boldsymbol{h}}^1)^\top, (\bar{\boldsymbol{h}}^2)^\top, \ldots, (\bar{\boldsymbol{h}}^q)^\top, \\ \cdots, (\bar{\boldsymbol{h}}^1)^\top, (\bar{\boldsymbol{h}}^2)^\top, \ldots, (\bar{\boldsymbol{h}}^q)^\top \Big]^\top \in \mathbb{R}^{(\bar{M}+1)\times q},$$

where the sequence $\bar{\boldsymbol{h}}^1, \bar{\boldsymbol{h}}^2, \ldots, \bar{\boldsymbol{h}}^q$ is repeated $\bar{M}+1$ times. For $M = (\bar{M}+1) \times q$, this leads to a problem in the form of Problem $(P3)$ where we just need to determine the switching times t_i, $i = 1, \ldots, (\bar{M}+1) \times q - 1$. Note that any possible order of the $\bar{\boldsymbol{h}}^j$, $j = 1, \ldots, q$ with respect to $\bar{M}$ switches can be parametrized in this way if any of the successive switching times are allowed to coalesce. In other words, all possible combinations of control height sequences of Problem $(P5)$ with up to $\bar{M}$ switches are contained in the vector $\bar{\boldsymbol{h}}$. Optimization of the resulting Problem $(P3)$ (via the time scaling transformation leading to the equivalent Problem $(TP3)$) will then lead to many of the switching times coalescing and leave us with an optimal switching sequence [122, 126]. While this is an effective heuristic scheme, there are several issues with this approach.

(i) It is possible to get more than the assumed $\bar{M}$ switches. For many practical problems, this is not a serious issue, as the number of necessary switches is generally not known in the first place. If a limited number of switches does need to be strictly adhered to, additional constraints can be imposed along with the time scaling transformation for this purpose [297].
(ii) The introduction of many potentially unnecessary switchings creates a large mathematical programming problem. Numerical experience indicates that a lot of locally optimal solutions exist for this problem and it is easy to get trapped in these. This is particularly well illustrated by a complex problem requiring the calculation of an optimal path by a submarine through a sensor field [38].

If the problem under consideration also includes continuous valued control functions and the number of switches for the discrete valued controls is small compared to the size of the partition required for the continuous valued controls, a modified time scaling transformation proposed in [67] can be used. This involves a coarse partition of the time horizon for the discrete valued control components. Within each interval of this partition, a much finer partition is set up for the continuous valued controls.

A special class of optimal control problems which often occurs in practice is one where the Hamiltonian function turns out to be linear in the control variables and where the controls have constant upper and lower bounds. In this case, application of the Maximum Principle leads to an optimal control which is made up of a combination of bang or singular arcs only, as demonstrated for several basic examples in Chapter 6. If there are only bang arcs, the problem can be considered in the form of Problem $(P5)$. If the formulae for the singular control during singular intervals are known and do not depend on the costate, the problem can be considered in the form of Problem $(P6)$ (see [269] for an example). If the singular control can only be expressed in terms of the costate, time scaling can still be used by formulating an auxiliary problem where additional dynamics are included for the costate determination [228].

Another approach to determine optimal switching sequences is a graph-based semi analytical method proposed in [118]. For a class of optimal control problems with a single state and multiple controls, an equivalence between the search for the optimal solution to the problem and the search for the shortest path in a specially constructed graph is established. The graph is based on analytical properties derived from applying the Maximum Principle. Thus the problem of finding optimal sequence for the control policies, which is part of finding the optimal control, is reduced to the problem of finding the shortest path in the corresponding graph. While the method is capable of finding exact global optimal solutions, it is not readily applicable to complex optimal control problems which are difficult to treat analytically or to those where part of the problem formulation depends on experimental data rather than on well defined functions. Nevertheless, the method has been shown to applicable to a broad class of machine scheduling problems in [119]. Note that a related approach can be used for many practical problems which allow some analytical analysis. An instance of this is given in [95].

The approaches described above are mostly limited to problems where the switches in the dynamics depend on time only. There is another class of problems where dynamic switches are driven by changes in the state of the system. Typically, a change in dynamics is triggered when the state trajectory moves from one particular region in the state space to another. This will be elaborated in Chapter 10.2. See also [27, 128, 153].

There are many practical examples of discrete valued optimal control problems. In addition to some of the application mentioned above, these include optimizing driving strategies for trains power by diesel electric locomotives [96, 126, 310], battery re-charge scheduling for conventional submarines [213], operation of hybrid power systems [217], optimal operation of vehicles [66, 128] and sensor scheduling [56, 124].

7.5 Time-Lag System

In this section, we consider a class of optimal control problems with time-delay. The main reference of the section is Chapter 12 of [253].

Consider a process described by the following system of differential equations defined on the fixed time interval $(0, T]$:

$$\frac{d\boldsymbol{x}(t)}{dt} = \boldsymbol{f}(t, \boldsymbol{x}(t), \boldsymbol{x}(t-h), \boldsymbol{\zeta}), \tag{7.5.1a}$$

where

$$\boldsymbol{x} = [x_1, \ldots, x_n]^\top \in \mathbb{R}^n, \quad \boldsymbol{\zeta} = [\zeta_1, \ldots, \zeta_s]^\top \in \mathbb{R}^s$$

are, respectively, the state and system parameter vectors, $\boldsymbol{f} : [0, T] \times \mathbb{R}^{2n} \times \mathbb{R}^s \mapsto \mathbb{R}^n$, $\boldsymbol{f} = [f_1, \ldots, f_n]^\top \in \mathbb{R}^n$, and h is the time delay satisfying $0 < h < T$.

For the sake of simplicity, we have confined our analysis to the case of a single time-delay. Nevertheless, all the results can be extended in a straightforward manner to the case of multiple time delays. The initial function for the state vector is

$$\boldsymbol{x}(t) = \boldsymbol{\phi}(t), \;\; t \in [-h, 0); \quad \boldsymbol{x}(0) = \boldsymbol{x}^0, \tag{7.5.1b}$$

where

$$\boldsymbol{\phi}(t) = [\phi_1(t), \ldots, \phi_n(t)]^\top$$

is a given piecewise continuous function mapping from $[-h, 0)$ to $\mathbb{R}^n$, and $\boldsymbol{x}^0$ is a given vector in $\mathbb{R}^n$.

Let $\mathcal{Z}$ be a compact and convex subset of $\mathbb{R}^s$. For each $\boldsymbol{\zeta} \in \mathcal{Z}$, let $\boldsymbol{x}(\cdot \mid \boldsymbol{\zeta})$ be the corresponding vector-valued function which is absolutely continuous on $(0, T]$ and satisfies the differential equation (7.5.1a) almost everywhere on $(0, T]$ and the initial condition (7.5.1b) everywhere on $[-h, 0]$. This function is called the solution of the system (7.5.1) corresponding to the system parameter vector $\boldsymbol{\zeta} \in \mathcal{Z}$.

We may now state an optimal parameter selection problem for the time-delay system as follows:

Problem $(P7)$**:** Given the system (7.5.1), find a system parameter vector $\boldsymbol{\zeta} \in \mathcal{Z}$ such that the cost functional

$$g_0(\boldsymbol{\zeta}) = \Phi_0(\boldsymbol{x}(T \mid \boldsymbol{\zeta})) + \int_0^T \mathcal{L}_0(t, \boldsymbol{x}(t \mid \boldsymbol{\zeta}), \boldsymbol{x}(t-h \mid \boldsymbol{\zeta}), \boldsymbol{\zeta})\, dt \tag{7.5.2}$$

is minimized over $\mathcal{Z}$ and subject to the equality constraints (in canonical form):

$$g_i(\boldsymbol{\zeta}) = \Phi_i(\boldsymbol{x}(T \mid \boldsymbol{\zeta})) + \int_0^T \mathcal{L}_i(t, \boldsymbol{x}(t \mid \boldsymbol{\zeta}), \boldsymbol{x}(t-h \mid \boldsymbol{\zeta}), \boldsymbol{\zeta})\, dt = 0,$$
$$i = 1, \ldots, N_e, \tag{7.5.3a}$$

and inequality constraints (in canonical form):

$$g_i(\boldsymbol{\zeta}) = \Phi_i(\boldsymbol{x}(T \mid \boldsymbol{\zeta})) + \int_0^T \mathcal{L}_i(t, \boldsymbol{x}(t \mid \boldsymbol{\zeta}), \boldsymbol{x}(t-h \mid \boldsymbol{\zeta}), \boldsymbol{\zeta})\, dt \leq 0,$$
$$i = N_e + 1, \ldots, N, \tag{7.5.3b}$$

where $\Phi_i : \mathbb{R}^n \mapsto \mathbb{R}$ and $\mathcal{L}_i : [0, T] \times \mathbb{R}^{2n} \times \mathbb{R}^s \mapsto \mathbb{R}$, $i = 0, 1, \ldots, N$, are given real valued functions.

We assume that the following conditions are satisfied:

Assumption 7.5.1 *For each compact subset* V *of* $\mathbb{R}^s$*, there exists a positive constant* K *such that*

$$|\boldsymbol{f}(t, \boldsymbol{x}, \boldsymbol{y}, \boldsymbol{\zeta})| \leq K(1 + |\boldsymbol{x}| + |\boldsymbol{y}|),$$

and, for each $i = 0, \ldots, N$,

$$|\mathcal{L}_i(t, \boldsymbol{x}, \boldsymbol{y}, \boldsymbol{\zeta})| \leq K(1 + |\boldsymbol{x}| + |\boldsymbol{y}|)$$

for all $(t, \boldsymbol{x}, \boldsymbol{y}, \boldsymbol{\zeta}) \in [0, T] \times \mathbb{R}^{2n} \times V$.

Assumption 7.5.2 *$\boldsymbol{f}(t, \boldsymbol{x}, \boldsymbol{y}, \boldsymbol{\zeta})$ and $\mathcal{L}_i(t, \boldsymbol{x}, \boldsymbol{y}, \boldsymbol{\zeta})$, $i = 0, 1, \ldots, N$, are piecewise continuous on $[0, T]$ for each $(\boldsymbol{x}, \boldsymbol{y}, \boldsymbol{\zeta}) \in \mathbb{R}^{2n} \times \mathbb{R}^s$. Furthermore, $\boldsymbol{f}(t, \boldsymbol{x}, \boldsymbol{y}, \boldsymbol{\zeta})$ and $\mathcal{L}_i(t, \boldsymbol{x}, \boldsymbol{y}, \boldsymbol{\zeta})$, $i = 0, 1, \ldots, N$, are continuously differentiable with respect to each of the components of $\boldsymbol{x}, \boldsymbol{y}$, and $\boldsymbol{\zeta}$ for each fixed $t \in [0, T]$.*

Assumption 7.5.3 *Φ_i, $i = 0, 1, \ldots, N$, are continuously differentiable with respect to $\boldsymbol{x}$ and $\boldsymbol{\zeta}$.*

Assumption 7.5.4 *$\boldsymbol{\phi}$ is piecewise continuous on $[-h, 0)$.*

Remark 7.5.1 *For each $\boldsymbol{\zeta} \in \mathbb{R}^s$ (and hence for each $\boldsymbol{\zeta} \in \mathcal{Z}$), there exists a unique absolutely continuous vector-valued function $\boldsymbol{x}(\cdot \mid \boldsymbol{\zeta})$ which satisfies the system (7.5.1a)–(7.5.1b). This is done as follows.*
Let $k =$ integer of (T/h). Then, we subdivide the interval $[0, T]$ into k subintervals if $\frac{T}{h} = k$ and $k + 1$ subintervals if $\frac{T}{h} > k$. Without loss of generality, we shall only consider the latter case. Clearly, these $k + 1$ subintervals may be written as

$$[(l-1)h, lh], \quad l = 1, \ldots, k, \quad [kh, T].$$

Finding the unique solution of the system (7.5.1a)–(7.5.1b) is the same as finding the unique solution of the system (7.5.1a)–(7.5.1b) on each of these subintervals successively with appropriate boundary conditions. The existence of a unique solution follows from repeated applications of well-known results on ordinary differential equations.

7.5.1 Gradient Formulae

For each $i = 0, 1, \ldots, N$, and for each system parameter vector $\boldsymbol{\zeta}$, consider the system

$$\begin{aligned}\frac{d\boldsymbol{\lambda}^i(t)}{dt} = &- \left[\frac{\partial H_i\left(t, \boldsymbol{x}(t \mid \boldsymbol{\zeta}), \boldsymbol{y}(t, \boldsymbol{\zeta}), \boldsymbol{\lambda}^i(t)\right)}{\partial \boldsymbol{x}}\right]^\top \\ &- \left[\frac{\partial \hat{H}_i\left(t, \boldsymbol{z}(t), \boldsymbol{x}(t \mid \boldsymbol{\zeta}), \boldsymbol{\zeta}, \hat{\boldsymbol{\lambda}}^i(t)\right)}{\partial \boldsymbol{x}}\right]^\top, t \in [0, T] \end{aligned} \tag{7.5.4a}$$

with the boundary conditions

$$(\boldsymbol{\lambda}^i(T))^\top = \frac{\partial \Phi_i(\boldsymbol{x}(T \mid \boldsymbol{\zeta}))}{\partial \boldsymbol{x}}, \tag{7.5.4b}$$

$$\boldsymbol{\lambda}^i(t) = \boldsymbol{0}, \quad t > T, \tag{7.5.4c}$$

where

$$\boldsymbol{y}(t) = \boldsymbol{x}(t - h \mid \boldsymbol{\zeta}), \tag{7.5.4d}$$

$$z(t) = x(t + h \mid \zeta), \tag{7.5.4e}$$
$$\hat{\lambda}^i(t) = \lambda^i(t + h), \tag{7.5.4f}$$

$$H_i(t, x, y, \zeta, \lambda) = \mathcal{L}_i(t, x, y, \zeta) + (\lambda)^\top f(t, x, y, \zeta), \tag{7.5.5a}$$
$$\begin{aligned}\hat{H}_i(t, z, x, \zeta, \hat{\lambda}) &= \mathcal{L}_i(t + h, z, x, \zeta)e(T - t - h), \\ &\quad +(\hat{\lambda})^\top f(t + h, z, x, \zeta)e(T - t - h),\end{aligned} \tag{7.5.5b}$$

and $e(\cdot)$ is the unit step function. To continue, we set

$$z(t) = \mathbf{0}, \qquad \text{for all } t \in [T - h, T]. \tag{7.5.6}$$

For each $i = 0, 1, \ldots, N$, the system (7.5.4) is again known as the corresponding costate system. Furthermore, let $\lambda^i(\cdot \mid \zeta)$ denote the solution of the costate system corresponding to $\zeta \in \mathcal{Z}$. It is solved backward in time from $t = T$ to $t = 0$, using a similar idea to that described in Remark 7.5.1.

Theorem 7.5.1 *Consider Problem (P7). For each $i = 0, 1, \ldots, N$, the gradient of the functional g_i is given by*

$$\frac{\partial g_i(\zeta)}{\partial \zeta} = \int_0^T \left\{ \frac{\partial H_i\left(t, x(t \mid \zeta), x(t - h \mid \zeta), \zeta, \lambda^i(t \mid \zeta)\right)}{\partial \zeta} \right\} dt. \tag{7.5.7}$$

Proof. Let $\zeta \in \mathcal{Z}$ be any system parameter vector and let ρ be any perturbation about ζ. Define

$$\zeta(\varepsilon) = \zeta + \varepsilon\rho. \tag{7.5.8}$$

For brevity, let $x(\cdot)$ and $x(\cdot; \varepsilon)$ denote the solutions of the system (7.5.1) corresponding to ζ and $\zeta(\varepsilon)$, respectively. Let $y(\cdot)$, $z(\cdot)$, $y(\cdot; \varepsilon)$, $z(\cdot; \varepsilon)$ be as defined according to (7.5.4d)–(7.5.4e). Clearly, from (7.5.1), we have

$$x(t) = x^0 + \int_0^t f(s, x(s), y(s), \zeta)\, ds \tag{7.5.9}$$

and

$$x(t; \varepsilon) = x^0 + \int_0^t f(s, x(s; \varepsilon), y(s; \varepsilon), \zeta(\varepsilon))\, ds. \tag{7.5.10}$$

Thus,

$$\begin{aligned}\triangle x(t) &= \left.\frac{dx(t; \varepsilon)}{d\varepsilon}\right|_{\varepsilon=0} \\ &= \int_0^t \left\{ \frac{\partial f(s, x(s), y(s), \zeta)}{\partial x} \triangle x(s) + \frac{\partial f(s, x(s), y(s), \zeta)}{\partial y} \triangle y(s) \right. \\ &\quad \left. + \frac{\partial f(s, x(s), y(s), \zeta)}{\partial \zeta} \rho \right\} ds.\end{aligned} \tag{7.5.11}$$

Clearly,

$$\frac{d(\triangle \boldsymbol{x}(t))}{dt} = \frac{\partial \boldsymbol{f}(t, \boldsymbol{x}(t), \boldsymbol{y}(t), \boldsymbol{\zeta})}{\partial \boldsymbol{x}} \triangle \boldsymbol{x}(t) + \frac{\partial \boldsymbol{f}(t, \boldsymbol{x}(t), \boldsymbol{y}(t), \boldsymbol{\zeta})}{\partial \boldsymbol{y}} \triangle \boldsymbol{y}(t) + \frac{\partial \boldsymbol{f}(t, \boldsymbol{x}(t), \boldsymbol{y}(t), \boldsymbol{\zeta})}{\partial \boldsymbol{\zeta}} \boldsymbol{\rho} \tag{7.5.12a}$$

$$\triangle \boldsymbol{x}(t) = 0, \text{ for } t \leq 0. \tag{7.5.12b}$$

Now, by (7.5.3), we have

$$g_i(\boldsymbol{\zeta}(\varepsilon)) = \Phi_i(\boldsymbol{x}(T;\varepsilon)) + \int_0^T \mathcal{L}_i(t, \boldsymbol{x}(t;\varepsilon), \boldsymbol{y}(t;\varepsilon), \boldsymbol{\zeta}(\varepsilon))\, dt. \tag{7.5.13}$$

Define

$$\bar{\mathcal{L}}_i = \mathcal{L}_i(t, \boldsymbol{x}(t), \boldsymbol{y}(t), \boldsymbol{\zeta}), \tag{7.5.14a}$$

$$\hat{\mathcal{L}}_i = \mathcal{L}_i(t+h, \boldsymbol{z}(t), \boldsymbol{x}(t), \boldsymbol{\zeta}), \tag{7.5.14b}$$

$$\bar{\boldsymbol{f}} = \boldsymbol{f}(t, \boldsymbol{x}(t), \boldsymbol{y}(t), \boldsymbol{\zeta}), \tag{7.5.14c}$$

$$\hat{\boldsymbol{f}} = \boldsymbol{f}(t+h, \boldsymbol{z}(t), \boldsymbol{x}(t), \boldsymbol{\zeta}), \tag{7.5.14d}$$

$$\bar{H}_i = H_i(t, \boldsymbol{x}(t), \boldsymbol{y}(t), \boldsymbol{\zeta}, \boldsymbol{\lambda}^i(t)), \tag{7.5.14e}$$

and

$$\hat{H}_i = \hat{H}_i\left(t, \boldsymbol{z}(t), \boldsymbol{x}(t), \boldsymbol{\zeta}, \hat{\boldsymbol{\lambda}}^i(t)\right), \tag{7.5.14f}$$

where $\boldsymbol{\lambda}^i(t)$ is the solution of the costate system (7.5.4) corresponding to the system parameter vector $\boldsymbol{\zeta}$, and $\hat{\boldsymbol{\lambda}}^i(t)$ is defined by (7.5.4f). From (7.5.13), we have

$$\begin{aligned}\triangle g_i(\boldsymbol{\zeta}) &= \left.\frac{dg_i(\boldsymbol{\zeta}(\varepsilon))}{d\varepsilon}\right|_{\varepsilon=0} \\ &= \frac{\partial g_i(\boldsymbol{\zeta})}{\partial \boldsymbol{\zeta}} \boldsymbol{\rho} \\ &= \frac{\partial \Phi_i(\boldsymbol{x}(T))}{\partial \boldsymbol{x}} \triangle \boldsymbol{x}(T) + \int_0^T \left[\frac{\partial \bar{\mathcal{L}}_i}{\partial \boldsymbol{x}} \triangle \boldsymbol{x}(t) + \frac{\partial \bar{\mathcal{L}}_i}{\partial \boldsymbol{y}} \triangle \boldsymbol{y}(t) + \frac{\partial \bar{\mathcal{L}}_i}{\partial \boldsymbol{\zeta}} \boldsymbol{\rho} \right] dt. \end{aligned} \tag{7.5.15}$$

In view of (7.5.12b), we have

$$\int_0^T \left[\frac{\partial \bar{\mathcal{L}}_i}{\partial \boldsymbol{y}} \triangle \boldsymbol{y}(t) \right] dt = \int_0^T \left[e(T-t-h) \frac{\partial \hat{\mathcal{L}}_i}{\partial \boldsymbol{x}} \triangle \boldsymbol{x}(t) \right] dt. \tag{7.5.16}$$

Combining (7.5.15), (7.5.16) and (7.5.5), we have

$$
\begin{aligned}
&\triangle g_i(\boldsymbol{\zeta}) \\
&= \frac{\partial \Phi_i(\boldsymbol{x}(T))}{\partial \boldsymbol{x}} \triangle \boldsymbol{x}(T) + \int_0^T \left[\frac{\partial \bar{H}_i}{\partial \boldsymbol{x}} \triangle \boldsymbol{x}(t) + \frac{\partial \hat{H}_i}{\partial \boldsymbol{x}} \triangle \boldsymbol{x}(t) - \left(\boldsymbol{\lambda}^i(t)\right)^\top \frac{\partial \bar{\boldsymbol{f}}}{\partial \boldsymbol{x}} \triangle \boldsymbol{x}(t) \right. \\
&\quad \left. - \left(\hat{\boldsymbol{\lambda}}^i(t)\right)^\top \frac{\partial \hat{\boldsymbol{f}}}{\partial \boldsymbol{x}} \triangle \boldsymbol{x}(t) e(T-t-h) + \frac{\partial \bar{H}_i}{\partial \boldsymbol{\zeta}} \boldsymbol{\rho} - \left(\boldsymbol{\lambda}^i(t)\right)^\top \frac{\partial \bar{\boldsymbol{f}}}{\partial \boldsymbol{\zeta}} \boldsymbol{\rho} \right] dt.
\end{aligned}
\tag{7.5.17}
$$

In view of (7.5.6), (7.5.4c) and (7.5.12b), we have

$$
\int_0^T \left[\left(\hat{\boldsymbol{\lambda}}^i(t)\right)^\top \frac{\partial \hat{\boldsymbol{f}}}{\partial \boldsymbol{x}} \triangle \boldsymbol{x}(t) e(T-t-h) \right] dt = \int_0^T \left(\boldsymbol{\lambda}^i(t)\right)^\top \frac{\partial \bar{\boldsymbol{f}}}{\partial \boldsymbol{y}} \triangle \boldsymbol{y}(t)\, dt.
\tag{7.5.18}
$$

Thus, from (7.5.17), (7.5.18) and (7.5.12a), we get

$$
\begin{aligned}
\triangle g_i(\boldsymbol{\zeta}) = {} & \frac{\partial \Phi_i(\boldsymbol{x}(T))}{\partial \boldsymbol{x}} \triangle \boldsymbol{x}(T) + \int_0^T \left[\frac{\partial \bar{H}_i}{\partial \boldsymbol{x}} \triangle \boldsymbol{x}(t) + \frac{\partial \hat{H}_i}{\partial \boldsymbol{x}} \triangle \boldsymbol{x}(t) \right. \\
& \left. + \frac{\partial \bar{H}_i}{\partial \boldsymbol{\zeta}} \boldsymbol{\rho} - \left(\boldsymbol{\lambda}^i(t)\right)^\top \frac{d(\triangle \boldsymbol{x}(t))}{dt} \right] dt.
\end{aligned}
\tag{7.5.19}
$$

Using (7.5.4a), (7.5.4b), (7.5.12b), and integration by parts, it follows that

$$
\begin{aligned}
& \int_0^T \left[\frac{\partial \bar{H}_i}{\partial \boldsymbol{x}} + \frac{\partial \hat{H}_i}{\partial \boldsymbol{x}} \right] \triangle \boldsymbol{x}(t)\, dt \\
&= \int_0^T -\frac{d\left(\boldsymbol{\lambda}^i(t)\right)^\top}{dt} \triangle \boldsymbol{x}(t)\, dt \\
&= -\frac{\partial \Phi_i(\boldsymbol{x}(T))}{\partial \boldsymbol{x}} \triangle \boldsymbol{x}(T) + \int_0^T \left(\boldsymbol{\lambda}^i(t)\right)^\top \frac{d(\triangle \boldsymbol{x}(t))}{dt} dt.
\end{aligned}
\tag{7.5.20}
$$

From (7.5.19) and (7.5.20), we have

$$
\triangle g_i(\boldsymbol{\zeta}) = \frac{\partial g_i(\boldsymbol{\zeta})}{\partial \boldsymbol{\zeta}} \boldsymbol{\rho} = \left[\int_0^T \frac{\partial \bar{H}_i}{\partial \boldsymbol{\zeta}} dt \right] \boldsymbol{\rho}.
\tag{7.5.21}
$$

Since $\boldsymbol{\rho}$ is arbitrary, the conclusion of the theorem follows readily.

Remark 7.5.2 *The procedure for calculating the values of the cost functional and the constraint functionals are similar to that described in Algorithm 7.2.2. Their gradients can be computed by an algorithm similar to Algorithm 7.2.3, using the formulae presented in Theorem 7.5.1. Thus, we see that the time-lag optimal parameter selection problem can also be viewed and hence solved as a standard mathematical programming problem. However, we point out that the time-lag system and its costate system are solved successively over a finite* *number of subintervals as specified in Remark 7.5.1. This method is known as the* method of steps *in the literature [14, 84].*

7.6 Multiple Characteristic Time Points

In this section, we consider a class of optimal parameter selection problems with multiple characteristic time points in the cost and constraint functionals. The main references for this section are [180, 181]. Consider a process described by the following system of differential equations defined on the fixed time interval $(0, T]$:

$$\frac{d\boldsymbol{x}(t)}{dt} = \boldsymbol{f}(t, \boldsymbol{x}(t), \boldsymbol{\zeta}), \tag{7.6.1a}$$

where $\boldsymbol{x} = [x_1, \ldots, x_n]^\top \in \mathbb{R}^n$ is the state vector, $\boldsymbol{\zeta} = [\zeta_1, \ldots, \zeta_s]^\top \in \mathbb{R}^s$ is a vector of system parameters, $\boldsymbol{f} = [f_1, \ldots, f_n]^\top \in \mathbb{R}^n$, and $\boldsymbol{f} : [0, T] \times \mathbb{R}^n \times \mathbb{R}^s \mapsto \mathbb{R}^n$. The initial condition for (7.6.1a) is

$$\boldsymbol{x}(0) = \boldsymbol{x}^0, \tag{7.6.1b}$$

where $\boldsymbol{x}^0 \in \mathbb{R}^n$ is given.

For each $\boldsymbol{\zeta} \in \mathbb{R}^s$, let $\boldsymbol{x}(\cdot \mid \boldsymbol{\zeta})$ be the corresponding solution of the system (7.6.1). We may now state the optimal parameter selection problem as follows.

Problem $(P8)$: Given the system (7.6.1), find a system parameter vector $\boldsymbol{\zeta} \in \mathbb{R}^s$ such that the cost functional

$$g_0(\boldsymbol{\zeta}) = \Phi_0(\boldsymbol{x}(\tau_1 \mid \boldsymbol{\zeta}), \ldots, \boldsymbol{x}(\tau_M \mid \boldsymbol{\zeta})) + \int_0^T \mathcal{L}_0(t, \boldsymbol{x}(t \mid \boldsymbol{\zeta}), \boldsymbol{\zeta})\, dt \tag{7.6.2}$$

is minimized subject to the canonical inequality constraints

$$g_m(\boldsymbol{\zeta}) = \Phi_m(\boldsymbol{x}(\tau_1 \mid \boldsymbol{\zeta}), \ldots, \boldsymbol{x}(\tau_M \mid \boldsymbol{\zeta})) + \int_0^T \mathcal{L}_m(t, \boldsymbol{x}(t \mid \boldsymbol{\zeta}), \boldsymbol{\zeta})\, dt \leq 0,$$
$$m = 1, \ldots, N. \tag{7.6.3}$$

Here, $\Phi_m : \mathbb{R}^n \times \cdots \times \mathbb{R}^n \mapsto \mathbb{R}$, $m = 0, 1, \ldots, N$, $\mathcal{L}_m : [0, T] \times \mathbb{R}^n \times \mathbb{R}^s \mapsto \mathbb{R}$, $m = 0, 1, \ldots, N$, are given real-valued functions and the time points τ_i, $0 < \tau_i < T$, $i = 1, \ldots, M$, are referred to as the *characteristic times*. For standard optimal parameter selection problems such as those considered in previous sections, each canonical constraint (as well as the cost which corresponds to $m = 0$) depends only on one such time point. Here, however, there may be many such time points. For convenience, define $\tau_0 = 0$ and $\tau_{M+1} = T$.

We assume throughout that the following conditions are satisfied.

Assumption 7.6.1 *For any compact subset $V \subset \mathbb{R}^s$, there exists a positive constant K such that*

$$|\boldsymbol{f}(t, \boldsymbol{x}, \boldsymbol{\zeta})| \leq K(1 + |\boldsymbol{x}|)$$

for all $(t, \boldsymbol{x}, \boldsymbol{\zeta}) \in [0, T] \times \mathbb{R}^n \times V$.

Assumption 7.6.2 *$\boldsymbol{f}$ and $\mathcal{L}_m$, $m = 0, 1, \ldots, N$, together with their partial derivatives with respect to each of the components of $\boldsymbol{x}$ and $\boldsymbol{\zeta}$ are piecewise continuous on $[0, T]$ for each $(\boldsymbol{x}, \boldsymbol{\zeta}) \in \mathbb{R}^n \times \mathbb{R}^s$ and continuous on $\mathbb{R}^n \times \mathbb{R}^s$ for each $t \in [0, T]$.*

Assumption 7.6.3 *Φ_m, $m = 0, 1, \ldots, N$, are continuously differentiable on $\mathbb{R}^n \times \cdots \times \mathbb{R}^n$.*

To derive the gradient formulae for the cost and the constraint functionals given by (7.6.2) and (7.6.3), respectively, we consider the following system.

$$\frac{d\boldsymbol{\lambda}^m(t)}{dt} = -\left[\frac{\partial H_m(t, \boldsymbol{x}(t \mid \boldsymbol{\zeta}), \boldsymbol{\zeta}, \boldsymbol{\lambda}^m(t))}{\partial \boldsymbol{x}(t \mid \boldsymbol{\zeta})}\right]^\top, \tag{7.6.4a}$$

where $t \in (\tau_{k-1}, \tau_k)$ for $k = 1, \ldots, M + 1$, with the jump conditions

$$\boldsymbol{\lambda}^m\left(\tau_k^+\right) - \boldsymbol{\lambda}^m(\tau_k^-) = -\left[\frac{\partial \Phi_m}{\partial \boldsymbol{x}(\tau_k)}\right], \text{ for } k = 1, \ldots, M, \tag{7.6.4b}$$

and the terminal condition

$$\boldsymbol{\lambda}^m(T) = \mathbf{0}, \tag{7.6.4c}$$

where H_m is the Hamiltonian function defined by

$$H_m(t, \boldsymbol{x}(t \mid \boldsymbol{\zeta}), \boldsymbol{\zeta}, \boldsymbol{\lambda}^m(t)) = \mathcal{L}_m(t, \boldsymbol{x}(t \mid \boldsymbol{\zeta}), \boldsymbol{\zeta}) + (\boldsymbol{\lambda}^m(t))^\top \boldsymbol{f}(t, \boldsymbol{x}(t \mid \boldsymbol{\zeta}), \boldsymbol{\zeta}). \tag{7.6.5}$$

For each $m = 0, 1, \ldots, N$, the corresponding system (7.6.4) is called the costate system for g_m. Let $\boldsymbol{\lambda}^m(\cdot \mid \boldsymbol{\zeta})$ be the solution of the costate system (7.6.4) corresponding to $\boldsymbol{\zeta} \in \mathbb{R}^s$.

Remark 7.6.1 *For each $\boldsymbol{\zeta} \in \mathbb{R}^s$, the solution $\boldsymbol{\lambda}^m(\cdot \mid \boldsymbol{\zeta})$ is calculated as follows.*

Step 1. *Solve the system (7.6.1), yielding $\boldsymbol{x}(\cdot \mid \boldsymbol{\zeta})$ on $[0, T]$.*

Step 2. *Solve the costate differential equations (7.6.4a) with the terminal condition (7.6.4c) backward from $t = T$ to $t = \tau_M^+$, yielding $\lambda^m(\cdot \mid \boldsymbol{\zeta})$ on the subinterval $(\tau_M, T]$ and $\boldsymbol{\lambda}^m\left(\tau_M^+ \mid \boldsymbol{\zeta}\right)$.*

Step 3. *Note that $\boldsymbol{x}(\tau_m \mid \boldsymbol{\zeta})$ is known from Step 1 and that Φ_m is a given continuously differentiable function of $\boldsymbol{x}(\tau_k \mid \boldsymbol{\zeta})$, $k = 1, \ldots, M$. Calculate $\boldsymbol{\lambda}^m(\tau_M^- \mid \boldsymbol{\zeta})$ by using the jump condition (7.6.4b) with $k = M$.*

Step 4. *Solve the costate differential equations (7.6.4a) backward from $t = \tau_M^-$ to $t = \tau_{M-1}^+$ with the condition at $t = \tau_M^-$ being taken as $\boldsymbol{\lambda}^m(\tau_M^- \mid \boldsymbol{\zeta})$. This yields $\boldsymbol{\lambda}^m(\cdot \mid \boldsymbol{\zeta})$ on the subinterval $(\tau_{M-1}, \tau_M]$ and $\boldsymbol{\lambda}^m\left(\tau_{M-1}^+ \mid \boldsymbol{\zeta}\right)$.*

Step 5. *Use the jump condition* (7.6.4b) *with* $k = M - 1$ *to obtain* $\boldsymbol{\lambda}^m(\tau_{M-1}^- \mid \boldsymbol{\zeta})$.

Step 6. *Solve the costate system* (7.6.4a) *backward from* $t = \tau_{M-1}^-$ *to* τ_{M-2}^+ *with the condition at* $t = \tau_{M-1}^-$ *being taken as* $\boldsymbol{\lambda}^m(\tau_{M-1}^- \mid \boldsymbol{\zeta})$. *This yields* $\boldsymbol{\lambda}^m(\cdot \mid \boldsymbol{\zeta})$ *on the subinterval* $(\tau_{M-2}, \tau_{M-1}]$ *and* $\boldsymbol{\lambda}^m(\tau_{M-2}^+ \mid \boldsymbol{\zeta})$.

Step 7. *The process is continued until* $\boldsymbol{\lambda}^m(\cdot \mid \boldsymbol{\zeta})$ *is obtained on the subinterval* $[0, \tau_1]$, *which includes* $\boldsymbol{\lambda}^m(0 \mid \boldsymbol{\zeta})$ *at* $t = 0$.

The solution of the costate system (7.6.4a) *with the jump conditions* (7.6.4b) *and the terminal condition* (7.6.4c) *is thus obtained by combining* $\boldsymbol{\lambda}^m(\cdot \mid \boldsymbol{\zeta})$ *in* $[\tau_k, \tau_{k+1}]$, $k = 0, 1, \ldots, M$.

Theorem 7.6.1 *For each* $m = 0, 1, \ldots, N$, *the gradient of* g_m *is given by*

$$\frac{\partial g_m(\boldsymbol{\zeta})}{\partial \zeta_j} = \int_0^T \frac{\partial H_m(t, \boldsymbol{x}(t \mid \boldsymbol{\zeta}), \boldsymbol{\zeta}, \boldsymbol{\lambda}^m(t \mid \boldsymbol{\zeta}))}{\partial \zeta_j} \, dt, \tag{7.6.6}$$

where $j = 1, \ldots, s$.

Proof. The functions H_m and $\boldsymbol{\lambda}^m$ may have discontinuities at the characteristic time points τ_i, $i = 1, \ldots, M$. Note that we can write

$$\begin{aligned} g_m(\boldsymbol{\zeta}) &= \Phi_m(\boldsymbol{x}(\tau_1 \mid \boldsymbol{\zeta}), \ldots, \boldsymbol{x}(\tau_M \mid \boldsymbol{\zeta})) \\ &\quad + \sum_{k=1}^{M+1} \int_{\tau_{k-1}}^{\tau_k} \left[H_m(t, \boldsymbol{x}(t), \boldsymbol{\zeta}, \boldsymbol{\lambda}^m(t)) - (\boldsymbol{\lambda}^m(t))^\top \boldsymbol{f}(t, \boldsymbol{x}(t), \boldsymbol{\zeta}) \right] dt. \end{aligned} \tag{7.6.7}$$

Solving (7.6.1) gives

$$\boldsymbol{x}(t) = \boldsymbol{x}(t \mid \boldsymbol{\zeta}) = \boldsymbol{x}(0) + \int_0^t \boldsymbol{f}(s, \boldsymbol{x}(s \mid \boldsymbol{\zeta}), \boldsymbol{\zeta}) \, ds. \tag{7.6.8}$$

The partial derivative of $\boldsymbol{x}(t \mid \boldsymbol{\zeta})$ with respect to the $j-$th component, ζ_j, of the vector $\boldsymbol{\zeta}$, $j = 1, \ldots, s$, yields

$$\frac{\partial \boldsymbol{x}(t)}{\partial \zeta_j} = \int_0^t \left[\frac{\partial \boldsymbol{f}(s, \boldsymbol{x}(s), \boldsymbol{\zeta})}{\partial \boldsymbol{x}(t)} \frac{\partial \boldsymbol{x}(s)}{\partial \zeta_j} + \frac{\partial \boldsymbol{f}(s, \boldsymbol{x}(s), \boldsymbol{\zeta})}{\partial \zeta_j} \right] ds. \tag{7.6.9}$$

Note that

$$\frac{\partial}{\partial \zeta_j} \left(\frac{d\boldsymbol{x}(t)}{dt} \right) = \frac{d}{dt} \left(\frac{\partial \boldsymbol{x}(t)}{\partial \zeta_j} \right). \tag{7.6.10}$$

The gradient of g_m can be calculated as:

$$\frac{\partial g_m(\boldsymbol{\zeta})}{\partial \zeta_j} = \sum_{l=1}^{M} \frac{\partial \Phi_m(\boldsymbol{x}(\tau_1), \ldots, \boldsymbol{x}(\tau_M))}{\partial \boldsymbol{x}(\tau_l)} \frac{\partial \boldsymbol{x}(\tau_l)}{\partial \zeta_j}$$

$$+\sum_{k=1}^{M+1}\int_{\tau_{k-1}}^{\tau_k}\left[\frac{\partial H_m}{\partial \boldsymbol{x}}\frac{\partial \boldsymbol{x}}{\partial \zeta_j}+\frac{\partial H_m}{\partial \zeta_j}+\frac{\partial H_m}{\partial \boldsymbol{\lambda}^m}\frac{\partial \boldsymbol{\lambda}^m}{\partial \zeta_j}\right.$$
$$\left.-\frac{\partial (\boldsymbol{\lambda}^m)^\top}{\partial \zeta_j}\boldsymbol{f}(t,\boldsymbol{x},\boldsymbol{\zeta})-(\boldsymbol{\lambda}^m)^\top\left(\frac{d}{dt}\left(\frac{\partial \boldsymbol{x}(t)}{\partial \zeta_j}\right)\right)\right]dt. \tag{7.6.11}$$

From (7.6.5), we obtain

$$\frac{\partial H_m}{\partial \boldsymbol{\lambda}^m}\frac{\partial \boldsymbol{\lambda}^m}{\partial \zeta_j}=\frac{\partial (\boldsymbol{\lambda}^m)^\top}{\partial \zeta_j}\boldsymbol{f}(t,\boldsymbol{x},\boldsymbol{\zeta}). \tag{7.6.12}$$

Applying integration by parts to the last term of the right hand side of (7.6.11) yields

$$\int_{\tau_{k-1}}^{\tau_k}(\boldsymbol{\lambda}^m)^\top\frac{d}{dt}\left(\frac{\partial \boldsymbol{x}}{\partial \zeta_j}\right)dt=(\boldsymbol{\lambda}^m)^\top\frac{\partial \boldsymbol{x}}{\partial \zeta_j}\bigg|_{t=\tau_{k-1}^+}^{t=\tau_k^-}-\int_{\tau_{k-1}}^{\tau_k}\left(\frac{d\boldsymbol{\lambda}^m}{dt}\right)^\top\left(\frac{\partial \boldsymbol{x}}{\partial \zeta_j}\right)dt. \tag{7.6.13}$$

From (7.6.12) and (7.6.13), it follows from (7.6.11) that

$$\frac{\partial g_m(\boldsymbol{\zeta})}{\partial \zeta_j}=\sum_{l=1}^{M}\frac{\partial \Phi_m(\boldsymbol{x}(\tau_1),\ldots,\boldsymbol{x}(\tau_M))}{\partial \boldsymbol{x}(\tau_l)}\frac{\partial \boldsymbol{x}(\tau_l)}{\partial \zeta_j}-\sum_{k=1}^{M+1}(\boldsymbol{\lambda}^m)^\top\frac{\partial \boldsymbol{x}}{\partial \zeta_j}\bigg|_{t=\tau_{k-1}^+}^{t=\tau_k^-}$$
$$+\sum_{k=1}^{M+1}\int_{\tau_{k-1}}^{\tau_k}\left[\left(\frac{\partial H_m}{\partial \boldsymbol{x}}+\left(\frac{d\boldsymbol{\lambda}^m}{dt}\right)^\top\right)\frac{\partial \boldsymbol{x}}{\partial \zeta_j}+\frac{\partial H_m}{\partial \zeta_j}\right]dt. \tag{7.6.14}$$

Since the state and its gradient with respect to ζ_j are continuous in t on $[0,T]$, $\boldsymbol{x}(\tau_k^-)=\boldsymbol{x}\left(\tau_k^+\right)$ and $\dfrac{\partial \boldsymbol{x}(\tau_k^-)}{\partial \zeta_j}=\dfrac{\partial \boldsymbol{x}\left(\tau_k^+\right)}{\partial \zeta_j}$ for $k=1,\ldots,M$. Thus, we obtain

$$\sum_{k=1}^{M+1}(\boldsymbol{\lambda}^m)^\top\frac{\partial \boldsymbol{x}}{\partial \zeta_j}\bigg|_{t=\tau_{k-1}^+}^{t=\tau_k^-}$$
$$=\sum_{k=1}^{M}\left[\left(\boldsymbol{\lambda}^m(\tau_k^-)\right)^\top-\left(\boldsymbol{\lambda}^m\left(\tau_k^+\right)\right)^\top\right]\frac{\partial \boldsymbol{x}(\tau_k)}{\partial \zeta_j}-\left(\boldsymbol{\lambda}^m\left(\tau_0^+\right)\right)^\top\frac{\partial \boldsymbol{x}(\tau_0)}{\partial \zeta_j}$$
$$+\left(\boldsymbol{\lambda}^m(\tau_{M+1}^-)\right)^\top\frac{\partial \boldsymbol{x}(\tau_{M+1})}{\partial \zeta_j}$$
$$=\sum_{k=1}^{M}\left[\left(\boldsymbol{\lambda}^m(\tau_k^-)\right)^\top-\left(\boldsymbol{\lambda}^m\left(\tau_k^+\right)\right)^\top\right]\frac{\partial \boldsymbol{x}(\tau_k)}{\partial \zeta_j}+\left(\boldsymbol{\lambda}^m(T)\right)^\top\frac{\partial \boldsymbol{x}(T)}{\partial \zeta_j}. \tag{7.6.15}$$

Since $\boldsymbol{x}\left(\tau_0^+\right)=\boldsymbol{x}(0)=\boldsymbol{x}^0$, which is a fixed vector in $\mathbb{R}^n$, (7.6.14) becomes

$$
\begin{aligned}
&\frac{\partial g_m(\boldsymbol{\zeta})}{\partial \zeta_j} \\
&= \sum_{k=1}^{M} \left(\frac{\partial \Phi_m(\boldsymbol{x}(\tau_1), \ldots, \boldsymbol{x}(\tau_M))}{\partial \boldsymbol{x}(\tau_k)} - \left(\boldsymbol{\lambda}^m(\tau_k^-)\right)^\top + \left(\boldsymbol{\lambda}^m\left(\tau_k^+\right)\right)^\top \right) \frac{\partial \boldsymbol{x}(\tau_k)}{\partial \zeta_j} \\
&\quad - \left(\boldsymbol{\lambda}^m(T)\right)^\top \frac{\partial \boldsymbol{x}(T)}{\partial \zeta_j} + \sum_{k=1}^{M+1} \int_{\tau_{k-1}}^{\tau_k} \left[\left(\frac{\partial H_m}{\partial \boldsymbol{x}} + \left(\frac{d\boldsymbol{\lambda}^m}{dt} \right)^\top \right) \frac{\partial \boldsymbol{x}}{\partial \zeta_j} + \frac{\partial H_m}{\partial \zeta_j} \right] dt.
\end{aligned}
\tag{7.6.16}
$$

By virtue of the definition of the costate system corresponding to g_m given in (7.6.4a) with the jump conditions (7.6.4b) and terminal condition (7.6.4c), we obtain

$$
\frac{\partial g_m(\boldsymbol{\zeta})}{\partial \zeta_j} = \sum_{k=1}^{M+1} \int_{\tau_{k-1}}^{\tau_k} \frac{\partial H_m}{\partial \zeta_j} dt. \tag{7.6.17}
$$

This completes the proof.

Note that the costates are discontinuous at the characteristic time points. The sizes of the jumps are determined by the interior-point conditions given by (7.6.4b).

7.7 Exercises

7.7.1 *Consider Problem (P1). Write down the corresponding definition for a system parameter $\boldsymbol{\zeta}^*$ to be a regular point of the constraints (7.2.4) in the sense of Definition 3.1.3.*

7.7.2 *Consider Problem (P1). Write down the corresponding first order necessary conditions in the sense of Theorem 3.1.1.*

7.7.3 *Consider a process governed by the following scalar differential equation defined on a free terminal time interval $(0, \beta(\zeta)]$:*

$$
\frac{dx(t)}{dt} = f(t, x(t), \zeta)
$$

$$
x(0) = x^0(\zeta),
$$

where ζ is a system parameter yet to be determined, and both $\beta(\zeta)$ and $x^0(\zeta)$ are given scalar functions of the system parameter ζ. The problem is to find a system parameter $\zeta \in \mathbb{R}$ such that the following cost function:

$$
g(\zeta) = \int_0^{\beta(\zeta)} \mathcal{L}(t, x(t), \zeta) dt
$$

is minimized.

(a) *Reduce the problem to the one of fixed terminal time by rescaling the time with respect to* $\beta(\zeta)$, *i.e., by setting* $t = \beta(\zeta)\tau$. *With reference to the fixed terminal time problem, show that the gradient of the corresponding cost functional, again denoted by* $g(\zeta)$, *is given by*

$$\frac{\partial g}{\partial \zeta} = \lambda(0)\frac{\partial x^0(\zeta)}{\partial \zeta} + \int_0^1 \frac{\partial H}{\partial \zeta} d\tau,$$

where

$$H = \beta(\zeta)\left[\tilde{\mathcal{L}}(\tau, x(\tau), \zeta) + \lambda(\tau)\ \tilde{f}(\tau, x(\tau), \zeta)\right]$$

$$\frac{d\lambda(\tau)}{d\tau} = -\frac{\partial H}{\partial x}, \quad \lambda(1) = 0$$

$$\frac{dx(\tau)}{d\tau} = \beta(\zeta)\ \tilde{f}(\tau, x(\tau), \zeta), \quad x(0) = x^0(\zeta),$$

while

$$\tilde{\mathcal{L}}(\tau, x(\tau), \zeta) = \mathcal{L}(\beta(\zeta)\tau, x(\tau), \zeta)$$

and

$$\tilde{f}(\tau, x(\tau), \zeta) = f(\beta(\zeta)\tau, x(\tau), \zeta).$$

(b) *It is also possible to derive the gradient formula directly. By going through the steps given in the proof of Theorem 7.2.2, show that the gradient of the cost functional* $g(\zeta)$ *is given by*

$$\begin{aligned}
\frac{\partial g}{\partial \zeta} =& \mathcal{L}(\beta(\zeta), x(\beta(\zeta)), \zeta)\frac{\partial \beta(\zeta)}{\partial \zeta} + \lambda(0)\frac{\partial x^0(\zeta)}{\partial \zeta} + \int_0^{\beta(\zeta)} \frac{\partial H}{\partial \zeta} dt \\
\frac{d\lambda(t)}{dt} =& -\frac{\partial H}{\partial x}, \quad \lambda(\beta(\zeta)) = 0 \\
\frac{dx(t)}{dt} =& f(t, x(t), \zeta), \quad x(0) = x^0(\varsigma),
\end{aligned}$$

where

$$H = \mathcal{L}(t, x, \zeta) + \lambda f(t, x, \zeta).$$

(c) *Are the results given in part (a) and part (b) equivalent?*

7.7.4 *(Optimal Design of Suspended Cable [251].) Consider a cable with its own weight and a distributed load along its span. After appropriate statical analysis and normalization, the total (non-dimensional) weight of the cable is given by*

$$\Phi = \int_0^1 \frac{1}{\beta}\sqrt{(1+S^2)\left(1+\left(\frac{dy(x)}{dx}\right)^2\right)}dx$$

where

$$\frac{d^2y(x)}{dx^2} = \alpha\sqrt{1+\left(\frac{dy(x)}{dx}\right)^2}\sqrt{(1+S^2)}+\beta,$$

$y(0) = 0$, $\frac{dy(0)}{dx} = 0$, $\frac{dy(1)}{dx} = S$, $\alpha =$ *a given constant which relates the specific weight and the maximum permissible stress,* $\beta =$ *an adjustable parameter representing the ratio of total loading to horizontal tension in the cable, and* $S =$ *the maximum slope of the cable which is also adjustable. The optimal design problem is to determine* β *and* S *such that* Φ *is minimized.*

(a) Formulate the problem as an optimal parameter selection problem by setting $y(x) = y_1(x)$ *and* $\frac{dy_1(x)}{dx} = y_2(x)$.

(b) Show that the problem is equivalent to

$$\min\ (S-\beta)/(\alpha\beta)$$

subject to

$$\frac{dy_2(x)}{dx} = \alpha\sqrt{1+(y_2(x))^2}\sqrt{(1+S^2)}+\beta,$$

$y_2(0) = 0$, *and* $y_2(1) = S \Rightarrow g_1(\beta, S) = y_2(1) - S = 0$.

(c) Write down the necessary conditions for optimality for the reduced problem in (b).

(d) Determine the gradient formulae of the cost functional and the equality constraint function $g_1(\beta, S)$ *of the problem in (b) with respect to the decision variables* β *and* S.

7.7.5 *(Computation of Eigenvalues for Sturm-Liouville Boundary-Value Problems.) Consider the well-known Sturm-Liouville problem*

$$\frac{d}{dx}\left[p(x)\frac{dy(x)}{dx}\right] + q(x)y(x) + \lambda\omega(x)y(x) = 0 \tag{7.7.1a}$$

subject to the boundary conditions

$$\alpha_1 y(0) + \alpha_2\frac{dy(0)}{dx} = 0 \tag{7.7.1b}$$

$$\beta_1 y(1) + \beta_2\frac{dy(1)}{dx} = 0. \tag{7.7.1c}$$

The functions $p(x)$, $q(x)$ *and* $\omega(x)$ *are assumed to be continuous. In addition,* $p(x)$ *does not vanish in* $(0,1)$. *The problem is solvable only for a countable number of distinct values of* λ *known as the eigenvalues. For each* λ, *the corresponding solution* $y(x)$ *is known as an eigenfunction of the problem. Traditionally, the eigenvalues and eigenfunctions are obtained by using the Rayleigh-Ritz method. Later approaches use finite difference and finite element methods. This exercise illustrates how the problem can be solved, rather easily, by posing it as an optimal parameter selection problem.*

(a) *Express* (7.7.1a) *as a system of first order differential equations in* y *and* $\frac{dy}{dt}$.

(b) *Since an eigenfunction for a given eigenvalue is non-unique (it is only unique up to a multiplicative constant), one can normalize it by fixing either* $y(0)$ *or* $\frac{dy(0)}{dt}$. *With this in mind, write down the initial conditions for the differential equations obtained in (a).*

(c) *The end condition* (7.7.1c) *can be satisfied if and only if* λ *is an eigenvalue. Thus,* (7.7.1c) *can be viewed as a function of* λ, *i.e.,*

$$g_1(\lambda) = \beta_1 y(1) + \beta_2 \frac{dy(1)}{dt} = 0. \tag{7.7.2}$$

In principle, one can find the eigenvalues of the problem by solving for the zeros of (7.7.2). *Alternatively, one can formulate the solution of* (7.7.2) *as an optimal parameter selection problem. The following functions may be useful.*

$$\Phi(g) = g^2$$

or

$$\Phi_\varepsilon(g) = \begin{cases} 0 & \text{if } g < -\varepsilon \\ \frac{1}{2\varepsilon}(g+\varepsilon)^2 & \text{if } -\varepsilon \le g \le \varepsilon \\ g & \text{if } g > \varepsilon, \end{cases}$$

where ε *is a small positive constant.*

7.7.6 *Consider the optimal control problem involving the dynamical system* (7.2.1) *and the cost functional* (7.2.3). *Assume that the control* $\boldsymbol{u}$ *has the following special structure:*

$$\boldsymbol{u} = h(\boldsymbol{z}, \boldsymbol{x}),$$

where $\boldsymbol{z}$ *is some control law parameter to be determined. Formulate the corresponding optimal feedback control problem as an optimal parameter selection problem.*

7.7.7 *Consider the linear quadratic optimal regulator problem as described in Section 6.3 with* $S_f = 0$ *and with* A, B, Q *and* R *not depending on time* t. *Assume that the feedback control law takes the form:*

$$\boldsymbol{u} = -K\boldsymbol{x},$$

where K *is a* $r \times n$ *dimensional constant gain matrix. The problem is to find a constant gain matrix* K *such that the cost functional:*

$$g(K) = \int_0^T \frac{1}{2}(\boldsymbol{x}(t))^\top [Q + K^\top R K] \boldsymbol{x}(t)\, dt$$

is minimized subject to the dynamical system:

$$\frac{d\boldsymbol{x}(t)}{dt} = (A - BK)\boldsymbol{x}(t)$$

with the initial condition

$$\boldsymbol{x}(0) = \boldsymbol{x}^0.$$

(a) Use Theorem 7.2.2 to show that the necessary condition for optimality is

$$\int_0^T \frac{\partial H}{\partial K}\, dt = 0, \tag{7.7.3}$$

where

$$H = \frac{1}{2}\boldsymbol{x}^\top [Q + K^\top RK]\boldsymbol{x} + \boldsymbol{\lambda}^\top (A - BK)\boldsymbol{x}$$

and

$$\frac{d\boldsymbol{\lambda}(t)}{dt} = -\left(Q + K^\top RK\right)\boldsymbol{x}(t) - \left(A^\top - K^\top B^\top\right)\boldsymbol{\lambda}(t)$$

with

$$\boldsymbol{\lambda}(T) = \mathbf{0}.$$

(b) Prove that

$$\frac{\partial}{\partial X}[Tr(CX^\top DX)] = DXC + D^\top XC^\top$$

and

$$\frac{\partial}{\partial X}[Tr(EX)] = E^\top,$$

where C, D, E and X are matrices of appropriate dimensions.

(c) Making use of the result given in (b), show that (7.7.3) is equivalent to:

$$\int_0^T R\left[K\boldsymbol{x} - R^{-1}B^\top \boldsymbol{\lambda}(t)\right](\boldsymbol{x}(t))^\top\, dt = 0.$$

(d) Assuming that $\boldsymbol{\lambda}(t) = S(t)\boldsymbol{x}(t)$, show that

$$\frac{dS(t)}{dt} + S(t)(A - BK) + \left(A^\top - K^\top B^\top\right)S(t) + \left(Q + K^\top RK\right) = 0$$

and

$$\int_0^T R\left[K - R^{-1}B^\top S(t)\right]\boldsymbol{x}(t)(\boldsymbol{x}(t))^\top\, dt = 0.$$

7.7.8 *Consider Problem (P1) without the equality constraints (7.2.4a) and without the inequality constraints (7.2.4b). Let this be referred to as Problem (P). If $\boldsymbol{\zeta}^* \in \mathcal{Z}$ is an optimal parameter vector for Problem (P), use Theorem 7.2.2 to show that*

$$\left(\frac{\partial g_0(\boldsymbol{\zeta}^*)}{\partial \boldsymbol{\zeta}}\right)(\boldsymbol{\zeta} - \boldsymbol{\zeta}^*) \geq 0$$

for all $\boldsymbol{\zeta} \in \mathcal{Z}$.

Develop a two-phase method to solve the discrete valued optimal control problem formulated in Section 7.4.4, where the upper level is the simulated annealing algorithm (to determine the control sequence) and the lower level is an optimal parameter selection problem.

7.7.9 *Consider the time-lag optimal control problem described in Section 7.5, but without the canonical equality and inequality constraints (i.e., without* (7.5.3a) *and* (7.5.3b)*). The control* $\mathbf{u}$ *is assumed to take the structure given by* (7.3.2)*. Derive the gradient formulae for the cost functional with respect to the variable switching times.*

7.7.10 *Consider the optimal control problem involving multiple characteristic time points, where the system dynamic is given by* (7.3.1)*, where the control is assumed to take the form of* (7.3.2) *and the cost functional is of the form* (7.6.2)*. No canonical equality and inequality constraints are involved. Derive the gradient formula for the cost functional with respect to the variable switching times.*

7.7.11 *Give detailed proof of Theorem 7.3.1.*

7.7.12 *Give detailed proof of Theorem 7.3.2.*

7.7.13 *Give detailed proof of Theorem 7.4.2.*

7.7.14 *Give detailed proof of Theorem 7.4.5.*

7.7.15 *Show the validity of Equations (7.4.57) and (7.4.58).*

7.7.16 *Explain in detail the equivalence of Problem (P5) and Problem (P6).*

7.7.17 *Consider Problem (P7) with two time-delays. State and show the validity of the corresponding version of Theorem 7.5.1.*

7.7.18 *Can the time scaling transform be applied to Problem (P8)? Why?*

Chapter 8
Control Parametrization for Canonical Optimal Control Problems

8.1 Introduction

The methods reported in [75, 244, 245, 248, 249] and [253], as well as many papers cited in the references list, are developed based on the control parameterization technique for solving various classes of optimal control problems. Basically, the method partitions the time interval $[0, T]$ into several subintervals and the control variables are approximated by piecewise constant or piecewise linear functions with pre-fixed switching times. Through this process, the optimal control problem is approximated by a sequence of optimal parameter selection problems. Each of these optimal parameter selection problems can be viewed as a mathematical programming problem and is hence solvable by existing optimization techniques. The software package MISER3.3 (both FORTRAN and 'Visual Fortran' or 'Matlab version of MISER') [104] was developed by implementing the control parametrization method. The Visual MISER [295] is now available. Many practical problems have been solved using this approach. See relevant references in the reference list. Intuitively, the optimal parameter selection problem with a finer partition will yield a more accurate solution to the original optimal control problem. Convergence results are first obtained in [260] for a class of optimal control problems involving linear time-lag systems subject to linear control constraints. Subsequently, a number of control parametrization type algorithms with associated proof of convergence have been developed in [240, 244–246, 248, 253, 279, 280], and the relevant references cited therein. Many of these results are included in [253]. Although convergence analysis may or may not be of serious consequence for implementation purposes, it nevertheless provides important insight concerning the performance of an algorithm. Thus, it has become a widely accepted requirement for any new algorithmic development.

K. L. Teo et al., *Applied and Computational Optimal Control*, Springer Optimization and Its Applications 171,
https://doi.org/10.1007/978-3-030-69913-0_8

In practice, the accuracy of the optimal control obtained by the control parametrization method is not high, as it is impossible to know the precise switching times a priori. To obtain higher accuracy the switching times should also be regarded as decision variables. This can be accomplished by a time scaling transform, which was called the control parametrization enhancing transform (CPET) in [125, 126] and [215] but has been renamed as the time scaling transformation. Under this time scaling transform, the variable switching times of the control are mapped onto a set of fixed knots in a new time scale. The transformed problems then have the same structure as those obtained by the classical control parametrization technique. Thus, they can be solved by using software packages such as MISER3.3 (both FORTRAN and 'Visual Fortran' or 'Matlab version of MISER') [104] or Visual MISER [294]. The details will be covered in Section 8.9.

The main references for this chapter are [75, 89, 125, 148, 168, 215, 244, 245, 254] and Chapter 6 of [253].

8.2 Problem Statement

Consider the process described by the following system of nonlinear differential equations on the fixed time interval $(0, T]$:

$$\frac{d\boldsymbol{x}(t)}{dt} = \boldsymbol{f}(t, \boldsymbol{x}(t), \boldsymbol{u}(t)), \tag{8.2.1a}$$

where $\boldsymbol{x} = [x_1, \ldots, x_n]^\top \in \mathbb{R}^n$, $\boldsymbol{u} = [u_1, \ldots, u_r]^\top \in \mathbb{R}^r$, are, respectively, the state and control vectors, $\boldsymbol{f} = [f_1, \ldots, f_n]^\top \in \mathbb{R}^n$ and $\boldsymbol{f} : [0, T] \times \mathbb{R}^n \times \mathbb{R}^r \to \mathbb{R}^n$. The initial condition for the differential equation (8.2.1a) is

$$\boldsymbol{x}(0) = \boldsymbol{x}^0, \tag{8.2.1b}$$

where $\boldsymbol{x}^0$ is a given vector in $\mathbb{R}^n$.

Define

$$U_1 = \{\boldsymbol{v} = [v_1, \ldots, v_r]^\top \in \mathbb{R}^r : \left(E^i\right)^\top \boldsymbol{v} \leq b_i,\ i = 1, \ldots, q\}, \tag{8.2.2a}$$

where E^i, $i = 1, \ldots, q$, are r-vectors, and b_i, $i = 1, \ldots, q$, are real numbers; and

$$U_2 = \{\boldsymbol{v} = [v_1, \ldots, v_r]^\top \in \mathbb{R}^r : \alpha_i \leq v_i \leq \beta_i,\ i = 1, \ldots, r\}, \tag{8.2.2b}$$

where α_i, $i = 1, \ldots, r$, and β_i, $i = 1, \ldots, r$, are real numbers. Let

$$U = U_1 \cap U_2. \tag{8.2.3}$$

Clearly, U is a compact and convex subset of $\mathbb{R}^r$. A bounded measurable function $\boldsymbol{u} = [u_1, \ldots, u_r]^\top$ from $[0,T]$ into $\mathbb{R}^r$ is *said to be an admissible control* if $\boldsymbol{u}(t) \in U$ for almost all $t \in [0,T]$. Let $\mathcal{U}$ be the class of all such admissible controls.

For each $\boldsymbol{u} \in \mathcal{U}$, let $\boldsymbol{x}(\cdot|\boldsymbol{u})$ be the corresponding vector-valued function that is absolutely continuous on $(0,T]$ and satisfies the differential equation (8.2.1a) almost everywhere on $(0,T]$ and the initial condition (8.2.1b). This function is called the solution of the system (8.2.1) corresponding to $\boldsymbol{u} \in \mathcal{U}$. We may now state the canonical optimal control problem as follows.

Problem (P_1)**:** Given system (8.2.1), find a control $\boldsymbol{u} \in \mathcal{U}$ such that the cost functional

$$g_0(\boldsymbol{u}) = \Phi_0(\boldsymbol{x}(T|\boldsymbol{u})) + \int_0^T \mathcal{L}_0(t, \boldsymbol{x}(t|\boldsymbol{u}), \boldsymbol{u}(t))\, dt \tag{8.2.4}$$

is minimized over $\mathcal{U}$, subject to the equality constraints

$$g_i(\boldsymbol{u}) = \Phi_i(\boldsymbol{x}(\tau_i|\boldsymbol{u})) + \int_0^{\tau_i} \mathcal{L}_i(t, \boldsymbol{x}(t|\boldsymbol{u}), \boldsymbol{u}(t))\, dt = 0, \ i = 1, \ldots, N_e, \tag{8.2.5a}$$

and subject to the inequality constraints:

$$g_i(\boldsymbol{u}) = \Phi_i(\boldsymbol{x}(\tau_i|\boldsymbol{u})) + \int_0^{\tau_i} \mathcal{L}_i(t, \boldsymbol{x}(t|\boldsymbol{u}), \boldsymbol{u}(t))\, dt \le 0, \tag{8.2.5b}$$

$$i = N_e + 1, \ldots, N, \tag{8.2.5c}$$

where $\Phi_i : \mathbb{R}^n \mapsto \mathbb{R}$, $i = 0, 1, \ldots, N$, and $\mathcal{L}_i : [0,T] \times \mathbb{R}^n \times \mathbb{R}^r \mapsto \mathbb{R}$, $i = 0, 1, \ldots, N$, are given real-valued functions; and $0 < \tau_i \le T$ is referred to as the *characteristic time* for the i–th constraint, $i = 1, \ldots, N$, with $\tau_0 = T$ by convention.

Remark 8.2.1

(i) *Both the equality constraints* (8.2.5a) *and the inequality constraints* (8.2.5c) *are said to be in their canonical form.*

(ii) *Equations* (8.2.5a) *and* (8.2.5c) *reduce to terminal equality constraints and inequality constraints, respectively, if* $\tau_i = T$ *and* $\mathcal{L}_i = 0$.

(iii) *Similarly, the corresponding versions of* (8.2.5a) *and* (8.2.5c) *with* $0 < \tau_i < T$ *and* $\mathcal{L}_i = 0$ *are, respectively, equality interior point constraints and inequality interior point constraints.*

(iv) *The continuous inequality constraint:* $h(t, \boldsymbol{x}(t), \boldsymbol{u}(t)) \le 0$, $t \in [0,T]$, *is equivalent to* (8.2.5a) *with* $\tau_i = T$, $\Phi_i(\boldsymbol{x}(T)) = 0$, *and*

$$\mathcal{L}_i(t, \boldsymbol{x}(t), \boldsymbol{u}(t)) = [\max\{h(t, \boldsymbol{x}(t), \boldsymbol{u}(t)), 0\}]^2.$$

This constraint transcription was first introduced in [241]. For more details on this constraint transcription, see Remark 8.6.5.

We assume throughout that the following conditions are satisfied.

Assumption 8.2.1 *For any compact subset $V \subset \mathbb{R}^r$, there exists a positive constant K such that*

$$|\boldsymbol{f}(t,\boldsymbol{x},\boldsymbol{u})| \leq K(1+|\boldsymbol{x}|)$$

for all $(t,\boldsymbol{x},\boldsymbol{u}) \in [0,T] \times \mathbb{R}^n \times V$.

Assumption 8.2.2 *$\boldsymbol{f}$ and $\mathcal{L}_i$, $i = 0,1,\ldots,N$, together with their partial derivatives with respect to each of the components of $\boldsymbol{x}$ and $\boldsymbol{u}$ are piecewise continuous on $[0,T]$ for each $(\boldsymbol{x},\boldsymbol{u}) \in \mathbb{R}^n \times \mathbb{R}^r$ and continuous on $\mathbb{R}^n \times \mathbb{R}^r$ for each $t \in [0,T]$.*

Assumption 8.2.3 *Φ_i, $i = 0,1,\ldots,N$, are continuously differentiable with respect to $\boldsymbol{x}$.*

Remark 8.2.2 *From the theory of differential equations, we recall that the system* (8.2.1) *admits a unique solution, $\boldsymbol{x}(\cdot|\boldsymbol{u})$, corresponding to each $\boldsymbol{u} \in L_\infty([0,T],\mathbb{R}^r)$, and hence for each $\boldsymbol{u} \in \mathcal{U}$.*

8.3 Control Parametrization

In this section, we construct a sequence of approximate problems such that their solutions are progressively better approximations of an optimal solution to Problem (P). This is done through the discretization of the control space. The classical control parametrization technique approximates each control by a piecewise constant control as follows.

Consider a monotonically non-decreasing sequence $\{S^p\}_{p=1}^{\infty}$ of finite subsets of $[0,T]$. For each p, let n_p+1 points of S^p be denoted by $t_0^p, t_1^p, \ldots, t_{n_p}^p$. These points are chosen such that $t_0^p = 0$, $t_{n_p}^p = T$, and $t_{k-1}^p < t_k^p$, $k = 1,2,\ldots,n_p$. Associated with each S^p there is a partition $\mathcal{I}^p$ of $[0,T)$, defined by

$$\mathcal{I}^p = \{I_k^p : k = 1,\ldots,n_p\},$$

where $I_k^p = [t_{k-1}^p, t_k^p)$. We choose S^p such that the following two properties are satisfied.

Assumption 8.3.1 *S^{p+1} is a refinement of S^p.*

Assumption 8.3.2 $\lim_{p\to\infty} S^p$ *is dense in $[0,T]$, i.e.,*

$$\lim_{p\to\infty} \max_{k=1,\ldots,n_p} |I_k^p| = 0,$$

where $|I_k^p| = t_k^p - t_{k-1}^p$ is the length of the k-th interval.

Example 8.3.1 For each positive integer p, let the interval $[0,T]$ be partitioned into 2^p equal subintervals. Let Δ denote the length of each of these.

Then, the partition $\mathcal{I}^p$ of $[0,T]$ associated with the corresponding S^p is defined by

$$\mathcal{I}^p = \{[(k-1)\Delta, k\Delta] : k = 1, \ldots, 2^p\}.$$

Note that this form of $\mathcal{I}^p$ is most commonly used in practice.

Let $\mathcal{U}^p$ consist of all those elements from $\mathcal{U}$ that are piecewise constant and consistent with the partition $\mathcal{I}^p$. It is clear that each $\boldsymbol{u} \in \mathcal{U}^p$ can be written as

$$\boldsymbol{u}^p(t) = \sum_{k=1}^{n_p} \boldsymbol{\sigma}^{p,k} \chi_{I_k^p}(t), \tag{8.3.1a}$$

where $\boldsymbol{\sigma}^{p,k} \in U$ and χ_I denotes the indicator function of I, defined by

$$\chi_I(t) = \begin{cases} 1, & t \in I \\ 0, & \text{elsewhere.} \end{cases} \tag{8.3.1b}$$

Let

$$\boldsymbol{\sigma}^p = \left[\left(\boldsymbol{\sigma}^{p,1}\right)^\top, \ldots, \left(\boldsymbol{\sigma}^{p,n_p}\right)^\top \right]^\top, \tag{8.3.2a}$$

where

$$\boldsymbol{\sigma}^{p,k} = \left[\sigma_1^{p,k}, \ldots, \sigma_r^{p,k} \right]^\top. \tag{8.3.2b}$$

When restricting to $\mathcal{U}^p$, the control constraints defined in (8.2.2a) and (8.2.2b) become

$$\left(E^i\right)^\top \boldsymbol{\sigma}^{p,k} \le b_i,\ i = 1, \ldots, q,\ k = 1, \ldots, n_p \tag{8.3.3a}$$

and

$$\alpha_i \le \sigma_i^{p,k} \le \beta_i,\ i = 1, \ldots, r,\ \ k = 1, \ldots, n_p, \tag{8.3.3b}$$

respectively. Let Ξ^p be the set of all those $\boldsymbol{\sigma}^p$ vectors that satisfy the constraints (8.3.3). Clearly, for each control $\boldsymbol{u}^p \in \mathcal{U}^p$, there exists a unique control parameter vector $\boldsymbol{\sigma}^p \in \Xi^p$ such that the relation (8.3.1) is satisfied and vice versa.

With $\boldsymbol{u}^p \in \mathcal{U}^p$, system (8.2.1) takes the form:

$$\frac{d\boldsymbol{x}(t)}{dt} = \widetilde{\boldsymbol{f}}(t, \boldsymbol{x}(t), \boldsymbol{\sigma}^p), \qquad t \in [0,T] \tag{8.3.4a}$$

$$\boldsymbol{x}(0) = \boldsymbol{x}^0, \tag{8.3.4b}$$

where

$$\widetilde{\boldsymbol{f}}(t, \boldsymbol{x}(t), \boldsymbol{\sigma}^p) = \boldsymbol{f}\left(t, \boldsymbol{x}(t), \sum_{k=1}^{n_p} \boldsymbol{\sigma}^{p,k} \chi_{I_k^p}(t) \right). \tag{8.3.4c}$$

Let $\boldsymbol{x}(\cdot|\boldsymbol{\sigma}^p)$ be the solution of the system (8.3.4) corresponding to the control parameter vector $\boldsymbol{\sigma}^p \in \Xi^p$ in the sense that it satisfies the differential equation (8.3.4a) *a.e.* on $(0,T]$ and the initial condition (8.3.4b).

By restricting $\boldsymbol{u}$ in $\mathcal{U}^p$, the constraints (8.2.5a) and (8.2.5c) are reduced to

$$G_i(\boldsymbol{\sigma}^p) = \Phi_i(\boldsymbol{x}(\tau_i|\boldsymbol{\sigma}^p)) + \int_0^{\tau_i} \widetilde{\mathcal{L}}_i(t, \boldsymbol{x}(t|\boldsymbol{\sigma}^p), \boldsymbol{\sigma}^p)\, dt = 0,$$
$$i = 1, \ldots, N_e, \tag{8.3.5a}$$

and

$$G_i(\boldsymbol{\sigma}^p) = \Phi_i(\boldsymbol{x}(\tau_i|\boldsymbol{\sigma}^p)) + \int_0^{\tau_i} \widetilde{\mathcal{L}}_i(t, \boldsymbol{x}(t|\boldsymbol{\sigma}^p), \boldsymbol{\sigma}^p)\, dt \le 0,$$
$$i = N_e + 1, \ldots, N, \tag{8.3.5b}$$

respectively. Here, $\widetilde{\mathcal{L}}_i$, $i = 1, \ldots, N$, are obtained from $\mathcal{L}_i$, $i = 1, \ldots, N$, in the same manner as $\widetilde{\boldsymbol{f}}$ is obtained from $\boldsymbol{f}$ according to (8.3.4c).

Let Ω^p be the subset of Ξ^p such that the constraints (8.3.5) are satisfied. Furthermore, let $\mathcal{F}^p$ be the subset of $\mathcal{U}^p$, which consists of all those corresponding piecewise constant controls of the form (8.3.1a). We may now specify the approximate problem $(P_1(p))$ as follows.

Problem $(P_1(p))$**:** Find a control parameter vector $\boldsymbol{\sigma}^p \in \Xi^p$ such that the cost function

$$G_0(\boldsymbol{\sigma}^p) = \Phi_0(\boldsymbol{x}(T|\boldsymbol{\sigma}^p)) + \int_0^T \widetilde{\mathcal{L}}_0(t, \boldsymbol{x}(t|\boldsymbol{\sigma}^p), \boldsymbol{\sigma}^p)\, dt \tag{8.3.6}$$

is minimized over Ω^p, where $\widetilde{\mathcal{L}}_0$ is obtained from $\mathcal{L}_0$ in the same way as $\widetilde{\boldsymbol{f}}$ is obtained from $\boldsymbol{f}$ according to (8.3.4c).

Note that for each p, the approximate Problem $(P_1(p))$ is an optimal parameter selection problem. It is effectively a finite dimensional optimization problem, i.e., a mathematical programming problem. This is the main theme behind the control parametrization technique: *to approximate an optimal control problem by a sequence of appropriate mathematical programming problems.* The discussion of the computational aspects of this approach will be given in Section 8.5.

Remark 8.3.1 *In this section, we have assumed that the partition points t_j^p, $j = 0, \ldots, n_p$, are the same for each component of the control. This assumption is merely for the sake of brevity and not a rigid requirement.*

8.4 Four Preliminary Lemmas

In this section, we present four lemmas that will be used to support the convergence results in the next section.

Lemma 8.4.1 *For each $\boldsymbol{u} \in \mathcal{U}$ and each p, let*

$$\boldsymbol{u}^p(t) = \sum_{k=1}^{n_p} \boldsymbol{\sigma}^{p,k} \chi_{I_k^p}(t), \tag{8.4.1}$$

where $\chi_{I_k^p}$ is the indicator function defined by (8.3.1b),

$$\boldsymbol{\sigma}^{p,k} = \frac{1}{|I_k^p|} \int_{I_k^p} \boldsymbol{u}(s)\, ds$$

and $|I_k^p| = \left| t_k^p - t_{k-1}^p \right|$. Then as $p \to \infty$,

$$\boldsymbol{u}^p \to \boldsymbol{u} \tag{8.4.2a}$$

almost everywhere in $[0, T]$ and

$$\lim_{p \to \infty} \int_0^T |\boldsymbol{u}^p(t) - \boldsymbol{u}(t)|\, dt = 0. \tag{8.4.2b}$$

Proof. Let t_1 be a regular point of $\boldsymbol{u}$. Then clearly, there exists a sequence of intervals $\{I_{k(p)}^p\}_{p=1}^{\infty}$ such that

$$t_1 \in I_{k(p+1)}^{p+1} \subset I_{k(p)}^p, \quad \forall\, p,$$

and

$$\left| I_{k(p)}^p \right| \to 0 \text{ as } p \to \infty.$$

Then,

$$\{t_1\} = \bigcap_{p=1}^{\infty} \bar{I}_{k(p)}^p,$$

where '$-$' denotes closure, and so, by Theorem A.1.14,

$$\boldsymbol{u}(t_1) = \lim_{p \to \infty} \frac{1}{\left| \bar{I}_{k(p)}^p \right|} \int_{\bar{I}_{k(p)}^p} \boldsymbol{u}(s) ds = \lim_{p \to \infty} \frac{1}{\left| I_{k(p)}^p \right|} \int_{I_{k(p)}^p} \boldsymbol{u}(s) ds = \lim_{p \to \infty} \boldsymbol{u}^p(t_1).$$

Since almost all points of $\boldsymbol{u}(t)$ are regular points, we conclude that $\boldsymbol{u}^p \to \boldsymbol{u}$ *a.e.* in $[0, T]$. To prove (8.4.2b), we note that $\boldsymbol{u}$ is a bounded measurable function in $[0, T]$. Thus, it is clear from the construction of $\boldsymbol{u}^p$ that $\{\boldsymbol{u}^p\}_{p=1}^{\infty}$ are uniformly bounded. Thus, the result follows from Theorem A.1.10.

Remark 8.4.1 *Note that the second part of Lemma 8.4.1 remains valid if we take $\{\boldsymbol{u}^p\}_{p=1}^{\infty}$ to be any bounded sequence of functions in $L_\infty([0, T], \mathbb{R}^r)$ that converges to $\boldsymbol{u}$ a.e. in $[0, T]$ as $p \to \infty$.*

In the next three lemmas, $\{\boldsymbol{u}^p\}_{p=1}^{\infty}$ is assumed to be a bounded sequence of functions in $L_\infty([0, T], \mathbb{R}^r)$. In particular, these results are valid if $\{\boldsymbol{u}^p\}_{p=1}^{\infty}$

and $\boldsymbol{u}$ are as defined in Lemma 8.4.1. The details are left to the reader as exercises.

Lemma 8.4.2 *Let $\{\boldsymbol{u}^p\}_{p=1}^{\infty}$ be a bounded sequence of functions in $L_\infty([0,T], \mathbb{R}^r)$. Then, the sequence $\{\boldsymbol{x}(\cdot|\boldsymbol{u}^p)\}_{p=1}^{\infty}$ of the corresponding solutions of the system* (8.2.1) *is also bounded in $L_\infty([0,T],\mathbb{R}^n)$.*

Proof. From (8.2.1a), we have

$$\boldsymbol{x}(t|\boldsymbol{u}^p) = \boldsymbol{x}^0 + \int_0^t \boldsymbol{f}(s, \boldsymbol{x}(s|\boldsymbol{u}^p), \boldsymbol{u}^p(s))\, ds, \tag{8.4.3}$$

for all $t \in [0,T]$. Then, applying Assumption 8.2.1 to (8.4.3), we get

$$|\boldsymbol{x}(t|\boldsymbol{u})| \le N_0 + K \int_0^t |\boldsymbol{x}(s|\boldsymbol{u}^p)|\, ds, \tag{8.4.4}$$

where $N_0 = \left|\boldsymbol{x}^0\right| + KT$. Applying the Gronwall-Bellman lemma (see Theorem A.1.19), we obtain

$$|\boldsymbol{x}(t|\boldsymbol{u}^p)| \le N_0 \, \exp(KT),$$

for all $t \in [0,T]$. Thus, the proof is complete.

Lemma 8.4.3 *Let $\{\boldsymbol{u}^p\}_{p=1}^{\infty}$ be a bounded sequence of functions in $L_\infty([0,T], \mathbb{R}^r)$ that converges to a function $\boldsymbol{u}$ a.e. in $[0,T]$. Then,*

$$\lim_{p\to\infty} \|\boldsymbol{x}(\cdot|\boldsymbol{u}^p) - \boldsymbol{x}(\cdot|\boldsymbol{u})\|_\infty = 0,$$

and, for each $t \in [0,T]$,

$$\lim_{p\to\infty} |\boldsymbol{x}(t|\boldsymbol{u}^p) - \boldsymbol{x}(t|\boldsymbol{u})| = 0.$$

Proof. Let the bound of the sequence $\{\|\boldsymbol{u}^p\|_\infty\}_{p=1}^{\infty}$ be denoted by N_0. It follows from Lemma 8.4.2 that there exists a constant $N_1 > 0$ such that

$$\|\boldsymbol{x}(\cdot|\boldsymbol{u}^p)\|_\infty \le N_1,$$

for all integers $p \ge 1$. From (8.2.1a), we get

$$|\boldsymbol{x}(t|\boldsymbol{u}^p) - \boldsymbol{x}(t|\boldsymbol{u})| \le \int_0^t |\boldsymbol{f}(s, \boldsymbol{x}(s|\boldsymbol{u}^p), \boldsymbol{u}^p(s)) - \boldsymbol{f}(s, \boldsymbol{x}(s|\boldsymbol{u}), \boldsymbol{u}(s))|\, ds.$$

From (8.2.2), the partial derivatives of $\boldsymbol{f}(t, \boldsymbol{x}, \boldsymbol{u})$ with respect to each component of $\boldsymbol{x}$ and $\boldsymbol{u}$ are piecewise continuous in $[0,T]$ for each $(\boldsymbol{x}, \boldsymbol{u}) \in B \times V$ and continuous in $B \times V$ for each $t \in [0,T]$, where $B = \{\boldsymbol{y} \in \mathbb{R}^n : |\boldsymbol{y}| \le N_1\}$ and $V = \{\boldsymbol{z} \in \mathbb{R}^r : |\boldsymbol{z}| \le N_0\}$. Thus, there exists a constant $N_2 > 0$ such that

$$|\boldsymbol{x}(t|\boldsymbol{u}^p) - \boldsymbol{x}(t|\boldsymbol{u})| \leq N_2 \int_0^t \{|\boldsymbol{x}(s|\boldsymbol{u}^p) - \boldsymbol{x}(s|\boldsymbol{u})| + |\boldsymbol{u}^p(s) - \boldsymbol{u}(s)|\}\, ds.$$

Applying Theorem A.1.19 once more, we obtain

$$|\boldsymbol{x}(t|\boldsymbol{u}^p) - \boldsymbol{x}(t|\boldsymbol{u})| \leq N_2 \left(\int_0^t |\boldsymbol{u}^p(s) - \boldsymbol{u}(s)|\, ds \right) \exp(N_2 T).$$

Thus, both the conclusions of the lemma follow easily from Remark 8.4.1.

Lemma 8.4.4 *Let* $\{\boldsymbol{u}^p\}_{p=1}^{\infty}$ *denote the bounded sequence of functions in* $L_\infty([0,T], \mathbb{R}^r)$ *that converges to a function* $\boldsymbol{u}$ *a.e. in* $[0,T]$. *Then*

$$\lim_{p \to \infty} g_0(\boldsymbol{u}^p) = g_0(\boldsymbol{u}).$$

Proof. From (8.2.4), we have

$$|g_0(\boldsymbol{u}^p) - g_0(\boldsymbol{u})| \leq |\Phi_0(\boldsymbol{x}(T|\boldsymbol{u}^p)) - \Phi_0(\boldsymbol{x}(T|\boldsymbol{u}))| + \int_0^T |\mathcal{L}_0(t, \boldsymbol{x}(t|\boldsymbol{u}^p), \boldsymbol{u}^p(t)) - \mathcal{L}_0(t, \boldsymbol{x}(t|\boldsymbol{u}), \boldsymbol{u}(t))|\, dt.$$

The conclusion of this lemma then follows from Lemma 8.4.3, Remark 8.4.1, Lemma 8.4.2, Assumptions 8.2.2 and 8.2.3 and Theorem A.1.10.

8.5 Some Convergence Results

In this section, we present some convergence properties of the sequence of approximate optimal controls. To be more precise, for each $p = 1, 2, \ldots$, let $\boldsymbol{\sigma}^{p,*}$ be an optimal control parameter vector of Problem $(P_1(p))$, which is a finite dimensional optimization problem. Furthermore, let $\{\boldsymbol{u}^{p,*}\}$ be the corresponding sequence of piecewise constant controls. Thus, for each $p = 1, 2, ..$, $\boldsymbol{u}^{p,*}$ is referred to as the optimal piecewise constant control of Problem $(P_1(p))$. In view of Assumption 8.3.1 given in Section 8.3, we see that each of these controls is suboptimal for the original Problem (P_1) and

$$g_0\left(\boldsymbol{u}^{p+1,*}\right) \leq g_0\left(\boldsymbol{u}^{p,*}\right)$$

for all $p = 1, 2, \ldots$. Now, two obvious questions to ask are as follows:

(i) Does $g_0(\boldsymbol{u}^{p,*})$ converge to the true optimal cost?
(ii) Does $\boldsymbol{u}^{p,*}$ converge to the true optimal control in some sense?

We can provide partial answers to the first question if an additional assumption is satisfied. For the second question, it can be shown that $\boldsymbol{u}^{p,*}$ converges to the true optimal control in the weak* topology of $L_\infty([0,T],\mathbb{R}^r)$ if the dynamical system is linear and the cost functional is convex. For further detail, see Chapter 8 of [253]. To state the required additional assumption, we need the following preliminary definition.

Definition 8.5.1 *A control parameter vector* $\boldsymbol{\sigma}^p \in \Xi^p$ *is said to be* ε-*tolerated feasible if it satisfies the following* ε-*tolerated constraints:*

$$-\varepsilon \le G_i(\boldsymbol{\sigma}^p) \le \varepsilon, \quad i = 1, \ldots, N_e, \tag{8.5.1a}$$

$$G_i(\boldsymbol{\sigma}^p) \le \varepsilon, \quad i = N_e + 1, \ldots, N, \tag{8.5.1b}$$

where G_i *is defined by* (8.3.5).

Let $\Omega^{p,\varepsilon}$ be the subset of Ξ^p such that the ε-tolerated constraints (8.5.1) are satisfied; and furthermore, let $\mathcal{F}^{p,\varepsilon}$ be the subset of $\mathcal{U}^{p,\varepsilon}$, which consists of all those corresponding piecewise constant controls of the form (8.3.1a). Clearly, $\Omega^p \subset \Omega^{p,\varepsilon}$ (and hence $\mathcal{F}^p \subset \mathcal{F}^{p,\varepsilon}$) for any $\varepsilon > 0$. We now consider the ε-tolerated version of the approximate Problem $(P_1(p))$ as follows.

Problem $(P_{1,\varepsilon}(p))$**:** Find a control parameter vector $\boldsymbol{\sigma}^p \in \Omega^{p,\varepsilon}$ such that the cost functional (8.3.6) is minimized over $\Omega^{p,\varepsilon}$.

Since $\Omega^p \subset \Omega^{p,\varepsilon}$ for any $\varepsilon > 0$, it follows that

$$G_0(\boldsymbol{\sigma}^{p,\varepsilon,*}) \le G_0(\boldsymbol{\sigma}^{p,*})$$

for any $\varepsilon > 0$, where $\boldsymbol{\sigma}^{p,\varepsilon,*}$ and $\boldsymbol{\sigma}^{p,*}$ are optimal control parameter vectors of Problems $(P_{1,\varepsilon}(p))$ and $(P_1(p))$, respectively. Furthermore, let $\boldsymbol{u}^{p,\varepsilon,*}$ and $\boldsymbol{u}^{p,*}$ be the corresponding piecewise constant controls in the form of (8.3.1a) with $\boldsymbol{\sigma}^p$ replaced by $\boldsymbol{\sigma}^{p,\varepsilon,*}$ and $\boldsymbol{\sigma}^{p,*}$, respectively. They are referred to as optimal piecewise constant controls of Problems $(P_{1,\varepsilon}(p))$ and $(P_1(p))$, respectively. We can now specify the additional required assumption mentioned earlier.

Assumption 8.5.1 *There exists an integer* p_0 *such that*

$$\lim_{\varepsilon \to 0} G_0(\boldsymbol{\sigma}^{p,\varepsilon,*}) = G_0(\boldsymbol{\sigma}^{p,*})$$

uniformly with respect to $p \ge p_0$.

Note that Assumption 8.5.1 is not really restrictive from the practical viewpoint. Indeed, a real practical problem is most likely solved numerically. The problem formulation would clearly be in doubt if this assumption was not satisfied. We are now in a position to present the convergence results in the next two theorems.

Theorem 8.5.1 *Let* $\boldsymbol{u}^{p,*}$ *be an optimal piecewise constant control of the approximate Problem* $(P_1(p))$. *Suppose that the original Problem* (P_1) *has an optimal control* $\boldsymbol{u}^*$. *Then*

$$\lim_{p\to\infty} g_0(\boldsymbol{u}^{p,*}) = g_0(\boldsymbol{u}^*).$$

Proof. Let $\boldsymbol{u}^{p,\varepsilon,*}$ be an optimal piecewise constant control of Problem $(P_{1,\varepsilon}(p))$. Then, it is clear from Assumption 8.5.1 that for any $\delta > 0$, there exists an $\varepsilon_0 > 0$ such that

$$g_0(\boldsymbol{u}^{p,\varepsilon,*}) > g_0(\boldsymbol{u}^{p,*}) - \delta \tag{8.5.2}$$

for any ε, $0 < \varepsilon < \varepsilon_0$, uniformly with respect to $p > p_0$. Let $\boldsymbol{u}^{*,p}$ be the control defined from $\boldsymbol{u}^*$ by (8.4.1). Then, for any ε, $0 < \varepsilon < \varepsilon_0$, it follows from Lemmas 8.4.1 and 8.4.3 and Assumptions 8.2.2 and 8.2.3 that there exists an integer $p_1 > p_0$ such that

$$\boldsymbol{u}^{*,p} \in \mathcal{F}^{p,\varepsilon} \tag{8.5.3}$$

for all $p \geq p_1$. Thus,

$$g_0(\boldsymbol{u}^{p,\varepsilon,*}) \leq g_0(\boldsymbol{u}^{*,p}) \tag{8.5.4}$$

for all $p \geq p_1$. Combining (8.5.2) and (8.5.4), we get

$$g_0(\boldsymbol{u}^{*,p}) > g_0(\boldsymbol{u}^{p,*}) - \delta \tag{8.5.5}$$

for all $p \geq p_1$. On the other hand, by virtue of Lemmas 8.4.1 and 8.4.4, we have

$$\lim_{p\to\infty} g_0(\boldsymbol{u}^{*,p}) = g_0(\boldsymbol{u}^*). \tag{8.5.6}$$

Hence, it follows from (8.5.5) and (8.5.6) that

$$\delta + g_0(\boldsymbol{u}^*) \geq \lim_{p\to\infty} g_0(\boldsymbol{u}^{p,*}). \tag{8.5.7}$$

Since $\delta > 0$ is arbitrary and $\boldsymbol{u}^*$ is an optimal control, we conclude that

$$\lim_{p\to\infty} g_0(\boldsymbol{u}^{p,*}) = g_0(\boldsymbol{u}^*).$$

This completes the proof.

Theorem 8.5.2 *Let $\boldsymbol{u}^*$ be an optimal control of Problem (P_1), and let $\boldsymbol{u}^{p,*}$ be an optimal piecewise constant control of the approximate Problem $(P_1(p))$. Suppose that*

$$\boldsymbol{u}^{p,*} \to \bar{\boldsymbol{u}},$$

a.e. on $[0,T]$. Then, $\bar{\boldsymbol{u}}$ is also an optimal control of Problem (P_1).

Proof. Since $\boldsymbol{u}^{p,*} \to \bar{\boldsymbol{u}}$ *a.e.* in $[0,T]$, it follows from Lemma 8.4.4 that

$$\lim_{p\to\infty} g_0(\boldsymbol{u}^{p,*}) = g_0(\bar{\boldsymbol{u}}). \tag{8.5.8}$$

Next, it is easy to verify from Remark 8.4.1, Lemma 8.4.3 and Assumptions 8.2.2 and 8.2.3 that $\bar{\boldsymbol{u}}$ is also a feasible control of Problem (P). On the other hand, it follows from Theorem 8.5.1 that

$$\lim_{p\to\infty} g_0(\boldsymbol{u}^{p,*}) = g_0(\boldsymbol{u}^*). \tag{8.5.9}$$

Hence, the conclusion of the theorem follows easily from (8.5.8) and (8.5.9).

8.6 A Unified Computational Approach

After the application of control parametrization technique, the constrained optimal control Problem (P) is approximated by a sequence of optimal parameter selection Problems $(P_1(p))$. For each positive integer p, the optimal parameter selection Problem $(P_1(p))$ can be viewed as the following nonlinear optimization problem.

Minimize

$$G_0(\boldsymbol{\sigma}^p) \tag{8.6.1a}$$

subject to

$$G_i(\boldsymbol{\sigma}^p) = 0, \ i = 1, \ldots, N_e, \tag{8.6.1b}$$
$$G_i(\boldsymbol{\sigma}^p) \le 0, \ i = N_e + 1, \ldots, N, \tag{8.6.1c}$$
$$F_i\left(\boldsymbol{\sigma}^{p,k}\right) = \left(\boldsymbol{E}^i\right)^\top \boldsymbol{\sigma}^{p,k} - b_i \le 0, \ i = 1, \ldots, q, \ k = 1, \ldots, n_p, \tag{8.6.1d}$$
$$\alpha_i \le \sigma_i^{p,k} \le \beta_i, \ i = 1, \ldots, r, \ k = 1, \ldots, n_p, \tag{8.6.1e}$$

where G_0 is defined by (8.3.6), and G_i, $i = 1, \ldots, N$, are defined by (8.3.5). (8.6.1c)–(8.6.1e) are the constraints that specify the set Ξ^p. The constraints in (8.6.1e) are known as boundedness constraints in nonlinear mathematical programming. Let Λ^p be the set that consists of all those $\boldsymbol{\sigma}^p \in \mathbb{R}^{rn_p}$ such that the boundedness constraints (8.6.1e) are satisfied.

This nonlinear mathematical programming problem in terms of the control parameters can be solved by using any suitable nonlinear optimization technique, such as sequential quadratic programming (SQP) (see Section 3.5). Like most nonlinear optimization techniques, SQP requires an initial control parameter vector $(\boldsymbol{\sigma}^p)^{(0)} \in \Lambda^p$ to start up the iterative search for an optimal solution. For each iterate $(\boldsymbol{\sigma}^p)^{(i)} \in \Lambda^p$ generated during the optimization, the values of the cost function (8.6.1a) and the constraints (8.6.1b)–(8.6.1c) as well as their respective gradients are required in order to generate the next iterate $(\boldsymbol{\sigma}^p)^{(i+1)}$. Consequently, like many others, the technique gives rise to a sequence of control parameter vectors. The optimal control parameter vector obtained by the SQP routine is then taken as an optimal control parameter vector of Problem $(P_1(p))$. In what follows, we shall explain how to calculate, for each $\boldsymbol{\sigma}^p \in \Lambda^p$, the values of the cost function $G_0(\boldsymbol{\sigma}^p)$ and the constraint functions $G_i(\boldsymbol{\sigma}^p)$, $i = 1, \ldots, N$, and $\mathcal{F}_i\left(\boldsymbol{\sigma}^{p,k}\right)$, $i = 1, \ldots, q$; $k = 1, \ldots, n_p$, as well as their respective gradients.

Remark 8.6.1 *For each $\boldsymbol{\sigma}^p \in \Lambda^p$, the corresponding values of the constraint functions $\mathcal{F}_i\left(\boldsymbol{\sigma}^{p,k}\right)$, $i = 1, \ldots, q$; $k = 1, \ldots, n_p$, are straightforward to calculate via* (8.6.1c). *Their gradients are given by*

$$\frac{\partial \mathcal{F}_i\left(\boldsymbol{\sigma}^{p,k}\right)}{\partial \boldsymbol{\sigma}^{p,k}} = \left(\boldsymbol{E}^i\right)^\top, \quad i = 1, \ldots, q, \ k = 1, \ldots, n_p. \tag{8.6.2}$$

To calculate the values of the cost functional (8.6.1a) and the constraint functionals (8.6.1b)–(8.6.1c) corresponding to each $\boldsymbol{\sigma}^p \in \Lambda^p$, the first task is to calculate the solution of the system (8.3.4) corresponding to each $\boldsymbol{\sigma}^p \in \Lambda^p$. This is presented as an algorithm for future reference.

Algorithm 8.6.1 *For each given $\boldsymbol{\sigma}^p \in \Lambda^p$, compute the solution $\boldsymbol{x}(\cdot|\boldsymbol{\sigma}^p)$ of the system* (8.3.4) *by solving the differential equations* (8.3.4a) *forward in time from $t = 0$ to $t = T$ with the initial condition* (8.3.4b).

With the information obtained in Algorithm 8.6.1, the values of G_i corresponding to each $\boldsymbol{\sigma}^p \in \Lambda^p$ can be easily calculated by the following simple algorithm.

Algorithm 8.6.2

Step 1. Use Algorithm 8.6.1 to solve for $\boldsymbol{x}(\cdot|\boldsymbol{\sigma}^p)$. Thus, $\boldsymbol{x}(t|\boldsymbol{\sigma}^p)$ is known for each $t \in [0, T]$. This implies that

(a) $\Phi_i(\boldsymbol{x}(\tau_i|\boldsymbol{\sigma}^p))$, $i = 0, 1, \ldots, N$, are known; and
(b) $\widetilde{\mathcal{L}}_i(t, \boldsymbol{x}(t|\boldsymbol{\sigma}^p), \boldsymbol{\sigma}^p)$, $i = 0, 1, \ldots, N$, are known for each $t \in [0, \tau_i]$.

Hence, their integrals

$$\int_0^{\tau_i} \widetilde{\mathcal{L}}_i(t, \boldsymbol{x}(t|\boldsymbol{\sigma}^p), \boldsymbol{\sigma}^p)\, dt, \ \ i = 0, 1, \ldots, N,$$

can be obtained readily using Simpson's rule.

Step 2. Calculate the values of the cost functional ($i = 0$) and the constraint functionals, $i = 1, \ldots, N$, according to

$$G_i(\boldsymbol{\sigma}^p) = \Phi_i(\boldsymbol{x}(\tau_i|\boldsymbol{\sigma}^p)) + \int_0^{\tau_i} \widetilde{\mathcal{L}}_i(t, \boldsymbol{x}(t|\boldsymbol{\sigma}^p), \boldsymbol{\sigma}^p)\, dt, i = 0, 1, \ldots, N.$$

In view of Section 7.3, we see that the derivations of the gradient formulae for the cost functional and the canonical constraint functionals are the same. For each $i = 0, 1, \ldots, N$, the gradient of the corresponding G_i may be computed using the following algorithm.

Algorithm 8.6.3 *Given $\boldsymbol{\sigma}^p \in \Lambda^p$, proceed as follows.*

Step 1. Solve the costate differential equation

$$\frac{d\boldsymbol{\lambda}^i(t)}{dt} = -\left[\frac{\partial \widetilde{H}_i\left(t, \boldsymbol{x}(t|\boldsymbol{\sigma}^p), \boldsymbol{\sigma}^p, \boldsymbol{\lambda}^i(t)\right)}{\partial \boldsymbol{x}}\right]^\top \tag{8.6.3a}$$

with the boundary condition

$$\boldsymbol{\lambda}^i(\tau_i) = \left[\frac{\partial \Phi_i(\boldsymbol{x}(\tau_i|\boldsymbol{\sigma}^p))}{\partial \boldsymbol{x}}\right]^\top \tag{8.6.3b}$$

backward in time from $t = \tau_i$ *to* $t = 0$*, where* $\boldsymbol{x}(\cdot|\boldsymbol{\sigma}^p)$ *is the solution of the system* (8.3.4) *corresponding to* $\boldsymbol{\sigma}^p \in \Xi^p$*; and* $\widetilde{H}_i$*, the corresponding Hamiltonian function for the cost function if* $i = 0$ *and the* i*-th constraint function if* $i = 1, \ldots, N$*, defined by*

$$\widetilde{H}_i(t, \boldsymbol{x}, \boldsymbol{\sigma}^p, \boldsymbol{\lambda}) = \widetilde{\mathcal{L}}_i(t, \boldsymbol{x}, \boldsymbol{\sigma}^p) + \boldsymbol{\lambda}^\top \widetilde{\boldsymbol{f}}(t, \boldsymbol{x}, \boldsymbol{\sigma}^p). \tag{8.6.4}$$

Let $\boldsymbol{\lambda}^i(\cdot|\boldsymbol{\sigma}^p)$ *be the solution of the costate system* (8.6.3).

Step 2. The gradient of G_i *is computed as*

$$\frac{\partial G_i(\boldsymbol{\sigma}^p)}{\partial \boldsymbol{\sigma}^p} = \int_0^{\tau_i} \frac{\partial \widetilde{H}_i\left(t, \boldsymbol{x}(t|\boldsymbol{\sigma}^p), \boldsymbol{\sigma}^p, \boldsymbol{\lambda}^i(t|\boldsymbol{\sigma}^p)\right)}{\partial \boldsymbol{\sigma}^p} dt. \tag{8.6.5}$$

Remark 8.6.2 *During actual computation, very often the control parametrization is carried out on a uniform partition of the interval* $[0, T]$*, i.e.,*

$$u_j^p(t) = \sum_{k=1}^{n_p} \sigma_j^{p,k} \chi_k(t), \quad j = 1, \ldots, r, \tag{8.6.6}$$

where χ_k *is the indicator function given by*

$$\chi_k(t) = \begin{cases} 1, & (k-1)\Delta < t < k\Delta, \\ 0, & \text{otherwise}, \end{cases} \tag{8.6.7}$$

$\boldsymbol{u}^p(t) = [u_1^p, \ldots, u_r^p]^\top$, $\boldsymbol{\sigma}^{p,k} = \left[\sigma_1^{p,k}, \ldots, \sigma_r^{p,k}\right]^\top$, n_p *is the number of equal subintervals and* $\Delta = \frac{T}{n_p}$ *is the uniform interval length. In this case, each component of the gradient formula* (8.6.5) *can be written in a more specific form*

$$\frac{\partial G_i(\boldsymbol{\sigma}^p)}{\partial \sigma_j^{p,k}} = \begin{cases} \displaystyle\int_{(k-1)\Delta}^{k\Delta} \frac{\partial H_i}{\partial u_j^p} dt, & k \le l_i, \\ \displaystyle\int_{l_i\Delta}^{\tau_i} \frac{\partial H_i}{\partial u_j^p} dt, & k = l_i + 1, \\ 0 & k > l_i + 1, \end{cases} \tag{8.6.8}$$

$i = 0, 1, \ldots, N$, $j = 1, \ldots, r$, $k = 1, \ldots, n_p$,

$$H_i(t, \boldsymbol{x}, \boldsymbol{u}, \boldsymbol{\lambda}) = \mathcal{L}_i(t, \boldsymbol{x}, \boldsymbol{u}) + \boldsymbol{\lambda}^\top (\boldsymbol{f}(t, \boldsymbol{x}, \boldsymbol{u})); \tag{8.6.9a}$$

$$l_i = \left\lfloor \frac{\mathcal{T}_i}{\Delta} \right\rfloor, \tag{8.6.9b}$$

and $\left\lfloor \frac{\mathcal{T}_i}{\Delta} \right\rfloor$ *denotes the largest integer smaller than or equal to* $\frac{\mathcal{T}_i}{\Delta}$.

Note that a uniform partition for the parametrization is not a strict requirement but rather a computational convenience. At times when it is known a priori that the control changes rapidly over certain intervals and changes slowly over others, it will be more effective to use a nonuniform partition.

Remark 8.6.3 *Note that the gradient formulae for the cost and constraint functionals can also be obtained using the variational approach as detailed in Section 7.3.*

Remark 8.6.4 *Clearly, when n_p increases, the computational time required to solve the corresponding approximate problem will increase with some power of n_p. To overcome this difficulty, we propose to solve any given problem as follows.*

Let q be a small positive integer. To begin, we solve the approximate Problem $(P(1))$ with $n_1 = q$. Let $\boldsymbol{\sigma}^{1,*}$ be the optimal solution so obtained, and let $\boldsymbol{u}^{1,*}$ be the corresponding control. Then, it is clear that $\boldsymbol{u}^{1,*}$ is a suboptimal control for the original Problem (P). Next, we choose $n_2 = 2q$, and let $\boldsymbol{\sigma}_0^2$ denote the parameter vector that describes $\boldsymbol{u}^{1,*}$ over a new partition with $n_2 = 2q$ intervals. Clearly, $\boldsymbol{\sigma}_0^2$ is a feasible parameter vector for Problem $(P(2))$. We then solve Problem $(P(2))$ using $\boldsymbol{\sigma}_0^2$ as the initial guess. The process continues until the cost reduction becomes negligible. Computational experience indicates that the reduction in the cost value appears to be insignificant for $n_p > 20$ in many problems. Also, by virtue of the construction of the subsequent initial guess, the increase in CPU time is seldom drastic.

Remark 8.6.5 *Note that the constraint transcription introduced in [241] is used in the transformation of the continuous constraints in Remark 8.2.1(iv). However, this constraint transcription has a serious disadvantage because the canonical equality state constraint so obtained does not satisfy any constraint qualification (see Remark 3.1.1). In particular, let us consider the constraint specified in Remark 8.2.1(iv). Then,*

$$\frac{\partial G(\boldsymbol{\sigma}^p)}{\partial \boldsymbol{\sigma}^p} = 0$$

if the parameter vector $\boldsymbol{\sigma}^p$ is such that

$$\max_{0 \le t \le T} h(t, \boldsymbol{x}(t|\boldsymbol{\sigma}^p), \boldsymbol{\sigma}^p) = 0,$$

where the function h is defined in Remark 8.2.1(iv). In this situation, the linear approximation of the constraint G is equal to zero for all search directions. This, in turn, implies that the search direction, which is obtained from the *corresponding quadratic programming subproblem, may be one along which the corresponding constraint will always be violated. Two better transformation methods have been presented in Chapter 4—one is based on the constraint transcription method and the other one is based on the exact penalty function method that will be introduced for the continuous inequality constraints in Chapter 9.3.*

8.7 Illustrative Examples

To illustrate the simple and yet efficient solution procedure outlined in the previous sections, we now present the numerical results of applying this procedure to several examples. Note that the procedure has been implemented in the software package MISER 3.3 (see [104]) that is used to generate the results presented here and also those in later sections and chapters of the text.

Example 8.7.1 (Bitumen Pyrolysis) This problem has been widely studied in the literature, more recently in the context of determining global optimal solutions of nonlinear optimal control problems (see [53] and the references cited therein). The task is to find an optimal temperature profile in a plug flow reactor with the following reactions involving components A_i, $i = 1, \ldots, 4$, with reactions rates k_j, $j = 1, \ldots, 5$.

$$A_1 \xrightarrow{k_1} A_2, \quad A_2 \xrightarrow{k_2} A_3, \quad A_1 + A_2 \xrightarrow{k_3} A_2 + A_2,$$

$$A_1 + A_2 \xrightarrow{k_4} A_3 + A_2, \quad A_1 + A_2 \xrightarrow{k_5} A_4 + A_2.$$

While a more comprehensive model of the problem [176] involves all of the components, the version commonly presented in the literature (and which we adopt here) includes components A_1 and A_2 only [53]. The aim is to determine an optimal temperature profile to maximize the final amount of A_2 subject to bound constraints on the temperature. In standard form, this problem can be stated as follows. Minimize

$$g_0(u) = -x_2(T)$$

subject to

$$\begin{aligned}
\frac{dx_1(t)}{dt} &= -k_1 x_1(t) - (k_3 + k_4 + k_5) x_1(t) x_2(t), \\
\frac{dx_2(t)}{dt} &= k_1 x_1(t) - k_2 x_2(t) + k_3 x_1(t) x_2(t), \\
x_1(0) &= 1, \\
x_2(0) &= 0,
\end{aligned}$$

where

$$k_i = a_i \exp\left[-\frac{b_i/R}{u}\right], \ i = 1, \ldots, 5,$$

and the values of a_i, b_i/R, $i = 1, \ldots, 5$, are given in Table 8.7.1. We also have the control constraint

$$698.15 \leq u(t) \leq 748.15, \ t \in [0, T].$$

We solve the problem with $T = 10$ and assume that u is piecewise constant over a uniform partition with 10 subintervals.

Table 8.7.1: Data for the bitumen pyrolysis example

i	$\ln a_i$	b_i/R
1	8.86	10,215.4
2	24.25	18,820.5
3	23.67	17,008.9
4	18.75	14,190.8
5	20.70	15,599.8

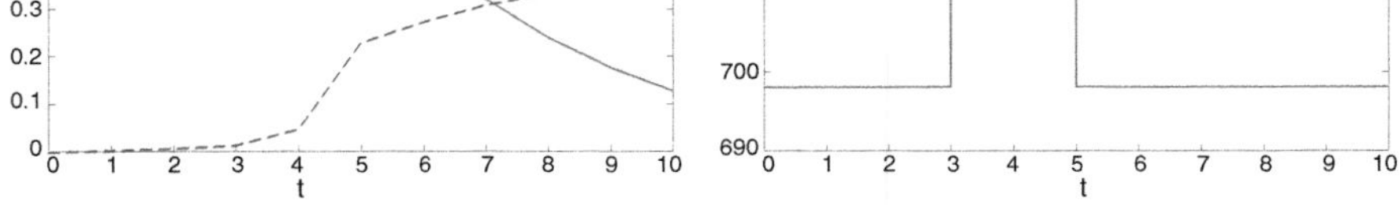

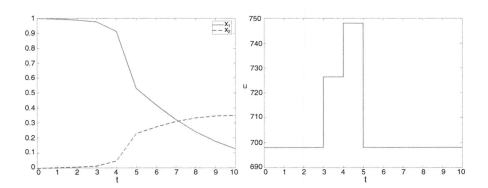

Fig. 8.7.1: Computed optimal solution for Example 8.7.1 with 10 subintervals

The optimal solution obtained yields an objective function value of −0.353434883 when a constant initial guess for the control was chosen halfway between the upper and lower bounds. As noted in [53], the problem has several local minima. While the global solution is obtained with the initial guess we choose here, other initial guesses for the control can lead to one of the local optimal solutions. To illustrate the effect of the size of the partition for the piecewise constant control, we also solved the problem assuming a uniform piecewise constant partition with intervals for the control. A slightly improved objective function value of -0.353717047 is obtained with this refined partition, although the shape of the optimal control function is now more well-defined (compare Figures 8.7.1 and 8.7.2).

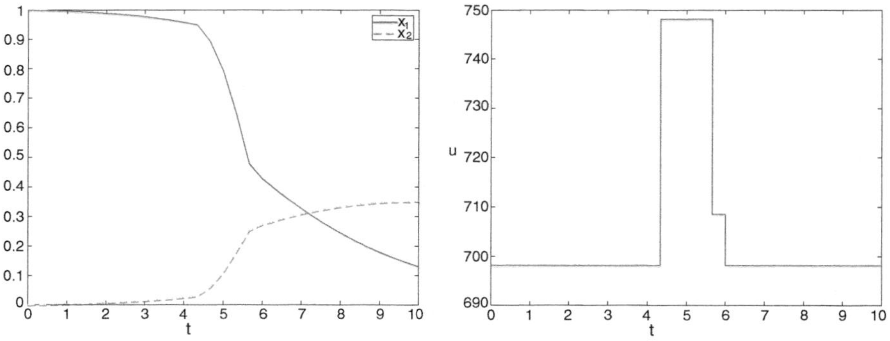

Fig. 8.7.2: Computed optimal solution for Example 8.7.1 with 30 subintervals

Example 8.7.2 (Student Problem) Consider the Student Problem presented in Chapter 1. Choosing $b = 0.5$, $c = 0.1$, $k_0 = 15$, $T = 15$ and $\bar{w} = 15$, the problem may be written in standard form as follows. Minimize

$$g_0(u) = \int_0^{15} u(t)\, dt$$

subject to

$$\frac{dx(t)}{dt} = 0.5u(t) - 0.1x(t), \ x(0) = 0,$$

with the control constraint

$$0 \leq u(t) \leq 15, \ t \in [0, 15].$$

We assume that u is piecewise constant over a uniform partition with 20 subintervals.

The analytic optimal solution for this problem can be shown to involve a pure bang-bang control. The approximate solution calculated here (see Figure 8.7.3b) is therefore clearly suboptimal. This is due to the fixed partition we have chosen for the control function, because the exact switching time

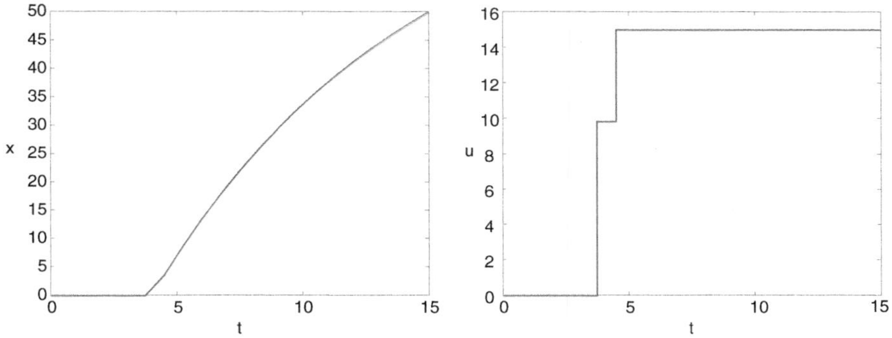

Fig. 8.7.3: Computed optimal solution for Example 8.7.2

for the optimal control does not happen to coincide with one of the fixed knot points in the chosen partition. While an increasingly finer partition can yield improved solutions, a much better way to solve bang-bang and related optimal control problems numerically is discussed in Section 8.9.

8.8 Combined Optimal Control and Optimal Parameter Selection Problems

In the previous sections, a unified and efficient computational scheme is developed for solving a general class of fixed terminal time optimal control problems in canonical form. However, there are many practical problems that do not belong to this general class of optimal control problems. Examples include:

(i) optimal parameter selection problems;
(ii) free terminal time optimal control problems, including minimum-time problems;
(iii) minimax optimal control problems (problems with Chebyshev performance index);
(iv) boundary-value control problems, including problems with periodic and interrelated boundary conditions and
(v) combined optimal control and optimal parameter selection problems.

The aim of this section is to extend the results presented in the previous sections to a more general class of optimization problems that covers all the important problems mentioned above as special cases. Since the extension is rather straightforward, the following exposition is relatively brief. To begin, let us consider a process described by the following system of differential equations on the fixed time interval $(0, T]$,

$$\frac{d\boldsymbol{x}(t)}{dt} = \boldsymbol{f}(t, \boldsymbol{x}(t), \boldsymbol{\zeta}, \boldsymbol{u}(t)), \tag{8.8.1a}$$

where $\boldsymbol{x} = [x_1, \ldots, x_n]^\top \in \mathbb{R}^n$, $\boldsymbol{\zeta} = [\zeta_1, \ldots, \zeta_s]^\top \in \mathbb{R}^s$, $\boldsymbol{u} = [u_1, \ldots, u_r]^\top \in \mathbb{R}^r$, are, respectively, the state, system parameter and control parameter vectors. We have $\boldsymbol{f} = [f_1, \ldots, f_n]^\top \in \mathbb{R}^n$ with $\boldsymbol{f} : \mathbb{R} \times \mathbb{R}^n \times \mathbb{R}^s \times \mathbb{R}^r \mapsto \mathbb{R}^n$. The initial condition for the differential equation (8.8.1a) is

$$\boldsymbol{x}(0) = \boldsymbol{x}^0(\boldsymbol{\zeta}), \tag{8.8.1b}$$

where $\boldsymbol{x}^0 = \left[x_1^0, \ldots, x_n^0\right]^\top$ is a given function of the system parameters $\boldsymbol{\zeta}$. Define

$$\mathcal{Z} = \{\boldsymbol{\zeta} = [\zeta_1, \ldots, \zeta_s]^\top \in \mathbb{R}^s : \alpha_i \leq \zeta_i \leq \beta_i, \;\; i = 1, \ldots, s\}, \tag{8.8.2}$$

where α_i and β_i, $i = 1, \ldots, s$, are given real numbers. Clearly, $\mathcal{Z}$ is a compact and convex subset of $\mathbb{R}^s$. Let $\mathcal{U}$ be as defined in Section 8.2.

For each $(\boldsymbol{\zeta}, \boldsymbol{u}) \in \mathbb{R}^s \times L_\infty([0, T], \mathbb{R}^r)$, let $\boldsymbol{x}(\cdot|\boldsymbol{\zeta}, \boldsymbol{u})$ be the corresponding solution of the system (8.8.1). We may then state the combined optimal control and optimal parameter selection problem as follows.

Problem ($\boldsymbol{Q}$): Given system (8.8.1), find a combined element $(\boldsymbol{\zeta}, \boldsymbol{u}) \in \mathcal{Z} \times \mathcal{U}$ such that the cost function

$$g_0(\boldsymbol{\zeta}, \boldsymbol{u}) = \Phi_0(\boldsymbol{x}(T|\boldsymbol{\zeta}, \boldsymbol{u}), \boldsymbol{\zeta}) + \int_0^T \mathcal{L}_0(t, \boldsymbol{x}(t|\boldsymbol{\zeta}, \boldsymbol{u}), \boldsymbol{\zeta}, \boldsymbol{u}(t))\, dt \tag{8.8.3}$$

is minimized over $\mathcal{Z} \times \mathcal{U}$, subject to the equality constraints

$$g_i(\boldsymbol{\zeta}, \boldsymbol{u}) = \Phi_i(\boldsymbol{x}(T|\boldsymbol{\zeta}, \boldsymbol{u}), \boldsymbol{\zeta}) + \int_0^{\tau_i} \mathcal{L}_i(t, \boldsymbol{x}(t|\boldsymbol{\zeta}, \boldsymbol{u}), \boldsymbol{\zeta}, \boldsymbol{u}(t))\, dt = 0,$$
$$i = 1, \ldots, N_e, \tag{8.8.4a}$$

and subject to the inequality constraints

$$g_i(\boldsymbol{\zeta}, \boldsymbol{u}) = \Phi_i(\boldsymbol{x}(T|\boldsymbol{\zeta}, \boldsymbol{u}), \boldsymbol{\zeta}) + \int_0^{\tau_i} \mathcal{L}_i(t, \boldsymbol{x}(t|\boldsymbol{\zeta}, \boldsymbol{u}), \boldsymbol{\zeta}, \boldsymbol{u}(t))\, dt \leq 0,$$
$$i = N_e + 1, \ldots, N, \tag{8.8.4b}$$

where $\Phi_i : \mathbb{R}^n \times \mathbb{R}^s \mapsto \mathbb{R}$, $i = 0, 1, \ldots, N$, and $\mathcal{L}_i : \mathbb{R} \times \mathbb{R}^n \times \mathbb{R}^s \times \mathbb{R}^r \mapsto \mathbb{R}$, $i = 0, 1, \ldots, N$, are given real-valued functions. As before, $\tau_i \leq T$ is referred to as the characteristic time for the i-th constraint.

We assume throughout this section that the corresponding versions of Assumptions (8.2.1)–(8.5.1) are satisfied. We now apply the concept of control parametrization to Problem (Q). Thus, system (8.8.1) takes the form

$$\frac{d\boldsymbol{x}(t)}{dt} = \tilde{\boldsymbol{f}}(t, \boldsymbol{x}(t), \boldsymbol{\zeta}, \boldsymbol{\sigma}^p), \tag{8.8.5a}$$

$$\boldsymbol{x}(0) = \boldsymbol{x}^0(\boldsymbol{\zeta}), \tag{8.8.5b}$$

where

$$\tilde{\boldsymbol{f}}(t, \boldsymbol{x}(t), \boldsymbol{\zeta}, \boldsymbol{\sigma}^p) = \boldsymbol{f}\left(t, \boldsymbol{x}(t), \boldsymbol{\zeta}, \sum_{k=1}^{n_p} \boldsymbol{\sigma}^{p,k} \chi_{I_k^p}(t)\right), \tag{8.8.5c}$$

$\chi_{I_k^p}(t)$ is defined by (8.3.1b) and I_k^p, $k = 1, \ldots, n_p$, are as defined in Section 8.3. Let $\boldsymbol{x}(\cdot|\boldsymbol{\zeta}, \boldsymbol{\sigma}^p)$ be the solution of the system (8.8.5) corresponding to the combined vector $(\boldsymbol{\zeta}, \boldsymbol{\sigma}^p) \in \mathcal{Z} \times \Xi^p$, where Ξ^p is defined in Section 8.3. The constraints (8.8.4a) and (8.8.4b) are reduced to

$$G_i(\boldsymbol{\zeta}, \boldsymbol{\sigma}^p) = \Phi_i(\boldsymbol{x}(T|\boldsymbol{\zeta}, \boldsymbol{\sigma}^p), \boldsymbol{\zeta}) + \int_0^T \tilde{\mathcal{L}}_i(t, \boldsymbol{x}(t|\boldsymbol{\zeta}, \boldsymbol{\sigma}^p), \boldsymbol{\zeta}, \boldsymbol{\sigma}^p)\, dt = 0,$$
$$i = 1, \ldots, N_e, \tag{8.8.6a}$$

and

$$G_i(\boldsymbol{\zeta}, \boldsymbol{\sigma}^p) = \Phi_i(\boldsymbol{x}(T|\boldsymbol{\zeta}, \boldsymbol{\sigma}^p), \boldsymbol{\zeta}) + \int_0^T \tilde{\mathcal{L}}_i(t, \boldsymbol{x}(t|\boldsymbol{\zeta}, \boldsymbol{\sigma}^p), \boldsymbol{\zeta}, \boldsymbol{\sigma}^p)\, dt \leq 0,$$
$$i = N_e + 1, \ldots, N, \tag{8.8.6b}$$

respectively, where

$$\tilde{\mathcal{L}}_i(t, \boldsymbol{x}, \boldsymbol{\zeta}, \boldsymbol{\sigma}^p) = \mathcal{L}_i\left(t, \boldsymbol{x}, \boldsymbol{\zeta}, \sum_{k=1}^{n_p} \boldsymbol{\sigma}^{p,k} \chi_{I_k^p}(t)\right). \tag{8.8.6c}$$

Let $\mathcal{D}^p$ be the set that consists of all those combined vectors $(\boldsymbol{\zeta}, \boldsymbol{\sigma}^p)$ in $\mathcal{Z} \times \Xi^p$ that satisfy the constraints (8.8.6a) and (8.8.6b). Furthermore, let $\mathcal{B}^p$ be the corresponding subset of $\mathcal{Z} \times \mathcal{U}^p$. We are now in a position to specify an approximate version of Problem (Q) as follows.

Problem $(Q(p))$: Subject to system (8.8.5), find a combined vector $(\boldsymbol{\zeta}, \boldsymbol{\sigma}^p) \in \mathcal{D}^p$ such that the cost function

$$G_0(\boldsymbol{\zeta}, \boldsymbol{\sigma}^p) = \Phi_0(\boldsymbol{x}(T|\boldsymbol{\zeta}, \boldsymbol{\sigma}^p), \boldsymbol{\zeta}) + \int_0^T \tilde{\mathcal{L}}_0(t, \boldsymbol{x}(t|\boldsymbol{\zeta}, \boldsymbol{\sigma}^p), \boldsymbol{\zeta}, \boldsymbol{\sigma}^p)\, dt \tag{8.8.7}$$

is minimized over $\mathcal{D}^p$.

Problem $(Q(p))$ can also be stated in the following form.
Minimize

$$G_0(\boldsymbol{\zeta}, \boldsymbol{\sigma}^p) \tag{8.8.8a}$$

subject to

$$G_i(\boldsymbol{\zeta}, \boldsymbol{\sigma}^p) = 0,\ i = 1, \ldots, N_e, \tag{8.8.8b}$$
$$G_i(\boldsymbol{\zeta}, \boldsymbol{\sigma}^p) \leq 0,\ i = N_e + 1, \ldots, N, \tag{8.8.8c}$$

$$\digamma_i\left(\boldsymbol{\sigma}^{p,k}\right) = \left(\boldsymbol{E}^i\right)^\top \boldsymbol{\sigma}^{p,k} - b_i \leq 0,\ i = 1, \ldots, q,\ k = 1, \ldots, n_p, \tag{8.8.8d}$$

$$\alpha_i \leq \sigma_i^{p,k} \leq \beta_i,\ \ i = 1, \ldots, r,\ k = 1, \ldots, n_p, \tag{8.8.8e}$$

$$a_i \leq \zeta_i \leq b_i,\ i = 1, \ldots, s, \tag{8.8.8f}$$

where G_0 is defined by (8.8.7), G_i, $i = 1, \ldots, N$, are defined by (8.8.6a) and (8.8.6b), while (8.8.8d)–(8.8.8f) are the constraints that specify the set $\mathcal{Z} \times \Xi^p$. The constraints (8.8.8e)–(8.8.8f) are known as the boundedness constraints in nonlinear optimization programming. Let Λ^p be the set that consists of all those $(\boldsymbol{\zeta}, \boldsymbol{\sigma}^p) \in \mathbb{R}^s \times \mathbb{R}^{rn_p}$ such that the boundedness constraints (8.8.8e)–(8.8.8f) are satisfied.

This nonlinear mathematical programming problem in the control parameter vectors can be solved by using any nonlinear optimization technique, such as the sequential quadratic programming (SQP) approach (see Section 3.5). In solving the nonlinear optimization problem (8.8.8) via SQP, we choose an initial control parameter vector $(\boldsymbol{\zeta}, \boldsymbol{\sigma}^p)^{(0)} \in \Lambda^p$ to initialize the SQP process. Then, for each $(\boldsymbol{\zeta}, \boldsymbol{\sigma}^p)^{(i)} \in \Lambda^p$, the values of the cost function (8.8.8a) and the constraints (8.8.8b)–(8.8.8d) as well as their respective gradients are required by SQP to generate the next iterate $(\boldsymbol{\zeta}, \boldsymbol{\sigma}^p)^{(i+1)}$. Consequently, it gives rise to a sequence of combined vectors. The optimal combined vector obtained by the SQP process is then regarded as an approximate optimal combined vector of Problem $(Q(p))$.

For each $(\zeta, \boldsymbol{\sigma}^p) \in \Lambda^p$, the values of the cost function $G_0(\boldsymbol{\zeta}, \boldsymbol{\sigma}^p)$ and the constraint functions $G_i(\boldsymbol{\zeta}, \boldsymbol{\sigma}^p)$, $i = 1, \ldots, N$, and $\digamma_i\left(\boldsymbol{\sigma}^{p,k}\right)$, $i = 1, \ldots, q$, $k = 1, \ldots, n_p$, can be calculated in a manner similar to the corresponding components of Algorithm 8.6.1 and Algorithm 8.6.2, and Remark 8.6.1. The gradients of $\digamma_i\left(\boldsymbol{\sigma}^{p,k}\right)$, $i = 1, \ldots, q$, $k = 1, \ldots, n_p$, are given in Remark 8.6.1. A procedure similar to that described in Algorithm 8.6.3 is given below for computing the gradient of $G_i(\boldsymbol{\sigma}^p)$ for each $i = 0, 1, \ldots, N$.

Algorithm 8.8.1 *Let $(\boldsymbol{\zeta}, \boldsymbol{\sigma}^p) \in \Lambda^p$ be given.*

Step 1. Solve the costate system (8.6.3) with $\boldsymbol{\sigma}^p$ replaced by $(\boldsymbol{\zeta}, \boldsymbol{\sigma}^p)$ backward in time from $t = \tau_i$ to $t = 0$ (again $\tau_0 = T$ by convention). Let the corresponding solution be denoted by $\widetilde{\boldsymbol{\lambda}}^i(\cdot|\boldsymbol{\zeta}, \boldsymbol{\sigma}^p)$.

Step 2. The gradient is computed from

$$\frac{\partial G_i(\boldsymbol{\zeta}, \boldsymbol{\sigma}^p)}{\partial \boldsymbol{\sigma}} = \int_0^{\tau_i} \frac{\partial \widetilde{H}_i\left(t, \boldsymbol{x}(t|\boldsymbol{\zeta}, \boldsymbol{\sigma}^p), \boldsymbol{\zeta}, \boldsymbol{\sigma}^p, \widetilde{\boldsymbol{\lambda}}^i(t|\boldsymbol{\zeta}, \boldsymbol{\sigma}^p)\right)}{\partial \boldsymbol{\sigma}} dt \tag{8.8.9a}$$

and

$$\frac{\partial G_i(\boldsymbol{\zeta}, \boldsymbol{\sigma}^p)}{\partial \boldsymbol{\zeta}} = (\widetilde{\boldsymbol{\lambda}}^i(0|\boldsymbol{\zeta}, \boldsymbol{\sigma}^p))^\top \frac{\partial \boldsymbol{x}^0(\boldsymbol{\zeta})}{\partial \boldsymbol{\zeta}} + \int_0^{T_i} \frac{\partial \widetilde{H}_i\left(t, \boldsymbol{x}(t|\boldsymbol{\zeta}, \boldsymbol{\sigma}^p), \boldsymbol{\zeta}, \boldsymbol{\sigma}^p, \widetilde{\lambda}^i(t|\boldsymbol{\zeta}, \boldsymbol{\sigma}^p)\right)}{\partial \boldsymbol{\zeta}} dt, \tag{8.8.9b}$$

where $\widetilde{H}_i$ is defined by an equation similar to (8.6.4).

Remark 8.8.1 *In the above algorithm, the solution $\boldsymbol{x}(\cdot|\boldsymbol{\zeta}, \boldsymbol{\sigma}^p)$ of the system (8.8.5) corresponding to each $(\boldsymbol{\zeta}, \boldsymbol{\sigma}^p) \in \Lambda^p$ can be computed by an algorithm similar to Algorithm 8.6.1.*

The convergence properties of the proposed method will be presented in the next two theorems. Their proofs are similar to those given for Theorems 8.5.1 and 8.5.2, respectively.

Theorem 8.8.1 *Let $(\boldsymbol{\zeta}^{p,*}, \boldsymbol{\sigma}^{p,*})$ be an optimal combined vector of Problem $(Q(p))$, and let $(\boldsymbol{\zeta}^{p,*}, \bar{\boldsymbol{u}}^{p,*})$ be the corresponding element in $\mathcal{Z} \times \mathcal{U}^p$. Suppose that the original problem (Q) has an optimal combined element $(\boldsymbol{\zeta}^*, \boldsymbol{u}^*)$. Then,*

$$\lim_{p\to\infty} g_0(\boldsymbol{\zeta}^{p,*}, \bar{\boldsymbol{u}}^{p,*}) = g_0(\boldsymbol{\zeta}^*, \boldsymbol{u}^*).$$

Theorem 8.8.2 *Let $(\boldsymbol{\zeta}^{p,*}, \bar{\boldsymbol{u}}^{p,*})$ be as defined in Theorem 8.8.1. Suppose that*

$$\lim_{p\to\infty} |\bar{\boldsymbol{u}}^{p,*}(t) - \boldsymbol{u}^*(t)| = 0, \qquad a.e. \text{ on } [0, T]$$

and

$$\lim_{p\to\infty} |\boldsymbol{\zeta}^{p,*} - \boldsymbol{\zeta}^*| = 0.$$

Then, $(\boldsymbol{\zeta}^*, \boldsymbol{u}^*)$ is an optimal combined system parameter vector and control.

8.8.1 Model Transformation

In this subsection, our aim is to show that many different classes of optimal control problems can be transformed into special cases of Problem (Q). The following is a list of some of these transformations, but it is by no means exhaustive. Readers are advised to exercise their ingenuity and initiative in applying these transformations and devising new ones.

Note that the references for Sections 8.8.1 and 8.8.2 are from Section 6.8.1 and Section 6.8.2 of [253], respectively.

(i) Free terminal time problems (including minimum-time problems).

$$\min_{\boldsymbol{u}(\cdot),T} g_0(\boldsymbol{u}, T),$$

where

$$g_0(\boldsymbol{u}, T) = \Phi_0(\boldsymbol{x}(T), T) + \int_0^T \mathcal{L}_0(t, \boldsymbol{x}(t), \boldsymbol{u}(t))\, dt$$

subject to the differential equation

$$\frac{d\boldsymbol{x}(t)}{dt} = \boldsymbol{f}(t, \boldsymbol{x}(t), \boldsymbol{u}(t)), \ t \in (0, T],$$

with the initial condition

$$\boldsymbol{x}(0) = \boldsymbol{x}^0$$

and terminal condition

$$g(\boldsymbol{x}(T)) = 0 \text{ (or } \leq 0).$$

This problem is not in the form of Problem (P) or Problem (Q), as the terminal time T is not fixed, but variable. The simple time scale transformation

$$t = T\tau$$

then converts the problem into

$$\min_{T, \hat{\boldsymbol{u}}(\cdot)} g_0(\hat{\boldsymbol{u}}, T),$$

where

$$g_0(\hat{\boldsymbol{u}}, T) = \Phi_0(\hat{\boldsymbol{x}}(1), T) + \int_0^1 T\mathcal{L}_0(\tau T, \hat{\boldsymbol{x}}(\tau), \hat{\boldsymbol{u}}(\tau))\, d\tau,$$

subject to the differential equation

$$\frac{d\hat{\boldsymbol{x}}(\tau)}{d\tau} = T\boldsymbol{f}(\tau T, \hat{\boldsymbol{x}}(\tau), \hat{\boldsymbol{u}}(\tau)),$$

the initial condition

$$\hat{\boldsymbol{x}}(0) = \boldsymbol{x}^0,$$

and the terminal condition

$$g(\hat{\boldsymbol{x}}(1)) = 0 \text{ (or } \leq 0).$$

Note that the transformed problem takes the form of Problem (Q) if we treat T as a system parameter.

(ii) Minimax optimal control problems.
The state dynamical equations are as in (8.2.1) but the cost functional to be minimized takes the form

$$g_0(\boldsymbol{u}) = \max_{0 \leq t \leq T} C(t, \boldsymbol{x}(t), \boldsymbol{u}(t)) + \widehat{\Phi}_0(\boldsymbol{x}(T)) + \int_0^T \mathcal{L}_0(t, \boldsymbol{x}(t), \boldsymbol{\zeta}, \boldsymbol{u}(t))\, dt$$

(often referred to as a Chebyshev performance index). If we introduce the additional parameter

$$S = \max_{0 \le t \le T} C(t, \boldsymbol{x}(t), \boldsymbol{u}(t)),$$

then the cost functional is equivalent to

$$\hat{g}_0(\boldsymbol{u}, S) = \Phi_0(\boldsymbol{x}(T), S) + \int_0^T \mathcal{L}_0(t, \boldsymbol{x}(t), \boldsymbol{\zeta}, \boldsymbol{u}(t))dt$$

subject to the continuous state constraint

$$C(t, \boldsymbol{x}(t), \boldsymbol{u}(t)) - S \le 0, \ \forall t \in [0, T],$$

where

$$\Phi_0(\boldsymbol{x}(T), S) = \widehat{\Phi}_0(\boldsymbol{x}(T)) + S.$$

The resulting problem, due to the additional continuous state constraint, is not exactly in the form of Problem (Q). However, the techniques to be introduced in Chapter 9.3 can be readily used to solve it.

(iii) Problems with periodic boundary conditions.
The cost functional and the state dynamical equations are as described by (8.2.4) and (8.2.1), but the initial and final state values are related by

$$h(\boldsymbol{x}(0), \boldsymbol{x}(T)) = 0. \tag{8.8.10}$$

In this case, we can introduce a system parameter vector $\boldsymbol{\zeta} \in \mathbb{R}^n$ and put

$$\boldsymbol{x}(0) = \boldsymbol{\zeta}.$$

Then the constraint (8.8.10) is equivalent to

$$h(\boldsymbol{\zeta}, \boldsymbol{x}(T)) = 0,$$

and, once again, we have a special case of Problem (Q).

8.8.2 Smoothness of Optimal Control

Normally, the control parametrization often uses piecewise constant approximation, and hence the resulting optimal control obtained is discontinuous. While this is reasonable for many applications and actually desirable for some, others may require a continuous control or even a control with a certain degree of smoothness. In fact, the control parametrization can also be in terms of piecewise linear functions, and hence the resulting optimal control obtained will be continuous. Alternatively, as it is detailed in Chapter 9 of [253], optimal controls with any degree of smoothness can be readily found with the control parametrization approach and some additional computational effort.

For example, if a piecewise linear continuous control is desired for Problem (Q), one only needs to introduce an additional set of differential equations,

$$\frac{d\boldsymbol{u}(t)}{dt} = \boldsymbol{v}(t)$$

with the initial conditions

$$\boldsymbol{u}(0) = \boldsymbol{\zeta}_u = [\zeta_{s+1}, \ldots, \zeta_{s+r}]^\top,$$

where $\zeta_{s+1}, \ldots, \zeta_{s+r}$ are additional system parameters to be optimized. In the context of Problem (Q), $\boldsymbol{u}$ is now effectively a state function rather than a control function and $\boldsymbol{v}$ is the new piecewise constant control function. Note that the resulting $\boldsymbol{u}$ is a piecewise linear and continuous function.

Similarly, if we wish to approximate the optimal control by a C^1 (i.e., smooth) function, we introduce two additional sets of differential equations,

$$\frac{d\boldsymbol{u}(t)}{dt} = \boldsymbol{v}(t),$$

$$\frac{d\boldsymbol{v}(t)}{dt} = \boldsymbol{w}(t),$$

together with the initial conditions

$$\boldsymbol{u}(0) = \boldsymbol{\zeta}_u,$$

$$\boldsymbol{v}(0) = \boldsymbol{\zeta}_v,$$

where $\boldsymbol{w}(t)$ is piecewise constant, $\boldsymbol{u}(t)$ and $\boldsymbol{v}(t)$ are effectively new state variables and $\boldsymbol{\zeta}_u$ and $\boldsymbol{\zeta}_v$ are additional system parameter vectors to be optimized. Clearly, we can extend this process further to generate piecewise polynomial optimal controls with any desired degree of smoothness.

8.8.3 Illustrative Examples

The efficiency and versatility of the proposed computational schemes will be illustrated by several numerical examples.

Example 8.8.1 In this example, we consider the planar motion of a free flying robot (FFR). Let x_1 and x_2 denote the coordinates of the FFR, x_4 and x_5 denote the corresponding velocities, x_3 denotes the direction of the thrust, x_6 denotes the angular velocity and u_1 and u_2 denote the thrusts of the two jets. Furthermore we assume that the robot moves at a constant height from the initial position to the final equilibrium position. The robot is controlled by the thrust of the two jets, i.e., u_1 and u_2, and these two control variables are subject to boundedness constraints. This problem was studied earlier in [218] and [270], where it was formulated and solved as an L_p-minimization problem in [270]. In [13], this problem is formulated as an L_2-minimization problem, which is described formally as given below.

Given the dynamical system

$$\begin{aligned}
\frac{dx_1(t)}{dt} &= x_4(t),\\
\frac{dx_2(t)}{dt} &= x_5(t),\\
\frac{dx_3(t)}{dt} &= x_6(t),\\
\frac{dx_4(t)}{dt} &= (u_1(t)+u_2(t))\cos x_3(t),\\
\frac{dx_5(t)}{dt} &= (u_1(t)+u_2(t))\sin x_3(t),\\
\frac{dx_6(t)}{dt} &= \alpha(u_1(t)-u_2(t)),
\end{aligned}$$

with initial and terminal conditions given, respectively, by

$$\begin{aligned}
x(0) &= (x_1(0), x_2(0), x_3(0), x_4(0), x_5(0), x_6(0))^\top\\
&= (-10, -10, \pi/2, 0, 0, 0)^\top,
\end{aligned}$$

and

$$x(T) = (x_1(T), x_2(T), x_3(T), x_4(T), x_5(T), x_6(T))^\top = (0,0,0,0,0,0)^\top,$$

find a control $\boldsymbol{u} = (u_1, u_2)^\top \in \mathcal{U}$ such that the following cost functional

$$g_0(\boldsymbol{u}) = \int_0^T \left\{(u_1(t))^2 + (u_2(t))^2\right\} dt$$

is minimized, where

$$\mathcal{U} = \left\{\boldsymbol{u} = (u_1, u_2)^\top : |u_1(t)| \leq 0.8,\ |u_2(t)| \leq 0.4,\ \forall t \in [0,T]\right\}.$$

Then, we consider the cases of $n_p = 20, 30$ and 40, where n_p is the number of partitions points. For each of these cases, the problem is solved using the MISER software[104]. The optimization method used within the MISER software is the sequential quadratic programming (SQP). The optimal costs obtained with different numbers n_p of partition points are given in Table 8.8.1.

From the results obtained in Table 8.8.1, we see the convergence of the suboptimal costs obtained using the control parametrization method.

Table 8.8.1: Optimal costs for Example 8.8.1 with different n_p

n_p	Optimal cost
20	6.28983430
30	6.20898660
40	6.18595308

Note that this problem was solved in [13], where it was first discretized using Euler discretization scheme. Then, both inexact resporation (IR) method [113, 115, 182] and Ipopt optimization software [277] are used to solve this discretized problem. The optimal cost obtained with $N = 1500$ is 6.154193, where N denotes the number of partition points used in the Euler discretization scheme. Comparing this cost value with that obtained for the case of $n_p = 40$ using the control parametrization method, we see that their costs are sufficiently close. The approximate optimal controls and approximate optimal state trajectories for the case of $n_p = 40$ obtained using the control parametrization method are shown in Figure 8.8.1. Their trends are similar to those obtained in [13].

Example 8.8.2 Consider a tubular chemical reactor of length L and plug flow capacity v. We wish to carry out the following set of parallel reactions

$$\begin{aligned} A &\xrightarrow{k_1} B \\ A &\xrightarrow{k_2} C, \end{aligned}$$

where both reactions are irreversible. Assuming that the reactions are of first order and the velocity constants are given by

$$k_i = A_i \exp(-E_i/RT), \quad i = 1, 2,$$

where A_i, E_i, $i = 1, 2$ and R are fixed constants, the material balance equations along the tube $0 \leq z \leq L$ are

$$\begin{aligned} v\frac{dA}{dz} &= -(k_1 + k_2)A, \quad && A(0) = A_f, \\ v\frac{dB}{dz} &= k_1 A, && B(0) = 0, \end{aligned}$$

where A_f is also a given constant. Due to physical limitations, we need to specify practical bounds on the temperature:

$$0 \leq T(z) \leq \bar{T}, \quad 0 \leq z \leq L.$$

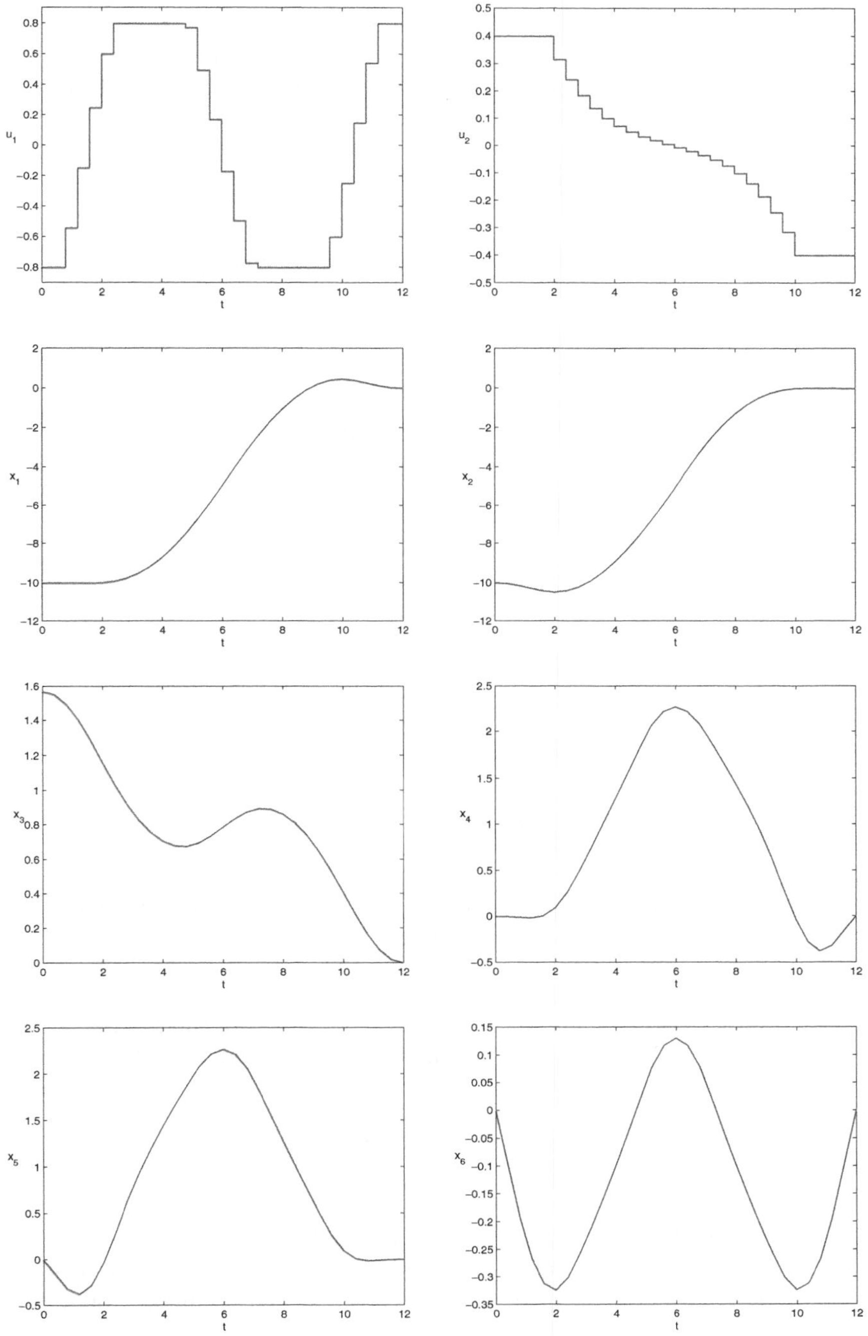

Fig. 8.8.1: Optimal control and optimal state trajectories for Example 8.8.1 with $n_p = 40$

To non-dimensionalize the problem, define $x_1 = A/A_f$, $x_2 = B/A_f$, $u = k_1L/v$, $y = z/L$, $p = E_2/E_1$, $\beta = LA_2/[v((L/v)A_1)^p]$, $\bar{u} = \left[\bar{k_1}L\right]/v$ and $\bar{k_1} = A_1 \exp\left(-E_1/R\bar{T}\right)$. It can be shown that the transformed problem is to minimize

$$g_0 = -x_2(1)$$

subject to the transformed state equations

$$\frac{dx_1(y)}{dy} = -(u(y) + \beta(u(y))^p)x_1(y),$$
$$\frac{dx_2(y)}{dy} = ux_1(y),$$

and

$$0 \leq u(y) \leq \bar{u}, \ 0 \leq y < 1.$$

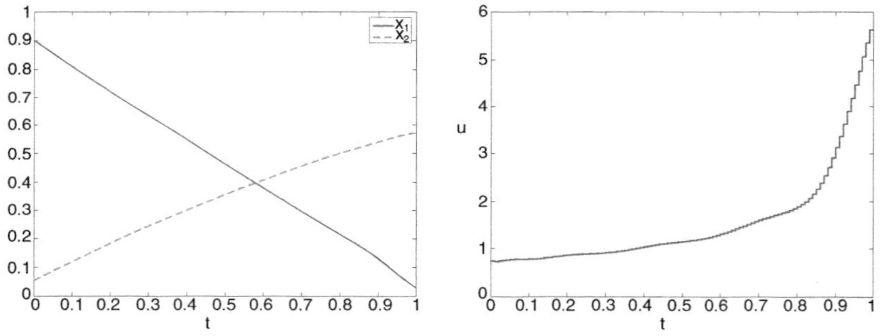

Fig. 8.8.2: Computed optimal solution for Example 8.8.2 with 100 subintervals

In practice, it is common to recycle a fraction γ of the product stream and mix it with fresh feed. This means the initial and final values of x_1 and x_2 are related by

$$x_1(0) = (1 - \gamma) + \gamma x_1(1)$$
$$x_2(0) = \gamma x_2(1).$$

We can deal with this aspect of the problem by putting $x_1(1) = \zeta_1$ and $x_2(1) = \zeta_2$ where ζ_1 and ζ_2 are system parameters, both of which are restricted to $[0, 1]$. Hence, we have two equality constraints as follows:

$$g_1 = x_1(1) - \zeta_1 = 0,$$
$$g_2 = x_2(1) - \zeta_2 = 0.$$

We solve this problem numerically for $\beta = 0.5$, $p = 2$, $\bar{u} = 6$ and $\gamma = 0.1$. We assume that the control is piecewise constant and choose the partition for u to contain $n_p = 100$ equally spaced intervals. The solution yields an optimal objective functional value of -0.57581 with $z_1 = 0.03066$ and $z_2 = 0.57581$. The corresponding optimal control and optimal state trajectories are shown in Figure 8.8.2.

Example 8.8.3 The design of optimal feedback controllers was considered for a general class of nonlinear systems in [252]. While the actual feedback control problem is defined over an infinite time horizon, an approximate problem is constructed over a finite time horizon as follows. Minimize

$$g_0 = \frac{1}{2} \int_0^{15} \left((x_1(t))^2 + 0.1(x_2(t))^2 + 0.2(u(t))^2 \right) dt$$

subject to

$$\begin{aligned} \frac{dx_1(t)}{dt} &= x_2(t), \; x_1(0) = -5, \\ \frac{dx_2(t)}{dt} &= -x_1(t) + \left(1.4 - 0.14(x_2(t))^2\right) x_2(t) + u(t), \; x_2(0) = -5. \end{aligned}$$

Instead of an open-loop optimal control, the aim is to construct state-dependent suboptimal feedback control of the form

$$u = \zeta_1 x_1 + \zeta_2 x_2 + \zeta_3 x_1^2 + \zeta_4 x_2^2.$$

Note the inclusion of the quadratic terms is in the hope of being able to address the nonlinearities in the state differential equations. Once the form of u has been substituted into the statement of the problem, we end up with a pure optimal parameter selection problem involving 4 system parameters. We choose upper and lower bounds of -10 and 10 for each of these 4 system parameters. The problem is solved using MISER Software [104]. The suboptimal feedback control obtained is

$$u = -2.9862x_1 - 2.3228x_2 - 0.2808(x_1)^2 + 0.0926(x_2)^2.$$

8.9 Control Parametrization Time Scaling Transform

In spite of the flexibility and efficiency of the control parametrization approach, there are several numerical difficulties associated with it. Consider the case in which the optimal control that we are seeking is a piecewise continuous function. Then, it has a finite number of discontinuity points. These discontinuity points are referred to as *switching times*. Clearly, the number of, as well as the locations of, these switching times are not known in advance. The accuracy of the classical control parametrization method thus depends greatly on the choice of knots distribution. The ideal knot distribution would be to have a knot placed exactly at the location of each switching time. However, before the optimal control problem is solved, we do not have any idea what the optimal control looks like. Thus, we have no insight of how the switching times are distributed. Thus, a usual practice is to choose a set of dense and evenly distributed knots in the hope that there would be a knot placed near each switching time. Hence, the number of parameters in the approximate optimal parameter selection problem would be very large if the approximate optimal control obtained is to be accurate. However, as the number of parameters increases, the optimization process quickly becomes much more expensive in terms of the computational time required. Intuitively, the control parametrization method would be more effective if the switching times could also be treated as decision variables to be optimized just like the control parameters. This could largely reduce the overall number of parameters used. However, there are a number of numerical difficulties associated with such a strategy as pointed out in the following remark.

Remark 8.9.1 *The gradient formulae of the cost functional and constraint functionals with respect to these switching times may not exist when two or more switching times coalesce (see Section 7.4 for details). Another difficulty is that the differential equations governing the dynamics of the problem would now be only piecewise continuous with the points of discontinuity (that is, the switching times) varying from one iteration to the next in the optimization process. Furthermore, the number of the decision variables would change when two or more switching times coalesce. Thus, the task of integrating the differential equations accurately can be very involved. For these reasons, the use of the gradient formulae presented in [89] and Theorem 7.4.3 is not as popular as those obtained based on time scaling transformation to be presented in the section.*

In this section, the time scaling transform introduced in Section 7.4.2 is used to enhance the classical control parametrization technique. This time scaling transform is earlier called the control parametrization enhancing transform (CPET). See, for example, [125, 126] and [215].

Consider the process described by the following system of nonlinear differential equations on the fixed time interval $(0, T]$:

$$\frac{d\boldsymbol{x}(t)}{dt} = \boldsymbol{f}(t, \boldsymbol{x}(t), \boldsymbol{u}(t)), \tag{8.9.1a}$$

where $\boldsymbol{x} = [x_1, \ldots, x_n]^\top \in \mathbb{R}^n$, $\boldsymbol{u} = [u_1, \ldots, u_r]^\top \in \mathbb{R}^r$ are, respectively, the state and control vectors; and $\boldsymbol{f} = [f_1, \ldots, f_n]^\top \in \mathbb{R}^n$. The initial condition for the system of differential equations (8.9.1a) is:

$$\boldsymbol{x}(0) = \boldsymbol{x}^0, \tag{8.9.1b}$$

where $\boldsymbol{x}^0$ is a given vector in $\mathbb{R}^n$. Let U_1, U_2 and U be as defined by (8.2.2a), (8.2.2b) and (8.2.3), respectively.

Definition 8.9.1 *A Borel measurable function* $\boldsymbol{u} = [u_1, \ldots, u_r]^\top$ *mapping from* $[0, T]$ *into* $\mathbb{R}^r$ *is said to be an* admissible control *if* $\boldsymbol{u}(t) \in U$ *for almost all* $t \in [0, T]$. *Let* $\mathcal{U}$ *be the class of all such admissible controls.*

For each $\boldsymbol{u} \in \mathcal{U}$, let $\boldsymbol{x}(\cdot \mid \boldsymbol{u})$ be the corresponding vector-valued function that is absolutely continuous on $(0, T]$ and satisfies the differential equations (8.9.1a) almost everywhere on $(0, T]$ and the initial condition (8.9.1b). This function is called *the solution* of system (8.9.1) corresponding to $\boldsymbol{u} \in \mathcal{U}$.

We may now state the canonical optimal control problem as follows:

Problem (P_2). Given system (8.9.1), find a control $\boldsymbol{u} \in \mathcal{U}$ such that the cost functional

$$g_0(\boldsymbol{u}) = \Phi_0(\boldsymbol{x}(T \mid \boldsymbol{u})) + \int_0^T \mathcal{L}_0(t, \boldsymbol{x}(t \mid \boldsymbol{u}), \boldsymbol{u}(t))dt \tag{8.9.2}$$

is minimized over $\mathcal{U}$ subject to the equality constraints

$$g_i(\boldsymbol{u}) = \Phi_i(\boldsymbol{x}(T \mid \boldsymbol{u})) + \int_0^T \mathcal{L}_i(t, \boldsymbol{x}(t \mid \boldsymbol{u}), \boldsymbol{u}(t))dt = 0, \ i = 1, \ldots, N_e, \tag{8.9.3}$$

and the inequality constraints

$$g_i(\boldsymbol{u}) = \Phi_i(\boldsymbol{x}(T \mid \boldsymbol{u})) + \int_0^T \mathcal{L}_i(t, \boldsymbol{x}(t \mid \boldsymbol{u}), \boldsymbol{u}(t))dt \leq 0, \ i = N_e + 1, \ldots, N, \tag{8.9.4}$$

where Φ_i, $i = 0, 1, \ldots, N$, and $\mathcal{L}_i$, $i = 0, 1, \ldots, N$, are given real-valued functions.

For the given functions $\boldsymbol{f}$, Φ_i, $i = 0, 1, \ldots, N$, and $\mathcal{L}_i$, $i = 0, 1, \ldots, N$, we assume that the following conditions are satisfied.

Assumption 8.9.1 *The relevant conditions of Assumption 8.2.1.*

Assumption 8.9.2 *The relevant conditions of Assumption 8.2.2.*

Assumption 8.9.3 *The relevant conditions of Assumption 8.2.3.*

8.9.1 Control Parametrization Time Scaling Transform

For each integer $p \geq 1$, let the planning horizon $[0, T]$ be partitioned into n_p subintervals with $n_p + 1$ partition points denoted by

$$\tau_0^p, \tau_1^p, \ldots, \tau_{n_p}^p,$$

where $\tau_0^p = 0$, $\tau_{n_p}^p = T$ and τ_k^p, $k = 1, \ldots, n_p$, are decision variables that are subject to the following conditions:

$$\tau_{k-1}^p \leq \tau_k^p, \; k = 1, 2, \ldots, n_p. \tag{8.9.5}$$

Let

$$\mathcal{S}^p = \left\{ \tau_0^p, \tau_1^p, \ldots, \tau_{n_p}^p \right\}. \tag{8.9.6}$$

For all $p > 1$, the partition points in $\mathcal{S}^p$ are chosen such that

$$\mathcal{S}^p \subset \mathcal{S}^{p+1}. \tag{8.9.7}$$

The control is now approximated in the form of piecewise constant function as

$$\boldsymbol{u}^p(t) = \sum_{k=1}^{n_p} \boldsymbol{\sigma}^{p,k} \chi_{[\tau_{k-1}^p, \tau_k^p)}(t), \tag{8.9.8}$$

where $\chi_{[\tau_{k-1}^p, \tau_k^p)}(\cdot)$ denotes the indicator function of the interval $[\tau_{k-1}^p, \tau_k^p)$ defined by (8.3.1b), and

$$\boldsymbol{\sigma}^{p,k} = \left[\sigma_1^{p,k}, \ldots, \sigma_r^{p,k} \right]^\top \tag{8.9.9a}$$

with

$$\boldsymbol{\sigma}^{p,k} \in U, \qquad k = 1, \ldots, n_p \tag{8.9.9b}$$

and the set U defined by (8.2.3). Let

$$\boldsymbol{\sigma}^p = \left[\left(\boldsymbol{\sigma}^{p,1}\right)^\top, \ldots, \left(\boldsymbol{\sigma}^{p,n_p}\right)^\top \right]^\top \tag{8.9.10}$$

and

$$\boldsymbol{\tau}^p = \left[\tau_1^p, \tau_2^p, \ldots, \tau_{n_p-1}^p \right]^\top, \tag{8.9.11}$$

where τ_i^p, $i = 1, 2, \ldots, n_p - 1$, are decision variables satisfying

$$0 = \tau_0^p \leq \tau_1^p \leq \tau_2^p \leq \cdots \leq \tau_{n_p}^p = T. \tag{8.9.12}$$

Definition 8.9.2 *Let Γ^p be the set of all switching vectors $\boldsymbol{\tau}^p$ defined by* (8.9.11) *such that* (8.9.12) *is satisfied.*

Definition 8.9.3 *Any piecewise constant control of the form of* (8.9.8) *satisfying the conditions* (8.9.9b) *and* (8.9.12) *is called an* admissible piecewise constant control. *Let $\widetilde{\mathcal{U}^p}$ be the set of all such admissible piecewise constant controls.*

Definition 8.9.4 *Let Ξ^p be the set of all control parameter vectors $\boldsymbol{\sigma}^p = \left[\left(\boldsymbol{\sigma}^{p,1}\right)^\top, \ldots, \left(\boldsymbol{\sigma}^{p,n_p}\right)^\top \right]^\top$, where $\boldsymbol{\sigma}^{p,k}$, $k = 1, \ldots, n_p$, are as defined by* (8.9.9).

Clearly, for each control $\boldsymbol{u}^p \in \widetilde{\mathcal{U}^p}$, there exists a unique combined control parameter vector and switching vector $(\boldsymbol{\sigma}^p, \boldsymbol{\tau}^p) \in \Xi^p \times \Gamma^p$ such that the relation (8.9.8) is satisfied. Conversely, there also exists a unique control $\boldsymbol{u}^p \in \widetilde{\mathcal{U}^p}$ corresponding to each combined control parameter vector and switching vector $(\boldsymbol{\sigma}^p, \boldsymbol{\tau}^p) \in \Xi^p \times \Gamma^p$.

Restricting the controls to $\widetilde{\mathcal{U}^p}$, system (8.9.1) takes the form

$$\frac{d\boldsymbol{x}(t)}{dt} = \widetilde{\boldsymbol{f}}(t, \boldsymbol{x}(t), \boldsymbol{\sigma}^p, \boldsymbol{\tau}^p) \tag{8.9.13a}$$

$$\boldsymbol{x}(0) = \boldsymbol{x}^0, \tag{8.9.13b}$$

where

$$\widetilde{\boldsymbol{f}}(t, \boldsymbol{x}(t), \boldsymbol{\sigma}^p, \boldsymbol{\tau}^p) = \boldsymbol{f}\left(t, \boldsymbol{x}(t), \sum_{k=1}^{n_p} \boldsymbol{\sigma}^{p,k} \chi_{[\tau_{k-1}^p, \tau_k^p)}(t)\right). \tag{8.9.13c}$$

Let $\boldsymbol{x}(\cdot \mid \boldsymbol{\sigma}^p, \boldsymbol{\tau}^p)$ be the solution of system (8.9.13) corresponding to the combined control parameter vector and the switching vector $(\boldsymbol{\sigma}^p, \boldsymbol{\tau}^p) \in \Xi^p \times \Gamma^p$.

Similarly, by restricting $\boldsymbol{u} \in \widetilde{\mathcal{U}^p}$, the constraints (8.9.3) and (8.9.4) are reduced, respectively, to

$$\widetilde{G}_i(\boldsymbol{\sigma}^p, \boldsymbol{\tau}^p) = \Phi_i(\boldsymbol{x}(T \mid \boldsymbol{\sigma}^p, \boldsymbol{\tau}^p)) + \int_0^T \widetilde{\mathcal{L}}_i(t, \boldsymbol{x}(t \mid \boldsymbol{\sigma}^p, \boldsymbol{\tau}^p), \boldsymbol{\sigma}^p, \boldsymbol{\tau}^p) dt = 0, \quad i = 1, \ldots, N_e, \tag{8.9.14a}$$

and

$$\widetilde{G}_i(\boldsymbol{\sigma}^p, \boldsymbol{\tau}^p) = \Phi_i(\boldsymbol{x}(T \mid \boldsymbol{\sigma}^p, \boldsymbol{\tau}^p)) + \int_0^T \widetilde{\mathcal{L}}_i(t, \boldsymbol{x}(t \mid \boldsymbol{\sigma}^p, \boldsymbol{\tau}^p), \boldsymbol{\sigma}^p, \boldsymbol{\tau}^p) dt \leq 0, \quad i = N_e + 1, \ldots, N, \tag{8.9.14b}$$

where, for $i = 1, \ldots, N$,

$$\widetilde{\mathcal{L}}_i(t, \boldsymbol{x}(t \mid \boldsymbol{\sigma}^p, \boldsymbol{\tau}^p), \boldsymbol{\sigma}^p, \boldsymbol{\tau}^p) = \mathcal{L}_i\left(t, \boldsymbol{x}(t), \sum_{k=1}^{n_p} \boldsymbol{\sigma}^{p,k} \chi_{[\tau_{k-1}^p, \tau_k^p)}(t)\right). \tag{8.9.15}$$

Definition 8.9.5 *Let $\widetilde{\Omega^p}$ be the subset of $\Xi^p \times \Gamma^p$ such that the constraints* (8.9.14) *are satisfied. Furthermore, let $\widetilde{\mathcal{F}^p}$ be the corresponding subset of $\widetilde{\mathcal{U}^p}$ uniquely defined by elements from $\widetilde{\Omega^p}$ via* (8.9.8).

We may now specify an approximate problem corresponding to Problem (P_2).

Problem $(P_2(p))$. Subject to the dynamical system (8.9.13), find a combined control parameter vector and switching vector $(\boldsymbol{\sigma}^p, \boldsymbol{\tau}^p) \in \widetilde{\Omega^p}$ such that the cost functional

$$\widetilde{G}_0(\boldsymbol{\sigma}^p, \boldsymbol{\tau}^p) = \Phi_0(\boldsymbol{x}(T \mid \boldsymbol{\sigma}^p, \boldsymbol{\tau}^p)) + \int_0^T \widetilde{\mathcal{L}}_0(t, \boldsymbol{x}(t \mid \boldsymbol{\sigma}^p, \boldsymbol{\tau}^p), \boldsymbol{\sigma}^p, \boldsymbol{\tau}^p)dt \tag{8.9.16}$$

is minimized over $\widetilde{\Omega^p}$, where $\widetilde{\mathcal{L}}_0$ is defined by (8.9.15) for $i = 0$.

Note that for each p, Problem $(P_2(p))$ is an optimal parameter selection problem. The gradient formulae of the cost functional and the constraint functionals with respect to the switching vector can be obtained readily as those given in either (7.4.15) or (7.4.18). However, there are deficiencies associated with the algorithms developed based on these gradient formulae as mentioned in Remarks 7.4.3 and 8.9.1. Thus, there is a need for a more effective approach for dealing with the problem $(P_2(p))$. This is the main motivation behind the control parametrization time scaling transform introduced below.

Consider the new time scale $s \in [0, 1]$. We wish to construct a transformation from $t \in [0, T]$ to $s \in [0, 1]$ that maps the variable knots

$$\tau_1^p, \tau_2^p, \ldots, \tau_{n_p-1}^p$$

to the fixed knots

$$\xi_k^p = \frac{k}{n_p}, \qquad k = 1, \ldots, n_p - 1. \tag{8.9.17a}$$

Clearly, these fixed knots satisfy

$$0 = \xi_0^p < \xi_1^p < \xi_2^p < \cdots < \xi_{n_p-1}^p < \xi_{n_p}^p = 1. \tag{8.9.17b}$$

The required transformation from $t \in [0, T]$ to $s \in [0, 1]$ can be defined by the following differential equation.

$$\frac{dt(s)}{ds} = v^p(s) \tag{8.9.18a}$$

with the initial condition

$$t(0) = 0. \tag{8.9.18b}$$

Definition 8.9.6 *A scalar function $v^p(s) \geq 0$ for all $s \in [0, 1]$ is called a* time scaling control *if it is a piecewise non-negative constant function with possible discontinuities at the fixed knots $\xi_0^p, \xi_1^p, \ldots, \xi_{n_p}^p$, that is,*

$$v^p(s) = \sum_{k=1}^{n_p} \theta_k^p \chi_{[\xi_{k-1}^p, \xi_k^p)}(s), \tag{8.9.19}$$

where $\theta_k^p \geq 0$, $k = 1, \ldots, n_p$, are decision variables, and $\chi_{[\xi_{k-1}^p, \xi_k^p)}(\cdot)$ is the indicator function on the interval $[\xi_{k-1}^p, \xi_k^p)$ defined by (8.3.1b). *Clearly, $v^p(s)$ depends on the choice of $\boldsymbol{\theta} = \left[\theta_1^p, \ldots, \theta_{n_p}^p\right]^\top$.*

Definition 8.9.7 *Let Θ^p be the set containing all those $\boldsymbol{\theta}^p = \left[\theta_1^p, \ldots, \theta_{n_p}^p\right]^\top$ with $\theta_i^p \geq 0$, $i = 1, \ldots, n_p$. Furthermore, let $\mathcal{V}^p$ be the set containing all the corresponding time scaling controls obtained by elements from Θ^p via* (8.9.19).

Clearly, each $\boldsymbol{\theta}^p \in \Theta^p$ uniquely defines a $v^p \in \mathcal{V}^p$, and vice versa. By (8.9.19), it is clear from (8.9.18) that

$$t^p(s) = \int_0^s v^p(\tau)d\tau = \sum_{j=1}^{k-1} \theta_j^p \left(\xi_j^p - \xi_{j-1}^p\right) + \theta_k^p \left(s - \xi_{k-1}^p\right), \quad s \in \left[\xi_{k-1}^p, \xi_k^p\right]. \tag{8.9.20}$$

In particular,

$$t^p(1) = \int_0^1 v^p(\tau)d\tau = \sum_{k=1}^{n_p} \theta_k^p \left(\xi_k^p - \xi_{k-1}^p\right). \tag{8.9.21}$$

Define

$$\boldsymbol{\omega}^p(s) = \boldsymbol{u}^p(t(s)), \qquad \hat{\boldsymbol{x}}(s) = \left[(\boldsymbol{x}(s))^\top, t(s)\right]^\top, \tag{8.9.22}$$

where, by abusing the notation, we write

$$\boldsymbol{x}(s) = \boldsymbol{x}(t(s)).$$

Clearly, the piecewise constant control $\boldsymbol{\omega}^p$ can be written as

$$\boldsymbol{\omega}^p(s) = \sum_{k=1}^{n_p} \boldsymbol{\sigma}^{p,k} \chi_{[\xi_{k-1}^p, \xi_k^p)}(s), \qquad s \in [0, 1], \tag{8.9.23}$$

where ξ_k^p, $k = 0, 1, \ldots, n_p$, are fixed knots chosen according to (8.9.17), $\chi_{[\xi_{k-1}^p, \xi_k^p)}(\cdot)$ denotes the indicator function of the interval $\left[\xi_{k-1}^p, \xi_k^p\right)$ defined by (8.3.1b) and $\boldsymbol{\sigma}^p$ as defined for $\boldsymbol{u}^p$ given in (8.9.8) is defined by (8.9.10).

Definition 8.9.8 *Let $\widehat{\mathcal{U}^p}$ be the set containing all piecewise constant controls given by* (8.9.23) *with $\boldsymbol{\sigma}^p \in \Xi^p$.*

Clearly, each $(\boldsymbol{\sigma}^p, \boldsymbol{\theta}^p) \in \Xi^p \times \Theta^p$ defines uniquely a $(\boldsymbol{\omega}^p, v^p) \in \widehat{\mathcal{U}^p} \times \mathcal{V}^p$, and vice versa.

Definition 8.9.9 *Let $\widehat{\Omega^p}$ be the set, which consists of all those elements from $\Xi^p \times \Theta^p$ such that the following constraints are satisfied.*

$$\widehat{G}_i(\boldsymbol{\sigma}^p, \boldsymbol{\theta}^p) = \Phi_i(\boldsymbol{x}(1 \mid \boldsymbol{\sigma}^p, \boldsymbol{\theta}^p)) \quad + \int_0^1 \widehat{\mathcal{L}}_i(\hat{\boldsymbol{x}}(s \mid \boldsymbol{\sigma}^p, \boldsymbol{\theta}^p), \boldsymbol{\sigma}^p, \boldsymbol{\theta}^p)ds = 0,$$
$$i = 1, \ldots, N_e, \tag{8.9.24a}$$

$$\widehat{G}_i(\boldsymbol{\sigma}^p, \boldsymbol{\theta}^p) = \Phi_i(\boldsymbol{x}(1 \mid \boldsymbol{\sigma}^p, \boldsymbol{\theta}^p)) \quad + \int_0^1 \widehat{\mathcal{L}}_i(\hat{\boldsymbol{x}}(s \mid \boldsymbol{\sigma}^p, \boldsymbol{\theta}^p), \boldsymbol{\sigma}^p, \boldsymbol{\theta}^p) ds \leq 0,$$

$$i = N_e + 1, \ldots, N, \tag{8.9.24b}$$

$$\sum_{k=1}^{n_p} \theta_k^p \left(\xi_k^p - \xi_{k-1}^p\right) = T, \tag{8.9.24c}$$

where

$$\widehat{\mathcal{L}}_i(\hat{\boldsymbol{x}}(s), \boldsymbol{\sigma}^p, \boldsymbol{\theta}^p) = v^p(s) \mathcal{L}_i\left(t(s), \boldsymbol{x}(s), \sum_{k=1}^{n_p} \boldsymbol{\sigma}_k^p \chi_{[\xi_{k-1}^p, \xi_k^p)}(s)\right). \tag{8.9.25}$$

Definition 8.9.10 *Let $\mathcal{A}^p$ be the subset of $\widehat{\mathcal{U}^p} \times \mathcal{V}^p$ uniquely defined by elements from $\widehat{\Omega^p}$.*

Clearly, each $(\boldsymbol{\omega}^p, v^p) \in \mathcal{A}^p$ can be written as

$$\boldsymbol{\omega}^p(s) = \sum_{k=1}^{n_p} \boldsymbol{\sigma}^{p,k} \chi_{[\xi_{k-1}^p, \xi_k^p)}(s), \qquad s \in [0, 1], \tag{8.9.26a}$$

$$v^p(s) = \sum_{i=1}^{n_p} \theta_i^p \chi_{[\xi_{i-1}^p, \xi_i^p)}(s), \qquad s \in [0, 1], \tag{8.9.26b}$$

with $(\boldsymbol{\sigma}^p, \boldsymbol{\theta}^p) \in \widehat{\Omega^p}$.

The equivalent transformed optimal parameter selection problem may now be stated as follows.

Problem $(P_3(p))$. Subject to the system of differential equations

$$\frac{d\hat{\boldsymbol{x}}(s)}{ds} = \hat{\boldsymbol{f}}(\hat{\boldsymbol{x}}(s), \boldsymbol{\sigma}^p, \boldsymbol{\theta}^p), \tag{8.9.27a}$$

where

$$\hat{\boldsymbol{f}}(\hat{\boldsymbol{x}}(s), \boldsymbol{\sigma}^p, \boldsymbol{\theta}^p) = v^p(s) \boldsymbol{f}\left(t(s), \boldsymbol{x}(s), \sum_{k=1}^{n_p} \boldsymbol{\sigma}_k^p \chi_{[\xi_{k-1}^p, \xi_k^p)}(s)\right) \tag{8.9.27b}$$

with initial condition

$$\hat{\boldsymbol{x}}(0) = \begin{bmatrix} \boldsymbol{x}^0 \\ 0 \end{bmatrix}, \tag{8.9.27c}$$

find a combined control parameter vector and switching vector $(\boldsymbol{\sigma}^p, \boldsymbol{\theta}^p) \in \widehat{\Omega^p}$ such that the cost functional

$$\widehat{G}_0(\boldsymbol{\sigma}^p, \boldsymbol{\theta}^p) = \Phi_0(\hat{\boldsymbol{x}}(1 \mid \boldsymbol{\sigma}^p, \boldsymbol{\theta}^p)) + \int_0^1 \widehat{\mathcal{L}}_0(\hat{\boldsymbol{x}}(s \mid \boldsymbol{\sigma}^p, \boldsymbol{\theta}^p), \boldsymbol{\sigma}^p, \boldsymbol{\theta}^p) ds \quad (8.9.28)$$

is minimized over $\widehat{\Omega^p}$.

Remark 8.9.2 *Note that in the transformed problem $(P_3(p))$, only the knots contribute to the discontinuities of the state differential equation. Thus, all locations of the discontinuities of the state differential equation are known and fixed during the optimization process. These locations will not change from one iteration to the next during the optimization process. Even when two or more of the original switching times coalesce, the number of these locations remains unchanged in the transformed problem. Furthermore, the gradient formulae of the cost function and constraint functions with respect to the original switching times in the new transformed problem are provided by the usual gradient formulae for the classical optimal parameter selection problem as given in Section 7.2.*

The basic idea behind the control parametrization time scaling transform is aiming to include the switching times as parameters to be optimized and at the same time, to avoid the numerical difficulties mentioned in Remark 8.9.1. The time scaling control captures the discontinuities of the optimal control if the number of knots in the partition of the new time horizon is greater than or equal to the number of discontinuities of the optimal control. Since the time scaling control parameters θ_k^p, $k = 1, \ldots, n_p$, are allowed to vary, the control parametrization time scaling transform technique gives rise to a larger search space and hence produces a better or at least equal approximate optimal cost. Clearly, if the optimal control is a piecewise constant function with discontinuities at $\bar{t}_1, \ldots, \bar{t}_M$, then, by solving the transformed problems with the number of knots greater or equal to M, and by using (8.9.23), we obtain the exact optimal control. For the general case, the convergence results are much harder to establish.

8.9.2 Convergence Analysis

Let us provide similar convergence results as those presented in Section 8.8. Similar to Definition 8.5.1, we need the following definitions.

Definition 8.9.11 *A combined vector $(\boldsymbol{\sigma}^p, \boldsymbol{\theta}^p) \in \Xi^p \times \Theta^p$ is said to be ε-tolerated feasible if it satisfies the following ε-tolerated constraints:*

$$-\varepsilon \leq \widehat{G}_i(\boldsymbol{\sigma}^p, \boldsymbol{\theta}^p) \leq \varepsilon, \qquad i = 1, \ldots, N_e \quad (8.9.29a)$$

$$\widehat{G}_i(\boldsymbol{\sigma}^p, \boldsymbol{\theta}^p) \leq \varepsilon, \qquad i = N_e + 1, \ldots, N. \quad (8.9.29b)$$

Definition 8.9.12 *Let $\widehat{\Omega^{p,\varepsilon}}$ be the subset of $\Xi^p \times \Theta^p$ such that the ε-tolerated constraints (8.9.29) are satisfied. Furthermore, let $\mathcal{A}^{p,\varepsilon}$ be the corresponding* *subset of* $\widehat{\mathcal{U}^p} \times \mathcal{V}^p$ *defined uniquely by elements from* $\widehat{\Omega^{p,\varepsilon}}$. *Clearly,* $\widehat{\Omega^p} \subset \widehat{\Omega^{p,\varepsilon}}$ *(and hence* $\mathcal{A}^p \subset \mathcal{A}^{p,\varepsilon}$ *) for any* $\varepsilon > 0$.

We now consider the following ε-tolerated version of the approximate problem $(P_{3,\varepsilon}(p))$.

Problem $(P_{3,\varepsilon}(p))$. Find a combined vector $(\boldsymbol{\sigma}^p, \boldsymbol{\theta}^p) \in \widehat{\Omega^{p,\varepsilon}}$ such that the cost functional (8.9.28) is minimized over $\widehat{\Omega^{p,\varepsilon}}$.

Since $\widehat{\Omega^p} \subset \widehat{\Omega^{p,\varepsilon}}$ for any $\varepsilon > 0$, it follows that

$$\widehat{G}_0(\boldsymbol{\sigma}^{p,\varepsilon,*}, \boldsymbol{\theta}^{p,\varepsilon,*}) \leq \widehat{G}_0(\boldsymbol{\sigma}^{p,*}, \boldsymbol{\theta}^{p,*}) \tag{8.9.30}$$

for any $\varepsilon > 0$, where $(\boldsymbol{\sigma}^{p,\varepsilon,*}, \boldsymbol{\theta}^{p,\varepsilon,*})$ and $(\boldsymbol{\sigma}^{p,*}, \boldsymbol{\theta}^{p,*})$ are optimal vectors to Problem $(P_{3,\varepsilon}(p))$ and Problem $(P_3(p))$, respectively.

Let the following additional condition be satisfied.

Assumption 8.9.4 *There exists an integer* p_0 *such that*

$$\lim_{\varepsilon \to 0} \widehat{G}_0(\boldsymbol{\sigma}^{p,\varepsilon,*}, \boldsymbol{\theta}^{p,\varepsilon,*}) = \widehat{G}_0(\boldsymbol{\sigma}^{p,*}, \boldsymbol{\theta}^{p,*}) \tag{8.9.31}$$

uniformly with respect to $p \geq p_0$.

Remark 8.9.3 *Note that each* $(\boldsymbol{\sigma}^p, \boldsymbol{\theta}^p) \in \widehat{\Omega^p}$ *uniquely defines a* $(\boldsymbol{\omega}^p, v^p) \in \mathcal{A}^p$ *via* (8.9.19) *and* (8.9.23), *and vice versa. We further note that for each* $(\boldsymbol{\omega}^p, v^p) \in \mathcal{A}^p$ *there exists a unique* $(\boldsymbol{u}^p, \boldsymbol{\tau}^p)$ *defined in the original time horizon* $[0, T]$ *such that*

$$\boldsymbol{u}^p(t) = \sum_{k=1}^{n_p} \boldsymbol{\sigma}^{p,k} \chi_{[\tau_{k-1}^p, \tau_k^p)}(t), \tag{8.9.32}$$

where $\boldsymbol{\sigma}^p = \left[\left(\boldsymbol{\sigma}^{p,1}\right)^\top, \ldots, \left(\boldsymbol{\sigma}^{p,n_p}\right)^\top \right]^\top$ *is the same as for* $\boldsymbol{\omega}^p$ *defined by* (8.9.23), *and* $\boldsymbol{\tau}^p = \left[\tau_1^p, \ldots, \tau_{n_p}^p\right]^\top$ *is determined uniquely by* $\boldsymbol{\theta}^p$ *via evaluating* (8.9.18) *at* $s = k/n_p$, $k = 1, \ldots, n_p$.

The functions $\boldsymbol{\omega}^{p,*}$ and $v^{p,*}$ are piecewise constant functions with possible discontinuities at $s = \frac{k}{n_p}$, $k = 1, 2, \ldots, n_p - 1$. We choose n_p such that

$$n_{p+1} > n_p \text{ for } p \geq 1.$$

In practice, we first choose an integer n_1 corresponding to $p = 1$. Then, by (8.9.18) with $v(s) = v^{1,*}(s)$, we obtain the corresponding t as a function of s. Thus, by (8.9.23), (8.9.22), (8.9.18) and (8.9.17a), we obtain $\boldsymbol{\omega}^{1,*}(s)$ and hence $\boldsymbol{u}^{1,*}(t)$, where the switching times τ_k^1, $k = 1, \ldots, n_1$, are given by

$$\tau_k^1 = t\left(\frac{k}{n_1}\right) = \int_0^{k/n_1} v^{1,*}(s)ds, \; k = 1, 2, \ldots, n_1. \tag{8.9.33}$$

Define

$$\mathcal{S}^1 = \left\{ \frac{k}{n_1} \ : \ k = 1, 2, \ldots, n_1 - 1 \right\}. \tag{8.9.34}$$

Then, we choose an integer n_2 such that $n_2 > n_1$, and let

$$\mathcal{S}^2 = \left\{ \frac{k}{n_2} \ : \ k = 1, 2, \ldots, n_2 - 1 \right\}. \tag{8.9.35}$$

This process is continued in such a way that the following condition is satisfied.

Assumption 8.9.5 *$\mathcal{S}^{p+1} \supset \mathcal{S}^p$ and $\lim_{p \to \infty} \mathcal{S}^p$ is dense in $[0, 1]$.*

The procedure for solving Problem (P_2) may be stated as follows. For each $p \geq 1$, we use the control parametrization time scaling transform technique to obtain Problem $(P_3(p))$. In what follows, we present a computational procedure to solve Problem $(P_3(p))$, giving an approximate optimal solution of Problem (P_2).

Algorithm 8.9.1

Step 1. Solve Problem $(P_3(p))$ as a standard optimal parameter selection problem by using a computational procedure similar to that described in Section 8.6. Let the optimal control vector obtained be denoted by $(\boldsymbol{\sigma}^{p,*}, \boldsymbol{\theta}^{p,*})$. Then, by Remark 8.9.3, we obtain the corresponding piecewise constant control $(\boldsymbol{\omega}^{p,*}, \nu^{p,*})$.

Step 2. If $n_p \geq M$, where M is a pre-specified positive constant, go to Step 3. Otherwise go to Step 1 with n_p increased to n_{p+1}.

Step 3. Stop. Construct $\boldsymbol{\tau}^{p,*} = \left[\tau_1^{p,*}, \ldots, \tau_{n_p}^{p,*}\right]^\top$ from $\boldsymbol{\theta}^{p,*}$. Then, obtain

$$\boldsymbol{u}^{p,*}(t) = \sum_{k=1}^{n_p} \boldsymbol{\sigma}^{p,k,*} \chi_{[\tau_{k-1}^{p,*}, \tau_k^{p,*})}(t), \tag{8.9.36}$$

where $\boldsymbol{\sigma}^{p,*} = \left[\left(\boldsymbol{\sigma}^{p,1,*}\right)^\top, \ldots, \left(\boldsymbol{\sigma}^{p,n_p,*}\right)^\top\right]^\top$. The piecewise constant control $\boldsymbol{u}^{p,*}$ obtained is an approximate optimal solution of Problem (P).

We are now in a position to present the convergence results in the next two theorems.

Theorem 8.9.1 *Let $(\boldsymbol{\sigma}^{p,*}, \boldsymbol{\theta}^{p,*})$ be an optimal parameter vector of Problem $(P_3(p))$, and let $(\boldsymbol{u}^{p,*}, \boldsymbol{\tau}^{p,*})$ be the corresponding piecewise constant optimal control and switching vector of Problem $(P_2(p))$ such that*

$$\boldsymbol{u}^{p,*}(t) = \sum_{k=1}^{n_p} \boldsymbol{\sigma}^{p,k,*} \chi_{[\tau_{k-1}^{p,*}, \tau_k^{p,*})}(t), \tag{8.9.37}$$

where $\boldsymbol{\sigma}^{p,*} = \left[\left(\boldsymbol{\sigma}^{p,1,*}\right)^\top, \ldots, \left(\boldsymbol{\sigma}^{p,n_p,*}\right)^\top\right]^\top$, *and* $\boldsymbol{\tau}^{p,*} = \left[\tau_1^{p,*}, \ldots, \tau_{n_p}^{p,*}\right]^\top$. *Suppose that Problem (P) has an optimal control* $\boldsymbol{u}^*$. *Then*

$$\lim_{p\to\infty} g_0(\boldsymbol{u}^{p,*}) = g_0(\boldsymbol{u}^*). \tag{8.9.38}$$

Proof. Let $(\boldsymbol{\omega}^{p,*}, \nu^{p,*})$ be determined uniquely by $(\boldsymbol{\sigma}^{p,*}, \boldsymbol{\theta}^{p,*})$ via (8.9.23) and (8.9.19). More precisely, solve (8.9.18) to obtain $t^*(s)$, $s \in [0,1]$. Then, by evaluating $t^*(s)$ at $s = k/n_p$, $k = 0, 1, \ldots, n_p$, we obtain $\boldsymbol{\tau}^{p,*} = \left[\tau_1^{p,*}, \ldots, \tau_{n_p-1}^{p,*}\right]^\top$. Now, by (8.9.22), $\boldsymbol{u}^{p,*}$ can be written as

$$\boldsymbol{u}^{p,*}(t) = \sum_{k=1}^{n_p} \boldsymbol{\sigma}^{p,k,*} \chi_{[\tau_{k-1}^{p,*}, \tau_k^{p,*})}(t), \tag{8.9.39}$$

where $\boldsymbol{\sigma}^{p,*} = \left[\left(\boldsymbol{\sigma}^{p,1,*}\right)^\top, \ldots, \left(\boldsymbol{\sigma}^{p,n_p,*}\right)^\top\right]^\top$, $\tau_0^{p,*} = 0$, $\tau_{n_p}^{p,*} = T$ and $\boldsymbol{\tau}^{p,*} = \left[\tau_1^{p,*}, \ldots, \tau_{n_p-1}^{p,*}\right]^\top$.

Consider Problem $(P_3(p))$ but with the switching vector $\boldsymbol{\tau}^p = \left[\tau_1^p, \ldots, \tau_{n_p-1}^p\right]^\top$ taken as fixed. Let this problem be referred to as Problem $(P_4(p))$. Clearly, Problem $(P_4(p))$ can be considered as the type obtained in Section 8.3 through the application of the control parametrization technique. Let $\bar{\boldsymbol{u}}^{p,*}$ be an optimal control of Problem $(P_4(p))$.

Note that the switching vector in Problem $(P_4(p))$ is taken as fixed, while Problem $(P_3(p))$ treats the switching vector as part of the decision vector to be optimized over. Thus, it is clear that

$$g_0(\boldsymbol{u}^*) \le g_0(\boldsymbol{u}^{p,*}) \le g_0(\bar{\boldsymbol{u}}^{p,*}) \tag{8.9.40}$$

for all $p \ge 1$. Now, by Theorem 8.5.1, it follows that for any $\delta > 0$, there exists a positive integer $\bar{p}$ such that

$$g_0(\boldsymbol{u}^*) \le g_0(\bar{\boldsymbol{u}}^{p,*}) < g_0(\boldsymbol{u}^*) - \delta, \tag{8.9.41}$$

for all $p \ge \bar{p}$. Combining (8.9.40) and (8.9.41), we obtain that, for all $p \ge \bar{p}$,

$$g_0(\boldsymbol{u}^*) \le g_0(\boldsymbol{u}^{p,*}) \le g_0(\boldsymbol{u}^*) - \delta. \tag{8.9.42}$$

Taking the limit as $p \to \infty$ and noting that $\delta > 0$ is arbitrary, the conclusion of the theorem follows readily.

Theorem 8.9.2 *Let* $(\boldsymbol{\sigma}^{p,*}, \boldsymbol{\theta}^{p,*})$ *and* $(\boldsymbol{u}^{p,*}, \boldsymbol{\tau}^{p,*})$ *be as defined in Theorem 8.9.1. Suppose that*

$$\lim_{p\to\infty} \boldsymbol{u}^{p,*}(t) = \hat{\boldsymbol{u}}(t), \text{ almost everywhere in } [0,T]. \tag{8.9.43}$$

Then, $\hat{\boldsymbol{u}}$ is also an optimal control of Problem (P_2).

Proof. Since $\boldsymbol{u}^{p,*} \to \hat{\boldsymbol{u}}$ almost everywhere in $[0, T]$, it follows from Lemma 8.4.4 that

$$\lim_{p\to\infty} g_0(\boldsymbol{u}^{p,*}) = g_0(\hat{\boldsymbol{u}}). \tag{8.9.44}$$

Next, it is easy to verify from Remark 8.4.1, Lemma 8.4.3, Assumptions 8.9.2 and 8.9.3 that $\hat{\boldsymbol{u}}$ is also a feasible control of Problem (P_2). On the other hand, it follows from Theorem 8.9.1 that

$$\lim_{p\to\infty} g_0(\boldsymbol{u}^{p,*}) = g_0(\boldsymbol{u}^*). \tag{8.9.45}$$

Hence, the conclusion of the theorem follows easily from (8.9.44) and (8.9.45).

8.10 Examples

Example 8.10.1 Considers the free flying robot (FFR) problem as described in Example 8.8.1. We shall solve the problem again for the cases of $n_p = 20, 30$ and 40, where n_p is the number of partitions points. For each of these cases, the problem is solved using the MISER Software [104]. The optimal costs obtained with different numbers n_p of partition points are given in Table 8.10.1.

Table 8.10.1: Approximate optimal costs for Example 8.10.1 with different n_p and solved using time scaling

n_p	Optimal cost with time scaling
20	6.22527918
30	6.19719836
40	6.18444730

From the results listed in Table 8.10.1, we see the convergence of the approximate optimal costs. The solutions obtained with the time scaling transform are better than those obtained without time scaling transform (see Table 8.8.1). Figure 8.10.1 shows the plots of the approximate optimal controls and approximate optimal state trajectories obtained for the case of $n_p = 40$. Their trends are similar to those obtained in [13] for $N = 1000$, where N denotes the number of partition points used in the Euler discretization scheme. Furthermore, the cost of 6.18444730 obtained for the case of $n_p = 40$ is close to the cost of 6.154193 obtained in [13] for $N = 1000$.

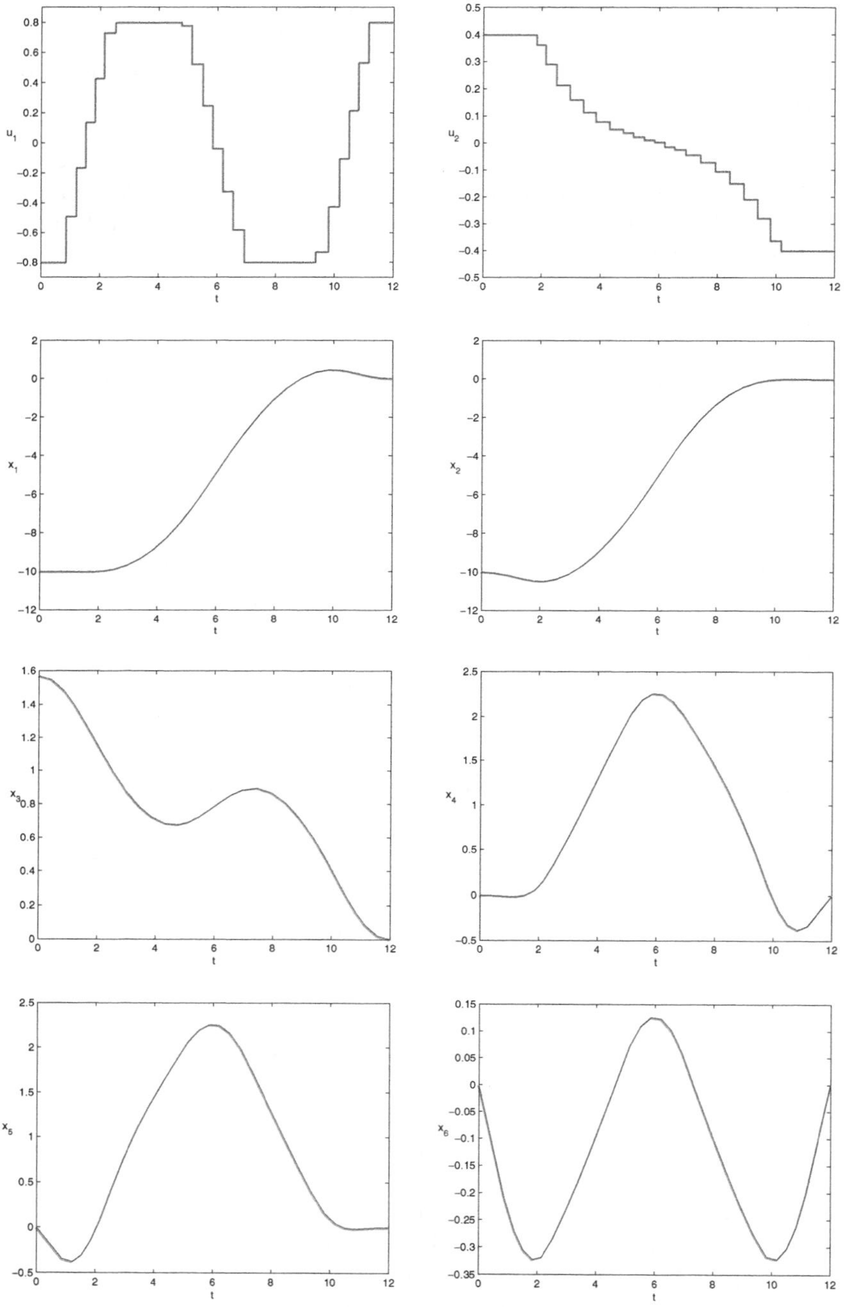

Fig. 8.10.1: Optimal control and optimal state trajectories for Example 8.10.1 with $n_p = 40$ and solved using time scaling

Example 8.10.2 Consider the tubular chemical reactor problem as described in Example 8.8.2. We solve the problem again using the control parametrization method with time scaling transformation being utilized. The number of partition points of the control function is $n_p = 20$. For this problem, it is solved using the MISER software [104] with time scaling transform being applied. The successful termination of the optimization software indicates that the solution obtained is such that the KKT conditions are satisfied. The optimal cost obtained is $-5.75817152 \times 10^{-1}$. The optimal cost of $-5.75805641 \times 10^{-1}$ is obtained in Example 8.8.2 without the use of the time scaling transform for which $n_p = 100$. Figure 8.10.2 shows the plots of the approximate optimal control and the optimal state trajectories.

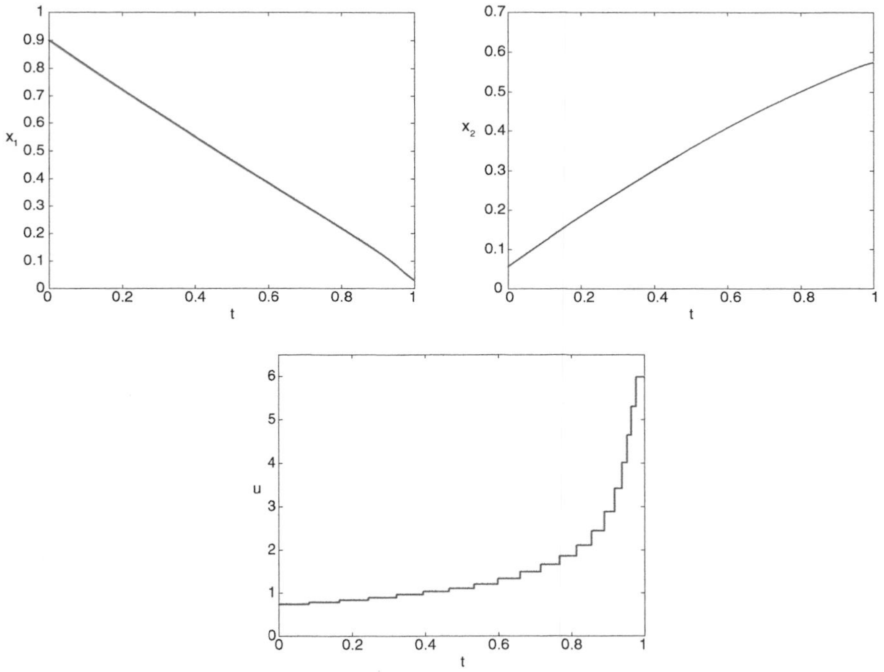

Fig. 8.10.2: Optimal state trajectories and control for Example 8.10.2 with $n_p = 20$ and solved using time scaling

Example 8.10.3 In Example 8.10.3, the control is in the form of a feedback control. In this example, we consider the same optimal control problem but the control is now an open-loop control subject to the following boundedness constraints:

$$\underline{u} \leq u(t) \leq \overline{u}, \quad \forall t \in [0, 15],$$

where $\underline{u} = -0.5$ and $\overline{u} = 0.5$. The control is parametrized as a piecewise constant function with the time horizon $[0, 15]$ being subdivided into n_p subintervals. We solve the problem for the cases of $n_p = 10$ and 20. The respective costs obtained are 80.2451689 and 79.7608063. Figure 8.10.3 shows the plots of the approximate optimal control and the approximate optimal state trajectories.

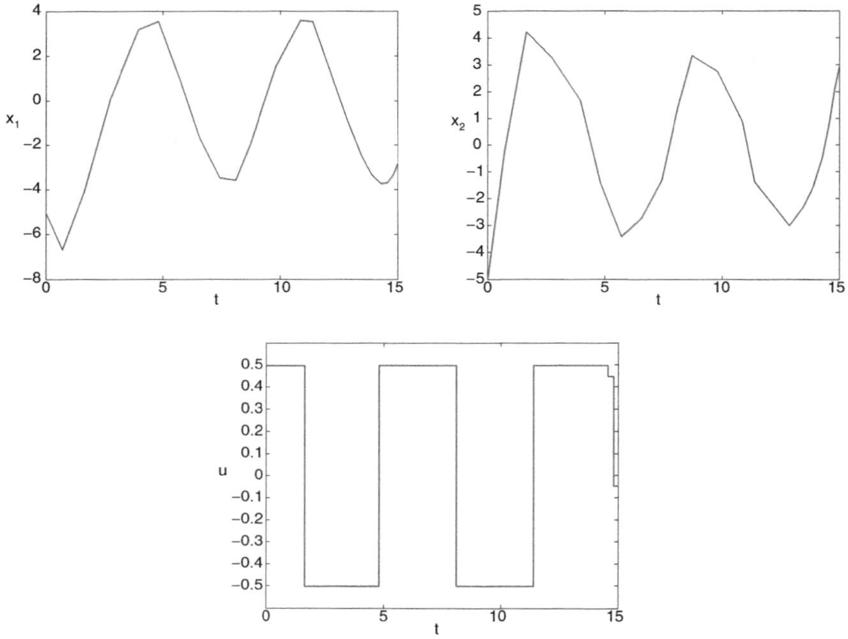

Fig. 8.10.3: Optimal state trajectories and optimal control for Example 8.10.3 with $n_p = 20$ and solved using time scaling

8.11 Exercises

8.11.1 *With reference to Remark 8.6.5, we consider the constraint specified in Remark 8.2.1(iv). Show that*

$$\frac{\partial G(\boldsymbol{\sigma}^p)}{\partial \boldsymbol{\sigma}^p} = 0$$

if $\boldsymbol{\sigma}^p$ *is such that*

$$\max_{0 \le t \le T} h(t, \boldsymbol{x}(t|\boldsymbol{\sigma}^p), \boldsymbol{\sigma}^p) = 0.$$

8.11.2 *Show that*

$$\int_0^T \frac{\partial \widetilde{H}_i}{\partial \sigma_j^{p,k}} dt = \int_{I_k^p} \frac{\partial H_i}{\partial u_j^p} dt,$$

where $\widetilde{H}_i$ *and* H_i *are defined by* (8.6.4) *and* (8.6.9a), *respectively.*

8.11.3 *Prove Theorem 8.8.1.*

8.11.4 *Prove Theorem 8.8.2.*

8.11.5 *Consider the optimal control problem, where the cost functional* (8.2.4) *is to be minimized subject to the dynamical system* (8.2.1). *Let* $\mathcal{U}$ *be the class of admissible controls that consists of all* C^1 *functions from* $[0, T]$ *to* U, *where* U *is as defined by* (8.2.2b) *with* $\alpha_i = -1$, *and* $\beta_i = 1$, $i = 1, \ldots, r$. *Pose this problem in a form solvable by the usual control parametrization technique. Derive the gradient formula for the cost functional with respect to the control parameters and the resulting system parameters.*

8.11.6 *Consider a minimax optimal control problem, where the following cost functional*

$$g_0(\boldsymbol{u}) = \max_{0 \le t \le T} C(t, \boldsymbol{x}(t), \boldsymbol{u}(t))$$

is to be minimized subject to the dynamical system (8.2.1). *The class of admissible controls is as defined in Section 8.2. Following the idea of [134], we define*

$$\|\gamma\|_p = \left(\int_0^T |\gamma(t)|^p \, dt \right)^{1/p}$$

and

$$\|\gamma\|_\infty = \operatorname*{ess\,sup}_{0 \le t \le T} |\gamma(t)| \, .$$

Then, it is known (cf. [FMT1]) that

$$\|\gamma\|_p \uparrow \|\gamma\|_\infty , \qquad as \;\; p \to \infty.$$

(i) Use this fact to construct a sequence of optimal control problems so that each of them is solvable by the control parametrization technique.

(ii) Show that the optimal costs of these approximate problems converge monotonically from below to the true optimal cost of the original problem.

8.11.7 *Show that the result of Lemma 8.4.1 remains valid if* $\{\boldsymbol{u}^p\}_{p=1}^{\infty}$ *is a bounded sequence of controls in* $L_\infty([0, T], \mathbb{R}^r)$ *that converges to* $\boldsymbol{u}$ *a.e. in* $[0, T]$, *as* $p \to \infty$.

8.11.8 *Show that Lemmas 8.4.2–8.4.4 still hold if* $\{\boldsymbol{u}^p\}_{p=1}^{\infty}$ *and* $\boldsymbol{u}$ *are as defined in Lemma 8.4.1.*

8.11.9 *Can the control functions chosen in Definition 8.9.1 be just measurable, rather than Borel measurable functions?*

8.11.10 *Derive the gradient formulae given by* (8.6.5) *and show that it is equivalent to* (8.6.8).

8.11.11 *Show that* $\hat{\mathbf{u}}$ *in the proof of Theorem 8.9.2 is a feasible control of Problem* (P).

8.11.12 *Consider the problem* (P) *subject to additional terminal equality constraints*

$$q_i(\boldsymbol{x}(T \mid \boldsymbol{u})) = 0, \quad i = 1, \ldots, N_E, \tag{8.11.1}$$

where q_i, $i = 1, \ldots, N_E$, *are continuously differentiable functions. Let this optimal control problem be referred to as the Problem* (R). *Use the control parametrization time scaling transform technique to derive a computational method for solving Problem* (R). *State all the essential assumptions and then prove all the relevant convergence results.*

8.11.13 *Derive the gradient formulae for* G_i, $i = 0, 1, \ldots, N$, *in Algorithm 8.8.1 by using the variational approach (Hint: Please refer to Sections 7.2 and 7.3).*

Chapter 9
Optimal Control Problems with State and Control Constraints

9.1 Introduction

In real world, optimal control problems are often subject to constraints on the state and/or control. These constraints can be point constraints and/or continuous inequality constraints. The point constraints are expressed as functions of the states at the end point or some intermediate interior points of the time horizon. These point constraints can be handled without much difficulty. However, for the continuous inequality constraints, they are expressed as functions of the states and/or controls over the entire time horizon, and hence are very difficult to handle. This chapter is devoted to devise computational methods for solving optimal control problems subject to point and continuous constraints. It is divided into two sections. In Section 9.2, our focus is on optimal control problems subject to continuous state and/or control inequality constraints. Through the application of the control parametrization technique, an approximate optimal parameter selection problem with continuous state inequality constraints is obtained. Then, the constraint transcription method introduced in Section 4.3 is applied to construct a smooth approximate inequality canonical constraint for each continuous inequality state constraint. In Section 9.3, exact penalty function method introduced in Section 4.4 will be utilized to develop an effective computational method for solving the class of optimal control problems considered in Section 9.3. The main references for this Chapter are [103, 104, 133, 134, 143, 145, 146, 168, 171, 249, 253, 254, 259].

K. L. Teo et al., *Applied and Computational Optimal Control*, Springer Optimization and Its Applications 171,
https://doi.org/10.1007/978-3-030-69913-0_9

9.2 Optimal Control with Continuous State Inequality Constraints

Consider the process described by the following system of nonlinear differential equations on the fixed time interval $(0, T]$.

$$\frac{d\boldsymbol{x}(t)}{dt} = \boldsymbol{f}(t, \boldsymbol{x}(t), \boldsymbol{u}(t)), \tag{9.2.1a}$$

where $\boldsymbol{x} = [x_1, \ldots, x_n]^\top \in \mathbb{R}^n$, $\boldsymbol{u} = [u_1, \ldots, u_r]^\top \in \mathbb{R}^r$ are, respectively, the state and control vectors; and $\boldsymbol{f} = [f_1, \ldots, f_n]^\top \in \mathbb{R}^n$.

The initial condition for the system of differential equations (9.2.1a) is

$$\boldsymbol{x}(0) = \boldsymbol{x}^0, \tag{9.2.1b}$$

where $\boldsymbol{x}^0$ is a given vector in $\mathbb{R}^n$.

Let U_1, U_2 and U be as defined in (8.2.2a) and (8.2.2b) in Section 8.2.

Definition 9.2.1 *A piecewise continuous function $\boldsymbol{u} = [u_1, \ldots, u_r]^\top$ from $[0, T]$ into $\mathbb{R}^r$ is said to be an admissible control if $\boldsymbol{u}(t) \in U$ for almost all $t \in [0, T]$ and it is continuous from the right.*

Let $\mathcal{U}$ be the class of all such admissible controls.

For each $\boldsymbol{u} \in \mathcal{U}$, let $\boldsymbol{x}(\cdot \mid \boldsymbol{u})$ be the corresponding vector-valued function which is absolutely continuous on $(0, T]$ and satisfies the differential equations (9.2.1a) almost everywhere on $(0, T]$ and the initial condition (9.2.1b). This function is called *the solution* of system (9.2.1) corresponding to $\boldsymbol{u} \in \mathcal{U}$.

Inequality terminal state constraints and inequality continuous state and control constraints are imposed, respectively, as follows.

$$\Phi_i(\boldsymbol{x}(T \mid \boldsymbol{u})) \leq 0, \quad i = 1, \ldots, N_T, \tag{9.2.2}$$

where Φ_i, $i = 1, \ldots, N_T$, are real-valued functions defined on $\mathbb{R}^n$, and

$$h_i(t, \boldsymbol{x}(t \mid \boldsymbol{u}), \boldsymbol{u}(t)) \leq 0, \text{ a.e. in } [0, T], \ i = 1, \ldots, N_S, \tag{9.2.3}$$

where h_i, $i = 1, \ldots, N_S$, are real-valued functions defined on $[0, T] \times \mathbb{R}^n \times \mathbb{R}^r$.

Let $\mathcal{F}$ be the set that consists of all those elements from $\mathcal{U}$ such that the constraints (9.2.2) and (9.2.3) are satisfied. Elements from $\mathcal{F}$ are called *feasible controls* and $\mathcal{F}$ is called *the class of feasible controls.*

We may now state the following optimal control problem.

Problem (P) Given the system (9.2.1), find a control $\boldsymbol{u} \in \mathcal{F}$ such that the cost functional

$$g_0(\boldsymbol{u}) = \Phi_0(\boldsymbol{x}(T \mid \boldsymbol{u})) + \int_0^T \mathcal{L}_0(t, \boldsymbol{x}(t \mid \boldsymbol{u}), \boldsymbol{u}(t))dt, \tag{9.2.4}$$

is minimized over $\mathcal{F}$, where Φ_0 and $\mathcal{L}_0$ are given real-valued functions.

We assume throughout that the following conditions are satisfied.

Assumption 9.2.1 $\boldsymbol{f}$ *and* $\mathcal{L}_0$ *satisfy the relevant conditions appearing in Assumptions 8.2.1 and 8.2.2.*

Assumption 9.2.2 *For each* $i = 0, 1, \ldots, N_T$, $\Phi_i : \mathbb{R}^n \to \mathbb{R}$ *is continuously differentiable.*

Assumption 9.2.3 *For each* $i = 1, \ldots, N_S$, $h_i : [0, T] \times \mathbb{R}^n \times \mathbb{R}^n \to \mathbb{R}$ *is continuously differentiable.*

9.2.1 Time Scaling Transform

We shall apply the time scaling transform introduced in Section 7.4.2 to Problem (P). For each $p \geq 1$, let the planning horizon $[0, T]$ be partitioned into n_p subintervals with $n_p + 1$ partition points denoted by

$$\tau_0^p, \tau_1^p, \ldots, \tau_{n_p}^p, \tag{9.2.5}$$

where $\tau_0^p = 0$ and $\tau_{n_p}^p = T$, while τ_i^p, $i = 1, \ldots, n_p - 1$, are decision variables satisfying the following conditions:

$$\tau_{k-1}^p \leq \tau_k^p, \quad k = 1, 2, \ldots, n_p. \tag{9.2.6}$$

Let the number n_p of the partition points be chosen such that $n_{p+1} > n_p$. The control is now approximated in the form of piecewise constant function as:

$$\boldsymbol{u}^p(t) = \sum_{k=1}^{n_p} \boldsymbol{\sigma}^{p,k} \chi_{[\tau_{k-1}^p, \tau_k^p)}(t), \tag{9.2.7}$$

where $(\boldsymbol{\sigma}^{p,k})^\top = [\sigma_1^{p,k}, \ldots, \sigma_r^{p,k}]^\top$ with $\boldsymbol{\sigma}^{p,k} \in U$, $k = 1, \ldots, n_p$, and the set U is as defined in Section 8.2. Let

$$\boldsymbol{\sigma}^p = \left[\left(\boldsymbol{\sigma}^{p,1}\right)^\top, \ldots, (\boldsymbol{\sigma}^{p,n_p})^\top \right]^\top \tag{9.2.8}$$

and

$$\boldsymbol{\tau}^p = \left[\tau_1^p, \tau_2^p, \ldots, \tau_{n_p-1}^p \right]^\top, \tag{9.2.9}$$

where $\tau_0^p = 0$, $\tau_{n_p}^p = T$, while τ_i^p, $i = 1, \ldots, n_p - 1$, are decision variables satisfying

$$0 = \tau_0^p \leq \tau_1^p \leq \tau_2^p \leq \cdots \leq \tau_{n_p}^p = T. \tag{9.2.10}$$

Let Ξ^p be the set consisting all those $\boldsymbol{\sigma}^p$ and let Γ^p be the set consisting all those $\boldsymbol{\tau}^p$.

To apply the time scaling transform, we construct a transformation from $t \in [0, T]$ to a new time scale $s \in [0, 1]$. This transformation maps the knots:

$$\tau_0^p, \tau_1^p, \tau_2^p, \ldots, \tau_{n_p-1}^p, \tau_{n_p}^p$$

into the fixed knots:

$$\xi_k^p = \frac{k}{n_p}, \quad k = 0, 1, \ldots, n_p. \tag{9.2.11a}$$

Clearly, these fixed knots are such that

$$0 = \xi_0^p < \xi_1^p < \xi_2^p < \cdots < \xi_{n_p-1}^p < \xi_{n_p}^p = 1. \tag{9.2.11b}$$

The required transformation from $t \in [0, T]$ to $s \in [0, 1]$ can be defined by the following differential equation.

$$\frac{dt(s)}{ds} = v^p(s) \tag{9.2.12a}$$

with the initial condition:

$$t(0) = 0. \tag{9.2.12b}$$

In view of Definition 8.9.6, we note that $v^p(s)$ is called *a time scaling control* defined by (8.9.19), that is,

$$v^p(s) = \sum_{k=1}^{n_p} \theta_k^p \chi_{[\xi_{k-1}^p, \xi_k^p)}(s), \tag{9.2.13}$$

where $\theta_k^p \geq 0$, $k = 1, \ldots, n_p$, are decision variables. Define $\boldsymbol{\theta}^p = \left[\theta_1^p, \ldots, \theta_{n_p}^p\right]^\top \in \mathbb{R}^{n_p}$ with $\theta_k^p \geq 0$, $k = 1, \ldots, n_p$. Let Θ^p be the set consisting all those $\boldsymbol{\theta}^p$, and let V^p be the set consisting of all the piecewise constant functions in the form of (9.2.13) with $\boldsymbol{\theta}^p \in \Theta^p$. Clearly, each $\boldsymbol{\theta}^p \in \Theta^p$ defines uniquely a $v^p \in \mathcal{V}^p$, and vice versa. By (9.2.13), it is clear from (9.2.12) that, for $k = 1, \ldots, n_p$,

$$t^p(s) = \int_0^s v^p(\tau)d\tau = \sum_{j=1}^{k-1} \theta_j^p(\xi_j^p - \xi_{j-1}^p) + \theta_k^p(s - \xi_{k-1}^p), \quad \text{for } s \in [\xi_{k-1}^p, \xi_k^p]. \tag{9.2.14}$$

In particular,

$$t^p(1) = \int_0^1 v^p(\tau)d\tau = \sum_{k=1}^{n_p} \theta_k^p(\xi_k^p - \xi_{k-1}^p) = T. \tag{9.2.15}$$

and

$$t^p(k/n_p) = \int_0^{k/n_p} v^p(\tau)d\tau = \sum_{j=1}^{k} \theta_j^p(\xi_j^p - \xi_{j-1}^p) = \tau_k, \quad k = 1, \dots, n_p. \quad (9.2.16)$$

Define

$$\boldsymbol{\omega}^p(s) = \boldsymbol{u}^p(t(s)), \quad (9.2.17)$$

where $\boldsymbol{u}^p$ is in the form of (9.2.7). Clearly, $\boldsymbol{\omega}^p$ is a piecewise constant control which can be written as

$$\boldsymbol{\omega}^p(s) = \sum_{k=1}^{n_p} \boldsymbol{\sigma}^{p,k} \chi_{[\xi_{k-1}^p, \xi_k^p)}(s), \; s \in [0,1], \quad (9.2.18)$$

where ξ_k^p, $k = 0, 1, \dots, n_p$, are fixed knots chosen as specified by (9.2.11), $\chi_{[\xi_{k-1}^p, \xi_k^p)}(\cdot)$ denotes the indicator function of the interval $[\xi_{k-1}^p, \xi_k^p)$ defined by (7.3.3), and $\boldsymbol{\sigma}^p \in \Xi^p$.

Let $\widehat{\mathcal{U}^p}$ be the set containing all those piecewise constant controls given by (9.2.18) with $\boldsymbol{\sigma}^p \in \Xi^p$. Clearly, each $(\boldsymbol{\sigma}^p, \boldsymbol{\theta}^p) \in \Xi^p \times \Theta^p$ defines uniquely a $(\boldsymbol{\omega}^p, v^p) \in \widehat{\mathcal{U}^p} \times \mathcal{V}^p$, and vice versa.

With the transformation from $t \in [0, T]$ into $s \in [0, 1]$ via the introduction of (9.2.12), system (9.2.1) becomes

$$\frac{d\hat{\boldsymbol{x}}(s)}{ds} = \widehat{\boldsymbol{f}}(s, \hat{\boldsymbol{x}}(s), \boldsymbol{\sigma}^p, \boldsymbol{\theta}^p), \quad (9.2.19)$$

$$\hat{\boldsymbol{x}}(0) = \begin{bmatrix} \widetilde{\boldsymbol{x}}(0) \\ 0 \end{bmatrix} = \begin{bmatrix} \boldsymbol{x}^0 \\ 0 \end{bmatrix}, \quad (9.2.20)$$

where

$$\hat{\boldsymbol{x}}(s) = \begin{bmatrix} \widetilde{\boldsymbol{x}}(s) \\ t(s) \end{bmatrix} = \begin{bmatrix} \boldsymbol{x}(t(s)) \\ t(s) \end{bmatrix}, \quad (9.2.21a)$$

$$\widetilde{\boldsymbol{x}}(s) = \boldsymbol{x}(t(s)), \quad (9.2.21b)$$

$$\hat{\boldsymbol{f}}(s, \hat{\boldsymbol{x}}(s), \boldsymbol{\sigma}^p, \boldsymbol{\theta}^p) = v^p(s) \boldsymbol{f}\left(\widetilde{\boldsymbol{x}}(s), \sum_{k=1}^{n_p} \boldsymbol{\sigma}_k^p \chi_{[\xi_{k-1}^p, \xi_k^p)}(s) \right), \quad (9.2.21c)$$

and v^p is given by (9.2.13).

Let $\mathcal{B}^p$ be the set which consists of all those elements from $\Xi^p \times \Theta^p$ such that the following constraints are satisfied.

$$\Phi_i(\widetilde{\boldsymbol{x}}(1 \mid \boldsymbol{\sigma}^p, \boldsymbol{\theta}^p)) \le 0, \quad i = 1, \dots, N_T, \quad (9.2.22)$$

$$\hat{h}_i\left(s, \hat{\boldsymbol{x}}(s \mid \boldsymbol{\sigma}^p, \boldsymbol{\theta}^p), \boldsymbol{\sigma}^{p,k}\right) \le 0, \; s \in \overline{I_k}, \; i = 1, \dots, N_S, \; k = 1, \dots, n_p \quad (9.2.23)$$

and

$$\widehat{\Upsilon}(\boldsymbol{\theta}^p) = \sum_{k=1}^{n_p} \theta_k^p(\xi_k^p - \xi_{k-1}^p) - T = 0, \quad (9.2.24)$$

where

$$\hat{h}_i\left(s, \widehat{\boldsymbol{x}}(s), \boldsymbol{\sigma}^{p,k}\right) = h_i\left(t(s), \widetilde{\boldsymbol{x}}(s), \sum_{k=1}^{n_p} \boldsymbol{\sigma}^{p,k} \chi_{[\xi_{k-1}^p, \xi_k^p)}(s)\right), \quad s \in \overline{I_k}, \quad (9.2.25)$$

and, for each $k = 1, \ldots, n_p$, $\overline{I_k}$ denotes the closure of I_k with $I_k = [\xi_{k-1}^p, \xi_k^p)$.

Let $\mathcal{D}^p$ be the corresponding subset of $\widehat{\mathcal{U}^p} \times \mathcal{V}^p$ uniquely defined by elements from $\mathcal{B}^p$. Clearly, for each $(\boldsymbol{\omega}^p, \nu^p) \in \mathcal{D}^p$, there exists a unique $(\boldsymbol{\sigma}^p, \boldsymbol{\theta}^p) \in \mathcal{B}^p$ such that

$$\boldsymbol{\omega}^p(s) = \sum_{k=1}^{n_p} \boldsymbol{\sigma}^{p,k} \chi_{[\xi_{k-1}^p, \xi_k^p)}(s), \quad s \in [0, 1], \quad (9.2.26a)$$

$$v^p(s) = \sum_{k=1}^{n_p} \theta_k^p \chi_{[\xi_{k-1}^p, \xi_k^p)}(s), \quad s \in [0, 1]. \quad (9.2.26b)$$

The equivalent transformed optimal parameter selection problem may now be stated as follows:

Problem $(P(p))$ Subject to system (9.2.19)–(9.2.20), find a combined control parameter vector and switching vector $(\boldsymbol{\sigma}^p, \boldsymbol{\theta}^p) \in \mathcal{B}^p$ such that the cost functional

$$G_0(\boldsymbol{\sigma}^p, \boldsymbol{\theta}^p) = \Phi_0(\widetilde{\boldsymbol{x}}(1 \mid \boldsymbol{\sigma}^p, \boldsymbol{\theta}^p)) + \int_0^1 \widehat{\mathcal{L}}_0(s, \hat{\boldsymbol{x}}(s \mid \boldsymbol{\sigma}^p, \boldsymbol{\theta}^p), \boldsymbol{\sigma}^p, \boldsymbol{\theta}^p) ds \quad (9.2.27)$$

is minimized over $\mathcal{B}^p$, where

$$\widehat{\mathcal{L}}_0(s, \hat{\boldsymbol{x}}(s), \boldsymbol{\sigma}^p, \boldsymbol{\theta}^p) = v^p(s) \mathcal{L}_0(\widehat{\boldsymbol{x}}(s), \sum_{k=1}^{n_p} \boldsymbol{\sigma}_k^p \chi_{[\xi_{k-1}^p, \xi_k^p)}(s)). \quad (9.2.28)$$

9.2.2 Constraint Approximation

For each $i = 1, \ldots, N_S$, and $k = 1, \ldots, n_p$, the corresponding inequality continuous state constraint in (9.2.23) is equivalent to

$$G_{i,k}(\boldsymbol{\sigma}^p, \boldsymbol{\theta}^p) = \int_{\xi_{k-1}^p}^{\xi_k^p} \max\left\{\hat{h}_i\left(s, \widehat{\boldsymbol{x}}(s \mid \boldsymbol{\sigma}^p, \boldsymbol{\theta}^p), \boldsymbol{\sigma}^{p,k}\right), 0\right\} ds = 0. \quad (9.2.29)$$

However, the equality constraint (9.2.29) is non-differentiable at $(\boldsymbol{\sigma}^p, \boldsymbol{\theta}^p) \in \Xi^p \times \Theta^p$ such that $\hat{h}_i = 0$. Nevertheless, since for each $i = 1, \ldots, N_S$, and $k = 1, \ldots, n_p$, the continuous inequality constraint in (9.2.23) is equivalent to the equality constraint in (9.2.29), Problem $(P(p))$ with (9.2.23) replaced by (9.2.29) is again referred to as Problem $(P(p))$.

Let $\mathring{\mathcal{B}}^p$ be the interior of the set $\mathcal{B}^p$ in the sense that it consists of all those $(\boldsymbol{\sigma}^p, \boldsymbol{\theta}^p) \in \mathcal{B}^p$ such that

$$\Phi_i(\widetilde{\boldsymbol{x}}(1 \mid \boldsymbol{\sigma}^p, \boldsymbol{\theta}^p)) < 0, \quad i = 1, \ldots, N_T, \tag{9.2.30}$$

$$\max_{s \in \overline{I}_k} \left(\hat{h}_i \left(s, \widehat{\boldsymbol{x}}(s \mid \boldsymbol{\sigma}^p, \boldsymbol{\theta}^p), \boldsymbol{\sigma}^{p,k} \right) \right) < 0, \ i = 1, \ldots, N_S, \ k = 1, \ldots, n_p. \tag{9.2.31}$$

To continue, we assume that the following condition is satisfied.

Assumption 9.2.4 $\mathring{\mathcal{B}}^p \neq \varnothing$.

To handle the non-differentiable equality constraints (9.2.28), we replace $\max\left\{\hat{h}_i \left(s, \widehat{\boldsymbol{x}}(s \mid \boldsymbol{\sigma}^p, \boldsymbol{\theta}^p), \boldsymbol{\sigma}^{p,k} \right), 0\right\}$ by $\widehat{\mathcal{L}}_{i,\varepsilon} \left(s, \widehat{\boldsymbol{x}}(s \mid \boldsymbol{\sigma}^p, \boldsymbol{\theta}^p), \boldsymbol{\sigma}^{p,k} \right)$, where

$$\widehat{\mathcal{L}}_{i,\varepsilon} \left(s, \widehat{\boldsymbol{x}}(s \mid \boldsymbol{\sigma}^p, \boldsymbol{\theta}^p), \boldsymbol{\sigma}^{p,k} \right) = \begin{cases} 0, & \text{if } \hat{h}_i \left(s, \widehat{\boldsymbol{x}}(s \mid \boldsymbol{\sigma}^p, \boldsymbol{\theta}^p), \boldsymbol{\sigma}^{p,k} \right) < -\varepsilon \\ \dfrac{\left(\hat{h}_i \left(s, \widehat{\boldsymbol{x}}(s \mid \boldsymbol{\sigma}^p, \boldsymbol{\theta}^p), \boldsymbol{\sigma}^{p,k} \right) + \varepsilon\right)^2}{4\varepsilon}, & \text{if } -\varepsilon \leq \hat{h}_i \left(s, \widehat{\boldsymbol{x}}(s \mid \boldsymbol{\sigma}^p, \boldsymbol{\theta}^p), \boldsymbol{\sigma}^{p,k} \right) \leq \varepsilon \\ \hat{h}_i \left(s, \widehat{\boldsymbol{x}}(s \mid \boldsymbol{\sigma}^p, \boldsymbol{\theta}^p), \boldsymbol{\sigma}^{p,k} \right), & \text{if } \hat{h}_i \left(s, \widehat{\boldsymbol{x}}(s \mid \boldsymbol{\sigma}^p, \boldsymbol{\theta}^p), \boldsymbol{\sigma}^{p,k} \right) > \varepsilon. \end{cases} \tag{9.2.32}$$

This function is obtained by smoothing out the sharp corner of the function $\max\left\{\hat{h}_i \left(s, \widehat{\boldsymbol{x}}(s \mid \boldsymbol{\sigma}^p, \boldsymbol{\theta}^p), \boldsymbol{\sigma}^{p,k} \right), 0\right\}$ which is in the form as shown in Figure 4.2.1. For each $i = 1, \ldots, N_S$, and $k = 1, \ldots, n_p$, define

$$G_{i,\varepsilon}(\boldsymbol{\sigma}^p, \boldsymbol{\theta}^p) = \int_0^1 \sum_{k=1}^{n_p} \widehat{\mathcal{L}}_{i,\varepsilon} \left(s, \widehat{\boldsymbol{x}}(s \mid \boldsymbol{\sigma}^p, \boldsymbol{\theta}^p), \boldsymbol{\sigma}^{p,k} \right) \chi_{\left[\xi^p_{k-1}, \xi^p_k\right)}(s) ds. \tag{9.2.33}$$

We now define two related approximate problems, which will be referred to as Problem $(P_\varepsilon(p))$ and Problem $(P_{\varepsilon,\gamma}(p))$, respectively. The first approximate problem is:

Problem $(P_\varepsilon(p))$: The Problem $(P(p))$ with the continuous inequality constraints (9.2.23) replaced by

$$G_{i,\varepsilon}(\boldsymbol{\sigma}^p, \boldsymbol{\theta}^p) = 0, \quad i = 1, \ldots, N_S. \tag{9.2.34}$$

Let $\mathcal{B}^p_\varepsilon$ be the feasible region of Problem $(P_\varepsilon(p))$ containing all those $(\boldsymbol{\sigma}^p, \boldsymbol{\theta}^p) \in \Xi^p \times \Theta^p$ such that the constraints (9.2.22), (9.2.34) and (9.2.21) are satisfied. Clearly, for each $\varepsilon > 0$, $\mathcal{B}^p_\varepsilon \subset \mathcal{B}^p$.

Note that equality constraints (9.2.34) fail to satisfy the usual constraint qualification. To overcome this difficulty, we consider the second approximate problem as follows:

Problem ($P_{\varepsilon,\gamma}(p)$): The Problem $(P(p))$ with (9.2.23) replaced by

$$\widehat{G}_{i,\varepsilon,\gamma}(\boldsymbol{\sigma}^p) = -\gamma + \widehat{G}_{i,\varepsilon}(\boldsymbol{\sigma}^p) \leq 0, \quad i = 1, \ldots, N_S. \tag{9.2.35}$$

Note that constraints (9.2.35) are already in canonical form, i.e., in the form of (9.2.29), where the functions $\hat{G}_{i,\epsilon}$ in (9.2.35) are equal to the constant $-\gamma$ in the present case.

We assume that the following condition is satisfied.

Assumption 9.2.5 *For any combined vector $(\boldsymbol{\sigma}^p, \boldsymbol{\theta}^p)$ in $\mathcal{B}^p$, there exists a combined vector $(\boldsymbol{\sigma}^p, \boldsymbol{\theta}^p) \in \mathring{\mathcal{B}}^p$ such that*

$$\alpha(\boldsymbol{\sigma}^p, \boldsymbol{\theta}^p) + (1-\alpha)(\boldsymbol{\sigma}^p, \boldsymbol{\theta}^p) \in \mathring{\mathcal{B}}^p \text{ for all } \alpha \in (0,1].$$

To relate the solutions of Problems $(P(p))$, $(P_\varepsilon(p))$ and $(P_{\varepsilon,\gamma}(p))$ as $\varepsilon \to 0$, we have the following lemma.

Lemma 9.2.1 *For any $\varepsilon > 0$, there exists a $\gamma(\varepsilon) > 0$ such that for all γ, $0 < \gamma < \gamma(\varepsilon)$, if $(\boldsymbol{\sigma}^p_{\varepsilon,\gamma}, \boldsymbol{\theta}^p_{\varepsilon,\gamma}) \in \Xi^p \times \Theta^p$ satisfies the constraints of Problem $(P_{\varepsilon,\gamma}(p))$, i.e.,*

$$\Phi_i(\widetilde{\boldsymbol{x}}(1 \mid \boldsymbol{\sigma}^p_{\varepsilon,\gamma}, \boldsymbol{\theta}^p_{\varepsilon,\gamma})) \leq 0, \quad i = 1, \ldots, N_T, \tag{9.2.36a}$$

$$G_{i,k,\varepsilon,\gamma}(\boldsymbol{\sigma}^p_{\varepsilon,\gamma}, \boldsymbol{\theta}^p_{\varepsilon,\gamma}) \leq 0, \quad i = 1, \ldots, N_S;\ k = 1, \ldots, n_p, \tag{9.2.36b}$$

$$\widehat{\Upsilon}(\boldsymbol{\theta}^p) = \sum_{k=1}^{n_p} \theta^p_k(\xi^p_k - \xi^p_{k-1}) - T = 0, \tag{9.2.36c}$$

then it also satisfies the constraints of Problem $(P(p))$.

Proof. For each $i = 1, \ldots, N_S$ and any $(\boldsymbol{\sigma}^p, \boldsymbol{\theta}^p) \in \Xi^p \times \Theta^p$, we have

$$\begin{aligned}\frac{dh_i(t(s), \overline{\boldsymbol{x}}(s \mid \boldsymbol{\sigma}^p, \boldsymbol{\theta}^p))}{ds} \\ = \frac{\partial h_i(t(s), \overline{\boldsymbol{x}}(s \mid \boldsymbol{\sigma}^p, \boldsymbol{\theta}^p))}{\partial t} \frac{dt(s)}{ds} \\ + \sum_{j=1}^{n} \frac{\partial h_i(t(s), \overline{\boldsymbol{x}}(s \mid \boldsymbol{\sigma}^p, \boldsymbol{\theta}^p))}{\partial \overline{x}_j} \tilde{f}_j(t(s), \overline{\boldsymbol{x}}(s \mid \boldsymbol{\sigma}^p, \boldsymbol{\theta}^p), \boldsymbol{\sigma}^p, \boldsymbol{\theta}^p).\end{aligned} \tag{10.5.33}$$

Note that $t(s)$ depends on $\boldsymbol{\theta}^p \in \Theta^p$. Now, by Assumption 9.2.3 and Lemma 8.4.2, there exists a positive constant m_i such that, for all $(\boldsymbol{\sigma}^p, \boldsymbol{\theta}^p) \in \mathcal{A}^p$,

$$\left| \frac{dh_i(t(s), \overline{\boldsymbol{x}}(s \mid \sigma^p, \boldsymbol{\theta}^p))}{ds} \right| \leq m_i, \quad \forall s \in [0,1]. \tag{9.2.37}$$

Next, for any $\varepsilon > 0$, define

$$k_{i,\varepsilon} = \frac{\varepsilon}{16} \min\left\{T, \frac{\varepsilon}{2m_i}\right\}. \tag{9.2.38}$$

It suffices to show that, for each $i = 1, \ldots, N_s$,

$$\mathcal{B}^p_{i,\varepsilon,\gamma} \subset \mathcal{B}^p_i \tag{9.2.39}$$

for any γ such that $0 < \gamma < k_{i,\varepsilon}$, where

$$\mathcal{B}^p_{i,\varepsilon,\gamma} = \left\{(\sigma^p, \boldsymbol{\theta}^p) \in \mathcal{A}^p : -\gamma + \widehat{G}_{i,\varepsilon}(\sigma^p, \boldsymbol{\theta}^p) \leq 0\right\} \tag{9.2.40}$$

and

$$\mathcal{B}^p_i = \left\{(\sigma^p, \boldsymbol{\theta}^p) \in \mathcal{A}^p : \widehat{G}_i(\sigma^p, \boldsymbol{\theta}^p)) = 0\right\}, \tag{9.2.41}$$

where $\widehat{G}_i(\sigma^p, \boldsymbol{\theta}^p)$ is defined by (9.2.29). Assume the contrary. Then, there exists an $i \in \{1, \ldots, N_S\}$ and a $(\sigma^p, \boldsymbol{\theta}^p) \in \mathcal{A}^p$ such that

$$-\gamma + \widehat{G}_{i,\varepsilon}(\sigma^p, \boldsymbol{\theta}^p) \leq 0 \tag{9.2.42}$$

for any γ such that $0 < \gamma < k_{i,\varepsilon}$ but

$$\widehat{G}_i(\sigma^p, \boldsymbol{\theta}^p) > 0. \tag{9.2.43}$$

Since $h_i(t(s), \overline{\boldsymbol{x}}(s \mid \sigma^p, \boldsymbol{\theta}^p))$ is a continuous function of s in $[0, 1]$, (9.2.43) implies that there exists a $\widetilde{s} \in [0, 1]$ such that

$$h_i(t(\widetilde{s}), \overline{\boldsymbol{x}}(\widetilde{s} \mid \sigma^p, \boldsymbol{\theta}^p)) > 0. \tag{9.2.44}$$

Again by continuity, there exists an interval $I_i \subset [0, 1]$ containing $\widetilde{s}$ such that

$$h_i(t(\widetilde{s}), \overline{\boldsymbol{x}}(\widetilde{s} \mid \sigma^p)) > -\varepsilon/2, \qquad \forall t \in I_i. \tag{9.2.45}$$

For geometrical interpretation, please refer to Figure 4.3.1. Using (9.2.45), and Figure 4.3.1, we see that

$$m_i = \tan\theta = \frac{y}{z}. \tag{9.2.46}$$

Thus, the length $|I_i|$ of the interval I_i must satisfy

$$|I_i| \geq \min\{T, z\} = \min\left\{T, \frac{y}{m_i}\right\} \geq \min\left\{T, \frac{\varepsilon}{2m_i}\right\}. \tag{9.2.47}$$

From the definition of $\widehat{G}_{i,\varepsilon}(\sigma^p)$ and the fact that $\widehat{\mathcal{L}}_{i,\varepsilon}(t(s), \overline{\boldsymbol{x}}(s \mid \sigma^p))$ is nonnegative, it follows from (9.2.42) that

$$0 \geq -\gamma + \widehat{G}_{i,\varepsilon}(\sigma^p, \boldsymbol{\theta}^p) = -\gamma + \int_0^T \widehat{\mathcal{L}}_{i,\varepsilon}(t(s), \overline{\boldsymbol{x}}(s \mid \sigma^p, \boldsymbol{\theta}^p))ds$$
$$\geq -\gamma + \int_{I_i} \widehat{\mathcal{L}}_{i,\varepsilon}(t(s), \overline{\boldsymbol{x}}(s \mid \sigma^p, \boldsymbol{\theta}^p))ds \geq -\gamma$$
$$+ \min_{s \in I_i} \left\{ \widehat{\mathcal{L}}_{i,\varepsilon}(t(s), \overline{\boldsymbol{x}}(s \mid \sigma^p, \boldsymbol{\theta}^p)) \right\} |I_i| . \tag{9.2.48}$$

Now, by virtue of (9.2.45), we have

$$\min_{s \in I_i} \widehat{\mathcal{L}}_{i,\varepsilon}(t(s), \overline{\boldsymbol{x}}(s \mid \sigma^p, \boldsymbol{\theta}^p)) > \varepsilon/16. \tag{9.2.49}$$

Combining (9.2.47), (9.2.48), and (9.2.49), we obtain

$$0 \geq -\gamma + \widehat{G}_{i,\varepsilon}(\sigma^p, \boldsymbol{\theta}^p) > -\gamma + \frac{\varepsilon}{16} \min \left\{ T, \frac{\varepsilon}{2m_i} \right\} = -\gamma + k_{i,\varepsilon}.$$

This is a contradiction, because $\gamma < k_{i,\epsilon}$. Thus, the proof is complete. ■

Remark 9.2.1 *An alternative proof of Lemma 8.4.1 without the geometrical interpretation of Figure 4.3.1 is left as an exercise (use Lemma 4.2.1)*

9.2.3 A Computational Algorithm

To solve the constrained optimal control Problem $(P(p))$, we construct a sequence of approximate problems $(P_{\varepsilon,\gamma}(p))$ in ε and γ. For a given positive integer $p > 0$, and for each $\varepsilon > 0$ and $\gamma > 0$, let $\left(\boldsymbol{\sigma}^{p,*}_{\varepsilon,\gamma}, \boldsymbol{\theta}^{p,*}_{\varepsilon,\gamma}\right)$ be the solution of the problem $(P_{\varepsilon,\gamma}(p))$. The following algorithm can be used to generate a sequence of combined vectors in the feasible region of the Problem $(P(p))$.

Algorithm 9.2.1
Set $\varepsilon > 0$, $\gamma > 0$. *(In particular, we may choose* $\varepsilon = 10^{-1}$ *and* $\gamma = T\varepsilon/16$*).*
Step 1. *Solve* $(P_{\varepsilon,\gamma}(p))$ *to give* $\left(\boldsymbol{\sigma}^{p,*}_{\varepsilon,\gamma}, \boldsymbol{\theta}^{p,*}_{\varepsilon,\gamma}\right)$.
Step 2. *Check the feasibility of* $\hat{h}_i\left(s, \hat{\boldsymbol{x}}\left(s \mid \boldsymbol{\sigma}^{p,*}_{\varepsilon,\gamma}, \boldsymbol{\theta}^{p,*}_{\varepsilon,\gamma}\right), \boldsymbol{\theta}^{p,*}_{\varepsilon,\gamma}\right) \leq 0$ *for all* $s \in \overline{I_k}$ *and* $i = 1, \ldots, N_S$; $k = 1, \ldots, n_p$.
Step 3. *If* $\left(\boldsymbol{\sigma}^{p,*}_{\varepsilon,\gamma}, \boldsymbol{\theta}^{p,*}_{\varepsilon,\gamma}\right)$ *is feasible, go to* ***Step 5****. Otherwise, go to* ***Step 4****.*
Step 4. *Set* $\gamma = \gamma/2$ *and go to* ***Step 1****.*
Step 5. *Set* $\varepsilon = \varepsilon/10$, $\gamma = \gamma/10$ *and go to* ***Step 1****.*

Remark 9.2.2 *From Lemma 9.2.1, it is clear that the halving process of* γ *in Step 4 of Algorithm 9.2.1 needs only to be carried out a finite number of times. Let* $\tilde{\gamma}(\varepsilon)$ *be the parameter corresponding to each* $\varepsilon > 0$ *obtained in the halving process of* γ *in Step 4 of the algorithm. Clearly,* $\left(\boldsymbol{\sigma}^{p,*}_{\varepsilon,\tilde{\gamma}(\varepsilon)}, \boldsymbol{\theta}^{p,*}_{\varepsilon,\tilde{\gamma}(\varepsilon)}\right)$

satisfies the constraints of Problem $(P(p))$. The algorithm produces a sequence $\left\{\boldsymbol{\sigma}^{p,*}_{\varepsilon,\tilde{\gamma}(\varepsilon)}, \theta^{p,*}_{\varepsilon,\tilde{\gamma}(\varepsilon)}\right\}$ *in* $\varepsilon > 0$, *where each of these combined vectors* $\left\{\boldsymbol{\sigma}^{p,*}_{\varepsilon,\tilde{\gamma}(\varepsilon)}, \boldsymbol{\theta}^{p,*}_{\varepsilon,\tilde{\gamma}(\varepsilon)}\right\}$ *is in the feasible region of Problem $(P(p))$. Thus, it is a sequence of approximate optimal combined vectors of Problem $(P(p))$.*

Remark 9.2.3 *Let* $\left\{\boldsymbol{\sigma}^{p,*}_{\varepsilon,\tilde{\gamma}(\varepsilon)}, \boldsymbol{\theta}^{p,*}_{\varepsilon,\tilde{\gamma}(\varepsilon)}\right\}$ *be the sequence of approximate optimal combined vectors of Problem $(P(p))$ obtained by Algorithm 9.2.1 as explained in Remark 9.2.2. This gives rise to a corresponding sequence* $\left\{\boldsymbol{\omega}^{p,*}_{\varepsilon,\tilde{\gamma}(\varepsilon)}, \nu^{p,*}_{\varepsilon,\tilde{\gamma}(\varepsilon)}\right\}$ *of approximate piecewise constant controls, where* $\boldsymbol{\omega}^{p,*}_{\varepsilon,\tilde{\gamma}(\varepsilon)}$ *and* $\nu^{p,*}_{\varepsilon,\tilde{\gamma}(\varepsilon)}$ *are, respectively, given by* (9.2.26a) *and* (9.2.26b) *with* $\boldsymbol{\sigma}^p$ *and* $\boldsymbol{\theta}^p$ *taken as* $\boldsymbol{\sigma}^{p,*}_{\varepsilon,\tilde{\gamma}(\varepsilon)}$ *and* $\boldsymbol{\theta}^{p,*}_{\varepsilon,\tilde{\gamma}(\varepsilon)}$, *respectively. Then, by* (9.2.7) *and* (9.2.17), *we obtain a corresponding sequence* $\left\{\boldsymbol{u}^{p,*}_{\varepsilon,\tilde{\gamma}(\varepsilon)}\right\}$ *of approximate controls to Problem $(P(p))$, i.e.,*

$$\boldsymbol{u}^{p,*}_{\varepsilon,\tilde{\gamma}(\varepsilon)}(t) = \sum_{k=1}^{n_p} \boldsymbol{\sigma}^{p,*,k}_{\varepsilon,\tilde{\gamma}(\varepsilon)} \chi_{\left[\tau^{p,*}_{\varepsilon,\tilde{\gamma}(\varepsilon),k-1}, \tau^{p,*}_{\varepsilon,\tilde{\gamma}(\varepsilon),k}\right)}(t). \tag{9.2.50}$$

Here, we solve (9.2.12a) *and* (9.2.12b) *with* $\boldsymbol{\theta}^p$ *taken as* $\boldsymbol{\theta}^{p,*}_{\varepsilon,\tilde{\gamma}(\varepsilon)}$, *giving* $t^{p,*}_{\varepsilon,\tilde{\gamma}(\varepsilon)}(s)$ *defined by* (9.2.14). *Then, by evaluating* $t^{p,*}_{\varepsilon,\tilde{\gamma}(\varepsilon)}(s)$ *at* $s = k/n_p$, $k = 0, 1, \ldots, n_p$, *we obtain* $\tau^{p,*}_{\varepsilon,\tilde{\gamma}(\varepsilon),k}$, $k = 0, 1, \ldots, n_p$.

Remark 9.2.4 *By examining the proof of Lemma 9.2.1, we see that ε and γ are closely related to each other. At the solution of a particular problem, if a constraint is active over a large fraction of $[0, 1]$, then we should choose $\gamma = O(\varepsilon)$. On the other hand, if the constraint is active only over a very small fraction of $[0, 1]$, then $\gamma = O\left(\varepsilon^2\right)$.*

Remark 9.2.5 *In the actual implementation, Algorithm 9.2.1 is terminated when either of the following two conditions is satisfied.*

(i) In Step 4: If $\gamma < 10^{-10}$, then the algorithm is terminated as abnormal exit.
(ii) In Step 5: If $\varepsilon < 10^{-7}$, then the algorithm is terminated as successful exit.

Remark 9.2.6 *In Step 1 of Algorithm 9.2.1, we are required to solve Problem $(P_{\varepsilon,\gamma}(p))$. It can be solved as a nonlinear optimization problem. The details are given in the next section.*

9.2.4 Solving Problem $(P_{\varepsilon,\gamma}(p))$

Problem $(P_{\varepsilon,\gamma}(p))$ can be viewed as the following nonlinear optimization problem, which is again referred to as Problem $(P_{\varepsilon,\gamma}(p))$.

Minimize:

$$G_0(\boldsymbol{\sigma}^p, \boldsymbol{\theta}^p) \tag{9.2.51}$$

subject to

$$\Phi_i(\widetilde{\boldsymbol{x}}(1 \mid \boldsymbol{\sigma}^p, \boldsymbol{\theta}^p)) \le 0, \qquad i = 1, \ldots, N_T, \tag{9.2.52}$$

$$G_{i,k,\varepsilon,\gamma}(\boldsymbol{\sigma}^p, \boldsymbol{\theta}^p) \le 0, \qquad i = 1, \ldots, N_S;\ k = 1, \ldots, n_p, \tag{9.2.53}$$

$$\widehat{\Upsilon}(\boldsymbol{\theta}^p) = \sum_{k=1}^{n_p} \theta_k^p(\xi_k^p - \xi_{k-1}^p) - T = 0 \tag{9.2.54}$$

$$F_i(\boldsymbol{\sigma}^p) = \left(\boldsymbol{E}^i\right)^\top \boldsymbol{\sigma}^{p,k} - b_i \le 0, \ \ i = 1, \ldots, q;\ k = 1, \ldots, n_p \tag{9.2.55}$$

$$\alpha_i \le \boldsymbol{\sigma}_i^{p,k} \le \beta_i, \ \ i = 1, \ldots, r;\ k = 1, \ldots, n_p \tag{9.2.56}$$

$$\theta_k^p \ge 0, \ \ k = 1, \ldots, n_p. \tag{9.2.57}$$

The constraints (9.2.55)–(9.2.57) are to specify the sets Ξ^p and Θ^p, respectively. The constraints (9.2.56)–(9.2.57) are known as the boundedness constraints in nonlinear optimization programming. Let Λ be the set which consists of all those $(\boldsymbol{\sigma}^p, \boldsymbol{\theta}^p) \in \mathbb{R}^{rn_p} \times \mathbb{R}^{n_p}$ such that the constraints (9.2.56)–(9.2.57) are satisfied.

This nonlinear mathematical programming problem in the combined vectors can be solved by using any nonlinear optimization technique, such as the sequential quadratic programming (SQP) approximation routine with the active set strategy (see Section 3.5).

To use the SQP approximation routine with the active set strategy to solve the nonlinear optimization problem (9.2.51)–(9.2.57), we choose an initial combined vector $\left((\boldsymbol{\sigma}^p)^{(0)}, (\boldsymbol{\theta}^p)^{(0)}\right) \in \Lambda$ to start up the SQP approximation routine. The SQP approximation routine will use, for each $\left((\boldsymbol{\sigma}^p)^{(i)}, (\boldsymbol{\theta}^p)^{(i)}\right) \in \Lambda$, the values of the cost function (9.2.51) and the constraint functions (9.2.52)–(9.2.55) as well as their respective gradients to generate the next iterate $\left((\boldsymbol{\sigma}^p)^{(i+1)}, (\boldsymbol{\theta}^p)^{(i+1)}\right)$. Consequently, it gives rise to a sequence of combined vectors. The combined vector obtained by the SQP approximation routine is regarded as an optimal combined vector of Problem $(P(p))$.

In what follows, we shall explain how the values of the cost functional (9.2.51) and the constraint functionals (9.2.52)–(9.2.55), as well as their respective gradients are calculated corresponding to each $(\boldsymbol{\sigma}^p, \boldsymbol{\theta}^p) \in \Lambda$.

Remark 9.2.7 *For each $(\boldsymbol{\sigma}^p, \boldsymbol{\theta}^p) \in \Lambda$, the corresponding values of the constraint functions $F_i(\boldsymbol{\sigma}^{p,k})$, $i = 1, \ldots, n_p$; $k = 1, \ldots, n_p$ and $\widehat{\Upsilon}\ (\boldsymbol{\theta}^p)$ are straightforward to calculate via (9.2.54)–(9.2.55). Their gradients are*

$$\frac{\partial F_i\left(\boldsymbol{\sigma}^{p,k}\right)}{\partial \boldsymbol{\sigma}^{p,k}} = \left(\boldsymbol{E}^i\right)^\top, \ \ i = 1, \ldots, q;\ k = 1, \ldots, n_p \tag{9.2.58}$$

$$\frac{\partial \widehat{\Upsilon}(\boldsymbol{\theta}^p)}{\partial \theta_k^p} = (\xi_k^p - \xi_{k-1}^p), \ \ k = 1, \ldots, n_p. \tag{9.2.59}$$

To calculate, for each $(\boldsymbol{\sigma}^p, \boldsymbol{\theta}^p) \in \Lambda$, *the values of the cost functional* (9.2.51) *and the constraint functionals* (9.2.52) *and* (9.2.53), *the first task is to use the following algorithm to calculate the solution of system* (9.2.19)–(9.2.20) *corresponding to each* $(\boldsymbol{\sigma}^p, \boldsymbol{\theta}^p) \in \Lambda$.

Algorithm 9.2.2
For each given $(\boldsymbol{\sigma}^p, \boldsymbol{\theta}^p) \in \Lambda$, *compute the solution* $\hat{\boldsymbol{x}}(\cdot \mid \boldsymbol{\sigma}^p, \boldsymbol{\theta}^p)$ *of system* (9.2.19)–(9.2.20) *by solving the differential equations* (9.2.19) *forward in time from* $s = 0$ *to* $s = 1$ *with the initial condition* (9.2.20).

With the solution $\hat{\boldsymbol{x}}(\cdot \mid \boldsymbol{\sigma}^p, \boldsymbol{\theta}^p)$ of the system (9.2.19)–(9.2.20) corresponding to the $(\boldsymbol{\sigma}^p, \boldsymbol{\theta}^p) \in \Lambda$ obtained, the values of the cost functional $G_0(\boldsymbol{\sigma}^p, \boldsymbol{\theta}^p)$ and the constraint functionals Φ_i, $i = 1, \ldots, N_T$, and $G_{i,\varepsilon,\gamma}$, $i = 1, \ldots, N_S$; $k = 1, \ldots, n_p$, can be easily calculated using the following simple algorithm.

Algorithm 9.2.3
For a given $(\boldsymbol{\sigma}^p, \boldsymbol{\theta}^p) \in \Lambda$,
Step 1. *Use Algorithm 9.2.2 to solve for* $\hat{\boldsymbol{x}}(\cdot \mid \boldsymbol{\sigma}^p, \boldsymbol{\theta}^p)$. *Thus,* $\hat{\boldsymbol{x}}(s \mid \boldsymbol{\sigma}^p, \boldsymbol{\theta}^p)$ *is known for each* $s \in [0, 1]$. *This implies that:*

(a). $\Phi_0(\widetilde{\boldsymbol{x}}(1 \mid \boldsymbol{\sigma}^p, \boldsymbol{\theta}^p))$, *and* $\Phi_i(\widetilde{\boldsymbol{x}}(1 \mid \boldsymbol{\sigma}^p, \boldsymbol{\theta}^p))$, $i = 1, \ldots, N_T$, *are known; and*
(b). $\widehat{\mathcal{L}}_0(s, \hat{\boldsymbol{x}}(s \mid \boldsymbol{\sigma}^p, \boldsymbol{\theta}^p), \boldsymbol{\sigma}^p, \boldsymbol{\theta}^p)$, *and* $\widehat{\mathcal{L}}_{i,\varepsilon}(s, \hat{\boldsymbol{x}}(s \mid \boldsymbol{\sigma}^p, \boldsymbol{\theta}^p), \boldsymbol{\sigma}^p, \boldsymbol{\theta}^{p,k})$, $i = 1, \ldots, N_S$; $k = 1, \ldots, n_p$, *are known for each* $s \in [0, 1]$. *Hence, their integrals:*

$$\int_0^1 \widehat{\mathcal{L}}_0(s, \hat{\boldsymbol{x}}(s \mid \boldsymbol{\sigma}^p, \boldsymbol{\theta}^p), \boldsymbol{\sigma}^p, \boldsymbol{\theta}^p)ds \tag{9.2.60}$$

$$\int_{\xi_{k-1}^p}^{\xi_{k-1}^p} \widehat{\mathcal{L}}_{i,\varepsilon}\left(s, \hat{\boldsymbol{x}}(s \mid \boldsymbol{\sigma}^p, \boldsymbol{\theta}^p), \boldsymbol{\sigma}^p, \boldsymbol{\theta}^{p,k}\right) ds, \quad i = 1, \ldots, N_S;\ k = 1, \ldots, n_p, \tag{9.2.61}$$

can be obtained readily.

Step 2. *The value of the cost functional:*

$$G_0(\boldsymbol{\sigma}^p, \boldsymbol{\theta}^p) = \Phi_0(\widetilde{\boldsymbol{x}}(1 \mid \boldsymbol{\sigma}^p, \boldsymbol{\theta}^p)) + \int_0^1 \widehat{\mathcal{L}}_0(s, \hat{\boldsymbol{x}}(s \mid \boldsymbol{\sigma}^p, \boldsymbol{\theta}^p), \boldsymbol{\sigma}^p, \boldsymbol{\theta}^p)ds \tag{9.2.62}$$

and the values of the constraint functionals:

$$\Phi_i(\widetilde{\boldsymbol{x}}(1 \mid \boldsymbol{\sigma}^p, \boldsymbol{\theta}^p)), \quad i = 1, \ldots, N_T, \tag{9.2.63}$$

and

$$G_{i,k,\varepsilon,\gamma}(\boldsymbol{\sigma}^p, \boldsymbol{\theta}^p) = -\gamma + \int_{\xi_{k-1}^p}^{\xi_k^p} \widehat{\mathcal{L}}_{i,\varepsilon}(s, \hat{\boldsymbol{x}}(s \mid \boldsymbol{\sigma}^p, \boldsymbol{\theta}^p), \boldsymbol{\sigma}^p, \boldsymbol{\theta}^{p,k})ds,$$

$$i = 1, \ldots, N_S;\ k = 1, \ldots, n_p \qquad (9.2.64)$$

are calculated.

Let us now move to present an algorithm for calculating the gradients of these functionals for each given $(\boldsymbol{\sigma}^p, \boldsymbol{\theta}^p) \in \Lambda$. For this, we note from Section 7.3 that the derivations of the gradient formulae for the cost functional and the constraint functionals are similar. These gradients can be computed using the following algorithm:

Algorithm 9.2.4
For a given $(\boldsymbol{\sigma}^p, \boldsymbol{\theta}^p) \in \Lambda$,
Step 1. *Solve the costate systems corresponding to the cost functional* (9.2.62), *the terminal constraint functionals* (9.2.63), *and the inequality state constraint functionals* (9.2.64), *respectively, as follows:*

(i). For the cost functional (9.2.62):

Solve the following costate system of differential equations:

$$\frac{d\hat{\boldsymbol{\lambda}}^0(s)}{ds} = -\left(\frac{\partial \widehat{H}_0\left(s, \hat{\boldsymbol{x}}(s \mid \boldsymbol{\sigma}^p, \boldsymbol{\theta}^p), \boldsymbol{\sigma}^p, \boldsymbol{\theta}^p, \hat{\boldsymbol{\lambda}}^0(s)\right)}{\partial \hat{\boldsymbol{x}}}\right)^\top \qquad (9.2.65)$$

with the boundary condition:

$$\hat{\boldsymbol{\lambda}}^0(1) = \left(\frac{\partial \Phi_0(\overline{\boldsymbol{x}}(1 \mid \boldsymbol{\sigma}^p, \boldsymbol{\theta}^p))}{\partial \overline{\boldsymbol{x}}}\right)^\top \qquad (9.2.66)$$

backward in time from $s = 1$ *to* $s = 0$, *where* $\hat{\boldsymbol{x}}(\cdot \mid \boldsymbol{\sigma}^p, \boldsymbol{\theta}^p)$ *is the solution of system* (9.2.19)–(9.2.20) *corresponding to* $(\boldsymbol{\sigma}^p, \boldsymbol{\theta}^p) \in \Lambda$; *and the Hamiltonian function* $\widehat{H}_0$ *is defined by*

$$\widehat{H}_0(s, \hat{\boldsymbol{x}}, \boldsymbol{\sigma}^p, \boldsymbol{\theta}^p, \boldsymbol{\lambda}) = \widehat{\mathcal{L}}_0(s, \hat{\boldsymbol{x}}, \boldsymbol{\sigma}^p, \boldsymbol{\theta}^p) + \boldsymbol{\lambda}^\top \widehat{\boldsymbol{f}}(s, \hat{\boldsymbol{x}}, \boldsymbol{\sigma}^p, \boldsymbol{\theta}^p). \qquad (9.2.67)$$

Let $\hat{\boldsymbol{\lambda}}^0(\cdot \mid \boldsymbol{\sigma}^p, \boldsymbol{\theta}^p)$ *be the solution of the costate system* (9.2.65)–(9.2.66).

(ii) For the i-th terminal constraint functional given in (9.2.63):

Solve the following costate system of differential equations:

$$\frac{d\hat{\boldsymbol{\lambda}}^i(s)}{ds} = -\left(\frac{\partial \widehat{H}_i\left(s, \hat{\boldsymbol{x}}(s \mid \boldsymbol{\sigma}^p, \boldsymbol{\theta}^p), \boldsymbol{\sigma}^p, \boldsymbol{\theta}^p, \hat{\boldsymbol{\lambda}}^i(s)\right)}{\partial \hat{\boldsymbol{x}}}\right)^\top \qquad (9.2.68)$$

with the boundary condition:

$$\hat{\boldsymbol{\lambda}}^i(1) = \left(\frac{\partial \Phi_i(\overline{\boldsymbol{x}}(1 \mid \boldsymbol{\sigma}^p, \boldsymbol{\theta}^p))}{\partial \overline{\boldsymbol{x}}}\right)^\top \qquad (9.2.69)$$

backward in time from $s = 1$ *to* $s = 0$, *where* $\hat{\boldsymbol{x}}(\cdot \mid \boldsymbol{\sigma}^p, \boldsymbol{\theta}^p)$ *is the solution of system* (9.2.19) *and* (9.2.20) *corresponding to* $(\boldsymbol{\sigma}^p, \boldsymbol{\theta}^p) \in \Lambda$; *and* $\widehat{H}_i$, *the corresponding Hamiltonian function for the i-th terminal constraint functional, is defined by*

$$\widehat{H}_i(s, \hat{\boldsymbol{x}}, \boldsymbol{\sigma}^p, \boldsymbol{\theta}^p, \boldsymbol{\lambda}) = \boldsymbol{\lambda}^\top \widehat{\boldsymbol{f}}(s, \hat{\boldsymbol{x}}, \boldsymbol{\sigma}^p, \boldsymbol{\theta}^p). \tag{9.2.70}$$

Let $\hat{\boldsymbol{\lambda}}^i(\cdot \mid \boldsymbol{\sigma}^p, \boldsymbol{\theta}^p)$ *be the solution of the costate system* (9.2.68) *and* (9.2.69) .

(iii) For the (i,k)*-th inequality state constraint functional given in* (9.2.64)*:*

Solve the following costate system of differential equations:

$$\frac{d\hat{\boldsymbol{\lambda}}^{i,,k,\varepsilon,\gamma}(s)}{ds} = -\left(\frac{\partial \widehat{H}_{i,k,\varepsilon,\gamma}\left(s, \hat{\boldsymbol{x}}(s \mid \boldsymbol{\sigma}^p, \boldsymbol{\theta}^p), \boldsymbol{\sigma}^p, \boldsymbol{\theta}^p, \hat{\boldsymbol{\lambda}}^{i,k,\varepsilon,\gamma}(s)\right)}{\partial \hat{\boldsymbol{x}}}\right)^\top \tag{9.2.71}$$

with the boundary condition:

$$\hat{\boldsymbol{\lambda}}^{i,k,\varepsilon,\gamma}(1) = \mathbf{0} \tag{9.2.72}$$

backward in time from $s = 1$ *to* $s = 0$, *where* $\hat{\boldsymbol{x}}(\cdot \mid \boldsymbol{\sigma}^p, \boldsymbol{\theta}^p)$ *is the solution of the system* (9.2.19)–(9.2.20) *corresponding to* $(\boldsymbol{\sigma}^p, \boldsymbol{\theta}^p) \in \Lambda$; *and the Hamiltonian function* $\widehat{H}_{i,k,\varepsilon,\gamma}$ *is defined by*

$$\widehat{H}_{i,k,\varepsilon,\gamma}(s, \hat{\boldsymbol{x}}, \boldsymbol{\sigma}^p, \boldsymbol{\theta}^p, \boldsymbol{\lambda}) = \widehat{\mathcal{L}}_{i,k,\varepsilon,\gamma}(s, \hat{\boldsymbol{x}}, \boldsymbol{\sigma}^p, \boldsymbol{\theta}^p) + \boldsymbol{\lambda}^\top \widehat{\boldsymbol{f}}(s, \hat{\boldsymbol{x}}, \boldsymbol{\sigma}^p, \boldsymbol{\theta}^p). \tag{9.2.73}$$

Let $\hat{\boldsymbol{\lambda}}^{i,\varepsilon}(\cdot \mid \boldsymbol{\sigma}^p, \boldsymbol{\theta}^p)$ *be the solution of the costate system* (9.2.71) *and* (9.2.72).
Step 2. *The gradients of the cost functional* (9.2.62), *the terminal inequality constraint functionals* (9.2.63), *and the inequality constraint functionals* (9.2.64) *are computed, respectively, as follows:*

(i) For the cost functional (9.2.62)*:*

$$\frac{\partial G_0(\boldsymbol{\sigma}^p, \boldsymbol{\theta}^p)}{\partial \boldsymbol{\sigma}^p} = \int_0^1 \frac{\partial \widehat{H}_0\left(s, \hat{\boldsymbol{x}}(s \mid \boldsymbol{\sigma}^p, \boldsymbol{\theta}^p), \boldsymbol{\sigma}^p, \boldsymbol{\theta}^p, \hat{\boldsymbol{\lambda}}^0(s \mid \boldsymbol{\sigma}^p, \boldsymbol{\theta}^p)\right)}{\partial \boldsymbol{\sigma}^p} ds \tag{9.2.74a}$$

$$\frac{\partial G_0(\boldsymbol{\sigma}^p, \boldsymbol{\theta}^p)}{\partial \boldsymbol{\theta}^p} = \int_0^1 \frac{\partial \widehat{H}_0\left(s, \hat{\boldsymbol{x}}(s \mid \boldsymbol{\sigma}^p, \boldsymbol{\theta}^p), \boldsymbol{\sigma}^p, \boldsymbol{\theta}^p, \hat{\boldsymbol{\lambda}}^0(s \mid \boldsymbol{\sigma}^p, \boldsymbol{\theta}^p)\right)}{\partial \boldsymbol{\theta}^p} ds. \tag{9.2.74b}$$

(ii) For the i-th terminal constraint functional given in (9.2.63)*:*

$$\frac{\partial G_i(\boldsymbol{\sigma}^p, \boldsymbol{\theta}^p)}{\partial \boldsymbol{\sigma}^p} = \int_0^1 \frac{\partial \widehat{H}_i\left(s, \hat{\boldsymbol{x}}(s \mid \boldsymbol{\sigma}^p, \boldsymbol{\theta}^p), \boldsymbol{\sigma}^p, \boldsymbol{\theta}^p, \hat{\boldsymbol{\lambda}}^i(s \mid \boldsymbol{\sigma}^p, \boldsymbol{\theta}^p)\right)}{\partial \boldsymbol{\sigma}^p} ds \tag{9.2.75}$$

$$\frac{\partial G_i(\boldsymbol{\sigma}^p,\boldsymbol{\theta}^p)}{\partial \boldsymbol{\theta}^p}=\int_0^1 \frac{\partial \widehat{H}_i\left(s,\hat{\boldsymbol{x}}(s\mid\boldsymbol{\sigma}^p,\boldsymbol{\theta}^p),\boldsymbol{\sigma}^p,\boldsymbol{\theta}^p,\hat{\boldsymbol{\lambda}}^i(s\mid\boldsymbol{\sigma}^p,\boldsymbol{\theta}^p)\right)}{\partial \boldsymbol{\theta}^p}ds. \tag{9.2.76}$$

(iii) For the (i,k)-th inequality constraint functional given in (9.2.64):

$$\frac{\partial G_{i,\varepsilon}(\boldsymbol{\sigma}^p,\boldsymbol{\theta}^p)}{\partial \boldsymbol{\sigma}^p}=\int_0^1 \frac{\partial \widehat{H}_{i,\varepsilon}\left(s,\hat{\boldsymbol{x}}(s\mid\boldsymbol{\sigma}^p,\boldsymbol{\theta}^p),\boldsymbol{\sigma}^p,\boldsymbol{\theta}^p,\hat{\boldsymbol{\lambda}}^{i,\varepsilon}(s\mid\boldsymbol{\sigma}^p,\boldsymbol{\theta}^p)\right)}{\partial \boldsymbol{\sigma}^p}ds \tag{9.2.77}$$

$$\frac{\partial G_{i,\varepsilon}(\boldsymbol{\sigma}^p,\boldsymbol{\theta}^p)}{\partial \boldsymbol{\theta}^p}=\int_0^1 \frac{\partial \widehat{H}_{i,\varepsilon}\left(s,\hat{\boldsymbol{x}}(s\mid\boldsymbol{\sigma}^p,\boldsymbol{\theta}^p),\boldsymbol{\sigma}^p,\boldsymbol{\theta}^p,\hat{\boldsymbol{\lambda}}^{i,\varepsilon}(s\mid\boldsymbol{\sigma}^p,\boldsymbol{\theta}^p)\right)}{\partial \boldsymbol{\theta}^p}ds. \tag{9.2.78}$$

9.2.5 Some Convergence Results

In this section, we shall investigate some convergence properties of the sequences of approximate optimal controls obtained in Sections 9.2.3 and 9.2.4.

Theorem 9.2.1 *Let $\left\{\boldsymbol{\sigma}^{p,*}_{\varepsilon,\tilde{\gamma}(\varepsilon)},\boldsymbol{\theta}^{p,*}_{\varepsilon,\tilde{\gamma}(\varepsilon)}\right\}$ be a sequence in ε of the approximate optimal combined vectors produced by Algorithm 9.2.1, where $\tilde{\gamma}(\varepsilon)$ is determined as described in Remark 9.2.1 so as to ensure that the constraints of Problem $(P(p))$ are satisfied. Then,*

$$G_0\left(\boldsymbol{\sigma}^{p,*}_{\varepsilon,\tilde{\gamma}(\varepsilon)},\boldsymbol{\theta}^{p,*}_{\varepsilon,\tilde{\gamma}(\varepsilon)}\right)\to G_0(\boldsymbol{\sigma}^{p,*},\boldsymbol{\theta}^{p,*}),$$

as $\varepsilon\to 0$, where $(\boldsymbol{\sigma}^{p,*},\boldsymbol{\theta}^{p,*})$ is an optimal combined vector of Problem $(P(p))$. Furthermore, any accumulation point of $\left\{\boldsymbol{\sigma}^{p,*}_{\varepsilon,\tilde{\gamma}(\varepsilon)},\boldsymbol{\theta}^{p,*}_{\varepsilon,\tilde{\gamma}(\varepsilon)}\right\}$ is a solution of Problem $(P(p))$.

Proof. The proof is similar to that given for Theorem 4.3.1.

Remark 9.2.8 *Let $(\boldsymbol{\sigma}^{p,*},\boldsymbol{\theta}^{p,*})$ be an optimal combined vector of the approximate Problem $(P(p))$. Then, $(\boldsymbol{\sigma}^{p,*},\boldsymbol{\theta}^{p,*})$ defines uniquely a $\{\boldsymbol{\omega}^{p,*},\nu^{p,*}\}\in\mathcal{B}^p$ via (9.2.12a) and (9.2.18), and vice versa. Furthermore, corresponding to $\{\boldsymbol{\omega}^{p,*},\nu^{p,*}\}\in\mathcal{B}^p$, there exists a unique $(\boldsymbol{u}^{p,*},\boldsymbol{\tau}^{p,*})$ defined on the original time horizon $[0,T]$ such that*

$$\boldsymbol{u}^{p,*}(t)=\sum_{k=1}^{n_p}\boldsymbol{\sigma}^{p,*,k}\chi_{[\tau^{p,*}_{k-1},\tau^{p,*}_k)}(t). \tag{9.2.79}$$

Here, $\boldsymbol{\tau}^{p,*}=[\tau^{p,*}_1,\ldots,\tau^{p,*}_{n_p-1}]^\top$, $\tau^{p,*}_0=0$ and $\tau^{p,*}_{n_p}=T$, are determined uniquely by $\boldsymbol{\theta}^{p,*}$ in the same way as $\boldsymbol{u}^{p,*}_{\varepsilon,\tilde{\gamma}(\varepsilon)}$ is obtained from $\boldsymbol{\theta}^{p,*}_{\varepsilon,\tilde{\gamma}(\varepsilon)}$ as described in Remark 9.2.3 . Note that $\boldsymbol{u}^{p,*}$ is an approximate optimal control of Problem (P).

Theorem 9.2.2 *Let $\boldsymbol{u}^{p,*}$ be as defined in Remark 9.2.3. Suppose that the original Problem (P) has an optimal control $\boldsymbol{u}^*$. Then,*

$$\lim_{p \to \infty} g_0(\boldsymbol{u}^{p,*}) = g_0(\boldsymbol{u}^*). \tag{9.2.80}$$

Proof. The proof is similar to that given for Theorem 8.9.1.

Theorem 9.2.3 *Let $\boldsymbol{u}^*$ be an optimal control of Problem (P), and let $\boldsymbol{u}^{p,*}$ be as defined in Remark 9.2.3 Suppose that*

$$\boldsymbol{u}^{p,*} \to \bar{\boldsymbol{u}}, \tag{9.2.81}$$

a.e. in $[0, 1]$. *Then, $\bar{\boldsymbol{u}}$ is an optimal control of Problem (P).*

Proof. The proof is similar to that given for Theorem 8.9.2.

9.2.6 Illustrative Examples

Example 9.2.1 A non-trivial test problem is taken from Example 6.7.2 of [253]:

$$\min\ g_0, \tag{9.2.82}$$

where

$$g_0 = \int_0^1 \left\{ (x_1(t))^2 + (x_2(t))^2 + 0.005\,(u(t))^2 \right\} dt \tag{9.2.83}$$

subject to

$$\frac{dx_1(t)}{dt} = x_2(t) \tag{9.2.84a}$$

$$\frac{dx_2(t)}{dt} = -x_2(t) + u(t) \tag{9.2.84b}$$

with initial conditions

$$x_1(0) = 0,\ x_2(0) = -1 \tag{9.2.84c}$$

and the continuous state inequality constraint

$$h = -8(t - 0.5)^2 + 0.5 + x_2(t) \le 0,\ \forall t \in [0, 1] \tag{9.2.85}$$

together with the control constraints

$$-20 \le u(t) \le 20,\ \forall t \in [0, 1]. \tag{9.2.86}$$

The constraint transcription of Section 9.2.2 is used to handle the continuous state inequality constraint (9.2.85). Let $\mathcal{L}_\varepsilon$ be constructed from h

according to (9.2.32). Then, we consider the cases of $n_p = 10, 20, 30$ and 40, where n_p is the number of partitions points of the parametrized control as a piecewise constant function. For each of these cases, the problem is solved using the MISER software[104], first without using time scaling transform and then using time scaling transform. The optimization method used within the MISER software is the sequential quadratic programming (SQP). Note that for each case, the constraint transcription method is used to ensure that the continuous inequality constraint (9.2.85) is satisfied, where $\mathcal{L}_\varepsilon$ is constructed from h according to (9.2.32). Take the case of $N = 40$, the numerical results obtained from the use of the constraint transcription method are summarized in Table 9.2.1.

Table 9.2.1: Numerical results for example 9.2.1 with $n_p = 40$ and solved using the constraint transcription method but without using time scaling

ε	γ	$g_0(u)$	$\int_0^1 \mathcal{L}_\varepsilon dt$	$\int_0^1 \max\{h, 0\}dt$	Reason for termination
10^{-2}	0.25×10^{-2}	0.1709	-0.196×10^{-2}	-0.15×10^{-2}	Normal
10^{-2}	0.79×10^{-3}	0.1732	-0.744×10^{-3}	-0.84×10^{-4}	Normal
10^{-2}	0.25×10^{-3}	0.1751	-0.218×10^{-4}	0	Normal
10^{-3}	0.25×10^{-4}	0.1727	-0.205×10^{-5}	-0.13×10^{-6}	Normal
10^{-4}	0.25×10^{-5}	0.1727	-0.335×10^{-6}	-0.10×10^{-6}	Zero derivative

The cost values obtained without using time scaling transform and using time scaling transformation for the cases of $n_p = 10, 20, 30$ and 40 are summarized in Table 9.2.2 and Table 9.2.3, respectively.

Table 9.2.2: Approximate optimal costs for Example 9.2.1 with different n_p and solved without using time scaling

n_p	g_0	Terminated successfully
10	$1.81473135 \times 10^{-1}$	Yes
20	$1.73092969 \times 10^{-1}$	Yes
30	$1.71593913 \times 10^{-1}$	Yes
40	$1.70814337 \times 10^{-1}$	Yes

For the case of $n_p = 40$, Figures 9.2.1 and 9.2.2 show, respectively, the approximate optimal control and approximate optimal state trajectories. For this example, the optimal solution has been obtained in [189]. Compared the optimal solution obtained in [189] with the solution obtained using our method with $n_p = 40$, we see that their trends are similar. In particular, their cost values are basically the same. This confirms the convergence of the approximate optimal costs of our method.

Table 9.2.3: Approximate optimal costs for different n_p and solved using time scaling

n_p	g_0	Terminated successfully
10	$1.78540476 \times 10^{-1}$	Yes
20	1.7275432×10^{-1}	Yes
30	$1.71188006 \times 10^{-1}$	Yes
40	$1.70634166 \times 10^{-1}$	Yes

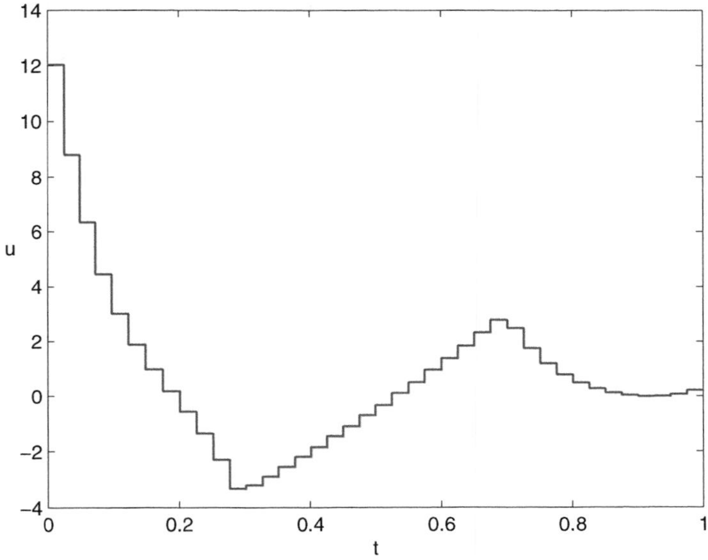

Fig. 9.2.1: Optimal control for Example 9.2.1 with $n_p = 40$ and solved using time scaling

Example 9.2.2 A realistic and complex problem of transferring containers from a ship to a cargo truck at the port of Kobe. It is taken from Example 6.7.3 of [253]. The containers crane is driven by a hoist motor and a trolley drive motor. For safety reason, the objective is to minimize the swing during and at the end of the transfer. See Example 1.2.2 of Chapter 1 for details. Here the problem is summarized after appropriate normalization as follows:

$$\min \left\{ g_0 = 4.5 \int_0^1 \left[(x_3(t))^2 + (x_6(t))^2 \right] dt \right\} \tag{9.2.87}$$

subject to

$$\frac{dx_1(t)}{dt} = 9x_4(t) \tag{9.2.88a}$$

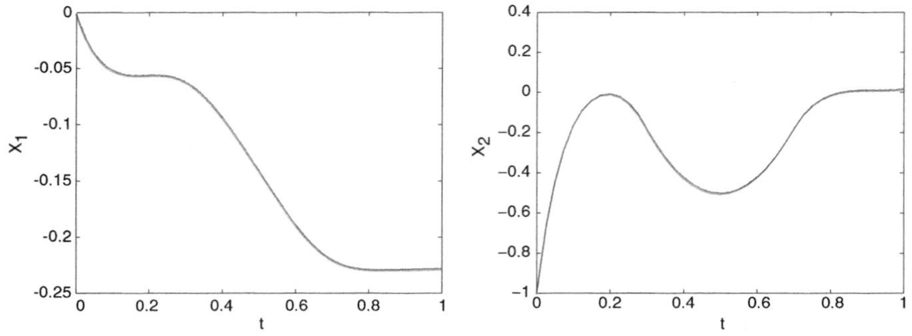

Fig. 9.2.2: Optimal state trajectories for Example 9.2.1 with $n_p = 40$ and solved using time scaling

$$\frac{dx_2(t)}{dt} = 9x_5(t) \tag{9.2.88b}$$

$$\frac{dx_3(t)}{dt} = 9x_6(t) \tag{9.2.88c}$$

$$\frac{dx_4(t)}{dt} = 9(u_1(t) + 17.2656x_3(t)) \tag{9.2.88d}$$

$$\frac{dx_5(t)}{dt} = 9u_2(t) \tag{9.2.88e}$$

$$\frac{dx_6(t)}{dt} = -\frac{9}{x_2(t)}[u_1(t) + 27.0756x_3(t) + 2x_5(t)x_6(t)], \tag{9.2.88f}$$

where

$$\mathbf{x}(0) = [0, 22, 0, 0, -1, 0]^\top \tag{9.2.89a}$$

$$\mathbf{x}(1) = [10, 14, 0, 2.5, 0, 0]^\top \tag{9.2.89b}$$

and

$$|u_1(t)| \leq 2.83374 \tag{9.2.90}$$

$$-0.80865 \leq u_2(t) \leq 0.71265, \ \forall t \in [0, 1]. \tag{9.2.91}$$

with continuous state inequality constraints

$$|x_4(t)| \leq 2.5, \ \forall t \in [0, 1], \tag{9.2.92}$$

$$|x_5(t)| \leq 1.0, \ \forall t \in [0, 1]. \tag{9.2.93}$$

The bounds of the states can be formulated as the continuous inequality constraints as follows:

$$h_1 = x_4(t) - 2.5 \leq 0 \tag{9.2.94}$$

$$h_2 = -x_4(t) - 2.5 \leq 0 \tag{9.2.95}$$

$$h_3 = x_5(t) - 1.0 \leq 0 \tag{9.2.96}$$
$$h_4 = -x_5(t) - 1.0 \leq 0 \tag{9.2.97}$$

$$\mathcal{L}_\varepsilon(t, \boldsymbol{x}(t)) = \sum_{i=1}^{4} \mathcal{L}_{i,\varepsilon}(t, \boldsymbol{x}(t)),$$

where for each $i = 1, \ldots, 4$, $\mathcal{L}_{i,\varepsilon}(t, \boldsymbol{x}(t))$ is constructed from $h_i(t, \boldsymbol{x}(t))$ according to (9.2.32), and $h_i(t, \boldsymbol{x}(t))$, $i = 1, \ldots, 4$, are defined by (9.2.94)–(9.2.97), respectively.

Take the case of $n_p = 40$, the numerical results obtained from the use of the constraint transcription method are summarized in Table 9.2.4.

Table 9.2.4: Numerical results for Example 9.2.2 using constraint transformation method but without using time scaling

ε	γ	$g_0(u)$	$\int_0^1 \mathcal{L}_\varepsilon dt$	$\int_0^1 \max\{h, 0\}dt$	Reason for termination
10^{-2}	0.25×10^{-2}	0.55×10^{-2}	-0.25×10^{-2}	-0.799×10^{-5}	Normal
10^{-3}	0.79×10^{-3}	0.53×10^{-2}	-0.25×10^{-3}	-0.795×10^{-13}	Normal
10^{-4}	0.25×10^{-4}	0.53×10^{-2}	-0.25×10^{-4}	-0.365×10^{-8}	Normal

The optimal cost values obtained without using time scaling and those obtained using time scaling for the cases of $n_p = 20, 30$ and 40 are summarized in the following tables. The reason for the termination of the optimization software is normal for each of these cases, showing that the solution obtained for each of the cases is such that the KKT conditions are satisfied, and so are the continuous inequality constraints.

Table 9.2.5: Approximate optimal costs for Example 9.2.2 with different n_p and solved using time scaling and without using time scaling

n_p	g_0 without time scaling	g_0 with time scaling
20	5.6549084×10^{-3}	$5.50843808 \times 10^{-3}$
30	$5.54097657 \times 10^{-3}$	$5.18506216 \times 10^{-3}$
40	$5.30287028 \times 10^{-3}$	$5.17354239 \times 10^{-3}$

Figures 9.2.3 and 9.2.4 show, respectively, the approximate optimal controls and approximate optimal state trajectories obtained using time scaling transform.

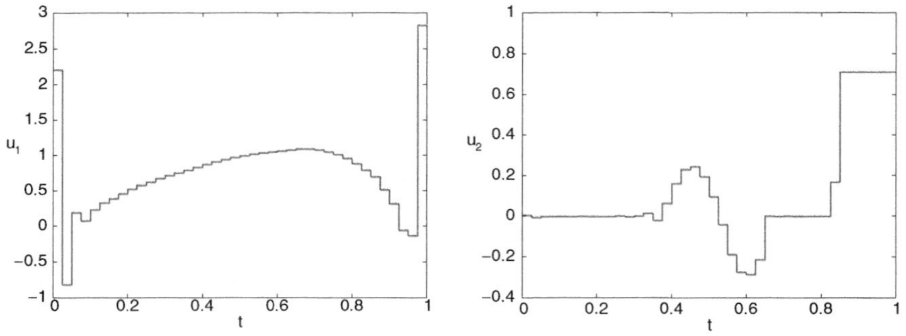

Fig. 9.2.3: Optimal controls for Example 9.2.2 with $n_p = 40$ and solved using time scaling

From Table 9.2.5, we see that for each approximate problem, the optimization software is unable to reduce the cost value further due to the smallness of the cost value. We thus multiply the cost functional with a weighting factor of 10^3 to give $\widetilde{g}_0 = 10^3 \times g_0$. Then, we redo the calculation. The results obtained are listed in Table 9.2.6. From which, we can see the convergence of the approximate optimal controls.

Table 9.2.6: Approximate optimal costs for Example 9.2.2 with different n_p and solved with a weighting factor of 1000 and using time scaling and without using time scaling

n_p	g_0 without time scaling	g_0 with time scaling
20	$5.25532980 \times 10^{-3}$	$5.15123152 \times 10^{-3}$
30	$5.17674340 \times 10^{-3}$	$5.15078395 \times 10^{-3}$
40	$5.15527749 \times 10^{-3}$	$5.15059955 \times 10^{-3}$

Figures 9.2.5 and 9.2.6 show the approximate optimal state trajectories, and approximate optimal controls for the case of $n_p = 40$.

Based on Euler discretization scheme, an algorithm is developed using iterative restoration method in [13]. The modeling language AMPL [F2] is then used to implement the algorithm for constrained optimal control problems, where the optimization software Ipopt [277] is used. This example is solved by using the algorithm developed in [13], where the optimal control obtained is shown to satisfy the optimality conditions. The optimal cost obtained in [13] is 0.005139 with $N = 1000$, where N denotes number of grid points used in Euler discretization. From Table 9.2.6, we see that the difference between the optimal cost obtained in [13] and that obtained by our method for the case of $n_p = 40$ is insignificant. In view of Figures 9.2.5 and 9.2.6, we see that

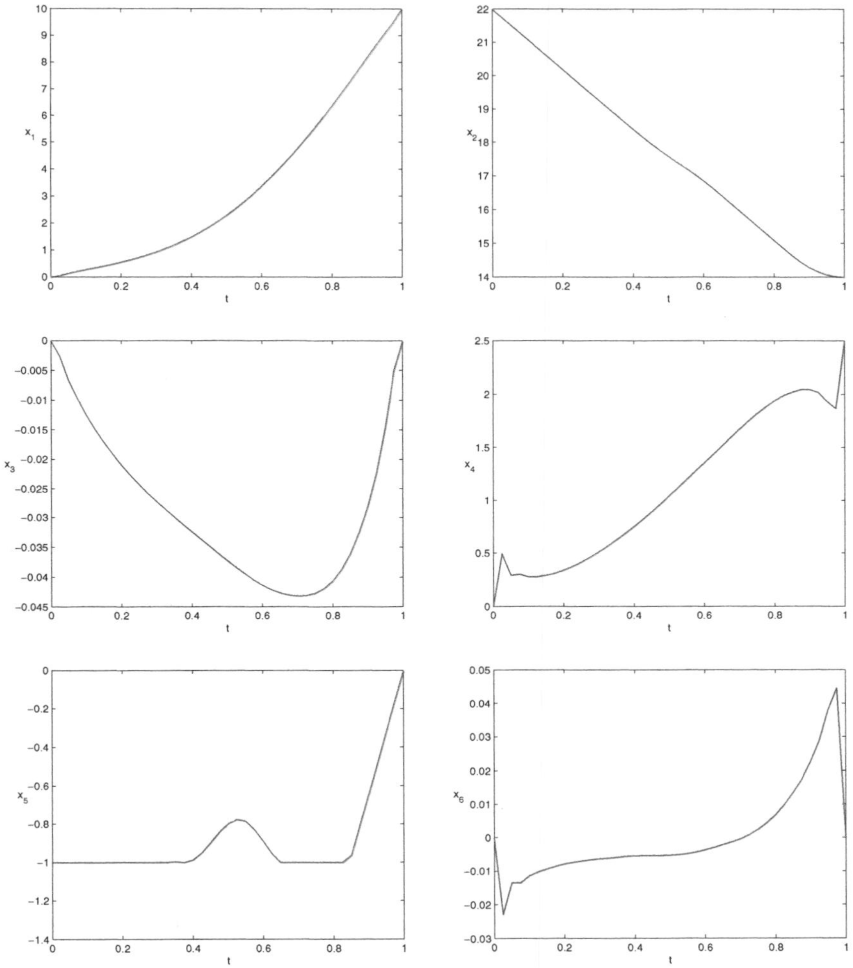

Fig. 9.2.4: Optimal state trajectories for Example 9.2.2 with $n_p = 40$ and solved with a weighting factor of 1000, solved with a weighting factor but without time scaling

the trends of the approximate optimal state trajectories and approximate optimal controls are similar but not identical to those obtained in [13].

In many optimal control problems, a relatively large change in the control parameters near the optimal solution often produces only a small change in the value of the cost functional g_0. In other words, the exact value of the optimal control does not matter much in some problems. This phenomenon, while may be valuable to the system designer, produces ill-conditioning in the computation of optimal control.

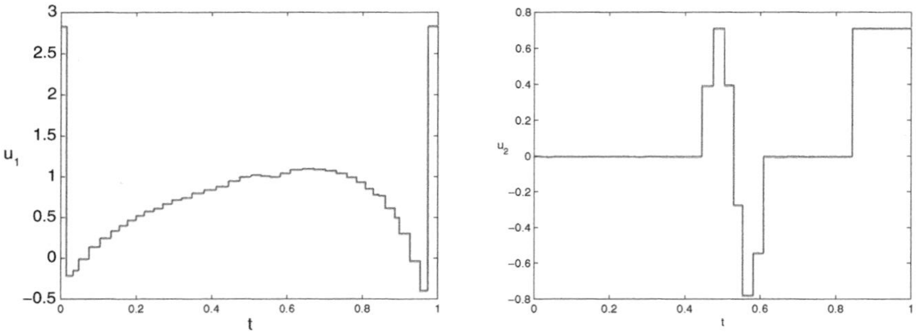

Fig. 9.2.5: Optimal control for Example 9.2.2 with $n_p = 40$ and solved with a weighting factor of 1000 and using time scaling

9.3 Exact Penalty Function Approach

In this section, the exact penalty function approach detailed in Section 4.4 will be used to develop a computational method for solving a class of optimal control problems to be described below. The main references for this section are [134, 147, 300, 301].

Consider the system of differential equations given by (9.2.1a) with initial condition (9.2.1b) and terminal equality constraint given by

$$\boldsymbol{x}(T) = \boldsymbol{x}^f, \tag{9.3.1}$$

where T is the terminal time, $\boldsymbol{x} = [x_1, \ldots, x_n]^\top \in \mathbb{R}^n$ and $\boldsymbol{u} = [u_1, \ldots u_r]^\top \in \mathbb{R}^r$are, respectively, state and control vectors, and $\boldsymbol{f} = [f_1, \ldots, f_n]^\top \in \mathbb{R}^n$ is a given functional.

Assumption 9.3.1 *The function $\boldsymbol{f}$ satisfies the relevant conditions appearing in Assumptions 8.2.1 and 8.2.2.*

Define

$$U = \{\boldsymbol{\nu} = [v_1, \ldots v_r]^\top \in \mathbb{R}^r : \alpha_i \le v_i \le \beta_i, \ i = 1, \ldots, r\}, \tag{9.3.2}$$

where α_i, $i = 1, \ldots, r$, and β_i, $i = 1, \ldots, r$, are given real numbers. A piecewise continuous function $\boldsymbol{u}$ is said to be an admissible control if $\boldsymbol{u}(t) \in U$ for all $t \in [0, T]$. Let $\mathcal{U}$ be the class of all such admissible controls. Furthermore, let $\boldsymbol{x}(\cdot \mid \boldsymbol{u})$ denote the solution of system (9.2.1a)–(9.2.1b) corresponding to $\boldsymbol{u} \in \mathcal{U}$.

Consider the following continuous state inequality constraints (9.2.3).

$$h_i(t, \boldsymbol{x}(t \mid \boldsymbol{u}), \boldsymbol{u}(t)) \le 0, \ \forall t \in [0, T], \quad i = 1, \ldots, N. \tag{9.3.3}$$

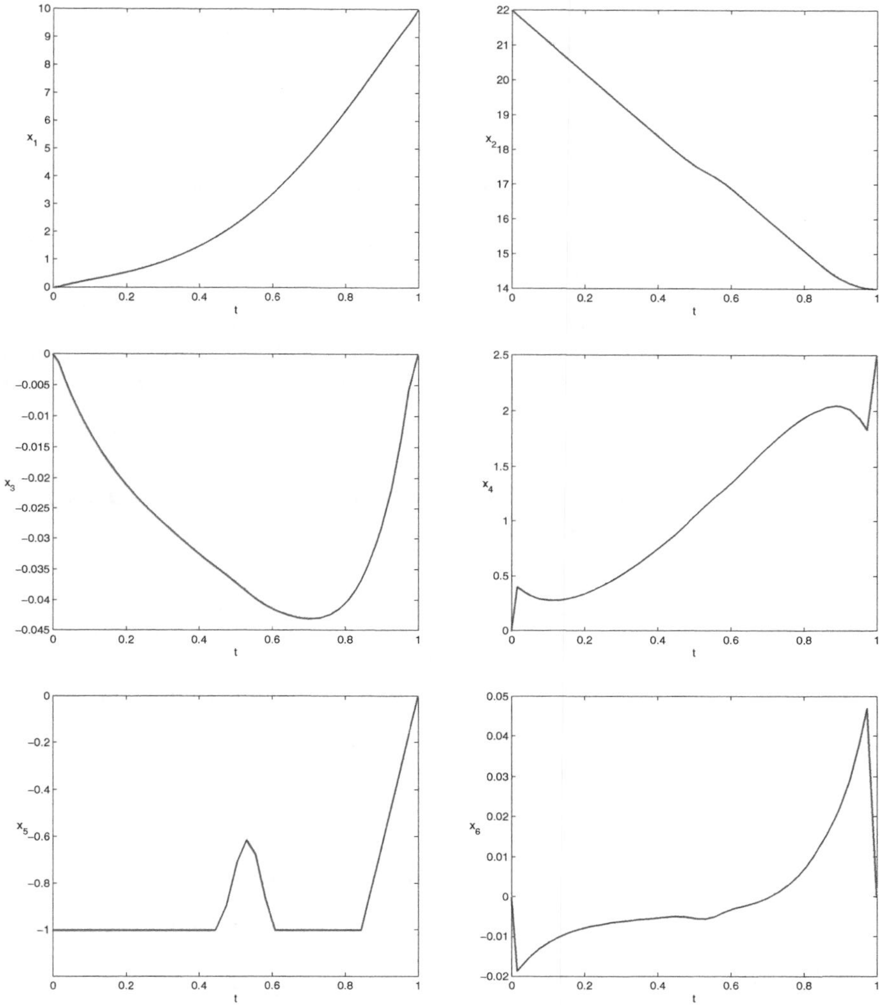

Fig. 9.2.6: Optimal state trajectories for Example 9.2.2 with $n_p = 40$ and solved with a weighting factor of 1000 and using time scaling

We now state the corresponding optimal control problem formally as follows:

Problem ($\hat{\boldsymbol{P}}$) Given the dynamical system (9.2.1a) and (9.2.1b) subject to the terminal constraint (9.3.1) and the continuous inequality constraints (9.3.3), find a control $\boldsymbol{u} \in \mathcal{U}$ such that the cost function

$$g_0(\boldsymbol{u}) = \Phi_0(\boldsymbol{x}(T \mid \boldsymbol{u})) + \int_0^T \mathcal{L}_0(t, \boldsymbol{x}(t \mid \boldsymbol{u}), \boldsymbol{u}(t))dt \tag{9.3.4}$$

is minimized.

We assume that the following conditions are satisfied.

Assumption 9.3.2 *Φ_0 is continuously differentiable with respect to x.*

Assumption 9.3.3 *h_i, $i = 1, \ldots, N$ and $\mathcal{L}_0$ are continuously differentiable with respect to all their arguments.*

Remark 9.3.1 *By Assumption 9.3.1 and the definition of $\mathcal{U}$, it follows from an argument similar to that given for the proof of Lemma 8.4.2 that for all $\boldsymbol{u} \in \mathcal{U}$, $x(t|\boldsymbol{u}) \in X$ holds, and for all $t \in [0, T]$, $X \subset \mathbb{R}^n$ is a compact subset.*

9.3.1 Control Parametrization and Time Scaling Transformation

To solve Problem $(\hat{P})$, we shall apply the control parametrization scheme together with a time scaling transform. The time horizon $[0, T]$ is partitioned with a sequence $\boldsymbol{\tau} = \{\tau_0, \ldots, \tau_p\}$ of time points τ_i, $i = 1, \ldots, p-1$. Then, the control is approximated by a piecewise constant function as follows.

$$\boldsymbol{u}^p(t \mid \boldsymbol{\sigma}, \boldsymbol{\tau}) = \sum_{j=1}^{p} \boldsymbol{\sigma}^j \chi_{[\tau_{j-1}, \tau_j)}(t), \tag{9.3.5}$$

where $\tau_{j-1} \le \tau_j$, $j = 1, \ldots, p$, with $\tau_0 = 0$ and $\tau_p = T$, $\boldsymbol{\sigma}^j = \left[\sigma_1^j, \ldots, \sigma_r^j\right]^\top \in \mathbb{R}^r$, $j = 1, \ldots, p$, $\boldsymbol{\sigma} = \left[\left(\boldsymbol{\sigma}^1\right)^\top, \ldots, \left(\boldsymbol{\sigma}^p\right)^\top\right]^\top \in \mathbb{R}^{pr}$ and χ_I is the indicator function of I defined by (7.3.3).

As $\boldsymbol{u}^p \in \mathcal{U}$, $\boldsymbol{\sigma}^j \in U$ for $j = 1, \ldots, p$. Let Ξ be the set of all those $\boldsymbol{\sigma} = \left[\left(\boldsymbol{\sigma}^1\right)^\top, \ldots, \left(\boldsymbol{\sigma}^p\right)^\top\right]^\top \in \mathbb{R}^{pr}$ such that $\boldsymbol{\sigma}^j \in U$ for $j = 1, \ldots, p$.

The switching times τ_j, $1 \le j \le p-1$, are also regarded as decision variables. The time scaling transform is employed to map these switching times into a set of fixed time points $\frac{k}{p}$, $k = 1, \ldots, p-1$, on a new time horizon $[0, 1]$. This is easily achieved by the following differential equation

$$\frac{dt(s)}{ds} = \upsilon^p(s), \quad s \in [0, 1], \tag{9.3.6a}$$

with initial condition

$$t(0) = 0, \tag{9.3.6b}$$

where

$$\upsilon^p(s) = \sum_{j=1}^{p} \theta_j \chi_{\left[\frac{j-1}{p}, \frac{j}{p}\right)}(s). \tag{9.3.7}$$

Here, $\theta_j \geq 0,\ j = 1, \ldots p$. Let $\boldsymbol{\theta} = [\theta_1, \ldots, \theta_p]^\top \in \mathbb{R}^p$ and let Θ be the set containing all such $\boldsymbol{\theta}$.

Taking integration of (9.3.6a) with initial condition (9.3.6b), it is easy to see that, for $s \in \left[\frac{k-1}{p}, \frac{k}{p}\right)$, $k = 1, \ldots, p$,

$$t(s) = \sum_{j=1}^{k-1} \frac{\theta_j}{p} + \frac{\theta_k}{p}(ps - k + 1), \tag{9.3.8}$$

where $k = 1, \ldots, p$. Clearly, for $k = 1, \ldots, p-1$,

$$\tau_k = \sum_{j=1}^{k} \frac{\theta_j}{p} \tag{9.3.9}$$

and

$$t(1) = \sum_{j=1}^{p} \frac{\theta_j}{p} = T. \tag{9.3.10}$$

The approximate control given by (9.3.5) in the new time horizon $[0,1]$ becomes

$$\tilde{\boldsymbol{u}}^p(s) = \boldsymbol{u}^p(t(s)) = \sum_{j=1}^{p} \boldsymbol{\sigma}^j \chi_{\left[\frac{j-1}{p}, \frac{j}{p}\right)}(s), \tag{9.3.11}$$

which has fixed switching times at $s = \frac{1}{p}, \ldots, \frac{p-1}{p}$. Now, by using the time scaling transform (9.3.6a) and (9.3.6b), the dynamic system (9.2.1a) and (9.2.1b) is transformed into

$$\frac{d\boldsymbol{y}(s)}{ds} = \theta_k \boldsymbol{f}\left(t(s), \boldsymbol{y}(s), \boldsymbol{\sigma}^k\right), \quad s \in \mathcal{J}_k,\ k = 1, \ldots, p \tag{9.3.12a}$$

$$\frac{dt(s)}{ds} = \upsilon^p(s) \tag{9.3.12b}$$

$$\boldsymbol{y}(0) = \boldsymbol{x}^0 \quad \text{and} \quad t(0) = 0, \tag{9.3.12c}$$

where

$$\boldsymbol{y}(s) = [y_1(s), \ldots, y_n(s)]^\top,$$

and

$$\boldsymbol{x}^0 = \left[x_1^0, \ldots, x_n^0\right]^\top$$

.

The terminal conditions (9.3.1) and (9.3.10) become

$$\boldsymbol{y}(1) = \boldsymbol{x}^f \quad \text{and} \quad t(1) = T, \tag{9.3.13a}$$

respectively, where $\boldsymbol{y}(s) = \boldsymbol{x}(t(s))$ and

$$\mathcal{J}_k = \begin{cases} \left[\frac{k-1}{p}, \frac{k}{p}\right), & \text{if } k = 1, \\ \left(\frac{k-1}{p}, \frac{k}{p}\right), & \text{if } k \in \{2, \ldots, p-1\}, \\ \left(\frac{k-1}{p}, \frac{k}{p}\right], & \text{if } k = p. \end{cases} \tag{9.3.14}$$

We then rewrite system (9.3.12a)–(9.3.12c) as follows.

$$\frac{d\tilde{\boldsymbol{y}}(s)}{ds} = \tilde{\boldsymbol{f}}(s, \tilde{\boldsymbol{y}}(s), \boldsymbol{\sigma}, \boldsymbol{\theta}), \quad s \in [0, 1] \tag{9.3.15a}$$

$$\tilde{\boldsymbol{y}}(0) = \tilde{\boldsymbol{y}}^0 \tag{9.3.15b}$$

with the terminal conditions

$$\tilde{\boldsymbol{y}}(1) = \tilde{\boldsymbol{y}}^f, \tag{9.3.15c}$$

where

$$\tilde{\boldsymbol{y}}(s) = [\widetilde{y}_1(s), \ldots, \widetilde{y}_n(s), \widetilde{y}_{n+1}(s)]^\top \tag{9.3.16}$$

with

$$\widetilde{y}_i(s) = y_i(s), \;\; i = 1, \ldots, n, \;\; \widetilde{y}_{n+1}(s) = t(s);$$

$$\tilde{\boldsymbol{f}}(s, \tilde{\boldsymbol{y}}(s), \boldsymbol{\sigma}, \boldsymbol{\theta}) = \begin{bmatrix} \sum_{k=1}^{p} \theta_k \boldsymbol{f}(t(s), \boldsymbol{y}(s), \boldsymbol{\sigma}^k) \chi_{\mathcal{J}_k}(s) \\ \upsilon^p(s) \end{bmatrix}; \tag{9.3.17}$$

$$\tilde{\boldsymbol{y}}^0 = \left[\widetilde{y}_1^0, \ldots, \widetilde{y}_n^0, \widetilde{y}_{n+1}^0\right]^\top \tag{9.3.18}$$

with

$$\widetilde{y}_i^0 = x_i^0, \; i = 1, \ldots, n, \; \widetilde{y}_{n+1}^0 = 0;$$

and

$$\tilde{\boldsymbol{y}}^f = \left[\widetilde{y}_1^f, \ldots, \widetilde{y}_n^f, \widetilde{y}_{n+1}^f\right]^\top \tag{9.3.19}$$

with

$$\widetilde{y}_i^f = x_i^f, \; i = 1, \ldots, n, \; \widetilde{y}_{n+1}^f = T.$$

To proceed further, let $\tilde{\boldsymbol{y}}(\cdot \mid \boldsymbol{\sigma}, \boldsymbol{\theta})$ denote the solution of system (9.3.15a)–(9.3.15b) corresponding to $(\boldsymbol{\sigma}, \boldsymbol{\theta}) \in \Xi \times \Theta$. Similarly, applying the time scaling transform to the continuous inequality constraints (9.3.3) and the cost functional (9.3.4) yields

$$h_i\left(\tilde{\boldsymbol{y}}(s \mid \boldsymbol{\sigma}, \boldsymbol{\theta}), \boldsymbol{\sigma}^k\right) \le 0, \;\; \forall s \in \mathcal{J}_k, \;\; k = 1, \ldots, p; \; i = 1, \ldots, N \tag{9.3.20}$$

and

$$\widetilde{g}_0(\boldsymbol{\sigma},\boldsymbol{\theta}) = \Phi_0(\boldsymbol{y}(1 \mid \boldsymbol{\sigma},\boldsymbol{\theta})) + \int_0^1 \bar{\mathcal{L}}_0(s,\tilde{\boldsymbol{y}}(s \mid \boldsymbol{\sigma},\boldsymbol{\theta}),\boldsymbol{\sigma},\boldsymbol{\theta})ds, \tag{9.3.21}$$

respectively, where

$$\bar{\mathcal{L}}_0(s,\tilde{\boldsymbol{y}}(s \mid \boldsymbol{\sigma},\boldsymbol{\theta}),\boldsymbol{\sigma},\boldsymbol{\theta}) = v^p(s)\mathcal{L}_0(t(s),\boldsymbol{y}(s),\tilde{\boldsymbol{u}}^p(s)). \tag{9.3.22}$$

Remark 9.3.2 *Assumption 9.3.1, Assumption 9.3.2 and Assumption 9.3.3, it follows from Remark 9.3.1 that there exits constants $K_1 > 0$ and $K_2 > 0$ such that*

$$|\Upsilon(s,\tilde{\boldsymbol{y}}(s \mid \boldsymbol{\sigma},\boldsymbol{\theta}),\boldsymbol{\sigma},\boldsymbol{\theta})| \leq K_1, \;\; s \in \mathcal{J}_k, \;\; k = 1,\ldots,p; \; (\boldsymbol{\sigma},\boldsymbol{\theta}) \in \Xi \times \Theta$$

$$\left|\frac{\partial \Upsilon(s,\tilde{\boldsymbol{y}}(s \mid \boldsymbol{\sigma},\boldsymbol{\theta}),\boldsymbol{\sigma},\boldsymbol{\theta})}{\partial \boldsymbol{\sigma}}\right| \leq K_2, \;\; s \in \mathcal{J}_k, \;\; k = 1,\ldots,p; \; (\boldsymbol{\sigma},\boldsymbol{\theta}) \in \Xi \times \Theta$$

$$\left|\frac{\partial \Upsilon(s,\tilde{\boldsymbol{y}}(s \mid \boldsymbol{\sigma},\boldsymbol{\theta}),\boldsymbol{\sigma},\boldsymbol{\theta})}{\partial \boldsymbol{\theta}}\right| \leq K_2, \;\; s \in \mathcal{J}_k, \;\; k = 1,\ldots,p; \; (\boldsymbol{\sigma},\boldsymbol{\theta}) \in \Xi \times \Theta$$

$$\left|\frac{\partial \Upsilon(s,\tilde{\boldsymbol{y}}(s \mid \boldsymbol{\sigma},\boldsymbol{\theta}),\boldsymbol{\sigma},\boldsymbol{\theta})}{\partial \widetilde{\boldsymbol{y}}}\right| \leq K_2, \;\; s \in \mathcal{J}_k, \;\; k = 1,\ldots,p; \; (\boldsymbol{\sigma},\boldsymbol{\theta}) \in \Xi \times \Theta,$$

where Υ is used to denote $\widetilde{\boldsymbol{f}}_i$, $i = 1,\ldots,n$; $h_i(t(s),\tilde{\boldsymbol{y}}(s \mid \boldsymbol{\sigma},\boldsymbol{\theta}),\boldsymbol{\sigma}^k)$, $i = 1,\ldots,N$; $k = 1,\ldots,p$; and $\bar{\mathcal{L}}_0$.

The approximate problem to Problem $(\hat{P})$ may now be stated formally as follows.

Problem $(\hat{\boldsymbol{P}}(\boldsymbol{p}))$ Given system (9.3.15a) and (9.3.15b), find a $(\boldsymbol{\sigma},\boldsymbol{\theta}) \in \Xi \times \Theta$ such that the cost functional (9.3.21) is minimized subject to (9.3.15c) and (9.3.20).

Thus, Problem $(\hat{P}(p))$ becomes an optimization problem subject to both the equality constraints (9.3.15c) and the continuous inequality constraints (9.3.20). To solve this problem, an exact penalty function method introduced in Section 4.4 is used.

First, we define

$$\mathcal{F}_\varepsilon = \{(\boldsymbol{\sigma},\boldsymbol{\theta},\varepsilon) \in \Xi \times \Theta \times \mathbb{R}_+ : h_i\left(\tilde{\boldsymbol{y}}(s \mid \boldsymbol{\sigma},\boldsymbol{\theta}),\boldsymbol{\sigma}^k\right) \leq \varepsilon^\gamma W_i,$$
$$\forall\, s \in \mathcal{J}_k, \; k = 1,\ldots,p; \; i = 1, \;\ldots\;, N\}, \tag{9.3.23}$$

where $\mathbb{R}_+ = \{\alpha \in \mathbb{R} : \; \alpha \geq 0\}$, $W_i \in (0,1)$, $i = 1, \;\ldots\;, N$, are fixed constants and γ is a positive real number. In particularly, when $\varepsilon = 0$,

$$\mathcal{F}_0 = \{(\boldsymbol{\sigma}, \boldsymbol{\theta}) \in \Xi \times \Theta : h_i\left(\tilde{\boldsymbol{y}}(s \mid \boldsymbol{\sigma}, \boldsymbol{\theta}), \boldsymbol{\sigma}^k\right) \le 0, \\ \forall\ s \in \mathcal{J}_k,\ k = 1, \ldots, p;\ i = 1,\ \ldots\ ,\ N\}. \tag{9.3.24}$$

Similarly, we define

$$\Omega_\varepsilon = \left\{(\boldsymbol{\sigma}, \boldsymbol{\theta}, \varepsilon) \in \mathcal{F}_\varepsilon : \tilde{\boldsymbol{y}}(1 \mid \boldsymbol{\sigma}, \boldsymbol{\theta}) - \tilde{\boldsymbol{y}}^f = \mathbf{0}\right\} \tag{9.3.25}$$

and

$$\Omega_0 = \left\{(\boldsymbol{\sigma}, \boldsymbol{\theta}) \in \mathcal{F}_0 : \tilde{\boldsymbol{y}}(1 \mid \boldsymbol{\sigma}, \boldsymbol{\theta}) - \tilde{\boldsymbol{y}}^f = \mathbf{0}\right\}. \tag{9.3.26}$$

Clearly, Problem $(\hat{P}(p))$ is equivalent to the following problem, which is denoted as Problem $(\tilde{P}(p))$.

Problem $(\tilde{\boldsymbol{P}}(\boldsymbol{p}))$ Given system (9.3.15a) and (9.3.15b), find a $(\boldsymbol{\sigma}, \boldsymbol{\theta}) \in \Omega_0$ such that the cost functional (9.3.21) is minimized.

Then, by applying the exact penalty function introduced in Section 4.4, we obtain a new cost functional defined below.

$$\widetilde{g}_0^\delta(\boldsymbol{\sigma}, \boldsymbol{\theta}, \varepsilon) \\ = \begin{cases} \widetilde{g}_0(\boldsymbol{\sigma}, \boldsymbol{\theta}), & \text{if } \varepsilon = 0,\ h_i(\tilde{\boldsymbol{y}}(s), \boldsymbol{\sigma}^k) \le 0 \\ & \quad (s \in \mathcal{J}_k, k = 1, \ldots, p), \\ \widetilde{g}_0(\boldsymbol{\sigma}, \boldsymbol{\theta}) + \varepsilon^{-\alpha}(\Delta(\boldsymbol{\sigma}, \boldsymbol{\theta}, \varepsilon) + \Delta_1) + \delta\varepsilon^\beta, & \text{if } \varepsilon > 0, \\ +\infty, & \text{otherwise}, \end{cases} \tag{9.3.27}$$

where $\Delta(\boldsymbol{\sigma}, \boldsymbol{\theta}, \varepsilon)$, which is referred to as the continuous inequality constraint violation, is defined by

$$\Delta(\boldsymbol{\sigma}, \boldsymbol{\theta}, \varepsilon) = \sum_{i=1}^{N} \sum_{k=1}^{p} \int_{\mathcal{J}_k} \left[\max\left\{0, h_i\left(\tilde{\boldsymbol{y}}(s \mid \boldsymbol{\sigma}, \boldsymbol{\theta}), \boldsymbol{\sigma}^k\right) - \varepsilon^\gamma W_i\right\}\right]^2 ds, \tag{9.3.28}$$

α and γ are positive real numbers, $\beta > 2$, and $\delta > 0$ is a penalty parameter, while Δ_1, which is referred to as the equality constraints violation, is defined by

$$\Delta_1 = \left|\tilde{\boldsymbol{y}}(1 \mid \boldsymbol{\sigma}, \boldsymbol{\theta}) - \tilde{\boldsymbol{y}}^f\right|^2 = \sum_{i=1}^{n+1} \left(\tilde{y}_i(1 \mid \boldsymbol{\sigma}, \boldsymbol{\theta}) - \tilde{y}_i^f\right)^2, \tag{9.3.29}$$

and $|\cdot|$ denotes the usual Euclidian norm.

Remark 9.3.3 *Note that other types of equality constraints, such as interior point equality constraints* (8.3.5) *can be dealt with similarly by introducing appropriate equality constraint violation as defined by* (9.3.29).

We now introduce a surrogate optimal control problem, which is referred to as Problem $(\tilde{P}_\delta(p))$, as follows.

Problem ($\tilde{P}_\delta(p)$) Given system (9.3.15a)–(9.3.15b), find a $(\boldsymbol{\sigma}, \boldsymbol{\theta}, \varepsilon) \in \Xi \times \Theta \times [0, +\infty)$ such that the cost functional (9.3.27) is minimized.

Intuitively, during the process of minimizing $\widetilde{g}_0^\delta(\boldsymbol{\sigma}, \boldsymbol{\theta}, \varepsilon)$, if δ is increased, ε^β should be reduced, meaning that ε should be reduced as β is fixed. Thus $\varepsilon^{-\alpha}$ will be increased, and hence the constraint violation will be reduced. This means that the values of

$$\sum_{i=1}^{N}\sum_{k=1}^{p}\int_{\mathcal{J}_k}\left[\max\left\{0, h_i\left(\tilde{y}(s|\sigma,\theta),\sigma^k\right) - \epsilon^\gamma W_i\right\}\right]^2 \text{ and } \sum_{i=1}^{n+1}\left(\tilde{y}_i(1|\sigma,\theta) - \tilde{y}_i^f\right)^2$$

must go down such that the continuous inequality constraints (9.3.20) and the equality constraints (9.3.15c) are satisfied.

Before deriving the gradient of the cost functional of Problem ($\tilde{P}_\delta(p)$), we will rewrite the cost functional in the canonical form below.

$$\begin{aligned}
\widetilde{g}_0^\delta(\boldsymbol{\sigma}, \boldsymbol{\theta}, \varepsilon) &= \Phi_0(\boldsymbol{y}(1 \mid \boldsymbol{\sigma}, \boldsymbol{\theta})) + \int_0^1 \bar{\mathcal{L}}_0(s, \tilde{\boldsymbol{y}}(s \mid \boldsymbol{\sigma}, \boldsymbol{\theta}), \boldsymbol{\sigma}, \boldsymbol{\theta}) ds \\
&\quad + \varepsilon^{-\alpha}\left\{\sum_{i=1}^{N}\int_0^1 \left[\max\left\{0, \bar{h}_i(s, \tilde{\boldsymbol{y}}(s \mid \boldsymbol{\sigma}, \boldsymbol{\theta}), \boldsymbol{\sigma}) - \varepsilon^\gamma W_i\right\}\right]^2 ds\right. \\
&\quad \left. + \sum_{i=1}^{n+1}\left(\tilde{\boldsymbol{y}}_i(1 \mid \boldsymbol{\sigma}, \boldsymbol{\theta}) - \tilde{\boldsymbol{y}}_i^f\right)^2\right\} + \delta\varepsilon^\beta \\
&= \left\{\Phi_0(\boldsymbol{y}(1 \mid \boldsymbol{\sigma}, \boldsymbol{\theta})) + \varepsilon^{-\alpha}\sum_{i=1}^{n+1}\left(\tilde{\boldsymbol{y}}_i(1 \mid \boldsymbol{\sigma}, \boldsymbol{\theta}) - \tilde{\boldsymbol{y}}_i^f\right)^2 + \delta\varepsilon^\beta\right\} \\
&\quad + \int_0^1 \bar{\mathcal{L}}_0(s, \tilde{\boldsymbol{y}}(s \mid \boldsymbol{\sigma}, \boldsymbol{\theta}), \boldsymbol{\sigma}, \boldsymbol{\theta}) ds \\
&\quad + \varepsilon^{-\alpha}\sum_{i=1}^{N}\int_0^1 \left[\max\left\{0, \bar{h}_i(s, \tilde{\boldsymbol{y}}(s \mid \boldsymbol{\sigma}, \boldsymbol{\theta}), \boldsymbol{\sigma}) - \varepsilon^\gamma W_i\right\}\right]^2 ds,
\end{aligned} \tag{9.3.30}$$

where

$$\bar{h}_i(s, \tilde{\boldsymbol{y}}(s \mid \boldsymbol{\sigma}, \boldsymbol{\theta}), \boldsymbol{\sigma}) = h_i(\tilde{\boldsymbol{y}}(s \mid \boldsymbol{\sigma}, \boldsymbol{\theta}), \tilde{\boldsymbol{u}}^p(s)), \ i = 1, \ldots, N \tag{9.3.31}$$

and $\tilde{\boldsymbol{u}}^p(s)$ is defined by (9.3.11).

Let

$$\tilde{\Phi}_0(\tilde{\boldsymbol{y}}(1 \mid \boldsymbol{\sigma}, \boldsymbol{\theta}), \varepsilon) = \Phi_0(\boldsymbol{y}(1 \mid \boldsymbol{\sigma}, \boldsymbol{\theta})) + \varepsilon^{-\alpha}\sum_{i=1}^{n+1}\left(\tilde{\boldsymbol{y}}_i(1 \mid \boldsymbol{\sigma}, \boldsymbol{\theta}) - \tilde{\boldsymbol{y}}_i^f\right)^2 + \delta\varepsilon^\beta \tag{9.3.32}$$

and

$$\tilde{\mathcal{L}}_0(s, \tilde{\boldsymbol{y}}(s \mid \boldsymbol{\sigma}, \boldsymbol{\theta}), \boldsymbol{\sigma}, \theta, \varepsilon) = \bar{\mathcal{L}}_0(s, \tilde{\boldsymbol{y}}(s \mid \boldsymbol{\sigma}, \boldsymbol{\theta}), \boldsymbol{\sigma}, \boldsymbol{\theta})$$

$$+\varepsilon^{-\alpha}\sum_{i=1}^{N}\int_0^1 \left[\max\left\{0,\bar{h}_i(s,\tilde{\boldsymbol{y}}(s\mid\boldsymbol{\sigma},\boldsymbol{\theta}),\boldsymbol{\sigma})-\varepsilon^{\gamma}W_i\right\}\right]^2. \tag{9.3.33}$$

We then substitute (9.3.32) and (9.3.33) into (9.3.30) to give

$$\widetilde{g}_0^{\delta}(\sigma,\theta,\epsilon)=\tilde{\Phi}_0(\tilde{\boldsymbol{y}}(1\mid\boldsymbol{\sigma},\boldsymbol{\theta}),\varepsilon)+\int_0^1\tilde{\mathcal{L}}_0(s,\tilde{\boldsymbol{y}}(s\mid\boldsymbol{\sigma},\boldsymbol{\theta}),\boldsymbol{\sigma},\boldsymbol{\theta},\varepsilon)ds. \tag{9.3.34}$$

Now, the cost functional of Problem $(\tilde{P}_\delta(p))$ is in canonical form. As derived for the proof of Theorem 7.2.2, the gradient formulas of the cost functional (9.3.34) are given by the following theorem.

Theorem 9.3.1 *The gradients of the cost functional $\widetilde{g}_0^{\delta}(\boldsymbol{\sigma},\boldsymbol{\theta},\varepsilon)$ with respect to $\boldsymbol{\sigma}$, $\boldsymbol{\theta}$, and ε are*

$$\frac{\partial\widetilde{g}_0^{\delta}(\boldsymbol{\sigma},\boldsymbol{\theta},\varepsilon)}{\partial\boldsymbol{\sigma}}=\int_0^1\frac{\partial H_0\left(s,\tilde{\boldsymbol{y}}(s\mid\boldsymbol{\sigma},\boldsymbol{\theta}),\boldsymbol{\sigma},\boldsymbol{\theta},\varepsilon,\boldsymbol{\lambda}^0(s\mid\boldsymbol{\sigma},\boldsymbol{\theta},\varepsilon)\right)}{\partial\boldsymbol{\sigma}}ds \tag{9.3.35}$$

$$\frac{\partial\widetilde{g}_0^{\delta}(\boldsymbol{\sigma},\boldsymbol{\theta},\varepsilon)}{\partial\boldsymbol{\theta}}=\int_0^1\frac{\partial H_0\left(s,\tilde{\boldsymbol{y}}(s\mid\boldsymbol{\sigma},\boldsymbol{\theta}),\boldsymbol{\sigma},\boldsymbol{\theta},\varepsilon,\boldsymbol{\lambda}^0(s\mid\boldsymbol{\sigma},\boldsymbol{\theta},\varepsilon)\right)}{\partial\boldsymbol{\theta}}ds \tag{9.3.36}$$

$$\begin{aligned}\frac{\partial\widetilde{g}_0^{\delta}(\boldsymbol{\sigma},\boldsymbol{\theta},\varepsilon)}{\partial\varepsilon}&=-\alpha\varepsilon^{-\alpha-1}\Bigg\{\sum_{i=1}^{N}\int_0^1\left[\max\left\{0,\bar{h}_i(s,\tilde{\boldsymbol{y}}(s\mid\boldsymbol{\sigma},\boldsymbol{\theta}),\boldsymbol{\sigma})-\varepsilon^{\gamma}W_i\right\}\right]^2ds\\
&\quad+\sum_{i=1}^{n+1}\left(\tilde{y}_i(1\mid\boldsymbol{\sigma},\boldsymbol{\theta})-\tilde{y}_i^f\right)^2\Bigg\}\\
&\quad-2\gamma\varepsilon^{\gamma-\alpha-1}\sum_{i=1}^{N}\int_0^1\max\left\{0,\bar{h}_i(s,\tilde{\boldsymbol{y}}(s\mid\boldsymbol{\sigma},\boldsymbol{\theta}),\boldsymbol{\sigma})-\varepsilon^{\gamma}W_i\right\}W_ids\\
&\quad+\delta\beta\varepsilon^{\beta-1}\\
&=\varepsilon^{-\alpha-1}\Bigg\{-\alpha\sum_{i=1}^{N}\int_0^1\left[\max\left\{0,\bar{h}_i(s,\tilde{\boldsymbol{y}}(s\mid\boldsymbol{\sigma},\boldsymbol{\theta}),\boldsymbol{\sigma})-\varepsilon^{\gamma}W_i\right\}\right]^2ds\\
&\quad+2\gamma\sum_{i=1}^{N}\int_0^1\max\left\{0,\bar{h}_i(s,\tilde{\boldsymbol{y}}(s\mid\boldsymbol{\sigma},\boldsymbol{\theta}),\boldsymbol{\sigma})-\varepsilon^{\gamma}W_i\right\}(-\varepsilon^{\gamma}W_i)ds\\
&\quad-\alpha\sum_{i=1}^{n+1}\left(\tilde{y}_i(1\mid\boldsymbol{\sigma},\boldsymbol{\theta})-\tilde{y}_i^f\right)^2\Bigg\}+\delta\beta\varepsilon^{\beta-1},\end{aligned} \tag{9.3.37}$$

respectively, where $H_0\left(s,\tilde{\boldsymbol{y}}(s\mid\boldsymbol{\sigma},\boldsymbol{\theta}),\boldsymbol{\sigma},\boldsymbol{\theta},\varepsilon,\boldsymbol{\lambda}^0(s\mid\boldsymbol{\sigma},\boldsymbol{\theta},\varepsilon)\right)$ is the Hamiltonian function for the cost functional (9.3.34) *given by*

$$\begin{aligned}&H_0\big(s,\tilde{\boldsymbol{y}}(s\mid\boldsymbol{\sigma},\boldsymbol{\theta}),\boldsymbol{\sigma},\boldsymbol{\theta},\varepsilon,\boldsymbol{\lambda}^0(s\mid\boldsymbol{\sigma},\boldsymbol{\theta},\varepsilon)\big)\\
&\quad=\tilde{\mathcal{L}}_0(s,\tilde{\boldsymbol{y}}(s\mid\boldsymbol{\sigma},\boldsymbol{\theta}),\boldsymbol{\sigma},\boldsymbol{\theta},\varepsilon)+\left[\boldsymbol{\lambda}^0(s\mid\boldsymbol{\sigma},\boldsymbol{\theta},\varepsilon)\right]^{\top}\tilde{f}(s,\tilde{\boldsymbol{y}}(s\mid\boldsymbol{\sigma},\boldsymbol{\theta}),\boldsymbol{\sigma},\boldsymbol{\theta})\end{aligned} \tag{9.3.38}$$

and $\boldsymbol{\lambda}^0(\cdot|\sigma,\theta,\epsilon)$ *is the solution of the following system of costate differential equations*

$$\frac{d\boldsymbol{\lambda}^0(s)}{ds} = -\left[\frac{\partial H_0\left(s, \tilde{\boldsymbol{y}}(s \mid \boldsymbol{\sigma},\boldsymbol{\theta}), \boldsymbol{\sigma}, \boldsymbol{\theta}, \varepsilon, \boldsymbol{\lambda}^0(s)\right)}{\partial \tilde{\boldsymbol{y}}}\right]^\top \tag{9.3.39a}$$

with the boundary condition

$$\boldsymbol{\lambda}^0(1) = \left[\frac{\partial \tilde{\Phi}_0(\tilde{\boldsymbol{y}}(1 \mid \boldsymbol{\sigma},\boldsymbol{\theta}), \varepsilon)}{\partial \tilde{\boldsymbol{y}}}\right]^\top . \tag{9.3.39b}$$

Remark 9.3.4 *By Assumptions 9.3.1, 9.3.2, and 9.3.3, Remarks 9.3.1 and 9.3.2, it follows from arguments similar to those given for the proof of Lemma 8.4.2 that there exits a compact set* $\mathcal{Z} \subset \mathbb{R}^n$ *such that* $\boldsymbol{\lambda}^0(s \mid \boldsymbol{\sigma},\boldsymbol{\theta},\varepsilon) \in \mathcal{Z}$ *for all* $s \in [0,1]$, $(\boldsymbol{\sigma},\boldsymbol{\theta}) \in \Xi \times \Theta$ *and* $\varepsilon \geq 0$.

9.3.2 Some Convergence Results

In this section, we shall show that, under some mild assumptions, if the parameter δ_k is sufficient large ($\delta_k \to +\infty$ as $k \to +\infty$) and $\left(\boldsymbol{\sigma}^{(k),*}, \boldsymbol{\theta}^{(k),*}, \varepsilon^{(k),*}\right)$ is a local minimizer of Problem $(\tilde{P}_\delta(p))$, then $\varepsilon^{(k),*} \to \varepsilon^* = 0$, and $\left(\boldsymbol{\sigma}^{(k),*}, \boldsymbol{\theta}^{(k),*}\right) \to (\boldsymbol{\sigma}^*, \boldsymbol{\theta}^*)$ with $(\boldsymbol{\sigma}^*, \boldsymbol{\theta}^*)$ being a local minimizer of Problem $(\tilde{P}(p))$.

For every positive integer k, let $\left(\boldsymbol{\sigma}^{(k),*}, \boldsymbol{\theta}^{(k),*}\right)$ be a local minimizer of Problem $(\tilde{P}_\delta(p))$. To obtain our main result, we need

Lemma 9.3.1 *Let* $\left(\boldsymbol{\sigma}^{(k),*}, \boldsymbol{\theta}^{(k),*}, \varepsilon^{(k),*}\right)$ *be a local minimizer of Problem* $(\tilde{P}_\delta(p))$. *Suppose that* $\widetilde{g}_0^{\delta_k}\left(\boldsymbol{\sigma}^{(k),*}, \boldsymbol{\theta}^{(k),*}, \varepsilon^{(k),*}\right)$ *is finite and that* $\varepsilon^{(k),*} > 0$. *Then*

$$\left(\boldsymbol{\sigma}^{(k),*}, \boldsymbol{\theta}^{(k),*}, \varepsilon^{(k),*}\right) \notin \Omega_{\varepsilon_k},$$

where Ω_{ε_k} *is as defined by* (9.3.25).

Proof. Since $(\boldsymbol{\sigma}^{(k),*}, \boldsymbol{\theta}^{(k),*}, \varepsilon^{(k),*})$ is a local minimizer of Problem $(\tilde{P}_\delta(p))$ and $\varepsilon^{(k),*} > 0$, we have

$$\frac{\partial \widetilde{g}_0^{\delta_k}\left(\boldsymbol{\sigma}^{(k),*}, \boldsymbol{\theta}^{(k),*}, \varepsilon^{(k),*}\right)}{\partial \varepsilon} = 0. \tag{9.3.40}$$

On the contrary, we assume that the conclusion of the lemma is false. Then, we have

$$h_i\left(\tilde{\boldsymbol{y}}(s \mid \boldsymbol{\sigma}^{(k),*}, \boldsymbol{\theta}^{(k),*}), \boldsymbol{\sigma}^{(k),*}\right) \leq \left(\varepsilon^{(k),*}\right)^\gamma W_i,$$

$$\forall\ s \in \mathcal{J}_j,\ j = 1, \ldots, p;\ i = 1,\ \ldots\ , N, \tag{9.3.41}$$

and

$$\tilde{\boldsymbol{y}}\left(1 \mid \boldsymbol{\sigma}^{(k),*}, \boldsymbol{\theta}^{(k),*}\right) - \tilde{\boldsymbol{y}}^f = \boldsymbol{0}. \tag{9.3.42}$$

Thus, by (9.3.41), (9.3.43), (9.3.27), and (9.3.40), we obtain

$$0 = \frac{\partial \widetilde{g}_0^{\delta_k}\left(\boldsymbol{\sigma}^{(k),*}, \boldsymbol{\theta}^{(k),*}, \varepsilon^{(k),*}\right)}{\partial \varepsilon} = \beta \delta_k \varepsilon^{\beta - 1} > 0$$

This is a contradiction, and hence completing the proof.

Before we introduce the definition of the constraint qualification, we first define

$$\phi_i(\tilde{\boldsymbol{y}}(1 \mid \boldsymbol{\sigma}, \boldsymbol{\theta})) = \tilde{y}_i(1 \mid \boldsymbol{\sigma}, \boldsymbol{\theta}) - \tilde{y}_i^f\ ,\ i = 1,\ \ldots\ , n+1. \tag{9.3.43}$$

Assumption 9.3.4 *The constraint qualification in the sense of Definition 4.1 is satisfied for the continuous inequality constraints* (9.3.20) *at* $(\sigma, \theta) = (\bar{\sigma}, \bar{\theta})$

Theorem 9.3.2 *Suppose that* $\left(\boldsymbol{\sigma}^{(k),*}, \boldsymbol{\theta}^{(k),*}, \varepsilon^{(k),*}\right)$ *is a local minimizer of Problem* $\left(\tilde{P}_{\delta_k}(p)\right)$ *such that* $\widetilde{g}_0^{\delta_k}\left(\boldsymbol{\sigma}^{(k),*}, \boldsymbol{\theta}^{(k),*}, \varepsilon^{(k),*}\right)$ *is finite and* $\varepsilon^{(k),*} > 0$*. If* $\left(\boldsymbol{\sigma}^{(k),*}, \boldsymbol{\theta}^{(k),*}, \varepsilon^{(k),*}\right) \to \left(\boldsymbol{\sigma}^*, \boldsymbol{\theta}^*, \varepsilon^*\right)$ *as* $k \to +\infty$*, and the constraint qualification is satisfied for the continuous inequality constraints* (9.3.20) *at* $(\boldsymbol{\sigma}, \boldsymbol{\theta}) = (\boldsymbol{\sigma}^*, \boldsymbol{\theta}^*)$*, then* $\varepsilon^* = 0$ *and* $(\boldsymbol{\sigma}^*, \boldsymbol{\theta}^*) \in \Omega_0$.

Proof. From Lemma 9.2.1, it follows that $(\boldsymbol{\sigma}^{(k),*}, \boldsymbol{\theta}^{(k),*}, \varepsilon^{(k),*}) \notin \Omega_{\varepsilon^{(k),*}}$. Furthermore, in terms of (9.3.27), we have

$$\begin{aligned}
&\frac{\partial \widetilde{g}_0^{\delta}\left(\boldsymbol{\sigma}^{(k),*}, \boldsymbol{\theta}^{(k),*}, \varepsilon^{(k),*}\right)}{\partial \boldsymbol{\sigma}} \\
&= \int_0^1 \frac{\partial H_0\left(s, \tilde{\boldsymbol{y}}\left(s \mid \boldsymbol{\sigma}^{(k),*}, \boldsymbol{\theta}^{(k),*}\right), \boldsymbol{\sigma}, \boldsymbol{\theta}, \varepsilon, \boldsymbol{\lambda}^0\left(s \mid \boldsymbol{\sigma}^{(k),*}, \boldsymbol{\theta}^{(k),*}, \varepsilon^{(k),*}\right)\right)}{\partial \boldsymbol{\sigma}} ds \\
&= \int_0^1 \frac{\partial \bar{\mathcal{L}}_0\left(s, \tilde{y}\left(s \mid \boldsymbol{\sigma}^{(k),*}, \boldsymbol{\theta}^{(k),*}\right), \boldsymbol{\sigma}^{(k),*}, \boldsymbol{\theta}^{(k),*}\right)}{\partial \boldsymbol{\sigma}} ds + 2(\varepsilon^{(k),*})^{-\alpha}. \\
&\sum_{i=1}^{N} \int_0^1 \max\left\{0, \bar{h}_i\left(s, \tilde{\boldsymbol{y}}\left(s \mid \boldsymbol{\sigma}^{(k),*}, \boldsymbol{\theta}^{(k),*}\right), \boldsymbol{\sigma}^{(k),*}\right) - \left(\varepsilon^{(k),*}\right)^{\gamma} W_i\right\} \cdot \\
&\frac{\partial \bar{h}_i\left(s, \tilde{\boldsymbol{y}}\left(s \mid \boldsymbol{\sigma}^{(k),*}, \boldsymbol{\theta}^{(k),*}\right), \boldsymbol{\sigma}^{(k),*}\right)}{\partial \boldsymbol{\sigma}} ds + \int_0^1 \boldsymbol{\lambda}^0\left(s \mid \boldsymbol{\sigma}^{(k),*}, \boldsymbol{\theta}^{(k),*}, \varepsilon^{(k),*}\right) \cdot \\
&\frac{\partial \tilde{\boldsymbol{f}}\left(s, \tilde{\boldsymbol{y}}\left(s \mid \boldsymbol{\sigma}^{(k),*}, \boldsymbol{\theta}^{(k),*}\right), \boldsymbol{\sigma}^{(k),*}, \boldsymbol{\theta}^{(k),*}\right)}{\partial \boldsymbol{\sigma}} ds \\
&= 0
\end{aligned} \tag{9.3.44}$$

$$
\begin{aligned}
&\frac{\partial \tilde{g}_0^\delta\left(\boldsymbol{\sigma}^{(k),*},\boldsymbol{\theta}^{(k),*},\varepsilon^{(k),*}\right)}{\partial \varepsilon}\\
&=(\varepsilon^{(k),*})^{-\alpha-1}\Bigg\{-\alpha\sum_{i=1}^{N}\int_0^1\Big[\max\Big\{0,\bar{h}_i\Big(s,\tilde{\boldsymbol{y}}\Big(s\mid\boldsymbol{\sigma}^{(k),*},\boldsymbol{\theta}^{(k),*}\Big),\boldsymbol{\sigma}^{(k),*}\Big)\\
&\quad-(\varepsilon)^{\gamma}W_i\Big\}\Big]^2ds+2\gamma\sum_{i=1}^{N}\int_0^1\max\Big\{0,\bar{h}_i\Big(s,\tilde{\boldsymbol{y}}\Big(s\mid\boldsymbol{\sigma}^{(k),*},\boldsymbol{\theta}^{(k),*}\Big),\boldsymbol{\sigma}^{(k),*}\Big)\\
&\quad-\Big(\varepsilon^{(k),*}\Big)^{\gamma}W_i\Big\}(-(\varepsilon^{(k),*})^{\gamma}W_i)ds-\alpha\sum_{i=1}^{n+1}\Big(\tilde{y}_i\Big(1\mid\boldsymbol{\sigma}^{(k),*},\boldsymbol{\theta}^{(k),*}\Big)-\tilde{y}_i^f\Big)^2\Bigg\}\\
&\quad+\delta_k\beta\Big(\varepsilon^{(k),*}\Big)^{\beta-1}\\
&=0. \qquad (9.3.45)
\end{aligned}
$$

Suppose that $\varepsilon^{(k),*}\to\varepsilon^*\neq 0$. Then, by (9.3.45), it can be shown by using Remarks 9.3.1 and 9.3.2 and Theorem A.1.10 (Lebesgue dominated convergence theorem) that its first term tends to a finite value, while the last term tends to infinity as $\delta_k\to+\infty$, when $k\to+\infty$. This is impossible for the validity of (9.3.45). Thus, $\varepsilon^*=0$.

Now, by (9.3.44), we obtain

$$
\begin{aligned}
&\int_0^1\frac{\partial\bar{\mathcal{L}}_0\left(s,\tilde{\boldsymbol{y}}\left(s\mid\boldsymbol{\sigma}^{(k),*},\boldsymbol{\theta}^{(k),*}\right),\boldsymbol{\sigma}^{(k),*},\boldsymbol{\theta}^{(k),*}\right)}{\partial\boldsymbol{\sigma}}ds+2\Big(\varepsilon^{(k),*}\Big)^{-\alpha}\cdot\\
&\sum_{i=1}^{N}\int_0^1\max\Big\{0,\bar{h}_i\Big(s,\tilde{\boldsymbol{y}}\Big(s\mid\boldsymbol{\sigma}^{(k),*},\boldsymbol{\theta}^{(k),*}\Big),\boldsymbol{\sigma}^{(k),*\cdot}\Big)-\Big(\varepsilon^{(k),*}\Big)^{\gamma}W_i\Big\}\cdot\\
&\frac{\partial\bar{h}_i\left(s,\tilde{\boldsymbol{y}}\left(s\mid\boldsymbol{\sigma}^{(k),*},\boldsymbol{\theta}^{(k),*}\right),\boldsymbol{\sigma}^{(k),*}\right)}{\partial\boldsymbol{\sigma}}ds+\int_0^1\boldsymbol{\lambda}_0\Big(s\mid\boldsymbol{\sigma}^{(k),*},\boldsymbol{\theta}^{(k),*},\varepsilon^{(k),*}\Big)\cdot\\
&\frac{\partial\tilde{\boldsymbol{f}}\left(\tilde{\boldsymbol{y}}\left(s\mid\boldsymbol{\sigma}^{(k),*},\boldsymbol{\theta}^{(k),*}\right),\boldsymbol{\sigma}^{(k),*},\boldsymbol{\theta}^{(k),*}\right)}{\partial\boldsymbol{\sigma}}ds=0
\end{aligned}
$$

Thus,

$$
\begin{aligned}
&\lim_{k\to+\infty}\Bigg\{\int_0^1\frac{\partial\bar{\mathcal{L}}_0\left(s,\tilde{\boldsymbol{y}}\left(s\mid\boldsymbol{\sigma}^{(k),*},\boldsymbol{\theta}^{(k),*}\right),\boldsymbol{\sigma}^{(k),*},\boldsymbol{\theta}^{(k),*}\right)}{\partial\boldsymbol{\sigma}}ds+2\Big(\varepsilon^{(k),*}\Big)^{-\alpha}\cdot\\
&\sum_{i=1}^{N}\int_0^1\max\Big\{0,\bar{h}_i\Big(s,\tilde{\boldsymbol{y}}\Big(s\mid\boldsymbol{\sigma}^{(k),*},\boldsymbol{\theta}^{(k),*}\Big),\boldsymbol{\sigma}^{(k),*}\Big)-\Big(\varepsilon^{(k),*}\Big)^{\gamma}W_i\Big\}\cdot\\
&\frac{\partial\bar{h}_i\left(s,\tilde{\boldsymbol{y}}\left(s\mid\boldsymbol{\sigma}^{(k),*},\boldsymbol{\theta}^{(k),*}\right),\boldsymbol{\sigma}^{(k),*}\right)}{\partial\boldsymbol{\sigma}}ds+\int_0^1\lambda_0\Big(s\mid\boldsymbol{\sigma}^{(k),*},\boldsymbol{\theta}^{(k),*},\varepsilon^{(k),*}\Big)\cdot\\
&\frac{\partial\tilde{\boldsymbol{f}}\left(s,\tilde{\boldsymbol{y}}\left(s\mid\boldsymbol{\sigma}^{(k),*},\boldsymbol{\theta}^{(k),*}\right),\boldsymbol{\sigma}^{(k),*},\boldsymbol{\theta}^{(k),*}\right)}{\partial\boldsymbol{\sigma}}ds\Bigg\}=0. \qquad (9.3.46)
\end{aligned}
$$

Then, by Remarks 9.3.2 and 9.3.3, it follows from Theorem A.1.10 that the first and third terms appeared on the right hand side of (9.3.46) converge to finite values. On the other hand, the second term tends to infinity, which is impossible. Thus,

$$\sum_{i=1}^{N}\int_0^1 \max\left\{0, \bar{h}_i(s, \tilde{\boldsymbol{y}}(s \mid \boldsymbol{\sigma}^*, \boldsymbol{\theta}^*), \boldsymbol{\sigma}^*)\right\} \frac{\partial \bar{h}_i(s, \tilde{\boldsymbol{y}}(s \mid \boldsymbol{\sigma}^*, \boldsymbol{\theta}^*), \boldsymbol{\sigma}^*)}{\partial \boldsymbol{\sigma}} ds = 0. \tag{9.3.47}$$

Since the constraint qualification is satisfied for the continuous inequality constraints (9.3.20) at $(\boldsymbol{\sigma}, \boldsymbol{\theta}) = (\boldsymbol{\sigma}^*, \boldsymbol{\theta}^*)$, it follows that, for each $i = 1, \ldots, N$,

$$\max\left\{0, \bar{h}_i(s, \tilde{\boldsymbol{y}}(s \mid \boldsymbol{\sigma}^*, \boldsymbol{\theta}^*), \boldsymbol{\sigma}^*)\right\} = 0$$

for each $s \in [0, 1]$. This, in turn, implies that, for each $i = 1, \ldots, N$

$$\bar{h}_i(s, \tilde{\boldsymbol{y}}(s \mid \boldsymbol{\sigma}^*, \boldsymbol{\theta}^*), \boldsymbol{\sigma}^*) \leq 0 \tag{9.3.48}$$

for each $s \in [0, 1]$. Next, from (9.3.45) and (9.3.48), it is easy to see that when $k \to +\infty$, for each $i = 1, \ldots, n+1$,

$$\tilde{y}_i(1 \mid \boldsymbol{\sigma}^*, \boldsymbol{\theta}^*) - \tilde{y}_i^f = 0. \tag{9.3.49}$$

The proof is complete.

Corollary 9.3.1 *Suppose that* $\left(\boldsymbol{\sigma}^{(k),*}, \boldsymbol{\theta}^{(k),*}\right) \to (\boldsymbol{\sigma}^*, \boldsymbol{\theta}^*) \in \Omega_0$ *and that* $\varepsilon^{(k),*} \to \varepsilon^* = 0$. *Then,* $\Delta\left(\boldsymbol{\sigma}^{(k),*}, \boldsymbol{\theta}^{(k),*}, \varepsilon^{(k),*}\right) \to \Delta(\sigma^*, \boldsymbol{\theta}^*, \varepsilon^*) = 0$, *and* $\Delta_1 \to 0$.

Proof. The conclusion follows readily from the definitions of $\Delta(\boldsymbol{\sigma}, \boldsymbol{\theta}, \varepsilon)$ and Δ_1, and the continuity of h_i and $\tilde{\boldsymbol{f}}$.

In what follows, we shall turn our attention to the exact penalty function constructed in (9.3.27). We shall see that, under some mild conditions, $\widetilde{g}_0^\delta(\boldsymbol{\sigma}, \boldsymbol{\theta}, \varepsilon)$ is continuously differentiable with continuous limit. For this, we need the following lemmas.

Lemma 9.3.2 *Assume that*

$$\left|\left(\boldsymbol{\sigma}^{(k),*}, \boldsymbol{\theta}^{(k),*}\right) - (\boldsymbol{\sigma}^*, \theta^*)\right| = o\left(\left(\varepsilon^{(k),*}\right)^\xi - (\varepsilon^*)^\xi\right),$$

then

$$\left\|\tilde{\boldsymbol{y}}\left(s \mid \boldsymbol{\sigma}^{(k),*}, \boldsymbol{\theta}^{(k),*}\right) - \tilde{\boldsymbol{y}}(s \mid \boldsymbol{\sigma}^*, \boldsymbol{\theta}^*)\right\|_\infty = o\left(\left(\varepsilon^{(k),*}\right)^\xi - (\varepsilon^*)^\xi\right), \tag{9.3.50}$$

where

$$\|\tilde{\boldsymbol{y}}(\cdot \mid \boldsymbol{\sigma}, \boldsymbol{\theta})\|_\infty = \operatorname*{ess\,sup}_{s\in[0,1]} |\tilde{\boldsymbol{y}}(\cdot \mid \boldsymbol{\sigma}, \boldsymbol{\theta})|. \tag{9.3.51}$$

Proof. Note that

$$\tilde{\boldsymbol{y}}\left(s \mid \boldsymbol{\sigma}^{(k),*}, \boldsymbol{\theta}^{(k),*}\right) = \tilde{\boldsymbol{y}}(0) + \int_0^s \tilde{\boldsymbol{f}}\left(\tau, \tilde{\boldsymbol{y}}(\tau), \boldsymbol{\sigma}^{(k),*}, \boldsymbol{\theta}^{(k),*}\right) d\tau \tag{9.3.52}$$

for any $s \in [0,1]$. Thus,

$$\begin{aligned}&\left|\tilde{\boldsymbol{y}}\left(s \mid \boldsymbol{\sigma}^{(k),*}, \boldsymbol{\theta}^{(k),*}\right) - \tilde{\boldsymbol{y}}\left(s \mid \boldsymbol{\sigma}^{*}, \boldsymbol{\theta}^{*}\right)\right| \\ &= \left|\int_0^s \tilde{\boldsymbol{f}}\left(\tau, \tilde{\boldsymbol{y}}(\tau), \boldsymbol{\sigma}^{(k),*}, \boldsymbol{\theta}^{(k),*}\right) - \tilde{\boldsymbol{f}}\left(\tau, \tilde{\boldsymbol{y}}(\tau), \boldsymbol{\sigma}^{*}, \boldsymbol{\theta}^{*}\right) d\tau\right|\end{aligned} \tag{9.3.53}$$

for any $s \in [0,1]$. By Assumption 10.1.1, there exists a constant $N_2 > 0$ such that

$$\begin{aligned}&\left|\tilde{\boldsymbol{y}}\left(s \mid \boldsymbol{\sigma}^{(k),*}, \boldsymbol{\theta}^{(k),*}\right) - \tilde{\boldsymbol{y}}(s \mid \boldsymbol{\sigma}^{*}, \boldsymbol{\theta}^{*})\right| \\ \le& N_2 \int_0^s \left\{\left|\tilde{\boldsymbol{y}}\left(\tau \mid \boldsymbol{\sigma}^{(k),*}, \boldsymbol{\theta}^{(k),*}\right) - \tilde{\boldsymbol{y}}(\tau \mid \boldsymbol{\sigma}^{*}, \boldsymbol{\theta}^{*})\right| \right. \\ &\left. + \left|\left(\boldsymbol{\sigma}^{(k),*}, \boldsymbol{\theta}^{(k),*}\right) - (\boldsymbol{\sigma}^{*}, \boldsymbol{\theta}^{*})\right| \right\} d\tau\end{aligned} \tag{9.3.54}$$

for any $s \in [0,1]$. By applying Gronwall–Bellman's lemma, we have

$$\begin{aligned}&\left|\tilde{\boldsymbol{y}}\left(s \mid \boldsymbol{\sigma}^{(k),*}, \boldsymbol{\theta}^{(k),*}\right) - \tilde{\boldsymbol{y}}(s \mid \boldsymbol{\sigma}^{*}, \boldsymbol{\theta}^{*})\right| \\ &\le N_2 \left\{\int_0^1 \left|\left(\boldsymbol{\sigma}^{(k),*}, \boldsymbol{\theta}^{(k),*}\right) - (\boldsymbol{\sigma}^{*}, \boldsymbol{\theta}^{*})\right| d\tau\right\} exp(N_2) \\ &= N_2 \left|\left(\boldsymbol{\sigma}^{(k),*}, \boldsymbol{\theta}^{(k),*}\right) - (\boldsymbol{\sigma}^{*}, \boldsymbol{\theta}^{*})\right| exp(N_2)\end{aligned} \tag{9.3.55}$$

for any $s \in [0,1]$. Let $N_3 = N_2 exp(N_2)$, (9.3.55) becomes

$$\begin{aligned}\left|\tilde{\boldsymbol{y}}\left(s \mid \boldsymbol{\sigma}^{(k),*}, \boldsymbol{\theta}^{(k),*}\right) - \tilde{\boldsymbol{y}}(s \mid \boldsymbol{\sigma}^{*}, \boldsymbol{\theta}^{*})\right| &\le N_3 \left|\left(\boldsymbol{\sigma}^{(k),*}, \boldsymbol{\theta}^{(k),*}\right) - (\boldsymbol{\sigma}^{*}, \boldsymbol{\theta}^{*})\right| \\ &= N_3 o\left(\left(\varepsilon^{(k),*}\right)^{\xi} - (\varepsilon^{*})^{\xi}\right)\end{aligned} \tag{9.3.56}$$

for any $s \in [0,1]$. Since this is valid for all $s \in [0,1]$, it completes the proof.

We assume that the following conditions are satisfied.

Assumption 9.3.5

$$\begin{aligned}\bar{h}_i\left(s, \tilde{\boldsymbol{y}}\left(s \mid \boldsymbol{\sigma}^{(k),*}, \boldsymbol{\theta}^{(k),*}\right), \boldsymbol{\sigma}^{(k),*}\right) = o\left(\left(\varepsilon^{(k),*}\right)^{\xi}\right), \\ \xi > 0,\ s \in [0,1],\ i = 1,\ \ldots,\ N.\end{aligned} \tag{9.3.57}$$

Assumption 9.3.6

$$\phi_i\left(\tilde{\boldsymbol{y}}\left(1 \mid \boldsymbol{\sigma}^{(k),*}, \boldsymbol{\theta}^{(k),*}\right)\right) = o\left(\left(\varepsilon^{(k),*}\right)^{\xi'}\right), \quad \xi' > 0,\ i = 1,\ \ldots,\ n+1. \tag{9.3.58}$$

Theorem 9.3.3 *Suppose that* $\gamma > \alpha,\quad \xi > \alpha,\quad \xi' > \alpha,\quad -\alpha - 1 + 2\xi > 0,\quad -\alpha - 1 + 2\xi' > 0,\quad 2\gamma - \alpha - 1 > 0$. *Then, as* $\varepsilon^{(k),*} \to \varepsilon^* = 0$ *and* $\left(\boldsymbol{\sigma}^{(k),*}, \boldsymbol{\theta}^{(k),*}\right) \to (\boldsymbol{\sigma}^*, \boldsymbol{\theta}^*) \in \Omega_0$, *it holds that*

$$\widetilde{g}_0^{\delta_k}\left(\boldsymbol{\sigma}^{(k),*}, \boldsymbol{\theta}^{(k),*}, \varepsilon^{(k),*}\right) \to \widetilde{g}_0^{0}(\boldsymbol{\sigma}^*, \boldsymbol{\theta}^*, 0) = \widetilde{g}_0(\boldsymbol{\sigma}^*, \boldsymbol{\theta}^*) \tag{9.3.59}$$

$$\begin{aligned}\nabla_{(\boldsymbol{\sigma},\boldsymbol{\theta},\varepsilon)}\widetilde{g}_0^{\delta_k}\left(\boldsymbol{\sigma}^{(k),*}, \boldsymbol{\theta}^{(k),*}, \varepsilon^{(k),*}\right) &\to \nabla_{(\boldsymbol{\sigma},\boldsymbol{\theta},\varepsilon)}\widetilde{g}_0^{0}(\boldsymbol{\sigma}^*, \boldsymbol{\theta}^*, 0) \\ &= (\nabla_{(\boldsymbol{\sigma},\boldsymbol{\theta})}\widetilde{g}_0(\boldsymbol{\sigma}^*, \boldsymbol{\theta}^*), 0).\end{aligned} \tag{9.3.60}$$

Proof. For notational brevity, the following abbreviations will be used through the proof of this theorem and that of Theorem 9.3.5

$$\bar{h}_i^{(k),*}(\cdot) = \bar{h}_i\left(\cdot, \tilde{y}\left(\cdot|\sigma^{(k),*}, \theta^{(k),*}\right), \sigma^{(k),*}\right) \tag{9.3.61}$$

$$\bar{\mathcal{L}}_0^{(k),*}(\cdot) = \bar{\mathcal{L}}_0\left(\cdot, \tilde{y}\left(\cdot|\sigma^{(k),*}, \theta^{(k),*}\right), \sigma^{(k),*}, \theta^{(k),*}\right) \tag{9.3.62}$$

$$\tilde{\mathcal{L}}_{0,\varepsilon}^{(k),*}(\cdot) = \tilde{\mathcal{L}}_0\left(\cdot, \tilde{y}\left(\cdot|\sigma^{(k),*}, \theta^{(k),*}\right), \sigma^{(k),*}, \theta^{(k),*}, \varepsilon\right) \tag{9.3.63}$$

$$\tilde{y}^{(k),*}(\cdot) = \tilde{y}\left(\cdot|\sigma^{(k),*}, \theta^{(k),*}\right) \tag{9.3.64}$$

$$\tilde{y}^{*}(\cdot) = \tilde{y}\left(\cdot|\sigma^{*}, \theta^{*}\right) \tag{9.3.65}$$

$$\tilde{y}_i^{(k),*}(\cdot) = \tilde{y}_i\left(\cdot|\sigma^{(k),*}, \theta^{(k),*}\right) \tag{9.3.66}$$

$$\tilde{f}^{(k),*}(\cdot) = \tilde{f}\left(\cdot, \tilde{y}\left(\cdot|\sigma^{(k),*}, \theta^{(k),*}\right), \sigma^{(k),*}, \theta^{(k),*}\right) \tag{9.3.67}$$

$$\bar{\lambda}^{0,(k),*}(\cdot) = \bar{\lambda}\left(\cdot|\sigma^{(k),*}, \theta^{(k),*}\right) \tag{9.3.68}$$

$$\lambda^{0,(k),*,\varepsilon}(\cdot) = \lambda^0\left(\cdot|\sigma^{(k),*}, \theta^{(k),*}, \varepsilon^{(k),*}\right). \tag{9.3.69}$$

Now, based on the conditions of the theorem, we can show that, for $\varepsilon \neq 0$,

$$\begin{aligned}&\lim_{\substack{\varepsilon^{(k),*}\to\varepsilon^*=0 \\ \left(\boldsymbol{\sigma}^{(k),*},\boldsymbol{\theta}^{(k),*}\right)\to(\boldsymbol{\sigma}^*,\boldsymbol{\theta}^*)\in\Omega_0}} \widetilde{g}_0^{\delta_k}\left(\boldsymbol{\sigma}^{(k),*}, \boldsymbol{\theta}^{(k),*}, \varepsilon^{(k),*}\right) \\ &= \lim_{\substack{\varepsilon^{(k),*}\to\varepsilon^*=0 \\ \left(\boldsymbol{\sigma}^{(k),*},\boldsymbol{\theta}^{(k),*}\right)\to(\boldsymbol{\sigma}^*,\boldsymbol{\theta}^*)\in\Omega_0}} \left\{\widetilde{g}_0\left(\boldsymbol{\sigma}^{(k),*}, \boldsymbol{\theta}^{(k),*}\right)\right. \\ &+ \left(\varepsilon^{(k),*}\right)^{-\alpha}\sum_{i=1}^{N}\int_0^1 \left[\max\left\{0, \bar{h}_i^{(k),*}(s) - (\varepsilon)^{\gamma} W_i\right\}\right]^2 ds\end{aligned}$$

$$+\left(\varepsilon^{(k),*}\right)^{-\alpha}\sum_{i=1}^{n+1}\left(\tilde{y}_i^{(k),*}(1)-\tilde{y}_i^f\right)^2+\delta_k\left(\varepsilon^{(k),*}\right)^{\beta}\Big\}. \tag{9.3.70}$$

It is easy to see that when $(\boldsymbol{\sigma}^{(k),*},\boldsymbol{\theta}^{(k),*})\to(\boldsymbol{\sigma}^*,\boldsymbol{\theta}^*)$,

$$\tilde{y}^{(k),*}(s)\to\tilde{y}^*(s) \tag{9.3.71}$$

for each $s\in[0,1]$. By (9.3.71) and (9.3.70), we obtain

$$\lim_{\substack{\varepsilon^{(k),*}\to\varepsilon^*=0\\ \left(\boldsymbol{\sigma}^{(k),*},\boldsymbol{\theta}^{(k),*}\right)\to(\boldsymbol{\sigma}^*,\boldsymbol{\theta}^*)\in\Omega_0}}\widetilde{g}_0(\boldsymbol{\sigma}^{(k),*},\boldsymbol{\theta}^{(k),*})=\widetilde{g}_0\left(\boldsymbol{\sigma}^*,\boldsymbol{\theta}^*\right) \tag{9.3.72}$$

Substituting (9.3.72) into (9.3.70), we have

$$\begin{aligned}
&\lim_{\substack{\varepsilon^{(k),*}\to\varepsilon^*=0\\ \left(\boldsymbol{\sigma}^{(k),*},\boldsymbol{\theta}^{(k),*}\right)\to(\boldsymbol{\sigma}^*,\boldsymbol{\theta}^*)\in\Omega_0}}\widetilde{g}_0^{\delta_k}\left(\boldsymbol{\sigma}^{(k),*},\boldsymbol{\theta}^{(k),*},\varepsilon^{(k),*}\right)\\
&=\tilde{g}_0(\sigma^*,\theta^*)\\
&\quad+\lim_{\substack{\varepsilon^{(k),*}\to\varepsilon^*=0\\ \left(\boldsymbol{\sigma}^{(k),*},\boldsymbol{\theta}^{(k),*}\right)\to(\boldsymbol{\sigma}^*,\boldsymbol{\theta}^*)\in\Omega_0}}\frac{\int_0^1\left[\max\left\{0,\bar{h}_i^{(k),*}(s)-(\varepsilon^{(k),*})^{\gamma}W_i\right\}\right]^2ds}{(\varepsilon^{(k),*})^{\alpha}}\\
&\quad+\lim_{\substack{\varepsilon^{(k),*}\to\varepsilon^*=0\\ \left(\boldsymbol{\sigma}^{(k),*},\boldsymbol{\theta}^{(k),*}\right)\to(\boldsymbol{\sigma}^*,\boldsymbol{\theta}^*)\in\Omega_0}}\frac{\sum_{i=1}^{n+1}\left(\tilde{y}_i^{(k),*}(1)-\tilde{y}_i^f\right)^2}{(\varepsilon^{(k),*})^{\alpha}}.
\end{aligned} \tag{9.3.73}$$

For the second term and the third term of (9.3.73), it is clear from Lemma 9.2.1 that

$$\begin{aligned}
&\lim_{\substack{\varepsilon^{(k),*}\to\varepsilon^*=0\\ \left(\boldsymbol{\sigma}^{(k),*},\boldsymbol{\theta}^{(k),*}\right)\to(\boldsymbol{\sigma}^*,\boldsymbol{\theta}^*)\in\Omega_0}}\frac{\sum_{i=1}^{N}\int_0^1\left[\max\left\{0,\bar{h}_i^{(k),*}(s)-\left(\varepsilon^{(k),*}\right)^{\gamma}W_i\right\}\right]^2ds}{(\varepsilon^{(k),*})^{\alpha}}\\
&=\lim_{\substack{\varepsilon^{(k),*}\to\varepsilon^*=0\\ \left(\boldsymbol{\sigma}^{(k),*},\boldsymbol{\theta}^{(k),*}\right)\to(\boldsymbol{\sigma}^*,\boldsymbol{\theta}^*)\in\Omega_0}}\sum_{i=1}^{N}\int_0^1\left[\left(\varepsilon^{(k),*}\right)^{-\frac{\alpha}{2}}\bar{h}_i^{(k),*}(s)-\left(\varepsilon^{(k),*}\right)^{\gamma-\frac{\alpha}{2}}W_i\right]^2ds\,.
\end{aligned}$$

Since $\xi>\alpha,\ \gamma>\alpha$, it follows from Assumption 9.3.5, for any $s\in[0,1]$,

$$\lim_{\substack{\varepsilon^{(k),*}\to\varepsilon^*=0\\ \left(\boldsymbol{\sigma}^{(k),*},\boldsymbol{\theta}^{(k),*}\right)\to(\boldsymbol{\sigma}^*,\boldsymbol{\theta}^*)\in\Omega_0}}\left|\left(\varepsilon^{(k),*}\right)^{-\frac{\alpha}{2}}\bar{h}_i^{(k),*}(s)\right|=0. \tag{9.3.74}$$

Thus, we obtain

$$\lim_{\substack{\varepsilon^{(k),*}\to\varepsilon^*=0\\ \left(\boldsymbol{\sigma}^{(k),*},\boldsymbol{\theta}^{(k),*}\right)\to(\boldsymbol{\sigma}^*,\boldsymbol{\theta}^*)\in\Omega_0}} \sum_{i=1}^{N}\int_0^1 \left[\left(\varepsilon^{(k),*}\right)^{-\frac{\alpha}{2}}\bar{h}_i^{(k),*}(s)-\left(\varepsilon^{(k),*}\right)^{\gamma-\frac{\alpha}{2}}W_i\right]^2 ds$$
$$=\sum_{i=1}^{N}\int_0^1 \lim_{\substack{\varepsilon^{(k),*}\to\varepsilon^*=0\\ \left(\boldsymbol{\sigma}^{(k),*},\boldsymbol{\theta}^{(k),*}\right)\to(\boldsymbol{\sigma}^*,\boldsymbol{\theta}^*)\in\Omega_0}} \left[\left(\varepsilon^{(k),*}\right)^{-\frac{\alpha}{2}}\bar{h}_i^{(k),*}(s)-\left(\varepsilon^{(k),*}\right)^{\gamma-\frac{\alpha}{2}}W_i\right]^2 ds$$
$$=0. \tag{9.3.75}$$

Similarly, for the third term of (9.3.73), we have

$$\lim_{\substack{\varepsilon^{(k),*}\to\varepsilon^*=0\\ \left(\boldsymbol{\sigma}^{(k),*},\boldsymbol{\theta}^{(k),*}\right)\to(\boldsymbol{\sigma}^*,\boldsymbol{\theta}^*)\in\Omega_0}} \frac{\sum_{i=1}^{n+1}\left(\tilde{y}_i^{(k),*}(1)-\tilde{y}_i^f\right)^2}{(\varepsilon^{(k),*})^{\alpha}}=0$$

. Combining (9.3.73), (9.3.74) and (9.3.75) gives

$$\lim_{\substack{\varepsilon^{(k),*}\to\varepsilon^*=0\\ \left(\boldsymbol{\sigma}^{(k),*},\boldsymbol{\theta}^{(k),*}\right)\to(\boldsymbol{\sigma}^*,\boldsymbol{\theta}^*)\in\Omega_0}} \widetilde{g}_0^{\delta_k}\left(\boldsymbol{\sigma}^{(k),*},\boldsymbol{\theta}^{(k),*},\varepsilon^{(k),*}\right)=\widetilde{g}_0^{\delta_k}(\boldsymbol{\sigma}^*,\boldsymbol{\theta}^*,0)=\widetilde{g}_0(\boldsymbol{\sigma}^*,\boldsymbol{\theta}^*). \tag{9.3.76}$$

For the second part of the theorem, we need gradient formulas of $\widetilde{g}_0(\boldsymbol{\sigma},\boldsymbol{\theta})$, which can be derived in the same way as that for Theorem 7.2.2. These gradient formulas are given as follows.

$$\frac{\partial\widetilde{g}_0(\boldsymbol{\sigma},\boldsymbol{\theta})}{\partial\boldsymbol{\sigma}}=\int_0^1\frac{\partial\bar{H}_0\left(s,\tilde{\boldsymbol{y}}(s\mid\boldsymbol{\sigma},\boldsymbol{\theta}),\boldsymbol{\sigma},\boldsymbol{\theta},\bar{\boldsymbol{\lambda}}^0(s\mid\boldsymbol{\sigma},\boldsymbol{\theta})\right)}{\partial\boldsymbol{\sigma}}ds \tag{9.3.77}$$

$$\frac{\partial\widetilde{g}_0(\boldsymbol{\sigma},\boldsymbol{\theta})}{\partial\boldsymbol{\theta}}=\int_0^1\frac{\partial\bar{H}_0\left(s,\tilde{\boldsymbol{y}}(s\mid\boldsymbol{\sigma},\boldsymbol{\theta}),\boldsymbol{\sigma},\boldsymbol{\theta},\bar{\boldsymbol{\lambda}}^0(s\mid\boldsymbol{\sigma},\boldsymbol{\theta})\right)}{\partial\boldsymbol{\theta}}ds, \tag{9.3.78}$$

where $\bar{H}_0(s,\tilde{\boldsymbol{y}}(s\mid\boldsymbol{\sigma},\boldsymbol{\theta}),\boldsymbol{\sigma},\boldsymbol{\theta},\bar{\boldsymbol{\lambda}}^0(s\mid\boldsymbol{\sigma},\boldsymbol{\theta}))$ is the Hamiltonian function defined by

$$\begin{aligned}&\bar{H}_0\big(s,\tilde{\boldsymbol{y}}(s\mid\boldsymbol{\sigma},\boldsymbol{\theta}),\boldsymbol{\sigma},\boldsymbol{\theta},\bar{\boldsymbol{\lambda}}^0(s\mid\boldsymbol{\sigma},\boldsymbol{\theta})\big)\\&\quad=\bar{\mathcal{L}}_0(s,\tilde{\boldsymbol{y}}(s\mid\boldsymbol{\sigma},\boldsymbol{\theta}),\boldsymbol{\sigma},\boldsymbol{\theta})+\left[\bar{\boldsymbol{\lambda}}^0(s\mid\boldsymbol{\sigma},\boldsymbol{\theta})\right]^{\top}\tilde{\boldsymbol{f}}(s,\tilde{\boldsymbol{y}}(s\mid\boldsymbol{\sigma},\boldsymbol{\theta}),\boldsymbol{\sigma},\boldsymbol{\theta}),\end{aligned} \tag{9.3.79}$$

and $\bar{\boldsymbol{\lambda}}^0(\cdot\mid\boldsymbol{\sigma},\boldsymbol{\theta})$ is the solution of the following system of costate differential equations corresponding to $(\boldsymbol{\sigma},\boldsymbol{\theta})\in\Xi\times\Theta$.

$$\frac{d\bar{\boldsymbol{\lambda}}^0(s)}{dt}=-\left[\frac{\partial\bar{H}_0\left(s,\tilde{\boldsymbol{y}}(s\mid\boldsymbol{\sigma},\boldsymbol{\theta}),\boldsymbol{\sigma},\boldsymbol{\theta},\bar{\boldsymbol{\lambda}}^0(s)\right)}{\partial\tilde{\boldsymbol{y}}}\right]^{\top} \tag{9.3.80a}$$

with the boundary condition

$$\bar{\boldsymbol{\lambda}}^0(1) = \left[\frac{\partial \Phi_0(\boldsymbol{y}(1 \mid \boldsymbol{\sigma}, \boldsymbol{\theta}))}{\partial \boldsymbol{y}}\right]^\top . \tag{9.3.80b}$$

By (9.3.79), we can rewrite (9.3.80a) as:

$$\frac{d\bar{\boldsymbol{\lambda}}^0(s)}{dt} = -\left[\frac{\partial \bar{\mathcal{L}}_0(s, \tilde{\boldsymbol{y}}(s \mid \boldsymbol{\sigma}, \boldsymbol{\theta}), \boldsymbol{\sigma}, \boldsymbol{\theta})}{\partial \tilde{\boldsymbol{y}}}\right]^\top - \left[\frac{\partial \tilde{\boldsymbol{f}}(s, \tilde{\boldsymbol{y}}(s \mid \boldsymbol{\sigma}, \boldsymbol{\theta}), \boldsymbol{\sigma})}{\partial \tilde{\boldsymbol{y}}}\right]^\top \bar{\boldsymbol{\lambda}}^0(s) \tag{9.3.81}$$

(9.3.81) can be written as:

$$\begin{aligned}\bar{\boldsymbol{\lambda}}^0(s \mid \boldsymbol{\sigma}, \boldsymbol{\theta}) =& S(s,1)\bar{\boldsymbol{\lambda}}^0(1 \mid \boldsymbol{\sigma}, \boldsymbol{\theta}) \\ &+ \int_1^s S(s,\omega)\left(-\frac{\partial \bar{\mathcal{L}}_0(\omega, \tilde{\boldsymbol{y}}(\omega \mid \boldsymbol{\sigma}, \boldsymbol{\theta}), \boldsymbol{\sigma}, \boldsymbol{\theta})}{\partial \tilde{\boldsymbol{y}}}\right) d\omega,\end{aligned} \tag{9.3.82}$$

where $S(s, s')$ is the fundamental matrix of the differential equation (9.3.81). Similarly, (9.3.80a), (9.3.80b) and (9.3.81) with (σ, θ) replaced by $(\sigma, \theta, \epsilon)$, we have

$$\begin{aligned}\boldsymbol{\lambda}^0(s \mid \boldsymbol{\sigma}, \boldsymbol{\theta}, \varepsilon) =& S(s,1)\boldsymbol{\lambda}^0(1 \mid \boldsymbol{\sigma}, \boldsymbol{\theta}, \varepsilon) \\ &+ \int_1^s S(s,\omega)\left(-\frac{\partial \tilde{\mathcal{L}}_0(\omega, \tilde{\boldsymbol{y}}(\omega \mid \boldsymbol{\sigma}, \boldsymbol{\theta}), \boldsymbol{\sigma}, \boldsymbol{\theta}, \varepsilon)}{\partial \tilde{\boldsymbol{y}}}\right) d\omega.\end{aligned} \tag{9.3.83}$$

Thus, by (9.3.82) and (9.3.83), we obtain for each $s \in [0,1]$,

$$\begin{aligned}&\lim_{\substack{\varepsilon^{(k),*} \to \varepsilon^* = 0 \\ \left(\boldsymbol{\sigma}^{(k),*}, \boldsymbol{\theta}^{(k),*}\right) \to (\boldsymbol{\sigma}^*, \boldsymbol{\theta}^*) \in \Omega_0}} \left|\bar{\boldsymbol{\lambda}}^{0,(k),*}(s) - \boldsymbol{\lambda}^{0,(k),*}(s)\right| \\ &= \lim_{\substack{\varepsilon^{(k),*} \to \varepsilon^* = 0 \\ \left(\boldsymbol{\sigma}^{(k),*}, \boldsymbol{\theta}^{(k),*}\right) \to (\boldsymbol{\sigma}^*, \boldsymbol{\theta}^*) \in \Omega_0}} \left|S(s,1)\left[\bar{\boldsymbol{\lambda}}^{0,(k),*}(1) - \boldsymbol{\lambda}^{0,(k),*}(1)\right]\right. \\ &\quad \left. + \int_1^s S(s,\omega)\left\{-\frac{\partial \bar{\mathcal{L}}_0^{(k),*}(\omega)}{\partial \tilde{\boldsymbol{y}}} + \frac{\partial \tilde{\mathcal{L}}_{0,\varepsilon}^{(k),*}(\omega)}{\partial \tilde{\boldsymbol{y}}}\right\} d\omega\right| \\ &\le \lim_{\substack{\varepsilon^{(k),*} \to \varepsilon^* = 0 \\ \left(\boldsymbol{\sigma}^{(k),*}, \boldsymbol{\theta}^{(k),*}\right) \to (\boldsymbol{\sigma}^*, \boldsymbol{\theta}^*) \in \Omega_0}} |S(s,1)| \left|\bar{\boldsymbol{\lambda}}^{0,(k),*}(1) - \boldsymbol{\lambda}^{0,(k),*}(1)\right| \\ &\quad + \int_1^0 |S(s,\omega)| d\omega \int_1^0 \left|-\frac{\partial \bar{\mathcal{L}}_0^{(k),*}(\omega)}{\partial \tilde{\boldsymbol{y}}} + \frac{\partial \tilde{\mathcal{L}}_{0,\varepsilon}^{(k),*}(\omega)}{\partial \tilde{\boldsymbol{y}}}\right| d\omega.\end{aligned} \tag{9.3.84}$$

By (9.3.39b) and (9.3.80b), Assumption 9.3.5 and $\xi' > \alpha$, we have

$$\lim_{\substack{\varepsilon^{(k),*} \to \varepsilon^* = 0 \\ \left(\boldsymbol{\sigma}^{(k),*}, \boldsymbol{\theta}^{(k),*}\right) \to (\boldsymbol{\sigma}^*, \boldsymbol{\theta}^*) \in \Omega_0}} \left|\bar{\boldsymbol{\lambda}}^{0,(k),*}(1) - \boldsymbol{\lambda}^{0,(k),*}(1)\right|$$

$$= \lim_{\substack{\varepsilon^{(k),*}\to\varepsilon^*=0 \\ \left(\sigma^{(k),*},\theta^{(k),*}\right)\to(\sigma^*,\theta^*)\in\Omega_0}} \left| \left(\varepsilon^{(k),*}\right)^{-\alpha} \frac{\partial}{\partial \boldsymbol{y}} \sum_{i=1}^{n+1} \left(\tilde{y}_i^{(k),*}(1) - \tilde{y}_i^f\right)^2 \right|$$
$$= 0. \tag{9.3.85}$$

On the other hand, by (9.3.33), $\xi > \alpha$ and $\gamma > \alpha$, it follows from Assumption 9.3.6 that, for each $s \in [0,1]$,

$$\lim_{\substack{\varepsilon^{(k),*}\to\varepsilon^*=0 \\ \left(\sigma^{(k),*},\theta^{(k),*}\right)\to(\sigma^*,\theta^*)\in\Omega_0}} \int_1^0 \left| \left[-\frac{\partial \bar{\mathcal{L}}_0^{(k),*}(\omega)}{\partial \tilde{\boldsymbol{y}}} + \frac{\partial \tilde{\mathcal{L}}_{0,\varepsilon}^{(k),*}(\omega)}{\partial \tilde{\boldsymbol{y}}} \right] \right| d\omega$$
$$= \lim_{\substack{\varepsilon^{(k),*}\to\varepsilon^*=0 \\ \left(\sigma^{(k),*},\theta^{(k),*}\right)\to(\sigma^*,\theta^*)\in\Omega_0}} \int_1^0 2\varepsilon^{(k),*} \sum_{i=1}^{N} \left| \left[\max\left\{0, \bar{h}_i^{(k),*}(\omega) - \varepsilon^{(k),*} W_i\right\}\right] \cdot \right.$$
$$\left. \frac{\partial \bar{h}_i^{(k),*}(\omega)}{\partial \tilde{\boldsymbol{y}}} \right| d\omega$$
$$= 0. \tag{9.3.86}$$

We then substitute (9.3.85), (9.3.86) into (9.3.84) to give, for each $s \in [0,1]$,

$$\lim_{\substack{\varepsilon^{(k),*}\to\varepsilon^*=0 \\ \left(\sigma^{(k),*},\theta^{(k),*}\right)\to(\sigma^*,\theta^*)\in\Omega_0}} \left| \bar{\boldsymbol{\lambda}}^{0,(k),*}(s) - \boldsymbol{\lambda}^{0,(k),*}(s) \right| = 0. \tag{9.3.87}$$

Then we have

$$\lim_{\substack{\varepsilon^{(k),*}\to\varepsilon^*=0 \\ \left(\sigma^{(k),*},\theta^{(k),*}\right)\to(\sigma^*,\theta^*)\in\Omega_0}} \nabla_{\sigma} \tilde{g}_0^{\delta_k} \left(\boldsymbol{\sigma}^{(k),*}, \boldsymbol{\theta}^{(k),*}, \varepsilon^{(k),*}\right)$$
$$= \lim_{\substack{\varepsilon^{(k),*}\to\varepsilon^*=0 \\ \left(\sigma^{(k),*},\theta^{(k),*}\right)\to(\sigma^*,\theta^*)\in\Omega_0}} \left[\int_0^1 \frac{\partial \bar{\mathcal{L}}_0^{(k),*}(s)}{\partial \sigma} ds + 2\left(\varepsilon^{(k),*}\right)^{-\alpha} \cdot \right.$$
$$\sum_{i=1}^{N} \int_0^1 \max\left\{0, \bar{h}_i^{(k),*}(s) - \left(\varepsilon^{(k),*}\right)^{\gamma} W_i\right\} \frac{\partial \bar{h}_i^{(k),*}(s)}{\partial \boldsymbol{\sigma}} ds$$
$$\left. + \int_0^1 \boldsymbol{\lambda}^{0,(k),*}(s) \frac{\partial \tilde{f}^{(k),*}(s)}{\partial \boldsymbol{\sigma}} ds \right]$$
$$= \lim_{\substack{\varepsilon^{(k),*}\to\varepsilon^*=0 \\ \left(\sigma^{(k),*},\theta^{(k),*}\right)\to(\sigma^*,\theta^*)\in\Omega_0}} \left[\int_0^1 \frac{\partial \bar{\mathcal{L}}_0^{(k),*}(s)}{\partial \boldsymbol{\sigma}} ds + \int_0^1 \boldsymbol{\lambda}^0\left(s \mid \boldsymbol{\sigma}^{(k),*}, \boldsymbol{\theta}^{(k),*}\right) \cdot \right.$$
$$\left. \frac{\partial \tilde{\boldsymbol{f}}^{(k),*}(s)}{\partial \boldsymbol{\sigma}} ds \right] + \lim_{\substack{\varepsilon^{(k),*}\to\varepsilon^*=0 \\ \left(\sigma^{(k),*},\theta^{(k),*}\right)\to(\sigma^*,\theta^*)\in\Omega_0}} \left[\sum_{i=1}^{N} \int_0^1 \left\{ \left(\varepsilon^{(k),*}\right)^{-\alpha} \bar{h}_i^{(k),*}(s) \right. \right.$$

$$- \left(\varepsilon^{(k),*}\right)^{\gamma-\alpha} W_i\} \cdot \frac{\partial \bar{h}_i^{(k),*}(s)}{\partial \boldsymbol{\sigma}} ds \Bigg]. \tag{9.3.88}$$

Note that $\partial \bar{\mathcal{L}}_0 / \partial \boldsymbol{\sigma}$, $\boldsymbol{\lambda}^0$ and $\partial \tilde{\boldsymbol{f}} / \partial \boldsymbol{\sigma}$ are all bounded. Thus, it follows from (9.3.87) and Theorem A.1.10 that

$$\begin{aligned}
&\lim_{\substack{\varepsilon^{(k),*} \to \varepsilon^* = 0 \\ \left(\boldsymbol{\sigma}^{(k),*}, \boldsymbol{\theta}^{(k),*}\right) \to (\boldsymbol{\sigma}^*, \boldsymbol{\theta}^*) \in \Omega_0}} \Bigg[\int_0^1 \frac{\partial \bar{\mathcal{L}}_0^{(k),*}(s)}{\partial \boldsymbol{\sigma}} ds \\
&+ \int_0^1 \boldsymbol{\lambda}^0 \left(s \mid \boldsymbol{\sigma}^{(k),*}, \boldsymbol{\theta}^{(k),*}, \varepsilon^{(k),*}\right) \cdot \frac{\partial \tilde{\boldsymbol{f}}^{(k),*}(s)}{\partial \boldsymbol{\sigma}} ds \Bigg] \\
= &\int_0^1 \lim_{\substack{\varepsilon^{(k),*} \to \varepsilon^* = 0 \\ \left(\boldsymbol{\sigma}^{(k),*}, \boldsymbol{\theta}^{(k),*}\right) \to (\boldsymbol{\sigma}^*, \boldsymbol{\theta}^*) \in \Omega_0}} \frac{\partial \bar{\mathcal{L}}_0^{(k),*}(s)}{\partial \boldsymbol{\sigma}} ds \\
&+ \int_0^1 \lim_{\substack{\varepsilon^{(k),*} \to \varepsilon^* = 0 \\ \left(\boldsymbol{\sigma}^{(k),*}, \boldsymbol{\theta}^{(k),*}\right) \to (\boldsymbol{\sigma}^*, \boldsymbol{\theta}^*) \in \Omega_0}} \boldsymbol{\lambda}^0 \left(s \mid \boldsymbol{\sigma}^{(k),*}, \boldsymbol{\theta}^{(k),*}, \varepsilon^{(k),*}\right) \frac{\partial \tilde{\boldsymbol{f}}^{(k),*}(s)}{\partial \boldsymbol{\sigma}} ds \\
= &\int_0^1 \frac{\partial \bar{\mathcal{L}}_0^{(k),*}(s)}{\partial \boldsymbol{\sigma}} ds + \int_0^1 \bar{\boldsymbol{\lambda}}^0 \left(s \mid \boldsymbol{\sigma}^*, \boldsymbol{\theta}^*\right) \frac{\partial \tilde{\boldsymbol{f}}^{(k),*}(s)}{\partial \boldsymbol{\sigma}} ds \\
= &\nabla_{\boldsymbol{\sigma}} \widetilde{g}_0(\boldsymbol{\sigma}^*, \boldsymbol{\theta}^*).
\end{aligned} \tag{9.3.89}$$

Similarly, $(\varepsilon^{(k),*})^{-\alpha} g_i$, $\partial g_i / \partial \boldsymbol{\sigma}$ are all bounded, and $\xi > \alpha$, $\gamma > \alpha$. It follows from Assumption 9.3.5 that

$$\begin{aligned}
&\lim_{\substack{\varepsilon^{(k),*} \to \varepsilon^* = 0 \\ \left(\boldsymbol{\sigma}^{(k),*}, \boldsymbol{\theta}^{(k),*}\right) \to (\boldsymbol{\sigma}^*, \boldsymbol{\theta}^*) \in \Omega_0}} \Bigg[2 \sum_{i=1}^{N} \int_0^1 \left\{ \left(\varepsilon^{(k),*}\right)^{-\alpha} \bar{h}_i^{(k),*}(s) - \left(\varepsilon^{(k),*}\right)^{\gamma-\alpha} W_i \right\} \cdot \\
&\qquad \frac{\partial \bar{h}_i^{(k),*}(s)}{\partial \boldsymbol{\sigma}} ds \Bigg] \\
= &2 \sum_{i=1}^{N} \int_0^1 \lim_{\substack{\varepsilon^{(k),*} \to \varepsilon^* = 0 \\ \left(\boldsymbol{\sigma}^{(k),*}, \boldsymbol{\theta}^{(k),*}\right) \to (\boldsymbol{\sigma}^*, \boldsymbol{\theta}^*) \in \Omega_0}} \Bigg[\left\{ \left(\varepsilon^{(k),*}\right)^{-\alpha} \bar{h}_i^{(k),*}(s) - \left(\varepsilon^{(k),*}\right)^{\gamma-\alpha} W_i \right\} \cdot \\
&\frac{\partial \bar{h}_i^{(k),*}(s)}{\partial \boldsymbol{\sigma}} \Bigg] ds \\
= &0.
\end{aligned} \tag{9.3.90}$$

We substitute (9.3.89) and (9.3.90) into (9.3.88) to give

$$\lim_{\substack{\epsilon^{(k),*} \to \epsilon^* = 0 \\ \left(\sigma^{(k),*}, \theta^{(k),*}\right) \to (\sigma^*, \theta^*) \in \Omega_0}} \nabla_{\sigma} \widetilde{g}_0^{\delta_k} \left(\boldsymbol{\sigma}^{(k),*}, \boldsymbol{\theta}^{(k),*}, \varepsilon^{(k),*}\right) = \nabla_{\sigma} \widetilde{g}_0(\boldsymbol{\sigma}^*, \boldsymbol{\theta}^*). \tag{9.3.91}$$

Similarly, we can show that

$$\lim_{\substack{\varepsilon^{(k),*}\to\varepsilon^*=0\\ \left(\boldsymbol{\sigma}^{(k),*},\boldsymbol{\theta}^{(k),*}\right)\to(\boldsymbol{\sigma}^*,\boldsymbol{\theta}^*)\in\Omega_0}} \nabla_{\boldsymbol{\theta}}\widetilde{g}_0^{\delta_k}\left(\boldsymbol{\sigma}^{(k),*},\boldsymbol{\theta}^{(k),*},\varepsilon^{(k),*}\right) = \nabla_{\theta}\widetilde{g}_0(\boldsymbol{\sigma}^*,\boldsymbol{\theta}^*). \tag{9.3.92}$$

On the other hand, we note that

$$\begin{aligned}
&\lim_{\substack{\varepsilon^{(k),*}\to\varepsilon^*=0\\ \left(\boldsymbol{\sigma}^{(k),*},\boldsymbol{\theta}^{(k),*}\right)\to(\boldsymbol{\sigma}^*,\boldsymbol{\theta}^*)\in\Omega_0}} \nabla_{\varepsilon}\widetilde{g}_0^{\delta_k}\left(\boldsymbol{\sigma}^{(k),*},\boldsymbol{\theta}^{(k),*},\varepsilon^{(k),*}\right)\\
&= \lim_{\substack{\varepsilon^{(k),*}\to\varepsilon^*=0\\ \left(\boldsymbol{\sigma}^{(k),*},\boldsymbol{\theta}^{(k),*}\right)\to(\boldsymbol{\sigma}^*,\boldsymbol{\theta}^*)\in\Omega_0}} \Bigg[\left(\varepsilon^{(k),*}\right)^{-\alpha-1}\Bigg[-\alpha\cdot\\
&\sum_{i=1}^{N}\int_0^1\left[\max\left\{0,\bar{h}_i^{(k),*}(s)-\left(\varepsilon^{(k),*}\right)^{\gamma}W_i\right\}\right]^2 ds\\
&\quad + 2\gamma\sum_{i=1}^{N}\int_0^1\max\left\{0,\bar{h}_i^{(k),*}(s)-\left(\varepsilon^{(k),*}\right)^{\gamma}W_i\right\}\left(\left(-\varepsilon^{(k),*}\right)^{\gamma}W_i\right)ds\\
&\quad + \sum_{i=1}^{n+1}\left(\phi_i\left(\tilde{\boldsymbol{y}}_i^{(k),*}(1)\right)\right)^2\Bigg] + \sigma_k\beta\left(\varepsilon^{(k),*}\right)^{\beta-1}\Bigg]\\
&= \lim_{\substack{\varepsilon^{(k),*}\to\varepsilon^*=0\\ \left(\boldsymbol{\sigma}^{(k),*},\boldsymbol{\theta}^{(k),*}\right)\to(\boldsymbol{\sigma}^*,\boldsymbol{\theta}^*)\in\Omega_0}} \Bigg\{-\alpha\sum_{i=1}^{N}\int_0^1\Bigg[\bar{h}_i^{(k),*}(s)\left(\varepsilon^{(k),*}\right)^{-\frac{\alpha+1}{2}}\\
&\quad -\left(\varepsilon^{(k),*}\right)^{\gamma-\frac{\alpha+1}{2}}W_i\Bigg]^2 ds + 2\gamma\sum_{i=1}^{N}\int_0^1\left[\bar{h}_i^{(k),*}(s)-\left(\varepsilon^{(k),*}\right)^{\gamma}W_i\right]\cdot\\
&\left[\left(-\varepsilon^{(k),*}\right)^{\gamma}W_i\right]\left(\varepsilon^{(k),*}\right)^{-\alpha-1}ds + \sum_{i=1}^{n+1}\left(\phi_i\left(\tilde{\boldsymbol{y}}_i^{(k),*}(1)(\varepsilon^{(k),*})^{-\frac{\alpha+1}{2}}\right)\right)^2\Bigg\}
\end{aligned} \tag{9.3.93}$$

As all the terms in (9.3.93) are bounded and $-\alpha-1+2\xi>0$, $-\alpha-1+2\xi'>0$, $2\gamma-\alpha-1>0$, we have

$$\begin{aligned}
&\lim_{\substack{\varepsilon^{(k),*}\to\varepsilon^*=0\\ \left(\boldsymbol{\sigma}^{(k),*},\boldsymbol{\theta}^{(k),*}\right)\to(\boldsymbol{\sigma}^*,\boldsymbol{\theta}^*)\in\Omega_0}} \nabla_{\varepsilon}g_0^{\delta_k}\left(\boldsymbol{\sigma}^{(k),*},\boldsymbol{\theta}^{(k),*},\varepsilon^{(k),*}\right)\\
&=\Bigg\{-\alpha\sum_{i=1}^{N}\int_0^1 \lim_{\substack{\varepsilon^{(k),*}\to\varepsilon^*=0\\ \left(\boldsymbol{\sigma}^{(k),*},\boldsymbol{\theta}^{(k),*}\right)\to(\boldsymbol{\sigma}^*,\boldsymbol{\theta}^*)\in\Omega_0}}\Bigg[\bar{h}_i^{(k),*}(s)\left(\varepsilon^{(k),*}\right)^{-\frac{\alpha+1}{2}}\\
&\quad -\left(\varepsilon^{(k),*}\right)^{\gamma-\frac{\alpha+1}{2}}W_i\Bigg]^2 ds
\end{aligned} \tag{9.3.94}$$

$$+ 2\gamma \sum_{i=1}^{N} \int_0^1 \lim_{\substack{\varepsilon^{(k),*} \to \varepsilon^* = 0 \\ \left(\sigma^{(k),*}, \theta^{(k),*}\right) \to (\sigma^*, \theta^*) \in \Omega_0}} \left(\bar{h}_i^{(k),*}(s) - \left(\varepsilon^{(k),*}\right)^{\gamma} W_i \right) \cdot$$

$$\left(\left(-\varepsilon^{(k),*}\right)^{\gamma} W_i \right) \left(\varepsilon^{(k),*}\right)^{-\alpha-1} ds + \sum_{i=1}^{n+1} \left(\phi_i \left(\tilde{\boldsymbol{y}}_i^{(k),*}(1) \left(\varepsilon^{(k),*}\right)^{-\frac{\alpha+1}{2}} \right) \right)^2 \Bigg\}$$

$$=0. \tag{9.3.95}$$

Thus, the proof is complete.

Theorem 9.3.4 *Let* $\varepsilon^{(k),*} \to \varepsilon^* = 0$ *and* $\left(\boldsymbol{\sigma}^{(k),*}, \boldsymbol{\theta}^{(k),*}\right) \to (\boldsymbol{\sigma}^*, \boldsymbol{\theta}^*) \in \Omega_0$ *be such that* $\widetilde{g}_0^{\delta_k}\left(\sigma^{(k),*}, \theta^{(k),*}, \epsilon^{(k),*}\right)$ *is finite. Then,* $(\boldsymbol{\sigma}^*, \boldsymbol{\theta}^*)$ *is a local minimizer of Problem* $(\tilde{P}(p))$.

Proof. On the contrary, assume that $(\boldsymbol{\sigma}^*, \boldsymbol{\theta}^*)$ is not a local minimizer of Problem $(\tilde{P}(p))$. Then, there must exist a feasible point $\left(\hat{\boldsymbol{\sigma}}^*, \hat{\boldsymbol{\theta}}^*\right) \in \mathcal{N}_\delta(\boldsymbol{\sigma}^*, \boldsymbol{\theta}^*)$ of Problem $(\tilde{P}(p))$ such that

$$\widetilde{g}_0\left(\hat{\boldsymbol{\sigma}}^*, \hat{\boldsymbol{\theta}}^*\right) < \widetilde{g}_0(\boldsymbol{\sigma}^*, \boldsymbol{\theta}^*) \tag{9.3.96}$$

where $\mathcal{N}_\delta(\boldsymbol{\sigma}^*, \boldsymbol{\theta}^*)$ is a δ neighbourhood of $\left(\hat{\boldsymbol{\sigma}}^*, \hat{\boldsymbol{\theta}}^*\right)$ in Ω_0 for some $\bar{\delta} > 0$. Since $\left(\boldsymbol{\sigma}^*, \theta^*, \varepsilon^{(k),*}\right)$ is a local minimizer of Problem $(\tilde{P}_\delta(p))$, there exists a sequence $\{\xi^k\}$, such that

$$\widetilde{g}_0^{\delta_k}\left(\boldsymbol{\sigma}, \boldsymbol{\theta}, \varepsilon^{(k),*}\right) \geq \widetilde{g}_0^{\delta_k}\left(\boldsymbol{\sigma}^{(k),*}, \boldsymbol{\theta}^{(k),*}, \varepsilon^{(k),*}\right)$$

for any $\boldsymbol{x} \in \mathcal{N}_{\delta^k}\left(\boldsymbol{\sigma}^{(k),*}, \boldsymbol{\theta}^{(k),*}\right)$. Now, we construct a sequence $\left\{\hat{\boldsymbol{\sigma}}^{(k),*}, \hat{\boldsymbol{\theta}}^{(k),*}\right\}$ satisfying

$$\left|\left(\hat{\boldsymbol{\sigma}}^{(k),*}, \hat{\boldsymbol{\theta}}^{(k),*}\right) - \left(\boldsymbol{\sigma}^{(k),*}, \boldsymbol{\theta}^{(k),*}\right)\right| \leq \frac{\xi^k}{k}$$

. Clearly,

$$\widetilde{g}_0^{\delta_k}\left(\hat{\boldsymbol{\sigma}}^{(k),*}, \hat{\boldsymbol{\theta}}^{(k),*}, \varepsilon^{(k),*}\right) \geq \widetilde{g}_0^{\delta_k}\left(\boldsymbol{\sigma}^{(k),*}, \boldsymbol{\theta}^{(k),*}, \varepsilon^{(k),*}\right) \tag{9.3.97}$$

Letting $k \to +\infty$, we have

$$\begin{aligned}
&\lim_{k \to +\infty} \left|\left(\hat{\boldsymbol{\sigma}}^{(k),*}, \hat{\boldsymbol{\theta}}^{(k),*}\right) - \left(\hat{\boldsymbol{\sigma}}^*, \hat{\theta}^*\right)\right| \\
&\quad \leq \lim_{k \to +\infty} \left|\left(\hat{\boldsymbol{\sigma}}^{(k),*}, \hat{\boldsymbol{\theta}}^{(k),*}\right) - \left(\boldsymbol{\sigma}^{(k),*}, \boldsymbol{\theta}^{(k),*}\right)\right| \\
&\qquad + \lim_{k \to +\infty} \left|\left(\boldsymbol{\sigma}^{(k),*}, \boldsymbol{\theta}^{(k),*}\right) - (\boldsymbol{\sigma}^*, \boldsymbol{\theta}^*)\right| + \left|(\boldsymbol{\sigma}^*, \boldsymbol{\theta}^*) - \left(\hat{\boldsymbol{\sigma}}^*, \hat{\boldsymbol{\theta}}^*\right)\right| \\
&\quad \leq 0 + 0 + \bar{\delta}.
\end{aligned} \tag{9.3.98}$$

However, $\bar{\delta} > 0$ is arbitrary. Thus,

$$\lim_{k \to +\infty} \left(\hat{\boldsymbol{\sigma}}^{(k),*}, \hat{\boldsymbol{\theta}}^{(k),*} \right) = \left(\hat{\boldsymbol{\sigma}}^*, \hat{\boldsymbol{\theta}}^* \right). \tag{9.3.99}$$

Letting $k \to +\infty$ in (9.3.97), it follows from the continuity of $\tilde{g}_0$ and (9.3.99) that

$$\begin{aligned} &\lim_{k \to +\infty} \widetilde{g}_0^{\delta_k} \left(\hat{\boldsymbol{\sigma}}^{(k),*}, \hat{\boldsymbol{\theta}}^{(k),*}, \varepsilon^{(k),*} \right) \\ &= \widetilde{g}_0^{\delta_k} \left(\hat{\boldsymbol{\sigma}}^*, \hat{\boldsymbol{\theta}}^*, 0 \right) = \widetilde{g}_0 \left(\hat{\boldsymbol{\sigma}}^*, \hat{\boldsymbol{\theta}}^* \right) \geq \lim_{k \to +\infty} \widetilde{g}_0^{\delta_k} \left(\boldsymbol{\sigma}^{(k),*}, \boldsymbol{\theta}^{(k),*}, \varepsilon^{(k),*} \right) \\ &= \widetilde{g}_0^{\delta_k} \left(\boldsymbol{\sigma}^*, \boldsymbol{\theta}^*, 0 \right) = \widetilde{g}_0 \left(\boldsymbol{\sigma}^*, \boldsymbol{\theta}^* \right). \end{aligned} \tag{9.3.100}$$

This is a contradiction to (9.3.96), and hence it completes the proof.

Theorem 9.3.5 *Let $-\alpha - \beta + 2\xi > 0$, $-\alpha - \beta + 2\xi' > 0$ and $-\alpha - \beta + 2\gamma > 0$, then there exists a $k_0 > 0$, such that $\varepsilon^{(k),*} = 0$, $\left(\boldsymbol{\sigma}^{(k),*}, \boldsymbol{\theta}^{(k),*} \right)$ is local minimizer of Problem $(\tilde{P}(p))$, for $k \geq k_0$.*

Proof. On the contrary, we assume that the conclusion is false. Then, there exists a subsequence of $\left\{ \left(\boldsymbol{\sigma}^{(k),*}, \boldsymbol{\theta}^{(k),*}, \varepsilon^{(k),*} \right) \right\}$, which is denoted by the original sequence, such that for any $k_0 > 0$, there exists a $k' > k_0$ satisfying $\varepsilon^{(k'),*} \neq 0$. By Theorem 9.3.2, we have

$$\varepsilon^{(k),*} \to \varepsilon^* = 0, \quad \left(\boldsymbol{\sigma}^{(k),*}, \boldsymbol{\theta}^{(k),*} \right) \to \left(\boldsymbol{\sigma}^*, \boldsymbol{\theta}^* \right) \in \Omega_0, \quad \text{as} \quad k \to +\infty$$

. Since $\varepsilon^{(k),*} \neq 0$ for all k, it follows from dividing (9.3.45) by $(\varepsilon^{(k),*})^{\beta-1}$ that

$$\begin{aligned} &\left(\varepsilon^{(k),*} \right)^{-\alpha-\beta} \left\{ -\alpha \sum_{i=1}^{N} \int_0^1 \left[\max \left\{ 0, \bar{h}_i^{(k),*}(s) - \left(\varepsilon^{(k),*} \right)^\gamma W_i \right\} \right]^2 ds \right. \\ &+ 2\gamma \sum_{i=1}^{N} \int_0^1 \max \left\{ 0, \bar{h}_i^{(k),*}(s) - \left(\varepsilon^{(k),*} \right)^\gamma W_i \right\} \left(\left(-\varepsilon^{(k),*} \right)^\gamma W_i \right) ds \\ &\left. - \alpha \sum_{i=1}^{n+1} \left(\tilde{y}_i^{(k),*}(1) - \tilde{y}_i^f \right)^2 \right\} + \delta_k \beta = 0. \end{aligned} \tag{9.3.101}$$

This is equivalent to

$$\begin{aligned} &\left(\varepsilon^{(k),*} \right)^{-\alpha-\beta} \left\{ -\alpha \sum_{i=1}^{N} \int_0^1 \left[\max \left\{ 0, \bar{h}_i^{(k),*}(s) - \left(\varepsilon^{(k),*} \right)^\gamma W_i \right\} \right]^2 ds \right. \\ &+ 2\gamma \sum_{i=1}^{N} \int_0^1 \left\{ \max \left\{ 0, \bar{h}_i^{(k),*}(s) - \left(\varepsilon^{(k),*} \right)^\gamma W_i \right\} \left(\left(-\varepsilon^{(k),*} \right)^\gamma W_i \right) \right. \end{aligned}$$

$$+\max\left\{0,\bar{h}_i^{(k),*}(s)-\left(\varepsilon^{(k),*}\right)^{\gamma}W_i\right\}\bar{h}_i^{(k),*}(s)$$
$$-\max\left\{0,\bar{h}_i^{(k),*}(s)-\left(\varepsilon^{(k),*}\right)^{\gamma}W_i\right\}\bar{h}_i^{(k),*}(s)\Bigg\}ds$$
$$-\alpha\sum_{i=1}^{n+1}\left(\tilde{y}_i^{(k),*}(1)-\tilde{y}_i^f\right)^2\Bigg\}+\delta_k\beta=0 \tag{9.3.102}$$

Rearranging (9.3.102) yields

$$\left(\varepsilon^{(k),*}\right)^{-\alpha-\beta}\Bigg\{(2\gamma-\alpha)\sum_{i=1}^{N}\int_0^1\left[\max\left\{0,\bar{h}_i^{(k),*}(s)-\left(\varepsilon^{(k),*}\right)^{\gamma}W_i\right\}\right]^2ds$$
$$-\alpha\sum_{i=1}^{n+1}\left(\tilde{y}_i^{(k),*}(1)-\tilde{y}_i^f\right)^2\Bigg\}+\delta_k\beta$$
$$=2\gamma\left(\varepsilon^{(k),*}\right)^{-\alpha-\beta}\sum_{i=1}^{N}\int_0^1\max\left\{0,\bar{h}_i^{(k),*}(s)-\left(\varepsilon^{(k),*}\right)^{\gamma}W_i\right\}\times\bar{h}_i^{(k),*}(s)ds. \tag{9.3.103}$$

Let $k\to+\infty$ in (9.3.102), and note that $-\alpha-\beta+2\xi>0$ and $-\alpha-\beta+2\xi'>0$. Then, it follows that the left hand side of (9.3.103) yields

$$\left(\varepsilon^{(k),*}\right)^{-\alpha-\beta}\Bigg\{(2\gamma-\alpha)\sum_{i=1}^{N}\int_0^1\left[\max\left\{0,\bar{h}_i^{(k),*}(s)-\left(\varepsilon^{(k),*}\right)^{\gamma}W_i\right\}\right]^2ds$$
$$-\alpha\sum_{i=1}^{n+1}\left(\tilde{y}_i^{(k),*}(1)-\tilde{y}_i^f\right)^2\Bigg\}+\delta_k\beta\to\infty. \tag{9.3.104}$$

However, under the same conditions and $-\alpha-\beta+2\gamma>0$, the right hand side of (9.3.103) gives

$$2\gamma\left(\varepsilon^{(k),*}\right)^{-\alpha-\beta}\sum_{i=1}^{N}\int_0^1\max\left\{0,\bar{h}_i^{(k),*}(s)-\left(\varepsilon^{(k),*}\right)^{\gamma}W_i\right\}\bar{h}_i^{(k),*}(s)ds\to 0. \tag{9.3.105}$$

This is a contradiction. Thus, the proof is complete.

Theorem 9.3.6 *Let* $\mathbf{u}^{p,*}$ *be an optimal control of the approximate Problem* $(\hat{P}(p))$. *Suppose that* $\mathbf{u}^*$ *is an optimal control of the Problem* $(\tilde{P})$. *Then,*

$$\lim_{p\to+\infty}g_0(\boldsymbol{u}^{p,*})=g_0(\boldsymbol{u}^*). \tag{9.3.106}$$

Proof. The proof is similar to that given for Theorem 9.2.2.

Theorem 9.3.7 *Let $\boldsymbol{u}^{p,*}$ be an optimal control of the approximate Problem $(\tilde{P}(p))$, and $\boldsymbol{u}^*$ be an optimal control of the Problem $(\hat{P})$. Suppose that*

$$\lim_{p\to+\infty} \boldsymbol{u}^{p,*} = \bar{\boldsymbol{u}}, \quad a.e. \text{ on } [0,T]. \tag{9.3.107}$$

Then, $\bar{\mathbf{u}}$ is an optimal control of the Problem $(\hat{P})$

$$\lim_{p\to+\infty} g_0(\boldsymbol{u}^{p,*}) = g_0(\boldsymbol{u}^*). \tag{9.3.108}$$

Proof. The proof is similar to that given for Theorem 9.2.3.

9.3.3 Computational Algorithm

In this section, we are in the position to present the computational algorithm for solving Problem $(\tilde{P}(p))$ as follows.

Algorithm 9.3.1
Step 1 *set $\delta^{(1)} = 10$, $\varepsilon^{(1)} = 0.1$, $\varepsilon^* = 10^{-9}$, $\beta > 2$, choose an initial point $(\boldsymbol{\sigma}^0, \boldsymbol{\theta}^0, \varepsilon^0)$, the iteration index $k = 0$. The values of γ and α are chosen depending on the specific structure of Problem $(\tilde{P}(p))$ concerned.*
Step 2 *Solve Problem $(\tilde{P}_{\delta_k}(p))$, and let $\left(\boldsymbol{\sigma}^{(k),*}, \boldsymbol{\theta}^{(k),*}, \varepsilon^{(k),*}\right)$ be the minimizer obtained.*
Step 3 If *$\varepsilon^{(k),*} > \varepsilon^*$, $\delta^{(k)} < 10^8$,*
set $\delta^{(k+1)} = 10 \times \sigma^{(k)}$, $k = k+1$. Go to ***Step 2*** *with $\left(\boldsymbol{\sigma}^{(k),*}, \boldsymbol{\theta}^{(k),*}, \varepsilon^{(k),*}\right)$ as the new initial point in the new optimization process*
Else *set $\varepsilon^{(k),*} = \varepsilon^*$, then go to* ***Step 4***
Step 4 *Check the feasibility of $\left(\boldsymbol{\sigma}^{(k),*}, \boldsymbol{\theta}^{(k),*}\right)$ (i.e., check whether or not $\max\limits_{1\le i\le N} \max\limits_{s\in[0,1]} h_i\left(\boldsymbol{y}(s), \boldsymbol{\sigma}^{(k),*}\right) \le 0$).*
If *$\left(\boldsymbol{\sigma}^{(k),*}, \boldsymbol{\theta}^{(k),*}\right)$ is feasible, then it is a local minimizer of Problem $(\tilde{P}(p))$. Exit.*
Else *go to* ***Step 5***
Step 5*: Adjust the parameters α, β and γ such that the conditions of Lemma 9.3.1 are satisfied. Set $\delta^{(k+1)} = 10\delta^{(k)}$, $\varepsilon^{(k+1)} = 0.1\varepsilon^{(k)}$, $k := k+1$. Go to* ***Step 2****.*

Remark 9.3.5 *In Step 3, if $\varepsilon^{(k),*} > \varepsilon^*$, it follows from Theorem 9.3.2 and Theorem 9.3.3 that $\left(\boldsymbol{\sigma}^{(k),*}, \boldsymbol{\theta}^{(k),*}\right)$ cannot be a feasible point, meaning that the penalty parameter δ may not be chosen large enough. Thus we need to increase δ. If $\delta_k > 10^8$, but still $\varepsilon^{(k),*} > \varepsilon^*$, then we should adjust the value of α, β and γ, such that the conditions of Theorem 9.3.3 are satisfied. Then, go to Step 2.*

Remark 9.3.6 *Clearly, we cannot check the feasibility of $h_i(\boldsymbol{y}(s), \boldsymbol{\sigma}) \leq 0$, $i = 1, \ldots, N$, for every $s \in [0, 1]$. In practice, we choose a set which contains a dense enough of points in $[0, 1]$. Check the feasibility of $h_i(\boldsymbol{y}(s), \boldsymbol{\sigma}) \leq 0$ over this set for each $i = 1, \ldots, N$.*

Remark 9.3.7 *Although we have proved that a local minimizer of the exact penalty function optimization Problem $(\tilde{P}_{\delta_k}(p))$ will converge to a local minimizer of the original Problem $(\tilde{P}(p))$, we need, in actual computation, set a lower bound $\varepsilon^* = 10^{-9}$ for $\varepsilon^{(k),*}$ so as to avoid the situation of being divided by $\varepsilon^{(k),*} = 0$, leading to infinity.*

9.3.4 Examples

Example 9.3.1 In this example, we revisit Example 9.2.2 in Section 9.2.6 by using the exact penalty function method. In this problem, we set $p = 20$, $\gamma = 3$ and $W_1 = 0.3$. The result is shown below. The optimal objective function value $g_0^* = 1.75101803 \times 10^{-1}$, where $\delta = 1.0 \times 10^6$ and $\varepsilon = 1.89531e \times 10^{-5}$.

As shown in Figure 9.3.1, the continuous inequality constraints (9.2.85) is satisfied for all t$\in [0, 1]$. Although the minimum value of the cost functional is almost the same (which is 1.727×10^{-1} for Example 9.2.1) comparing with the results obtained for Example 9.2.1, (9.2.85) in Example 9.2.1 is violated at some $t \in [0, 1]$. The optimal state strategies and optimal control are plotted in Figures 9.3.2 and 9.3.3, respectively.

Example 9.3.2 In this example, we revisit Example 9.2.2 in Section 9.2.6 by using the exact penalty function method. In this problem, we set $p = 20$, $\gamma = 3$ and $W_1 = W_2 = W_3 = W_4 = 0.3$.

The result obtained is shown below. The optimal cost function value is $g_0^* = 5.75921513 \times 10^{-3}$, where $\delta = 1.0 \times 10^5$ and $\varepsilon = 1.00057 \times 10^{-7}$. All the continuous inequality constraints are satisfied for all $t \in [0, 1]$. Comparing with the results obtained for Example 9.2.2, our minimum value of the cost functional is slightly larger (which is 5.3×10^{-3} for Example 9.2.2). However, the continuous inequality constraints (9.2.94)–(9.2.97) in Example 9.2.2 are not satisfied at all $t \in [0, 1]$. The continuous inequality constraints are shown in Figure 9.3.4, and the optimal state trajectories and the optimal control are shown in Figures 9.3.5 and 9.3.6, respectively.

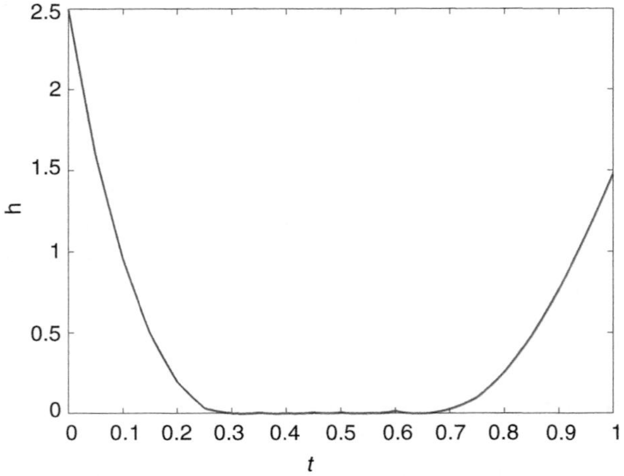

Fig. 9.3.1: $h(t)$ for Example 9.3.1

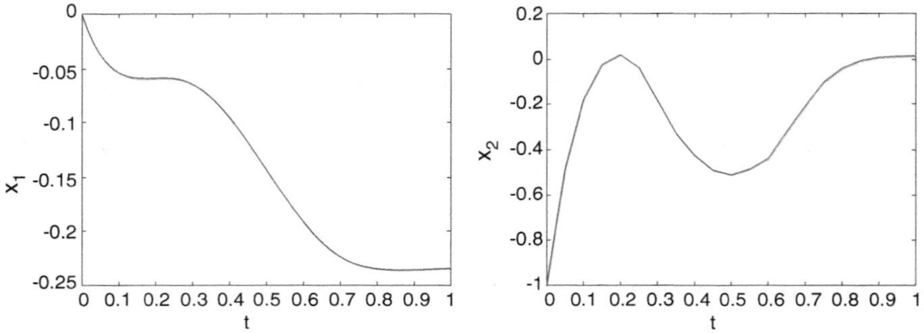

Fig. 9.3.2: Optimal state trajectories $x_1(t)$ and $x_2(t)$ for Example 9.3.1

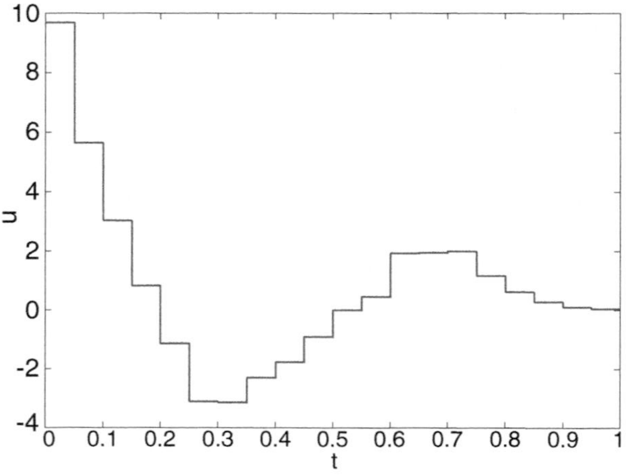

Fig. 9.3.3: Optimal control $u(t)$ for Example 9.3.1

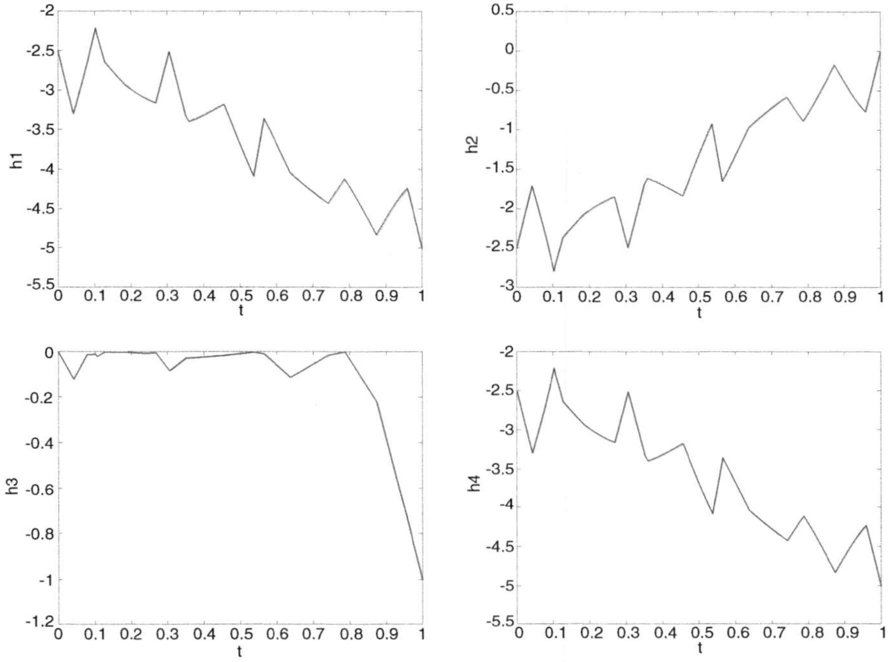

Fig. 9.3.4: Continuous inequality constraints for Example 9.3.2

Example 9.3.3 The following problem is taken from [68]: Find a control $u : [0, 4.5] \to \mathbb{R}$ that minimizes the cost functional

$$\int_0^{4.5} \left\{ (u(t))^2 + (x_1(t))^2 \right\} dt \tag{9.3.109}$$

subject to the following dynamical equations

$$\frac{dx_1(t)}{dt} = x_2(t) \tag{9.3.110}$$

$$\frac{dx_2(t)}{dt} = -x_1(t) + x_2(t) \left(1.4 - 0.14(x_2(t))^2(t)\right) + 4u(t) \tag{9.3.111}$$

with the initial conditions $x_1(0) = -5$ and $x_2(0) = -5$, and the continuous inequality constraint

$$h = -u(t) - \frac{1}{6}x_1(t) \geq 0, \ t \in [0, 4.5]. \tag{9.3.112}$$

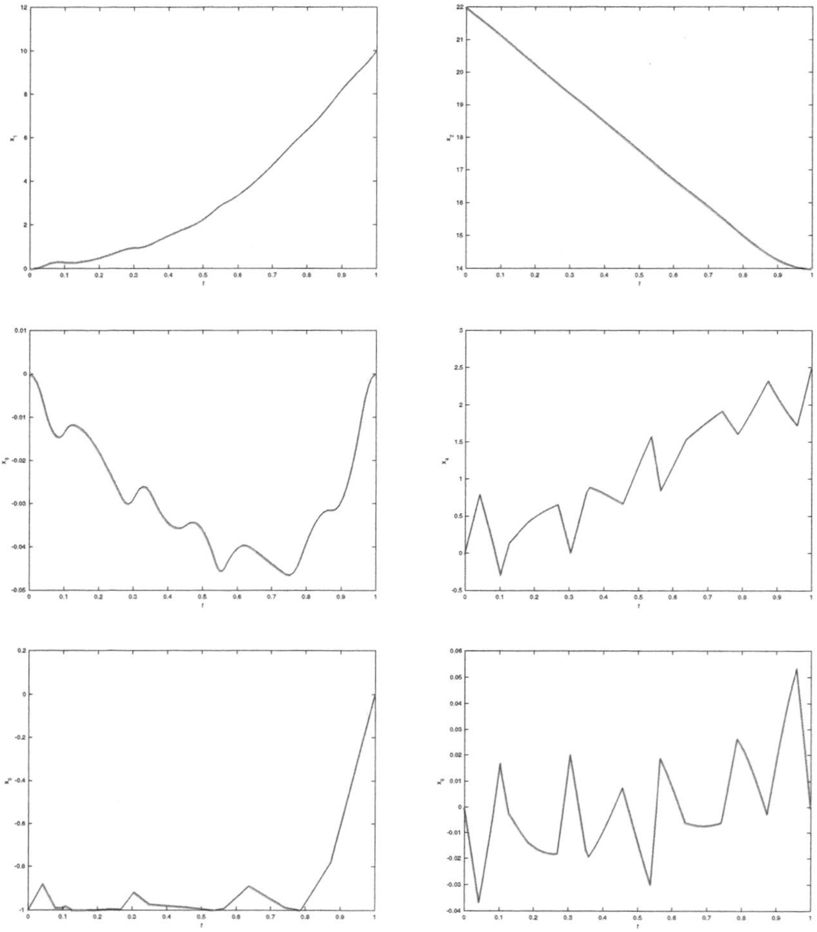

Fig. 9.3.5: Optimal state trajectories for Example 9.3.2

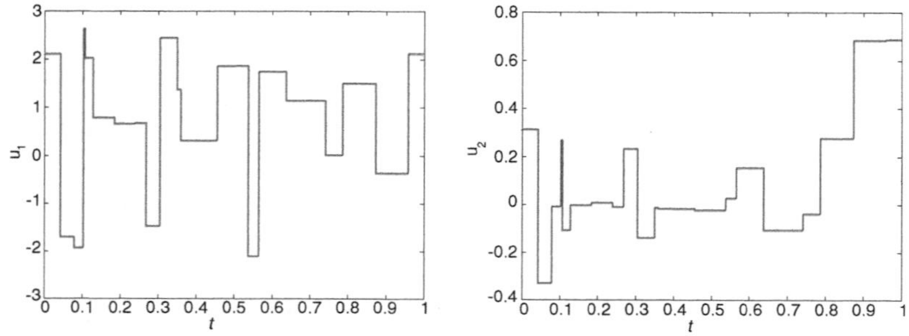

Fig. 9.3.6: Optimal controls for Example 9.3.2

In this problem, we set $p = 10$, $\gamma = 3$ and $W_1 = 0.3$. The result is shown below. The optimal objective function value obtained is $g_0^* = 4.58048380e \times 10^1$, where $\delta = 1.0 \times 10^4$ and $\varepsilon = 9.99998 \times 10^{-5}$. The continuous inequality constraint (9.3.112) is satisfied for all $t \in [0, 4.5]$. The continuous inequality constraints are shown in Figure 9.3.7, and the optimal states and the optimal control are shown in Figures 9.3.8 and 9.3.9, respectively.

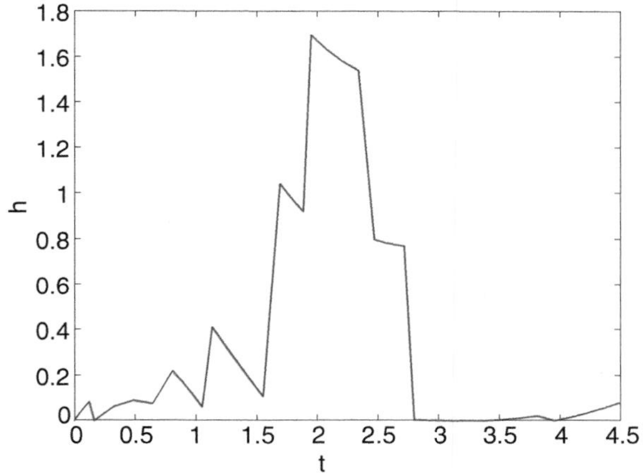

Fig. 9.3.7: Continuous inequality constraint for Example 9.3.3.

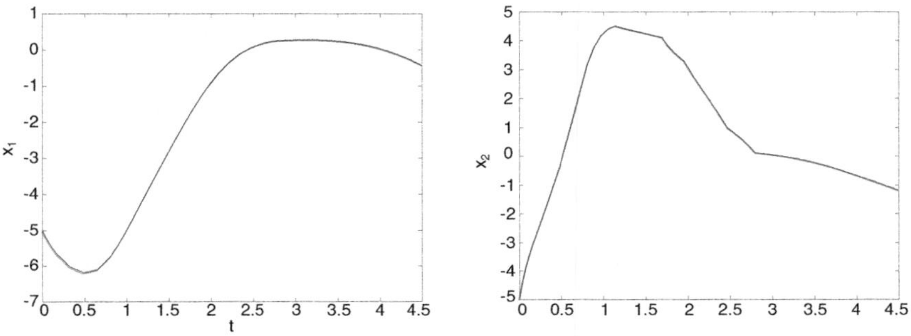

Fig. 9.3.8: Optimal state trajectories for Example 9.3.3.

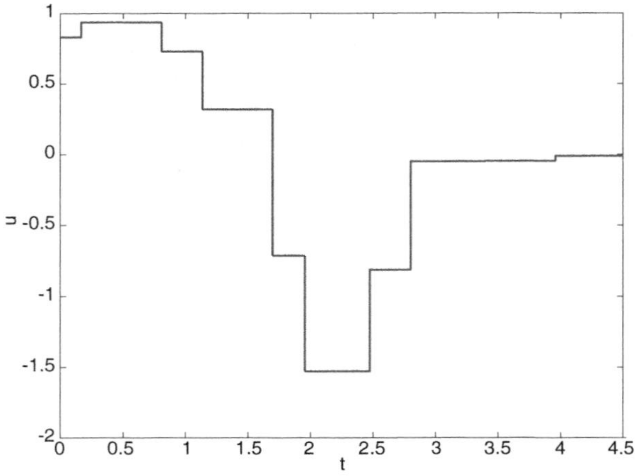

Fig. 9.3.9: Optimal control for Example 9.3.3.

9.4 Exercises

9.4.1 *Can the control functions chosen in Definition 9.2.1 be just measurable, rather than Borel measurable functions?*

9.4.2 *Derive the gradient formulae given by (9.3.32) and (9.3.33).*

9.4.3 *Consider the problem (P) subject to additional terminal equality constraints:*

$$q_i(\boldsymbol{x}(T \mid \boldsymbol{u})) = 0, \quad i = 1, \ldots, N_E, \tag{9.4.1}$$

where q_i, $i = 1, \ldots, N_E$, are continuously differentiable functions. Let this optimal control problem be referred to as the problem (R). Use the control parametrization time scaling transform technique to derive a computational method for solving Problem (R). State all the essential assumptions and then prove all the relevant convergence results.

9.4.4 *Provide detailed derivations of the gradient formulae appeared in Algorithm 9.2.4.*

9.4.5 *Prove Theorem 9.2.1.*

9.4.6 *Prove Theorem 9.2.2.*

9.4.7 *Prove Theorem 9.2.3.*

9.4.8 *Show the validity of Remark 9.3.2.*

9.4.9 *Construct the equality constraint violation as defined by (9.3.29) fort the interior equality constraint (8.9.14a).*

9.4.10 *Provide detailed derivations of the gradient formulae presented in Theorem 9.3.1.*

9.4.11 *Show the validity of the statement made in Remark 9.3.4.*

9.4.12 *Provide detailed derivations of the gradient formulae which is given by* (9.3.35), (9.3.36) *and* (9.3.37).

9.4.13 *Give detailed proof of Corollary 9.3.1.*

9.4.14 *Prove Theorem 9.3.6.*

9.4.15 *Prove Theorem 9.3.7.*

Chapter 10
Time-Lag Optimal Control Problems

In this chapter, we consider three types of optimal control problems: (1) Time-lag optimal control problems. The main reference is [289, 290, 298]. (2) Optimal control problems with state-dependent switched time-delayed systems. The main reference is [153]. (3) Min-max optimal control of linear continuous dynamical systems with uncertainty and quadratic terminal constraints problems. The main reference is [287].

10.1 Time-Lag Optimal Control

10.1.1 Introduction

A time-delay system is a dynamic system, which evolves depending not only on the current state and/or control variables but also on the state and/or control variables at some past time instants. Such systems arise in plethora of real-world applications in science and engineering, including epidemiological modeling [16], vehicle suspension design [129] and spacecraft attitude control [47]. In this section, we consider a general class of time-delayed optimal control problems where the cost functional is to be minimized subject to canonical constraints. The control parametrization method together with a hybrid time-scaling transformation strategy is used to devise a computational algorithm for solving this general class of time-delayed optimal control problems. For illustration, several numerical examples are solved using the proposed algorithm. The main references of this section are [289], [290] and [298].

K. L. Teo et al., *Applied and Computational Optimal Control*, Springer Optimization and Its Applications 171,
https://doi.org/10.1007/978-3-030-69913-0_10

10.1.2 Problem Formulation

Consider the following time-delay system, defined on the fixed time interval $(-\infty, T]$:

$$\frac{d\boldsymbol{x}}{dt} = \boldsymbol{f}(\boldsymbol{x}(t), \bar{\boldsymbol{x}}(t), \boldsymbol{u}(t), \bar{\boldsymbol{u}}(t)),\ t \in [0, T], \tag{10.1.1}$$

$$\boldsymbol{x}(t) = \boldsymbol{\phi}(t),\ t \leq 0, \tag{10.1.2}$$

$$\boldsymbol{u}(t) = \boldsymbol{\varphi}(t),\ t < 0, \tag{10.1.3}$$

where $\boldsymbol{x}(t) = [x_1(t), x_2(t), \ldots, x_n(t)]^\top \in \mathbb{R}^n$ is the state vector; $\boldsymbol{u}(t) = [u_1(t), u_2(t), \ldots, u_r(t)]^\top \in \mathbb{R}^r$ is the control vector; $\bar{\boldsymbol{x}}(t) = [x_1(t-h_1), x_2(t-h_2), \ldots, x_n(t-h_n)]^\top$ and $\bar{\boldsymbol{u}}(t) = [u_1(t-h_{n+1}), u_2(t-h_{n+2}), \ldots, u_r(t-h_{n+r})]^\top$, in which $h_q > 0, q = 1, \ldots, n+r$, are given time-delays; $\boldsymbol{f} : \mathbb{R}^n \times \mathbb{R}^n \times \mathbb{R}^r \times \mathbb{R}^r \to \mathbb{R}^n$ and $\boldsymbol{\phi}(t) = [\phi_1(t), \ldots, \phi_n(t)]^\top$ are given continuously differentiable functions; and $\boldsymbol{\varphi}(t) = [\varphi_1(t), \ldots, \varphi_r(t)]^\top$ is a given function. A Borel measurable function $\boldsymbol{u}(t) : (-\infty, T] \to \mathbb{R}^r$ is said to be an admissible control if $\boldsymbol{u}(t) \in U$ for almost all $t \in [0, T]$ and $\boldsymbol{u}(t) = \boldsymbol{\varphi}(t)$ for all $t < 0$, where U is a compact and convex subset of $\mathbb{R}^r$. Let $\mathcal{U}$ denote the class of all such admissible controls. For simplicity, we use $\boldsymbol{u}$ to denote $\boldsymbol{u}(t)$ for the rest of the section. Let $\boldsymbol{x}(\cdot)$ denote the solution of (10.1.1)–(10.1.3) corresponding to each $\boldsymbol{u} \in \mathcal{U}$. It is an absolutely continuous function that satisfies the dynamic (10.1.1) almost everywhere on $[0, T]$, and the initial condition (10.1.2) everywhere on $(-\infty, 0]$. Our optimization problem is defined formally as follows: Given the dynamic system (10.1.1)–(10.1.3), choose an admissible control $\boldsymbol{u} \in \mathcal{U}$ to minimize the following cost functional:

$$\Phi_0(\boldsymbol{x}(T)) + \int_0^T \mathcal{L}_0(\boldsymbol{x}(t), \bar{\boldsymbol{x}}(t), \boldsymbol{u})dt, \tag{10.1.4}$$

subject to the canonical constraints

$$\Phi_k(\boldsymbol{x}(T)) + \int_0^T \mathcal{L}_k(\boldsymbol{x}(t), \bar{\boldsymbol{x}}(t), \boldsymbol{u})dt = 0, \quad k = 1, \ldots, N_e, \tag{10.1.5}$$

$$\Phi_k(\boldsymbol{x}(T)) + \int_0^T \mathcal{L}_k(\boldsymbol{x}(t), \bar{\boldsymbol{x}}(t), \boldsymbol{u})dt \geq 0, \quad k = N_e + 1, \ldots, N_e + N_m, \tag{10.1.6}$$

where $\Phi_k : \mathbb{R}^n \to \mathbb{R},\ k = 0, 1, \ldots, N_e + N_m$, and $\mathcal{L}_k : \mathbb{R}^n \times \mathbb{R}^n \times \mathbb{R}^r \to \mathbb{R},\ k = 0, 1, \ldots, N_e + N_m$, are given real-valued functions. We denote this problem as Problem (P_1).

We assume that the following conditions are satisfied throughout this section.

Assumption 10.1.1 *$\mathcal{L}_k : \mathbb{R}^n \times \mathbb{R}^n \times \mathbb{R}^r \to \mathbb{R}$, $k = 0, 1, \ldots, N_e + N_m$ and $\Phi_k : \mathbb{R}^n \to \mathbb{R}$, $k = 0, 1, \ldots, N_e + N_m$ are continuously differentiable with respect to each of their arguments.*

Assumption 10.1.2 *$\boldsymbol{f}$ is twice continuously differentiable.*

Assumption 10.1.3 *There exists a real number $L_1 > 0$ such that $|\boldsymbol{f}(\boldsymbol{\alpha}, \boldsymbol{e}, \boldsymbol{\nu}, \boldsymbol{\omega})| \leq L_1(1 + |\boldsymbol{\alpha}| + |\boldsymbol{e}|)$, $(\boldsymbol{\alpha}, \boldsymbol{e}, \boldsymbol{\nu}, \boldsymbol{\omega}) \in \mathbb{R}^n \times \mathbb{R}^n \times U \times U$.*

Assumptions 10.1.1–10.1.3 ensure the existence and uniqueness of the solution of the dynamic system (10.1.1)–(10.1.3).

10.1.3 Control Parametrization

In this section, we shall propose a numerical method to solve Problem (P_1). We subdivide the planning horizon $[0, T]$ into $p \geq 1$ subintervals $[t_i, t_{i+1}), i = 0, 1, \ldots, p-1$, where t_i, $i = 0, 1, \ldots, p$, are the partition points that satisfying

$$0 = t_0 \leq t_1 \leq t_2 \leq \ldots \leq t_{p-1} \leq t_p = T. \tag{10.1.7}$$

Let Ξ denote the set of all vectors $\boldsymbol{\sigma} = [t_1, \ldots, t_p]^\top$ such that (10.1.7) is satisfied. Then the control $\boldsymbol{u}$ is approximated as follows:

$$\boldsymbol{u} \approx \sum_{i=1}^{p} \boldsymbol{\delta}^{(i)} \chi_{[t_{i-1}, t_i)}(t), \ t \in [0, T], \tag{10.1.8}$$

where $\boldsymbol{\delta}^{(i)} = \left[\delta_1^{(i)}, \ldots, \delta_r^{(i)}\right]^\top$ is the value of the control on the ith subinterval and $\chi_{[t_{i-1}, t_i)}(t)$ is the indicator function defined by

$$\chi_{[t_{i-1}, t_i)}(t) = \begin{cases} 1, & \text{if } t \in [t_{i-1}, t_i), \\ 0, & \text{otherwise.} \end{cases}$$

Let Δ denote the set of all such vectors $\boldsymbol{\delta} = \left[(\boldsymbol{\delta}^{(1)})^\top, \ldots, (\boldsymbol{\delta}^{(p)})^\top\right]^\top$. Substituting (10.1.8) into (10.1.1), the time-delay system defined on the subinterval $[t_{i-1}, t_i)$ becomes

$$\frac{d\boldsymbol{x}}{dt} = \boldsymbol{f}\left(\boldsymbol{x}(t), \bar{\boldsymbol{x}}(t), \boldsymbol{\theta}^{(i)}, \bar{\boldsymbol{\theta}}(t)\right), \tag{10.1.9}$$

where $\bar{\boldsymbol{\theta}}(t) = \left[\bar{\theta}_1(t), \ldots, \bar{\theta}_r(t)\right]^\top$. For $\bar{\theta}_m(t)$, $m = 1, \ldots, r$, there are two cases: (1) if $t - h_{n+m} < 0$, then $\bar{\theta}_m(t) = \varphi_m(t - h_{n+m})$; and (2) if $t - h_{n+m} \geq 0$, then there exist q $(q \leq p)$ distinct partition points (let these points be denoted as $t_{i_l}, l = 1, \ldots, q$,) such that

$$0 < t_{i_1} < t_{i_2} < \cdots < t_{i_q} = T.$$

Clearly,

$$[0,T] = [0,t_{i_1}) \cup [t_{i_1},t_{i_2}) \ldots \cup \left[t_{i_{q-1}},t_{i_q}\right], \tag{10.1.10}$$

and for any $k, l \in \{1,2,\ldots,q\}, k \neq l$,

$$\left[t_{i_{k-1}},t_{i_k}\right) \cap \left[t_{i_{l-1}},t_{i_l}\right) = \emptyset. \tag{10.1.11}$$

Therefore, we can find a unique $j \in \{1,\ldots,p\}$ such that $t - h_{n+m} \in [t_{j-1},t_j)$ and $\bar{\delta}_m(t) = \delta_m^{(j)}$.

Now, $\bar{\delta}_m(t),\ m = 1,\ldots,r$, can be expressed as

$$\bar{\delta}_m(t) = \begin{cases} \delta_m^{(j)}, & \text{if } t \in [t_{j-1}+h_{n+m}, t_j + h_{n+m}) \\ & \text{for some } j \in \{1,\ldots,p\}, \\ \varphi_m(t-h_{n+m}), & \text{if } t < h_{n+m}. \end{cases}$$

Let $\boldsymbol{x}(\cdot \mid \boldsymbol{\sigma},\boldsymbol{\delta})$ denote the solution of (10.1.9) corresponding to $(\boldsymbol{\sigma},\boldsymbol{\delta}) \in \Xi \times \Delta$. The original time-delay problem (P_1) is now approximated as: Given system (10.1.9), choose a $(\boldsymbol{\sigma},\boldsymbol{\delta}) \in \Xi \times \Delta$ such that the cost functional

$$g_0(\boldsymbol{\sigma},\boldsymbol{\delta}) = \Phi_0(\boldsymbol{x}(T \mid \boldsymbol{\sigma},\boldsymbol{\delta})) + \sum_{i=1}^{p} \int_{t_{i-1}}^{t_i} \mathcal{L}_0\left(\boldsymbol{x}(t \mid \boldsymbol{\sigma},\boldsymbol{\delta}), \bar{\boldsymbol{x}}(t \mid \boldsymbol{\sigma},\boldsymbol{\delta}), \boldsymbol{\delta}^{(i)}\right) dt, \tag{10.1.12}$$

is minimized subject to the following constraints:

$$g_k(\boldsymbol{\sigma},\boldsymbol{\delta}) = \Phi_k(\boldsymbol{x}(T \mid \boldsymbol{\sigma},\boldsymbol{\delta})) + \sum_{i=1}^{p} \int_{t_{i-1}}^{t_i} \mathcal{L}_k\left(\boldsymbol{x}(t \mid \boldsymbol{\sigma},\boldsymbol{\delta}), \bar{\boldsymbol{x}}(t \mid \boldsymbol{\sigma},\boldsymbol{\delta}), \boldsymbol{\delta}^{(i)}\right) dt = 0, \quad k = 1,\ldots,N_e, \tag{10.1.13}$$

$$g_k(\boldsymbol{\sigma},\boldsymbol{\delta}) = \Phi_k(\boldsymbol{x}(T \mid \boldsymbol{\sigma},\boldsymbol{\delta})) + \sum_{i=1}^{p} \int_{t_{i-1}}^{t_i} \mathcal{L}_k\left(\boldsymbol{x}(t \mid \boldsymbol{\sigma},\boldsymbol{\delta}), \bar{\boldsymbol{x}}(t \mid \boldsymbol{\sigma},\boldsymbol{\delta}), \boldsymbol{\delta}^{(i)}\right) dt \geq 0, \quad k = N_e+1,\ldots,N_e+N_m, \tag{10.1.14}$$

where $\bar{\boldsymbol{x}}(t \mid \boldsymbol{\sigma},\boldsymbol{\delta}) = [x_1(t-h_1 \mid \boldsymbol{\sigma},\boldsymbol{\delta}),\ldots,x_n(t-h_n \mid \boldsymbol{\sigma},\boldsymbol{\delta})]^\top$. A pair $(\boldsymbol{\sigma},\boldsymbol{\delta}) \in \Xi \times \Delta$ is said to be a feasible pair if it satisfies constraints (10.1.13)–(10.1.14). Let $\mathcal{F}$ consist of all such feasible pairs. To proceed further, let this approximate problem be referred to as Problem $(P_1(p))$.

10.1.4 The Time-Scaling Transformation

For time-delayed optimal control problems with variable switching times, the conventional time scaling transformation fails to work. This is because the conventional time scaling transformation will map variable switching times into fixed switching times in a new time horizon. However, the time-delays will become variable delays. As a consequence, the transformed problem will be even harder to solve. In this section, we shall develop a novel time-scaling transformation to transform Problem $(P(p))$ into an equivalent problem in which the switching times are fixed.

For any $\boldsymbol{\sigma} = [t_1, \ldots, t_p]^\top \in \Xi$, define a vector $\boldsymbol{\theta} = [\theta_1, \ldots, \theta_p]^\top \in \mathbb{R}^p$ where $\theta_i = t_i - t_{i-1}, i = 1, \ldots, p$. Clearly, $\theta_i \geq 0$ is the duration between two consecutive switching times for the control vector $\boldsymbol{\delta}^{(i)}$ and $\theta_1 + \cdots + \theta_p = T$. Let Θ denote the set of all such vectors $\boldsymbol{\theta} \in \mathbb{R}^p$, where vectors in Θ are called admissible duration vectors.

Now, we introduce a new time variable $\boldsymbol{s}$ in a new time horizon $(-\infty, p)$. For each admissible duration vector $\boldsymbol{\theta} \in \Theta$ and a time instant s in the new time horizon, define the corresponding time-scaling function as follows:

$$\mu(s \mid \boldsymbol{\theta}) = \sum_{i=1}^{\lfloor s \rfloor} \theta_i + \theta_{\lfloor s \rfloor + 1}\{s - \max(\lfloor s \rfloor, 0)\}, \ s \in (-\infty, p], \tag{10.1.15}$$

where $\lfloor \cdot \rfloor$ denotes the floor function. θ_{p+1} is arbitrary and $\theta_i = 1$ for $i \leq 0$. The form of the time-scaling function is depicted in Figure 10.1.1 below:

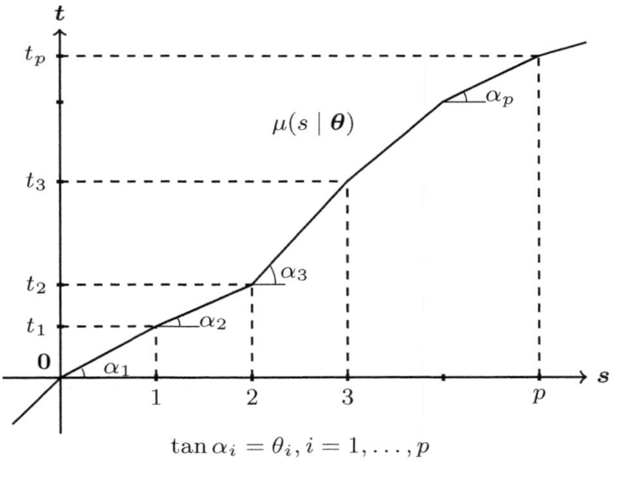

Fig. 10.1.1: Time-scaling function

It is clear from Figure 10.1.1 that the time scaling function (10.1.15) is a continuous piecewise linear non-decreasing function, which maps $s \in [i-1, i)$ in the new time horizon to $[t_{i-1}, t_i)$ in the original time horizon. The switching times for the control vector are fixed integer points $1, 2, \ldots, p-1$, in the new time horizon. Moreover, $\mu(\cdot \mid \boldsymbol{\theta})$ is strictly increasing on $[i-1, i)$ if and only if $\theta_i > 0$.

By the nature of the time-scaling function, we see that a fixed time-delay h will become a variable in the new time horizon. It is clearly important to find the explicit formula for the variable time-delay in the new time horizon.

For each $\boldsymbol{\theta} \in \Theta$, delay $h \in \{h_1, \ldots, h_{n+r}\}$, and any $s \in (-\infty, p]$, define the variable delay in the new time horizon as follows:

$$\zeta(s \mid \boldsymbol{\theta}) = \sup\{\eta(s \mid \boldsymbol{\theta}) \in (-\infty, p] : \mu(\eta(s \mid \boldsymbol{\theta}) \mid \boldsymbol{\theta}) = \mu(s \mid \boldsymbol{\theta}) - h\}. \quad (10.1.16)$$

For simplicity, let $\eta(s)$, $\zeta(s)$ and $\mu(s)$ denote $\eta(s \mid \boldsymbol{\theta})$, $\zeta(s \mid \boldsymbol{\theta})$ and $\mu(s \mid \boldsymbol{\theta})$, respectively, for the rest of the section.

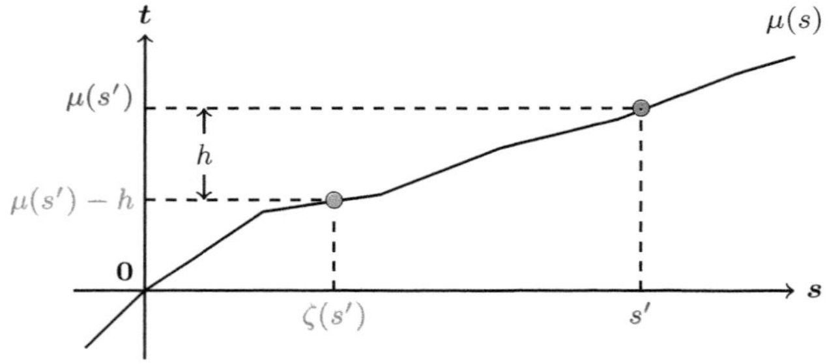

Case 1: Only one point in the new time scale

satisfies $\mu(\eta) = \mu(s') - h$

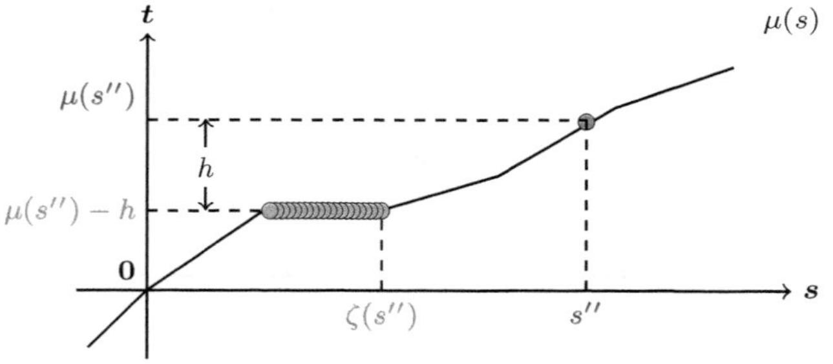

Case 2: Infinite many points in the new time horizon

satisfy $\mu(\eta) = \mu(s'') - h$

Fig. 10.1.2: Two cases for finding the delay in the new time horizon

Note that, for the duration vector $\boldsymbol{\theta}$, it is possible that $\theta_i = 0$ for some $i \in \{1, \ldots, p\}$. This means that for some time s in the new time horizon, there may exist infinite many $\eta(s)$ satisfying

$$\mu(\eta(s)) = \mu(s) - h.$$

By the definition of $\zeta(s)$, it follows that the delay time in the new time horizon is unique. The process for finding the delay time in the new time horizon is illustrated in Figure 10.1.2.

To proceed, we need the following lemma.

Lemma 10.1.1 *For any given $t \in [0, T)$ and $\boldsymbol{\theta} \in \Theta$, there exists a unique $m \in \{1, \ldots, p\}$ such that $\theta_m > 0$ and $t \in [\mu(m-1), \mu(m))$.*

Proof. By the definition of the time-scaling function, we have $t_i = \mu(i)$. For any $t' \in [0, T]$, it is clear from (10.1.10) and (10.1.11) that there exists a unique $j \in \{1, 2, \ldots, q\}$, such that $t' \in [t_{i_{j-1}}, t_{i_j})$. This indicates that there exists a unique $j \in \{1, \ldots, q\} \subset \{1, \ldots, p\}$ such that $\theta_{i_j} = t_{i_j} - t_{i_{j-1}} > 0$ and $t' \in [\mu(i_j - 1), \mu(i_j))$.

Now, we are in the position to give an explicit formula for $\zeta(s)$. This is presented as a theorem below:

Theorem 10.1.1 *Let $\boldsymbol{\theta} \in \Theta$. Then, for each $s \in (-\infty, p]$, if $\mu(s) - h < 0$, then*

$$\zeta(s) = \mu(s) - h.$$

Otherwise, let $\kappa(s \mid \boldsymbol{\theta})$ denote the unique integer such that $\theta_{\kappa(s|\boldsymbol{\theta})+1} > 0$ and

$$\mu(s) - h \in \left[\sum_{i=0}^{\kappa(s|\boldsymbol{\theta})} \theta_i, \sum_{i=0}^{\kappa(s|\boldsymbol{\theta})+1} \theta_i \right). \tag{10.1.17}$$

Then, the following equation holds:

$$\zeta(s) = \kappa(s \mid \boldsymbol{\theta}) + \sum_{l=\kappa(s|\boldsymbol{\theta})+1}^{\lfloor s \rfloor} \theta_{\kappa(s|\boldsymbol{\theta})+1}^{-1} \theta_l + \theta_{\kappa(s|\boldsymbol{\theta})+1}^{-1} \theta_{\lfloor s \rfloor + 1}(s - \lfloor s \rfloor) - h\theta_{\kappa(s|\boldsymbol{\theta})+1}^{-1}.$$

Proof. For simplicity, we omit the argument $\boldsymbol{\theta}$ in $\kappa(s \mid \boldsymbol{\theta})$. Suppose first that $\mu(s) - h < 0$. Then $\mu(\zeta(s)) < 0$. Thus, $\mu(\zeta(s)) = \zeta(s)$. Combining this equation with equation (10.1.15) gives

$$\zeta(s) = \mu(\zeta(s)) = \mu(s) - h. \tag{10.1.18}$$

Suppose now that $\mu(s) - h \in [0, T)$, it follows from (10.1.17) that $\kappa(s) \geq 0$, and $\zeta(s) \in [\kappa(s), \kappa(s) + 1)$. That is, $\kappa(s) = \lfloor \zeta(s) \rfloor$, and hence it follows from (10.1.15) and (10.1.18) that

$$\mu(\zeta(s)) = \sum_{l=1}^{\kappa(s)} \theta_l + \theta_{\kappa(s)+1}(\zeta(s) - \kappa(s)) = \sum_{l=1}^{\lfloor s \rfloor} \theta_l + \theta_{\lfloor s \rfloor + 1}(s - \lfloor s \rfloor) - h. \tag{10.1.19}$$

Since $\zeta(s) \leq s$, we have $0 \leq \kappa(s) \leq \lfloor s \rfloor$. Thus, (10.1.19) can be rearranged to give

$$\zeta(s) = \kappa(s) + \sum_{l=\kappa(s)+1}^{\lfloor s \rfloor} \theta_{\kappa(s)+1}^{-1}\theta_l + \theta_{\kappa(s)+1}^{-1}\theta_{\lfloor s \rfloor+1}(s - \lfloor s \rfloor) - h\theta_{\kappa(s)+1}^{-1}.$$

The proof is complete.

In the new time horizon, we consider the following new time-delay system with fixed switching time, defined on the subinterval $[i-1, i),\ i = 1, \ldots, p$:

$$\frac{d\boldsymbol{y}}{ds} = \theta_i \boldsymbol{f}\left(\boldsymbol{y}(s), \bar{\boldsymbol{y}}(s), \boldsymbol{\delta}^{(i)}, \bar{\boldsymbol{\delta}}(s)\right), \tag{10.1.20}$$

$$\boldsymbol{y}(s) = \boldsymbol{\phi}(s), s \leq 0, \tag{10.1.21}$$

where $\boldsymbol{\theta} \in \Theta,\ \boldsymbol{\delta} \in \Delta$,

$$\begin{aligned} \boldsymbol{y}(s) &:= [y_1(s), y_2(s), \ldots, y_n(s)]^\top \in \mathbb{R}^n \\ \bar{\boldsymbol{y}}(s) &:= [y_1(\bar{s}_1), y_2(\bar{s}_2), \ldots, y_n(\bar{s}_n)]^\top \in \mathbb{R}^n, \\ \bar{\boldsymbol{\delta}}(s) &= \left[\bar{\delta}_1(\bar{s}_{n+1}), \ldots, \bar{\delta}_r(\bar{s}_{n+r})\right]^\top, \end{aligned}$$

where $\bar{s}_q(s \mid \boldsymbol{\theta}) := \zeta(s)$ when $h = h_q,\ q = 1, \ldots, n+r$ and $\bar{s}_q(s \mid \boldsymbol{\theta})$ is denoted as $\bar{s}_q$ for simplicity. For each $m = 1, \ldots, r$,

$$\bar{\delta}_m(\bar{s}_{n+m}) = \begin{cases} \delta_m^{(j)}, & \text{if } \bar{s}_{n+m} \in [j-1, j) \\ & \text{for some } j \in \{1, \ldots, p\}, \\ \varphi_m(\bar{s}_{n+m}), & \text{if } \bar{s}_{n+m} < 0. \end{cases}$$

$\boldsymbol{f} : \mathbb{R}^n \times \mathbb{R}^n \times \mathbb{R}^r \times \mathbb{R}^r \to \mathbb{R}^n$, $\boldsymbol{\phi} : \mathbb{R} \to \mathbb{R}^n$ and $\boldsymbol{\varphi} : \mathbb{R} \to \mathbb{R}^r$ are as defined above. It is easy to see that $\zeta(s) < s$. Hence, $\bar{\boldsymbol{y}}(s)$ is a delay term in (10.1.20). Note that for the case of $\bar{s}_q < 0,\ q = 1, \ldots, n,\ y_q(\bar{s}_q) = \phi_q(\bar{s}_q)$. By Assumptions 10.1.1 and 10.1.2, system (10.1.20) have a unique solution for each admissible pair $(\boldsymbol{\theta}, \boldsymbol{\delta})$. Let $\boldsymbol{y}(\cdot \mid \boldsymbol{\theta}, \boldsymbol{\delta})$ denote the solution of (10.1.20).

Now, for each admissible duration vector $\boldsymbol{\theta} \in \Theta$, denote

$$\boldsymbol{v}(\boldsymbol{\theta}) = [\mu(1), \ldots, \mu(p-1)]^\top.$$

Since $\mu(\cdot)$ is non-decreasing, $\mu(i-1) \leq \mu(i)$ for each $i = 1, \ldots, p$. Thus, $\boldsymbol{v}(\boldsymbol{\theta})$ is an admissible switching time vector for Problem $(P(p))$, which means that the state trajectory $\boldsymbol{x}(\cdot \mid \boldsymbol{v}(\boldsymbol{\theta}), \boldsymbol{\delta})$ is well-defined. Note that $\boldsymbol{x}(\cdot \mid \boldsymbol{v}(\boldsymbol{\theta}), \boldsymbol{\delta})$ is a function of the original time variable $t \in (-\infty, T]$. For each $(\boldsymbol{\theta}, \boldsymbol{\delta}) \in \Theta \times \Delta$, let $\boldsymbol{x}(\cdot \mid \boldsymbol{v}(\boldsymbol{\theta}), \boldsymbol{\delta})$ be the corresponding state trajectory of system (10.1.9). We can show that

$$\boldsymbol{x}(t \mid \boldsymbol{v}(\boldsymbol{\theta}), \boldsymbol{\delta}) \mid_{t=\mu(s)} = \boldsymbol{y}(s \mid \boldsymbol{\theta}, \boldsymbol{\delta}), \; s \in (-\infty, p]. \tag{10.1.22}$$

We now consider the following constraints:

$$\begin{aligned} \tilde{g}_k(\boldsymbol{\theta}, \boldsymbol{\delta}) &= \Phi_k(\boldsymbol{y}(p \mid \boldsymbol{\theta}, \boldsymbol{\delta})) + \sum_{i=1}^{p} \int_{i-1}^{i} \theta_i \mathcal{L}_k \left(\boldsymbol{y}(s \mid \boldsymbol{\theta}, \boldsymbol{\delta}), \bar{\boldsymbol{y}}(s \mid \boldsymbol{\theta}, \boldsymbol{\delta}), \boldsymbol{\delta}^{(i)} \right) ds \\ &= 0, \quad k = 1, \ldots, N_e, \end{aligned} \tag{10.1.23}$$

$$\begin{aligned} \tilde{g}_k(\boldsymbol{\theta}, \boldsymbol{\delta}) &= \Phi_k(\boldsymbol{y}(p \mid \boldsymbol{\theta}, \boldsymbol{\delta})) + \sum_{i=1}^{p} \int_{i-1}^{i} \theta_i \mathcal{L}_k \left(\boldsymbol{y}(s \mid \boldsymbol{\theta}, \boldsymbol{\delta}), \bar{\boldsymbol{y}}(s \mid \boldsymbol{\theta}, \boldsymbol{\delta}), \boldsymbol{\delta}^{(i)} \right) ds \\ &\geq 0, \quad k = N_e + 1, \ldots, N_e + N_m, \end{aligned} \tag{10.1.24}$$

where the functions $\Phi_k : \mathbb{R}^n \to \mathbb{R}, \; k = 1, \ldots, N_e + N_m$ and $\mathcal{L}_k : \mathbb{R}^n \times \mathbb{R}^n \times \mathbb{R}^r \to \mathbb{R}, \; k = 1, \ldots, N_e + N_m$ are as defined in Section 10.1.2. Let $\tilde{\mathcal{F}}$ denote the set of all pairs $(\boldsymbol{\theta}, \boldsymbol{\delta}) \in \Theta \times \Delta$ which satisfy (10.1.23)–(10.1.24).

The new problem may now be stated formally as follows:
Given system (10.1.20)–(10.1.21), choose a feasible pair $(\boldsymbol{\theta}, \boldsymbol{\delta}) \in \tilde{\mathcal{F}}$ such that the following cost functional:

$$\tilde{g}_0(\boldsymbol{\theta}, \boldsymbol{\delta}) = \Phi_0(\boldsymbol{y}(p \mid \boldsymbol{\theta}, \boldsymbol{\delta})) + \sum_{i=1}^{p} \int_{i-1}^{i} \theta_i \mathcal{L}_0 \left(\boldsymbol{y}(s \mid \boldsymbol{\theta}, \boldsymbol{\delta}), \bar{\boldsymbol{y}}(s \mid \boldsymbol{\theta}, \boldsymbol{\delta}), \boldsymbol{\delta}^{(i)} \right) ds \tag{10.1.25}$$

is minimized over $\tilde{\mathcal{F}}$, where $\Phi_0 : \mathbb{R}^n \to \mathbb{R}$ and $\mathcal{L}_0 : \mathbb{R}^n \times \mathbb{R}^n \times \mathbb{R}^r \to \mathbb{R}$ are as defined in Section 10.1.2. Let this problem be denoted as Problem $(Q_1(p))$. It is easy to prove the equivalence of Problem $(P_1(p))$ and Problem $(Q_1(p))$.

10.1.5 Gradient Computation

To solve Problem $(Q_1(p))$ using the gradient-based nonlinear optimization algorithms, we require the gradients of the cost and constraint functionals with respect to each of their variables. We first rewrite $\tilde{g}_k(\boldsymbol{\theta}, \boldsymbol{\delta}), k = 0, \ldots, N_e + N_m$, in the following forms:

$$\tilde{g}_k(\boldsymbol{\theta}, \boldsymbol{\delta}) = \Phi_k(\boldsymbol{y}(p \mid \boldsymbol{\theta}, \boldsymbol{\delta})) + \int_0^p \frac{\partial \mu(s)}{\partial s} \hat{\mathcal{L}}_k \left(\boldsymbol{y}(s \mid \boldsymbol{\theta}, \boldsymbol{\delta}), \bar{\boldsymbol{y}}(s \mid \boldsymbol{\theta}, \boldsymbol{\delta}), \boldsymbol{\delta} \right) ds, \tag{10.1.26}$$

where

$$\hat{\mathcal{L}}_k(\boldsymbol{y}(s \mid \boldsymbol{\theta}, \boldsymbol{\delta}), \bar{\boldsymbol{y}}(s \mid \boldsymbol{\theta}, \boldsymbol{\delta}), \boldsymbol{\delta}) = \sum_{i=1}^{p} \mathcal{L}_k \left(\boldsymbol{y}(s \mid \boldsymbol{\theta}, \boldsymbol{\delta}), \bar{\boldsymbol{y}}(s \mid \boldsymbol{\theta}, \boldsymbol{\delta}), \boldsymbol{\delta}^{(i)} \right) \chi_{[i-1,i)}(s).$$

Then we consider the derivative of $\zeta(\cdot)$ with respect to θ_i, $i = 1, \ldots, p$. Let $\mathcal{S}'$ denote the set of points s such that $\zeta(s) \in \{0, 1, \ldots, p-1\}$. For all $s \notin \mathcal{S}'$, according to $\mu(\zeta(s)) = \mu(s) - h$, $h \in \{h_1, \ldots, h_{n+r}\}$, we obtain

$$\frac{\partial \zeta(s)}{\partial \theta_i} = \theta_{\kappa(s)+1}^{-1} \left\{ \frac{\partial \mu(s)}{\partial \theta_i} - \frac{\partial \mu(\zeta(s))}{\partial \theta_i} \right\}, i = 1, \ldots, p, \tag{10.1.27}$$

where $\kappa(s)$ is defined in Theorem 10.1.1.

Next, the gradient of the state with respect to the duration vector $\boldsymbol{\theta}$ is given as a theorem stated below.

Theorem 10.1.2 *For each pair $(\boldsymbol{\theta}, \boldsymbol{\delta}) \in \Theta \times \Delta$,*

$$\frac{\partial \boldsymbol{y}(s \mid \boldsymbol{\theta}, \boldsymbol{\delta})}{\partial \boldsymbol{\theta}} = \bar{\boldsymbol{\Lambda}}(s \mid \boldsymbol{\theta}, \boldsymbol{\delta}), \ s \in [0, p]. \tag{10.1.28}$$

Here, $\bar{\boldsymbol{\Lambda}}(\cdot \mid \boldsymbol{\theta}, \boldsymbol{\delta})$ is the solution of the following auxiliary dynamic on each $[i-1, i)$:

$$\begin{aligned} \frac{d\bar{\boldsymbol{\Lambda}}}{ds} =& \frac{\partial \boldsymbol{f}^i(\boldsymbol{y}(s \mid \boldsymbol{\theta}, \boldsymbol{\delta}), \bar{\boldsymbol{y}}(s \mid \boldsymbol{\theta}, \boldsymbol{\delta}), \boldsymbol{\theta}, \boldsymbol{\delta})}{\partial \boldsymbol{y}} \bar{\boldsymbol{\Lambda}}(s) \\ &+ \frac{\partial \boldsymbol{f}^i(\boldsymbol{y}(s \mid \boldsymbol{\theta}, \boldsymbol{\delta}), \bar{\boldsymbol{y}}(s \mid \boldsymbol{\theta}, \boldsymbol{\delta}), \boldsymbol{\theta}, \boldsymbol{\delta})}{\partial \bar{\boldsymbol{y}}} \frac{\partial \bar{\boldsymbol{y}}(s \mid \boldsymbol{\theta}, \boldsymbol{\delta})}{\partial \boldsymbol{\theta}} \\ &+ \frac{\partial \boldsymbol{f}^i(\boldsymbol{y}(s \mid \boldsymbol{\theta}, \boldsymbol{\delta}), \bar{\boldsymbol{y}}(s \mid \boldsymbol{\theta}, \boldsymbol{\delta}), \boldsymbol{\theta}, \boldsymbol{\delta})}{\partial \boldsymbol{\theta}} \end{aligned} \tag{10.1.29}$$

with

$$\bar{\boldsymbol{\Lambda}}(s) = \boldsymbol{0}, \ s \leq 0, \tag{10.1.30}$$

where

$$\frac{\partial \bar{\boldsymbol{y}}(s \mid \boldsymbol{\theta}, \boldsymbol{\delta})}{\partial \boldsymbol{\theta}} = \begin{bmatrix} \bar{\boldsymbol{\Lambda}}_1(\bar{s}_1 \mid \boldsymbol{\theta}, \boldsymbol{\delta}) + \frac{\partial y_1(\bar{s}_1 \mid \boldsymbol{\theta}, \boldsymbol{\delta})}{\partial s} \frac{\partial \bar{s}_1}{\partial \boldsymbol{\theta}} \\ \vdots \\ \bar{\boldsymbol{\Lambda}}_n(\bar{s}_n \mid \boldsymbol{\theta}, \boldsymbol{\delta}) + \frac{\partial y_n(\bar{s}_n \mid \boldsymbol{\theta}, \boldsymbol{\delta})}{\partial s} \frac{\partial \bar{s}_n}{\partial \boldsymbol{\theta}} \end{bmatrix},$$

and for $s \in [i-1, i)$,

$$\frac{d\boldsymbol{f}^i(\boldsymbol{y}(s \mid \boldsymbol{\theta}, s), \tilde{\boldsymbol{y}}(s \mid \boldsymbol{\theta}, s), \boldsymbol{\theta}, s)}{ds} = \theta_i \boldsymbol{f} \left(\boldsymbol{y}(s \mid \boldsymbol{\theta}, \boldsymbol{\delta}), \bar{\boldsymbol{y}}(s \mid \boldsymbol{\theta}, \boldsymbol{\delta}), \boldsymbol{\delta}^{(i)}, \bar{\boldsymbol{\delta}}(s) \right).$$

Proof. Let $\boldsymbol{\delta}$ and $s \in \{1, \cdots, p\}$ be arbitrary but fixed, and let $\boldsymbol{e}^r$ be the rth unit vector in $\mathbb{R}^p$. Then,

$$\frac{\partial \boldsymbol{y}(s)}{\partial \theta_r} = \lim_{\xi \to 0} \frac{\boldsymbol{y}\left(s|\boldsymbol{\delta}, \boldsymbol{\theta}^{\xi}\right) - \boldsymbol{y}(s|\boldsymbol{\theta}, \boldsymbol{\delta})}{\xi},$$

where $\boldsymbol{\theta}^{\xi} = \boldsymbol{\theta} + \xi \boldsymbol{e}^r$.

Now, we will prove the theorem in the following steps:

Step 1: Preliminaries

For each real number $\xi \in \mathbb{R}$, let $\boldsymbol{y}^{\xi}$ denote the function $\boldsymbol{y}\left(\cdot|\boldsymbol{\delta}, \boldsymbol{\theta}^{\xi}\right)$. Then, it follows from dynamic system that, for each $\xi \in \mathbb{R}$,

$$\boldsymbol{y}^{\xi}(s) = \boldsymbol{y}^{\xi}(0) + \int_0^s \boldsymbol{F}^{\xi}(t)dt, \quad s \in [0, p],$$

where $\boldsymbol{F}^{\xi}$ is defined as follows:

$$\boldsymbol{F}^{\xi}(s) = \theta^{\xi}_{\lfloor s \rfloor + 1} \boldsymbol{f}\left(\boldsymbol{y}\left(s|\boldsymbol{\delta}, \boldsymbol{\theta}^{\xi}\right), \bar{\boldsymbol{y}}\left(s^{\xi}|\boldsymbol{\delta}, \boldsymbol{\theta}^{\xi}\right), \bar{\boldsymbol{\delta}}\left(s^{\xi}\right)\right),$$

while $\bar{\boldsymbol{y}}\left(s^{\xi}|\boldsymbol{\delta}, \boldsymbol{\theta}^{\xi}\right) = \left[y_1\left(\bar{s}_1\left(s|\theta^{\xi}\right)\right), \ldots, y_n\left(\bar{s}_n\left(s|\theta^{\xi}\right)\right)\right]^{\top}$, $\bar{\boldsymbol{\delta}}\left(s^{\xi}\right) = \left[\bar{\delta}_1\left(\bar{s}^{\xi}_{n+1}\right), \ldots, \bar{\delta}_r\left(\bar{s}^{\xi}_{n+r}\right)\right]^{\top}$, and for each $m = 1, \ldots, r$,

$$\bar{\delta}_m\left(\bar{s}^{\xi}_{n+m}\right) = \begin{cases} \delta^j_m, & \text{if } \bar{s}^{\xi}_{n+m} \in [j-1, j), \\ & \text{for some } j \in \{1, \ldots, p\}, \\ \varphi_m\left(\bar{s}_{n+m}\left(s|\theta^{\xi}\right)\right), & \text{if } \bar{s}^{\xi}_{n+m} < 0. \end{cases}$$

Define

$$\Gamma^{\xi}(s) = \boldsymbol{y}\left(s|\boldsymbol{\delta}, \boldsymbol{\theta}^{\xi}\right) - \boldsymbol{y}(s|\boldsymbol{\delta}, \boldsymbol{\theta}) = \int_0^s \left(\boldsymbol{F}^{\xi} - \boldsymbol{F}^0\right) dt. \tag{10.1.31}$$

Applying the mean value theorem, we have, for $s \in [0, p]$,

$$\begin{aligned} &\boldsymbol{F}^{\xi}(s) - \boldsymbol{F}^0(s) \\ &= \int_0^1 \frac{\partial \boldsymbol{f}\left(\boldsymbol{y} + \eta \Gamma^{\xi}(s), \bar{\boldsymbol{y}} + \eta\left(\bar{\boldsymbol{y}}^{\xi} - \bar{\boldsymbol{y}}\right), \boldsymbol{\theta} + \eta \xi \boldsymbol{e}^r, \boldsymbol{\delta}\right)}{\partial \boldsymbol{y}} \Gamma^{\xi}(s) d\eta \\ &\quad + \int_0^1 \frac{\partial \boldsymbol{f}\left(\boldsymbol{y} + \eta \Gamma^{\xi}(s), \bar{\boldsymbol{y}} + \eta\left(\bar{\boldsymbol{y}}^{\xi} - \bar{\boldsymbol{y}}\right), \boldsymbol{\theta} + \eta \xi \boldsymbol{e}^r, \boldsymbol{\delta}\right)}{\partial \bar{\boldsymbol{y}}} \left(\bar{\boldsymbol{y}}^{\xi} - \bar{\boldsymbol{y}}\right) d\eta \\ &\quad + \int_0^p \frac{\partial \boldsymbol{f}\left(\boldsymbol{y} + \eta \Gamma^{\xi}(s), \bar{\boldsymbol{y}} + \eta\left(\bar{\boldsymbol{y}}^{\xi} - \bar{\boldsymbol{y}}\right), \boldsymbol{\theta} + \eta \xi \boldsymbol{e}^r, \boldsymbol{\delta}\right)}{\partial \theta_r} \xi d\eta \end{aligned} \tag{10.1.32}$$

and

$$\begin{aligned}\bar{\boldsymbol{y}}^{\xi}-\bar{\boldsymbol{y}}^{0} &= \bar{\boldsymbol{y}}\left(s^{\xi}|\boldsymbol{\delta},\boldsymbol{\theta}^{\xi}\right)-\bar{\boldsymbol{y}}(s|\boldsymbol{\delta},\boldsymbol{\theta}) \\ &= \bar{\boldsymbol{y}}\left(s^{\xi}|\boldsymbol{\delta},\boldsymbol{\theta}^{\xi}\right)-\bar{\boldsymbol{y}}\left(s|\boldsymbol{\delta},\boldsymbol{\theta}^{\xi}\right)+\bar{\boldsymbol{y}}\left(s|\boldsymbol{\delta},\boldsymbol{\theta}^{\xi}\right)-\bar{\boldsymbol{y}}(s|\boldsymbol{\delta},\boldsymbol{\theta}) \\ &= \bar{\boldsymbol{y}}\left(s^{\xi}|\boldsymbol{\delta},\boldsymbol{\theta}^{\xi}\right)-\bar{\boldsymbol{y}}\left(s|\boldsymbol{\delta},\boldsymbol{\theta}^{\xi}\right)+\Gamma^{\xi}(\bar{s}),\end{aligned}$$

where $\bar{s}$ is the corresponding delayed time point in the new time horizon.

From Assumption 10.1.3, it follows that the state set $\left\{\boldsymbol{y}^{\xi}(s):\xi\in[-a,a]\right\}$ is equibounded on $[0,p]$, where $a>0$ is a fixed small real number. Hence, there exists a real number $C_1>0$ such that for each $\xi\in[-a,a]$,

$$\boldsymbol{y}^{\xi}(s)\in\mathcal{N}_n(C_1),\quad s\in[0,p],$$

where $\mathcal{N}_n(C_1)$ denotes the closed ball in $\mathbb{R}^n$ of radius C_1 centered at the origin. Thus, for each $\xi\in[-a,a]$,

$$\boldsymbol{y}(s)+\eta\Gamma^{\xi}(s)\in\mathcal{N}_n(C_1),\quad s\in[0,p],\eta\in[0,1].$$

Furthermore, it is easy to see that for each $\xi\in[-a,a]$,

$$\boldsymbol{\theta}+\eta\xi\boldsymbol{e}^{r}\in\mathcal{N}_p(C_2),\eta\in[0,1],$$

where $C_2=|\boldsymbol{\theta}|_p+a$, $\mathcal{N}_p(C_2)$ denotes the closed ball in $\mathbb{R}^p$ of radius C_2 centered at the origin. Recall from Assumptions 10.1.1 and 10.1.2 that $\partial\boldsymbol{f}/\partial\boldsymbol{y}$ and $\partial\boldsymbol{f}/\partial\theta_r$ are continuous. Hence, it follows from the compactness of $[0,T]$, $\mathcal{V}$, $\mathcal{N}_n(C_1)$ and $\mathcal{N}_p(C_2)$ and the definitions of $z(s|\boldsymbol{\theta})$ and $\boldsymbol{\phi}$ that there exists a real number $C_3>0$ such that, for each $\xi\in[-a,a]$,

$$\left|\frac{\partial\boldsymbol{f}_{\eta}^{\xi}}{\partial\boldsymbol{y}}\right|_{n\times n}\le C_3,\quad s\in[0,p],\ \eta\in[0,1],$$

$$\left|\frac{\partial\boldsymbol{f}_{\eta}^{\xi}}{\partial\bar{\boldsymbol{y}}}\right|_{n\times n}\le C_3,\quad s\in[0,p],\ \eta\in[0,1],$$

$$\left|\frac{\partial\boldsymbol{f}_{\eta}^{\xi}}{\partial\theta_r}\right|_{n}\le C_3,\quad s\in[0,p],\ \eta\in[0,1],$$

$$\left|\frac{\partial\boldsymbol{\phi}_{\eta}^{\xi}}{\partial t}\right|_{n}\le C_3,\quad s\in[0,p],\ \eta\in[0,1],$$

where $\boldsymbol{f}_{\eta}^{\xi}$ denotes $\boldsymbol{f}\left(\boldsymbol{y}+\eta\Gamma^{\xi}(t),\bar{\boldsymbol{y}}+\eta\left(\bar{\boldsymbol{y}}^{\xi}-\bar{\boldsymbol{y}}\right),\boldsymbol{\theta}+\eta\xi\boldsymbol{e}^{r},\boldsymbol{\delta}\right)$, and $\boldsymbol{\phi}_{\eta}^{\xi}$ denotes $\boldsymbol{\phi}(\mu(t|\boldsymbol{\theta}+\eta\xi\boldsymbol{e}^{r})-h)$, and $|\cdot|$ denotes the Euclidian norm.

Step 2: The Function $\Gamma^{\xi}(s)$ Is of Order ξ

Let $\xi \in [-a, a]$ be arbitrary. When $\bar{s} < 0$, taking the norm of both sides of (10.1.31) and applying the definition of C_3 gives

$$\left|\Gamma^{\xi}(s)\right|_n = \left| \int_0^s \int_0^1 \left\{ \frac{\partial \boldsymbol{f}_\eta^\xi}{\partial \boldsymbol{y}} \Gamma^{\xi}(t) + \frac{\partial \boldsymbol{f}_\eta^\xi}{\partial \theta_r} \xi + \frac{\partial \boldsymbol{f}_\eta^\xi}{\partial \bar{\boldsymbol{y}}} \left(\boldsymbol{\phi}_\eta^\xi - \boldsymbol{\phi}_\eta^0 \right) \right\} d\eta dt \right|_n$$

where

$$\begin{aligned} \boldsymbol{\phi}_\eta^\xi - \boldsymbol{\phi}_\eta^0 &= \boldsymbol{\phi}(\mu(t|\boldsymbol{\theta} + \eta\xi \boldsymbol{e}^r) - h) - \boldsymbol{\phi}(\mu(t|\boldsymbol{\theta}) - h) \\ &= \xi \frac{\partial \boldsymbol{\phi}(\mu(t|\boldsymbol{\theta} + \eta\xi \boldsymbol{e}^r) - h)}{\partial t} \frac{\partial \mu(t|\boldsymbol{\theta} + \eta\xi \boldsymbol{e}^r)}{\partial \theta_r}, \quad \eta \in [0,1]. \end{aligned}$$

Thus, we have

$$\left|\Gamma^{\xi}(s)\right|_n \leq C_3|\xi| + C_3^2 T|\xi| + \int_0^s C_3 \left|\Gamma^{\xi}(t)\right|_n dt, \quad \bar{s} < 0.$$

Applying Theorem A.1.19 (Gronwall-Bellman Lemma) gives

$$\left|\Gamma^{\xi}(s)\right|_n \leq (C_3 + C_3^2 T) exp(C_3 p)|\xi|, \quad s \in [0, \alpha_1],$$

where α_1 is a time point such that

$$\mu(\alpha_1|\boldsymbol{\theta}) = h.$$

When $\bar{s} \geq 0$,

$$\begin{aligned} \left|\Gamma^{\xi}(s)\right|_n =& \left| \int_0^s \int_0^1 \left\{ \frac{\partial \boldsymbol{f}_\eta^\xi}{\partial \boldsymbol{y}} \Gamma^{\xi}(t) + \frac{\partial \boldsymbol{f}_\eta^\xi}{\partial \theta_r} \xi + \frac{\partial \boldsymbol{f}_\eta^\xi}{\partial \bar{\boldsymbol{y}}} \left(\bar{\boldsymbol{y}}\left(s^\xi|\boldsymbol{\delta}, \boldsymbol{\theta}^\xi\right) - \bar{\boldsymbol{y}}\left(s|\boldsymbol{\delta}, \boldsymbol{\theta}^\xi\right) \right. \right. \right. \\ &\left. \left. \left. + \Gamma^{\xi}(\bar{s}) \right) \right\} d\eta dt \right|_n \\ \leq& \left| \int_0^s \int_0^1 \frac{\partial \boldsymbol{f}_\eta^\xi}{\partial \boldsymbol{y}} \Gamma^{\xi}(t) d\eta dt \right|_n + \left| \int_0^s \int_0^1 \frac{\partial \boldsymbol{f}_\eta^\xi}{\partial \theta_r} \xi d\eta dt \right|_n \\ &+ \left| \int_0^s \int_0^1 \frac{\partial \boldsymbol{f}_\eta^\xi}{\partial \bar{\boldsymbol{y}}} \Gamma^{\xi}(\bar{s}) d\eta dt \right|_n \\ &+ \left| \int_0^s \int_0^1 \frac{\partial \boldsymbol{f}_\eta^\xi}{\partial \bar{\boldsymbol{y}}} \left(\bar{\boldsymbol{y}}\left(s^\xi|\boldsymbol{\delta}, \boldsymbol{\theta}^\xi\right) - \bar{\boldsymbol{y}}\left(s|\boldsymbol{\delta}, \boldsymbol{\theta}^\xi\right) \right) d\eta dt \right|_n \\ \leq& (C_3 + C_3^2) exp(C_3 p)|\xi| + \left| \int_{\alpha_1}^s \int_0^1 \frac{\partial \boldsymbol{f}_\eta^\xi}{\partial \boldsymbol{y}} \Gamma^{\xi}(t) d\eta dt \right|_n \end{aligned}$$

$$+\left|\int_{\alpha_1}^{s}\int_0^1 \frac{\partial \boldsymbol{f}_\eta^\xi}{\partial \theta_r}\xi d\eta dt\right|_n + \left|\int_{\alpha_1}^{s}\int_0^1 \frac{\partial \boldsymbol{f}_\eta^\xi}{\partial \bar{\boldsymbol{y}}}\Gamma^\xi(\bar{s}) d\eta dt\right|_n$$
$$+\left|\int_{\alpha_1}^{s}\int_0^1 \frac{\partial \boldsymbol{f}_\eta^\xi}{\partial \bar{\boldsymbol{y}}}\left(\bar{\boldsymbol{y}}\left(s^\xi|\boldsymbol{\delta},\boldsymbol{\theta}^\xi\right) - \bar{\boldsymbol{y}}\left(s|\boldsymbol{\delta},\boldsymbol{\theta}^\xi\right)\right) d\eta dt\right|_n$$

Since $\bar{s} \geq 0$, it follows from the definitions of $\bar{s}$ that

$$\left|\int_{\alpha_1}^{s}\int_0^1 \frac{\partial \boldsymbol{f}_\eta^\xi}{\partial \bar{\boldsymbol{y}}}\Gamma^\xi(\bar{s}) d\eta dt\right|_n \leq \int_{\alpha_1}^{s} C_3 \left|\Gamma^\xi(\bar{s})\right|_n dt \leq \int_0^s C_3 \left|\Gamma^\xi(t)\right|_n dt$$

and by the mean value theorem

$$\left|\int_{\alpha_1}^{s}\int_0^1 \frac{\partial \boldsymbol{f}_\eta^\xi}{\partial \bar{\boldsymbol{y}}}\left(\bar{\boldsymbol{y}}\left(s^\xi|\boldsymbol{\delta},\boldsymbol{\theta}^\xi\right) - \bar{\boldsymbol{y}}\left(s|\boldsymbol{\delta},\boldsymbol{\theta}^\xi\right)\right) d\eta dt\right|_n$$
$$\leq C_3 \int_{\alpha_1}^{s}\int_0^1 \left|\frac{\partial \boldsymbol{y}(\bar{s}(t|\boldsymbol{\theta}+l\xi \boldsymbol{e}^r)|\boldsymbol{\delta},\boldsymbol{\theta}+l\xi\boldsymbol{e}^r)}{\partial \bar{s}}\frac{\partial \bar{s}}{\partial \theta_r}\xi\right|_n dl dt \leq C_3^3 p|\xi|,$$

where $l \in [0,1]$. Again, by applying Theorem A.1.19 (Gronwall-Bellman Lemma), we have

$$\left|\Gamma^\xi(s)\right|_n \leq (C_3 + C_3^2 T)\exp(C_3 p)|\xi| + C_3|\xi| + C_3^3 p|\xi| + \int_0^s 2C_3\left|\Gamma^\xi(t)\right|_n dt$$
$$\leq (C_3 + C_3^2 T \exp(C_3 p|\xi|) + (C_3 + C_3^3 p)\exp(2C_3 p)|\xi|. \quad (10.1.33)$$

Since $\xi \in [-a, a]$ is arbitrary, the function $\Gamma^\xi(s)$ is of order ξ.

Step 3: The Definition of ρ and Its Properties

For each $\xi \in [-a,a] \in \mathbb{R}$, define the corresponding functions $\lambda^{1,\xi} : [0,p] \to \mathbb{R}^n$, $\lambda^{2,\xi} : [0,p] \to \mathbb{R}^n$, $\lambda^{3,\xi} : [0,p] \to \mathbb{R}^n$ as follows:

$$\lambda^{1,\xi}(t) = \int_0^1 \left\{\frac{\partial \boldsymbol{f}(\boldsymbol{y}+\eta\Gamma^\xi(s), \bar{\boldsymbol{y}}+\eta\left(\bar{\boldsymbol{y}}^\xi - \bar{\boldsymbol{y}}\right), \boldsymbol{\theta}+\eta\xi\boldsymbol{e}^r, \boldsymbol{\delta})}{\partial \boldsymbol{y}} - \frac{\partial \boldsymbol{f}(\boldsymbol{y},\bar{\boldsymbol{y}},\boldsymbol{\theta},\boldsymbol{\delta})}{\partial \boldsymbol{y}}\right\}\Gamma^\xi(t) d\eta$$
$$\lambda^{2,\xi}(t) = \int_0^1 \left\{\frac{\partial \boldsymbol{f}(\boldsymbol{y}+\eta\Gamma^\xi(s), \bar{\boldsymbol{y}}+\eta\left(\bar{\boldsymbol{y}}^\xi - \bar{\boldsymbol{y}}\right), \boldsymbol{\theta}+\eta\xi\boldsymbol{e}^r, \boldsymbol{\delta})}{\partial \bar{\boldsymbol{y}}} - \frac{\partial \boldsymbol{f}(\boldsymbol{y},\bar{\boldsymbol{y}},\boldsymbol{\theta},\boldsymbol{\delta})}{\partial \bar{\boldsymbol{y}}}\right\}\left(\bar{\boldsymbol{y}}^\xi - \bar{\boldsymbol{y}}\right) d\eta$$

$$\lambda^{3,\xi}(t) = \int_0^1 \left\{ \frac{\partial \boldsymbol{f}(\boldsymbol{y} + \eta \Gamma^{\xi}(s), \bar{\boldsymbol{y}} + \eta\left(\bar{\boldsymbol{y}}^{\xi} - \bar{\boldsymbol{y}}\right), \boldsymbol{\theta} + \eta\xi \boldsymbol{e}^r, \boldsymbol{\delta})}{\partial \theta_r} \right.$$
$$\left. - \frac{\partial \boldsymbol{f}(\boldsymbol{y}, \bar{\boldsymbol{y}}, \boldsymbol{\theta}, \boldsymbol{\delta})}{\partial \theta_r} \right\} \xi d\eta.$$

In addition, let the function $\rho : [-a, 0) \cup (0, a] \to \mathbb{R}$ be defined as follows:

$$\rho(\xi) = |\xi|^{-1} \int_0^p \left\{ \left|\lambda^{1,\xi}(t)\right|_n + \left|\lambda^{2,\xi}(t)\right|_n + \left|\lambda^{3,\xi}(t)\right|_n \right\} dt. \tag{10.1.34}$$

Since the function $\Gamma^{\xi}(s)$ is of order ξ, it follows that

$$\boldsymbol{y} + \eta \Gamma^{\xi}(t) \to \boldsymbol{y}, \text{ as } \xi \to 0 \tag{10.1.35}$$

$$\bar{\boldsymbol{y}} + \eta\left(\bar{\boldsymbol{y}}^{\xi} - \bar{\boldsymbol{y}}\right) \to \bar{\boldsymbol{y}}, \text{ as } \xi \to 0 \tag{10.1.36}$$

uniformly with respect to $t \in [0, p]$ and $\eta \in [0, 1]$. Meanwhile, it is obvious that

$$\boldsymbol{\theta} + \eta\xi \boldsymbol{e}^r \to \boldsymbol{\theta}, \text{ as } \xi \to 0 \tag{10.1.37}$$

uniformly with respect to $\eta \in [0, 1]$. Since the convergences in (10.1.35) and (10.1.36) take place inside the ball $\mathcal{N}_n(C_1)$, the convergence in (10.1.37) takes place inside the ball $\mathcal{N}_n(C_2)$, $\partial \boldsymbol{f}/\partial \boldsymbol{y}$, $\partial \boldsymbol{f}/\partial \boldsymbol{y}_z$ and $\partial \boldsymbol{f}/\partial \theta_r$ are uniformly continuous on the compact set $[0, p] \times \mathcal{N}_n(C_1) \times \mathcal{N}_n(C_1) \times \mathcal{V} \times \mathcal{N}_n(C_2)$,

$$\frac{\partial \boldsymbol{f}\left(\boldsymbol{y} + \eta \Gamma^{\xi}(s), \bar{\boldsymbol{y}} + \eta\left(\bar{\boldsymbol{y}}^{\xi} - \bar{\boldsymbol{y}}\right), \boldsymbol{\theta} + \eta\xi \boldsymbol{e}^r, \boldsymbol{\delta}\right)}{\partial \boldsymbol{y}} \to \frac{\partial \boldsymbol{f}(\boldsymbol{y}, \bar{\boldsymbol{y}}, \boldsymbol{\theta}, \boldsymbol{\delta})}{\partial \boldsymbol{y}}, \text{ as } \xi \to 0,$$
$$\frac{\partial \boldsymbol{f}\left(\boldsymbol{y} + \eta \Gamma^{\xi}(s), \bar{\boldsymbol{y}} + \eta\left(\bar{\boldsymbol{y}}^{\xi} - \bar{\boldsymbol{y}}\right), \boldsymbol{\theta} + \eta\xi \boldsymbol{e}^r, \boldsymbol{\delta}\right)}{\partial \bar{\boldsymbol{y}}} \to \frac{\partial \boldsymbol{f}(\boldsymbol{y}, \bar{\boldsymbol{y}}, \boldsymbol{\theta}, \boldsymbol{\delta})}{\partial \bar{\boldsymbol{y}}}, \text{ as } \xi \to 0,$$
$$\frac{\partial \boldsymbol{f}\left(\boldsymbol{y} + \eta \Gamma^{\xi}(s), \bar{\boldsymbol{y}} + \eta\left(\bar{\boldsymbol{y}}^{\xi} - \bar{\boldsymbol{y}}\right), \boldsymbol{\theta} + \eta\xi \boldsymbol{e}^r, \boldsymbol{\delta}\right)}{\partial \theta_r} \to \frac{\partial \boldsymbol{f}(\boldsymbol{y}, \bar{\boldsymbol{y}}, \boldsymbol{\theta}, \boldsymbol{\delta})}{\partial \theta_r}, \text{ as } \xi \to 0,$$
$$\frac{\partial \bar{\boldsymbol{y}}\left(s^{l,\xi}|\boldsymbol{\delta}, l\xi \boldsymbol{e}^r\right)}{\partial \bar{s}} \to \frac{\partial \bar{\boldsymbol{y}}(s|\boldsymbol{\delta}, \boldsymbol{\theta})}{\partial \bar{s}}, \text{ as } \xi \to 0,$$
$$\frac{\partial \bar{s}^{l,\xi}}{\partial \theta_r} \to \frac{\partial \bar{s}}{\partial \theta_r}, \text{ as } \xi \to 0,$$

uniformly with respect to $t \in [0, p]$, $\eta \in [0, 1]$ and $l \in [0, 1]$,where $\bar{s}^{l,\xi}$ is the corresponding delayed time of the control $\boldsymbol{\theta} + l\xi \boldsymbol{e}^r$. These results together with (10.1.33) imply that $|\xi|^{-1}\lambda^{1,\xi} \to 0$, $|\xi|^{-1}\lambda^{2,\xi} \to 0$, $|\xi|^{-1}\lambda^{3,\xi} \to 0$ uniformly on $[0, p]$ as $\xi \to 0$. Thus,

$$\lim_{\xi\to 0}\rho(\xi)=0.$$

Step 4: The Final Step

Let $\xi \in [-a,0)\cup(0,a]$ be arbitrary but fixed. Then, it follows from (10.1.31) that

$$\begin{aligned}\Gamma^{\xi}(s)=&\int_0^s\left[\lambda^{1,\xi}(t)+\lambda^{2,\xi}(t)+\lambda^{3,\xi}(t)\right]dt+\int_0^s\frac{\partial \boldsymbol{f}(\boldsymbol{y},\bar{\boldsymbol{y}},\boldsymbol{\theta},\boldsymbol{\delta})}{\partial \boldsymbol{y}}\Gamma^{\xi}(t)dt\\&+\int_0^s\frac{\partial \boldsymbol{f}(\boldsymbol{y},\bar{\boldsymbol{y}},\boldsymbol{\theta},\boldsymbol{\delta})}{\partial \bar{\boldsymbol{y}}}\left[\Gamma^{\xi}(\bar{s})+\frac{\partial \bar{\boldsymbol{y}}(s^{l,\xi}|\boldsymbol{\delta},\boldsymbol{\theta})}{\partial \bar{s}}\frac{\partial s^{l,\xi}}{\partial\theta_r}\xi\right]dt\\&+\int_0^s\frac{\partial \boldsymbol{f}(\boldsymbol{y},\bar{\boldsymbol{y}},\boldsymbol{\theta},\boldsymbol{\delta})}{\partial\theta_r}\xi dt.\end{aligned}\tag{10.1.38}$$

Furthermore, integrating the auxiliary system gives

$$\begin{aligned}\bar{\boldsymbol{\Lambda}}^r(s)=&\int_0^s\frac{\partial \boldsymbol{f}(\boldsymbol{y}(t|\boldsymbol{\theta},\boldsymbol{\delta},\bar{\boldsymbol{y}}(t|\boldsymbol{\theta},\boldsymbol{\delta}),\boldsymbol{\theta},\boldsymbol{\delta}))}{\partial \boldsymbol{y}}\bar{\boldsymbol{\Lambda}}^r(t)dt\\&+\int_0^s\frac{\partial \boldsymbol{f}(\boldsymbol{y}(s|\boldsymbol{\theta},\boldsymbol{\delta},\bar{\boldsymbol{y}}(s|\boldsymbol{\theta},\boldsymbol{\delta}),\boldsymbol{\theta},\boldsymbol{\delta}))}{\partial \bar{\boldsymbol{y}}}\left[\bar{\boldsymbol{\Lambda}}(\bar{s}|\boldsymbol{\delta},\boldsymbol{\theta})+\frac{\partial \bar{\boldsymbol{y}}(s|\boldsymbol{\delta},\boldsymbol{\theta})}{\partial \bar{s}}\frac{\partial \bar{s}}{\partial\theta_r}\right]dt\\&+\int_0^s\frac{\partial \boldsymbol{f}(\boldsymbol{y}(t|\boldsymbol{\theta},\boldsymbol{\delta},\bar{\boldsymbol{y}}(t|\boldsymbol{\theta},\boldsymbol{\delta}),\boldsymbol{\theta},\boldsymbol{\delta}))}{\partial\boldsymbol{\theta}_r}dt.\end{aligned}\tag{10.1.39}$$

Multiplying (10.1.38) by ξ^{-1}, and subtracting it from (10.1.39) yields

$$\begin{aligned}&\xi^{-1}\Gamma^{\xi}(s)-\bar{\boldsymbol{\Lambda}}^r(s)\\&\quad=\xi^{-1}\int_0^s\left[\lambda^{1,\xi}(t)+\lambda^{2,\xi}(t)+\lambda^{3,\xi}(t)\right]dt\\&\qquad+\int_0^s\frac{\partial \boldsymbol{f}(\boldsymbol{y},\bar{\boldsymbol{y}},\boldsymbol{\theta},\boldsymbol{\delta})}{\partial \boldsymbol{y}}(\xi^{-1}\Gamma^{\xi}(t)-\bar{\boldsymbol{\Lambda}}^r(t))dt\\&\qquad+\int_0^s\frac{\partial \boldsymbol{f}(\boldsymbol{y},\bar{\boldsymbol{y}},\boldsymbol{\theta},\boldsymbol{\delta})}{\partial \bar{\boldsymbol{y}}}\left[(\xi^{-1}\Gamma^{\xi}(\bar{s})-\bar{\boldsymbol{\Lambda}}^r(\bar{s}))\right]dt\\&\qquad+\int_0^s\frac{\partial \boldsymbol{f}(\boldsymbol{y},\bar{\boldsymbol{y}},\boldsymbol{\theta},\boldsymbol{\delta})}{\partial \bar{\boldsymbol{y}}}\left[\frac{\partial \bar{\boldsymbol{y}}(s^{l,\xi}|\boldsymbol{\delta},l\xi\boldsymbol{e}^r)}{\partial \bar{s}}\frac{\partial s^{l,\xi}}{\partial\theta_r}-\frac{\partial \bar{\boldsymbol{y}}(s|\boldsymbol{\delta},\boldsymbol{\theta})}{\partial \bar{s}}\frac{\partial \bar{s}}{\partial\theta_r}\right]dt.\end{aligned}$$

Let

$$\bar{\rho}(\xi)=\rho(\xi)+\int_0^s C_3\left|\frac{\partial \bar{\boldsymbol{y}}(s^{l,\xi}|\boldsymbol{\delta},l\xi\boldsymbol{e}^r)}{\partial \bar{s}}\frac{\partial s^{l,\xi}}{\partial\theta_r}-\frac{\partial \bar{\boldsymbol{y}}(s|\boldsymbol{\delta},\boldsymbol{\theta})}{\partial \bar{s}}\frac{\partial \bar{s}}{\partial\theta_r}\right|_n$$

Then, it is easy to see that $\bar{\rho}(\xi)\to 0$, as $\xi\to 0$. Therefore,

$$\left|\xi^{-1}\Gamma^{\xi}(s)-\bar{\boldsymbol{\Lambda}}^r(s)\right|_n$$

$$
\begin{aligned}
&\leq \rho(\xi) + \int_0^s C_3 \left|\xi^{-1}\Gamma^{\xi}(t) - \bar{\boldsymbol{\Lambda}}^r(t)\right|_n dt \\
&\quad + \int_0^s C_3 \left|\xi^{-1}\Gamma^{\xi}(\bar{s}) - \bar{\boldsymbol{\Lambda}}^r(\bar{s})\right|_n \\
&\quad + \int_0^s C_3 \left[\frac{\partial \bar{\boldsymbol{y}}(s^{l,\xi}|\boldsymbol{\delta}, l\xi \boldsymbol{e}^r)}{\partial \bar{s}} \frac{\partial s^{l,\xi}}{\partial \theta_r} - \frac{\partial \bar{\boldsymbol{y}}(s|\boldsymbol{\delta},\boldsymbol{\theta})}{\partial \bar{s}} \frac{\partial \bar{s}}{\partial \theta_r}\right]_n dt \\
&\leq \bar{\rho}(\xi) + \int_0^s 2C_3 \left|\xi^{-1}\Gamma^{\xi}(t) - \bar{\boldsymbol{\Lambda}}^r(t)\right|_n dt.
\end{aligned}
$$

By Theorem A.1.19 (Gronwall-Bellman Lemma), we obtain

$$
\left|\xi^{-1}\Gamma^{\xi}(s) - \bar{\boldsymbol{\Lambda}}^r(s)\right|_n \leq \bar{\rho}(\xi) exp(2C_3 p), \quad s \in [0, p]. \tag{10.1.40}
$$

Noting that $\xi \in [-a, 0) \cup (0, a]$ is arbitrary, we can take the limit as $\xi \to 0$ in (10.1.40). Since $\lim_{\xi \to 0} \rho(\xi) = 0$, it follows that

$$
\lim_{\xi \to 0} \xi^{-1}\Gamma^{\xi}(s) = \bar{\Lambda}^r(s|\boldsymbol{\delta}, \boldsymbol{\theta}), \ s \in [0, p].
$$

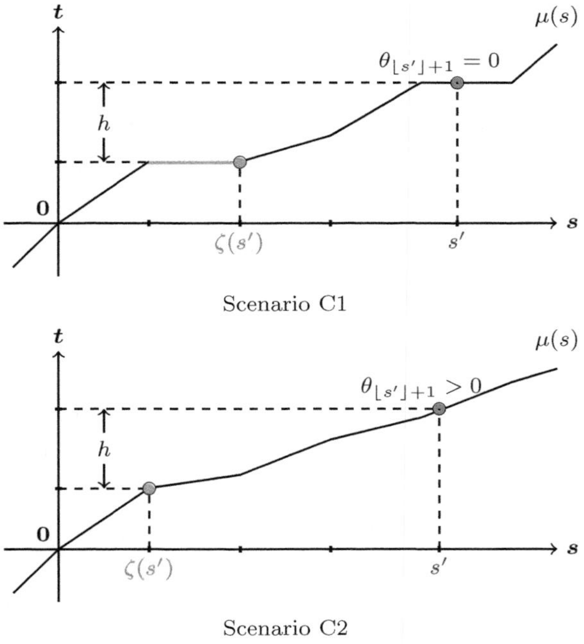

Fig. 10.1.3: A demonstration for two scenarios of $\theta_{\lfloor s \rfloor + 1}$

Note that Theorem 10.1.2 is valid only under the condition that the derivative of $\zeta(\cdot)$ with respect to θ_k exists. However, for the case of $s \in \mathcal{S}'$ ($\zeta(s) \in \{0, 1, \ldots, p-1\}$), there are two scenarios to consider:

(i) $\theta_{\lfloor s \rfloor + 1} = 0$.
(ii) $\theta_{\lfloor s \rfloor + 1} > 0$.

If scenarios (i) holds, clearly, the derivative of $\zeta(s)$ with respect to θ_i does not exist for all $s \in [j'-1, j')$ for some j'. However, this will not affect solving the auxiliary system (10.1.29), since $\partial\zeta(\cdot)/\partial\theta_i$ is always associated with $\partial \boldsymbol{f}^{j'}(\boldsymbol{y}(s \mid \boldsymbol{\theta}, \boldsymbol{\delta}), \bar{\boldsymbol{y}}(s \mid \boldsymbol{\theta}, \boldsymbol{\delta}), \boldsymbol{\theta}, \boldsymbol{\delta})/\partial\bar{\boldsymbol{y}}$, which takes the value of 0, when $\theta_{\lfloor s \rfloor + 1} = 0$.

As for scenario (ii), the derivative of $\mu(\zeta(s))$ with respect to s does not exist for only a finite number of time instant (less than or equal to the number of elements of $\mathcal{S}'$), which implies that, the value of $\partial\zeta(s)/\partial\theta_i$ does not exist only at these time instants. For this case, the auxiliary dynamic system (10.1.29) is still numerically solvable.

Remark 10.1.1 *Figure 10.1.3 depicts the situations for the two scenarios. Note that, in Scenario (C1), we have $\theta_{\zeta(s')-1} = 0$, while in Scenario (C2), we have $\theta_{\zeta(s')-1} > 0$. By the definition of $\zeta(\cdot)$ in (10.1.16), it is clear that $\theta_{\zeta(s')}$ is greater than 0 regardless of the value of $\theta_{\zeta(s')-1}$. Hence, the auxiliary dynamic system (10.1.29) is numerically solvable either in the case of $\theta_{\zeta(s')-1} > 0$ in Scenario (C1) or $\theta_{\zeta(s')-1} = 0$ in Scenario (C2).*

The gradient of $\tilde{g}_k(\boldsymbol{\theta}, \boldsymbol{\delta})$, $k = 0, 1, \ldots, N_e + N_m$, with respect to the duration vector $\boldsymbol{\theta}$ is given as a theorem stated below..

Theorem 10.1.3 *The gradient of $\tilde{g}_k(\boldsymbol{\theta}, \boldsymbol{\delta})$ for each $k = 0, 1, \ldots, N_e + N_m$ with respect to $\boldsymbol{\theta}$ is given by*

$$\begin{aligned}
\frac{\partial \tilde{g}_k(\boldsymbol{\theta}, \boldsymbol{\delta})}{\partial \boldsymbol{\theta}} = & \frac{\partial \Phi_k(\boldsymbol{y}(p \mid \boldsymbol{\theta}, \boldsymbol{\delta}))}{\partial \boldsymbol{y}} \cdot \frac{\partial \boldsymbol{y}(p \mid \boldsymbol{\theta}, \boldsymbol{\delta})}{\partial \boldsymbol{\theta}} \\
& + \int_0^p \left\{ \left[\frac{\partial \hat{\mathcal{L}}_k(\boldsymbol{y}(s \mid \boldsymbol{\theta}, \boldsymbol{\delta}), \bar{\boldsymbol{y}}(s \mid \boldsymbol{\theta}, \boldsymbol{\delta}), \boldsymbol{\delta})}{\partial \boldsymbol{y}} \cdot \frac{\partial \boldsymbol{y}(s \mid \boldsymbol{\theta}, \boldsymbol{\delta})}{\partial \boldsymbol{\theta}} \right.\right. \\
& + \left. \frac{\partial \hat{\mathcal{L}}_k(\boldsymbol{y}(s \mid \boldsymbol{\theta}, \boldsymbol{\delta}), \bar{\boldsymbol{y}}(s \mid \boldsymbol{\theta}, \boldsymbol{\delta}), \boldsymbol{\delta})}{\partial \bar{\boldsymbol{y}}} \cdot \frac{\partial \bar{\boldsymbol{y}}(s \mid \boldsymbol{\theta}, \boldsymbol{\delta})}{\partial \boldsymbol{\theta}} \right] \frac{\partial \mu(s)}{\partial s} \\
& + \left. \hat{\mathcal{L}}_k(\boldsymbol{y}(s \mid \boldsymbol{\theta}, \boldsymbol{\delta}), \bar{\boldsymbol{y}}(s \mid \boldsymbol{\theta}, \boldsymbol{\delta}), \boldsymbol{\delta}) \cdot \frac{\partial}{\partial \boldsymbol{\theta}} \left(\frac{\partial \mu(s)}{\partial s} \right) \right\} ds. \quad (10.1.41)
\end{aligned}$$

Proof. The proof follows from applying the chain rule to (10.1.26) and Theorem 10.1.2.

Finally, the gradients of the state and $\tilde{g}_k(\boldsymbol{\theta}, \boldsymbol{\delta})$ with respect to $\boldsymbol{\delta}$ are given below.

Theorem 10.1.4 *For each pair* $(\boldsymbol{\theta}, \boldsymbol{\delta}) \in \Theta \times \Delta$,

$$\frac{\partial \boldsymbol{y}(s \mid \boldsymbol{\theta}, \boldsymbol{\delta})}{\partial \delta} = \bar{\boldsymbol{\Upsilon}}(s \mid \boldsymbol{\theta}, \boldsymbol{\delta}), \; s \in [0, p], \tag{10.1.42}$$

where $\bar{\boldsymbol{\Upsilon}}(\cdot \mid \boldsymbol{\theta}, \boldsymbol{\delta})$ *is the solution of the following auxiliary dynamic system on each interval* $[i-1, i)$*:*

$$\begin{aligned}\frac{d\bar{\boldsymbol{\Upsilon}}(s)}{ds} = & \frac{\partial \boldsymbol{f}^i(\boldsymbol{y}(s \mid \boldsymbol{\theta}, \boldsymbol{\delta}), \bar{\boldsymbol{y}}(s \mid \boldsymbol{\theta}, \boldsymbol{\delta}), \boldsymbol{\theta}, \boldsymbol{\delta})}{\partial \boldsymbol{y}} \bar{\boldsymbol{\Upsilon}}(s) \\ & + \frac{\partial \boldsymbol{f}^i(\boldsymbol{y}(s \mid \boldsymbol{\theta}, \boldsymbol{\delta}), \bar{\boldsymbol{y}}(s \mid \boldsymbol{\theta}, \boldsymbol{\delta}), \boldsymbol{\theta}, \boldsymbol{\delta})}{\partial \bar{\boldsymbol{y}}} \frac{\partial \bar{\boldsymbol{y}}(s \mid \boldsymbol{\theta}, \boldsymbol{\delta})}{\partial \delta} \\ & + \frac{\partial \boldsymbol{f}^i(\boldsymbol{y}(s \mid \boldsymbol{\theta}, \boldsymbol{\delta}), \bar{\boldsymbol{y}}(s \mid \boldsymbol{\theta}, \boldsymbol{\delta}), \boldsymbol{\theta}, \boldsymbol{\delta})}{\partial \delta}\end{aligned} \tag{10.1.43}$$

with the initial condition

$$\bar{\boldsymbol{\Upsilon}}(s) = \mathbf{0}, \; s \leq 0, \tag{10.1.44}$$

where

$$\frac{\partial \bar{\boldsymbol{y}}(s \mid \boldsymbol{\theta}, \boldsymbol{\delta})}{\partial \delta} = \begin{bmatrix} \bar{\boldsymbol{\Upsilon}}_1(\bar{s}_1 \mid \boldsymbol{\theta}, \boldsymbol{\delta}) \\ \vdots \\ \bar{\boldsymbol{\Upsilon}}_n(\bar{s}_n \mid \boldsymbol{\theta}, \boldsymbol{\delta}) \end{bmatrix}.$$

Proof. The proof is similar to the proof of Theorem 10.1.2, and hence is omitted.

Theorem 10.1.5 *For each* $k = 0, 1, \ldots, N_e + N_m$*, the gradient of* $\tilde{g}_k(\boldsymbol{\theta}, \boldsymbol{\delta})$ *with respect to* $\boldsymbol{\delta}$ *is given by*

$$\begin{aligned}\frac{\partial \tilde{g}_k(\boldsymbol{\theta}, \boldsymbol{\delta})}{\partial \delta} = & \frac{\partial \Phi_k(\boldsymbol{y}(p \mid \boldsymbol{\theta}, \boldsymbol{\delta}))}{\partial \boldsymbol{y}} \cdot \frac{\partial \boldsymbol{y}(p \mid \boldsymbol{\theta}, \boldsymbol{\delta})}{\partial \delta} \\ & + \int_0^p \Big\{ \Big[\frac{\partial \hat{\mathcal{L}}_k(\boldsymbol{y}(s \mid \boldsymbol{\theta}, \boldsymbol{\delta}), \bar{\boldsymbol{y}}(s \mid \boldsymbol{\theta}, \boldsymbol{\delta}), \boldsymbol{\delta})}{\partial \boldsymbol{y}} \cdot \frac{\partial \boldsymbol{y}(s \mid \boldsymbol{\theta}, \boldsymbol{\delta})}{\partial \delta} \\ & + \frac{\partial \hat{\mathcal{L}}_k(\boldsymbol{y}(s \mid \boldsymbol{\theta}, \boldsymbol{\delta}), \bar{\boldsymbol{y}}(s \mid \boldsymbol{\theta}, \boldsymbol{\delta}), \boldsymbol{\delta})}{\partial \bar{\boldsymbol{y}}} \cdot \frac{\partial \bar{\boldsymbol{y}}(s \mid \boldsymbol{\theta}, \boldsymbol{\delta})}{\partial \delta} \\ & + \frac{\partial \hat{\mathcal{L}}_k(\boldsymbol{y}(s \mid \boldsymbol{\theta}, \boldsymbol{\delta}), \bar{\boldsymbol{y}}(s \mid \boldsymbol{\theta}, \boldsymbol{\delta}), \boldsymbol{\delta})}{\partial \delta} \Big] \frac{d\mu(s)}{ds} \Big\} ds.\end{aligned} \tag{10.1.45}$$

Proof. The proof follows from applying the chain rule to (10.1.26) and Theorem 10.1.2.

Note that Problem $(Q_1(p))$ is an optimal parameter selection problem. Theorems 10.1.3 and 10.1.5 give the gradients of the cost and constraint functionals in Problem $(Q_1(p))$ with respect to $\boldsymbol{\theta}$ and $\boldsymbol{\delta}$, respectively. On this basis, we can use existing nonlinear optimization software based on gradient descent techniques—for example, FMINCON in MATLAB or NLPQLP in

FORTRAN to solve Problem $(Q_1(p))$. In the next section, we will demonstrate the effectiveness of this approach with two numerical examples.

10.1.6 Numerical Examples

Example 10.1.1 (Optimal Control with State Delay)

Consider the following multiple time-delay optimal control problem give reference:

$$\min g_0(\boldsymbol{u}) = \frac{1}{2}\left(\boldsymbol{x}\left(t_f\right)\right)^\top S\boldsymbol{x}\left(t_f\right) + \frac{1}{2}\int_0^{t_f} \left\{\boldsymbol{x}(\boldsymbol{t})^\top Q\boldsymbol{x}(t) + (\boldsymbol{u}(t))^\top R\boldsymbol{u}(t)\right\} dt,$$

subject to the time-delay dynamic system

$$\frac{d\boldsymbol{x}}{dt} = A_1(t)\boldsymbol{x}(t) + A_2(t)\bar{\boldsymbol{x}}(t) + B(t)\boldsymbol{u}(t), \tag{10.1.46}$$

$$\boldsymbol{x}(t) = [1, 0]^\top,\ t \leq 0, \tag{10.1.47}$$

where

$$\bar{\boldsymbol{x}} = (x_1(t-1), x_2(t-0.5))$$

$$A_1(t) = \begin{bmatrix} 0 & 1 \\ -4\pi^2(a + c\cos 2\pi t) & 0 \end{bmatrix},$$

$$A_2(t) = \begin{bmatrix} 0 & 0 \\ -4\pi^2 b\cos 2\pi t & 0 \end{bmatrix},\ B(t) = \begin{bmatrix} 0 \\ 1 \end{bmatrix},$$

the parameters of the problem are in Table 10.1.1,

Table 10.1.1: Parameters in Example 10.1.1

a	b	c	t_f	Q	R	S
0.2	0.5	0.2	1.5	$I_{2\times 2}$	$I_{2\times 2}$	$10^4 I_{2\times 2}$

and the control constraints are

$$-3 \leq u_i \leq 4,\ t \in [0, t_f],\ i = 1, 2.$$

By choosing different partition number, i.e., $q = 5, 7, 10$, we obtain the corresponding optimal costs of $g_0(\boldsymbol{u}^*)$ by applying the new method and the conventional control parametrization method for each value of q. The detailed numerical results are listed in Table 10.1.2. It is clear from Table 10.1.2 that the optimal cost decreases when the partition number increases. Furthermore, since the variable switching times provide a larger flexibility for optimization, the proposed new method can always achieve a better cost when compared with the conventional control parametrization method for which the partition points are evenly distributed over the time horizon.

Note that the results obtained by applying the new method have similar cost values when compared with those obtained by applying the hybrid time-scaling transformation reported in [298]. However, the implementation of the new method is much simpler. In particular, it does not require the use of numerical interpolation to calculate the delay state values in the new time horizon. Consequently, the computational time requirement is much less.

Figure 10.1.4 shows optimal controls obtained by using the two different methods. Figures 10.1.5 and 10.1.6 depict, respectively, the two optimal state trajectories for the case of $q = 10$.

Table 10.1.2: Optimal costs for Example 10.1.1 using the two different methods

(a) New method

Number of subintervals	$g_0(\boldsymbol{u}^{q,*})$
$q = 10$	4.4355
$q = 7$	4.4610
$q = 5$	4.7630

(b) Conventional control parametrization method

Number of subintervals	$g_0(\boldsymbol{u}^{q,*})$
$q = 10$	7.2878
$q = 7$	7.4755
$q = 5$	8.1382

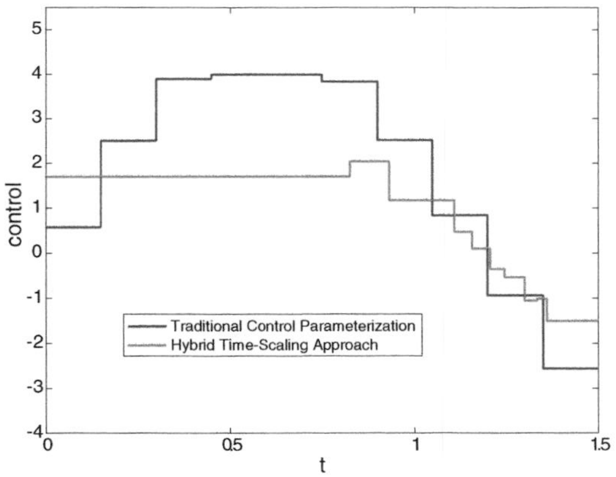

Fig. 10.1.4: Optimal controls obtained for Example 10.1.1 using the two different methods

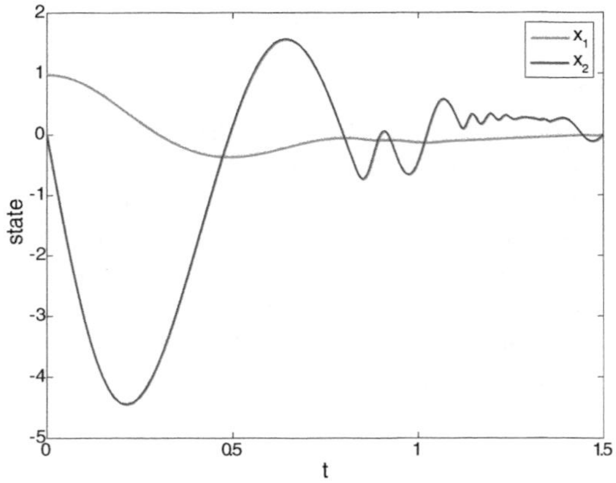

Fig. 10.1.5: Optimal state trajectory obtained for Example 10.1.1 using the new method

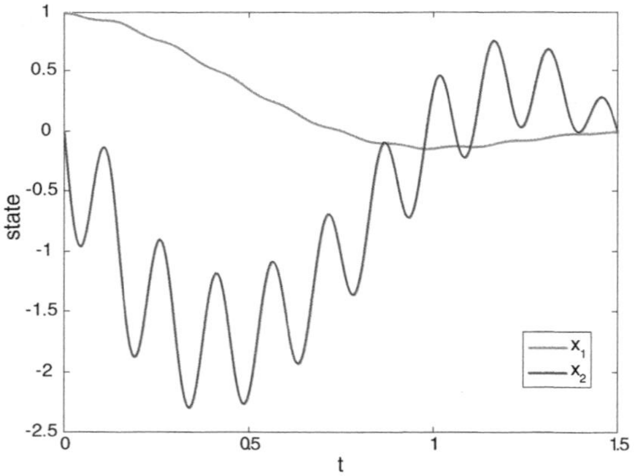

Fig. 10.1.6: Optimal state trajectory obtained for Example 10.1.1 using the conventional control parametrization method

Example 10.1.2 (Optimal Control with Multiple Time-Delay)

Consider the following time-delay optimal control problem give reference, which includes different time-delays in every state and control variables:

$$\min g_0 = \frac{1}{2}(\boldsymbol{x}(2))^\top S\boldsymbol{x}(2) + \frac{1}{2}\int_0^2 \left[(\boldsymbol{x}(t))^\top Q\boldsymbol{x}(t) + (\boldsymbol{u}(t))^\top R\boldsymbol{u}(t)\right] dt,$$

where

$$S = \begin{bmatrix} 1 & 2 & 0 & 0 \\ 2 & 1 & 0 & 0 \\ 0 & 0 & 1 & 2 \\ 0 & 0 & 1 & 1 \end{bmatrix}, \quad R = \begin{bmatrix} 1 & 0 & 0 & 0 \\ 0 & 1 & 0 & 0 \\ 0 & 0 & 1 & 0 \\ 0 & 0 & 0 & 1 \end{bmatrix},$$

$$Q = \begin{bmatrix} 1 & 0 & 0 & 0 \\ 0 & 2 & 0 & 0 \\ 0 & 0 & 1 & 0 \\ 0 & 0 & 0 & 2 \end{bmatrix}.$$

subject to the time-delay dynamic system

$$\begin{aligned} \frac{dx_1(t)}{dt} &= -2(x_1(t))^2 + x_1(t)x_2(t-0.2) + 2x_2(t) \\ &\quad - u_1(t)u_2(t-0.5), \\ \frac{dx_2(t)}{dt} &= -x_1(t-0.1) + 2x_3(t) + u_2(t), \\ \frac{dx_3(t)}{dt} &= -(x_3(t))^3 - x_1(t)x_2(t) - x_2(t-0.2)u_2(t) \\ &\quad + u_1(t-0.4) + 2u_3(t), \\ \frac{dx_4(t)}{dt} &= -(x_4(t))^2 + x_2(t)x_3(t) - 2x_3(t-0.3) + 2u_4(t), \end{aligned}$$

the initial conditions

$$\begin{aligned} x_1(t-0.1) &= 1,\ t \le 0.1; \quad x_2(t-0.2) = 1,\ t \le 0.2; \\ x_3(t-0.3) &= 1,\ t \le 0.3; \quad x_4(t-0.4) = 1,\ t \le 0.4; \\ u_1(t-0.5) &= 1,\ t < 0.5; \quad u_2(t-0.6) = 1,\ t < 0.6; \\ u_3(t-0.7) &= 1,\ t < 0.7; \quad u_4(t-0.8) = 1,\ t < 0.8, \end{aligned}$$

the terminal inequality constraints

$$\begin{aligned} g_1(\boldsymbol{u}) &= 4 - (x_1(2))^2 - (x_2(2))^2 - (x_3(2))^2 - (x_4(2))^2 \ge 0, \\ g_1(\boldsymbol{u}) &= (x_1(2))^2 + (x_2(2))^2 + (x_3(2))^2 + (x_4(2))^2 - 0.002 \ge 0, \end{aligned}$$

and the control constraints

$$-0.9 \le u_i(t) \le 1,\ t \in [0, 2],\ i = 1, \ldots, 4.$$

By choosing $q = 10$, we obtain a cost of $g_0(\mathbf{u}^*) = 2.1172$. However, the conventional control parametrization technique fails to solve this problem. The results obtained by applying the new method are again very similar to that obtained by applying the method reported in [298]. However, the implementation of the new method is much simpler and the required computational time is less. The obtained optimal controls are as shown in Figures 10.1.7

and 10.1.8, and the corresponding state trajectories are displayed in Figure 10.1.9.

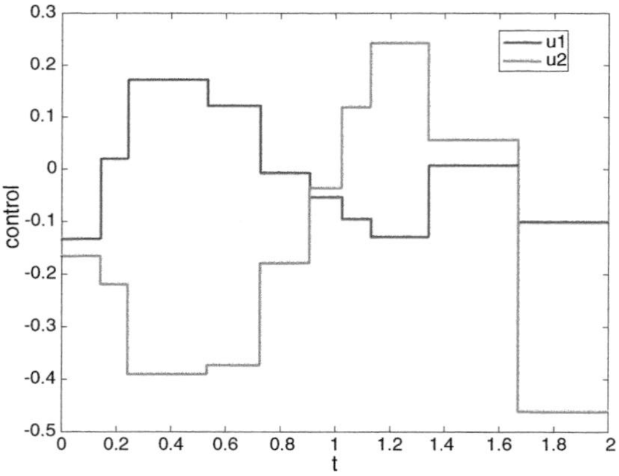

Fig. 10.1.7: Optimal controls u_1 and u_2 obtained for Example 10.1.2 using the new method

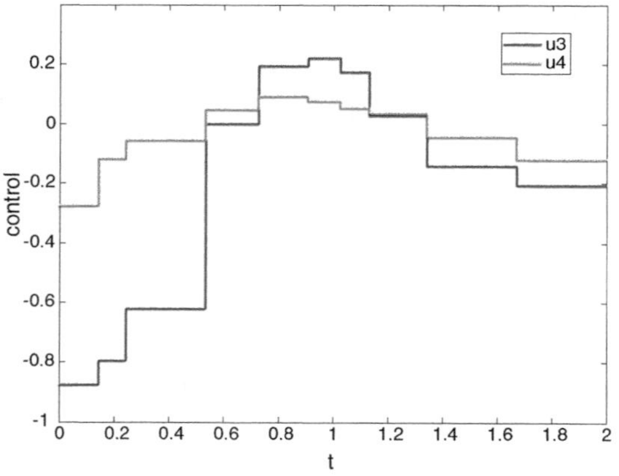

Fig. 10.1.8: Optimal controls u_3 and u_4 obtained for Example 10.1.2 using the new method

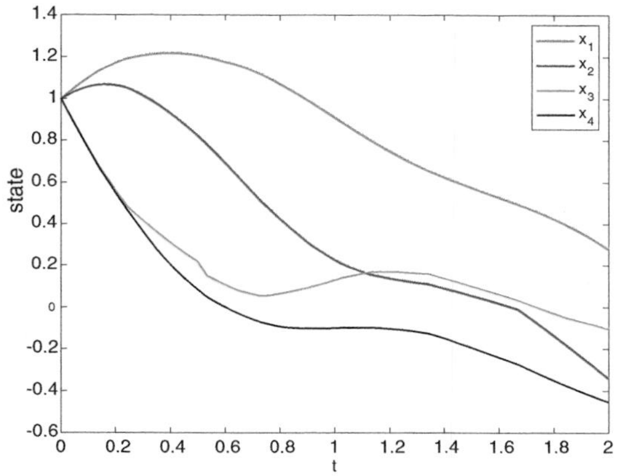

Fig. 10.1.9: Optimal state trajectories obtained for Example 10.1.2 using the new method

10.2 Time-Lag Optimal Control with State-Dependent Switched System

10.2.1 Introduction

In this section, we consider a switched system which consists of a number of sub-systems (or modes) and a switching law. It operates by switching among different sub-systems and the order of the switching sequence and the times at which the changes of the sub-systems take place are determined by the switching law. Switched systems are optimized by changing the switching sequence, switching times, and some input parameters in the dynamics of the sub-systems. The switching times and input parameters are normally continuous valued and hence they can be determined using gradient-based optimization techniques [142, 143, 148, 155]. On the other hand, optimizing the switching sequence is a discrete optimization problem, and it is a much harder optimization problem. Our focus is on the class of switching systems for which the switching sequence is pre-fixed (for a survey paper, see [148]). Furthermore, time-delays are assumed to appear in the state of each of the sub-systems and the switching mechanism is activated automatically when the state of the system satisfies the switching law. The system parameters appearing in the initial state functions are the decision variables. The main reference for this section is [152].

10.2.2 Problem Statement

This section is from [152]. Consider the following switched time-delay system, which consists of N sub-systems operating in succession, over the time horizon $[0, T]$:

$$\frac{d\boldsymbol{x}(t)}{dt} = \boldsymbol{f}^i(\boldsymbol{x}(t), \boldsymbol{x}(t-\gamma_1), \ldots, \boldsymbol{x}(t-\gamma_r)),$$
$$t \in (\tau_{i-1}, \tau_i),\ i = 1, \ldots, N, \tag{10.2.1a}$$
$$\boldsymbol{x}(t) = \boldsymbol{\phi}(t, \boldsymbol{\zeta}), \quad t \leq 0, \tag{10.2.1b}$$

where $\boldsymbol{x}(t) \in \mathbb{R}^n$ is the state; γ_j, $j = 1, \ldots, r$, are given time-delays; $\boldsymbol{\zeta} = [\zeta_1, \ldots, \zeta_s]^\top \in \mathbb{R}^s$ is a system parameter vector; τ_i, $i = 1, \ldots, N-1$, are switching times appeared in increasing order, with $\tau_0 = 0$ and $\tau_N = \infty$; and $\boldsymbol{f}^i : \mathbb{R}^{(m+1)\times n} \to \mathbb{R}^n$, $i = 1, \ldots, N$, and $\boldsymbol{\phi} : \mathbb{R} \times \mathbb{R}^r \to \mathbb{R}^n$ are given continuously differentiable functions.

The following linear growth condition is assumed throughout this section.

Assumption 10.2.1

$$\left|\boldsymbol{f}^i(\boldsymbol{y}^0, \ldots, \boldsymbol{y}^r)\right| \leq K_1\left(1 + |\boldsymbol{y}^0| + \cdots + |\boldsymbol{y}^r|\right),$$
$$(\boldsymbol{y}^0, \ldots, \boldsymbol{y}^r) \in \mathbb{R}^{(r+1)\times n},\ i = 1, \ldots, N, \tag{10.2.2}$$

where $K_1 > 0$ is a real constant and $|\cdot|$ stands for the Euclidean norm.

For $i = 1, \ldots, N$, the system switches from sub-system $i-1$ to sub-system i at the time $t = \tau_i$ defined by

$$\tau_i = \inf\left\{t > \tau_{i-1} : h_i(\boldsymbol{x}(t)) = 0\right\}, \tag{10.2.3}$$

where $h_i : \mathbb{R}^n \to \mathbb{R}$, $i = 1, \ldots, N$, are given continuously differentiable functions such that $h_i(\boldsymbol{x}(t)) = 0$ for all $t > \tau_{i-1}$ with $\tau_0 = 0$ and $\tau_N = \infty$. (10.2.3) is referred to as the switching law.

The evolution of the switched system (10.2.1) with its switching times determined by the switching law (10.2.3) is as follows:

Given a system parameter vector $\boldsymbol{\zeta}$, the system starts from the initial state $\boldsymbol{x}(0 \mid \boldsymbol{\zeta}) = \boldsymbol{\phi}(0, \boldsymbol{\zeta})$ at $t = 0$ with $\boldsymbol{x}(t \mid \boldsymbol{\zeta}) = \boldsymbol{\phi}(t, \boldsymbol{\zeta})$ for $t \leq 0$ and evolves smoothly according to (10.2.1a) with $i = 1$ until $t = \tau_1$. Then, the system switches to sub-system $i = 2$ and evolves according to (10.2.1a) with $i = 2$ until $t = \tau_2$. This process continues until $t = T$.

System (10.2.1) is influenced by the choice of the system parameter vector $\boldsymbol{\zeta} \in \mathcal{Z}$, where $\mathcal{Z}$ is defined by

$$\mathcal{Z} = \left\{\boldsymbol{\zeta} \in \mathbb{R}^s : a_k \leq \zeta_k \leq b_k,\ k = 1, \ldots, s\right\}, \tag{10.2.4}$$

where a_k, $k = 1, \ldots, s$ and b_k, $k = 1, \ldots, s$, are given constants such that $a_k < b_k$. Any vector $\boldsymbol{\zeta} \in \mathcal{Z}$ is called a feasible system parameter vector, and $\mathcal{Z}$ is called the set of feasible system parameter vectors.

Remark 10.2.1 *For (10.2.1), we note that the system parameter vector appears only in the initial function $\boldsymbol{\phi}$ (see (10.2.1b)). However, this is not a strict restriction but for the sake of brevity of presentation. The results can be easily extended to the case for which the system parameter vector appears also in the right hand side of (10.2.1a) and/or the switching law (10.2.3).*

Definition 10.2.1 *For each feasible parameter $\boldsymbol{\zeta} \in \mathcal{Z}$, $\boldsymbol{x}(\cdot \mid \boldsymbol{\zeta})$ is said to be a solution of system (10.2.1) with switching times determined by (10.2.3), if it satisfies the dynamics defined by (10.2.1a) almost everywhere on $[0, \infty)$, and the initial condition (10.2.1b) everywhere on $(-\infty, 0]$, where the switching times τ_i, $i = 1, \ldots, m$, are determined by the switching law (10.2.3).*

Theorem 10.2.1 *For each feasible system parameter vector $\boldsymbol{\zeta} \in \mathcal{Z}$, system (10.2.1) with switching times determined by (10.2.3) has a unique solution.*

Proof. For a given $\boldsymbol{\zeta} \in \mathcal{Z}$, define a set of auxiliary systems recursively as follows:

$$\frac{d\boldsymbol{\xi}^i(t)}{dt} = \boldsymbol{f}^i\left(\boldsymbol{\xi}^i(t), \boldsymbol{\xi}^i(t-\gamma_1), \ldots, \boldsymbol{\xi}^i(t-\gamma_r)\right), \quad t > \rho_{i-1}, \tag{10.2.5a}$$

$$\boldsymbol{\xi}^i(t) = \boldsymbol{\xi}^{i-1}(t), \quad t \leq \rho_{i-1}, \tag{10.2.5b}$$

where

$$\rho_i = \begin{cases} \inf\{\, t > \rho_{i-1} : h_i(\boldsymbol{\xi}^i(t)) = 0 \,\}, & \text{if } i \leq N-1, \\ \infty, & \text{if } i = N, \end{cases} \tag{10.2.6}$$

where $\boldsymbol{\xi}^0(t) = \boldsymbol{\phi}(t, \boldsymbol{\zeta})$ and $\rho_0 = 0$. Given ρ_{i-1} and $\boldsymbol{\xi}^{i-1}(\cdot)$ for some $i \in Z_N$, with

$$Z_N = \{1, \ldots, N\}.$$

The existence of a unique solution to (10.2.5) can be established as follows. Divide $[\rho_{i-1}, \infty)$ into consecutive subintervals of length $\min\{\gamma_1, \ldots, \gamma_r\}$. Then, consider the system on each of the subintervals consecutively. This gives rise to a set of consecutive non-delay systems on each of these subintervals in sequential order. Since the functions $\boldsymbol{\phi}$ and $\boldsymbol{f}^i$, $i = 1, \ldots, N$, are continuous and differentiable and Assumption 10.2.1 is satisfied, the well-known existence and uniqueness results for non-delay systems [3] can be applied to each of these consecutive systems one by one in a sequential order. More specifically, starting from $\boldsymbol{\xi}^0(t) = \boldsymbol{\phi}(t, \boldsymbol{\zeta})$ and $\rho_0 = 0$, we can show by induction that $\boldsymbol{\xi}^i(\cdot)$ and ρ_i are well-defined for each $i \in Z_N$. Then, we see that $\boldsymbol{\xi}^N(\cdot)$ satisfies (10.2.5) with $\rho_i = \tau_i$, $i = 0, \ldots, N$. Since each $\boldsymbol{\xi}^i(\cdot)$ is unique, it is clear that $\boldsymbol{\xi}^N(\cdot)$ is the only solution of (10.2.5). This completes the proof.

Now, consider the cost functional g_0 defined by

$$g_0(\boldsymbol{\zeta}) = \Phi(\boldsymbol{x}(T \mid \boldsymbol{\zeta})), \tag{10.2.7}$$

where $T > 0$ is a given terminal time, and $\Phi : \mathbb{R}^n \to \mathbb{R}$ is a given continuously differentiable function.

We may now formally state the optimal control problem under consideration in this section as follows.

Problem (P_2). Given the switched system (10.2.1) equipped with the switching law defined by (10.2.3), *find a feasible system parameter vector* $\boldsymbol{\zeta} \in \mathcal{Z}$ *such that the cost functional (10.2.7) is minimized.*

There are two special features for Problem (P_2): (1) the switchings among the sub-systems of the switched system (10.2.1) are influenced by multiple state-delays; and (2) the switchings among the sub-systems of the switched system are governed by a state-dependent switching law. We shall develop a gradient-based computational method to solve Problem (P_2).

10.2.3 Preliminaries

Some preliminary results are needed for the development of a gradient-based computational algorithm for solving Problem (P_2). To begin, let $\boldsymbol{e}^k$ denote the kth unit vector in $\mathbb{R}^r$. Furthermore, let $\partial \boldsymbol{x}(t \mid \boldsymbol{\zeta})/\partial \boldsymbol{\zeta}$ denote the $n \times s$ state variation matrix with its kth column defined by

$$\frac{\partial \boldsymbol{x}(t \mid \boldsymbol{\zeta})}{\partial \zeta_k} = \lim_{\varepsilon \to 0} \frac{\boldsymbol{x}\left(t \mid \boldsymbol{\zeta} + \varepsilon \boldsymbol{e}^k\right) - \boldsymbol{x}(t \mid \boldsymbol{\zeta})}{\varepsilon}. \tag{10.2.8}$$

Here, we assume that the limit on the right hand side is well-defined. For systems governed by ordinary differential equations, this assumption will always be satisfied. However, this assumption is not necessarily valid for the switched system (10.2.1). We shall show that even for the case where all the functions involved in the switched system (10.2.1) are smooth, the state variation matrix for the system does not exist in some situations due to the presence of time-delays and state-dependent switching law.

Now, suppose that the state variation matrix does exist at $t = T$. Then, we can take partial differentiation of the cost functional (10.2.7). In this way, it follows from the use of the chain rule that

$$\frac{\partial g_0(\boldsymbol{\zeta})}{\partial \boldsymbol{\zeta}} = \frac{\partial \Phi(\boldsymbol{x}(T \mid \boldsymbol{\zeta}))}{\partial \boldsymbol{x}} \frac{\partial \boldsymbol{x}(T \mid \boldsymbol{\zeta})}{\partial \boldsymbol{\zeta}}, \tag{10.2.9}$$

where $\partial g_0(\boldsymbol{\zeta})/\partial \boldsymbol{\zeta}$ is an s-dimensional row vector with its kth element being the partial derivative $\partial g_0(\boldsymbol{\zeta})/\partial \zeta_k$ of g_0 with respect to the kth element of the system parameter vector $\boldsymbol{\zeta}$.

To continue, the following assumption is assumed through this section.

Assumption 10.2.2 *For any* $\boldsymbol{\zeta} \in \mathcal{Z}$,

$$h_i(\boldsymbol{x}(\tau_{i-1} \mid \boldsymbol{\zeta})) \neq 0, \quad i \in Z_{N-1}, \text{ where } \tau_{i-1} < \infty. \tag{10.2.10}$$

This condition is to ensure that there is no switching of any two subsystems to occur at the same time. Therefore, the switching law defined by (10.2.3) is well-defined and the switching times are distinct. More specifically, $\tau_{i-1} < \tau_i$ for each integer $i \in Z_{N-1}$ with $\tau_{i-1} < \infty$.

The following assumption is also assumed throughout.

Assumption 10.2.3 *For a given* $\boldsymbol{\zeta} \in \mathcal{Z}$,

$$\frac{\partial h_i(\boldsymbol{x}(\tau_i \mid \boldsymbol{\zeta}))}{\partial \boldsymbol{x}} \boldsymbol{f}^i(\boldsymbol{x}(\tau_i \mid \boldsymbol{\zeta}), \boldsymbol{x}(\tau_i - \gamma_1 \mid \boldsymbol{\zeta}), \ldots, \boldsymbol{x}(\tau_i - \gamma_r \mid \boldsymbol{\zeta})) \neq 0,$$
$$i \in Z_{N-1}, \text{ where } \tau_i < \infty. \tag{10.2.11}$$

In this assumption, it is assumed that the scalar product of $\partial h_i/\partial \boldsymbol{x}$ (which is orthogonal to the switching surface $h_i = 0$) and $\boldsymbol{f}^i$ (which is tangent to the state trajectory) is non-zero at the ith switching time. This means that the state trajectory does not approach to the switching surfaces at a tangential direction. In the literature, there are similar assumptions being made. See, for example, [33, 190].

We are now ready to present formulae for both the state variation matrix and the partial derivatives of the switching times with respect to the system parameter vector. Throughout, we use the notation $\partial \tilde{\boldsymbol{x}}^j$ to denote the partial differentiation with respect to $\boldsymbol{x}(t - \gamma_j)$, with $\partial \tilde{\boldsymbol{x}}^0$ denoting the partial differentiation with respect to $\boldsymbol{x}(t)$ (that is, $\gamma_0 = 0$).

To continue, let $\boldsymbol{\zeta} \in \mathcal{Z}$ and $k \in \{1, \ldots, s\}$ be arbitrary. Consider the perturbed system parameter vector $\boldsymbol{\zeta} + \varepsilon \boldsymbol{e}^k$, where $\varepsilon \in [a_k - \zeta_k, b_k - \zeta_k]$, such that $\boldsymbol{\zeta} + \varepsilon \boldsymbol{e}^k \in \mathcal{Z}$. Let $\boldsymbol{\xi}^{i,\varepsilon}(\cdot)$, $i \in Z_N$ denote the trajectories obtained by solving (10.2.5) recursively corresponding to the perturbed system parameter vector $\boldsymbol{\zeta} + \varepsilon \boldsymbol{e}^k$, starting from $\boldsymbol{\xi}^{0,\varepsilon}(t) = \boldsymbol{\phi}\left(t, \boldsymbol{\zeta} + \varepsilon \boldsymbol{e}^k\right)$ for $t \leq 0$ and $\rho_0 = 0$. Following arguments similar to those used in the proof of Theorem 10.2.1, we can show that ρ_i defined by (10.2.6) for $\boldsymbol{\zeta} + \varepsilon \boldsymbol{e}^k$ is equal to $\tau_i^\varepsilon = \tau_i\left(\boldsymbol{\zeta} + \varepsilon \boldsymbol{e}^k\right)$, and $\boldsymbol{\xi}^{i,\varepsilon}(t) = \boldsymbol{x}\left(t \mid \boldsymbol{\zeta} + \varepsilon \boldsymbol{e}^k\right)$ for all $t \leq \tau_i^\varepsilon$.

Consider a function ψ of ε. The following notations are used. (1) If there exists a real number $M > 0$ and a positive integer k such that $|\psi(\varepsilon)| \leq M |\varepsilon|^k$ for all ε of sufficiently small magnitude, then $\psi(\varepsilon) = O\left(\varepsilon^k\right)$; (2) $\psi(\varepsilon) = \theta(\varepsilon)$ if $\psi(\varepsilon) \to 0$ as $\varepsilon \to 0$; and (3) $\psi(\varepsilon) = O(1)$ means that ψ is uniformly bounded with respect to ε.

Let

$$\bar{\gamma} = \max\{\gamma_1, \ldots, \gamma_r\}$$

and let

$$\boldsymbol{\mu}^{i,\varepsilon}(t) = \boldsymbol{\xi}^{i,\varepsilon}(t) - \boldsymbol{\xi}^{i,0}(t), \quad i = 0, \ldots, N. \tag{10.2.12}$$

Now, consider the following variational system:

$$\frac{d\Lambda_k(t)}{dt} = \sum_{j=0}^{m} \frac{\partial \boldsymbol{f}^i(\boldsymbol{x}(t), \boldsymbol{x}(t-\gamma_1), \ldots, \boldsymbol{x}(t-\gamma_m))}{\partial \tilde{\boldsymbol{x}}^j} \Lambda_k(t-\gamma_j),$$
$$t \in (\tau_{i-1}, \tau_i),\ i \in Z_N, \quad (10.2.13a)$$

with initial conditions

$$\Lambda_k(t) = \frac{\partial \boldsymbol{\phi}(t, \boldsymbol{\zeta})}{\partial \zeta_k}, \quad t \leq 0, \quad (10.2.13b)$$

$$\Lambda_k(0^+) = \frac{\partial \boldsymbol{\phi}(0, \boldsymbol{\zeta})}{\partial \zeta_k}, \quad (10.2.13c)$$

and intermediate jump conditions

$$\Lambda_k\left(\tau_i^+\right) = \Lambda_k(\tau_i^-) + \frac{\partial \tau_i(\boldsymbol{\zeta})}{\partial \zeta_k}\{\boldsymbol{f}^i(\boldsymbol{x}(\tau_i), \boldsymbol{x}(\tau_i - \gamma_1), \ldots, \boldsymbol{x}(\tau_i - \gamma_r))$$
$$- \boldsymbol{f}^{i+1}(\boldsymbol{x}(\tau_i), \boldsymbol{x}(\tau_i - \gamma_1), \ldots, \boldsymbol{x}(\tau_i - \gamma_r))\},\ i \in Z_{N-1} \text{ and } \tau_i < \infty. \quad (10.2.13d)$$

For each $i \in Z_N$, let Λ_k be the solution to the variational system governed by the differential equations (10.2.13a) with initial condition (10.2.13b)–(10.2.13c) and jump condition (10.2.13d). We need the following two lemmas:

Lemma 10.2.1 *For each $i \in Z_N$, it holds that*

$$\max_{t \in [-\bar{\gamma}, T_{\max}]} \left|\boldsymbol{\xi}^{i,\varepsilon}(t)\right| = O(1) \textit{ for every } T_{\max} > 0, \quad (10.2.14)$$

$$\max_{t \in [-\bar{\gamma}, T_{\max}]} \left|\boldsymbol{\mu}^{i,\varepsilon}(t)\right| = O(\varepsilon) \textit{ for every } T_{\max} > 0, \quad (10.2.15)$$

$$\lim_{\varepsilon \to 0} \varepsilon^{-1} \boldsymbol{\mu}^{i,\varepsilon}(t) = \Lambda_k(t^-), \quad t \in \left(-\infty, \tau_i^0\right] \setminus \bigcup_{k=0}^{i-1} \{\tau_k^0\}, \quad (10.2.16a)$$

$$\lim_{\varepsilon \to 0} \tau_i^\varepsilon = \tau_i^0, \quad (10.2.16b)$$

and, when τ_i^0 is finite,

$$\lim_{\varepsilon \to 0} \frac{\tau_i^\varepsilon - \tau_i^0}{\varepsilon}$$
$$= \begin{cases} 0 & \textit{if } i = 0. \\ -\dfrac{\partial h_i(\boldsymbol{\xi}^{i,0}(\tau_i^0))}{\partial \boldsymbol{x}} \Lambda_k(\tau_i^{0-}) \div \\ \left\{\dfrac{\partial h_i(\boldsymbol{\xi}^{i,0}(\tau_i^0))}{\partial \boldsymbol{x}} \boldsymbol{f}^i\left(\boldsymbol{\xi}^{i,0}\left(\tau_i^0\right), \boldsymbol{\xi}^{i,0}\left(\tau_i^0 - \gamma_1\right), \ldots, \boldsymbol{\xi}^{i,0}\left(\tau_i^0 - \gamma_r\right)\right)\right\}, & \textit{if } i \geq 1. \end{cases}$$
$$(10.2.16c)$$

Before we give the proof of Lemma 10.2.1, some important observations are noted in the following remark.

Remark 10.2.2 *From (10.2.14), we see that the solution of (10.2.1) corresponding to $\boldsymbol{\zeta} + \varepsilon \mathbf{e}^k$ is uniformly bounded with respect to ε. By (10.2.15) and (10.2.16a), the solution is continuous and differentiable at $\boldsymbol{\zeta}$ with respect to the kth component of the system parameter vector $\boldsymbol{\zeta}$. By (10.2.16b) and (10.2.16c), the switching times are continuous and differentiable at $\boldsymbol{\zeta}$ with respect to the kth component of the system parameter vector $\boldsymbol{\zeta}$.*

Proof of Lemma 10.2.1 The proof is by induction. To start with, consider (10.2.14)–(10.2.15), (10.2.16a)–(10.2.16c) for $i = 0$. Since $\tau_0^\varepsilon = 0$ and $\boldsymbol{\xi}^{0,\varepsilon}(t) = \boldsymbol{\phi}(t, \boldsymbol{\zeta} + \varepsilon \mathbf{e}^k)$, it is clear that (10.2.14)–(10.2.15), (10.2.16a)–(10.2.16c) for $i = 0$ are valid. For (10.2.15), we note that $\boldsymbol{\phi}$ is continuously differentiable on $[-\bar{\gamma}, T_{\max}] \times \mathcal{Z}$. Thus,

$$
\begin{aligned}
\max_{t\in[-\bar{\gamma},T_{\max}]} \left|\boldsymbol{\mu}^{0,\varepsilon}(t)\right| &= \max_{t\in[-\bar{\gamma},T_{\max}]} \left|\boldsymbol{\phi}\left(t, \boldsymbol{\zeta} + \varepsilon \mathbf{e}^k\right) - \boldsymbol{\phi}(t, \boldsymbol{\zeta})\right| \\
&\leq |\varepsilon| \max_{t\in[-\bar{\gamma},T_{\max}]} \int_0^1 \left| \frac{\partial \boldsymbol{\phi}\left(t, \boldsymbol{\zeta} + \varepsilon\eta \mathbf{e}^k\right)}{\partial \zeta_k} \right| d\eta = O(\varepsilon).
\end{aligned}
\tag{10.2.17}
$$

Now, given the inductive hypothesis (i.e., (10.2.14)–(10.2.15), (10.2.16a)–(10.2.16c) are valid for each $i = 1, \ldots, q$, where $q \leq N - 1$), we shall show that (10.2.14)–(10.2.15), (10.2.16a)–(10.2.16c) for $i = q + 1$ are also valid.

Consider the case of $\tau_q^0 = \infty$. Since $\boldsymbol{\xi}^{q+1,\varepsilon}(t) = \boldsymbol{\xi}^{q,\varepsilon}(t)$ when $|\varepsilon|$ is small, it is clear that (10.2.14)–(10.2.15), (10.2.16a)–(10.2.16b) for $i = q + 1$ are also satisfied. Here, (10.2.16c) is irrelevant.

We now consider the case of $\tau_r^0 < \infty$. Since (10.2.16c) is irrelevant, it suffices to prove the validity of (10.2.14)–(10.2.15), (10.2.16a)–(10.2.16b) for $i = q + 1$. This is done one by one as detailed below.

Proof of (10.2.14)

For each $i = 0, \ldots, q + 1$, define

$$
\begin{aligned}
&\hat{\boldsymbol{f}}^{i,\varepsilon}(t, \eta) \\
&= \begin{cases} \boldsymbol{f}^i\big(\boldsymbol{\xi}^{i,0}(t - \gamma_0) + \eta \boldsymbol{\mu}^{i,\varepsilon}(t - \gamma_0), \ldots, \\ \qquad \boldsymbol{\xi}^{i,0}(t - \gamma_r) + \eta \boldsymbol{\mu}^{i,\varepsilon}(t - \gamma_r)\big), & \text{if } i \geq 1, \\ d\boldsymbol{\phi}\left(t, \boldsymbol{\zeta} + \varepsilon\eta \mathbf{e}^k\right)/dt, & \text{if } i = 0, \end{cases}
\end{aligned}
\tag{10.2.18}
$$

and, for $i \geq 1$, let $\partial \hat{\boldsymbol{f}}^{i,\varepsilon}(s, \eta)/\partial \tilde{\boldsymbol{x}}^j$ denote the respective partial derivatives.

From (10.2.5) for $i = q + 1$, we have

$$
\boldsymbol{\xi}^{q+1,\varepsilon}(t) = \begin{cases} \boldsymbol{\xi}^{q,\varepsilon}(\tau_q^\varepsilon) + \int_{\tau_q^\varepsilon}^t \hat{\boldsymbol{f}}^{q+1,\varepsilon}(\omega, 1) d\omega, & \text{if } t \in [\tau_q^\varepsilon, T_{\max}], \\ \boldsymbol{\xi}^{q,\varepsilon}(t), & \text{if } t \in [-\bar{\gamma}, \tau_q^\varepsilon], \end{cases}
\tag{10.2.19}
$$

where τ_q^ε is as defined in the proof of Theorem 10.2.1 Thus, for $t \in [-\bar{\gamma}, \tau_q^\varepsilon]$,

$$\left|\boldsymbol{\xi}^{q+1,\varepsilon}(t)\right| = \left|\boldsymbol{\xi}^{q,\varepsilon}(t)\right| \leq \max_{\omega \in [-\bar{\gamma}, T_{\max}]} \left|\boldsymbol{\xi}^{q,\varepsilon}(\omega)\right|, \tag{10.2.20}$$

and for $t \in [\tau_q^\varepsilon, T_{\max}]$, it is clear from Assumption 10.2.1 that

$$\begin{aligned}\boldsymbol{\xi}^{q+1,\varepsilon}(t) &\leq \left|\boldsymbol{\xi}^{q,\varepsilon}(\tau_q^\varepsilon)\right| + \int_{\tau_q^\varepsilon}^{t} \left|\hat{f}^{q+1,\varepsilon}(\omega, 1)\right| d\omega \\ &\leq \max_{\omega \in [-\bar{\gamma}, T_{\max}]} \left|\boldsymbol{\xi}^{q,\varepsilon}(\omega)\right| + K_1 T_{\max} + \sum_{j=0}^{r} \int_{\tau_q^\varepsilon}^{t} K_1 \left|\boldsymbol{\xi}^{q+1,\varepsilon}(\omega - \gamma_j)\right| d\omega \\ &\leq \max_{\omega \in [-\bar{\gamma}, T_{\max}]} \left|\boldsymbol{\xi}^{q,\varepsilon}(\omega)\right| + K_1 T_{\max} + \int_{-\bar{\gamma}}^{t} (r+1) K_1 \left|\boldsymbol{\xi}^{q+1,\varepsilon}(\omega)\right| d\omega.\end{aligned} \tag{10.2.21}$$

Combining (10.2.20) and (10.2.21), it follows from (10.2.14) for $i = q$ that

$$\left|\boldsymbol{\xi}^{q+1,\varepsilon}(t)\right| \leq O(1) + \int_{-\bar{\gamma}}^{t} (r+1) K_1 \left|\boldsymbol{\xi}^{q+1,\varepsilon}(\omega)\right| d\omega, \quad t \in [-\bar{\gamma}, T_{\max}]. \tag{10.2.22}$$

Finally, by Theorem A.1.19 (Gronwall-Bellman Lemma), we obtain

$$\left|\boldsymbol{\xi}^{q+1,\varepsilon}(t)\right| \leq O(1) \exp(K_1 (r+1)(T_{\max} + \bar{\gamma})) = O(1), \quad t \in [-\bar{\gamma}, T_{\max}]. \tag{10.2.23}$$

Therefore, (10.2.14) for $i = q + 1$ is established.

Proof of (10.2.15)

From (10.2.16c) for $i = q$ (valid by the induction hypothesis), it follows that

$$\left|\tau_q^\varepsilon - \tau_q^0\right| = |\varepsilon| \cdot \left| \frac{\tau_q^\varepsilon - \tau_q^0}{\varepsilon} - \lim_{\varepsilon \to 0} \frac{\tau_q^\varepsilon - \tau_q^0}{\varepsilon} + \lim_{\varepsilon \to 0} \frac{\tau_q^\varepsilon - \tau_q^0}{\varepsilon} \right| = O(\varepsilon). \tag{10.2.24}$$

There are four cases to be considered for $t \in [-\bar{\gamma}, T_{\max}]$:

(i) $t < \min\left(\tau_q^\varepsilon, \tau_q^0\right)$; (ii) $\tau_q^0 \leq t < \tau_q^\varepsilon$;
(iii) $\tau_q^\varepsilon \leq t < \tau_q^0$; (iv) $t \geq \max\left(\tau_q^\varepsilon, \tau_q^0\right)$.

Using (10.2.5) and the fundamental theorem of calculus, we can derive the following formula for $\boldsymbol{\mu}^{q+1,\varepsilon}(t) = \boldsymbol{\xi}^{q+1,\varepsilon}(t) - \boldsymbol{\xi}^{q+1,0}(t)$ for each of the four cases:

$$\boldsymbol{\mu}^{q+1,\varepsilon}(t) = \boldsymbol{\mu}^{q,\varepsilon}(t) + \alpha_{qt} \int_{\tau_q^\varepsilon}^{t} \left\{\hat{\boldsymbol{f}}^{q+1,\varepsilon}(\omega, 1) - \hat{\boldsymbol{f}}^{q,\varepsilon}(\omega, 1)\right\} d\omega$$

$$+ \beta_{qt} \int_{\tau_q^0}^{t} \left\{ \hat{\boldsymbol{f}}^{q,0}(\omega, 0) - \hat{\boldsymbol{f}}^{q+1,0}(\omega, 0) \right\} d\omega, \tag{10.2.25}$$

where α_{qt} and β_{qt} are binary parameters indicating whether or not $t \geq \tau_q^\varepsilon$ and $t \geq \tau_q^0$, respectively.

Consider cases (i)–(iii) (i.e., at most one of $t \geq \tau_q^\varepsilon$ and $t \geq \tau_q^0$ holds). Then, it follows from (10.2.25) that

$$\begin{aligned} \left|\boldsymbol{\mu}^{q+1,\varepsilon}(t)\right| &\leq \left|\boldsymbol{\mu}^{q,\varepsilon}(t)\right| + \left(\boldsymbol{f}_{\max}^{q} + \boldsymbol{f}_{\max}^{q+1}\right)\left(\max\left(\tau_q^\varepsilon, \tau_q^0\right) - \min\left(\tau_q^\varepsilon, \tau_q^0\right)\right) \\ &= \left|\boldsymbol{\mu}^{q,\varepsilon}(t)\right| + \left(\boldsymbol{f}_{\max}^{q} + \boldsymbol{f}_{\max}^{q+1}\right)\left|\tau_q^\varepsilon - \tau_q^0\right|, \end{aligned} \tag{10.2.26}$$

where $\boldsymbol{f}_{\max}^{q}$ and $\boldsymbol{f}_{\max}^{q+1}$ are upper bounds for the norms of $\hat{\boldsymbol{f}}^{q,\varepsilon}(s, \eta)$ and $\hat{\boldsymbol{f}}^{q+1,\varepsilon}(s, \eta)$, respectively. The existence of these upper bounds is ensured by the inductive hypothesis and (10.2.14), the uniform boundedness of $\boldsymbol{\xi}^{q,\varepsilon}(\cdot)$ and $\boldsymbol{\xi}^{q+1,\varepsilon}(\cdot)$ on $[-\bar{\gamma}, T_{\max}]$ with respect to ε, and the continuity of the functions $\boldsymbol{f}^q$ and $\boldsymbol{f}^{q+1}$.

Next, consider case (iv) (i.e., $t \geq \tau_q^\varepsilon$ and $t \geq \tau_q^0$). Then, (10.2.25) becomes

$$\begin{aligned} &\boldsymbol{\mu}^{q+1,\varepsilon}(t) \\ &\quad = \boldsymbol{\mu}^{q,\varepsilon}(t) + \int_{\tau_q^\varepsilon}^{t} \left\{ \hat{\boldsymbol{f}}^{q+1,\varepsilon}(\omega, 1) - \hat{\boldsymbol{f}}^{q,\varepsilon}(\omega, 1) \right\} d\omega \\ &\quad\quad + \int_{\tau_q^0}^{t} \left\{ \hat{\boldsymbol{f}}^{q,0}(\omega, 0) - \hat{\boldsymbol{f}}^{q+1,0}(\omega, 0) \right\} d\omega \\ &\quad = \boldsymbol{\xi}^{q,\varepsilon}(\tau_q^\varepsilon) - \boldsymbol{\xi}^{q,0}(\tau_q^0) + \int_{\tau_q^\varepsilon}^{t} \left\{ \hat{\boldsymbol{f}}^{q+1,\varepsilon}(\omega, 1) ds - \int_{\tau_q^0}^{t} \hat{\boldsymbol{f}}^{q+1,0}(\omega, 0) \right\} d\omega \\ &\quad = \boldsymbol{\mu}^{q,\varepsilon}(\tau_q^0) + \int_{\tau_q^0}^{\tau_q^\varepsilon} \left\{ \hat{\boldsymbol{f}}^{q,\varepsilon}(\omega, 1) - \hat{\boldsymbol{f}}^{q+1,\varepsilon}(\omega, 1) \right\} d\omega \\ &\quad\quad + \int_{\tau_q^0}^{t} \left\{ \hat{\boldsymbol{f}}^{q+1,\varepsilon}(\omega, 1) - \hat{\boldsymbol{f}}^{q+1,0}(\omega, 0) \right\} d\omega, \end{aligned}$$

provided that $\tau_{q-1}^\varepsilon < \tau_q^0$ when $q \geq 1$. Thus, by the mean value theorem, we obtain

$$\begin{aligned} \boldsymbol{\mu}^{q+1,\varepsilon}(t) = \boldsymbol{\mu}^{q,\varepsilon}(\tau_q^0) &+ \int_{\tau_q^0}^{\tau_q^\varepsilon} \left\{ \hat{\boldsymbol{f}}^{q,\varepsilon}(\omega, 1) - \hat{\boldsymbol{f}}^{q+1,\varepsilon}(\omega, 1) \right\} d\omega \\ &+ \sum_{j=0}^{r} \int_{\tau_q^0}^{t} \int_0^1 \frac{\partial \hat{\boldsymbol{f}}^{q+1,\varepsilon}(\omega, \eta)}{\partial \tilde{\boldsymbol{x}}^j} \boldsymbol{\mu}^{q+1,\varepsilon}(\omega - \gamma_j) d\eta d\omega. \end{aligned} \tag{10.2.27}$$

Taking the norm of both sides, we have

$$\left|\boldsymbol{\mu}^{q+1,\varepsilon}(t)\right| \leq \left|\boldsymbol{\mu}^{q,\varepsilon}(\tau_q^0)\right| + \left(\boldsymbol{f}_{\max}^q + \boldsymbol{f}_{\max}^{q+1}\right)\left|\tau_q^\varepsilon - \tau_q^0\right| + \sum_{j=0}^{r}\int_{\tau_q^0}^{t} \partial\boldsymbol{f}_{\max}^{q+1}\left|\boldsymbol{\mu}^{q+1,\varepsilon}(\omega-\gamma_j)\right| d\omega, \tag{10.2.28}$$

where $\partial\boldsymbol{f}_{\max}^{q+1}$ denotes an upper bound for the norm of $\partial\hat{\boldsymbol{f}}^{q+1,\varepsilon}(s,\eta)/\partial\tilde{\boldsymbol{x}}^j$ (again, the existence of such an upper bound is ensured by the uniform boundedness of $\boldsymbol{\xi}^{q+1,\varepsilon}(\cdot)$ and the continuous differentiability of $\boldsymbol{f}^{q+1}$). Recall that (10.2.28) is established under the condition:

$$\tau_{q-1}^\varepsilon < \tau_q^0 \text{ when } q \geq 1. \tag{10.2.29}$$

By the inductive hypothesis and Assumption 10.2.2, we have

$$\lim_{\varepsilon\to 0}\tau_{q-1}^\varepsilon = \tau_{q-1}^0 = \tau_{q-1}(\boldsymbol{\zeta}) < \tau_q(\zeta) = \tau_q^0. \tag{10.2.30}$$

Hence, when ε is of sufficiently small magnitude, $\tau_{q-1}^\varepsilon < \tau_q^0$, showing the validity of (10.2.29). Consequently, (10.2.28) is valid.

Now, we combine (10.2.26) and (10.2.28) and shift the time variable in the integral. Then, it can be shown that, for all $t \in [-\bar{\gamma}, T_{\max}]$,

$$\left|\boldsymbol{\mu}^{q+1,\varepsilon}(t)\right| \leq \max_{\omega\in[-\bar{\gamma},T_{\max}]}\left|\boldsymbol{\mu}^{q,\varepsilon}(\omega)\right| + \left(\boldsymbol{f}_{\max}^q + \boldsymbol{f}_{\max}^{q+1}\right)\left|\tau_q^\varepsilon - \tau_q^0\right| + \int_{-\bar{\gamma}}^{t}(r+1)\partial\boldsymbol{f}_{\max}^{q+1}\left|\boldsymbol{\mu}^{q+1,\varepsilon}(\omega)\right| d\omega. \tag{10.2.31}$$

Therefore, since $\boldsymbol{\mu}^{q,\varepsilon}(\omega) = O(\varepsilon)$ and $\tau_q^\varepsilon - \tau_q^0 = O(\varepsilon)$ from (10.2.15) and (10.2.24), respectively, we have

$$\left|\boldsymbol{\mu}^{q+1,\varepsilon}(t)\right| \leq O(\varepsilon) + \int_{-\bar{\gamma}}^{t}(r+1)\partial\boldsymbol{f}_{\max}^{q+1}\left|\boldsymbol{\mu}^{q+1,\varepsilon}(\omega)\right| d\omega, \quad t \in [-\bar{\gamma}, T_{\max}]. \tag{10.2.32}$$

Finally, by Theorem A.1.19 (Gronwall-Bellman Lemma), we obtain

$$\left|\boldsymbol{\mu}^{q+1,\varepsilon}(t)\right| \leq O(\varepsilon)\exp\left((r+1)(T_{\max}+\bar{\gamma})\partial\boldsymbol{f}_{\max}^{q+1}\right) = O(\varepsilon), \quad t \in [-\bar{\gamma}, T_{\max}]. \tag{10.2.33}$$

Thus, (10.2.15) for $i = q+1$ follows readily. This completes the proof of (10.2.15).

The proofs of (10.2.16a)–(10.2.16c) will depend on some auxiliary results to be established below. First, from the inductive hypothesis, we recall that (10.2.16a)–(10.2.16c) are valid for each $i = 1, \ldots, q$, where $q \leq N-1$.

We have already shown in the proofs of (10.2.14) and (10.2.15) that for any $T_{\max} > 0$, $\boldsymbol{\xi}^{q+1,\varepsilon}(\cdot)$ is uniformly bounded on $[-\bar{\gamma}, T_{\max}]$ with respect to ε, and $\boldsymbol{\xi}^{q+1,\varepsilon}(\cdot) \to \boldsymbol{\xi}^{q+1,0}(\cdot)$ uniformly on $[-\bar{\gamma}, T_{\max}]$ as $\varepsilon \to 0$. Thus, since $\boldsymbol{f}^{q+1}$ is a continuously differentiable function, the following limit holds uniformly with respect to $t \in [0, T_{\max}]$ and $\eta \in [0,1]$.

$$\lim_{\varepsilon\to 0}\frac{\partial \hat{\boldsymbol{f}}^{q+1,\varepsilon}(t,\eta)}{\partial \tilde{\mathbf{x}}^j}=\frac{\partial \hat{\boldsymbol{f}}^{q+1,0}(t,0)}{\partial \tilde{\boldsymbol{x}}^j},\quad j=0,\ldots,r.$$

This implies that, for any $T_{\max}>0$,

$$\max_{t\in[0,T_{\max}]}\int_0^1\left|\frac{\partial \hat{\boldsymbol{f}}^{q+1,\varepsilon}(t,\eta)}{\partial \tilde{\boldsymbol{x}}^j}-\frac{\partial \hat{\boldsymbol{f}}^{q+1,0}(t,0)}{\partial \tilde{\boldsymbol{x}}^j}\right|d\eta=\theta(\varepsilon),\quad j=0,\ldots,r. \tag{10.2.34}$$

Now, for $t\in[0,T_{\max}]$ and $q\geq 1$, it follows from the mean value theorem that

$$\begin{aligned}\left|\hat{\boldsymbol{f}}^{q,\varepsilon}(t,1)-\hat{\boldsymbol{f}}^{q,0}(t,0)\right| &\leq \sum_{j=0}^{r}\int_0^1\left|\frac{\partial \hat{\boldsymbol{f}}^{q,\varepsilon}(t,\eta)}{\partial \tilde{\boldsymbol{x}}^j}\right|\times\left|\boldsymbol{\mu}^{q,\varepsilon}(t-\gamma_j)\right|d\eta\\ &\leq (r+1)\partial \boldsymbol{f}^q_{\max}O(\varepsilon)=O(\varepsilon),\end{aligned} \tag{10.2.35}$$

where $\partial \boldsymbol{f}^q_{\max}$ is an upper bound for the norm of $\partial \hat{\boldsymbol{f}}^{q,\varepsilon}(s,\eta)/\partial \tilde{\boldsymbol{x}}^j$, $j=0,1,\ldots,r$. Furthermore, for $t',t''\in[0,T_{\max}]$ and $q\geq 1$,

$$\begin{aligned}\left|\hat{\boldsymbol{f}}^{q,0}(t'',0)-\hat{\boldsymbol{f}}^{q,0}(t',0)\right| &\leq \sum_{j=0}^{r}\int_{t'}^{t''}\left|\frac{\partial \hat{\boldsymbol{f}}^{q,0}(t,0)}{\partial \tilde{\boldsymbol{x}}^j}\right|\times\left|\frac{d\boldsymbol{\xi}^{q,0}(t-\gamma_j)}{dt}\right|dt\\ &\leq (r+1)\partial \boldsymbol{f}^q_{\max}\max_{i=0,\ldots,q}\boldsymbol{f}^i_{\max}\left|t''-t'\right|\\ &=O(1)\left|t''-t'\right|,\end{aligned} \tag{10.2.36}$$

where, for each $i=0,\ldots,q$, the existence of the upper bound $\boldsymbol{f}^i_{\max}$ for the norm of $\hat{\boldsymbol{f}}^{i,\varepsilon}(s,\eta)$ is ensured by the uniform boundedness of $\boldsymbol{\xi}^{i,\varepsilon}(\cdot)$ and the continuity of the functions $\boldsymbol{f}^i$ and $\boldsymbol{\phi}$. Choose $T_{\max}>\max\left(\tau_q^\varepsilon,\tau_q^0\right)$. Then, it follows from (10.2.24), (10.2.35) and (10.2.36) that

$$\begin{aligned}&\int_{\min(\tau_q^\varepsilon,\tau_q^0)}^{\max(\tau_q^\varepsilon,\tau_q^0)}\left|\hat{\boldsymbol{f}}^{q,\varepsilon}(t,1)-\hat{\boldsymbol{f}}^{q,0}\left(\tau_q^0,0\right)\right|dt\\ &\leq \int_{\min(\tau_q^\varepsilon,\tau_q^0)}^{\max(\tau_q^\varepsilon,\tau_q^0)}\left\{\left|\hat{\boldsymbol{f}}^{q,\varepsilon}(t,1)-\hat{\boldsymbol{f}}^{q,0}(t,0)\right|+\left|\hat{\boldsymbol{f}}^{q,0}(t,0)-\hat{\boldsymbol{f}}^{q,0}\left(\tau_q^0,0\right)\right|\right\}dt\\ &\leq\left\{O(\varepsilon)+O(1)\left|\tau_q^\varepsilon-\tau_q^0\right|\right\}\left|\tau_q^\varepsilon-\tau_q^0\right|=O(\varepsilon^2).\end{aligned} \tag{10.2.37}$$

Similarly, we can show that

$$\int_{\min(\tau_q^\varepsilon,\tau_q^0)}^{\max(\tau_q^\varepsilon,\tau_q^0)}\left|\hat{\boldsymbol{f}}^{q+1,\varepsilon}(t,1)-\hat{\boldsymbol{f}}^{q+1,0}\left(\tau_q^0,0\right)\right|dt=O(\varepsilon^2). \tag{10.2.38}$$

Finally, we have

$$\int_{-\bar{\gamma}}^{\max(\tau_q^\varepsilon,\tau_q^0)} \left|\varepsilon^{-1}\boldsymbol{\mu}^{q+1,\varepsilon}(t) - \Lambda_k(t)\right| dt \le \int_{-\bar{\gamma}}^{\tau_q^0} \left|\varepsilon^{-1}\boldsymbol{\mu}^{q+1,\varepsilon}(t) - \Lambda_k(t)\right| dt$$
$$+ \int_{\tau_q^0}^{\max(\tau_q^\varepsilon,\tau_q^0)} \left|\varepsilon^{-1}\boldsymbol{\mu}^{q+1,\varepsilon}(t) - \Lambda_k(t)\right| dt. \quad (10.2.39)$$

Now, choose $T_{\max} > \max(\tau_q^\varepsilon, \tau_q^0)$. Then, from (10.2.15), we have

$$\varepsilon^{-1}\boldsymbol{\mu}^{q+1,\varepsilon}(\cdot) = O(1) \text{ uniformly on } [-\bar{\gamma}, \max(\tau_q^\varepsilon, \tau_q^0)].$$

Furthermore, by (10.2.15) and (10.2.16a) for $i = q$, it follows that, for almost every $t \in [-\bar{\gamma}, \tau_q^0)$,

$$\varepsilon^{-1}\boldsymbol{\mu}^{q+1,\varepsilon}(t) = \varepsilon^{-1}\boldsymbol{\mu}^{q,\varepsilon}(t) \to \Lambda_k(t) \text{ as } \varepsilon \to 0. \quad (10.2.40)$$

Hence, by Theorem A.1.10 (Lebesgue dominated convergence theorem), the first integral on the right hand side of (10.2.39) converges to zero as $\varepsilon \to 0$. For the second integral, it follows from (10.2.24) that

$$\int_{-\bar{\gamma}}^{\max(\tau_q^\varepsilon,\tau_q^0)} \left|\varepsilon^{-1}\boldsymbol{\mu}^{q+1,\varepsilon}(t) - \Lambda_k(t)\right| dt$$
$$\le \theta(\varepsilon) + \left\{O(1) + \max_{t\in[-\bar{\gamma},T_{\max}]} \left|\Lambda_k(t)\right|\right\} \cdot \left|\tau_q^\varepsilon - \tau_q^0\right|$$
$$= \theta(\varepsilon) + O(\varepsilon) = \theta(\varepsilon). \quad (10.2.41)$$

Remark 10.2.3 *Note that (10.2.34), (10.2.38), (10.2.39) and (10.2.41) will be used in the proof of (10.2.16a) for $i = q+1$.*

Proof of (10.2.16a)

Consider the case of $t < \tau_q^0$. Then, by (10.2.16b) for $i = q$ (which is valid under induction hypothesis), we have $t < \tau_q^\varepsilon$ for all ε of sufficiently small magnitude and hence $\boldsymbol{\xi}^{q+1,\varepsilon}(t) = \boldsymbol{\xi}^{q,\varepsilon}(t)$. If $t \ne \tau_l^0$, $l = 0, \ldots, q-1$, then it is clear from (10.2.16a) for $i = q$ that

$$\lim_{\varepsilon\to 0} \frac{\boldsymbol{\xi}^{q+1,\varepsilon}(t) - \boldsymbol{\xi}^{q+1,0}(t)}{\varepsilon}$$
$$= \lim_{\varepsilon\to 0} \frac{\boldsymbol{\xi}^{q,\varepsilon}(t) - \boldsymbol{\xi}^{q,0}(t)}{\varepsilon} = \lim_{\varepsilon\to 0} \varepsilon^{-1}\boldsymbol{\mu}^{q,\varepsilon}(t) = \Lambda_k^q(t^-). \quad (10.2.42)$$

Therefore, (10.2.16a) for $i = q+1$ is established for the case of $t < \tau_q^0$.

We now consider the case of $t > \tau_q^0$. From (10.2.27), we have, for $t \ge \max(\tau_q^\varepsilon, \tau_q^0)$,

$$\boldsymbol{\mu}^{q+1,\varepsilon}(t) = \boldsymbol{\mu}^{q,\varepsilon}(\tau_q^0) + \int_{\tau_q^0}^{\tau_q^\varepsilon} \left\{ \hat{\boldsymbol{f}}^{q,\varepsilon}(\omega,1) - \hat{\boldsymbol{f}}^{q+1,\varepsilon}(\omega,1) \right\} d\omega$$
$$+ \sum_{j=0}^{r} \int_{\tau_q^0}^{t} \int_0^1 \frac{\partial \hat{\boldsymbol{f}}^{q+1,\varepsilon}(\omega,\eta)}{\partial \tilde{\boldsymbol{x}}^j} \boldsymbol{\mu}^{q+1,\varepsilon}(\omega - \gamma_j) d\eta d\omega, \qquad (10.2.43)$$

provided that $\tau_q^0 > \tau_{q-1}^\varepsilon$ when $q \geq 1$, which is satisfied when ε is of sufficiently small magnitude (see the proof of (10.2.15)). Now, choose an arbitrary $T_{\max} > \max(\tau_q^\varepsilon, \tau_q^0)$. Then, for $t \leq T_{\max}$, it follows from (10.2.34) and (10.2.15) that

$$\int_{\tau_q^0}^{t} \int_0^1 \frac{\partial \hat{\boldsymbol{f}}^{q+1,\varepsilon}(\omega,\eta)}{\partial \tilde{\boldsymbol{x}}^j} \boldsymbol{\mu}^{q+1,\varepsilon}(\omega - \gamma_j) d\eta d\omega$$
$$= \int_{\tau_q^0}^{t} \frac{\partial \hat{\boldsymbol{f}}^{q+1,0}(\omega,0)}{\partial \tilde{\boldsymbol{x}}^j} \boldsymbol{\mu}^{q+1,\varepsilon}(\omega - \gamma_j) d\omega + \theta(\varepsilon) O(\varepsilon). \qquad (10.2.44)$$

Now, by (10.2.37) and (10.2.38), we can show that

$$\int_{\tau_q^0}^{\tau_q^\varepsilon} \left\{ \hat{\boldsymbol{f}}^{q,\varepsilon}(\omega,1) - \hat{\boldsymbol{f}}^{q+1,\varepsilon}(\omega,1) \right\} d\omega$$
$$= \left(\tau_q^\varepsilon - \tau_q^0\right) \left\{ \hat{\boldsymbol{f}}^{q,0}\left(\tau_q^0, 0\right) - \hat{\boldsymbol{f}}^{q+1,0}\left(\tau_q^0, 0\right) \right\} + O(\varepsilon^2). \qquad (10.2.45)$$

Using (10.2.44) and (10.2.45), the expression for $\boldsymbol{\mu}^{q+1,\varepsilon}(t)$ can be simplified. More specifically, for all times t satisfying $\max\left(\tau_q^\varepsilon, \tau_q^0\right) \leq t \leq T_{\max}$, we can deduce that

$$\boldsymbol{\mu}^{q+1,\varepsilon}(t) = \boldsymbol{\mu}^{q,\varepsilon}\left(\tau_q^0\right) + \left(\tau_q^\varepsilon - \tau_q^0\right) \left\{ \hat{\boldsymbol{f}}^{q,0}\left(\tau_q^0, 0\right) - \hat{\boldsymbol{f}}^{q+1,0}\left(\tau_q^0, 0\right) \right\}$$
$$+ \sum_{j=0}^{r} \int_{\tau_q^0}^{t} \frac{\partial \hat{\boldsymbol{f}}^{q+1,0}(\omega,0)}{\partial \tilde{\boldsymbol{x}}^j} \boldsymbol{\mu}^{q+1,\varepsilon}(\omega - \gamma_j) d\omega + O(\varepsilon^2) + \boldsymbol{\theta}(\varepsilon) O(\varepsilon).$$
$$(10.2.46)$$

Now, the solution of the variational system (10.2.13) on $(\tau_q^0, \tau_{q+1}^0]$ can be expressed as

$$\Lambda_k(t^-) = \Lambda_k\left(\tau_q^{0+}\right) + \sum_{j=0}^{r} \int_{\tau_q^0}^{t} \frac{\partial \hat{\boldsymbol{f}}^{q+1,0}(\omega,0)}{\partial \tilde{\boldsymbol{x}}^j} \Lambda_k(\omega - \gamma_j) d\omega,$$
$$t \in \left(\tau_q^0, \tau_{q+1}^0\right]. \qquad (10.2.47)$$

Multiplying (10.2.46) by ε^{-1} and then subtracting (10.2.47), we obtain

$$\varepsilon^{-1} \boldsymbol{\mu}^{q+1,\varepsilon}(t) - \Lambda_k(t^-)$$
$$= \varepsilon^{-1} \boldsymbol{\mu}^{q,\varepsilon}\left(\tau_q^0\right) - \Lambda_k\left(\tau_q^{0+}\right) + \varepsilon^{-1}\left(\tau_q^\varepsilon - \tau_q^0\right) \left\{ \hat{\boldsymbol{f}}^{q,0}\left(\tau_q^0, 0\right) - \hat{\boldsymbol{f}}^{q+1,0}\left(\tau_q^0, 0\right) \right\}$$

$$+\sum_{j=0}^{r}\int_{\tau_q^0}^{t}\frac{\partial \hat{\boldsymbol{f}}^{q+1,0}(\omega,0)}{\partial \tilde{\boldsymbol{x}}^j}\left\{\varepsilon^{-1}\boldsymbol{\mu}^{q+1,\varepsilon}(\omega-\gamma_j)-\Lambda_k(\omega-\gamma_j)\right\}d\omega+\theta(\varepsilon),$$

$$t\in\left(\max\left(\tau_q^\varepsilon,\tau_q^0\right),\min\left(\tau_{q+1}^0,T_{\max}\right)\right], \qquad (10.2.48)$$

which holds when ε is of sufficiently small magnitude. Hence, by taking the norm of both sides and changing the variable of integration in the last integral, we deduce that

$$\begin{aligned}&\left|\varepsilon^{-1}\boldsymbol{\mu}^{q+1,\varepsilon}(t)-\Lambda_k(t^-)\right|\\&\le\lambda^{q,\varepsilon}+\int_{-\bar{\gamma}}^{t}(r+1)\partial\boldsymbol{f}_{\max}^{q+1}\left|\varepsilon^{-1}\boldsymbol{\mu}^{q+1,\varepsilon}(\omega)-\Lambda_k(\omega)\right|d\omega+\theta(\varepsilon),\\&\qquad t\in\left(\max\left(\tau_q^\varepsilon,\tau_q^0\right),\min\left(\tau_{q+1}^0,T_{\max}\right)\right], \qquad (10.2.49)\end{aligned}$$

where $\partial\boldsymbol{f}_{\max}^{q+1}$ is as defined in the proof of (10.2.15) and

$$\begin{aligned}\lambda^{q,\varepsilon}=&\Big|\varepsilon^{-1}\boldsymbol{\mu}^{q,\varepsilon}\left(\tau_q^0\right)-\Lambda_k\left(\tau_q^{0+}\right)\\&+\varepsilon^{-1}\left(\tau_q^\varepsilon-\tau_q^0\right)\left\{\hat{\boldsymbol{f}}^{q,0}\left(\tau_q^0,0\right)-\hat{\boldsymbol{f}}^{q+1,0}\left(\tau_q^0,0\right)\right\}\Big|. \qquad (10.2.50)\end{aligned}$$

Next, by virtue of (10.2.16a) and (10.2.16c) for $i=q$ and the jump condition (10.2.13d) in the variational system (10.2.13), we have

$$\begin{aligned}\lim_{\varepsilon\to0}\lambda^{q,\varepsilon}=\lim_{\varepsilon\to0}\Big|\varepsilon^{-1}\left|\mu^{q,\varepsilon}\left(\tau_q^0\right)-\Lambda_k\left(\tau_q^{0-}\right)\right|+\left\{\frac{\tau_q^\varepsilon-\tau_q^0}{\varepsilon}-\frac{\partial\tau_q(\boldsymbol{\zeta})}{\partial\zeta_k}\right\}\cdot\\\left\{\hat{\boldsymbol{f}}^{q,0}\left(\tau_q^0,0\right)-\hat{\boldsymbol{f}}^{q+1,0}\left(\tau_q^0,0\right)\right\}\Big|=0. \qquad (10.2.51)\end{aligned}$$

Then, from (10.2.41) and (10.2.51), it is clear that

$$\begin{aligned}&\left|\varepsilon^{-1}\boldsymbol{\mu}^{q+1,\varepsilon}(t)-\Lambda_k(t^-)\right|\\&\le\boldsymbol{\theta}(\varepsilon)+\int_{\max(\tau_q^\varepsilon,\tau_q^0)}^{t}(m+1)\partial\boldsymbol{f}_{\max}^{q+1}\left|\varepsilon^{-1}\boldsymbol{\mu}^{q+1,\varepsilon}(\omega)-\Lambda_k(\omega)\right|d\omega,\\&\qquad t\in\left(\max\left(\tau_q^\varepsilon,\tau_q^0\right),\min\left(\tau_{q+1}^0,T_{\max}\right)\right]. \qquad (10.2.52)\end{aligned}$$

Thus, by Theorem A.1.19 (Gronwall-Bellman Lemma), we obtain

$$\begin{aligned}\left|\varepsilon^{-1}\boldsymbol{\mu}^{q+1,\varepsilon}(t)-\Lambda_k(t^-)\right|\le\theta(\varepsilon)\exp\left((r+1)\partial\boldsymbol{f}_{\max}^{q+1}T_{\max}\right),\\t\in\left(\max\left(\tau_q^\varepsilon,\tau_q^0\right),\min\left(\tau_{q+1}^0,T_{\max}\right)\right], \qquad (10.2.53)\end{aligned}$$

which holds for all ε of sufficiently small magnitude. Finally, for any fixed time point $t\in\left(\tau_q^0,\tau_{q+1}^0\right]$, we can choose $T_{\max}>t$ so that $t\in\left(\tau_q^\varepsilon,\min\left(\tau_{q+1}^0,T_{\max}\right)\right]$ when the magnitude of ε is sufficiently small. Thus, by (10.2.53), we have

$$\varepsilon^{-1}\boldsymbol{\mu}^{q+1,\varepsilon}(t) \to \Lambda_k(t^-).$$

This completes the proof of (10.2.16a) for $i = q+1$.

Proof of (10.2.16b)

Consider the case of $q = N-1$. Then, $\tau_{q+1}^{\varepsilon} = \tau_{q+1}^{0} = \infty$. Clearly, (10.2.16b) for $i = q+1$ holds.

It remains to consider the case of $q < N-1$. By virtue of Assumption 10.2.2 and the definition of τ_{q+1}^0, there exists a $\delta > 0$ such that $\tau_q^0 - \delta < \tau_q^0 < \tau_{q+1}^0 - \delta$ and

$$h_{q+1}\left(\boldsymbol{\xi}^{q+1,0}(t)\right) \neq 0, \quad t \in \left[\tau_q^0 - \delta, \tau_{q+1}^0 - \delta\right], \tag{10.2.54}$$

where $\tau_{q+1}^0 - \delta = \infty$ if $\tau_{q+1}^0 = \infty$. For any $T_{\max} > \tau_q^0$, we recall that $\boldsymbol{\xi}^{q+1,\varepsilon}(t) \to \boldsymbol{\xi}^{q+1,0}(t)$ uniformly on $[-\bar{\gamma}, T_{\max}]$ as $\varepsilon \to 0$ (see the proof of (10.2.15)). Thus, it follows from (10.2.16b) that

$$h_{q+1}\left(\boldsymbol{\xi}^{q+1,\varepsilon}(t)\right) \neq 0, \quad t \in \left[\tau_q^{\varepsilon}, \min\left(\tau_{q+1}^0 - \delta, T_{\max}\right)\right], \tag{10.2.55}$$

when ε is of sufficiently small magnitude. If $\tau_{q+1}^0 = \infty$, then (10.2.55) becomes

$$h_{q+1}\left(\boldsymbol{\xi}^{q+1,\varepsilon}(t)\right) \neq 0, \quad t \in \left[\tau_q^{\varepsilon}, T_{\max}\right], \tag{10.2.56}$$

which implies $\tau_{q+1}^{\varepsilon} \geq T_{\max}$. Since $T_{\max}$ was chosen arbitrarily, we can take $T_{\max} \to \infty$ and hence $\tau_{q+1}^{\varepsilon} \to \infty$ as $\varepsilon \to 0$. Thus, for the case of $\tau_{q+1}^0 = \infty$, the validity of (10.2.16b) for $i = q+1$ is established.

We now consider the case of $\tau_{q+1}^0 < \infty$. Clearly,

$$h_{q+1}\left(\boldsymbol{\xi}^{q+1,0}\left(\tau_{q+1}^0\right)\right) = 0. \tag{10.2.57}$$

Next, by Assumption 10.2.3, we have

$$\begin{aligned}
&\frac{\partial h_{q+1}\left(\boldsymbol{\xi}^{q+1,0}\left(\tau_{q+1}^0\right)\right)}{\partial \boldsymbol{x}} \hat{\boldsymbol{f}}^{q+1,0}\left(\tau_{q+1}^0, 0\right) \\
&= \frac{\partial h_{q+1}\left(\boldsymbol{x}\left(\tau_{q+1}^0 \mid \boldsymbol{\zeta}\right)\right)}{\partial \boldsymbol{x}} \cdot \\
&\quad \boldsymbol{f}^{q+1}\left(\boldsymbol{x}\left(\tau_{q+1}^0 \mid \boldsymbol{\zeta}\right), \boldsymbol{x}\left(\tau_{q+1}^0 - \gamma_1 \mid \boldsymbol{\zeta}\right), \ldots, \boldsymbol{x}\left(\tau_{q+1}^0 - \gamma_r \mid \boldsymbol{\zeta}\right)\right) \\
&\neq 0.
\end{aligned} \tag{10.2.58}$$

Thus, by continuity, $\delta > 0$ in (10.2.56) may be chosen such that

$$\begin{aligned}
\frac{d}{dt}\left\{h_{q+1}\left(\boldsymbol{\xi}^{q+1,0}(t)\right)\right\} &= \frac{\partial h_{q+1}\left(\boldsymbol{\xi}^{q+1,0}(t)\right)}{\partial \boldsymbol{x}} \frac{d\boldsymbol{\xi}^{q+1,0}(t)}{dt} \\
&= \frac{\partial h_{q+1}\left(\boldsymbol{\xi}^{q+1,0}(t)\right)}{\partial \boldsymbol{x}} \hat{\boldsymbol{f}}^{q+1,0}(t, 0) \neq 0, \\
&\qquad t \in \left(\tau_{q+1}^0 - \delta, \tau_{q+1}^0 + \delta\right).
\end{aligned} \tag{10.2.59}$$

This shows that $h_{q+1}\left(\boldsymbol{\xi}^{q+1,0}(\cdot)\right)$ is either strictly increasing or strictly decreasing on $\left(\tau_{q+1}^0-\delta, \tau_{q+1}^0+\delta\right)$. Therefore, in view of (10.2.58) and (10.2.59), $h_{q+1}\left(\boldsymbol{\xi}^{q+1,0}(\cdot)\right)$ has different sign at $\tau_{q+1}^0-\delta$ and $\tau_{q+1}^0+\delta$. This implies that

$$h_{q+1}\left(\boldsymbol{\xi}^{q+1,0}(\tau_{q+1}^0-\delta)\right)\cdot h_{q+1}\left(\boldsymbol{\xi}^{q+1,0}(\tau_{q+1}^0+\delta)\right)<0. \tag{10.2.60}$$

Choose $T_{\max}>\tau_{q+1}^0+\delta$. Note that $\boldsymbol{\xi}^{q+1,\varepsilon}(t)\to\boldsymbol{\xi}^{q+1,0}(t)$ uniformly on $[-\bar{\gamma}, T_{\max}]$ as $\varepsilon\to 0$. Thus,

$$h_{q+1}\left(\boldsymbol{\xi}^{q+1,\varepsilon}(\tau_{q+1}^0-\delta)\right)\cdot h_{q+1}\left(\boldsymbol{\xi}^{q+1,\varepsilon}(\tau_{q+1}^0+\delta)\right)<0, \tag{10.2.61}$$

when ε is sufficiently small. This implies that $h_{q+1}\left(\boldsymbol{\xi}^{q+1,\varepsilon}(\cdot)\right)$, for example, $h_{q+1}\left(\boldsymbol{\xi}^{q+1,0}(\cdot)\right)$, has different sign at $\tau_{q+1}^0-\delta$ and $\tau_{q+1}^0+\delta$ when ε is sufficiently small. Combining (10.2.55) and (10.2.61), it gives

$$\tau_{q+1}^0-\delta<\tau_{q+1}^\varepsilon<\tau_{q+1}^0+\delta.$$

Thus, the conclusion of (10.2.16b) follows readily by taking $\delta\to 0$.

Proof of (10.2.16c)

Since (10.2.16c) is not applicable when $i=N$, it suffices to consider the case of $q<N-1$. To begin, define

$$\widehat{h}_{q+1}^\varepsilon(\eta)=h_{q+1}\left(\eta\boldsymbol{\xi}^{q+1,\varepsilon}\left(\tau_{q+1}^\varepsilon\right)+(1-\eta)\boldsymbol{\xi}^{q+1,0}\left(\tau_{q+1}^0\right)\right) \tag{10.2.62}$$

and let $\partial\widehat{h}_{q+1}^\varepsilon(\eta)/\partial\boldsymbol{x}$ denote the respective partial derivative. By Taylor's theorem, there exists a constant $\eta_\varepsilon\in(0,1)$ such that

$$\begin{aligned}0&=\widehat{h}_{q+1}^\varepsilon(1)-\widehat{h}_{q+1}^0(0)\\&=\frac{\partial\widehat{h}_{q+1}^\varepsilon(\eta_\varepsilon)}{\partial\boldsymbol{x}}\left\{\boldsymbol{\xi}^{q+1,\varepsilon}(\tau_{q+1}^\varepsilon)-\boldsymbol{\xi}^{q+1,0}(\tau_{q+1}^0)\right\}.\end{aligned} \tag{10.2.63}$$

Now, since $\tau_q^\varepsilon\to\tau_q^0<\tau_{q+1}^0$ as $\varepsilon\to 0$, we have $\tau_q^\varepsilon<\tau_{q+1}^0$ when ε is of sufficiently small magnitude. Thus,

$$\begin{aligned}&\boldsymbol{\xi}^{q+1,\varepsilon}\left(\tau_{q+1}^\varepsilon\right)-\boldsymbol{\xi}^{q+1,0}\left(\tau_{q+1}^0\right)\\&\quad=\boldsymbol{\xi}^{q+1,\varepsilon}\left(\tau_{q+1}^\varepsilon\right)-\boldsymbol{\xi}^{q+1,\varepsilon}\left(\tau_{q+1}^0\right)+\boldsymbol{\mu}^{q+1,\varepsilon}\left(\tau_{q+1}^0\right)\\&\quad=\int_{\tau_{q+1}^0}^{\tau_{q+1}^\varepsilon}\hat{\boldsymbol{f}}^{q+1,\varepsilon}(\omega,1)d\omega+\boldsymbol{\mu}^{q+1,\varepsilon}\left(\tau_{q+1}^0\right)\\&\quad=\left(\tau_{q+1}^\varepsilon-\tau_{q+1}^0\right)\int_0^1\hat{\boldsymbol{f}}^{q+1,\varepsilon}\left(\eta\tau_{q+1}^\varepsilon+(1-\eta)\tau_{q+1}^0,1\right)d\eta+\boldsymbol{\mu}^{q+1,\varepsilon}\left(\tau_{q+1}^0\right).\end{aligned} \tag{10.2.64}$$

Substituting (10.2.64) into (10.2.63) and rearranging, we obtain

$$\left(\tau_{q+1}^{\varepsilon}-\tau_{q+1}^{0}\right)\frac{\partial\widehat{h}_{q+1}^{\varepsilon}(\eta_{\varepsilon})}{\partial\boldsymbol{x}}\int_{0}^{1}\hat{\boldsymbol{f}}^{q+1,\varepsilon}\left(\eta\tau_{q+1}^{\varepsilon}+(1-\eta)\tau_{q+1}^{0},1\right)d\eta$$
$$=-\frac{\partial\widehat{h}_{q+1}^{\varepsilon}(\eta_{\varepsilon})}{\partial\boldsymbol{x}}\boldsymbol{\mu}^{q+1,\varepsilon}\left(\tau_{q+1}^{0}\right), \tag{10.2.65}$$

which holds for all ε of sufficiently small magnitude.

Clearly, since $\tau_{q+1}^{\varepsilon}\to\tau_{q+1}^{0}$ as $\varepsilon\to 0$, we can choose $T_{\max}>\max\left(\tau_{q+1}^{\varepsilon},\tau_{q+1}^{0}\right)$. Thus, by Assumption 10.2.3 and (10.2.15), we obtain

$$\lim_{\varepsilon\to 0}\frac{\partial\widehat{h}_{q+1}^{\varepsilon}(\eta_{\varepsilon})}{\partial\boldsymbol{x}}\int_{0}^{1}\hat{\boldsymbol{f}}^{q+1,\varepsilon}\left(\eta\tau_{q+1}^{\varepsilon}+(1-\eta)\tau_{q+1}^{0},1\right)d\eta$$
$$=\frac{\partial\widehat{h}_{q+1}^{0}(0)}{\partial\boldsymbol{x}}\hat{\boldsymbol{f}}^{q+1,0}\left(\tau_{q+1}^{0},0\right)\neq 0. \tag{10.2.66}$$

Therefore, (10.2.65) can be arranged such that

$$\tau_{q+1}^{\varepsilon}-\tau_{q+1}^{0}=-\frac{\partial\widehat{h}_{q+1}^{\varepsilon}(\eta_{\varepsilon})}{\partial\boldsymbol{x}}\boldsymbol{\mu}^{q+1,\varepsilon}(\tau_{q+1}^{0})$$
$$\div\left\{\frac{\partial\widehat{h}_{q+1}^{\varepsilon}(\eta_{\varepsilon})}{\partial\boldsymbol{x}}\int_{0}^{1}\hat{\boldsymbol{f}}^{q+1,\varepsilon}\left(\eta\tau_{q+1}^{\varepsilon}+(1-\eta)\tau_{q+1}^{0},1\right)d\eta\right\}. \tag{10.2.67}$$

Now, dividing both sides of (10.2.67) by ε, it gives

$$\frac{\tau_{q+1}^{\varepsilon}-\tau_{q+1}^{0}}{\varepsilon}=-\frac{\partial\widehat{h}_{q+1}^{\varepsilon}(\eta_{\varepsilon})}{\partial\boldsymbol{x}}\varepsilon^{-1}\boldsymbol{\mu}^{q+1,\varepsilon}\left(\tau_{q+1}^{0}\right)$$
$$\div\left\{\frac{\partial\widehat{h}_{q+1}^{\varepsilon}(\eta_{\varepsilon})}{\partial\boldsymbol{x}}\int_{0}^{1}\hat{\boldsymbol{f}}^{q+1,\varepsilon}\left(\eta\tau_{q+1}^{\varepsilon}+(1-\eta)\tau_{q+1}^{0},1\right)d\eta\right\}. \tag{10.2.68}$$

Thus, by (10.2.16a) and (10.2.16b), we have

$$\lim_{\varepsilon\to 0}\frac{\tau_{q+1}^{\varepsilon}-\tau_{q+1}^{0}}{\varepsilon}=-\frac{\partial\widehat{h}_{q+1}^{0}(0)}{\partial\boldsymbol{x}}\Lambda_{k}\left(\tau_{q+1}^{0-}\right)\div\left\{\frac{\partial\widehat{h}_{q+1}^{0}(0)}{\partial\boldsymbol{x}}\hat{\boldsymbol{f}}^{q+1,0}\left(\tau_{q+1}^{0},0\right)\right\}. \tag{10.2.69}$$

Therefore, the validity of (10.2.16c) for $i=q+1$ is established. Now, the proof of Lemma 10.2.1 is complete.

10.2.4 Main Results

We are now in a position to derive formulas for the state variation matrix and the partial derivatives of the switching times with respect to each of the components of the system parameter vector.

Theorem 10.2.2 *For each $\boldsymbol{\zeta} \in Z$ and for each $k = 1, \ldots, s$, it holds that*

$$\frac{\partial \boldsymbol{x}(t \mid \boldsymbol{\zeta})}{\partial \zeta_k} = \Lambda_k(t), \quad t \in (\tau_{i-1}, \tau_i), \quad i \in Z_N, \tag{10.2.70}$$

and

$$\begin{aligned}\frac{\partial \tau_i(\boldsymbol{\zeta})}{\partial \zeta_k} = &-\frac{\partial h_i(\boldsymbol{x}(\tau_i))}{\partial \boldsymbol{x}} \Lambda_k(\tau_i^-) \\ &\div \left\{ \frac{\partial h_i(\boldsymbol{x}(\tau_i))}{\partial \boldsymbol{x}} \boldsymbol{f}^i(\boldsymbol{x}(\tau_i), \boldsymbol{x}(\tau_i - \gamma_1), \ldots, \boldsymbol{x}(\tau_i - \gamma_m)) \right\}, \\ &\qquad i \in Z_{N-1} \text{ and } \tau_i < \infty,\end{aligned} \tag{10.2.71}$$

where $\Lambda_k(\cdot)$ satisfies the variational system described by the differential equations (10.2.13a) with initial condition (10.2.13b)–(10.2.13c) and jump condition (10.2.13d).

Proof. Note that (10.2.16a) and (10.2.16b) are valid. Thus, given any $t \in (\tau_{i-1}, \tau_i)$, $i \in Z_N$, we have $t < \tau_i^\varepsilon$ for all ε of sufficiently small magnitude, and hence $\boldsymbol{x}(t \mid \boldsymbol{\zeta} + \varepsilon \boldsymbol{e}^k) = \boldsymbol{\xi}^{i,\varepsilon}(t)$. This implies that for $t < \tau_i^\varepsilon$ and for all ε of sufficiently small magnitude,

$$\begin{aligned}\frac{\partial \boldsymbol{x}\,(t \mid \boldsymbol{\zeta})}{\partial \zeta_k} &= \lim_{\varepsilon \to 0} \frac{\boldsymbol{x}(t \mid \boldsymbol{\zeta} + \varepsilon \boldsymbol{e}^k) - \boldsymbol{x}(t \mid \boldsymbol{\zeta})}{\varepsilon} \\ &= \lim_{\varepsilon \to 0} \varepsilon^{-1} \boldsymbol{\mu}^{i,\varepsilon}(t) = \Lambda_k(t^-) = \Lambda_k(t).\end{aligned} \tag{10.2.72}$$

Therefore, the validity of (10.2.70) is established. Now, for each $i \in Z_{N-1}$ with τ_i being finite, we recall that (10.2.16c) is satisfied.

Since $\boldsymbol{\xi}^{i,0}\left(\tau_i^0 - \gamma_j\right) = \boldsymbol{x}(\tau_i - \gamma_j \mid \boldsymbol{\zeta})$, $j = 0, \ldots, r$, it follows that

$$\begin{aligned}\frac{\partial \tau_i(\boldsymbol{\zeta})}{\partial \zeta_k} &= \lim_{\varepsilon \to 0} \frac{\tau_i^\varepsilon - \tau_i^0}{\varepsilon} \\ &= -\frac{\partial h_i\left(\boldsymbol{\xi}^{i,0}\left(\tau_i^0\right)\right)}{\partial \boldsymbol{x}} \Lambda_k\left(\tau_i^{0-}\right) \\ &\quad \div \left\{ \frac{\partial h_i\left(\xi^{i,0}\left(\tau_i^0\right)\right)}{\partial \boldsymbol{x}} \boldsymbol{f}^i\left(\boldsymbol{\xi}^{i,0}\left(\tau_i^0\right), \boldsymbol{\xi}^{i,0}\left(\tau_i^0 - \gamma_1\right), \ldots, \boldsymbol{\xi}^{i,0}\left(\tau_i^0 - \gamma_r\right)\right) \right\} \\ &= -\frac{\partial h_i(\boldsymbol{x}(\tau_i))}{\partial \boldsymbol{x}} \Lambda_k\left(\tau_i^-\right)\end{aligned}$$

$$\div \left\{ \frac{\partial h_i(\boldsymbol{x}(\tau_i))}{\partial \boldsymbol{x}} \boldsymbol{f}^i(\boldsymbol{x}(\tau_i), \boldsymbol{x}(\tau_i - \gamma_1), \ldots, \boldsymbol{x}(\tau_i - \gamma_r)) \right\}. \tag{10.2.73}$$

Thus, (10.2.71) is established.

Therefore, the conclusions of the theorem follow readily from (10.2.16a)–(10.2.16c). This completes the proof.

The switching times $t = \tau_i$, $i \in Z_{N-1}$ are deliberately excluded from equation (10.2.70) in Theorem 10.2.2. This is because the state variation matrix only exists at the switching times in rare circumstances, as the next result shows.

Theorem 10.2.3 *Let* $\boldsymbol{\zeta} \in \mathcal{Z}$ *and let* $\Lambda_k(\cdot) = \left[\Lambda_k^1, \ldots, \Lambda_k^N\right]$ *denote the solution of the variational system governed by the differential equations (10.2.13a) with initial condition (10.2.13b)–(10.2.13c) and jump condition (10.2.13d) corresponding to* $\boldsymbol{\zeta}$*. Then, for each* $k = 1, \ldots, s$ *and each* $i \in Z_{N-1}$ *with* $\tau_i < \infty$*, one of the following scenarios holds:*

(i) Suppose that

$$\begin{aligned} &\boldsymbol{f}^i(\boldsymbol{x}(\tau_i), \boldsymbol{x}(\tau_i - \gamma_1), \ldots, x(\tau_i - \gamma_r)) \\ &\quad = \boldsymbol{f}^{i+1}(\boldsymbol{x}(\tau_i), \boldsymbol{x}(\tau_i - \gamma_1), \ldots, \boldsymbol{x}(\tau_i - \gamma_r)) \end{aligned} \tag{10.2.74a}$$

or

$$\partial \tau_i(\boldsymbol{\zeta})/\partial \zeta_k = 0. \tag{10.2.74b}$$

Then

$$\frac{\partial \boldsymbol{x}(\tau_i \mid \zeta)}{\partial \zeta_k} = \Lambda_k(\tau_i^+) = \Lambda_k(\tau_i^-). \tag{10.2.75}$$

(ii) Suppose that

$$\begin{aligned} &\boldsymbol{f}^i(\boldsymbol{x}(\tau_i), \boldsymbol{x}(\tau_i - \gamma_1), \ldots, \boldsymbol{x}(\tau_i - \gamma_m)) \\ &\quad \neq \boldsymbol{f}^{i+1}(\boldsymbol{x}(\tau_i), \boldsymbol{x}(\tau_i - \gamma_1), \ldots, \boldsymbol{x}(\tau_i - \gamma_m)) \end{aligned} \tag{10.2.76a}$$

and

$$\partial \tau_i(\boldsymbol{\zeta})/\partial \zeta_k > 0. \tag{10.2.76b}$$

Then,

$$\frac{\partial^{\pm} \boldsymbol{x}(\tau_i \mid \boldsymbol{\zeta})}{\partial \zeta_k} = \lim_{\varepsilon \to 0\pm} \frac{\boldsymbol{x}\left(\tau_i \mid \boldsymbol{\zeta} + \varepsilon \boldsymbol{e}^k\right) - \boldsymbol{x}(\tau_i \mid \boldsymbol{\zeta})}{\varepsilon} = \Lambda_k(\tau_i^{\mp}). \tag{10.2.77}$$

(iii) Suppose that

$$\begin{aligned} &\boldsymbol{f}^i(\boldsymbol{x}(\tau_i), \boldsymbol{x}(\tau_i - \gamma_1), \ldots, \boldsymbol{x}(\tau_i - \gamma_r)) \\ &\quad \neq \boldsymbol{f}^{i+1}(\boldsymbol{x}(\tau_i), \boldsymbol{x}(\tau_i - \gamma_1), \ldots, \boldsymbol{x}(\tau_i - \gamma_r)) \end{aligned} \tag{10.2.78a}$$

and

$$\partial\tau_i(\boldsymbol{\zeta})/\partial\zeta_k < 0. \tag{10.2.78b}$$

Then,

$$\frac{\partial^{\pm}\boldsymbol{x}(\tau_i \mid \boldsymbol{\zeta})}{\partial\zeta_k} = \lim_{\varepsilon\to 0\pm} \frac{\boldsymbol{x}\left(\tau_i \mid \boldsymbol{\zeta}+\varepsilon e^k\right) - \boldsymbol{x}(\tau_i \mid \boldsymbol{\zeta})}{\varepsilon} = \Lambda_k(\tau_i^{\pm}). \tag{10.2.79}$$

Proof. Consider $i \in Z_{N-1}$ with $\tau_i < \infty$ and let $k \in \{1, \ldots, s\}$. Then, from the auxiliary system (10.2.5), we have

$$\begin{aligned} &\boldsymbol{x}\left(\tau_i^0 \mid \boldsymbol{\zeta}+\varepsilon \boldsymbol{e}^k\right) - \boldsymbol{x}\left(\tau_i^0 \mid \boldsymbol{\zeta}\right) \\ &= \boldsymbol{\xi}^{i,\varepsilon}\left(\tau_i^0\right) - \boldsymbol{\xi}^{i,0}\left(\tau_i^0\right) + \int_{\min(\tau_i^\varepsilon,\tau_i^0)}^{\tau_i^0} \left\{\hat{\boldsymbol{f}}^{i+1,\varepsilon}(\omega,1) - \hat{\boldsymbol{f}}^{i,\varepsilon}(\omega,1)\right\} d\omega \\ &= \boldsymbol{\mu}^{i,\varepsilon}\left(\tau_i^0\right) + \int_{\min\left(\tau_i^\varepsilon,\tau_i^0\right)}^{\tau_i^0} \left\{\hat{\boldsymbol{f}}^{i+1,\varepsilon}(\omega,1) - \hat{\boldsymbol{f}}^{i,\varepsilon}(\omega,1)\right\} d\omega, \end{aligned} \tag{10.2.80}$$

when $|\varepsilon|$ is sufficiently small such that $\tau_i^0 < \tau_{i+1}^\varepsilon$.

Now, by (10.2.37) and (10.2.38), which were established in the proof of (10.2.15), we obtain

$$\begin{aligned} &\boldsymbol{x}\left(\tau_i^0 \mid \boldsymbol{\zeta}+\varepsilon e^k\right) - \boldsymbol{x}\left(\tau_i^0 \mid \boldsymbol{\zeta}\right) \\ &= \boldsymbol{\mu}^{i,\varepsilon}\left(\tau_i^0\right) + \left\{\tau_i^0 - \min\left(\tau_i^\varepsilon,\tau_i^0\right)\right\} \cdot \left\{\hat{\boldsymbol{f}}^{i+1,0}\left(\tau_i^0,0\right) - \hat{\boldsymbol{f}}^{i,0}\left(\tau_i^0,0\right)\right\} + O(\varepsilon^2). \end{aligned} \tag{10.2.81}$$

Hence,

$$\begin{aligned} &\frac{\boldsymbol{x}\left(\tau_i^0 \mid \boldsymbol{\zeta}+\varepsilon \boldsymbol{e}^k\right) - \boldsymbol{x}\left(\tau_i^0 \mid \boldsymbol{\zeta}\right)}{\varepsilon} \\ &= \varepsilon^{-1}\boldsymbol{\mu}^{i,\varepsilon}\left(\tau_i^0\right) + \frac{\tau_i^0 - \min\left(\tau_i^\varepsilon,\tau_i^0\right)}{\varepsilon} \cdot \left\{\hat{\boldsymbol{f}}^{i+1,0}\left(\tau_i^0,0\right) - \hat{\boldsymbol{f}}^{i,0}\left(\tau_i^0,0\right)\right\} + O(\varepsilon). \end{aligned} \tag{10.2.82}$$

If $\partial\tau_i/\partial\zeta_k > 0$, then $\tau_i^\varepsilon \to \tau_i^{0\pm}$ as $\varepsilon \to 0^{\pm}$. Thus,

$$\lim_{\varepsilon\to 0^+} \frac{\tau_i^0 - \min\left(\tau_i^\varepsilon,\tau_i^0\right)}{\varepsilon} = 0, \qquad \lim_{\varepsilon\to 0^-} \frac{\tau_i^0 - \min(\tau_i^\varepsilon,\tau_i^0)}{\varepsilon} = -\frac{\partial\tau_i}{\partial\zeta_k}. \tag{10.2.83}$$

Therefore, since $\varepsilon^{-1}\boldsymbol{\mu}^{i,\varepsilon}\left(\tau_i^0\right) \to \Lambda_k\left(\tau_i^{0-}\right)$ by (10.2.16a), we have

$$\lim_{\varepsilon\to 0^+} \frac{\boldsymbol{x}\left(\tau_i^0 \mid \boldsymbol{\zeta}+\varepsilon \boldsymbol{e}^k\right) - \boldsymbol{x}\left(\tau_i^0 \mid \boldsymbol{\zeta}\right)}{\varepsilon} = \Lambda_k\left(\tau_i^{0-}\right) \tag{10.2.84}$$

and

$$\lim_{\varepsilon \to 0^-} \frac{\boldsymbol{x}\left(\tau_i^0 \mid \boldsymbol{\zeta} + \varepsilon \boldsymbol{e}^k\right) - \boldsymbol{x}\left(\tau_i^0 \mid \boldsymbol{\zeta}\right)}{\varepsilon}$$
$$= \Lambda_k\left(\tau_i^{0-}\right) + \frac{\partial \tau_i(\boldsymbol{\zeta})}{\partial \zeta_k} \cdot \left\{\hat{\boldsymbol{f}}^{i,0}\left(\tau_i^0, 0\right) - \hat{\boldsymbol{f}}^{i+1,0}\left(\tau_i^0, 0\right)\right\} = \Lambda_k\left(\tau_i^{0+}\right). \tag{10.2.85}$$

Similarly, if $\partial \tau_i / \partial \zeta_k < 0$, then

$$\lim_{\varepsilon \to 0^+} \frac{\boldsymbol{x}\left(\tau_i^0 \mid \boldsymbol{\zeta} + \varepsilon \boldsymbol{e}^k\right) - \boldsymbol{x}\left(\tau_i^0 \mid \boldsymbol{\zeta}\right)}{\varepsilon}$$
$$= \Lambda_k\left(\tau_i^{0-}\right) + \frac{\partial \tau_i(\boldsymbol{\zeta})}{\partial \zeta_k} \cdot \left\{\hat{\boldsymbol{f}}^{i,0}\left(\tau_i^0, 0\right) - \hat{\boldsymbol{f}}^{i+1,0}\left(\tau_i^0, 0\right)\right\} = \Lambda_k\left(\tau_i^{0+}\right) \tag{10.2.86}$$

and

$$\lim_{\varepsilon \to 0^-} \frac{\boldsymbol{x}\left(\tau_i^0 \mid \zeta + \varepsilon \boldsymbol{e}^k\right) - \boldsymbol{x}\left(\tau_i^0 \mid \boldsymbol{\zeta}\right)}{\varepsilon} = \Lambda_k\left(\tau_i^{0-}\right). \tag{10.2.87}$$

Finally, suppose that either $\partial \tau_i / \partial \zeta_k = 0$ or $\hat{\boldsymbol{f}}^{i+1,0}\left(\tau_i^0, 0\right) = \hat{\boldsymbol{f}}^{i,0}\left(\tau_i^0, 0\right)$ is satisfied. Then,

$$\Lambda_k\left(\tau_i^{0+}\right) = \Lambda_k\left(\tau_i^{0-}\right)$$

and

$$\frac{\partial \boldsymbol{x}\left(\tau_i^0 \mid \boldsymbol{\zeta}\right)}{\partial \zeta_k} = \lim_{\varepsilon \to 0} \frac{\boldsymbol{x}\left(\tau_i^0 \mid \boldsymbol{\zeta} + \varepsilon \boldsymbol{e}^k\right) - \boldsymbol{x}\left(\tau_i^0 \mid \boldsymbol{\zeta}\right)}{\varepsilon}$$
$$= \Lambda_k\left(\tau_i^{0-}\right) = \Lambda_k\left(\tau_i^{0+}\right). \tag{10.2.88}$$

The proof is complete.

Remark 10.2.4 *In the first scenario of Theorem 10.2.3, the state variation exists at $t = \tau_i$. In the last two scenarios (the more likely scenarios), the state variation does not exist at $t = \tau_i$ due to the facts that the left and right partial derivatives of the state with respect to the kth component of the system parameter are different, indicating that $\Lambda_k(\cdot)$ is discontinuous at the ith switching time.*

Let $\boldsymbol{\zeta} \in \mathcal{Z}$ be an arbitrary parameter vector. Then, under Assumptions 10.2.2 and 10.2.3, it follows from (10.2.15) and (10.2.16b) that $\boldsymbol{\xi}^i(\cdot)$, $i \in Z_N$, and $\tau_i(\cdot)$, $i \in Z_{N-1}$, are continuous with respect to system parameter vector at $\boldsymbol{\zeta}$ along each coordinate axis of the space $\mathbb{R}^s$. However, this does not necessarily imply the continuity (in the space $\mathbb{R}^s$) at $\boldsymbol{\zeta}$. For continuity, it is to be shown as follows. Consider a perturbed parameter vector $\boldsymbol{\zeta} + \boldsymbol{\sigma} \in \mathcal{Z}$ and let $\boldsymbol{\xi}^{i,\boldsymbol{\sigma}}(\cdot)$, $\tau_i^{\boldsymbol{\sigma}}(\cdot)$, and $\boldsymbol{\mu}^{i,\boldsymbol{\sigma}}(\cdot)$ denote the analogues of $\boldsymbol{\xi}^{i,\varepsilon}(\cdot)$, $\tau_i^{\varepsilon}(\cdot)$, and $\boldsymbol{\mu}^{i,\varepsilon}(\cdot)$ with $\boldsymbol{\sigma}$ replacing $\varepsilon \boldsymbol{e}^k$. Then, the proofs for (10.2.14), (10.2.15) and (10.2.16c) can be easily modified to prove that the following versions of (10.2.14), (10.2.15) and (10.2.16b) for each i are valid.

$$\max_{t\in[-\bar{\gamma},T_{\max}]} \left|\boldsymbol{\xi}^{i,\boldsymbol{\sigma}}(t)\right| = O(1) \text{ for every } T_{\max} > 0, \tag{10.2.89}$$

$$\max_{t\in[-\bar{\gamma},T_{\max}]} \left|\boldsymbol{\mu}^{i,\boldsymbol{\sigma}}(t)\right| = \theta(\boldsymbol{\sigma}) \text{ for every } T_{\max} > 0, \tag{10.2.90}$$

$$\lim_{\boldsymbol{\sigma}\to 0} \tau_i^{\boldsymbol{\sigma}} = \tau_i^0, \tag{10.2.91}$$

where $O(1)$ in (10.2.89) means that the left hand side is uniformly bounded with respect to $\boldsymbol{\sigma}$, and $\theta(\boldsymbol{\sigma})$ in (10.2.90) means that the left hand side converges to zero as $\boldsymbol{\sigma} \to \mathbf{0}$. In (10.2.90) and (10.2.91), the convergence $\sigma \to \mathbf{0}$ can be along any path to the origin, but in the original (10.2.15) and (10.2.16b), the convergence is restricted to be along one of the coordinate axes. Choose $T_{\max} > T$. Then, by virtue of (10.2.90) for $i = N$, we have

$$\begin{aligned}
&|\boldsymbol{x}(T \mid \boldsymbol{\zeta} + \boldsymbol{\sigma}) - \boldsymbol{x}(T \mid \boldsymbol{\zeta})| \\
&\quad = \left|\boldsymbol{\xi}^{N,\boldsymbol{\sigma}}(T) - \boldsymbol{\xi}^{N,\mathbf{0}}(T)\right| \\
&\quad = \left|\boldsymbol{\mu}^{N,\boldsymbol{\sigma}}(T)\right| \le \max_{t\in[-\bar{\gamma},T_{\max}]} \left|\mu^{N,\boldsymbol{\sigma}}(t)\right| = \theta(\boldsymbol{\sigma}).
\end{aligned} \tag{10.2.92}$$

This shows that $\boldsymbol{x}(T \mid \boldsymbol{\zeta} + \boldsymbol{\sigma}) \to \boldsymbol{x}(T \mid \boldsymbol{\zeta})$ as $\boldsymbol{\sigma} \to \mathbf{0}$. On this basis, since Φ is continuous, we have

$$g_0(\boldsymbol{\zeta} + \boldsymbol{\sigma}) = \Phi(\boldsymbol{x}(T \mid \boldsymbol{\zeta} + \sigma)) \to \Phi(\boldsymbol{x}(T \mid \boldsymbol{\zeta})) = g_0(\boldsymbol{\zeta}) \text{ as } \boldsymbol{\sigma} \to \mathbf{0}. \tag{10.2.93}$$

This shows that the cost functional g_0 is continuous. Note that the proof of (10.2.90) must be carried out simultaneously with (10.2.89) and (10.2.91) via induction as for the proof of Lemma 10.2.1.

We are now in a position to present the right and left partial derivatives of the cost functional g_0 with respect to the system parameter vector $\boldsymbol{\zeta}$ as given in the following theorem:

Theorem 10.2.4 *For each $\boldsymbol{\zeta} \in \mathcal{Z}$, it holds that*

$$\boldsymbol{x}\left(T \mid \boldsymbol{\zeta} + \varepsilon \boldsymbol{e}^k\right) \to \boldsymbol{x}(T \mid \boldsymbol{\zeta}) \quad \text{as} \quad \varepsilon \to 0^{\pm}, \tag{10.2.94}$$

and, for each $k = 1, \ldots, s$,

$$\begin{aligned}
&\frac{\partial^{\pm} g_0(\boldsymbol{\zeta})}{\partial \zeta_k} \\
&\quad = \lim_{\varepsilon\to 0^{\pm}} \frac{g_0\left(\boldsymbol{\zeta} + \varepsilon \boldsymbol{e}^k\right) - g_0(\boldsymbol{\zeta})}{\varepsilon} \\
&\quad = \frac{\partial \Phi(\boldsymbol{x}(T \mid \boldsymbol{\zeta}))}{\partial \boldsymbol{x}} \lim_{\varepsilon\to 0^{\pm}} \frac{\boldsymbol{x}\left(T \mid \zeta + \varepsilon \boldsymbol{e}^k\right) - \boldsymbol{x}(T \mid \boldsymbol{\zeta})}{\varepsilon}
\end{aligned}$$

$$= \frac{\partial \Phi(\boldsymbol{x}(T \mid \boldsymbol{\zeta}))}{\partial \boldsymbol{x}} \begin{cases} \Lambda_k(T \mid \boldsymbol{\zeta}), & \textit{if } \tau_i(\boldsymbol{\zeta}) \neq T,\ i = 1, \ldots, N-1, \\ \Lambda_k(T^{\mp} \mid \boldsymbol{\zeta}), & \textit{if } \tau_i(\boldsymbol{\zeta}) = T \textit{ and } \partial \tau_i(\boldsymbol{\zeta})/\partial \zeta_k \geq 0, \\ \Lambda_k(T^{\pm} \mid \boldsymbol{\zeta}), & \textit{if } \tau_i(\boldsymbol{\zeta}) = T \textit{ and } \partial \tau_i(\boldsymbol{\zeta})/\partial \zeta_k \leq 0, \end{cases} \tag{10.2.95}$$

where $\Lambda_k(\cdot \mid \boldsymbol{\zeta})$ is the solution of the variational system (10.2.13) corresponding to $\boldsymbol{\zeta}$.

Proof. By Theorems 10.2.2 and 10.2.3, we can derive the left and right partial derivatives of the cost functional g_0. First, by Taylor's theorem, there exists, for each $\varepsilon \neq 0$ and each $k = 1, \ldots, s$, a constant $\eta_{\varepsilon,k} \in (0, 1)$ such that

$$g_0(\boldsymbol{\zeta} + \varepsilon \mathbf{e}^k) - g_0(\boldsymbol{\zeta}) = \frac{\partial \Phi\left((1 - \eta_{\varepsilon,k})\boldsymbol{x}(T \mid \boldsymbol{\zeta}) + \eta_{\varepsilon,k}\boldsymbol{x}\left(T \mid \boldsymbol{\zeta} + \varepsilon \mathbf{e}^k\right)\right)}{\partial x} \cdot \left\{\boldsymbol{x}\left(T \mid \boldsymbol{\zeta} + \varepsilon \mathbf{e}^k\right) - \boldsymbol{x}(T \mid \boldsymbol{\zeta})\right\}. \tag{10.2.96}$$

Collectively, by Theorems 10.2.2 and 10.2.3, the existence of the right and left partial derivatives of the system state with respect to each component of the system parameter is assured under Assumptions 10.2.2 and 10.2.3 Thus, (10.2.94) and (10.2.95) are valid. The proof is complete.

Note from (10.2.95) that the left and right partial derivatives of g_0 exist at all $\boldsymbol{\zeta} \in \mathcal{Z}$ under Assumptions 10.2.2 and 10.2.3. In practice, these assumptions can be easily checked for a given $\boldsymbol{\zeta} \in \mathcal{Z}$ by numerically solving the switched time-delay system (10.2.1). Note that if T coincides with a switching time satisfying one of the last two scenarios in Theorem 10.2.3, then the left and right partial derivatives of g_0 with respect to ζ_k may differ, since in this case $\Lambda_k(T^-) \neq \Lambda_k(T^+)$.

Since g_0 has well-defined left and right partial derivatives (under Assumptions 10.2.2 and 10.2.3), it is continuous under Assumptions 10.2.2 and 10.2.3. If these assumptions hold at every point in the compact set $\mathcal{Z}$, then Problem (P_2) is guaranteed to admit an optimal solution. This result is summarized below.

Theorem 10.2.5 *Problem (P_2) admits an optimal solution.*

The left and right partial derivatives of g_0, as defined in (10.2.95), can be used to identify search directions at a given system parameter vector $\boldsymbol{\zeta}$ during the optimization process. Indeed, if $\partial^+ g_0(\boldsymbol{\zeta})/\partial \zeta_k < 0$, then $\boldsymbol{e}^k$ is a descent direction of g_0 at $\boldsymbol{\zeta}$, and if $\partial^- g_0(\boldsymbol{\zeta})/\partial \zeta_k > 0$, then $-\boldsymbol{e}^k$ is a descent direction of g_0 at $\boldsymbol{\zeta}$. Performing a line search along a descent direction will yield an improved point with lower cost.

If none of the switching times coincide with the terminal time, or if the conditions for the first scenario in Theorem 10.2.3 are satisfied at the terminal time, then the left and right partial derivatives of g_0 derived above become the full partial derivatives as shown in (10.2.9). We now present the following line search optimization algorithm for solving Problem (P_2).

Algorithm 10.2.1

---

1. *Choose an initial point* $\boldsymbol{\zeta} \in \mathcal{Z}$.
2. *Form an expanded switched time-delay system by combining the state system (10.2.1) with the variational system (10.2.13) for each* $k = 1, \ldots, s$.
3. *Solve the expanded system sub-system by sub-system, checking Assumptions 10.2.2 and 10.2.3 at the start and end of each sub-system. If these assumptions are violated at any stage, then stop with error.*
4. *Use* $\boldsymbol{x}(\cdot \mid \boldsymbol{\zeta})$ *and* $\Lambda_k(\cdot \mid \boldsymbol{\zeta})$, $k = 1, \ldots, s$, *to determine the left and right partial derivatives of* g_0 *according to (10.2.95).*
5. *Use* $\partial^{\pm} g_0(\boldsymbol{\zeta})/\partial \zeta_k$, $k = 1, \ldots, s$, *to check local optimality conditions at* $\boldsymbol{\zeta}$. *If the local optimality conditions hold, then stop; otherwise, continue to Step 6.*
6. *Use* $\partial^{\pm} g_0(\boldsymbol{\zeta})/\partial \zeta_k$, $k = 1, \ldots, s$, *to define a search direction.*
7. *Perform a line search along the direction from Step 6 to determine a new point* $\boldsymbol{\zeta}' \in \mathcal{Z}$.
8. *Set* $\boldsymbol{\zeta}' \to \boldsymbol{\zeta}$ *and return to Step 2.*

In most cases, the partial derivatives of g_0 will exist and Steps 5–7 can be implemented using well-known methods in nonlinear optimization (see Chapter 2). If any of the full partial derivatives of g_0 does not exist (i.e., one of the last two scenarios in Theorem 10.2.3 occurred at the terminal time), then the signs of the left and right partial derivatives can be used to identify an appropriate descent direction along one of the coordinate axes.

10.2.5 Numerical Example

We consider a fed-batch fermentation process for converting glycerol to 1,3-propanediol (1,3-PD). This process switches between two modes: batch mode (during which there is no input feed) and feeding mode (during which glycerol and alkali are added continuously to the fermentor). The switching of mode occurs when the concentration of glycerol reaches certain lower and upper thresholds. Moreover, since nutrient metabolization does not immediately lead to the production of new biomass, the fermentation process involves a time-delay.

The model is based on the work in [151, 197]. Let $\boldsymbol{x}(t) = [x_1(t), x_2(t), x_3(t), x_4(t)]^\top$, where t is time (hours). Here, $x_1(t)$ is the biomass concentration ($\mathrm{g\,L^{-1}}$), $x_2(t)$ is the glycerol concentration ($\mathrm{mmol\,L^{-1}}$), $x_3(t)$ is the 1,3-PD concentration ($\mathrm{mmol\,L^{-1}}$) and $x_4(t)$ is the fluid volume (L). The process dynamics due to natural fermentation are described by

$$\begin{bmatrix} \frac{dx_1(t)}{dt} \\ \frac{dx_2(t)}{dt} \\ \frac{dx_3(t)}{dt} \\ \frac{dx_4(t)}{dt} \end{bmatrix} = \begin{bmatrix} \mu(x_2(t), x_3(t))x_1(t-\gamma_1) \\ -q_2(x_2(t), x_3(t))x_1(t-\gamma_1) \\ q_3(x_2(t), x_3(t))x_1(t-\gamma_1) \\ 0 \end{bmatrix} = f^{\mathrm{ferm}}(x(t), x_1(t-\gamma_1)), \tag{10.2.97}$$

where $\gamma_1 = 0.1568$ is the time-delay; $\mu(\cdot,\cdot)$ is the cell growth rate; $q_2(\cdot,\cdot)$ is the substrate consumption rate; and $q_3(\cdot,\cdot)$ is the 1,3-PD formation rate. The process dynamics due to the input feed are

$$\begin{bmatrix} \frac{dx_1(t)}{dt} \\ \frac{dx_2(t)}{dt} \\ \frac{dx_3(t)}{dt} \\ \frac{dx_4(t)}{dt} \end{bmatrix} = \frac{u(t)}{x_4(t)} \begin{bmatrix} -x_1(t) \\ rc_{s0} - x_2(t) \\ -x_3(t) \\ x_4(t) \end{bmatrix} := f^{\mathrm{feed}}(x(t), u(t)), \tag{10.2.98}$$

where $u(t)$ is the input feeding rate ($\mathrm{L\,h^{-1}}$); $r = 0.5714$ is the proportion of glycerol in the input feed and $c_{s0} = 10762\,\mathrm{mmol\,L^{-1}}$ is the concentration of glycerol in the input feed. The functions $\mu(\cdot,\cdot)$, $q_2(\cdot,\cdot)$ and $q_3(\cdot,\cdot)$ in (10.2.97) are given by

$$\mu(x_2(t), x_3(t)) = \frac{\Delta_1 x_2(t)}{x_2(t) + k_1}\left(1 - \frac{x_2(t)}{x_2^*}\right)\left(1 - \frac{x_3(t)}{x_3^*}\right)^3, \tag{10.2.99}$$

$$q_2(x_2(t), x_3(t)) = m_1 + Y_1\mu(x_2(t), x_3(t)) + \frac{\Delta_2 x_2(t)}{x_2(t) + k_2}, \tag{10.2.100}$$

$$q_3(x_2(t), x_3(t)) = -m_2 + Y_2\mu(x_2(t), x_3(t)) + \frac{\Delta_3 x_2(t)}{x_2(t) + k_3}, \tag{10.2.101}$$

where $x_2^* = 2039\,\mathrm{mmol\,L^{-1}}$ and $x_3^* = 1036\,\mathrm{mmol\,L^{-1}}$ are, respectively, the critical concentrations of glycerol and 1,3-PD, and the values of the other parameters are given in Table 10.2.1.

Table 10.2.1: The other parameters in system

Δ_1	k_1	m_1	Y_1	Δ_2	k_2	m_2	Y_2	Δ_3	k_3
0.8037	0.4856	0.2977	144.9120	7.8367	9.4632	12.2577	80.8439	20.2757	38.75

Let N_{feed} be an upper bound for the number of feeding modes. Since the process starts and finishes in batch mode, the total number of potential modes is $N = 2N_{\text{feed}} + 1$ (N_{feed} feeding modes and $N_{\text{feed}} + 1$ batch modes). During batch mode, there is no input feed and the process is only governed by (10.2.97). On the other hand, the process is governed by both (10.2.97) and (10.2.98) during feeding mode. Thus,

$$\frac{d\boldsymbol{x}(t)}{dt} = \begin{cases} f^{\text{ferm}}(\boldsymbol{x(t)}, x_1(t-\gamma_1)), & \text{for batch mode,} \\ f^{\text{ferm}}(\boldsymbol{x(t)}, x_1(t-\gamma_1)) + f^{\text{feed}}(\boldsymbol{x(t)}, \zeta_i), & \text{for } i\text{th feeding mode,} \end{cases} \tag{10.2.102}$$

where ζ_i is the feeding rate during the ith feeding mode subject to the following boundedness constraints:

$$0043 \leq \zeta_i \leq 1.9266. \tag{10.2.103}$$

During the growth phase of the biomass, glycerol is being consumed. Since no new glycerol is added during the batch mode, the glycerol concentration will reduce and eventually it will become too low, and hence a switch into feeding mode is necessary. The corresponding switching condition is

$$x_2(t) - \zeta_{N_{\text{feed}}+1} = 0, \tag{10.2.104}$$

where $\zeta_{N_{\text{feed}}+1}$ is the lower switching concentration. This parameter is a decision parameter which is to be optimized. On the other hand, when the glycerol concentration becomes too high during feeding mode, cell growth is inhibited. Thus, the process must switch back into batch mode. The corresponding switching condition is

$$x_2(t) - \zeta_{N_{\text{feed}}+2} = 0, \tag{10.2.105}$$

where $\zeta_{N_{\text{feed}}+2}$ is the upper switching concentration. This is another parameter to be optimized.

The bounded constraints on $\zeta_{N_{\text{feed}}+1}$ and $\zeta_{N_{\text{feed}}+2}$ are

$$50 \leq \zeta_{N_{\text{feed}}+1} \leq 260, \quad 300 \leq \zeta_{N_{\text{feed}}+2} \leq 600. \tag{10.2.106}$$

Note that the system parameters in this example appear explicitly in the dynamics and switching conditions. Thus, to apply Theorem 10.2.2, we replace the system parameters with auxiliary state variables $x_{4+k}(t)$, $k = 1, \ldots, N_{\text{feed}} + 2$, where

$$\frac{dx_{4+k}(t)}{dt} = 0, t > 0, \tag{10.2.107}$$

and

$$x_{4+k}(t) = \zeta_k, t \leq 0. \tag{10.2.108}$$

Let δ_{ki} denote the Kronecker delta function and let $\partial \boldsymbol{x}$ and $\partial \tilde{x}_1$ denote the partial differentiation with respect to $\boldsymbol{x}(t)$ and $x_1(t-\gamma_1)$, respectively. Then, the variational system corresponding to ζ_k is

$$\frac{d\Lambda_k(t)}{dt} = \begin{cases} \frac{\partial f^{\text{ferm}}}{\partial x}\Lambda_k(t) + \frac{\partial f^{\text{ferm}}}{\partial \tilde{x}_1}\Lambda_{k1}(t-\gamma_1), \text{ batch mode}, \\ \frac{\partial f^{\text{ferm}}}{\partial x}\Lambda_k(t) + \frac{\partial f^{\text{ferm}}}{\partial \tilde{x}_1}\Lambda_{k1}(t-\gamma_1) \\ +\frac{\partial f^{\text{feed}}}{\partial x}\Lambda_k(t) + \delta_{ki}\frac{\partial f^{\text{feed}}}{\partial u}, \qquad i\text{th, feeding mode}, \end{cases} \tag{10.2.109a}$$

with jump conditions

$$\Lambda_k\left(\tau_i^+\right) = \begin{cases} \Lambda_k(\tau_i^-) - \frac{\partial \tau_i}{\partial \zeta_k} f^{\text{feed}}(\boldsymbol{x}(\tau_i), \zeta_i), \text{ if mode } i = \text{ batch} \\ \Lambda_k(\tau_i^-) + \frac{\partial \tau_i}{\partial \zeta_k} f^{\text{feed}}(\boldsymbol{x}(\tau_i), \zeta_i), \text{ if mode } i = \text{ feeding}. \end{cases} \tag{10.2.109b}$$

Furthermore, for a switch from batch mode to feeding mode,

$$\frac{\partial \tau_i}{\partial \zeta_k} = \begin{cases} (1-\Lambda_{k2}(\tau_i^-)) \div f_2^{\text{ferm}}(\boldsymbol{x}(\tau_i), x_1(\tau_i-\gamma_1)), \text{ if } k = N_{\text{feed}}+1, \\ -\Lambda_{k2}(\tau_i^-) \div f_2^{\text{ferm}}(\boldsymbol{x}(\tau_i), x_1(\tau_i-\gamma_1)), \qquad \text{otherwise}. \end{cases} \tag{10.2.110}$$

Similarly, for a switch from feeding mode to batch mode,

$$\frac{\partial \tau_i}{\partial \zeta_k} = \begin{cases} (1-\Lambda_{k2}(\tau_i^-)) \\ \div\{f_2^{\text{ferm}}(x(\tau_i), x_1(\tau_i-\gamma_1)) + f_2^{\text{feed}}(x(\tau_i), \zeta_i)\}, \text{ if } k = N_{\text{feed}}+2, \\ -\Lambda_{k2}(\tau_i^-) \\ \div\{f_2^{\text{ferm}}(x(\tau_i), x_1(\tau_i-\gamma_1)) + f_2^{\text{feed}}(x(\tau_i), \zeta_i)\}, \text{ otherwise}. \end{cases} \tag{10.2.111}$$

From the boundedness constraints specified in (10.2.106), we note that $\zeta_{N_{\text{feed}}+1} < \zeta_{N_{\text{feed}}+2}$. Thus, Assumption 10.2.2 is satisfied at all feasible points. For Assumption 10.2.3, we require

$$0 \neq \begin{cases} -q_2(x_2(\tau_i), x_3(\tau_i))x_1(\tau_i-\gamma_1), \text{ if mode } i = \text{batch}, \\ -q_2(x_2(\tau_i), x_3(\tau_i))x_1(\tau_i-\gamma_1) \\ +\frac{\zeta_i(rc_{s0}-x_2(\tau_i))}{x_4(\tau_i)}, \qquad \text{if mode } i = \text{ feeding}. \end{cases} \tag{10.2.112}$$

This condition is clearly satisfied with reasons given below: For batch mode, the right hand side of (10.2.112) is always non-zero because in practice both q_2 and x_1 are non-zero. For feeding mode, the right hand side of (10.2.112) is also non-zero because, during feeding mode, the glycerol loss from natural fermentation (first term) is dominated by the glycerol addition from the input feed (second term).

Since x_4 is non-decreasing and, for biologically meaningful trajectories, $\mu(\cdot,\cdot)$ is bounded, the linear growth assumption is also clearly valid.

The initial function ϕ for the dynamics (10.2.102) was obtained by applying cubic spline interpolation to the experimental data reported in [197]. As in [151], the terminal time for the fermentation process is taken as $T = 24.16$ hours. The upper bound for the number of feeding modes is chosen as $N_{\text{feed}} = 48$. Our goal is to maximize the concentration of 1,3-PD at the terminal time. Thus, the dynamic optimization problem is: Choose the parameters ζ_k, $k = 1, \ldots, N_{\text{feed}} + 2$, such that the cost functional $-x_3(T)$ is minimized subject to the boundedness constraints (10.2.103) and (10.2.104).

This dynamic optimization problem is solved using a FORTRAN program that implements the gradient-based optimization procedure in Section 10.2.4. In this program, NLPQLP [223] is used to perform the optimization iterations (optimality check and line search), and LSODAR [92] is used to solve the differential equations. Our gradient-based optimization strategy generates critical points satisfying local optimality conditions. However, the solution obtained is not necessarily a global optimal solution. Thus, it is necessary to repeat the optimization process from different starting points so that a better estimate of the global solution is obtained. We performed 100 test runs, where each run starts from a different randomly selected initial point. The average optimal cost over all runs is: -977.12854, and the best result, which is obtained on run 73, is: -986.16815. For this control strategy, there are 8 switches (5 batch modes and 4 feeding modes). The control parameters and the respective mode durations are listed in Table 10.2.2. The optimal state trajectories are shown in Figure 10.2.1 Due to the dilution effect from the new input feed, the concentrations of biomass and 1,3-PD decrease during the feeding modes. The control strategy listed in Table 10.2.2 is essentially a state feedback strategy. It produces more 1,3-PD (an increase of 5.789%) when compared with the time-dependent switching strategy reported in [151]. Furthermore, it requires far fewer switches. For the method reported in [151], it requires over 1000 switches.

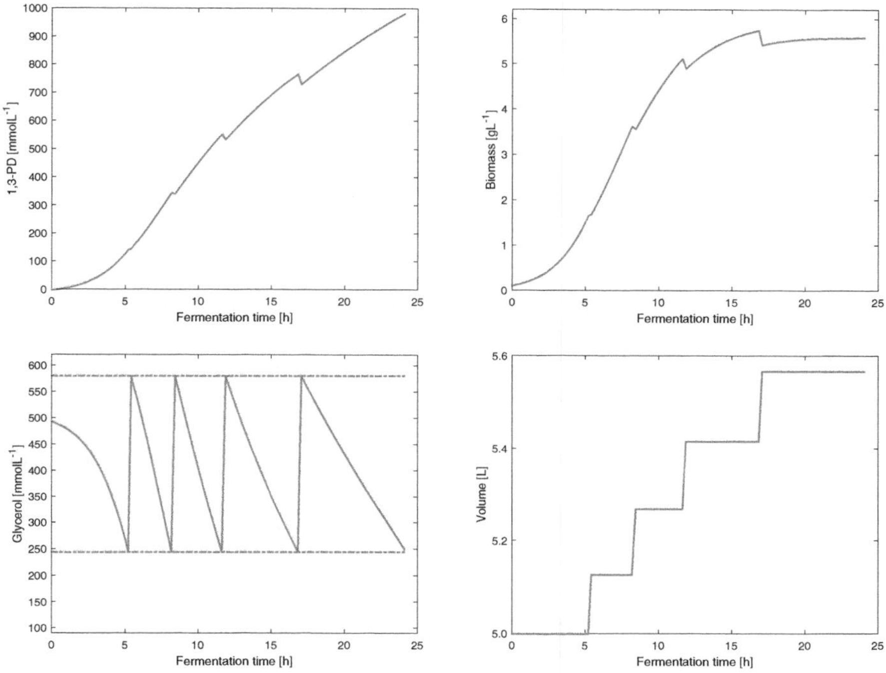

Fig. 10.2.1: Optimal state trajectory obtained for numerical example

Table 10.2.2: Optimal control parameters

Parameter	ζ_1	ζ_2	ζ_3	ζ_4	ζ_{49}	ζ_{50}
Optimal values	1.62662	1.36951	1.45283	1.64830	245.76512	581.35390

Remark 10.2.5 *$\zeta_1, \ldots, \zeta_4$ are the optimal feeding rates and ζ_{49} and ζ_{50} are the optimal switching concentrations. The optimal values of $\zeta_5, \ldots, \zeta_{48}$ are irrelevant because they represent the feeding rates after the terminal time.*

10.3 Min-Max Optimal Control

In this section, a class of min-max optimal control of linear continuous dynamical systems with uncertainty and quadratic terminal constraints is considered. It is shown that this min-max optimal control problem is transformed into a form, which can be solved via solving a sequence of semi-definite programming problems. This section is basically from [287].

10.3.1 Problem Statement

Consider the continuous linear uncertain dynamical system defined on the time horizon $[0, T]$ given below:

$$\frac{d\boldsymbol{x}(t)}{dt} = A(t)\,\boldsymbol{x}(t) + B(t)\,\boldsymbol{u}(t) + C(t)\,\boldsymbol{w}(t)\,,\ t \in [0, T]\,, \qquad (10.3.1)$$
$$\boldsymbol{x}(0) = \boldsymbol{x}^{o},$$

where $\boldsymbol{x}(t) \in \mathbb{R}^n$ is the state vector, $\boldsymbol{u}(t) \in \mathbb{R}^m$ is the control vector, $\boldsymbol{w}(t) \in \mathbb{R}^r$ is the disturbance, and $A(t) = [a_{i,j}(t)]$, $B(t) = [b_{i,j}(t)]$, $C(t) = [c_{i,j}(t)]$ are matrices with appropriate dimensions, and $\boldsymbol{x}^{o}$ is a given initial condition.

The cause of the disturbance $\boldsymbol{w}(t)$ in (10.3.1) can be due to the changes in external environment or errors in measurement. As in [23, 30], we assume that the disturbance $\boldsymbol{w}(t) \in \mathcal{W}_\rho$, where $\mathcal{W}_\rho$ is a L_2-norm bounded set defined by

$$\mathcal{W}_\rho = \left\{\boldsymbol{w} \in L_2([0, T], \mathbb{R}^r) : \|\boldsymbol{w}\|^2 \le \rho^2\right\}, \qquad (10.3.2)$$

where $\|\boldsymbol{w}\|^2 = \int_0^T (\boldsymbol{w}(t))^\top \boldsymbol{w}(t)\,dt$ and $\rho > 0$ is a given positive constant. Furthermore, the control $\boldsymbol{u}$ is restricted to be chosen from $\mathcal{U}_\delta$ which is a L_2-norm bounded set defined by

$$\mathcal{U}_\delta = \left\{\boldsymbol{u} \in L_2([0, T], \mathbb{R}^m) : \|\boldsymbol{u}\|^2 \le \delta^2\right\}, \qquad (10.3.3)$$

where $\|\boldsymbol{u}\|^2 = \int_0^T (\boldsymbol{u}(t))^\top \boldsymbol{u}(t)\,dt$ and $\delta > 0$ is a given positive constant.

The cost functional is considered to be a quadratic function given below:

$$J(\boldsymbol{u}, \boldsymbol{w}) = \int_0^T \left\{(\boldsymbol{x}(t))^\top Q(t)\,\boldsymbol{x}(t) + (\boldsymbol{u}(t))^\top R(t)\,\boldsymbol{u}(t)\right\} dt, \qquad (10.3.4)$$

where $Q(t) = [q_{i,j}(t)]$ and $R(t) = [r_{i,j}(t)]$, $t \in [0, T]$ are matrices with appropriate dimensions. It is assumed that the following terminal state constraint is satisfied:

$$\|\boldsymbol{x}(T) - \boldsymbol{x}^*\| \le \gamma,\ \forall\, \boldsymbol{w} \in \mathcal{W}_\rho, \qquad (10.3.5)$$

where $\boldsymbol{x}^*$ is the desired terminal state and $\gamma > 0$ is a given constant. Any element $\boldsymbol{u} \in \mathcal{U}_\delta$ is called a feasible control if the terminal state constraint (10.3.5) is satisfied.

We may now state our optimal control problem formally as follows.

Problem (P_3). Given the dynamical system (10.3.1) and the terminal state constraint (10.3.5), find a control $\boldsymbol{u} \in \mathcal{U}_\delta$ such that the worst-case performance $J(\boldsymbol{u}, \boldsymbol{w})$ is minimized over $\mathcal{U}_\delta$, i.e., finding a control $\boldsymbol{u} \in \mathcal{U}_\delta$ such that it solves the following min-max optimal control problem:

$$\min_{\boldsymbol{u} \in \mathcal{U}_\delta} \max_{\boldsymbol{w} \in \mathcal{W}_\rho} J(\boldsymbol{u}, \boldsymbol{w}). \qquad (10.3.6)$$

Clearly, without the presence of the disturbance $\boldsymbol{w}$ in (10.3.1), Problem (P_3) is easy to solve by existing optimal control methods, such as the control parametrization method used in conjunction with the time scaling transformation presented in earlier chapters. However, in the presence of disturbances, Problem (P_3) becomes much more complicated. In this section, we shall develop a computational scheme for solving this min-max optimal control problem. To continue, assume that the matrices $A(t)$, $B(t)$, $C(t)$, $Q(t)$ and $R(t)$ are all continuous on $[0,T]$, i.e., each of their elements is a continuous function in $[0,T]$. Furthermore, let $Q(t)$ and $R(t)$ be, respectively, positive semi-definite and positive definite for each $t \in [0,T]$.

10.3.2 Some Preliminary Results

To continue, we need some preliminary results, particularly, the following two lemmas, known as the S-Lemma and the Schur complement of a block matrix. The S-Lemma was developed independently in several different contexts [205, 263] and it has applications in control theory, linear algebra and mathematical optimization. It gives conditions under which a particular quadratic inequality is a consequence of another quadratic inequality. The statement of the lemma is given below.

Lemma 10.3.1 *Let M_1 and M_2 be symmetric matrices, $\boldsymbol{v}^1$ and $\boldsymbol{v}^2$ be vectors and α_1 and α_2 be real numbers. Suppose that there is a vector $\boldsymbol{x}^0$ such that the following strict inequality:*

$$\left(\boldsymbol{x}^0\right)^\top M_1\boldsymbol{x}^0 + 2(\boldsymbol{v}^1)^\top\boldsymbol{x}^0 + \alpha_1 < 0$$

holds. Then, the following implication:

$$(\boldsymbol{x})^\top M_1\boldsymbol{x} + 2(\boldsymbol{v}^1)^\top\boldsymbol{x} + \alpha_1 \le 0 \Rightarrow (\boldsymbol{x})^\top M_2\boldsymbol{x} + 2(\boldsymbol{v}^2)^\top\boldsymbol{x} + \alpha_2 \le 0$$

holds if and only if there exists a non-negative number λ such that

$$\lambda\begin{bmatrix} M_1 & v^1 \\ \left(v^1\right)^\top & \alpha_1 \end{bmatrix} - \begin{bmatrix} M_2 & v^2 \\ \left(v^2\right)^\top & \alpha_2 \end{bmatrix}$$

is positive semi-definite.

The following lemma is called the Schur complement of a block matrix.

Lemma 10.3.2 *Suppose that A, B, C, D are, respectively, $n\times n$, $n\times m$, $m\times p$ and $m\times m$ matrices, and that D is invertible. Let*

$$M = \begin{bmatrix} A & B \\ C & D \end{bmatrix}$$

such that M is a $(n+m)\times\ (p+m)$ matrix. Then, the Schur complement of the block D of the matrix M is the $n\times\ n$ matrix defined by

$$A - B\,(D)^{-1}\,C.$$

If A is invertible, then the Schur complement of the block A of the matrix M is the $m\times\ m$ matrix defined by

$$D - C\,(A)^{-1}\,B.$$

Furthermore, suppose that either A or D is singular. Then, by replacing a generalized inverse for the inverses of $D - C\,(A)^{-1}\,B$ and $A - B\,(D)^{-1}\,C$ gives rise to the generalized Schur complement.

To close this subsection, we briefly introduce some basic concepts and results related to the controllability of linear systems. Consider the following linear time invariant system:

$$\frac{d\boldsymbol{x}(t)}{dt} = A\boldsymbol{x}(t) + B\boldsymbol{u}(t)$$

$$\boldsymbol{y}(t) = C\boldsymbol{x}(t) + D\boldsymbol{u}(t),$$

where A, B, C and D are, respectively, $n\times n$, $n\times r$, $p\times n$ and $p\times r$ matrices.

We say that the linear time invariant system is controllable if and only if the pair (A, B) is controllable, namely the $n\times nr$ controllability matrix

$$C = \begin{bmatrix} B\ AB\ A^2B\ \cdots\ A^{n-1}B \end{bmatrix}$$

has rank n. The following statements are equivalent:

1. The pair (A, B) is controllable.
2. The $n\times n$ matrix

$$\begin{aligned} W_c(t) &= \int_0^t \exp\{A\tau\}BB^\top \exp\left\{A^\top \tau\right\} d\tau \\ &= \int_0^t \exp\{A(t-\tau)\}BB^\top \exp\left\{A^\top (t-\tau)\right\} d\tau \end{aligned}$$

   is non-singular for any $t > 0$.
3. The $n\times nr$ controllability matrix

$$C = \begin{bmatrix} B\ AB\ A^2B\ \cdots\ A^{n-1}B \end{bmatrix}$$

   has rank n.
4. The $n\times (n+r)$ matrix

$$[A - \lambda I\ B]$$

   has full row rank at every eigenvalue λ of A.

Now suppose that all the eigenvalues of A have negative real parts (A is stable), and that the unique solution of the Lyapunov equation

$$AW_c + W_c (A)^\top = -B (B)^\top$$

is positive definite. Then, the system is controllable. The solution is called the Controllability Gramian and can be expressed as

$$W_c = \int_0^t \exp\{A\tau\} BB^\top \exp\left\{A^\top \tau\right\} d\tau.$$

We now consider the following linear time varying system:

$$\frac{d\boldsymbol{x}(t)}{dt} = A(t)\boldsymbol{x}(t) + B(t)\boldsymbol{u}(t)$$

$$\boldsymbol{y}(t) = C(t)\boldsymbol{x}(t)$$

where A, B and C are, respectively, $n \times n$, $n \times r$, $p \times n$ matrices. Then the system $(A(t), B(t))$ is controllable at time t_0 if and only if there exists a finite time $t_1 > t_0$ such that the $n \times n$ matrix, also known as the controllability Gramian, defined by

$$W_c(t_0, t_1) = \int_{t_0}^{t_1} \Phi(t_1, \tau) B(\tau) \left(B(\tau)\right)^\top \left(\Phi(t_1, \tau)\right)^\top d\tau$$

is non-singular, where $\Phi(t, \tau)$ is the state transition matrix of the following system:

$$\frac{d\boldsymbol{x}(t)}{dt} = A(t)\boldsymbol{x}(t).$$

Note that for the Controllability Gramian $W_c(t_0, t_1)$, it holds that

$$W_c(t_0, t_1) = W_c(t, t_1) + \Phi(t_1, t) W_c(t_0, t) \left(\Phi(t_1, t)\right)^\top.$$

10.3.3 Problem Approximation

Let $\Phi(t, \tau)$ be the transition matrix of (10.3.1). For each $\boldsymbol{u}$ and $\boldsymbol{w}$, define

$$\mathcal{T}_0(\boldsymbol{u}) = \Phi(T, 0)\, \boldsymbol{x}^0 + \int_0^T \Phi(T, \tau)\, B(\tau)\, \boldsymbol{u}(\tau)\, d\tau,$$

$$\mathcal{T}_1(\boldsymbol{x}) = \left(Q(t)\right)^{1/2} \Phi(t, 0)\, \boldsymbol{x}^0 + \int_0^t \left(Q(t)\right)^{1/2} \Phi(t, \tau)\, B(\tau)\, \boldsymbol{u}(\tau)\, d\tau,$$

$$\mathcal{F}_0(\boldsymbol{w}) = \int_0^T \Phi(T,\tau)\, C(\tau)\, \boldsymbol{w}(\tau)\, d\tau,$$

$$\mathcal{F}_1(\boldsymbol{w}) = \int_0^t (Q(t))^{1/2}\, \Phi(t,\tau)\, C(\tau)\, \boldsymbol{w}(\tau)\, d\tau.$$

When no confusion can arise, the same notation $\langle \cdot, \cdot \rangle$ is used as the inner product in L_2 as well as in $\mathbb{R}^n$. The cost functional (10.3.4) and the terminal state constraint (10.3.5) can be rewritten as

$$J(\boldsymbol{u}, \boldsymbol{w}) = \langle \mathcal{T}_1(\boldsymbol{u}) + \mathcal{F}_1(\boldsymbol{w}), \mathcal{T}_1(\boldsymbol{u}) + \mathcal{F}_1(\boldsymbol{w}) \rangle + \left\langle (R)^{\frac{1}{2}}\, \boldsymbol{u}, (R)^{\frac{1}{2}}\, \boldsymbol{u} \right\rangle, \quad (10.3.7)$$

and

$$\langle \mathcal{T}_0(\boldsymbol{u}) + \mathcal{F}_0(\boldsymbol{w}) - x^*, \mathcal{T}_0(\boldsymbol{u}) + \mathcal{F}_0(\boldsymbol{w}) - \boldsymbol{x}^* \rangle \leq \gamma^2, \ \forall \boldsymbol{w} \in \mathcal{W}_\rho. \quad (10.3.8)$$

We have the following theorem.

Theorem 10.3.1 *$\mathcal{T}_1$ is a linear bounded operator from $L_2([0,T], \mathbb{R}^m)$ to $L_2([0,T], \mathbb{R}^n)$. Suppose that $\{\boldsymbol{u}^n\} \subset \mathcal{U}_\delta$ and $\boldsymbol{u}^n \rightharpoonup \boldsymbol{u}$. Then, $\mathcal{T}_1(\boldsymbol{u}^n) \to \mathcal{T}_1(\boldsymbol{u})$, where $\rightharpoonup$ and $\to$ stand for convergence in the weak topology and strong topology in L_2 space, respectively.*

Proof. It is easy to show that $\mathcal{T}_1$ is a bounded linear operator from $L_2([0,T], \mathbb{R}^m)$ to $L_2([0,T], \mathbb{R}^n)$. Now, suppose that $\{\boldsymbol{u}^n\} \subset \mathcal{U}_\delta$ and $\boldsymbol{u}^n \rightharpoonup \boldsymbol{u}$. Define

$$\Phi(t,\tau) = \begin{cases} \Phi(t,\tau), & \text{if } \tau \leq t, \\ 0_{n\times n}, & \text{else.} \end{cases}$$

Clearly, $\Phi(t,\cdot)$ is a continuous function except at $\tau = t$. Then, for each given $t \in [0,T]$, we have

$$\begin{aligned} &\lim_{n\to\infty} (\mathcal{T}_1(\boldsymbol{u}^n) - \mathcal{T}_1(\boldsymbol{u}))(t) \\ &= \int_0^t (Q(t))^{1/2}\, \Phi(t,\tau)\, B(\tau)\, (\boldsymbol{u}^n(\tau) - \boldsymbol{u}(\tau))\, d\tau \\ &= \int_0^T (Q(t))^{1/2}\, \Phi(t,\tau)\, B(\tau)\, (\boldsymbol{u}^n(\tau) - \boldsymbol{u}(\tau))\, d\tau = 0. \end{aligned}$$

On the other hand, since $\{\boldsymbol{u}^n\} \subset \mathcal{U}_\delta$, and $(Q(t))^{1/2}\, \Phi(t,\tau)\, B(\tau)$ is continuous with respect to $(t,\tau) \in [0,T] \times [0,t]$, we can easily show that there exists a constant K_1 such that

$$\left| \int_0^t \left[(Q(t))^{1/2}\, \Phi(t,\tau)\, B(\tau)\, \boldsymbol{u}^n(\tau) \right]_i d\tau \right| \leq K_1$$

for each $t \in [0,T]$, where $\int_0^t \left[(Q(t))^{1/2} \Phi(t,\tau) B(\tau) \boldsymbol{u}^n(\tau) \right]_i d\tau$ denotes the i-th element of $\int_0^t (Q(t))^{1/2} \Phi(t,\tau) B(\tau) \boldsymbol{u}^n(\tau) d\tau$. Now, by Theorem A.1.10 (Lebesgue Dominated Convergence Theorem), it follows that $\mathcal{T}_1(\boldsymbol{u}^n) \to \mathcal{T}_1(\boldsymbol{u})$.

Theorem 10.3.2 *$\mathcal{F}_1$ is a linear bounded operator from $L_2([0,T],\mathbb{R}^r)$ to $L_2([0,T],\mathbb{R}^n)$. Suppose that $\{\boldsymbol{w}^n\} \subset \mathcal{W}_\rho$ and $\boldsymbol{w}^n \rightharpoonup \boldsymbol{w}$. Then, $\mathcal{F}_1(\boldsymbol{w}^n) \to \mathcal{F}_1(\boldsymbol{w})$.*

Proof. The proof is similar to that given for Theorem 10.3.1.

Note that since both $\mathcal{T}_1$ and $\mathcal{F}_1$ are bounded operators, it follows readily that there exists a constant $K_2 > 0$, such that $0 \le J(\boldsymbol{u},\boldsymbol{w}) < K_2$, $\forall(\boldsymbol{u},\boldsymbol{w}) \in \mathcal{U}_\delta \times \mathcal{W}_\rho$. For a given $\boldsymbol{u} \in \mathcal{U}_\delta$, let $\{\boldsymbol{w}^n\} \subset \mathcal{W}_\rho \subset L_2([0,T],\mathbb{R}^r)$ be a maximizing sequence, meaning that $J(\boldsymbol{u},\boldsymbol{w}^n) \to \sup J(\boldsymbol{u},\boldsymbol{w})$. Since $L_2([0,T],\mathbb{R}^r)$ space is reflexive, $\mathcal{W}_\rho$, which is a ball with radius ρ, is weakly sequentially compact (see Remark A.1.4), there exists a subsequence of the sequence $\{\boldsymbol{w}^n\}$, which is denoted by the original sequence, such that $\boldsymbol{w}^n \rightharpoonup \boldsymbol{w}(u) \in \mathcal{W}_\rho$. Thus, by Theorem 10.3.1, it follows that $\mathcal{F}_1(\boldsymbol{w}^n) \to \mathcal{F}_1(\boldsymbol{w}(\boldsymbol{u}))$, and hence $J(\boldsymbol{u},\boldsymbol{w}^n) \to J(\boldsymbol{u},\boldsymbol{w}(\boldsymbol{u}))$. Clearly, $J(\boldsymbol{u},\boldsymbol{w}(\boldsymbol{u})) = \sup J(\boldsymbol{u},\boldsymbol{w})$. More precisely,

$$J(\boldsymbol{u},\boldsymbol{w}(\boldsymbol{u})) = \max_{\boldsymbol{w}\in\mathcal{W}_\rho} J(\boldsymbol{u},\boldsymbol{w}).$$

Clearly, $\boldsymbol{w}(\boldsymbol{u})$ may be not unique. However, they share the same cost function value $\max_{\boldsymbol{w}\in\mathcal{W}_\rho} J(\boldsymbol{u},\boldsymbol{w})$. For Problem (P), we have the following theorem.

Theorem 10.3.3 *Problem (P) has a unique solution $\boldsymbol{u}^* \in \mathcal{U}_\delta$ such that $\boldsymbol{u}^*$ satisfies (10.3.5) and*

$$J(\boldsymbol{u}^*,\boldsymbol{w}(\boldsymbol{u}^*)) = \min_{\boldsymbol{u}\in\mathcal{U}_\delta} \max_{\boldsymbol{w}\in\mathcal{W}_\rho} J(\boldsymbol{u},\boldsymbol{w}). \tag{10.3.9}$$

Proof. Suppose that $\boldsymbol{u}^n \rightharpoonup \boldsymbol{u}$. Since $R(t)$ is positive definite, then $\int_0^T \boldsymbol{u}(t) R(t) \boldsymbol{u}(t)\, dt$ is strictly convex. Hence,

$$\int_0^T (\boldsymbol{u}^n(t))^\top R(t) \boldsymbol{u}^n(t)\, dt \ge \int_0^T (\boldsymbol{u}(t))^\top R(t) \boldsymbol{u}(t)\, dt$$
$$+ 2\int_0^T (\boldsymbol{u}(t))^\top R(t) (\boldsymbol{u}^n(t) - \boldsymbol{u}(t))\, dt. \tag{10.3.10}$$

Thus,

$$\int_0^T (\boldsymbol{u}(t))^\top R(t) \boldsymbol{u}(t)\, dt \le \underline{\lim}_{n\to\infty} \int_0^T (\boldsymbol{u}^n(t))^\top R(t) \boldsymbol{u}^n(t)\, dt.$$

Note that, for any $\boldsymbol{u}^n$,

$$\begin{aligned} &\langle \mathcal{T}_1(\boldsymbol{u}^n) + \mathcal{F}_1(w(\boldsymbol{u}^n)), \mathcal{T}_1(\boldsymbol{u}^n) + \mathcal{F}_1(\boldsymbol{w}(\boldsymbol{u}^n))\rangle & (10.3.11) \\ &\quad = \max_{\boldsymbol{w}\in\mathcal{W}_\rho} \langle \mathcal{T}_1(\boldsymbol{u}^n) + \mathcal{F}_1(\boldsymbol{w}), \mathcal{T}_1(\boldsymbol{u}^n) + \mathcal{F}_1(w)\rangle \\ &\quad \geq \langle \mathcal{T}_1(\boldsymbol{u}^n) + \mathcal{F}_1(w(\boldsymbol{u})), \mathcal{T}_1(\boldsymbol{u}^n) + \mathcal{F}_1(\boldsymbol{w}(u))\rangle. & (10.3.12) \end{aligned}$$

By Theorem 10.3.1, $\mathcal{T}_1(\boldsymbol{u}^n) \to \mathcal{T}_1(\boldsymbol{u})$ when $\boldsymbol{u}^n \rightharpoonup \boldsymbol{u}$. Taking limit inferior on both sides of (10.3.12), it gives

$$\begin{aligned} &\underline{\lim}_{n\to\infty} \langle \mathcal{T}_1(\boldsymbol{u}^n) + \mathcal{F}_1(\boldsymbol{w}(\boldsymbol{u}^n)), \mathcal{T}_1(\boldsymbol{u}^n) + \mathcal{F}_1(\boldsymbol{w}(\boldsymbol{u}^n))\rangle & (10.3.13) \\ &\quad \geq \langle \mathcal{T}_1(\boldsymbol{u}) + \mathcal{F}_1(\boldsymbol{w}(\boldsymbol{u})), \mathcal{T}_1(\boldsymbol{u}) + \mathcal{F}_1(\boldsymbol{w}(\boldsymbol{u}))\rangle. \end{aligned}$$

By (10.3.10) and (10.3.13), it shows that $J(\boldsymbol{u}, \boldsymbol{w}(\boldsymbol{u}))$ is weakly sequentially lower semicontinuous, i.e.,

$$J(\boldsymbol{u}, \boldsymbol{w}(u)) \leq \underline{\lim}_{n\to\infty} J(\boldsymbol{u}^n, \boldsymbol{w}(\boldsymbol{u}^n)) \text{ as } \boldsymbol{u}^n \rightharpoonup \boldsymbol{u}.$$

Furthermore, we can show that $\max_{\boldsymbol{w}\in\mathcal{W}_\rho} \langle \mathcal{T}_1(\boldsymbol{u}) + \mathcal{F}_1(\boldsymbol{w}), \mathcal{T}_1(\boldsymbol{u}) + \mathcal{F}_1(\boldsymbol{w})\rangle$ is convex in $\boldsymbol{u}$. Since (10.3.18) is convex in $\boldsymbol{u}$ and $\int_0^T (\boldsymbol{u}(t))^\top R(t)\, \boldsymbol{u}(t)\, dt$ is strictly convex in $\boldsymbol{u}$, it follows that Problem (P) is strictly convex. Thus, the conclusion of the theorem holds.

Theorem 10.3.4 *Define*

$$S = \int_0^T \Phi(T,\tau)C(\tau)\,(C(\tau))^\top\,(\Phi(T,\tau))^\top\,d\tau \tag{10.3.14}$$

and let $\lambda_{\max}(S)$ be the largest eigenvalue of S. Suppose that S is invertible. If Problem (P) has a feasible control, then

$$\lambda_{\max}(S)\rho^2 \leq \gamma^2. \tag{10.3.15}$$

Furthermore, the terminal state constraint (10.3.8) is equivalent to the following constraint:

$$\begin{pmatrix} I & \mathcal{T}_0(\boldsymbol{u}) - x^* & I \\ (\mathcal{T}_0(\boldsymbol{u}) - x^*)^\top & \gamma^2 - \rho^2\varsigma & 0 \\ I & 0 & \varsigma\,(S)^{-1} \end{pmatrix} \succeq 0, \tag{10.3.16}$$

whereA $\succeq 0$ means that the matrix A is constrained to be positive semi-definite.

Proof. We first show that

$$\mathcal{F}_0(\mathcal{W}_\rho) = H, \tag{10.3.17}$$

where $\mathcal{F}_0(\mathcal{W}_\rho) = \{\mathcal{F}_0(\boldsymbol{w}) : \boldsymbol{w} \in \mathcal{W}_\rho\}$ and $H = \{\boldsymbol{h} \in \mathbb{R}^n : (\boldsymbol{h})^\top (S)^{-1} \boldsymbol{h} \le \rho^2\}$.

For notational simplicity, let $G(\tau) = \Phi(T,\tau) C(\tau)$ and $(S)^{-1/2} G(\tau) = \left[(\boldsymbol{g}^1(\tau))^\top, \ldots, (\boldsymbol{g}^n(\tau))^\top\right]^\top$. Clearly, $\boldsymbol{g}^i \in L_2([0,T], \mathbb{R}^m)$, $i = 1, \ldots, n$. Then, for any $\boldsymbol{w} \in \mathcal{W}_\rho$,

$$\begin{aligned}(\mathcal{F}_0(\boldsymbol{w}))^\top (S)^{-1} \mathcal{F}_0(\boldsymbol{w}) &= \sum_{i=1}^n \langle \boldsymbol{g}^i, \boldsymbol{w}\rangle^2 \le \sum_{i=1}^n \|\boldsymbol{g}^i\|^2 \|\boldsymbol{w}\|^2 \\ &= \mathrm{Tr}\left(\int_0^T (S)^{-1/2} \Phi(T,\tau) C(\tau) (C(\tau))^\top (\Phi(T,\tau))^\top (S)^{-1/2} dt\right) \|\boldsymbol{w}\|^2 \le \rho^2.\end{aligned}$$

Thus, $\mathcal{F}_0(\mathcal{W}_\rho) \subset H$. On the other hand, for any $\boldsymbol{h} \in H$, define $\boldsymbol{w}(t) = G(t)(S)^{-1}\boldsymbol{h}$. Then,

$$\begin{aligned}\int_0^T (\boldsymbol{w}(t))^\top \boldsymbol{w}(t) dt &= \int_0^T (\boldsymbol{h})^\top (S)^{-1} (G(t))^\top G(t) (S)^{-1} \boldsymbol{h} dt \\ &= (\boldsymbol{h})^\top (S)^{-1} \boldsymbol{h} \le \rho^2,\end{aligned}$$

and

$$\int_0^T G(t) \boldsymbol{w}(t) dt = \int_0^T G^\top(t) G(t) (S)^{-1} \boldsymbol{h} dt = \boldsymbol{h}.$$

Thus, $H \subset \mathcal{F}_0(\mathcal{W}_\rho)$. Therefore, $\mathcal{F}_0(\mathcal{W}_\rho) = H$. In light of (10.3.17), the constraint (10.3.5) is equivalent to the constraint

$$\langle \mathcal{T}_0(\boldsymbol{u}) + \boldsymbol{h} - \boldsymbol{x}^*, \mathcal{T}_0(\boldsymbol{u}) + \boldsymbol{h} - \boldsymbol{x}^*\rangle \le \gamma^2, \ \forall\, \boldsymbol{h} \in H. \tag{10.3.18}$$

For the simplicity of symbol, let $V_{\boldsymbol{u}} = \mathcal{T}_0(\boldsymbol{u}) - \boldsymbol{u}^*$. Inequality constraint (10.3.18) can be written as follows:

$$\begin{aligned}&(\boldsymbol{h})^\top \boldsymbol{h} + 2t (V_{\boldsymbol{u}})^\top \boldsymbol{h} + t^2 (V_{\boldsymbol{u}})^\top V_{\boldsymbol{u}} \\ &\quad \le \gamma^2 t^2, \ \forall\ (\boldsymbol{h}, t) \in \left\{(\boldsymbol{h}, t) : (\boldsymbol{h})^\top (S)^{-1} \boldsymbol{h} \le \rho^2 t^2\right\}.\end{aligned}$$

The above inequality can be rewritten as

$$\begin{aligned}&\begin{pmatrix} t \\ \boldsymbol{h}\end{pmatrix}^\top \begin{pmatrix} \gamma^2 - (V_{\boldsymbol{u}})^\top V_{\boldsymbol{u}} & -(V_{\boldsymbol{u}})^\top \\ -V_{\boldsymbol{u}} & -I\end{pmatrix} \begin{pmatrix} t \\ \boldsymbol{h}\end{pmatrix} \ge 0, \\ &\quad \forall (t, \boldsymbol{h}) : \begin{pmatrix} t \\ \boldsymbol{h}\end{pmatrix}^\top \begin{pmatrix} \rho^2 & 0 \\ 0 & -(S)^{-1}\end{pmatrix} \begin{pmatrix} t \\ \boldsymbol{h}\end{pmatrix} \ge 0.\end{aligned} \tag{10.3.19}$$

By Lemma 10.3.1, (10.3.3) holds if and only if there exists a $\varsigma \ge 0$ such that

$$\begin{pmatrix} \gamma^2 - (V_{\boldsymbol{u}})^\top V_{\boldsymbol{u}} - \varsigma\rho^2 & -(V_{\boldsymbol{u}})^\top \\ -V_{\boldsymbol{u}} & \varsigma (S)^{-1} - I \end{pmatrix}$$

$$= \begin{pmatrix} \gamma^2 - \varsigma\rho^2 & 0 \\ 0 & \varsigma (S)^{-1} \end{pmatrix} - \begin{pmatrix} (V_{\boldsymbol{u}})^\top \\ I \end{pmatrix} \begin{pmatrix} V_{\boldsymbol{u}} & I \end{pmatrix} \succeq 0 \tag{10.3.20}$$

which can be equivalently rewritten as (10.3.16) by Lemma 10.3.2. The inequality (10.3.3) implies that $\varsigma S^{-1} \succeq I$ and $\gamma^2 - \varsigma\rho^2 \geq 0$. Thus, $\lambda_{\max}(S)\rho^2 \leq \gamma^2$. This completes the proof.

The matrix S is the Controllability Gramian of the pair $(A(\cdot), C(\cdot))$. Thus, S is invertible if and only if the pair $(A(\cdot), C(\cdot))$ is controllable. If the system (10.3.1) is time-invariant, then S is invertible if and only if $(C, CA, \ldots, CA^{n-1})$ is a full rank matrix. In what follows, we assume that S is invertible.

By Theorem 10.3.4, (10.3.8) and (10.3.16) are equivalent. Problem (P_3) is equivalent to the problem defined by (10.3.6) and (10.3.16). Clearly, the problem defined by (10.3.6) and (10.3.16) is a convex infinite dimensional optimization problem. Although the maximization with respect to $\boldsymbol{w} \in \mathcal{W}_\rho$ is required to be carried out only in $J(\boldsymbol{u}, \boldsymbol{w})$ without involving constraint (10.3.16), the problem defined by (10.3.6) and (10.3.16) is still much too complicated to be solved analytically. It is inevitable to resort to numerical methods.

Suppose that $\{\boldsymbol{\gamma}^i\}_{i=1}^\infty$ and $\{\boldsymbol{\psi}^i\}_{i=1}^\infty$ are orthonormal bases (OB) of $L_2([0,T], \mathbb{R}^m)$ and $L_2([0,T], \mathbb{R}^r)$, respectively. Now we approximate $\boldsymbol{u}$ and $\boldsymbol{w}$ by the truncated OB as $\boldsymbol{u}(t) = \Gamma_N(t)\boldsymbol{\theta}$ and $\boldsymbol{w}(t) = \Psi_N(t)\boldsymbol{\vartheta}$, where N is the truncated number, $\Gamma_N(t) = [\boldsymbol{\gamma}^1(t), \boldsymbol{\gamma}^2(t), \ldots, \boldsymbol{\gamma}^N(t)]$, $\Psi_N(t) = [\boldsymbol{\psi}^1(t), \boldsymbol{\psi}^2(t), \ldots, \boldsymbol{\psi}^N(t)]$, $\boldsymbol{\theta} = [\theta_1, \theta_2, \ldots, \theta_N]^T \in \mathbb{R}^N$ and $\boldsymbol{\vartheta} = [\vartheta_1, \vartheta_2, \ldots, \vartheta_N]^T \in \mathbb{R}^N$. Denote $\boldsymbol{\Xi}_N = \{\boldsymbol{\theta} \in \mathbb{R}^N : \|\boldsymbol{\theta}\| \leq \delta\}$, $\mathcal{U}_N = \{\Gamma_N(t)\boldsymbol{\theta} : \boldsymbol{\theta} \in \boldsymbol{\Xi}_N\}$, $\boldsymbol{\Pi}_N = \{\boldsymbol{\vartheta} \in \mathbb{R}^N : \|\boldsymbol{\vartheta}\| \leq \rho\}$ and $\mathcal{W}_N = \{\Psi_N(t)\boldsymbol{\vartheta} : \boldsymbol{\theta} \in \boldsymbol{\Pi}_N\}$. Then, the parametrized finite dimensional optimization problem can be stated as: Find a control $\boldsymbol{u} \in \mathcal{U}_\delta \cap \mathcal{U}_N$ such that the cost function $\max_{\boldsymbol{w} \in \mathcal{W}_\rho \cap \mathcal{W}_N} J(\boldsymbol{u}, \boldsymbol{w})$ is minimized subject to the constraint (10.3.16). Let this problem be referred to as Problem (P_3^N). Following a similar proof as that given for Theorem 10.3.4, we have the following theorem.

Theorem 10.3.5 *Problem (P_3^N) is equivalent to the following semi-definite programming problem:*

$$\min_{\boldsymbol{\theta} \in \boldsymbol{\Xi}_N, t_1, t_2, \varsigma_1 \geq 0, \varsigma_2 \geq 0} t_1 + t_2 + 2(q_N)^\top \boldsymbol{\theta} + \mu_0 \tag{10.3.21}$$

subject to

$$\begin{pmatrix} I & (P_N)^{1/2}\boldsymbol{\theta} \\ (\boldsymbol{\theta})^\top (P_N)^{1/2} & t_2 \end{pmatrix} \succeq 0, \tag{10.3.22}$$

$$\begin{pmatrix} t_1 - \varsigma_1\rho^2 & -(\boldsymbol{\theta})^\top Q_N - \boldsymbol{r}_N \\ -Q_N\boldsymbol{\theta} - (\boldsymbol{r}_N)^\top & \varsigma_1 I - R_N \end{pmatrix} \succeq 0, \tag{10.3.23}$$

$$\begin{pmatrix} I & V\boldsymbol{x}^\circ - \boldsymbol{x}^* + V_N\boldsymbol{\theta} & I \\ (V\boldsymbol{x}^\circ - \boldsymbol{x}^* + V_N\boldsymbol{\theta})^\top & \gamma^2 - \rho^2\varsigma_2 & 0 \\ I & 0 & \varsigma_2 (S)^{-1} \end{pmatrix} \succeq 0, \tag{10.3.24}$$

where the explicit expressions of P_N, Q_N, R_N, $\boldsymbol{q}_N$, $\boldsymbol{r}_N$, V, V_N and μ_0 are given as below:

$$\begin{aligned} P_N = &\int_0^T \int_0^t (\Phi_{B,\Gamma}(t,\tau))^\top d\tau Q(t) \int_0^t \Phi_{B,\Gamma}(t,\tau) d\tau dt \\ &+ \int_0^T (\Gamma_N(t))^\top R(t) \Gamma_N(t)\, dt, \end{aligned}$$

$$\begin{aligned} Q_N &= \int_0^T \int_0^t (\Phi_{B,\Gamma}(t,\tau))^\top d\tau Q(t) \int_0^t \Phi_{C,\Psi}(t,\tau) d\tau dt, \\ R_N &= \int_0^T \left\{ \left[\int_0^t (\Phi_{C,\Psi}(t,\tau))^\top d\tau \right] Q(t) \left[\int_0^t \Phi F_{C,\Psi}(t,\tau) d\tau \right] \right\} dt, \\ V_N &= (P)^{1/2} \int_0^T \Phi(T,t) B(t) \Gamma_N(t)\, dt, \ V = (P)^{1/2} \Phi(T,0), \end{aligned}$$

$$\begin{aligned} \boldsymbol{q}_N &= \int_0^T \left\{ \left[\int_0^t (\Gamma_N(\tau))^\top B^\top(\tau) (\Phi(t,\tau))^\top d\tau \right] Q(t) \left[F(t,0)\boldsymbol{x}^\circ \right] \right\} dt, \\ \boldsymbol{r}_N &= \int_0^T \left\{ \left[\int_0^t (\Psi_N(\tau))^\top C^\top(\tau) (\Phi(t,\tau))^\top d\tau \right] Q(t) \left[\Phi(t,0)\boldsymbol{x}^\circ \right] \right\} dt, \\ \mu_0 &= (\boldsymbol{x}^\circ)^\top \left(\int_0^T (\Phi(t,0))^\top Q(t) \Phi(t,0)\, dt \right) \boldsymbol{x}^\circ, \end{aligned}$$

$$\begin{aligned} \Phi_{B,\Gamma}(t,\tau) &= \Phi(t,\tau) B(\tau) \Gamma_N(\tau) \\ \Phi_{C,\Psi}(t,\tau) &= \Phi(t,\tau) C(\tau) \Psi_N(\tau). \end{aligned}$$

Proof. Clearly,

$$J(\boldsymbol{u}^N, \boldsymbol{w}^N) = (\boldsymbol{\theta})^\top P_N\boldsymbol{\theta} + 2(\boldsymbol{\theta})^\top Q_N\boldsymbol{\vartheta} + (\boldsymbol{\vartheta})^\top R_N\boldsymbol{\vartheta} + 2(q_N)^\top\boldsymbol{\theta} + 2(\boldsymbol{r}_N)^\top\boldsymbol{\vartheta} + \mu_0.$$

Then, $\min_{\boldsymbol{\theta}\in\boldsymbol{\Xi}_N} \max_{\boldsymbol{\vartheta}\in\boldsymbol{\Pi}_N} J(u^N, w^N)$ can be equivalently rewritten as

$$\min \quad t_1 + t_2 + 2(q_N)^\top \boldsymbol{\theta} + \mu_0 \tag{10.3.25}$$

$$\text{subject to} \quad (\boldsymbol{\theta})^\top P_N\boldsymbol{\theta} \le t_2 \tag{10.3.26}$$

$$(\boldsymbol{\vartheta})^\top R_N\boldsymbol{\vartheta} + 2\left[(\boldsymbol{\theta})^\top Q_N + \boldsymbol{r}_N\right]^\top \boldsymbol{\vartheta} \le t_1, \ \forall \boldsymbol{\vartheta} \in \boldsymbol{\Pi}_N. \tag{10.3.27}$$

Using a similar argument as in the proof of Theorem 10.3.4, it follows that (10.3.27) is equivalent to (10.3.23). Thus, the conclusion of the theorem follows readily.

In view of Theorem 10.3.5 the solution of Problem (P_3^N) can be obtained through solving a SDP problem defined by (10.3.21)–(10.3.24).

For the solution procedure of SDP problems, the readers are referred to [163, 263]. The next theorem shows the relation between Problem (P_3) and Problem (P_3^N).

Theorem 10.3.6 *Let $\boldsymbol{u}^* \in U_\delta$ and $\boldsymbol{\theta}_N^* \in \Xi_N$ be the optimal solution of Problem (P_3) and the optimal solution of Problem (P_3^N), respectively. Let $\boldsymbol{u}_N^*(t) = \Gamma_N(t)\boldsymbol{\theta}_N^*$. Suppose that $\omega(\boldsymbol{u}^*)$ and $\boldsymbol{\omega}(\boldsymbol{u}_N^*)$ are such that*

$$J(\boldsymbol{u}^*, \omega(\boldsymbol{u}^*)) = \max_{\boldsymbol{\omega} \in \mathcal{W}_\rho} J(\boldsymbol{u}^*, \boldsymbol{\omega}), \tag{10.3.28}$$

and

$$J(\boldsymbol{u}_N^*, \omega(\boldsymbol{u}_N^*)) = \max_{\boldsymbol{\omega} \in \mathcal{W}_\rho \cap V_N} J(\boldsymbol{u}_N^*, \boldsymbol{\omega}),$$

respectively. Let $\boldsymbol{\omega}^* \in \mathcal{W}_\rho$ and $\boldsymbol{\omega}_N^* \in \mathcal{W}_\rho \cap V_N$ be such that

$$J(\boldsymbol{u}^*, \omega^*) = J(\boldsymbol{u}^*, \boldsymbol{\omega}(\boldsymbol{u}^*)), \tag{10.3.29}$$

and

$$J(\boldsymbol{u}_N^*, \omega_N^*) = J(\boldsymbol{u}_N^*, \boldsymbol{\omega}(\boldsymbol{u}_N^*)), \tag{10.3.30}$$

respectively, Then,

(i) $\lim_{N\to\infty} J(\boldsymbol{u}_N^*, \boldsymbol{\omega}_N^*) = J(\boldsymbol{u}^*, \boldsymbol{\omega}(\boldsymbol{u}^*))$; *and*
(ii) $\boldsymbol{u}_N^* \rightharpoonup \boldsymbol{u}^*$ *as* $N \to \infty$.

Proof. Note that $\boldsymbol{u}^*$ is the optimal solution of problem (P_3) and $\boldsymbol{\theta}_N^* \in \Xi_N$ is the optimal solution of Problem (P_3^N). Let $\boldsymbol{u}_N^*(t) = \Gamma_N(t)\boldsymbol{\theta}_N^*$. Suppose that for $\boldsymbol{u}^* \in \mathcal{U}_\delta$, and let $\omega(\boldsymbol{u}^*)$ be such that it satisfies (10.3.28). Note that $\omega(\boldsymbol{u}^*)$ may not be unique, but gives rise to the same value of $\max_{\boldsymbol{\omega}\in\mathcal{W}_\rho} J(\boldsymbol{u}^*, \boldsymbol{\omega})$. Let $\boldsymbol{\omega}^*$ be one of these maximizers. Similarly, let $\boldsymbol{\omega}_N^*$ be one of the maximizers $\boldsymbol{\omega}(\boldsymbol{u}_N^*)$. Without loss of generality, we suppose that $\boldsymbol{u}_N^* \rightharpoonup \widehat{\boldsymbol{u}}$ and $\boldsymbol{\omega}_N^* \rightharpoonup \widehat{\boldsymbol{\omega}}$. Let $\boldsymbol{u}^{N,*}$ and $\boldsymbol{\omega}^{N,*}$ denote, respectively, the projection of $\boldsymbol{u}^*$ onto U_N and $\boldsymbol{\omega}^*$ onto V_N. Then, $\boldsymbol{u}^{N,*} \to \boldsymbol{u}^*$ and $\boldsymbol{\omega}^{N,*} \to \boldsymbol{\omega}^*$. Thus,

$$\begin{aligned} J(\boldsymbol{u}_N^*, \boldsymbol{\omega}_N^*) &= \min_{\boldsymbol{u} \in U_N \cap \mathcal{U}_\rho} J(\boldsymbol{u}, \boldsymbol{\omega}_N^*) \\ &\le J\left(\boldsymbol{u}^{N,*}, \boldsymbol{\omega}_N^*\right) \to J(\boldsymbol{u}^*, \widehat{\boldsymbol{\omega}}) \le J(\boldsymbol{u}^*, \boldsymbol{\omega}^*). \end{aligned} \tag{10.3.31}$$

On the other hand,

$$J(\boldsymbol{u}^*, \boldsymbol{\omega}^*) = \min_{\boldsymbol{u} \in \mathcal{U}_\rho} J(\boldsymbol{u}, \boldsymbol{\omega}^*) \le J(\widehat{\boldsymbol{u}}, \boldsymbol{\omega}^*) \le \varliminf_{N\to\infty} J\left(\boldsymbol{u}_N^*, \boldsymbol{\omega}^{N,*}\right)$$

$$\leq \varliminf_{N\to\infty} \max_{\omega\in V_N\cap\mathcal{W}_\rho} J(\boldsymbol{u}_N^*, \boldsymbol{\omega}) = \varliminf_{N\to\infty} J(\boldsymbol{u}_N^*, \boldsymbol{\omega}_N^*). \tag{10.3.32}$$

Therefore,

$$\lim_{N\to\infty} J(\boldsymbol{u}_N^*, \boldsymbol{\omega}_N^*) = J(\boldsymbol{u}^*, \boldsymbol{\omega}(\boldsymbol{u}^*)) \tag{10.3.33}$$

We shall show that $\boldsymbol{u}_N^* \rightharpoonup \boldsymbol{u}^*$ by contradiction. Suppose that it is false. Then, there exists a subsequence $\{\boldsymbol{u}_{N_k}^*\}$ of $\{\boldsymbol{u}_N^*\}$ and a subsequence $\{\boldsymbol{\omega}_{N_k}^*\}$ of $\{\boldsymbol{\omega}(\boldsymbol{u}_N^*)\}$ such that

$$\boldsymbol{u}_{N_k}^* \rightharpoonup \widetilde{\boldsymbol{u}} \neq \boldsymbol{u}^* \text{ and } \boldsymbol{\omega}_{N_k}^* \rightharpoonup \widetilde{\boldsymbol{\omega}}.$$

Let $\boldsymbol{\omega}_{\widetilde{\boldsymbol{u}}}$ be one of the maximizers $\boldsymbol{\omega}(\widetilde{\boldsymbol{u}})$. Then, by virtue of the uniqueness of the solution of Problem (P_3), it is clear that

$$J(\boldsymbol{u}^*, \boldsymbol{\omega}^*) < J(\widetilde{\boldsymbol{u}}, \boldsymbol{\omega}_{\widetilde{\boldsymbol{u}}}). \tag{10.3.34}$$

Let $\boldsymbol{\omega}^{N_k}$ be the projection of $\boldsymbol{\omega}_{\widetilde{\boldsymbol{u}}}$ onto V_{N_k}. Then, $\boldsymbol{\omega}^{N_k} \to \boldsymbol{\omega}_{\widetilde{\boldsymbol{u}}}$. Since $J(\boldsymbol{u}, \boldsymbol{\omega}(\boldsymbol{u}))$ is weakly sequentially lower semicontinuous, it follows from (10.3.33) that

$$\begin{aligned} J(\widetilde{\boldsymbol{u}}, \boldsymbol{\omega}_{\widetilde{\boldsymbol{u}}}) &\leq \varliminf_{k\to\infty} J\left(\boldsymbol{u}_{N_k}^*, \boldsymbol{\omega}^{N_k}\right) \leq \varliminf_{k\to\infty} \max_{\omega\in V_{N_k}\cap\mathcal{W}_\rho} J(\boldsymbol{u}_{N_k}^*, \boldsymbol{\omega}) \\ &= \varliminf_{k\to\infty} J(\boldsymbol{u}_{N_k}^*, \boldsymbol{\omega}_{N_k}^*) = J(\boldsymbol{u}^*, \boldsymbol{\omega}(\boldsymbol{u}^*), \end{aligned}$$

which is a contradiction to (10.3.34). Thus, $\boldsymbol{u}_N^* \rightharpoonup \boldsymbol{u}^*$.

Theorem 10.3.6 shows that the min-max optimal control problem (P_3) is approximated by a sequence of finite dimensional convex optimization problems (P_3^N). Then, an intuitive scheme to solve Problem (P_3) can be stated as: For a given tolerance $\varepsilon > 0$, we solve Problem (P_3^N) until $\left|J\left(\boldsymbol{u}^{N+1,*}\right) - J\left(\boldsymbol{u}^{N,*}\right)\right| \leq \varepsilon$.

Problem (P_3) with $\rho = 0$ is a standard optimal control problem without disturbance. Let it be referred to as Problem $(\overline{P_3})$. Similarly, we can solve Problem $(\overline{P_3})$ through solving a sequence of approximate optimal control problems, denoted by Problems $(\overline{P_3}^N)$, by restricting the feasible control $\boldsymbol{u}$ in $U_N \cap \mathcal{U}_\delta$. We have the following results.

Corollary 10.3.1 *Problem*$(\overline{P}^N)$ *is equivalent to the following SDP problem*

$$\min_{\theta\in\boldsymbol{\Xi}_N, t\geq 0} t + 2\left(\boldsymbol{q}^N\right)^\top \boldsymbol{\theta} + \mu_0 \tag{10.3.35}$$

subject to

$$\begin{pmatrix} I & (P_N)^{1/2}\boldsymbol{\theta} \\ (\boldsymbol{\theta})^\top (P_N)^{1/2} & t \end{pmatrix} \succeq 0, \tag{10.3.36}$$

$$\begin{pmatrix} I & V\boldsymbol{x}^{\mathrm{o}} + V_N\boldsymbol{\theta} - x^* \\ (V\boldsymbol{x}^{\mathrm{o}} + V_N\boldsymbol{\theta} - \boldsymbol{x}^*)^\top & \gamma \end{pmatrix} \succeq 0. \tag{10.3.37}$$

Remark 10.1. During the computation, both $\boldsymbol{u}$ and $\boldsymbol{w}$ are approximated by truncated orthonormal bases. Suppose that $\boldsymbol{u}_N^* = \Gamma_N(t)\boldsymbol{\theta}^*$. Then $\mathcal{I}_0\left(\boldsymbol{u}^{N,*}\right) = V\boldsymbol{x}^{\mathrm{o}} + V_N\boldsymbol{\theta}^*$, where $\boldsymbol{\theta}^*$ is the optimal solution of Problem (P_3^N). Since $\boldsymbol{\theta}^*$ satisfies the linear matrix inequality (10.3.24), $\boldsymbol{u}^{N,*}$ satisfies the linear matrix inequality (10.3.16). Thus, by Theorem 10.3.4, the terminal inequality constraint (10.3.5) holds for all $\boldsymbol{w} \in \mathcal{W}_\rho$. Thus, $\boldsymbol{u}^{N,*}$ is a feasible solution. This feature is not shared by the control parametrization method given in previous chapters. More specifically, if we directly approximate $\mathcal{U}_\delta$, $\mathcal{W}_\rho$ by $\mathcal{U}_N$ and $\mathcal{W}_N$, then we can also transform the approximated problem as a SDP which is different from that defined by (10.3.21)–(10.3.24). Let the solution obtained by this method be $\bar{\boldsymbol{u}}^{N,*}$. Then, the terminal inequality constraint (10.3.5) is only satisfied for those $\boldsymbol{w} \in \mathcal{W}_N \subset \mathcal{W}_\rho$, not for all $\boldsymbol{w} \in \mathcal{W}_\rho$. Thus, the approximate solution $\bar{\boldsymbol{u}}^{N,*}$ may be infeasible. For our proposed approach, the approximations of $\boldsymbol{u}$ and $\boldsymbol{w}$ only affect the computation of the cost function value (10.3.4). The feasibility of the terminal constraint (10.3.5) is maintained for all $\boldsymbol{w} \in \mathcal{W}_\rho$.

10.3.4 Illustrative Example

Consider a worst-case DC motor control. The mathematical model of a DC motor is expressed as two linear differential equations [65] as

$$\begin{aligned} V &= R_a i + L_a \frac{di}{dt} + C_e \omega \\ C_m i &= J_r \frac{d\omega}{dt} + \mu\omega + m, \end{aligned} \tag{10.3.38}$$

where V is the voltage applied to the rotor circuit, i is the current, ω is the rotation speed, m is the resistant torque reduced to the motor shaft, R_a and L_a are the resistance and the inductance of the circuit, J_r is the inertia moment, C_e and C_m are the constants of the motor, μ is the coefficient of viscous friction. Let $\boldsymbol{x}(t) = [x_1(t), x_2(t)]^T = [\omega(t), i(t)]^\top$, $u(t) = V(t)$, $w(t) = m(t)$. Then, (10.3.38) can be rewritten as

$$\frac{d\boldsymbol{x}(t)}{dt} = A\boldsymbol{x}(t) + Bu(t) + Cw(t),$$

where

$$A = \begin{bmatrix} -\frac{\mu}{J_r} & \frac{C_m}{J_r} \\ -\frac{C_e}{L_a} & -\frac{R_a}{L_a} \end{bmatrix}, \; B = \begin{bmatrix} 0 \\ \frac{1}{L_a} \end{bmatrix}, \; C = \begin{bmatrix} -\frac{1}{J_r} \\ 0 \end{bmatrix}.$$

Suppose that the initial condition is $\boldsymbol{x}(0) = [0,0]^\top$. We wish to find an optimal control to drive the system to a neighbourhood around the desired state $\boldsymbol{x}^*$ with reference to all disturbances $w \in \mathcal{W}_\rho$ such that the energy consumption is minimized. In this case, $Q = 0$ and $R = 1$ in (10.3.4). Suppose that the nominal parameters of the DC motor are given as: $\mu = 0.01$, $J_r = 0.028$, $C_m = 0.58$, $L_a = 0.16$, $C_e = 0.58$ and $R_a = 3$. Let $T = 1$, $\delta = 5$, $\rho = 0.01$, $\gamma = 0.2$, $x^* = [3,1]^\top$. The two orthonormal bases $\{\gamma_i\}_{i=1}^\infty$ and $\{\psi_i\}_{i=1}^\infty$ are taken as the normalized shifted Legendre polynomial, i.e.,

$$\gamma_i(t) = \psi_i(t) = \sqrt{2i+1}P_i(2t-1), \; i = 0,1,2,\ldots,$$

where $P_i(t)$ is the i-th order Legendre polynomial. During the simulation, SeDuMi [231] and YALMIP [163] are used to solve the SDP problem defined by (10.3.21)–(10.3.24) and the SDP problem defined by (10.3.35)–(10.3.37).

Note that system (10.3.38) is time invariant. By direct computation, it follows that the matrix $[C, CA]$ is of full rank. Thus, S in (10.3.14) is invertible. Using Simpson's Rule to compute it, we obtain

$$S = \begin{bmatrix} 176.8542 & -27.7387 \\ -27.7387 & 5.3628 \end{bmatrix}$$

and $\lambda_{\max}(S) = 181.2293$, which indicates that γ should be far larger than ρ.

We set the tolerance ε as 10^{-8} and $N = 5$ to start solving Problem (P_N). For $N = 10$, we have $\left|J\left(u^{N+1,*}\right) - J\left(u^{N,*}\right)\right| \le 10^{-8}$. So we stop the computation. Meanwhile, we have $\left|\bar{J}\left(u^{N+1,*}\right) - \bar{J}\left(u^{N,*}\right)\right| \le 10^{-8}$, where $\bar{J}\left(u^{N,*}\right)$ is the optimal cost function value of Problem $(\overline{P}^N)$. The cost function values obtained are given in Table 10.3.1, from which we see that the convergence for the case with disturbance and that without disturbance are very fast. Figure 10.3.1 depicts the nominal state $[x_1(t), x_2(t)]^T = [w(t), i(t)]^\top$ and Figure 10.3.2 shows that the optimal control u_{11}^* under worst case performance. The terminal constraint (10.3.5) holds for any $w \in \mathcal{W}_\rho$ is ensured by Theorem 10.3.4.

Table 10.3.1: The optimal cost of Problem $(\bar{P}_3^N)$ and the optimal cost of Problem (P_3^N)

	Optimal cost of Problem $(\bar{P}_3^N)$	Optimal cost of Problem (P_3^N)
$N = 5$	2.157374508	2.283417844
$N = 6$	2.157349390	2.283381916
$N = 7$	2.157334886	2.283363764
$N = 8$	2.157331918	2.283360165
$N = 9$	2.157331543	2.283359716
$N = 10$	2.157331514	2.283359680
$N = 11$	2.157331514	2.283359680

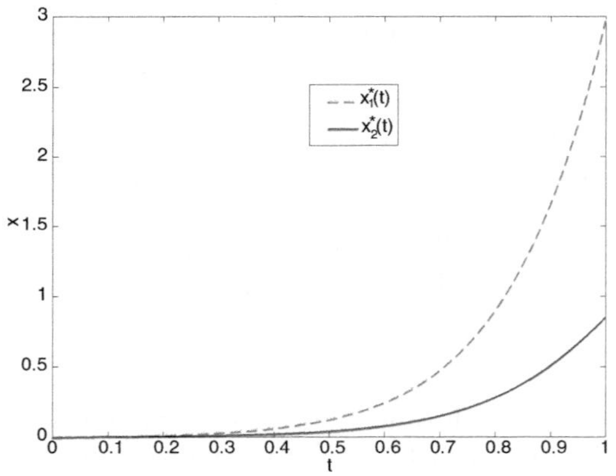

Fig. 10.3.1: The nominal state trajectories $[x_1^*(t), x_2^*(t)]^T$ of Problem $(\mathcal{P}^{11})$

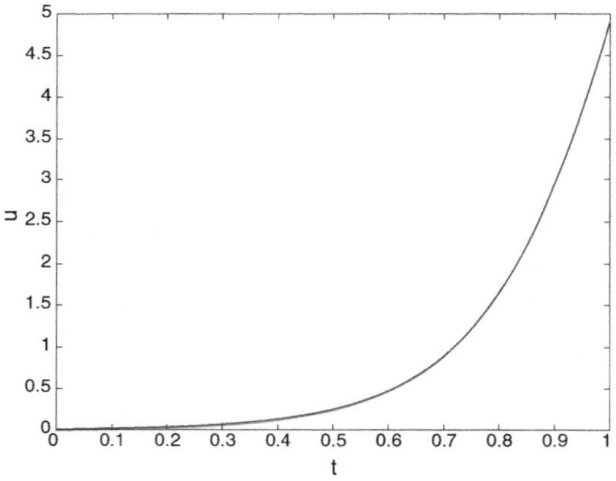

Fig. 10.3.2: The optimal control $\boldsymbol{u}^*$ of Problem $(\mathcal{P}^{11})$

10.4 Exercises

10.1. Show the validity of Equation (10.1.22).

10.2. Show the equivalence of Problem (P(p)) and Problem (Q(p)) (see Section 10.2.4).

10.3. Provide a proof of Theorem 10.2.2.

10.4. Provide a proof of Theorem 10.2.3.

10.5. Provide a proof of Theorem 10.2.4.

10.6. Show the validity of Equation (10.2.31).

10.7. Show the validity of Equation (10.2.28).

10.8. Show that $\mathcal{F}_1$ defined in Section 10.3.2 is bounded linear operator from $L_2([0,T],\mathbb{R}^r)$ to $L_2([0,T],\mathbb{R}^n)$.

10.9. In the proof of Theorem 10.3.1, show that there exists a constant $K > 0$, such that $0 \leq J(\boldsymbol{u},\boldsymbol{w}) < K$, $\forall(\boldsymbol{u},\boldsymbol{w}) \in \mathcal{U}_\delta \times \mathcal{W}_\rho$.

10.10. In the proof of Theorem 10.3.1, show that

$$\max_{\boldsymbol{w}\in\mathcal{W}_\rho} \langle \mathcal{T}_1(\boldsymbol{u}) + \mathcal{F}_1(\boldsymbol{w}), \mathcal{T}_1(\boldsymbol{u}) + \mathcal{F}_1(\boldsymbol{w})\rangle$$

is convex in $\boldsymbol{u}$.

10.11. Show the equivalence of Equations (10.3.17) and (10.3.18).

10.12. Show the validity of (10.3.3) by using S-Lemma.

10.13. Use Schur complement to show that (10.3.3) can be written as (10.3.16).

10.14. Give the proof of Theorem 10.3.4.

10.15. Give the proof of Corollary 10.3.1.

Chapter 11
Feedback Control

11.1 Introduction

In this chapter, we introduce two approaches to constructing suboptimal feedback controls for constrained optimal control problems. The first approach is known as the neighbouring extremals approach. The main references for this approach are [33, 107]. In this approach, we will present a solution method for constructing a first-order approximation of the optimal feedback control law for a class of optimal control problems governed by nonlinear continuous-time systems subject to continuous inequality constraints on the control and state. The control law constructed is in a state feedback form, and it is effective to small state perturbations caused by changes on initial conditions and/or modeling uncertainty. It has many potential applications, such as spacecraft guidance and control [140]. For illustration, a generalized Rayleigh problem with a mixed state and control constraint [24] is solved using the proposed method.

The second approach is to construct an optimal PID control for a class of optimal control problems subject to continuous inequality constraints and terminal equality constraint. The main reference for this approach is [132]. By applying the constraint transcription method and a local smoothing technique to these continuous inequality constraint functions, we construct the corresponding smooth approximate functions. We use the concept of the penalty function to append these smooth approximate functions to the cost function, forming a new cost function. Then, the constrained optimal PID control problem is approximated by a sequence of optimal parameter selection problems subject to only terminal equality constraint. Each of these optimal parameter selection problems can be viewed and hence solved as a nonlinear optimization problem. The gradient formulas of the new appended cost function and

K. L. Teo et al., *Applied and Computational Optimal Control*, Springer Optimization and Its Applications 171,
https://doi.org/10.1007/978-3-030-69913-0_11

the terminal equality constraint function are derived, and a reliable computation algorithm is given. The method proposed is used to solve a ship steering control problem.

11.2 Neighbouring Extremals

11.2.1 Problem Formulation

Consider a dynamic system governed by the following differential equations on the time horizon $(0, T]$:

$$\frac{d\boldsymbol{x}(t)}{dt} = \boldsymbol{f}(t, \boldsymbol{x}(t), \boldsymbol{u}(t)), \quad t \in (0, T], \qquad \boldsymbol{x}(0) = \boldsymbol{x}^0, \tag{11.2.1}$$

where $\boldsymbol{x}(t) \in \mathbb{R}^n$ and $\boldsymbol{u}(t) \in \mathbb{R}^r$ are, respectively, the state and control vectors; $\boldsymbol{f} : [0, T] \times \mathbb{R}^n \times \mathbb{R}^r \to \mathbb{R}^n$; T, $0 < T < \infty$, is the fixed terminal time and $\boldsymbol{x}^0 \in \mathbb{R}^n$ is a given vector.

For the control and state vectors, they are subject to the following continuous inequality constraints:

$$h_k(t, \boldsymbol{x}(t), \boldsymbol{u}(t)) \leq 0, \qquad t \in [0, T], \quad k = 1, \ldots, N \tag{11.2.2}$$

where $h_k : [0, T] \times \mathbb{R}^n \times \mathbb{R}^r \to \mathbb{R}$, $k = 1, \ldots, N$. Let

$$\boldsymbol{h} \triangleq [h_1, \ldots, h_N]^\top.$$

Furthermore, let $\mathcal{A}(t, \boldsymbol{x}, \boldsymbol{u}) \subseteq P \triangleq \{1, \ldots, N\}$ denote the index set of the active constraints in (11.2.2) at the point $(t, \boldsymbol{x}, \boldsymbol{u})$, that is,

$$\mathcal{A}(t, \boldsymbol{x}, \boldsymbol{u}) \triangleq \{k \in P : h_k(t, \boldsymbol{x}(t), \boldsymbol{u}(t)) = 0\}. \tag{11.2.3}$$

A measurable function $\boldsymbol{u} : [0, T] \to U \subset \mathbb{R}^r$ Satisfying (11.2.2) almost everywhere is called a feasible control. Let $\mathcal{F}$ denote the class of all such feasible controls.

Now, consider the following optimal control problem.

Problem (P_1) Subject to the dynamical system (11.2.1), find a feasible control $\boldsymbol{u} \in \mathcal{F}$ such that the cost functional

$$g_0(\boldsymbol{u}) \triangleq \Phi_0(\boldsymbol{x}(T)) + \int_0^T \mathcal{L}_0(t, \boldsymbol{x}(t), \boldsymbol{u}(t))dt \tag{11.2.4}$$

is minimized over $\mathcal{F}$, where $\Phi_0 : \mathbb{R}^n \to \mathbb{R}$ and $\mathcal{L}_0 : [0, T] \times \mathbb{R}^n \times \mathbb{R}^r \to \mathbb{R}$.

The following assumptions are assumed throughout the section.

Assumption 11.2.1 *$\boldsymbol{f}(t,\boldsymbol{x},\boldsymbol{u})$, $h_i(t,\boldsymbol{x},\boldsymbol{u})$, $i = 1,\ldots,N$, $\mathcal{L}_0(t,\boldsymbol{x},\boldsymbol{u})$ and $\Phi_0(\boldsymbol{x})$ are twice continuously differentiable with respect to each of their respective arguments.*

Assumption 11.2.2 *There exists a unique optimal solution $(\boldsymbol{x}^*,\boldsymbol{u}^*)$.*

Let H and L be, respectively, the Hamiltonian function and the augmented Hamiltonian function defined by

$$H(t,\boldsymbol{x},\boldsymbol{u},\boldsymbol{\lambda}) \triangleq \mathcal{L}_0(t,\boldsymbol{x},\boldsymbol{u}) + \boldsymbol{\lambda}^\top \boldsymbol{f}(t,\boldsymbol{x},\boldsymbol{u}), \tag{11.2.5}$$

$$L(t,\boldsymbol{x},\boldsymbol{u},\boldsymbol{\lambda},\boldsymbol{\rho}) \triangleq H(t,\boldsymbol{x},\boldsymbol{u},\boldsymbol{\lambda}) + \boldsymbol{\rho}^\top \boldsymbol{h}(t,\boldsymbol{x},\boldsymbol{u}), \tag{11.2.6}$$

where $\boldsymbol{\lambda}(t) \in \mathbb{R}^n$ is the costate and $\boldsymbol{\rho}(t) \in \mathbb{R}^p$ is the Lagrangian multiplier associated with constraints (11.2.2), where

$$\boldsymbol{\lambda} \triangleq [\lambda_1,\ldots,\lambda_n]^\top \quad and \quad \boldsymbol{\rho} \triangleq [\rho_1,\ldots,\rho_N]^\top.$$

Under Assumption 11.2.2, there exist multipliers $\boldsymbol{\lambda}^*(t) \in \mathbb{R}^n$ and $\boldsymbol{\rho}^*(t) \in \mathbb{R}^N$ such that the following necessary conditions are satisfied [33]:

$$\frac{d\boldsymbol{x}^*(t)}{dt} = \boldsymbol{f}(t,\boldsymbol{x}^*(t),\boldsymbol{u}^*(t)), \quad \boldsymbol{x}^*(0) = \boldsymbol{x}^0, \tag{11.2.7}$$

$$\left[\frac{d\lambda^*(t)}{dt}\right]^\top = -\frac{\partial L(t,\boldsymbol{x}^*(t),\boldsymbol{u}^*(t),\boldsymbol{\lambda}^*(t),\boldsymbol{\rho}^*(t))}{\partial \boldsymbol{x}}, \tag{11.2.8}$$

$$[\boldsymbol{\lambda}^*(T)]^\top = \frac{\partial \Phi_0(\boldsymbol{x}^*(T))}{\partial \boldsymbol{x}}, \tag{11.2.9}$$

$$\begin{aligned} 0 &= \frac{\partial L(t,\boldsymbol{x}^*(t),\boldsymbol{u}^*(t),\boldsymbol{\lambda}^*(t),\boldsymbol{\rho}^*(t))}{\partial \boldsymbol{u}} \\ &= \frac{\partial H(t,\boldsymbol{x}^*(t),\boldsymbol{u}^*(t),\boldsymbol{\lambda}^*(t))}{\partial \boldsymbol{u}} + [\boldsymbol{\rho}^*(t)]^\top \frac{\partial \boldsymbol{h}(t,\boldsymbol{x}^*(t),\boldsymbol{u}^*(t))}{\partial \boldsymbol{u}}, \end{aligned} \tag{11.2.10}$$

$$0 \geq h_i(t,\boldsymbol{x}^*(t),\boldsymbol{u}^*(t)); \qquad \rho_i^*(t) \geq 0, \quad i =,\ldots,N \tag{11.2.11}$$

$$0 = [\boldsymbol{\rho}^*(t)]^\top \boldsymbol{h}(t,\boldsymbol{x}^*(t),\boldsymbol{u}^*(t)). \tag{11.2.12}$$

In what follows, $(\boldsymbol{x}^*,\boldsymbol{u}^*)$ is also called the nominal solution, and a superscript '$*$' indicates that the corresponding function is evaluated along the nominal trajectory $(\boldsymbol{x}^*,\boldsymbol{u}^*)$.

Along this nominal solution, $(t_{k,1}^*,t_{k,2}^*) \subset [0,T]$, $k \in P$, is called an interior interval for the kth constraint if

$$h_k\left(t,\boldsymbol{x}^*(t),\boldsymbol{u}^*(t)\right) < 0 \; for \; all \; t \in (t_{k,1}^*,t_{k,2}^*)$$

and

$$h_k(t_{k,1}^*, \boldsymbol{x}^*(t_{k,1}^*), \boldsymbol{u}^*(t_{k,1}^*)) = h_k(t_{k,2}^*, \boldsymbol{x}^*(t_{k,2}^*), \boldsymbol{u}^*(t_{k,2}^*)) = 0.$$

$[t_{k,2}^*, t_{k,3}^*] \subset [0,T]$ is called a boundary interval if it is the maximal interval on which

$$h_k\left(t, \boldsymbol{x}^*(t), \boldsymbol{u}^*(t)\right) = 0 \ for \ all \ t \in [t_{k,2}^*, t_{k,3}^*]$$

$t_{k,1}^*$, $t_{k,2}^*$ and $t_{k,3}^*$ are called junction points. Let $\mathcal{T}_k^*$ denote the set of junction points $t_{k,j}^* \in [0,T]$ for $h_k(t, \boldsymbol{x}^*(t), \boldsymbol{u}^*(t)) \leq 0$. We assume that $(\boldsymbol{x}^*, \boldsymbol{u}^*)$ has the following regular structure.

Assumption 11.2.3 *The set $\mathcal{T}^* \triangleq \bigcup_{k \in P} \mathcal{T}_k^* = \{t_1^*, \ldots, t_M^*\}$ of all junction points is finite and $\mathcal{T}_k^* \bigcap \mathcal{T}_j^* = \emptyset$ for $k \neq j$, where $\emptyset$ denotes an empty set. Furthermore, there are no isolated touch points with the boundary for the nominal solution.*

In addition to this regular structure, we assume that the following strict complementarity condition and non-tangential junction condition are satisfied for $(\boldsymbol{x}^*, \boldsymbol{u}^*)$.

Assumption 11.2.4 *Let ρ_k^*, $k \in P$, denote the kth component of $\boldsymbol{\rho}^* \triangleq [\rho_1^*, \ldots, \rho_N^*]^\top$, and $\left[t_{k,j}^*, t_{k,j+1}^*\right] \subset [0,T]$ be any boundary interval for the constraint $h_k(t, \boldsymbol{x}^*, \boldsymbol{u}^*) \leq 0$. Then, $\rho_k^*(t) > 0$ for all $t \in \left(t_{k,j}^*, t_{k,j+1}^*\right)$.*

Assumption 11.2.5 *Let $\left[t_{k,j}^*, t_{k,j+1}^*\right] \subset [0,T]$, $k \in P$, be any boundary interval for the constraint $h_k(t, \boldsymbol{x}^*(t), \boldsymbol{u}^*(t)) \leq 0$. Then,*

$$\left.\frac{dh_k(t, \boldsymbol{x}(t), \boldsymbol{u}(t))}{dt}\right|_{t \to t_{k,j}^{*-}} \neq 0$$

and

$$\left.\frac{dh_k(t, \boldsymbol{x}(t), \boldsymbol{u}(t))}{dt}\right|_{t \to t_{k,j+1}^{*+}} \neq 0.$$

For convenience, let $\boldsymbol{u}\left(t_{k,j}^{*-}\right)$ and $\boldsymbol{u}\left(t_{k,j+1}^{*+}\right)$ denote, respectively, the limits of $\boldsymbol{u}^*(t)$ from the left at $t_{k,j}^*$ and right at $t_{k,j+1}^*$.

Let $\hat{\boldsymbol{h}}(t, \boldsymbol{x}^*(t), \boldsymbol{u}^*(t))$ and $\hat{\boldsymbol{\rho}}^*(t)$ denote, respectively, vectors composed of $h_k(t, \boldsymbol{x}^*(t), \boldsymbol{u}^*(t))$ and $\rho_k^*(t)$, where $k \in \mathcal{A}(t, \boldsymbol{x}^*(t), \boldsymbol{u}^*(t))$. Correspondingly, let $q(t) > 0$ be the number of the constraints in $\mathcal{A}(t, \boldsymbol{x}^*(t), \boldsymbol{u}^*(t))$. We have the following assumptions.

Assumption 11.2.6 *$\partial \hat{\boldsymbol{h}}(t, \boldsymbol{x}^*(t), \boldsymbol{u}^*(t))/\partial \boldsymbol{u}$ is of full row rank when*

$$\hat{\boldsymbol{h}}(t, \boldsymbol{x}^*(t), \boldsymbol{u}^*(t)) \neq \emptyset.$$

Assumption 11.2.7 *For all* $\boldsymbol{\gamma}(t) \in \ker(\partial\hat{\boldsymbol{h}}(t, \boldsymbol{x}^*(t), \boldsymbol{u}^*(t))/\partial\boldsymbol{u}) \setminus \{0\}$, *it holds that* $[\boldsymbol{\gamma}(t)]^\top (\partial^2 L^*(t, \boldsymbol{x}^*(t), \boldsymbol{u}^*(t)/\partial\boldsymbol{u}^2)\boldsymbol{\gamma}(t) > 0$. *where* $\ker(\cdot)$ *denotes the null space of a matrix, and*

$$\partial^2 L^*(t, \boldsymbol{x}^*(t), \boldsymbol{u}^*(t))/\partial\boldsymbol{u}^2 = (\partial/\partial\boldsymbol{u})^\top \left[\partial L^*(t, \boldsymbol{x}^*(t), \boldsymbol{u}^*(t))/\partial\boldsymbol{u}\right].$$

Now, treat $\boldsymbol{u}^*$, $\boldsymbol{\lambda}^*$ and $\boldsymbol{\rho}^*$ as functions of the nominal state $\boldsymbol{x}^*$. Let $\delta\boldsymbol{x}(t) \in \mathbb{R}^n$ be a perturbation of the nominal state $\boldsymbol{x}^*(t)$ such that $\delta\boldsymbol{x}(t) = \varepsilon\boldsymbol{\delta}(t)$ for some $\varepsilon \in \mathbb{R}^+$ and $\boldsymbol{\delta}(t) \in \mathcal{B}(n, 1)$, where $\mathcal{B}(n, s) \triangleq \{\boldsymbol{y} \in \mathbb{R}^n : |\boldsymbol{y}| \leq s\}$ and $|\cdot|$ denotes the usual Euclidean norm in $\mathbb{R}^n$.

We have the last assumption.

Assumption 11.2.8 $\boldsymbol{u}^*(\boldsymbol{x}^*)$, $\boldsymbol{\lambda}^*(\boldsymbol{x}^*)$ *and* $\boldsymbol{\rho}^*(\boldsymbol{x}^*)$ *are continuously differentiable with respect to* $\boldsymbol{x}^*$ *in a small neighbourhood of* $\boldsymbol{x}^*$.

Remark 11.2.1 *It is proved in [177] for optimal control problems depending on parameter* $\boldsymbol{\xi}$ *that if Assumptions 11.2.1–11.2.7, the controllability condition and the coercivity condition [177] are satisfied, then there exists a neighbourhood* G *of the nominal parameter* $\boldsymbol{\xi}^*$ *such that a local solution* $(\boldsymbol{x}, \boldsymbol{u})$ *and the associated Lagrangian multipliers* $\boldsymbol{\lambda}$ *and* $\boldsymbol{\rho}$ *exist for each* $\boldsymbol{\xi} \in G$. *All these functions* $\boldsymbol{x}$, $\boldsymbol{u}$, $\boldsymbol{\lambda}$ *and* $\boldsymbol{\rho}$ *are (Fréchet) differentiable with respect to* $\boldsymbol{\xi} \in G$ *satisfying* $\boldsymbol{x}(\boldsymbol{\xi}^*) = \boldsymbol{x}^*$, $\boldsymbol{u}(\boldsymbol{\xi}^*) = \boldsymbol{u}^*$, $\boldsymbol{\lambda}(\boldsymbol{\xi}^*) = \boldsymbol{\lambda}^*$ *and* $\boldsymbol{\rho}(\boldsymbol{\xi}^*) = \boldsymbol{\rho}^*$.

Consider neighbouring points $\boldsymbol{x} = \boldsymbol{x}^* + \varepsilon\boldsymbol{\delta}$, $\boldsymbol{u} = \boldsymbol{u}^*(\boldsymbol{x}^* + \varepsilon\boldsymbol{\delta})$, $\boldsymbol{\lambda} = \boldsymbol{\lambda}^*(\boldsymbol{x}^* + \varepsilon\boldsymbol{\delta})$ and $\boldsymbol{\rho} = \boldsymbol{\rho}^*(\boldsymbol{x}^* + \varepsilon\boldsymbol{\delta})$. For these neighbouring points to remain optimal, the following conditions are necessary [33]:

$$\frac{d\boldsymbol{x}(t)}{dt} = \boldsymbol{f}(t, \boldsymbol{x}(t), \boldsymbol{u}(t)), \qquad \boldsymbol{x}(0) = \boldsymbol{x}^0 + \varepsilon\boldsymbol{\delta}, \tag{11.2.13}$$

$$\left[\frac{d\boldsymbol{\lambda}(t)}{dt}\right]^\top = -\frac{\partial L(t, \boldsymbol{x}(t), \boldsymbol{u}(t), \boldsymbol{\lambda}(t), \boldsymbol{\rho}(t))}{\partial\boldsymbol{x}}, \quad [\boldsymbol{\lambda}(T)]^\top = \frac{\partial\Phi_0(\boldsymbol{x}(T))}{\partial\boldsymbol{x}}, \tag{11.2.14}$$

$$\begin{aligned} 0 &= \frac{\partial L(t, \boldsymbol{x}(t), \boldsymbol{u}(t), \boldsymbol{\lambda}(t), \boldsymbol{\rho}(t))}{\partial\boldsymbol{u}} \\ &= \frac{\partial H(t, \boldsymbol{x}(t), \boldsymbol{u}(t), \boldsymbol{\lambda}(t))}{\partial\boldsymbol{u}} + [\boldsymbol{\rho}(t)]^\top \frac{\partial\boldsymbol{h}(t, \boldsymbol{x}(t), \boldsymbol{u}(t))}{\partial\boldsymbol{u}}, \end{aligned} \tag{11.2.15}$$

$$0 \geq h_i(t, \boldsymbol{x}(t), \boldsymbol{u}(t)), \qquad \rho_i(t) \geq 0, \quad i = 1, \ldots, N \tag{11.2.16}$$

$$0 = [\boldsymbol{\rho}(t)]^\top \boldsymbol{h}(t, \boldsymbol{x}(t), \boldsymbol{u}(t)). \tag{11.2.17}$$

Now, the objective of this section can be stated formally as follows.

Problem ($\boldsymbol{P_{1F}}$) Given the optimal pair $(\boldsymbol{x}^*, \boldsymbol{u}^*)$ of **Problem ($\boldsymbol{P_1}$)**, construct a feedback control law expressed in the form of

$$\boldsymbol{u}(\boldsymbol{x}) \approx \boldsymbol{u}^* + \frac{\partial\boldsymbol{u}^*}{\partial\boldsymbol{x}}(\boldsymbol{x} - \boldsymbol{x}^*). \tag{11.2.18}$$

11.2.2 Construction of Suboptimal Feedback Control Law

Lemma 11.2.1 *Let $\boldsymbol{x} = \boldsymbol{x}^* + \varepsilon\boldsymbol{\delta}$ and $\boldsymbol{u} = \boldsymbol{u}^*(\boldsymbol{x}^* + \varepsilon\boldsymbol{\delta})$. If Assumptions 11.2.1, 11.2.3–11.2.5 and 11.2.8 are satisfied, then there exists an $\varepsilon_0 > 0$ such that for each $\varepsilon \in [0, \varepsilon_0]$,*

$$\frac{\partial \widehat{\boldsymbol{h}}^*}{\partial \boldsymbol{x}} + \frac{\partial \widehat{\boldsymbol{h}}^*}{\partial \boldsymbol{u}} \frac{\partial \boldsymbol{u}^*}{\partial \boldsymbol{x}} = 0. \tag{11.2.19}$$

Proof. From Assumptions 11.2.3–11.2.5, there exists a small $\varepsilon_1 > 0$ such that for each $\varepsilon \in [0, \varepsilon_1]$ the structure of the perturbed solution $(\boldsymbol{x}^* + \varepsilon\boldsymbol{\delta}, \boldsymbol{u}^*(\boldsymbol{x}^* + \varepsilon\boldsymbol{\delta}))$ is the same as that of $(\boldsymbol{x}^*, \boldsymbol{u}^*)$ [177]. Specifically, if the junction points of $(\boldsymbol{x}^*, \boldsymbol{u}^*)$ are such that $0 < t_1^* < t_2^* < \cdots < t_M^* < T$, then the solution $(\boldsymbol{x}^* + \varepsilon\boldsymbol{\delta}, \boldsymbol{u}^*(\boldsymbol{x}^* + \varepsilon\boldsymbol{\delta}))$ also has M junction points satisfying $0 < t_1 < t_2 < \cdots < t_M < T$ with t_i perturbed from t_i^*, $i = 1, \ldots, M$. Let $t_0^* = t_0 = 0$ and $t_{M+1}^* = t_{M+1} = T$. Then, if $\mathcal{A}^*(t) \triangleq \mathcal{A}^*(t, \boldsymbol{x}^*(t), \boldsymbol{u}^*(t)) = \mathcal{A}_i \subseteq P$ for all $t \in [t_i^*, t_{i+1}^*]$, $i = 0, \ldots, M$, it follows that $\mathcal{A}(t) = \mathcal{A}_i$ for all $t \in [t_i, t_{i+1}]$. Suppose that

$$\hat{\boldsymbol{h}}(t, \boldsymbol{x}^* + \varepsilon\boldsymbol{\delta}, \boldsymbol{u}^*(\boldsymbol{x}^* + \varepsilon\boldsymbol{\delta})) = 0, \quad \forall t \in [t_i, t_{i+1}].$$

From Assumption 11.2.8, there exists a small $\varepsilon_2 > 0$ such that, for each $\varepsilon \in [0, \varepsilon_2]$ and the perturbation $\delta\boldsymbol{x} = \varepsilon\boldsymbol{\delta}$, $\boldsymbol{u}^*(\boldsymbol{x}^*)$ is continuously differentiable with respect to $\boldsymbol{x}^*$. Let $\varepsilon_0 = \min\{\varepsilon_1, \varepsilon_2\}$. It follows from the continuity of $\hat{\boldsymbol{h}}$ at $\boldsymbol{x}^*$ and $\boldsymbol{u}^*(\boldsymbol{x}^*)$ that, for $\varepsilon \in [0, \varepsilon_0]$,

$$\left.\frac{d\hat{\boldsymbol{h}}\left(t, \boldsymbol{x}^*(t) + \varepsilon\boldsymbol{\delta}(t), \boldsymbol{u}^*(\boldsymbol{x}^*(t) + \varepsilon\boldsymbol{\delta}(t))\right)}{d\varepsilon}\right|_{\varepsilon=0} = 0.$$

Since $\boldsymbol{\delta h} \in \mathcal{B}(n, 1)$ is arbitrary, (11.2.19) follows.

Lemma 11.2.2 *Let $\boldsymbol{x} = \boldsymbol{x}^* + \varepsilon\boldsymbol{\delta}$ and $\boldsymbol{u} = \boldsymbol{u}^*(\boldsymbol{x}^* + \varepsilon\boldsymbol{\delta})$ with $\varepsilon \in [0, \varepsilon_0]$. If Assumptions 11.2.1–11.2.5 and 11.2.8 are satisfied, then*

$$\frac{\partial H^*}{\partial \boldsymbol{u}} \frac{\partial \boldsymbol{u}^*}{\partial \boldsymbol{x}} = \left[\boldsymbol{\rho}^*\right]^\top \frac{\partial \boldsymbol{h}^*}{\partial \boldsymbol{u}} \tag{11.2.20}$$

Proof. From the complementarity conditions (11.2.11) and (11.2.12), it follows that $\boldsymbol{\rho}_k^* = 0$ for $k \in P \backslash \mathcal{A}^*$. Then, from (11.2.19), we obtain

$$\left[\boldsymbol{\rho}^*\right]^\top \frac{\partial \boldsymbol{h}^*}{\partial \boldsymbol{x}} = -\left[\boldsymbol{\rho}^*\right]^\top \frac{\partial \boldsymbol{h}^*}{\partial \boldsymbol{u}} \frac{\partial \boldsymbol{u}^*}{\partial \boldsymbol{x}}. \tag{11.2.21}$$

Thus, the conclusion follows from (11.2.10).

Theorem 11.2.1 *Let* $\boldsymbol{x} = \boldsymbol{x}^* + \varepsilon\boldsymbol{\delta}$, $\boldsymbol{u} = \boldsymbol{u}^*(\boldsymbol{x}^* + \varepsilon\boldsymbol{\delta})$, $\boldsymbol{\lambda} = \boldsymbol{\lambda}^*(\boldsymbol{x}^* + \varepsilon\boldsymbol{\delta})$ *and* $\boldsymbol{\rho} = \boldsymbol{\rho}^*(\boldsymbol{x}^* + \varepsilon\boldsymbol{\delta})$ *with* $\varepsilon \in [0, \varepsilon_0]$. *If Assumptions 11.2.1–11.2.2 and 11.2.8 are satisfied, then*

$$0 = \left(\frac{\partial^2 H^*}{\partial \boldsymbol{u}^2} + \sum_{k\in\mathcal{A}^*} \rho_k^* \frac{\partial^2 h_k^*}{\partial \boldsymbol{u}^2}\right) \frac{\partial \boldsymbol{u}^*}{\partial \boldsymbol{x}} + \left[\frac{\partial \widehat{\boldsymbol{h}}^*}{\partial \boldsymbol{u}}\right]^\top \frac{\partial \widehat{\boldsymbol{\rho}}^*}{\partial \boldsymbol{x}} + \frac{\partial^2 H^*}{\partial \boldsymbol{x}\partial \boldsymbol{u}} + \sum_{k\in\mathcal{A}^*} \rho_k^* \frac{\partial^2 h_k^*}{\partial \boldsymbol{x}\partial \boldsymbol{u}} + \left[\frac{\partial \boldsymbol{f}^*}{\partial \boldsymbol{u}}\right]^\top \frac{\partial \boldsymbol{\lambda}^*}{\partial \boldsymbol{x}}. \tag{11.2.22}$$

Proof. For $(\boldsymbol{x}, \boldsymbol{u}, \boldsymbol{\lambda}, \boldsymbol{\rho})$ to remain optimal, Equation (11.2.15) holds with $\varepsilon \in [0, \varepsilon_0]$. Thus,

$$\begin{aligned}
0 &= \frac{d}{d\varepsilon}\left[\frac{\partial L}{\partial \boldsymbol{u}}\right]^\top\Bigg|_{\varepsilon=0} \\
&= \frac{d}{d\varepsilon}\left[\frac{\partial H}{\partial \boldsymbol{u}}\right]^\top\Bigg|_{\varepsilon=0} + \sum_{k=1}^{N}\left[\frac{d\rho_k}{d\varepsilon}\frac{\partial h_k}{\partial \boldsymbol{u}} + \rho_k \frac{d}{d\varepsilon}\left(\frac{\partial h_k}{\partial \boldsymbol{u}}\right)\right]^\top\Bigg|_{\varepsilon=0} \\
&= \Bigg\{\frac{\partial^2 H^*}{\partial \boldsymbol{x}\partial \boldsymbol{u}} + \frac{\partial^2 H^*}{\partial \boldsymbol{u}^2}\frac{\partial \boldsymbol{u}^*}{\partial \boldsymbol{x}} + \left[\frac{\partial \boldsymbol{f}^*}{\partial \boldsymbol{u}}\right]^\top \frac{\partial \boldsymbol{\lambda}^*}{\partial \boldsymbol{x}} \\
&\quad + \sum_{k=1}^{N}\left[\left[\frac{\partial h_k^*}{\partial \boldsymbol{u}}\right]^\top \frac{\partial \rho_k^*}{\partial \boldsymbol{x}} + \rho_k^* \frac{\partial^2 h_k^*}{\partial \boldsymbol{x}\partial \boldsymbol{u}} + \rho_k^* \frac{\partial^2 h_k^*}{\partial \boldsymbol{u}^2}\frac{\partial \boldsymbol{u}^*}{\partial \boldsymbol{x}}\right]\Bigg\}\boldsymbol{\delta}.
\end{aligned} \tag{11.2.23}$$

From Assumption 11.2.8, $\boldsymbol{\rho}$ is continuously differentiable with respect to $\boldsymbol{x}^*$ for $\varepsilon \in [0, \varepsilon_0]$. Then $\partial \rho_k^*/\partial \boldsymbol{x} = 0$ for $k \in P\backslash\mathcal{A}^*$. Thus, (11.2.23) holds for $\boldsymbol{\delta} \in \mathcal{B}(n, 1)$, which is arbitrary. Thus the validity of (11.2.22) follows readily.

Let $V^*(t, \boldsymbol{x}^*(t))$ be the optimal return function corresponding to $(\boldsymbol{x}^*, \boldsymbol{u}^*)$, which is defined by

$$V^*(t, \boldsymbol{x}^*(t)) \triangleq \Phi_0(\boldsymbol{x}^*(T)) + \int_t^T \mathcal{L}_0(t, \boldsymbol{x}^*(\tau), \boldsymbol{u}^*(\tau))d\tau. \tag{11.2.24}$$

By Assumption 11.2.2, it is known from [33] that V^* satisfies the Hamilton-Jacobi-Bellman equation:

$$\begin{aligned}
-\frac{\partial V^*}{\partial t} &= H(t, \boldsymbol{x}^*(t), \boldsymbol{u}^*(t), \boldsymbol{\lambda}^*(t)) \\
&= \mathcal{L}_0(t, \boldsymbol{x}^*(t), \boldsymbol{u}^*(t)) + [\boldsymbol{\lambda}^*(t)]^\top \boldsymbol{f}(t, \boldsymbol{x}^*(t), \boldsymbol{u}^*(t))
\end{aligned} \tag{11.2.25}$$

with $[\boldsymbol{\lambda}^*]^\top = \partial V^*/\partial \boldsymbol{x}$.

For the neighbouring points $\boldsymbol{x} = \boldsymbol{x}^* + \varepsilon\boldsymbol{\delta}$, $\boldsymbol{u} = \boldsymbol{u}^*(\boldsymbol{x}^* + \varepsilon\boldsymbol{\delta})$ and $\boldsymbol{\lambda} = \boldsymbol{\lambda}^*(\boldsymbol{x}^* + \varepsilon\boldsymbol{\delta})$ to remain optimal, the following equation must also be satisfied:

$$-\frac{\partial V}{\partial t} = H(t, \boldsymbol{x}(t), \boldsymbol{u}(t), \boldsymbol{\lambda}(t)) = \mathcal{L}_0(t, \boldsymbol{x}(t), \boldsymbol{u}(t)) + [\boldsymbol{\lambda}(t)]^\top \boldsymbol{f}(t, \boldsymbol{x}(t), \boldsymbol{u}(t)), \tag{11.2.26}$$

where $[\boldsymbol{\lambda}]^\top = \partial V / \partial \boldsymbol{x}$ and

$$V(t, \boldsymbol{x}(t)) \triangleq \Phi_0(\boldsymbol{x}(T)) + \int_t^T \mathcal{L}_0(t, \boldsymbol{x}(\tau), \boldsymbol{u}(\tau)) d\tau.$$

Let

$$Q^*(t) \triangleq \frac{\partial \boldsymbol{\lambda}^*(t)}{\partial \boldsymbol{x}} = \frac{\partial^2 V^*(t, \boldsymbol{x}(t))}{\partial \boldsymbol{x}^2}. \tag{11.2.27}$$

Theorem 11.2.2 *Let* $\boldsymbol{x} = \boldsymbol{x}^* + \varepsilon\boldsymbol{\delta}$, $\boldsymbol{u} = \boldsymbol{u}^*(\boldsymbol{x}^* + \varepsilon\boldsymbol{\delta})$ *and* $\boldsymbol{\lambda} = \boldsymbol{\lambda}^*(\boldsymbol{x}^* + \varepsilon\boldsymbol{\delta})$ *with* $\varepsilon \in [0, \varepsilon_0]$. *If Assumptions 11.2.1–11.2.5 and 11.2.8 are satisfied, then* Q^* *satisfies the matrix differential equation*

$$\begin{aligned} \frac{dQ^*(t)}{dt} = &- \frac{\partial^2 H^*}{\partial \boldsymbol{x}^2} - \left[\frac{\partial \boldsymbol{u}^*}{\partial \boldsymbol{x}}\right]^\top \frac{\partial^2 H^*}{\partial \boldsymbol{u}^2} \frac{\partial \boldsymbol{u}^*}{\partial \boldsymbol{x}} - \left[\frac{\partial \boldsymbol{f}^*}{\partial \boldsymbol{u}}\right]^\top Q^*(t) \\ &- Q^*(t) \frac{\partial \boldsymbol{f}^*}{\partial \boldsymbol{x}} - Q^*(t) \frac{\partial \boldsymbol{f}^*}{\partial \boldsymbol{u}} \frac{\partial \boldsymbol{u}^*}{\partial \boldsymbol{x}} - \left[\frac{\partial \boldsymbol{f}^*}{\partial \boldsymbol{u}} \frac{\partial \boldsymbol{u}^*}{\partial \boldsymbol{x}}\right]^\top Q^*(t) \\ &- \frac{\partial^2 H^*}{\partial \boldsymbol{u} \partial \boldsymbol{x}} \frac{\partial \boldsymbol{u}^*}{\partial \boldsymbol{x}} - \left[\frac{\partial \boldsymbol{u}^*}{\partial \boldsymbol{x}}\right]^\top \frac{\partial^2 H^*}{\partial \boldsymbol{x} \partial \boldsymbol{u}} \end{aligned} \tag{11.2.28}$$

with

$$Q^*(T) = \left. \frac{\partial^2 \Phi_0^*}{\partial \boldsymbol{x}^2} \right|_{t=T}. \tag{11.2.29}$$

Proof. Expanding $\boldsymbol{\lambda}$ into the first order in ε, it follows that

$$\boldsymbol{\lambda} = \boldsymbol{\lambda}^* + \varepsilon Q^* \boldsymbol{\delta} + o(\varepsilon^2).$$

Then, it can be derived by expanding V to the second order in ε that

$$\begin{aligned} V(x) &= V^* + \varepsilon \frac{\partial V^*}{\partial \boldsymbol{x}} \boldsymbol{\delta} + \frac{\varepsilon^2}{2} \boldsymbol{\delta}^\top \frac{\partial^2 V^*}{\partial \boldsymbol{x}^2} \boldsymbol{\delta} + o(\varepsilon^3) \\ &= V^* + \varepsilon [\boldsymbol{\lambda}^*]^\top \boldsymbol{\delta} + \frac{\varepsilon^2}{2} \boldsymbol{\delta}^\top Q^* \boldsymbol{\delta} + o(\varepsilon^3). \end{aligned} \tag{11.2.30}$$

Hence,

$$\begin{aligned} H + \frac{\partial V}{\partial t} = \mathcal{L}_0 + \frac{dV}{dt} = \mathcal{L}_0 + \frac{dV^*}{dt} + \varepsilon \left[\frac{d\boldsymbol{\lambda}^*(t)}{dt}\right]^\top \boldsymbol{\delta} + \varepsilon [\boldsymbol{\lambda}^*(t)]^\top \frac{d\boldsymbol{\delta}}{dt} \\ + \frac{\varepsilon^2}{2} \boldsymbol{\delta}^\top \frac{dQ^*(t)}{dt} \boldsymbol{\delta} + \varepsilon^2 \boldsymbol{\delta}^\top Q^* \frac{d\boldsymbol{\delta}}{dt} + o(\varepsilon^3). \end{aligned} \tag{11.2.31}$$

Thus, from

$$\mathcal{L}_0^* + \frac{dV^*}{dt} = H^* + \frac{\partial V^*}{\partial t} = 0,$$

and

$$\left[\frac{d\boldsymbol{\lambda}^*(t)}{dt}\right]^\top = -\frac{\partial L(t, \boldsymbol{x}^*(t), \boldsymbol{u}^*(t), \boldsymbol{\lambda}^*(t), \boldsymbol{\rho}^*(t))}{\partial \boldsymbol{x}},$$

it follows that

$$H + \frac{\partial V}{\partial t} = \mathcal{L}_0 - \mathcal{L}_0^* - \varepsilon \frac{\partial L^*}{\partial x}\delta + \underbrace{[\lambda^*]^\top (f - f^*) + \varepsilon \delta^\top Q^* (f - f^*)} \tag{11.2.32}$$

$$+ \frac{\varepsilon^2}{2}\delta^\top \frac{dQ^*(t)}{dt}\delta + o\left(\varepsilon^3\right) \tag{11.2.33}$$

$$= \mathcal{L}_0 - \mathcal{L}_0^* - \varepsilon \frac{\partial L^*}{\partial x}\delta + \lambda^\top (f - f^*) \tag{11.2.34}$$

$$+ \frac{\varepsilon^2}{2}\delta^\top \frac{dQ^*(t)}{dt}\delta + o\left(\varepsilon^3\right) \tag{11.2.35}$$

$$= \mathcal{L}_0 - \mathcal{L}_0^* - \varepsilon \left[\frac{\partial \mathcal{L}_0^*}{\partial x} + [\lambda^*]^\top \frac{\partial f^*}{\partial x} + [\rho^*]^\top \frac{\partial h^*}{\partial x}\right]\delta + \lambda^\top (f - f^*) \tag{11.2.36}$$

$$+ \frac{\varepsilon^2}{2}\delta^\top \frac{dQ^*(t)}{dt}\delta + o\left(\varepsilon^3\right) \tag{11.2.37}$$

Then, by expanding $\boldsymbol{\lambda}$ to first order in ε, and $\mathcal{L}_0$ and $\boldsymbol{f}$ to second order in ε, it gives

$$\begin{aligned}
H + \frac{\partial V}{\partial t}
&= \varepsilon \left(\frac{\partial \mathcal{L}_0^*}{\partial \boldsymbol{x}} + \frac{\partial \mathcal{L}_0^*}{\partial \boldsymbol{u}}\frac{\partial \boldsymbol{u}^*}{\partial \boldsymbol{x}}\right)\boldsymbol{\delta} \\
&\quad + \frac{\varepsilon^2}{2}\boldsymbol{\delta}^\top \left[\frac{\partial^2 \mathcal{L}_0^*}{\partial \boldsymbol{x}^2} + 2\frac{\partial^2 \mathcal{L}_0^*}{\partial \boldsymbol{u}\partial \boldsymbol{x}}\frac{\partial \boldsymbol{u}^*}{\partial \boldsymbol{x}} + \left[\frac{\partial \boldsymbol{u}^*}{\partial \boldsymbol{x}}\right]^\top \frac{\partial^2 \mathcal{L}_0^*}{\partial \boldsymbol{u}^2}\frac{\partial \boldsymbol{u}^*}{\partial \boldsymbol{x}}\right]\boldsymbol{\delta} \\
&\quad - \varepsilon \left[\frac{\partial \mathcal{L}_0^*}{\partial \boldsymbol{x}} + [\boldsymbol{\lambda}^*]^\top \frac{\partial \boldsymbol{f}^*}{\partial \boldsymbol{x}} + [\boldsymbol{\rho}^*]^\top \frac{\partial \boldsymbol{h}^*}{\partial \boldsymbol{x}}\right]\boldsymbol{\delta} \\
&\quad + \sum_{k=1}^{n}\left(\lambda_k^* + \varepsilon \frac{\partial \lambda_k^*}{\partial \boldsymbol{x}}\boldsymbol{\delta}\right)\left\{\varepsilon \left(\frac{\partial f_k^*}{\partial \boldsymbol{x}} + \frac{\partial f_k^*}{\partial \boldsymbol{u}}\frac{\partial \boldsymbol{u}^*}{\partial \boldsymbol{x}}\right)\boldsymbol{\delta} + \frac{\varepsilon^2}{2}\boldsymbol{\delta}^\top \left[\frac{\partial^2 f_k^*}{\partial \boldsymbol{x}^2}\right.\right. \\
&\quad \left.\left. + 2\frac{\partial^2 f_k^*}{\partial \boldsymbol{u}\partial \boldsymbol{x}}\frac{\partial \boldsymbol{u}^*}{\partial \boldsymbol{x}} + \left[\frac{\partial \boldsymbol{u}^*}{\partial \boldsymbol{x}}\right]^\top \frac{\partial^2 f_k^*}{\partial \boldsymbol{u}^2}\frac{\partial \boldsymbol{u}^*}{\partial \boldsymbol{x}}\right]\boldsymbol{\delta}\right\} + \frac{\varepsilon^2}{2}\boldsymbol{\delta}^\top \frac{dQ^*(t)}{dt}\boldsymbol{\delta} + o(\varepsilon^3) \\
&= \varepsilon \frac{\partial H^*}{\partial \boldsymbol{u}}\frac{\partial \boldsymbol{u}^*}{\partial \boldsymbol{x}}\boldsymbol{\delta} - \varepsilon [\boldsymbol{\rho}^*]^\top \frac{\partial \boldsymbol{h}^*}{\partial \boldsymbol{x}}\boldsymbol{\delta} + \frac{\varepsilon^2}{2}\boldsymbol{\delta}^\top \left[\frac{\partial^2 H^*}{\partial \boldsymbol{x}^2} + \frac{\partial^2 H^*}{\partial \boldsymbol{u}\partial \boldsymbol{x}}\frac{\partial \boldsymbol{u}^*}{\partial \boldsymbol{x}}\right.
\end{aligned}$$

$$
+ \left[\frac{\partial \boldsymbol{u}^*}{\partial \boldsymbol{x}}\right]^\top \frac{\partial^2 H^*}{\partial \boldsymbol{x}\partial \boldsymbol{u}} + \left[\frac{\partial \boldsymbol{u}^*}{\partial \boldsymbol{x}}\right]^\top \frac{\partial^2 H^*}{\partial \boldsymbol{u}^2}\frac{\partial \boldsymbol{u}^*}{\partial \boldsymbol{x}} + Q^*\frac{\partial \boldsymbol{f}^*}{\partial \boldsymbol{x}} + \left[\frac{\partial \boldsymbol{f}^*}{\partial \boldsymbol{x}}\right]^\top Q^*
$$

$$
\left. + Q^*\frac{\partial \boldsymbol{f}^*}{\partial \boldsymbol{u}}\frac{\partial \boldsymbol{u}^*}{\partial \boldsymbol{x}} + \left[\frac{\partial \boldsymbol{f}^*}{\partial \boldsymbol{u}}\frac{\partial \boldsymbol{u}^*}{\partial \boldsymbol{x}}\right]^\top Q^* + \frac{dQ^*(t)}{dt}\right] \boldsymbol{\delta} + o(\varepsilon^3). \qquad (11.2.38)
$$

From (11.2.20), the first two terms in the right hand side of (11.2.38) vanish. Then, (11.2.28) holds because $H + \partial V/\partial t = 0$ and $\boldsymbol{\delta} \in B(n,1)$ is arbitrary. Now, by expanding $\boldsymbol{\lambda}$ and $\partial \Phi_0/\partial \boldsymbol{x}$ to the first order in ε and using (11.2.9), (11.2.14) and (11.2.27), Equation (11.2.29) is obtained.

To continue, let

$$
A^* \triangleq \frac{\partial^2 H^*}{\partial \boldsymbol{u}^2} + \sum_{k\in\mathcal{A}^*} \rho_k^* \frac{\partial^2 h_k^*}{\partial \boldsymbol{u}^2}, \quad [B^*]^\top \triangleq \frac{\partial \hat{\boldsymbol{h}}^*}{\partial \boldsymbol{u}}, \qquad (11.2.39)
$$

$$
E^* \triangleq -\frac{\partial^2 H^*}{\partial \boldsymbol{x}\partial \boldsymbol{u}} - \sum_{k\in\mathcal{A}^*} \rho_k^* \frac{\partial^2 h_k^*}{\partial \boldsymbol{x}\partial \boldsymbol{u}} - \left[\frac{\partial \boldsymbol{f}^*}{\partial \boldsymbol{u}}\right]^\top Q^*, \qquad (11.2.40)
$$

and $F^* \triangleq -\partial \hat{\boldsymbol{h}}^*/\partial \boldsymbol{x}$.

We have the main theorem.

Theorem 11.2.3 *Suppose that the solution $(\boldsymbol{x}^*, \boldsymbol{u}^*, \boldsymbol{\lambda}^*, \boldsymbol{\rho}^*)$ satisfies the Assumptions 11.2.1–11.2.8. Then,*

$$
\frac{\partial \boldsymbol{u}^*}{\partial \boldsymbol{x}} = [I_r \quad o_{r\times q}] \begin{bmatrix} A^* & B^* \\ [B^*]^\top & 0_{q\times q} \end{bmatrix}^{-1} \begin{bmatrix} E^* \\ F^* \end{bmatrix}. \qquad (11.2.41)
$$

Proof. By (11.2.19), (11.2.22) and (11.2.27), we obtain

$$
\begin{bmatrix} A^* & B^* \\ [B^*]^\top & 0_{q\times q} \end{bmatrix} \begin{bmatrix} \partial \boldsymbol{u}^*/\partial \boldsymbol{x} \\ \partial \hat{\boldsymbol{\rho}}^*/\partial \boldsymbol{x} \end{bmatrix} = \begin{bmatrix} E^* \\ F^* \end{bmatrix}. \qquad (11.2.42)
$$

From Assumptions 11.2.6–11.2.7, the leftmost block matrix in (11.2.42) is non-singular. Thus, (11.2.41) holds.

Remark 11.2.2 *B^* in (11.2.41) will be an empty matrix when $\mathcal{A}^* = \emptyset$. In that case, (11.2.41) reduces from Assumption 11.2.7 to*

$$
\begin{aligned}
\frac{\partial \boldsymbol{u}^*}{\partial \boldsymbol{x}} &= [A^*]^{-1} E^* \qquad (11.2.43) \\
&= -\left[\frac{\partial^2 H^*}{\partial \boldsymbol{u}^2}\right]^{-1} \left[\frac{\partial^2 H^*}{\partial \boldsymbol{x}\partial \boldsymbol{u}} + \left[\frac{\partial \boldsymbol{f}^*}{\partial \boldsymbol{u}}\right]^\top Q^*\right].
\end{aligned}
$$

Now, either (11.2.41) or (11.2.43) is substituted into (11.2.28). Then, a differential Riccati equation for Q^* is derived with the terminal condition (11.2.29). Once Q^* is obtained, $\partial \boldsymbol{u}^*/\partial \boldsymbol{x}$ and hence

$$u(x) \approx u^* + \frac{\partial u^*}{\partial x}(x - x^*) \tag{11.2.44}$$

can be computed readily.

Remark 11.2.3 *Since the magnitude of the admissible perturbation ε_0 in Lemma 11.2.1 is hard to be determined, or the determined ε_0 is too small, the solution's structure may change after perturbations. Specifically, it is possible that a small boundary interval or a small interior interval along the nominal trajectory disappears after perturbations. In the first situation, the method proposed tries to keep the perturbed trajectory on the boundary, while in the second the perturbed trajectory may be infeasible in a small interval. As a remedy, we can modify the control law and project any infeasible point onto the boundary of the constraints. Suppose there are some constraints infeasible at time t, which satisfy that $h_k(t, x, u) > 0$ for $k \in \mathcal{K} \subseteq P$. Then, the control law (11.2.44) should be modified as*

$$u(x) = \overline{u}(x) = \{v \in U : h_k(t, x, v) = 0, \quad k \in \mathcal{K} \cup \mathcal{A}\}. \tag{11.2.45}$$

In this way, the perturbed solution is always feasible although some optimality may be lost.

The following algorithm gives the procedure to compute the feedback control (11.2.44) and its modification (11.2.45).

Algorithm 11.2.1

Step 1. *Solve Problem (P_1) to obtain $u^*(t)$ and $x^*(t)$ for $t \in [0, T]$. The expression of*

$$\rho^*(t) = \rho(t, x^*, u^*, \lambda^*)$$

can be solved from (11.2.10) to (11.2.12). Substituting the obtained ρ^* in (11.2.8), $\lambda^*(t)$ can be computed by integrating (11.2.8) backwards from $t = T$ to $t = 0$ with terminal condition (11.2.9). Then, $\rho^*(t)$ can be computed, and $\mathcal{A}^*(t)$ is also obtained.

Step 2. *Compute $Q^*(t)$, $t \in [0, T]$, by integrating (11.2.28) backwards in time with terminal condition (11.2.29), where $\partial u^*/\partial x$ is given by (11.2.41) or (11.2.43). Then, $\partial u^*/\partial x$ is obtained from (11.2.41) or (11.2.43) for each $t \in [0, T]$.*

Step 3. *For each neighbouring trajectory $x(t)$, the control $u(x)$ is given by (11.2.44). If any constraints are violated, $u(x)$ shall be modified as (11.2.45).*

11.2.3 Numerical Examples

Consider the following problem, which is generalized from the Rayleigh problem with a mixed state–control constraint [24].

For a given system

$$\frac{dx_1(t)}{dt} = x_2(t)\left(1 + \frac{t}{45}\right), \; x_1(0) = -5,$$
$$\frac{dx_2(t)}{dt} = -x_1(t) + x_2(t)\left(1.4 - p(x_2(t))^2\right) + 4\boldsymbol{u}(t), \; x_2(0) = -5,$$

with $p = 0.14$, find a control u that minimizes

$$g_0(u) = \int_0^{4.5} \left((u(t))^2 + (x_1(t))^2\right) dt$$

subject to a continuous inequality constraint

$$u(t) + \frac{x_1(t) + t}{6} \leq 0, \quad t \in [0, 4.5].$$

The Lagrangian L for this problem is

$$L = u^2 + x_1^2 + \lambda_1 x_2(1 + t/45) + \lambda_2\left[-x_1 + x_2\left(1.4 - px_2^2\right) + 4u\right] + \rho\left(u + (x_1 + t)/6\right).$$

From $\partial L/\partial u = 0$, ρ is solved as

$$\rho = -2u - 4\lambda_2.$$

Then, the dynamics of the costate $\boldsymbol{\lambda}$ is governed by the differential equations

$$\begin{aligned}\frac{d\lambda_1(t)}{dt} &= -\partial L/\partial x_1 = \lambda_2(t) - 2x_1(t) - \rho/6 \\ &= (5/3)\lambda_2(t) - 2x_1(t) + \boldsymbol{u}(t)/3, \lambda_1(T) = 0 \\ \frac{d\lambda_2(t)}{dt} &= -\partial L/\partial x_2 \\ &= 3p\lambda_2(t)(x_2(t))^2 - 1.4\lambda_2(t) - \lambda_1(t)(1 + t/45), \lambda_2(T) = 0.\end{aligned}$$

The nominal optimal pair $(\boldsymbol{x}^*, \boldsymbol{u}^*)$ of this problem can be computed by MISER software [104]. Then, for this nominal trajectory, it follows from (11.2.41) and (11.2.43) that

$$\frac{\partial u^*}{\partial x} = \begin{cases} \left[-\frac{1}{6}\ 0\right], & \text{if } u^* + (x_1^* + t)/6 = 0, \\ \left[0\ -2\right] Q^*, & \text{if } u^* + (x_1^* + t)/6 < 0, \end{cases} \tag{11.2.46}$$

where Q^* is the solution of the following differential equation:

$$\frac{dQ^*(t)}{dt} = F(Q^*), \quad Q^*(4.5) = 0_{2\times 2}.$$

Here,

$$F(Q^*) = \begin{bmatrix} -\frac{37}{18} & 0 \\ 0 & 0.84\lambda_2^* x_2^* \end{bmatrix} - \begin{bmatrix} 0 & -\frac{5}{3} \\ \frac{t+45}{45} & 1.4 - 0.42x_2^{*2} \end{bmatrix} Q^* \\ - Q^* \begin{bmatrix} 0 & \frac{t+45}{45} \\ -\frac{5}{3} & 1.4 - 0.42x_2^{*2} \end{bmatrix}$$

if $u^* + (x_1^* + t)/6 = 0$, while

$$F(Q^*) = \begin{bmatrix} -2 & 0 \\ 0 & 0.84\lambda_2^* x_2^* \end{bmatrix} - \begin{bmatrix} 0 & -1 \\ \frac{t+45}{45} & 1.4 - 0.42x_2^{*2} \end{bmatrix} Q^* \\ - Q^* \begin{bmatrix} 0 & \frac{t+45}{45} \\ -1 & 1.4 - 0.42x_2^{*2} \end{bmatrix} + Q^* \begin{bmatrix} 0 & 0 \\ 0 & 8 \end{bmatrix} Q^*$$

if $u^* + (x_1^* + t)/6 < 0$.

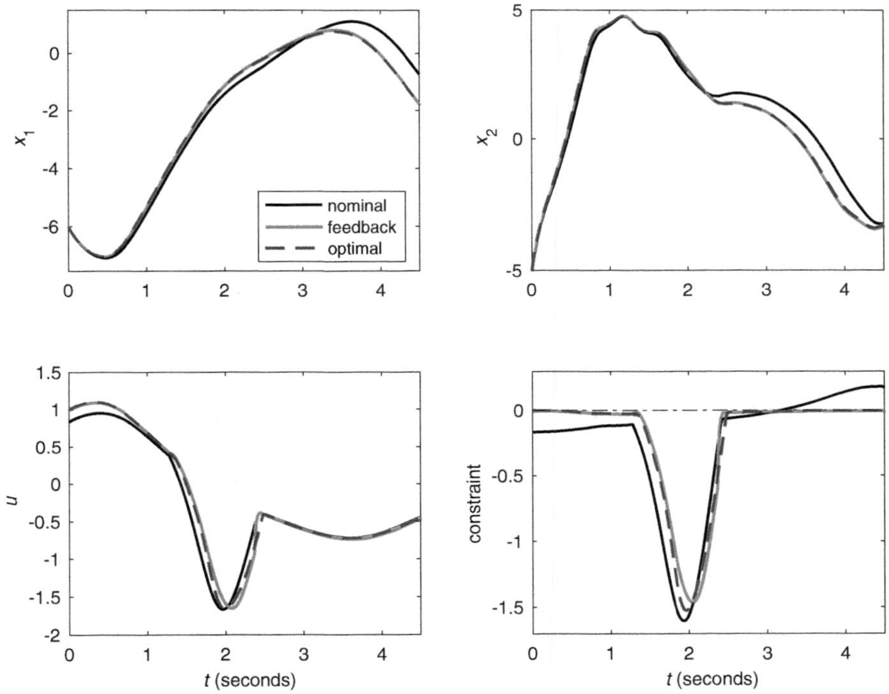

Fig. 11.2.1: Trajectories under different controls

Consider the case where $x_1(0)$ is perturbed to -6 and p is perturbed to $0.14[1 + 0.2\cos(40\pi t/9)]$. Figure 11.2.1 presents the trajectories of the perturbed system under three different controls, the nominal control u^*, the feedback control $u(x)$ of (11.2.44) and (11.2.45) and the optimal open-loop control for the perturbed problem. The respective trajectories of the state, control and constraint are, respectively, depicted by the black solid lines, the red solid lines and the blue dashed lines in Figure 11.2.1. It is seen that the errors between the system trajectories under the feedback control and those under the optimal open-loop control are relatively small, and the nominal control is infeasible for $t \in [3.18, 4.5]$. Since the perturbation is not small, the feedback control law (11.2.44) is infeasible in a small interval $t \in [1.29, 1.34]$, where the modified control law (11.2.45) is used instead. Under this modification, the feasibility of the feedback control is regained.

11.3 PID Control

11.3.1 Problem Statement

This section is from [132]. Consider the following dynamical system:

$$\frac{d\boldsymbol{x}(t)}{dt} = \boldsymbol{f}(\boldsymbol{x}(t), y(t), \boldsymbol{u}(t)), \ t \in (0, T] \tag{11.3.1}$$

$$\frac{dy(t)}{dt} = p(\boldsymbol{x}(t)) \tag{11.3.2}$$

$$\boldsymbol{x}(0) = \boldsymbol{x}^0 \tag{11.3.3}$$

$$y(0) = y^0 \tag{11.3.4}$$

where T is the terminal time, and $\boldsymbol{x} = [x_1, \ldots, x_n]^\top \in \mathbb{R}^n$, $\boldsymbol{u} = [u_1, \ldots, u_r]^\top \in \mathbb{R}^r$, $y \in \mathbb{R}$ are, respectively, state, control and output, while $\boldsymbol{f} = [f_1, \ldots, f_n]^\top \in \mathbb{R}^n$ and $p \in \mathbb{R}$ are, respectively, given continuously differentiable functions. $\boldsymbol{x}^0 \in \mathbb{R}^n$ and $y^0 \in \mathbb{R}$ are a given constant vector and a given scalar, respectively.

We assume that the following conditions are satisfied.

Assumption 11.3.1 *There exists a constant C_1 such that*

$$|\boldsymbol{f}(\boldsymbol{x}, y, \boldsymbol{u})| \leq C_1(1 + |\boldsymbol{x}| + |y| + |\boldsymbol{u}|)$$

for all $(\boldsymbol{x}, y, \boldsymbol{u}) \in \mathbb{R}^n \times \mathbb{R} \times \mathbb{R}^r$, where $|\cdot|$ denotes the usual Euclidean norm.

Assumption 11.3.2 *There exists a constant C_2 such that*

$$|p(\boldsymbol{x})| \leq C_2(1 + |\boldsymbol{x}|).$$

Remark 11.3.1 *Suppose that the output equations are algebraic equations given below rather than the output system (11.3.2) with the initial condition (11.3.4).*

$$y(t) = h(\hat{\boldsymbol{x}}(t)), \tag{11.3.5}$$

where, without loss of generality,

$$\hat{\boldsymbol{x}} = [x_1, \dots, x_s]^\top \tag{11.3.6}$$

with $s < n$. Furthermore, we assume that

$$\frac{d\hat{\boldsymbol{x}}(t)}{dt} = q(\boldsymbol{x}(t)), \tag{11.3.7}$$

where $\boldsymbol{q} = [q_1, \dots, q_s]^T$ is a continuously differentiable function. Then, it is easy to see that

$$\frac{dy(t)}{dt} = \sum_{i=1}^{s} \frac{\partial h(\hat{\boldsymbol{x}}(t))}{\partial x_i} \frac{dx_i(t)}{dt} = \sum_{i=1}^{s} \frac{\partial h(\hat{\boldsymbol{x}}(t))}{\partial x_i} q_i(\boldsymbol{x}(t)) \tag{11.3.8}$$

with initial condition

$$y(0) = h(\hat{\boldsymbol{x}}(0)). \tag{11.3.9}$$

Thus, the formulation of the output expressed in terms of differential equations given by (11.3.2) with initial condition (11.3.4) is rather general. Certainly it covers the ship steering problem to be considered later in this section as a special case.

The control u is assumed to take the form of a PID controller given below:

$$\begin{aligned} u(t) = &\sum_{j=1}^{N_1} k_{1,j}(y(t) - r(t))\chi_{I_{1,j}}(t) \\ &+ \sum_{j=1}^{N_2} k_{2,j} \int_0^t (y(s) - r(s))\chi_{I_{2,j}}(s)ds + \sum_{j=1}^{N_3} k_{3,j} \frac{dy(t)}{dt} \chi_{I_{3,j}}(t), \end{aligned} \tag{11.3.10}$$

where $r(t)$ denotes a given reference input, which is a piecewise continuous function defined on $[0, T]$,

$$I_{i,j} = [t_{i,j-1}, t_{i,j}), \ i = 1, 2, 3; \ j = 1, \dots, N_i, \tag{11.3.11}$$

while

$$0 = t_{i,0} < t_{i,1} < t_{i,2} < \dots < t_{i,N_i} < t_{i,N_{i+1}} = T, \ i = 1, 2, 3, \tag{11.3.12}$$

are the switching times for the proportional, integral and derivative control actions, respectively, and χ_I denotes the indicator function of I given by

$$\chi_I(t) = \begin{cases} 1, & t \in I, \\ 0, & \text{otherwise}. \end{cases} \tag{11.3.13}$$

Here, $\{k_{i,1}, \ldots, k_{i,N_i}\}$, $i = 1, 2, 3$, are respective gains for the proportional, integral and derivative terms of the PID controller.

The form of the PID controller is a generalized version of the conventional PID controller, particularly, the form of the integral control. For the conventional integral control, it performs the integral action over the whole period of the time horizon. Because of the accumulation effect, a large value of the gain for the integral control will cause huge overshoot. On the other hand, if the gain for the integral control is chosen to be very small, while the overshoot can become small, the steady state error will take a long time to reduce in the presence of constant disturbances. The generalized integral control is in the form for which it is re-set at appropriately chosen fixed switching time points so as to give a well-regulated control operation.

Remark 11.3.2 *Here, we assume that y and r are real-valued functions. It is straightforward to extend the results to the case where y and r are vector-valued functions at the expense of notational complexity.*

We now specify the region within which the output trajectory is allowed to move. This region is defined in terms of the following continuous inequality constraints, which arise due to practical requirements, such as constraints on the rise time and for avoiding overshoot. They may also arise due to engineering specification on the PID controller.

$$g_i(t, \boldsymbol{x}(t), y(t), u(t)) \leq 0, \ t \in [0, T], \ i = 1, \ldots, M. \tag{11.3.14}$$

For each $i = 1, \ldots, M$, the function g_i is continuously differentiable with respect to x, y and u while continuous with respect to t.

To ensure a satisfactory tracking of $r(t)$ by $y(t)$, the following terminal state constraint is imposed:

$$\Omega(y(T)) = y(T) - r(T) = 0. \tag{11.3.15}$$

The optimal control problem may now be stated below. Given system (11.3.1)–(11.3.4), design a PID controller in the form defined by (11.3.10) such that the output $y(t)$ of the corresponding closed loop system will move within the specified region defined by the continuous inequality constraints (11.3.14) and, at the same time, it will track the given reference input such that the terminal condition (11.3.15) is satisfied. Let this problem be referred to as Problem (P_2).

First, we formulate a cost functional below:

$$J(k) = \int_0^T \alpha_1 (y(t) - r(t))^2 + \alpha_2 \left[\frac{dy(t)}{dt} \right]^2 + \alpha_3 [u(t)]^2 dt, \tag{11.3.16}$$

where α_i, $i = 1, 2, 3$, are the weighting factors.

For the integral term of the PID controller given by (11.3.10), we define

$$z_j(t) = \int_0^t [y(s) - r(s)]\chi_{I_{2,j}}(s)ds, \; j = 1, \ldots, N_2. \tag{11.3.17}$$

Clearly, for each $j = 1, \ldots, N_2$, (11.3.17) is equivalent to

$$\frac{dz_j(t)}{dt} = (y(t) - r(t))\chi_{I_{2,j}}(t), \tag{11.3.18}$$

$$z_j(0) = 0. \tag{11.3.19}$$

Let $\boldsymbol{z}(t) = [z_1(t), \ldots, z_{N_2}(t)]^\top$ and $\boldsymbol{q}(t) = [q_1(t), \ldots, q_{N_2}(t)]^\top$, where

$$q_j(t) = (y(t) - r(t))\chi_{I_{2,j}}(t), \; j = 1, \ldots, N_2. \tag{11.3.20}$$

Then, system (11.3.18)–(11.3.19) become

$$\frac{d\boldsymbol{z}(t)}{dt} = \boldsymbol{q}(t), \tag{11.3.21}$$

$$\boldsymbol{z}(0) = \boldsymbol{0}. \tag{11.3.22}$$

Now, it follows from (11.3.21)–(11.3.22) that system (11.3.1)–(11.3.4) with $u(t)$ chosen as a PID controller given by (11.3.10) can be written as

$$\begin{cases} \dfrac{d\boldsymbol{x}(t)}{dt} = \overline{\boldsymbol{f}}(t, \boldsymbol{x}(t), y(t), \boldsymbol{z}(t), \boldsymbol{k}) \\ \dfrac{dy(t)}{dt} = p(\boldsymbol{x}(t)) \\ \dfrac{d\boldsymbol{z}(t)}{dt} = \boldsymbol{q}(t) \end{cases} \tag{11.3.23}$$

with initial conditions

$$\begin{cases} \boldsymbol{x}(0) = \boldsymbol{x}^0 \\ y(0) = y^0 \\ \boldsymbol{z}(0) = \boldsymbol{0} \end{cases} \tag{11.3.24}$$

where

$$\overline{\boldsymbol{f}}(t, \boldsymbol{x}(t), y(t), \boldsymbol{z}(t), \boldsymbol{k}) = \boldsymbol{f}(\boldsymbol{x}(t), y(t), u(t)), \tag{11.3.25}$$

while the PID controller $u(t)$ given by (11.3.10) becomes

$$\begin{aligned} u(t) = &\sum_{j=1}^{N_1} k_{1,j}(y(t) - r(t))\chi_{I_{1,j}}(t) \\ &+ \sum_{j=1}^{N_2} k_{2,j} z_j(t) + \sum_{j=1}^{N_3} k_{3,j} p(x(t))\chi_{I_{3,j}}(t). \end{aligned} \tag{11.3.26}$$

Here,

$$\boldsymbol{k} = [k_{1,1}, \ldots, k_{1,N_1}, k_{2,1}, \ldots, k_{2,N_2}, k_{3,1}, \ldots, k_{3,N_3}]^\top \tag{11.3.27}$$

is the vector containing the gains for the proportional, integral and derivative terms of the PID controller.

The specified region remains the same as given by (11.3.14). The cost functional (11.3.16) becomes

$$\begin{aligned}\overline{J}(k) = \int_0^T \Bigg\{&\alpha_1(y(t) - r(t))^2 + \alpha_2[p(x(t))]^2 \\ &+\alpha_3\Bigg[\sum_{j=1}^{N_1} k_{1,j}(y(t) - r(t))\chi_{I_{1,j}}(t) \\ &+\sum_{j=1}^{N_2} k_{2,j}z_j(t) + \sum_{j=1}^{N_3} k_{3,j}p(x(t))\chi_{I_{3,j}}(t)\Bigg]^2\Bigg\}dt. \end{aligned}\tag{11.3.28}$$

The problem may now be re-stated as: Given system (11.3.23) with initial condition (11.3.24), find a PID control parameter vector k such that the cost function (11.3.28) is minimized subject to the continuous inequality constraint (11.3.14) and the terminal equality constraint (11.3.15). Let this problem be referred to as **Problem** (Q_2). Clearly, Problem (Q_2) is an optimal parameter selection problem.

11.3.2 Constraint Approximation

The continuous inequality constraints (11.3.14) are handled by constraint transcription technique presented in Section 4.3. This leads to the following equivalent equality constraints:

$$\int_0^T \max\{g_i\left(t, \boldsymbol{x}(t), y(t), u(t)\right), 0\}dt = 0, \ i = 1, \ldots, M, \tag{11.3.29}$$

where $u(t)$ is given by (11.3.26). However, the integrands appeared under the integration in (11.3.29) are nonsmooth. Thus, for each $i = 1, \ldots, M$, we shall approximate the nonsmooth function $\max\{g_i\left(\boldsymbol{x}(t), y(t), u(t)\right), 0\}$ by a smooth function $\mathcal{L}_{i,\varepsilon}\left(t, \boldsymbol{x}(t), y(t), u(t)\right)$ given by

$$\begin{aligned}&\mathcal{L}_{i,\varepsilon}(t, \boldsymbol{x}(t), y(t), u(t)) \\ &= \begin{cases} 0, & \text{if } \ g_i(t, \boldsymbol{x}(t), y(t), u(t)) < -\varepsilon \\ \left(g_i(t, \boldsymbol{x}(t), y(t), u(t)) + \varepsilon\right)^2/4\varepsilon, & \text{if } -\varepsilon \le g_i(t, \boldsymbol{x}(t), y(t), u(t)) \le \varepsilon \\ g_i(t, \boldsymbol{x}(t), y(t), u(t)), & \text{if } \ g_i(t, \boldsymbol{x}(t), y(t), u(t)) > \varepsilon, \end{cases}\end{aligned} \tag{11.3.30}$$

where u is given by (11.3.26) and $\varepsilon > 0$ is an adjustable constant with small value. Then, for each $i = 1, \ldots, M$, we define

$$g_{i,\varepsilon}(k) = \int_0^T \bar{\mathcal{L}}_{i,\varepsilon}(t, \boldsymbol{x}(t), y(t), \boldsymbol{z}(t), \boldsymbol{k})dt, \tag{11.3.31}$$

where

$$\bar{\mathcal{L}}_{i,\varepsilon}(t, \boldsymbol{x}(t), y(t), \boldsymbol{z}(t), \boldsymbol{k}) = \mathcal{L}_{i,\varepsilon}(t, \boldsymbol{x}(t), y(t), u(t)) \tag{11.3.32}$$

and $u(t)$ is given by (11.3.26).

We now use the concept of the penalty function to append the functions $g_{i,\varepsilon}$ given by (11.3.31) to the cost functional (11.3.28), forming a new cost functional given below:

$$\begin{aligned}\overline{J}_{\varepsilon,\gamma}(\boldsymbol{k}) = &\int_0^T l(t, \boldsymbol{x}(t), y(t), \boldsymbol{z}(t), \boldsymbol{k})dt \\ &+ \gamma\sum_{i=1}^{M}\int_0^T \bar{\mathcal{L}}_{i,\varepsilon}(t, \boldsymbol{x}(t), y(t), \boldsymbol{z}(t), \boldsymbol{k})dt,\end{aligned}$$

where

$$l(t, \boldsymbol{x}, y, \boldsymbol{z}, \boldsymbol{k}) = \alpha_1\big(y - r\big)^2 + \alpha_2\big(p(\boldsymbol{x})\big)^2 + \alpha_3\big(u(t)\big)^2, \tag{11.3.33}$$

and $u(t)$ is given by (11.3.26) and $\gamma > 0$ is a penalty parameter.

We may now state the approximate problem for each $\varepsilon > 0$ and $\gamma > 0$ as follows. Given system (11.3.23) with initial condition (11.3.24) and terminal condition (11.3.15), find a PID control parameter vector $\boldsymbol{k}$ such that the cost functional

$$\overline{J}_{\varepsilon,\gamma}(\boldsymbol{k}) = \int_0^T \hat{\mathcal{L}}_{\varepsilon,\gamma}(t, \boldsymbol{x}, y, \boldsymbol{z}, \boldsymbol{k})dt \tag{11.3.34}$$

is minimized, where

$$\begin{aligned}\hat{\mathcal{L}}_{\varepsilon,\gamma}(t, \boldsymbol{x}, y, \boldsymbol{z}, \boldsymbol{k}) = &\ l(t, \boldsymbol{x}(t), y(t), \boldsymbol{z}(t), \boldsymbol{k}) \\ &+ \gamma\sum_{i=1}^{M}\bar{\mathcal{L}}_{i,\varepsilon}(t, \boldsymbol{x}(t), y(t), \boldsymbol{z}(t), \boldsymbol{k}).\end{aligned} \tag{11.3.35}$$

This problem is referred to as **Problem** $(Q_{2_{\varepsilon,\gamma}})$. The relationships between Problem $(Q_{2_{\varepsilon,\gamma}})$ and Problem (Q_2) are given in the following theorems. Their proofs are similar to those given for Theorems 2.1 and 2.2 in [259], respectively.

Theorem 11.3.1 *For any $\varepsilon > 0$, there exists a $\gamma(\varepsilon) > 0$ such that for all γ, $0 < \gamma < \gamma(\varepsilon)$, if $\boldsymbol{k}^*_{\varepsilon,\gamma}$ is an optimal solution of Problem $(Q_{2_{\varepsilon,\gamma}})$, then it satisfies the continuous inequality constraint (11.3.14) of Problem (Q_2).*

Theorem 11.3.2 *Let $\boldsymbol{k}^*$ and $\boldsymbol{k}^*_{\varepsilon,\overline{\gamma}(\varepsilon)}$ be, respectively, optimal solutions of Problem (Q_2) and Problem $(Q_{2\varepsilon,\gamma})$, where $\overline{\gamma}(\varepsilon)$ is chosen such that $\boldsymbol{k}^*_{\varepsilon,\overline{\gamma}(\varepsilon)}$ satisfies the continuous inequality constraint (11.3.14) of Problem (Q_2). Then,*

$$\lim_{\varepsilon \to 0} \overline{J}\left(\boldsymbol{k}^*_{\varepsilon,\overline{\gamma}(\varepsilon)}\right) = \overline{J}(\boldsymbol{k}^*), \tag{11.3.36}$$

where $\overline{J}$ is defined by (11.3.28).

On the basis of Theorems 11.3.1 and 11.3.2, Problem (P_2) can be solved through solving a sequence of optimal parameter selection problems $(Q_{2\varepsilon,\gamma})$ subject to only terminal equality condition (11.3.15). Each of these optimal parameter selection problems can be solved as a nonlinear optimization problem by using a gradient-based optimization method, such as the sequential quadratic programming approximation scheme. See Chapter 3 for details. Thus, the optimal control software, MISER, is applicable. Further details are given in the next section.

11.3.3 Computational Method

In this section, we will propose a reliable computational method for solving Problem (Q_2) via solving a sequence of Problems $(Q_{2\varepsilon,\gamma})$, where for each $\varepsilon > 0$ and $\gamma > 0$, Problem $(Q_{2\varepsilon,\gamma})$ is solved as a nonlinear optimization problem. For doing this, it is required to provide, for each $\boldsymbol{k}$, the value of the cost functional $\overline{J}_{\varepsilon,\gamma}(\boldsymbol{k})$, as well as its gradient $\partial\overline{J}_{\varepsilon,\gamma}(\boldsymbol{k})/\partial\boldsymbol{k}$. Furthermore, we also need the value of the terminal constraint function $\Omega(y(T|\boldsymbol{k}))$ and its gradient $\partial\Omega(y(T|\boldsymbol{k}))/\partial\boldsymbol{k}$. It is obvious that the values of the cost functional $\overline{J}_{\varepsilon,\gamma}(\boldsymbol{k})$ and the terminal constraint function $\Omega(y(T|\boldsymbol{k}))$ can be readily obtained after system (11.3.23) with initial condition (11.3.24) corresponding to $\boldsymbol{k}$ is solved. For the gradient formulas of the cost functional $\overline{J}_{\varepsilon,\gamma}(\boldsymbol{k})$ and the terminal constraint function $\Omega(y(T|u))$ corresponding to each $\boldsymbol{k}$, we have the following two theorems. Their proofs are similar to those given for Theorem 5.2.1 in [253].

Theorem 11.3.3 *The gradient formula for the cost function $\overline{J}_{\varepsilon,\gamma}(\boldsymbol{k})$ with respect to $\boldsymbol{k}$ is given by*

$$\frac{\partial\overline{J}_{\varepsilon,\gamma}(\boldsymbol{k})}{\partial\boldsymbol{k}} = \int_0^T \frac{\partial H_{\varepsilon,\gamma}(t,\boldsymbol{x}(t),y(t),\boldsymbol{z}(t),\boldsymbol{k},\boldsymbol{\lambda}_{\varepsilon,\gamma}(t))}{\partial\boldsymbol{k}}dt. \tag{11.3.37}$$

Here, $H_{\varepsilon,\gamma}(t,\boldsymbol{x},y,\boldsymbol{z},\boldsymbol{k},\boldsymbol{\lambda})$ is the Hamiltonian function given by

$$H_{\varepsilon,\gamma}(t,\boldsymbol{x},y,\boldsymbol{z},\boldsymbol{k},\boldsymbol{\lambda}) = \hat{\mathcal{L}}_{\varepsilon,\gamma}(t,\boldsymbol{x},y,\boldsymbol{z},\boldsymbol{k}) + \boldsymbol{\lambda}^\top_{\varepsilon,\gamma}\hat{f}(t,\boldsymbol{x},y,\boldsymbol{z},\boldsymbol{k}),$$

where $\hat{\mathcal{L}}_{\varepsilon,\gamma}$ is as defined by (11.3.35)

$$\hat{\boldsymbol{f}} = \left[(\overline{\boldsymbol{f}})^\top, p, \boldsymbol{q}^\top\right]^\top,$$

and $\boldsymbol{\lambda}_{\varepsilon,\gamma}$ *is the solution of following system of costate differential equations:*

$$\frac{d\boldsymbol{\lambda}(t)}{dt} = -\left[\frac{\partial H_{\varepsilon,\gamma}(t, \boldsymbol{x}(t), y(t), \boldsymbol{z}(t), \boldsymbol{k}, \boldsymbol{\lambda}(t))}{\partial \boldsymbol{x}}, \frac{\partial H_{\varepsilon,\gamma}(t, \boldsymbol{x}(t), y(t), \boldsymbol{z}(t), \boldsymbol{k}, \boldsymbol{\lambda}(t))}{\partial y}, \frac{\partial H_{\varepsilon,\gamma}(t, \boldsymbol{x}(t), y(t), \boldsymbol{z}(t), \boldsymbol{k}, \boldsymbol{\lambda}(t))}{\partial \boldsymbol{z}}\right]^\top \tag{11.3.38a}$$

with the boundary condition

$$\boldsymbol{\lambda}(T) = 0. \tag{11.3.38b}$$

Theorem 11.3.4 *The gradient formula for the terminal constraint function* $\Omega(y(T|\boldsymbol{k}))$ *with respect to* $\boldsymbol{k}$ *is given by*

$$\frac{\partial \Omega(y(T|\boldsymbol{k})}{\partial \boldsymbol{k}} = \int_0^T \frac{\partial \tilde{H}_{\varepsilon,\gamma}(t, \boldsymbol{x}(t), y(t), \boldsymbol{z}(t), \boldsymbol{k}, \tilde{\boldsymbol{\lambda}}_{\varepsilon,\gamma}(t))}{\partial \boldsymbol{k}} dt, \tag{11.3.39}$$

where $\tilde{H}_{\varepsilon,\gamma}(t, \boldsymbol{x}, y, \boldsymbol{z}, \boldsymbol{k}, \boldsymbol{\lambda})$ *is the Hamiltonian function given by*

$$\tilde{H}_{\varepsilon,\gamma}(t, \boldsymbol{x}, y, \boldsymbol{z}, \boldsymbol{k}, \boldsymbol{\lambda}) = \tilde{\boldsymbol{\lambda}}_{\varepsilon,\gamma}^\top \hat{\boldsymbol{f}}(t, \boldsymbol{x}, y, \boldsymbol{z}, \boldsymbol{k}). \tag{11.3.40}$$

Here, $\tilde{\boldsymbol{\lambda}}_{\varepsilon,\gamma}$ *is the solution of following system of costate differential equations:*

$$\frac{d\tilde{\boldsymbol{\lambda}}(t)}{dt} = -\left[\frac{\partial \tilde{H}_{\varepsilon,\gamma}(t, \boldsymbol{x}(t), y(t), \boldsymbol{z}(t), \boldsymbol{k}, \tilde{\boldsymbol{\lambda}}(t))}{\partial \boldsymbol{x}}, \frac{\partial \tilde{H}_{\varepsilon,\gamma}(t, \boldsymbol{x}(t), y(t), \boldsymbol{z}(t), \boldsymbol{k}, \tilde{\boldsymbol{\lambda}}(t))}{\partial y}, \frac{\partial \tilde{H}_{\varepsilon,\gamma}(t, \boldsymbol{x}(t), y(t), \boldsymbol{z}(t), \boldsymbol{k}, \tilde{\boldsymbol{\lambda}}(t))}{\partial \boldsymbol{z}}\right]^\top \tag{11.3.41a}$$

with the boundary condition

$$\tilde{\boldsymbol{\lambda}}(T) = \frac{d\Omega(y(T))}{dy}. \tag{11.3.41b}$$

For each $\varepsilon > 0$, $\gamma > 0$, Problem $(Q_{2_{\varepsilon,\gamma}})$ is to be solved as a nonlinear optimization problem using the gradient formulas given in Theorems 11.3.3 and 11.3.4. Details are reported in the following as an algorithm.

Algorithm 11.3.1

1. *Choose* $\varepsilon > 0$, $\gamma > 0$ *and* k.
2. *Solve Problem* $(Q_{2_{\varepsilon,\gamma}})$ *as a nonlinear optimization problem, yielding* $\boldsymbol{k}_{\varepsilon,\gamma}^*$.

3. *Check whether all the continuous inequality constraint (11.3.14) are satisfied or not. If they are satisfied, go to Step 4. Otherwise, increase* γ *to* 10γ *and go to Step 2 with* $\boldsymbol{k}^*_{\varepsilon,\gamma}$ *as the initial guess for the new optimization process.*
4. *If* ε *is small enough, say, less than or equal to a given small number, we have a successful exit. Else, decrease* ε *to* $\varepsilon/10$ *and go to Step 2, using* $\boldsymbol{k}^*_{\varepsilon,\gamma}$ *as the initial guess for the new optimization process.*

11.3.4 Application to a Ship Steering Control Problem

In this section, we apply the proposed method to a ship steering control problem. Our aim is to design a PID controller such that the heading angle, $y(t)$, of the ship will follow the change course set by the reference input signal $r(t)$. The control system is as shown in Figure 11.3.1. The ship motion can be described by the following differential equations defined on $[0, T]$ (see [17]). In this application, $T = 300$s.

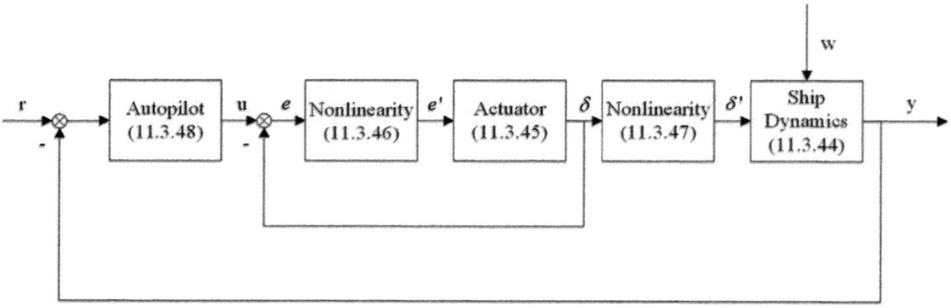

Fig. 11.3.1: The overall control system of the ship

$$\frac{d^3y(t)}{dt^3} + b_1\frac{d^2y(t)}{dt^2} + b_2\left(\left(a_1\frac{dy(t)}{dt}\right)^3 + a_2\frac{dy(t)}{dt}\right) = b_3\frac{d\delta'(t)}{dt} + b_2\delta'(t) + w, \tag{11.3.42}$$

where

$$w = b_3\frac{d(d)}{dt} + b_2 d,$$

$$\frac{d\delta(t)}{dt} = b_4 e'(t), \tag{11.3.43}$$

with

$$e' = \begin{cases} e, & \text{if } |e| \le e_{\max} \\ e_{\max}\text{sign}(e), & \text{if } |e| \ge e_{\max} \end{cases} \tag{11.3.44}$$

where $e = u - \delta$,

$$\delta' = \begin{cases} \delta, & \text{if } |\delta| \leq \delta_{\max} \\ \delta_{\max}\text{sign}(\delta), & \text{if } |\delta| \geq \delta_{\max} \end{cases} \tag{11.3.45}$$

The variable w is to account for sea disturbances acting on the ship with d a constant disturbance, u is the control that is chosen in the form of a PID controller defined by (11.3.10), δ is the rudder angle, e is the error as defined, e' and δ' are the real inputs to the actuator and ship dynamics, respectively, because of the saturation properties that are defined as (11.3.44) and (11.3.45). The ship model is in its full generality without resorting to simplification and linearization. This work develops further some previous studies of optimal ship steering strategies with time optimal control [257], phase advanced control [31], parameter self-turning [141], adaptive control [7] and constrained optimal model following [264].

For a ship steering problem, it has two phases: course changing and course keeping. During the course changing phase, it is required to manoeuvre the ship such that it moves quickly towards the desired course set by the command without violating the constraints arising from performance specifications and physical limitations on the controller. During the course keeping phase, the ship is required to move along the desired course. In this application, the PID controller of the form defined by (11.3.10) with $N_1 = N_2 = N_3 = 6$ is used. More specifically,

$$\begin{aligned} u(t) &= \sum_{i=1}^{6} k_{1,i}(y(t) - r(t))\chi_{[t_{i-1},t_i)}(t) \\ &+ \sum_{i=1}^{6} k_{2,i} \int_0^t (y(s) - r(s))\, \chi_{[t_{i-1},t_i)}(s)ds \\ &+ \sum_{i=1}^{6} k_{3,i} \frac{dy(t)}{dt} \chi_{[t_{i-1},t_i)}(t), \end{aligned} \tag{11.3.46}$$

where χ_I denotes the indicator function of I defined by (11.3.13), while t_i, $i = 1, \ldots, 5$, are fixed switching time points to be specified later.

Set

$$x_1(t) = y(t),\ x_2(t) = \frac{dy(t)}{dt},\ x_3(t) = \frac{d^2y(t)}{dt^2},\ x_4(t) = \delta(t) \tag{11.3.47}$$

and

$$x_{5,j}(t) = \int_0^t (y(s) - r(s))\, \chi_{[t_{j-1},t_j)}(s)ds,\ j = 1, \ldots, 6. \tag{11.3.48}$$

Then, the dynamics of the ship can be expressed as

$$\frac{dx_1(t)}{dt} = x_2(t) \tag{11.3.49}$$

$$\frac{dx_2(t)}{dt} = x_3(t) \tag{11.3.50}$$

$$\frac{dx_3(t)}{dt} = -b_1 x_3(t) - b_2\left(a_1 (x_2(t))^3 + a_2 x_2(t)\right) + b_3 b_4 e + b_2(x_4(t) + d) \tag{11.3.51}$$

$$\frac{dx_4(t)}{dt} = b_4 e \tag{11.3.52}$$

$$\frac{dx_{5,j}(t)}{dt} = x_1(t) - r(t)\chi_{[t_{j-1},t_j)}(t), \ j = 1, \ldots, 6 \tag{11.3.53}$$

with the initial condition

$$\boldsymbol{x}(0) = [0,\ 0,\ 0,\ 0,\ 0,\ 0,\ 0,\ 0,\ 0,\ 0]^\top, \tag{11.3.54}$$

where

$$\begin{aligned} \boldsymbol{x} &= [x_1, x_2, \ldots, x_{10}]^T \\ e &= u(t) - x_4(t) \end{aligned} \tag{11.3.55}$$

with

$$\begin{aligned} u(t) = &\sum_{i=1}^{6} k_{1,i}\left(x_1(t) - r(t)\right)\chi_{[t_{i-1},t_i)}(t) \\ &+ \sum_{i=1}^{6} k_{2,i} x_{5,j}(t) + \sum_{i=1}^{6} k_{3,i} x_2(t)\chi_{[t_{i-1},t_i)}(t). \end{aligned} \tag{11.3.56}$$

The values of the coefficients appeared in the equations are given in Table 11.3.1. The reference input signal $r(t)$ used in our example is $r(t) =$

Table 11.3.1: Coefficients for the ship model

a_1	a_2	b_1	b_2	b_3	b_4
−30.0	−5.6	0.1372	−0.0002014	−0.003737	0.5

$\pi/180$, for $t \in [0, 300s]$.

This ship steering problem is a special case of (11.3.1)–(11.3.4), where the output system is

$$\frac{dx_1(t)}{dt} = x_2(t)$$

with initial condition

$$x_1(0) = 0.$$

In practice, a large overshoot is undesirable. In this problem, the following constraint is imposed on the upper bound of the heading angle $x_1(t)$.

$$x_1(t) - 1.01r(t) \leq 0, \ t \in [0, 300s], \tag{11.3.57}$$

i.e., the heading angle should not go beyond 1% of the desired reference input $r(t)$. This constraint can be written as

$$g_1(t) = x_1(t) - 101\%r(t) \leq 0, \ t \in [0, 300s]. \tag{11.3.58}$$

We also impose constraint on the rise time of the heading angle such that the heading angle is constrained to reach at least 70% of the desired reference input in 30 seconds and 95% in 60 seconds, i.e.,

$$g_2(t) = h(t) - x_1(t) \leq 0, \ t \in [0, 300s], \tag{11.3.59}$$

where

$$h(t) = \begin{cases} 0, & t \in [0, 6) \\ 5.1 \times 10^{-4}t - 3.1 \times 10^{-3}, & t \in [6, 30) \\ 1.5 \times 10^{-4}t + 7.9 \times 10^{-3}, & t \in [30, 60) \\ 2.2 \times 10^{-6}t + 16.4 \times 10^{-3}, & t \in [60, 300]. \end{cases} \tag{11.3.60}$$

To cater for the saturation property of the actuator, it is equivalent to impose upper and lower bounds on $x_4(t)$, i.e.,

$$-\pi/6 \leq x_4(t) \leq \pi/6, \ t \in [0, 300s], \tag{11.3.61}$$

which are continuous inequality constraints. They can be rewritten as

$$g_3(t) = -x_4(t) - \pi/6 \leq 0, \ t \in [0, 300s] \tag{11.3.62}$$

and

$$g_4(t) = x_4(t) - \pi/6 \leq 0, \ t \in [0, 300s]. \tag{11.3.63}$$

Similarly, to cater for another saturation property, we have

$$-\pi/30 \leq x_4(t) - u(t) \leq \pi/30, \ t \in [0, 300s]. \tag{11.3.64}$$

They are again continuous inequality constraints, which can be rewritten as

$$g_5(t) = -x_4(t) + u(t) - \pi/30 \leq 0, \ t \in [0, 300s] \tag{11.3.65}$$

and

$$g_6(t) = x_4(t) - u(t) - \pi/30 \leq 0, \ t \in [0, 300s], \tag{11.3.66}$$

where $u(t)$ is given by (11.3.56).

The terminal equality constraint is

$$\begin{aligned}\Omega(x_1(300)) &= x_1(300) - r(300) \\ &= x_1(300) - \pi/180 = 0. \end{aligned} \tag{11.3.67}$$

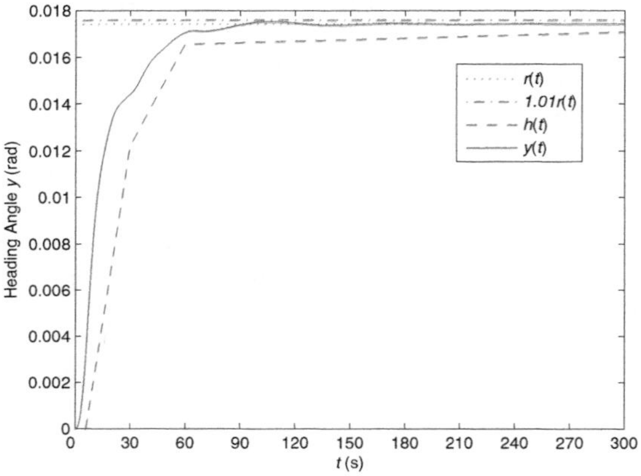

Fig. 11.3.2: The heading angle of the ship

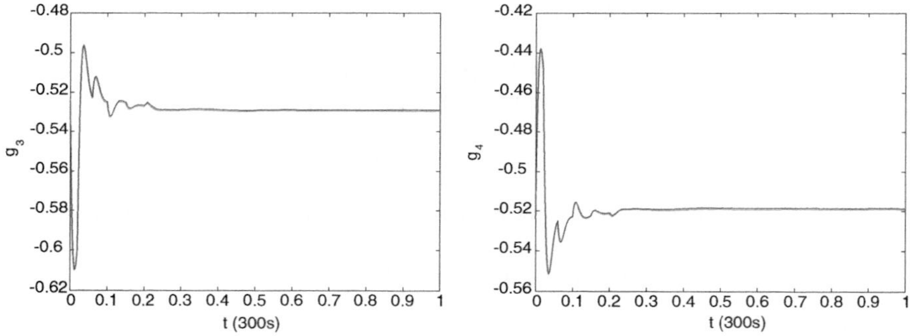

Fig. 11.3.3: The constraints for the saturation of the actuator

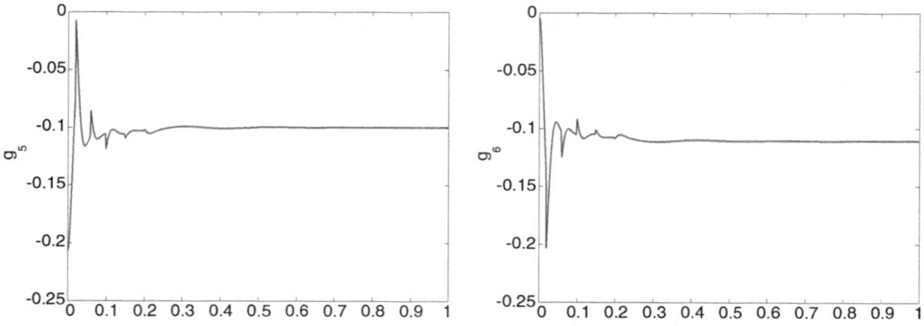

Fig. 11.3.4: The constraints for the saturation of the control

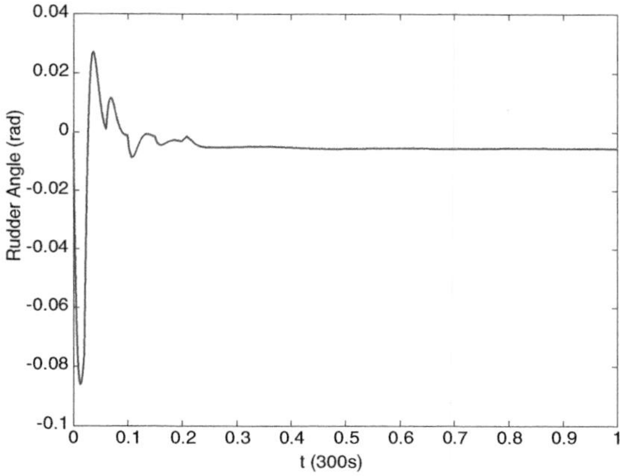

Fig. 11.3.5: The rudder angle of the ship

The optimal PID control problem may now be stated formally as follows.

Given system (11.3.49)–(11.3.54), find a PID control parameter vector $k = [(k^1)^\top, \ldots, (k^6)^\top]^\top$ with $k^i = [k_1^i, k_2^i, k_3^i]^\top$, $i = 1, 2, \ldots, 6$, such that the cost functional

$$J = \int_0^{300} \{\alpha_1(x_1(t) - r(t))^2 + \alpha_2 x_2^2(t) + \alpha_3 u^2(t)\}dt \tag{11.3.68}$$

is minimized subject to the continuous inequality constraints (11.3.62), (11.3.63), (11.3.65), (11.3.66) and (11.3.58) and the terminal condition (11.3.67), where

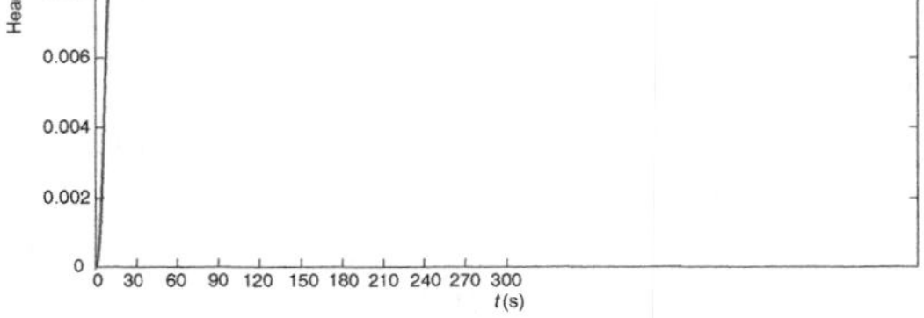

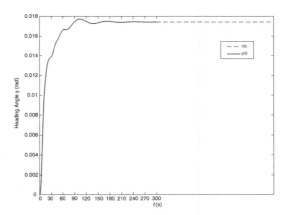

Fig. 11.3.6: The heading angle of the ship with a larger disturbance

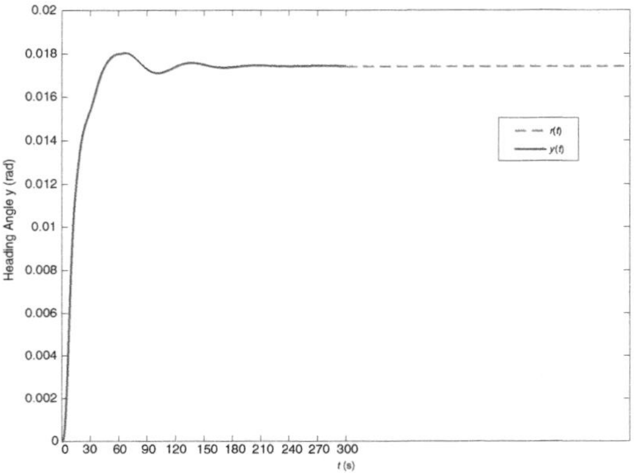

Fig. 11.3.7: The heading angle of the ship with a disturbance coming from the initial heading direction

$$u(t) = \sum_{i=1}^{6} k_{1,i}(x_1(t) - r(t))\chi_{[t_{i-1},t_i)}(t) + \sum_{i=1}^{6} k_{2,i}x_{5,j}(t) + \sum_{i=1}^{6} k_{3,i}x_2(t)\chi_{[t_{i-1},t_i)}(t). \quad (11.3.69)$$

Here, $t_0 = 0$, $t_6 = 300$ and t_i, $i = 1, 2, \ldots, 5$, are the switching time points that are chosen to be at the time points where the constraint function g_2 is non-differentiable. They are $t_1 = 6$, $t_2 = 18$, $t_3 = 30$, $t_4 = 45$ and $t_5 = 60$.

Let this problem be referred to as Problem (S), and it is solvable by the computational method developed in Section 11.3.3.

We then construct Problem $(P2_{\varepsilon,\gamma})$ according to the procedure as specified in Section 11.3.2, where the appended new cost functional is given by

$$\overline{J}_{\varepsilon,\gamma}(k) = \int_0^{300} \Big\{\alpha_1(x_1(t) - r(t))^2 + \alpha_2 x_2^2(t) + \alpha_3 u^2(t) + \gamma(g_{1,\varepsilon} + g_{2,\varepsilon} + g_{3,\varepsilon} + g_{4,\varepsilon} + g_{5,\varepsilon} + g_{6,\varepsilon})\Big\}dt, \quad (11.3.70)$$

where $g_{i,\varepsilon}$, $i = 1, \ldots, 6$, are obtained from g_i, $i = 1, \ldots, 6$, respectively, according to (11.3.31). In this problem, we set $\alpha_1 = 10$, $\alpha_2 = 400 and \alpha_3 = 0.05$. It is to be minimized subject to terminal equality constraint (11.3.67).

In real world, disturbances always exist and there are many kinds of disturbances. We consider the case, where the ship is encountered with a constant disturbance d. Assume that $d = 0.3\pi/180$.

Problem $(P2_{\varepsilon,\gamma})$ is solved by using Algorithm 3.3.1, where the final ε and γ are $\varepsilon = 0.01$ and $\gamma = 10$. The optimal parameters for the PID controller u obtained are

$$k^{1,*} = [5.78685,\ 7.27203,\ 2.62351,\ 6.98467,\ 9.76934,\ 7.39799]^\top$$
$$k^{2,*} = [1.03217,\ 0.00000,\ 0.00303,\ 0.00000,\ 0.24398,\ 0.84387]^\top$$
$$k^{3,*} = [99.81791,\ 100.52777,\ 100.67912,\ 100.13533,\ 99.75020,\ 99.70358]^\top.$$

The results obtained are shown in Figures 11.3.2, 11.3.3, 11.3.4, and 11.3.5. From the results obtained, we see that all the constraints are satisfied. The heading angle tracks the desired reference input with no steady state error after some small oscillation due to the constant disturbance. The overshooting of the heading angle above the reference input is less than 1%, and hence the constraint $g_1(t) \leq 0$, $t \in [0, 300]$, is satisfied. To test the robustness of this PID controller, we run the model with the optimal PID controller under the following environments: (1) The disturbance is much larger, more specifically, $d = 0.6 \times \pi/180$; and (2) the disturbance is coming from the initial heading direction, more specifically, $d = -0.3 \times \pi/180$. The results are shown in Figures 11.3.6 and 11.3.7. In both cases, we see that the heading angles track the desired reference input with no steady state error after some small oscillations.

11.4 Exercises

11.1. Provide a proof of Theorem 11.3.1.

11.2. Provide a proof of Theorem 11.3.2.

11.3. Provide a proof of Theorem 11.3.3.

11.4. Provide a proof of Theorem 11.3.4.

Chapter 12
On Some Special Classes of Stochastic Optimal Control Problems

12.1 Introduction

In this chapter, we consider two classes of stochastic optimal control problems. More specifically, in Section 12.2 we consider a class of combined optimal parameter selection and optimal control problems in which the dynamical system is governed by linear Ito stochastic differential equation involving a Wiener process. Both the control and system parameter vectors may, however, appear nonlinearly in the system dynamics. The cost functional is taken as an expected value of a quadratic function of the state vector, where the weighting matrices are time invariant but are allowed to be nonlinear in both the control and system parameters. Furthermore, certain realistic features such as probabilistic constraints on the state vector may also be included.

In Section 12.3, we consider a class of partially observed linear stochastic control problems described by three sets of stochastic differential equations: one for the system to be controlled, one for the observer (measurement) channel and one for the control channel driven by the observed process. The noise processes perturbing the system and observer dynamics are vector-valued Poisson processes.

For each of these stochastic optimal control problems, we show that it is equivalent to deterministic optimal control problems. These equivalent deterministic optimal control problems are further transformed into special cases of the form considered in Section 8.8. The main references of this chapter are [239] and [243].

K. L. Teo et al., *Applied and Computational Optimal Control*, Springer Optimization and Its Applications 171,
https://doi.org/10.1007/978-3-030-69913-0_12

12.2 A Combined Optimal Parameter and Optimal Control Problem

Consider a system described by the following system of linear Ito stochastic differential equations defined on the fixed time interval $(0, T]$:

$$d\boldsymbol{\xi}(t) = A(t, \boldsymbol{\delta}, \boldsymbol{u}(t))\boldsymbol{\xi}(t)dt + \boldsymbol{b}(t, \boldsymbol{\delta}, \boldsymbol{u}(t))dt + D(t, \boldsymbol{\delta}, \boldsymbol{u}(t))d\boldsymbol{w}(t) \quad (12.2.1a)$$

with the initial condition

$$\boldsymbol{\xi}(0) = \boldsymbol{\xi}^0, \quad (12.2.1b)$$

where $\boldsymbol{\xi}(t) = [\xi_1(t), \ldots, \xi_n(t)]^\top \in \mathbb{R}^n$ is the state vector, $\boldsymbol{\delta} = [\delta_1, \ldots, \delta_s] \in \mathbb{R}^s$ is the system parameter vector, $\boldsymbol{u} = [u_1, \ldots, u_r] \in \mathbb{R}^r$ is the control vector, $\boldsymbol{\xi}^0 = [\xi_1^0, \ldots, \xi_n^0]^\top \in \mathbb{R}^n$ is the initial state vector that is Gaussian distributed with mean $\boldsymbol{\mu}^0$ and covariance matrix M^0 and $\boldsymbol{w}(t) = [w_1(t), \ldots, w_m(t)]^\top \in \mathbb{R}^m$ is a Wiener process with zero mean and covariance matrix

$$\mathcal{E}\left\{\boldsymbol{w}(t)[\boldsymbol{w}(t)]^\top\right\} = \int_0^{\min\{t,\tau\}} \Theta(s)ds, \quad (12.2.2)$$

where $\Theta \in \mathbb{R}^{m\times m}$ is a symmetric positive definite matrix function, and $\mathcal{E}$ denotes the mathematical expectation.

We assume throughout this chapter that the following conditions are satisfied.

Assumption 12.2.1 *$A(t, \boldsymbol{\delta}, \boldsymbol{u}) \in \mathbb{R}^{n\times n}$, $\boldsymbol{b}(t, \boldsymbol{\delta}, \boldsymbol{u}) \in \mathbb{R}^n$ and $D(t, \boldsymbol{\delta}, \boldsymbol{u}) \in \mathbb{R}^{n\times m}$ are continuously differentiable with respect to all their arguments.*

Assumption 12.2.2 *The Wiener process $\boldsymbol{w}(t)$ and the initial random vector $\boldsymbol{\xi}^0$ are statistically independent.*

To continue, let $\mathcal{U}$ be the class of admissible controls as defined in Section 8.2. Furthermore, let

$$\Omega = \{\boldsymbol{\delta} \in \mathbb{R}^s : \eta_j(\boldsymbol{\delta}) \geq 0,\ j = 1, \ldots, M\}, \quad (12.2.3)$$

where η_j, $j = 1, \ldots, M$, are continuously differentiable functions of the parameter $\boldsymbol{\delta}$.

The next step is to introduce an important class of constraints on the state of the dynamical system (12.2.1). These constraints arise naturally in the situation when we wish to confine the state to be within a given acceptable region with a certain degree of confidence for all $t \in [0, T]$. These probabilistic state constraints may be stated as

$$\text{Prob}\left\{\alpha_i \leq \left(\boldsymbol{c}^i\right)^\top \boldsymbol{\xi}(t) \leq \beta_i, \text{ for all } t \in [0, T]\right\} > \varDelta_i, \quad i = 1, \ldots, N, \quad (12.2.4)$$

where $\boldsymbol{c}^i$, $i = 1, \ldots, N$, are n-vectors and α_i, β_i and Δ_i, $i = 1, \ldots, N$, are given real constants.

An element $(\boldsymbol{\delta}, \boldsymbol{u}) \in \Omega \times \mathcal{U}$ is said to be a feasible combined parameter and control if it satisfies the probabilistic state constraints specified in (12.2.4). Let $\mathcal{D}$ be the class of all such feasible combined parameters and controls. Hence $\mathcal{D}$ is called the class of feasible combined parameters and controls. We are now in a position to specify our problem formally as follows.

Subject to the dynamical system (12.2.1), find a feasible combined parameter and control $(\boldsymbol{\delta}, \boldsymbol{u}) \in \mathcal{D}$ such that the cost functional

$$g_0(\boldsymbol{\delta}, \boldsymbol{u}) = \mathcal{E}\Bigg\{[\boldsymbol{\xi}(T)]^\top S(\boldsymbol{\delta})\boldsymbol{\xi}(T) + [\boldsymbol{p}(\boldsymbol{\delta})]^\top \boldsymbol{\xi}(T) + \upsilon(\boldsymbol{\delta}) + \int_0^T \Big\{[\boldsymbol{q}(t, \boldsymbol{\delta}, \boldsymbol{u}(t))]^\top \boldsymbol{\xi}(t) + \vartheta(t, \boldsymbol{\delta}, \boldsymbol{u}(t)) + [\boldsymbol{\xi}(t)]^\top Q(t, \boldsymbol{\delta}, \boldsymbol{u}(t))\boldsymbol{\xi}(t)\Big\} dt\Bigg\} \tag{12.2.5}$$

is minimized over $\mathcal{D}$, where $S(\boldsymbol{\delta}) \in \mathbb{R}^{n\times n}$ and $Q(t, \boldsymbol{\delta}, \boldsymbol{u}) \in \mathbb{R}^{n\times n}$ are symmetric and positive semi-definite matrices continuously differentiable with respect to their respective arguments, while $\boldsymbol{q}(t, \boldsymbol{\delta}, \boldsymbol{u})$ and $\boldsymbol{p}(\boldsymbol{\delta})$ (respectively, $\vartheta(t, \boldsymbol{\delta}, \boldsymbol{u})$ and $\upsilon(\boldsymbol{\delta})$) are n vector-valued functions (respectively, real-valued functions) that are also continuously differentiable with respect to their arguments. For convenience, let this combined optimal parameter selection and optimal control problem be referred to as Problem (SP_1).

12.2.1 Deterministic Transformation

In this section, we wish to show that the combined optimal parameter selection and optimal control problem (SP_1) is equivalent to a deterministic one. To begin, we note that the solution of the system (12.2.1) corresponding to each $(\boldsymbol{\delta}, \boldsymbol{u})$ can be written as

$$\boldsymbol{\xi}(t \mid \boldsymbol{\delta}, \boldsymbol{u}) = \Phi(t, 0 \mid \boldsymbol{\delta}, \boldsymbol{u})\boldsymbol{\xi}^0 + \int_0^t \Phi(t, s \mid \boldsymbol{\delta}, \boldsymbol{u})\boldsymbol{b}(s, \boldsymbol{\delta}, \boldsymbol{u}(s))ds + \int_0^t \Phi(t, s \mid \boldsymbol{\delta}, \boldsymbol{u})D(s, \boldsymbol{\delta}, \boldsymbol{u}(s)))d\boldsymbol{w}(s), \tag{12.2.6}$$

where, for each $(\boldsymbol{\delta}, \boldsymbol{u})$, $\Phi(t, s \mid \boldsymbol{\delta}, \boldsymbol{u}) \in \mathbb{R}^{n\times n}$ is the principal solution matrix of the homogeneous system

$$\frac{\partial \Phi(t, \tau)}{\partial t} = A(t, \boldsymbol{\delta}, \boldsymbol{u}(t))\Phi(t, \tau), \quad t > \tau \tag{12.2.7a}$$

$$\Phi(\tau, \tau) = I, \tag{12.2.7b}$$

where I is the identity matrix.

Since a linear transformation of a Gaussian process is also Gaussian, the process $\{\boldsymbol{\xi}(t) : t > 0\}$ given by

$$\boldsymbol{\xi}_2(t) = \int_0^t \Phi(t, s \mid \boldsymbol{\delta}, \boldsymbol{u}) D(s, \boldsymbol{\delta}, \boldsymbol{u}(t)) d\boldsymbol{w}(s), \quad t > 0,$$

is Gaussian and

$$\boldsymbol{\xi}_1(t) = \Phi(t, 0 \mid \boldsymbol{\delta}, \boldsymbol{u}) \boldsymbol{\xi}^0$$

is also Gaussian if $\boldsymbol{\xi}^0$ is. Thus, for each $(\boldsymbol{\delta}, \boldsymbol{u})$, the process $\{\boldsymbol{\xi}(t) : t \geq 0\}$, given by (12.2.6), is a Gaussian Markov process with mean

$$\begin{aligned} \boldsymbol{\mu}(t \mid \boldsymbol{\delta}, \boldsymbol{u}) &= \mathcal{E}\{\boldsymbol{\xi}(t \mid \boldsymbol{\delta}, \boldsymbol{u})\} \\ &= \Phi(t, 0 \mid \boldsymbol{\delta}, \boldsymbol{u}) \mathcal{E}\{\boldsymbol{\xi}^0\} + \int_0^t \Phi(t, s \mid \boldsymbol{\delta}, \boldsymbol{u}) \boldsymbol{b}(s, \boldsymbol{\delta}, \boldsymbol{u}(s)) ds \\ &= \Phi(t, 0 \mid \boldsymbol{\delta}, \boldsymbol{u}) \boldsymbol{\mu}^0 + \int_0^t \Phi(t, s \mid \boldsymbol{\delta}, \boldsymbol{u}) \boldsymbol{b}(s, \boldsymbol{\delta}, \boldsymbol{u}(s)) ds \end{aligned} \tag{12.2.8}$$

and covariance matrix

$$\begin{aligned} \Psi(t \mid \boldsymbol{\delta}, \boldsymbol{u}) &= \Phi(t, 0 \mid \boldsymbol{\delta}, \boldsymbol{u}) M^0 [\Phi(t, 0 \mid \boldsymbol{\delta}, \boldsymbol{u})]^\top \\ &+ \int_0^t \Phi(t, \tau \mid \boldsymbol{\delta}, \boldsymbol{u}) D(\tau, \boldsymbol{\delta}, \boldsymbol{u}(\tau)) \Theta(\tau) [D(\tau, \boldsymbol{\delta}, \boldsymbol{u}(\tau))]^\top [\Phi(t, \tau \mid \boldsymbol{\delta}, \boldsymbol{u})]^\top d\tau. \end{aligned} \tag{12.2.9}$$

Differentiating (12.2.8) with respect to t and then using (12.2.7), we note that for each $(\boldsymbol{\delta}, \boldsymbol{u}) \in \mathcal{D}$, $\boldsymbol{\mu}(t \mid \boldsymbol{\delta}, \boldsymbol{u})$ is the corresponding solution of the following system of differential equations.

$$\frac{d\boldsymbol{\mu}(t)}{dt} = A(t, \boldsymbol{\delta}, \boldsymbol{u}(t)) \boldsymbol{\mu}(t) + \boldsymbol{b}(t, \boldsymbol{\delta}, \boldsymbol{u}(t)), \quad t > 0, \tag{12.2.10a}$$

$$\boldsymbol{\mu}(0) = \boldsymbol{\mu}^0. \tag{12.2.10b}$$

Differentiating (12.2.9) with respect to t and then using (12.2.7), it follows that for each $(\boldsymbol{\delta}, \boldsymbol{u}) \in \mathcal{D}$, $\boldsymbol{\Psi}(t \mid \boldsymbol{\delta}, \boldsymbol{u})$ is the corresponding solution of the following matrix differential equation.

$$\frac{d\Psi(t)}{dt} = A(t, \boldsymbol{\delta}, \boldsymbol{u}(t)) \Psi(t) + [\Psi(t)]^\top A(t, \boldsymbol{\delta}, \boldsymbol{u}(t)) + D(t, \boldsymbol{\delta}, \boldsymbol{u}(t)) \Theta(t) [D(t, \boldsymbol{\delta}, \boldsymbol{u}(t))]^\top \tag{12.2.11a}$$

with the initial condition

$$\Psi(0) = M^0. \tag{12.2.11b}$$

Note that $\Psi(t \mid \boldsymbol{\delta}, \boldsymbol{u})$ is symmetric. Thus, there are only $n(n+1)/2$ distinct differential equations in (12.2.11). The corresponding conditional joint probability density function for $\boldsymbol{\xi}(t)$ is given by

$$f(\boldsymbol{x},t\mid\boldsymbol{\delta},\boldsymbol{u})$$
$$=(2\pi)^{-\frac{n}{2}}[\det\Psi(t\mid\boldsymbol{\delta},\boldsymbol{u})]^{-\frac{1}{2}}\exp\left\{-\frac{[\boldsymbol{x}-\boldsymbol{\mu}(t\mid\boldsymbol{\delta},\boldsymbol{u})]^\top[\Psi(t\mid\boldsymbol{\delta},\boldsymbol{u})]^{-1}[\boldsymbol{x}-\boldsymbol{\mu}(t\mid\boldsymbol{\delta},\boldsymbol{u})]}{2}\right\}$$
$$. \tag{12.2.12}$$

Let us now turn our attention to the cost functional (12.2.5). First, we note that

$$\begin{aligned}
&\mathcal{E}\left\{[\boldsymbol{\xi}(t)]^\top Q(t,\boldsymbol{\delta},\boldsymbol{u}(t))\boldsymbol{\xi}(t)\right\}\\
&\quad=\mathcal{E}\left\{\mathrm{Tr}([\boldsymbol{\xi}(t)]^\top Q(t,\boldsymbol{\delta},\boldsymbol{u}(t))\boldsymbol{\xi}(t))\right\}\\
&\quad=\mathcal{E}\left\{\mathrm{Tr}(Q(t,\boldsymbol{\delta},\boldsymbol{u}(t))\boldsymbol{\xi}(t)[\boldsymbol{\xi}(t)]^\top)\right\}\\
&\quad=\mathrm{Tr}\left\{Q(t,\boldsymbol{\delta},\boldsymbol{u}(t))\Psi(t\mid\boldsymbol{\delta},\boldsymbol{u})+\boldsymbol{\mu}(t\mid\boldsymbol{\delta},\boldsymbol{u})[\boldsymbol{\mu}(t\mid\boldsymbol{\delta},\boldsymbol{u}))]^\top\right\},
\end{aligned}\tag{12.2.13}$$

where $\mathrm{Tr}(\cdot)$ denotes the trace of a matrix.

The first term of the cost functional (12.2.5) can be handled in a similar manner. The transformation of the second term into its equivalent deterministic form is obvious. The third term is already in deterministic form. Thus, we have the following lemma.

Lemma 12.2.1 *The cost functional* (12.2.5) *is equivalent to*

$$\begin{aligned}
g_0(\boldsymbol{\delta},\boldsymbol{u})=&\,\mathrm{Tr}\left\{S(\boldsymbol{\delta})\left[\Psi(T\mid\boldsymbol{\delta},\boldsymbol{u})+\boldsymbol{\mu}(T\mid\boldsymbol{\delta},\boldsymbol{u})[\boldsymbol{\mu}(T\mid\boldsymbol{\delta},\boldsymbol{u})]^\top\right]\right\}+[\boldsymbol{h}(\boldsymbol{\delta})]^\top\boldsymbol{\mu}(T)\\
&+\upsilon(\boldsymbol{\delta})+\int_0^T\left\{[\boldsymbol{q}(t,\boldsymbol{\delta},\boldsymbol{u}(t))]^\top\boldsymbol{\mu}(t\mid\boldsymbol{\delta},\boldsymbol{u})+\vartheta(t,\boldsymbol{\delta},\boldsymbol{u}(t))\right\}dt\\
&+\int_0^T\mathrm{Tr}\left\{Q(t,\boldsymbol{\delta},\boldsymbol{u}(t))\left[\boldsymbol{\Psi}(t\mid\boldsymbol{\delta},\boldsymbol{u})\right.\right.\\
&\left.\left.+\boldsymbol{\mu}(t,\boldsymbol{\delta},\boldsymbol{u}(t))[\boldsymbol{\mu}(t,\boldsymbol{\delta},\boldsymbol{u}(t))]^\top\right]\right\}dt,
\end{aligned}\tag{12.2.14}$$

where $\boldsymbol{\mu}(T\mid\boldsymbol{\delta},\boldsymbol{u})$ *and* $\boldsymbol{\Psi}(t\mid\boldsymbol{\delta},\boldsymbol{u})$ *are deterministic and determined, respectively, by* (12.2.10) *and* (12.2.11).

For the probabilistic state constraints specified in (12.2.4), we have the following lemma.

Lemma 12.2.2 *For each* $i=1,\ldots,N$, *the corresponding probabilistic constraint specified in* (12.2.4) *is equivalent to*

$$\mathrm{erf}\left\{\frac{b_i-\left(\boldsymbol{c}^i\right)^\top\boldsymbol{\mu}(t\mid\boldsymbol{\delta},\boldsymbol{u})}{\sqrt{2\pi\left(\boldsymbol{c}^i\right)^\top\Psi(t\mid\boldsymbol{\delta},\boldsymbol{u})\boldsymbol{c}^i}}\right\}-\mathrm{erf}\left\{\frac{a_i-\left(\boldsymbol{c}^i\right)^\top\boldsymbol{\mu}(t\mid\boldsymbol{\delta},\boldsymbol{u})}{\sqrt{2\pi\left(\boldsymbol{c}^i\right)^\top\Psi(t\mid\boldsymbol{\delta},\boldsymbol{u})\boldsymbol{c}^i}}\right\}\geq\Delta_i, \tag{12.2.15}$$

for all $t\in[0,T]$.

Proof. Since $\boldsymbol{\xi}(t)$ is Gaussian with mean $\boldsymbol{\mu}(t\mid\boldsymbol{\delta},\boldsymbol{u})$ and covariance $\Psi(t\mid\boldsymbol{\delta},\boldsymbol{u})$, it is clear that for each $i=1,\ldots,N$, the scalar process $\left(\boldsymbol{c}^i\right)^\top\boldsymbol{\xi}(t)$ is also

Gaussian with mean $\left(\boldsymbol{c}^i\right)^\top \boldsymbol{\mu}(t \mid \boldsymbol{\delta}, \boldsymbol{u})$ and covariance $\left(\boldsymbol{c}^i\right)^\top \Psi(t \mid \boldsymbol{\delta}, \boldsymbol{u})\boldsymbol{c}^i$. Thus, the corresponding constraint specified in (12.2.4) can be rewritten as

$$\int_{\alpha_i}^{\beta_i} \frac{1}{\sqrt{2\pi \left(\boldsymbol{c}^i\right)^\top \Psi(t \mid \boldsymbol{\delta}, \boldsymbol{u})\boldsymbol{c}^i}} \exp\left\{ \frac{1}{2} \frac{\left(y - \left(\boldsymbol{c}^i\right)^\top \boldsymbol{\mu}(t \mid \boldsymbol{\delta}, \boldsymbol{u})\right)^2}{\left(\boldsymbol{c}^i\right)^\top \Psi(t \mid \boldsymbol{\delta}, \boldsymbol{u})\boldsymbol{c}^i} \right\} dy \geq \Delta_i.$$

By carrying out the required integration, it is easy to show that this constraint is equivalent to (12.2.15). This completes the proof.

Remark 12.2.1 *Define*

$$g_i(t, \boldsymbol{\delta}, \boldsymbol{u}) = \operatorname{erf}\left\{ \frac{b_i - \left(\boldsymbol{c}^i\right)^\top \boldsymbol{\mu}(t \mid \boldsymbol{\delta}, \boldsymbol{u})}{\sqrt{2\pi \left(\boldsymbol{c}^i\right)^\top \Psi(t \mid \boldsymbol{\delta}, \boldsymbol{u})\boldsymbol{c}^i}} \right\} - \operatorname{erf}\left\{ \frac{a_i - \left(\boldsymbol{c}^i\right)^\top \boldsymbol{\mu}(t \mid \boldsymbol{\delta}, \boldsymbol{u})}{\sqrt{2\pi \left(\boldsymbol{c}^i\right)^\top \Psi(t \mid \boldsymbol{\delta}, \boldsymbol{u})\boldsymbol{c}^i}} \right\}. \tag{12.2.16}$$

Then, (12.2.15) *can be written as*

$$-\Delta_i + g_i(t, \boldsymbol{\delta}, \boldsymbol{u}) \geq 0, \quad \textit{for all } t \in [0, T], \quad i = 1, \ldots, N. \tag{12.2.17}$$

The constraints (12.2.17) are continuous inequality constraints. These continuous constraints can be approximated by a sequence of inequality constraints in canonical form by using the constraint transcription technique presented in Section 4.2.

Let $\boldsymbol{x}(t)$ be the vector formed from $\boldsymbol{\mu}(t)$ and the independent components of the matrix $\Psi(t)$, and let $\boldsymbol{f}$ be the corresponding vector obtained from the right hand side of (12.2.10) and (12.2.11). Furthermore, let $\mathcal{D}$ again denote the class of all feasible combined parameters and controls in the sense that each of its elements is in $\Omega \times \mathcal{U}$ and satisfies the constraints specified in (12.2.17). We can summarize the above analysis in the following theorem.

Theorem 12.2.1 *Problem* (SP_1) *is equivalent to the following deterministic combined optimal parameter selection and optimal control problem, denoted as Problem* (DP_1).

Subject to the dynamical system

$$\frac{d\boldsymbol{x}(t)}{dt} = \boldsymbol{f}(t, \boldsymbol{x}(t), \boldsymbol{\delta}, \boldsymbol{u}(t)) \tag{12.2.18a}$$

$$\boldsymbol{x}(0) = \boldsymbol{x}^0, \tag{12.2.18b}$$

where $\boldsymbol{x}^0$ *is formed by* $\boldsymbol{\mu}^0$ *and the components appearing in the upper triangular part of the covariance matrix* M^0*, find a combined parameter and control* $(\boldsymbol{\delta}, \boldsymbol{u}) \in \mathcal{D}$ *such that the cost functional*

$$g_0(\boldsymbol{\delta}, \boldsymbol{u}) = \Phi_0(\boldsymbol{x}(T \mid \boldsymbol{\delta}, \boldsymbol{u}), \boldsymbol{\delta}) + \int_0^T \mathcal{L}_0(t, \boldsymbol{x}(t \mid \boldsymbol{\delta}, \boldsymbol{u}), \boldsymbol{\delta}, \boldsymbol{u}(t))dt \tag{12.2.19}$$

is minimized over $\mathcal{D}$, where Φ_0 and $\mathcal{L}_0$ are obtained from the corresponding terms of (12.2.14) *in an obvious manner.*

Remark 12.2.2 *Note that the deterministic transformation has increased the dimension of the state from n to $\frac{1}{2}(n^2 + 3n)$.*

Problem (DP_1) can be solved by using the approach presented in Section 9.2. More specifically, the control parametrization method is first applied to approximate the control function in Problem (DP_1) by a piecewise constant function with its heights and switching times taken as decision variables. Then, the time scaling transformation (see Section 9.2.1) is applied to map the varying switching times into fixed switching times in a new time horizon with an additional control variable, called the time scaling control. Let the transformed problem be referred to as Problem $(DP_1(p))$. For the continuous inequality constraint (12.2.14), the constraint transcription technique (see Section 9.2.2) is applied to approximate it by a sequence of smooth inequality constraints. Consequently, Problem $(DP_1(p))$ is approximated by a sequence of optimal parameter selection problems $(DP_{1,\epsilon\gamma}(p))$, each of which can be solved by any gradient-based optimization methods as detailed in Sections 9.2.3 and 9.2.4. The convergence analysis of the approximation problems $(DP_{1,\epsilon\gamma}(p))$ to the original problem (DP_1) can be carried out as in Section 9.2.

Remark 12.2.3 *Note that if we consider Problem (SP_1) with the assumption that $\boldsymbol{\xi}^0$ is a deterministic vector rather than a Gaussian distributed random vector, similar results are also valid. In this situation, the initial conditions for* (12.3.10) *and* (12.2.11) *are, respectively, replaced by $\boldsymbol{x}^0$ and $\mathbf{0}$.*

12.2.2 A Numerical Example

Consider the optimal machine maintenance problem described in Example 6.4.1. It is a simple deterministic optimal control problem that can be solved easily by the Pontryagin Maximum Principle as shown in Section 6.4. However, from a practical view point, there is usually a multitude of other factors that contribute in a less significant manner to the deterioration of the machine's quality state. Since each of these factors may be assumed independent and random, the aggregated effect can be modeled as a random noise term superimposed on the deterministic part of the deterioration dynamic. The main reference of this section is [234].

The machine state $\xi(t)$ is thus governed by a stochastic differential equation:

$$d\xi(t) = -b\xi(t)dt + u(t)dt + dw(t) \tag{12.2.20a}$$

$$\xi(0) = x_0, \tag{12.2.20b}$$

where $w(t)$ is a Wiener process with zero mean and variance, i.e.,

$$\mathcal{E}\left\{w(t)w\left(t'\right)\right\} = \int_0^{\min\left(t,t'\right)} \theta(s)ds, \quad \theta(s) \geq 0, \text{ for all } s. \tag{12.2.21}$$

Here, we assume that the variance of the Wiener process is stationary, i.e.,

$$\theta(s) = \theta, \text{ for all } s \geq 0. \tag{12.2.22}$$

Furthermore, for satisfactory performance, we need to ensure that the machine maintains a relatively good quality state with a certain degree of confidence over the whole of its life span. This particular requirement is achieved by imposing the following probabilistic constraint over the whole time horizon $[0, T]$.

$$\text{Prob}\left\{\xi(t) \in [x_0 - \alpha, x_0 + \alpha]\right\} \geq \varepsilon, \text{ for all } t \in [0, T], \tag{12.2.23}$$

where $\alpha > 0$ and $\varepsilon > 0$. This constraint implies that we want to be at least $100\varepsilon\%$ confidence that the quality state of the machine is in the interval $[x_0 - \alpha, x_0 + \alpha]$ throughout its life span, where $\alpha > 0$ denotes the deviation of the quality state of the machine from its initial state. In practice, we would obviously like to make ε as close as possible to unity, while keeping δ as small as possible. This may, however, incur excessive control effort. The stochastic optimal maintenance problem may now be stated as follows.

Given the dynamical system (12.2.20), find an admissible control $u \in \mathcal{U}$ such that the expected return

$$g_0(u) = \mathcal{E}\left\{\exp(-rT)S\xi(T) + \int_0^T \exp(-rt)[\theta\xi(t) - u(t)]dt\right\} \tag{12.2.24}$$

is maximized subject to the probabilistic constraint (12.2.23), where $\mathcal{U}$ consists of all those controls that satisfy the constraints (6.4.8), while S, r, θ and T are, respectively, the salvage value per unit terminal quality, the interest rate, the productivity per unit quality and the sale date of the machine.

Let $\mu(t)$ and $\Psi(t)$ be the mean and variance of the state $\xi(t)$ determined, respectively, by the following differential equations:

$$\frac{d\mu(t)}{dt} = -b\mu(t) + u(t) \tag{12.2.25a}$$

$$\mu(0) = x_0 \tag{12.2.25b}$$

and

$$\frac{d\Psi(t)}{dt} = -2b\Psi(t) + \theta \tag{12.2.26a}$$

$$\Psi(0) = 0. \tag{12.2.26b}$$

Note that the variance $\Psi(t)$ does not depend on the control u. Thus, the differential equation (12.2.26) needs only to be solved once.

Now, by virtue of the theoretical results reported in Section 12.2.1, it follows readily that this stochastic optimal problem is equivalent to the following deterministic optimal control problem.

Given the dynamical system (12.2.25)–(12.2.26), find a control $u \in \mathcal{U}$ such that the cost functional

$$g_0(u) = h\mu(T)\exp\{-rT\} + \int_0^T \exp\{-rt\}[\theta\mu(t) - u(t)]dt \tag{12.2.27}$$

is minimized subject to the following continuous constraint

$$-2\varepsilon + g_1(t, \delta, u) \geq 0, \; t \in [0, T], \tag{12.2.28}$$

where

$$g_1(t, \delta, u) = \text{erf}\left(\frac{x_0 + \delta - \mu(t)}{\sqrt{2\pi\Psi(t)}}\right) - \text{erf}\left(\frac{x_0 - \delta - \mu(t)}{\sqrt{2\pi\Psi(t)}}\right). \tag{12.2.29}$$

To study the behavior of this optimal machine maintenance problem, the above deterministic optimal control problem with the following parameter values is solved by using the control parametrization technique, where the planning horizon $[0, T] = [0, 5]$ is partitioned into $n_p = 20$ subintervals with n_{p+1} switching time points in the partition. The results and the computational effectiveness can be improved if the time scaling technique is used. This task is left as an exercise for the reader. Furthermore, the following values are used for the various model constants.

$b = 0.2$, $s = 0.25$, $\bar{u} = 0.1$, $p = 0.6$, $x_0 = 1.0$, $r = 5\%$, $\delta = 0.5$ and $\theta = 0.04$.

Several cases of different values of ε were computed, and it was found that the optimal solutions are always of the bang-bang type, i.e.,

$$u^*(t) = \begin{cases} \bar{u} = 0.1, & 0 \leq t \leq t^* \\ 0, & t^* \leq t \leq .T \end{cases} \tag{12.2.30}$$

The optimal switching time, however, is substantially larger than that of the deterministic unconstrained case computed by (6.4.15). This is obvious due to the stringent state constraint specified in (12.2.28). The optimal switching times obtained for $\varepsilon = 0.5$, 0.6 and 0.7 are, respectively, $t^* = 3.350$, 3.697 and 4.680, with the corresponding return values of $g_0^* = 1.922$, 1.918 and 1.891. The unconstrained deterministic problem with the same set of parameter values has an optimal switching time of 3.284. The optimum solution is consistent with the intuitive notion that as the quality state requirement becomes more stringent, more control efforts are required thus resulting in a later switching time. It is expected that if ε becomes larger, full maintenance effort will be required, i.e., $u^*(t) = \bar{u}(t) = 0.1$, for all $t \in [0, T]$. However, if ε becomes too large, there may not be any feasible solution as the maximal control may still be insufficient to meet the quality requirement.

12.3 Optimal Feedback Control for Linear Systems Subject to Poisson Processes

The main reference for this section is [239]. Consider a system governed by the following stochastic differential equation over a finite time interval $(0, T]$.

$$d\boldsymbol{x}(t) = A(t)\boldsymbol{x}(t)dt + B(t)d\boldsymbol{u}(t) + G(t)d\boldsymbol{N}(t), \tag{12.3.1}$$

where $\boldsymbol{x}(t) \in \mathbb{R}^n$, $A(t) \in \mathbb{R}^{n\times n}$, $B(t) \in \mathbb{R}^{n\times r}$, $\boldsymbol{u}(t) \in \mathbb{R}^r$ is a control function that is of bounded variation and hence $d\boldsymbol{u}(t)$ is a measure, $\Gamma(t) \in \mathbb{R}^{n\times m}$, and $\boldsymbol{N}(t) \in \mathbb{R}^m$ is an m-dimensional Poisson process with mean intensity $\boldsymbol{\lambda}(t)$. We assume that the matrix-valued functions A, B and G are continuous on $[0.T]$.

Along with (12.3.1), suppose we have an observation system described by

$$d\boldsymbol{y}(t) = H(t)\boldsymbol{x}(t)dt + \Gamma^0(t)\left(d\boldsymbol{N}^0(t) - \boldsymbol{\lambda}^0(t)dt\right), \tag{12.3.2}$$

where $\boldsymbol{y}(t) \in \mathbb{R}^k$, $H(t) \in \mathbb{R}^{k\times n}$, $\Gamma^0(t) \in \mathbb{R}^{k\times q}$ and $\boldsymbol{N}^0(t) \in \mathbb{R}^q$ is a q-dimensional Poisson process with mean intensity $\boldsymbol{\lambda}^0(t)$.

It is assumed that all the components of the Poisson processes $\{\boldsymbol{N}(t), \boldsymbol{N}^0(t)\}$ are statistically mutually independent. Furthermore, we assume that all the components of their mean intensities, $\boldsymbol{\lambda}(t)$ and $\boldsymbol{\lambda}^0(t)$, are non-negative and bounded measurable functions.

Suppose that the control function $\boldsymbol{u}$ is such that the corresponding measure $d\boldsymbol{u}(t)$ is of the form

$$d\boldsymbol{u}(t) = K\boldsymbol{y}(t)dt + \hat{K}d\boldsymbol{y}(t) - C(t)\Gamma(t)\boldsymbol{\lambda}(t)dt, \tag{12.3.3}$$

where K, $\hat{K} \in \mathbb{R}^{r\times k}$, are constant matrices yet to be determined and

$$B(t)C(t)\Gamma(t)\boldsymbol{\lambda}(t) = \Gamma(t)\boldsymbol{\lambda}(t), \tag{12.3.4}$$

provided such a matrix $C(t)$ exists. In fact, if $B(t)$ has rank r and $n > r$, then $C(t)$ is just the right inverse of $B(t)$. Substituting (12.3.3) and (12.3.2) into (12.3.1), we obtain

$$\begin{aligned} d\boldsymbol{x}(t) = {} & \left(A(t) + B(t)\hat{K}H(t)\right)\boldsymbol{x}(t)dt + B(t)K\boldsymbol{y}(t)dt \\ & + B(t)\hat{K}\Gamma^0(t)\left(d\boldsymbol{N}^0(t) - \boldsymbol{\lambda}^0(t)dt\right) + \Gamma(t)\left(d\boldsymbol{N}(t) - \boldsymbol{\lambda}(t)dt\right). \end{aligned} \tag{12.3.5}$$

Define

$$\boldsymbol{\xi}(t) = \begin{bmatrix} \boldsymbol{x}(t) \\ \boldsymbol{y}(t) \end{bmatrix}.$$

Then, the system dynamics (12.3.5) together with the observation dynamics (12.3.2) can be jointly written as

$$d\boldsymbol{\xi}(t) = \tilde{A}(t,\boldsymbol{\kappa})\boldsymbol{\xi}(t)dt + \tilde{\Gamma}(t,\boldsymbol{\kappa})d\tilde{\boldsymbol{M}}(t), \tag{12.3.6}$$

where the vector $\boldsymbol{\kappa} \in \mathbb{R}^{2rk}$ is defined by

$$\boldsymbol{\kappa} = \left[K_1, \ldots, K_r, \hat{K}_1, \ldots, \hat{K}_r\right],$$

$$\tilde{A}(t,\boldsymbol{\kappa}) = \begin{bmatrix} A(t) + B(t)\hat{K}H(t) & B(t)K \\ H(t) & 0 \end{bmatrix},$$

$$\tilde{\Gamma}(t,\boldsymbol{\kappa}) = \begin{bmatrix} \Gamma(t) & B(t)\hat{K}\Gamma^0(t) \\ 0 & \Gamma^0(t) \end{bmatrix},$$

$$d\tilde{\boldsymbol{M}}(t) = \begin{bmatrix} d\boldsymbol{N}(t) - \boldsymbol{\lambda}(t)dt \\ d\boldsymbol{N}^0(t) - \boldsymbol{\lambda}^0(t)dt \end{bmatrix},$$

and for each $j = 1, \ldots, r$, K_j (respectively, $\hat{K}_j$) is the jth row of the matrix K (respectively, $\hat{K}$). For convenience, let the components of the vector $\boldsymbol{\kappa}$ be denoted as

$$\kappa_j, \; j = 1, \ldots, 2rk.$$

Note that $\tilde{\boldsymbol{M}}$ is a vector of zero-mean martingales.

The initial condition for the system dynamics may be deterministic or Gaussian, i.e.,

$$\boldsymbol{x}(0) = \boldsymbol{x}^0, \tag{12.3.7}$$

where $\boldsymbol{x}^0 \in \mathbb{R}^n$ is either a deterministic or a Gaussian vector. In the case when $\boldsymbol{x}^0$ is a Gaussian vector, let $\bar{\boldsymbol{x}}^0$ and P^0 be its mean and covariance, respectively. Furthermore, it is assumed that $\boldsymbol{x}^0$ is statistically independent of $\boldsymbol{N}$ and $\boldsymbol{N}^0$.

The initial condition for the observation dynamics is usually assumed to be

$$\boldsymbol{y}(0) = \boldsymbol{0}, \tag{12.3.8}$$

that is, no information is available at $t = 0$. Thus, in the notation of (12.3.6), we have

$$\boldsymbol{\xi}(0) = \begin{bmatrix} \boldsymbol{x}^0 \\ \boldsymbol{0} \end{bmatrix} = \boldsymbol{\xi}. \tag{12.3.9}$$

Consider the following homogeneous system.

$$\frac{\partial \tilde{\Phi}(t,\tau)}{\partial t} = \tilde{A}(t,\boldsymbol{\kappa})\tilde{\Phi}(t,\tau), \;\; 0 \le \tau \le t < \infty, \tag{12.3.10a}$$

$$\tilde{\Phi}(t,t) = I, \;\; \text{for any } t \in [0,\infty), \tag{12.3.10b}$$

where I denotes the identity matrix. For each $\boldsymbol{\kappa}$, let $\tilde{\Phi}(t,\tau \mid \boldsymbol{\kappa})$ be the corresponding solution of (12.3.10). Then, it is clear that for each $\boldsymbol{\kappa}$, the corresponding solution of the system (12.3.6) with the initial condition (12.3.9) can be written as

$$\boldsymbol{\xi}(t \mid \boldsymbol{\kappa}) = \tilde{\Phi}(t,0 \mid \boldsymbol{\kappa})\boldsymbol{\xi}^0 + \int_0^t \tilde{\Phi}(t,\tau \mid \boldsymbol{\kappa})\tilde{\Gamma}(\tau,\boldsymbol{\kappa})d\tilde{\boldsymbol{M}}(\tau). \tag{12.3.11}$$

Since $\boldsymbol{\xi}^0$ and $\tilde{\boldsymbol{M}}$ are independent, it follows from taking the expectation of (12.3.11) that

$$\bar{\boldsymbol{\xi}}(t \mid \boldsymbol{\kappa}) = \tilde{\Phi}(t,0 \mid \boldsymbol{\kappa})\bar{\boldsymbol{\xi}}^0, \tag{12.3.12}$$

where

$$\bar{\boldsymbol{\xi}}^0 = \begin{bmatrix} \bar{\boldsymbol{x}}^0 \\ \boldsymbol{0} \end{bmatrix}.$$

Define

$$\boldsymbol{\mu}(t \mid \boldsymbol{\kappa}) = \bar{\boldsymbol{\xi}}(t \mid \boldsymbol{\kappa}).$$

By differentiating (12.3.12) and using (12.3.10), the following theorem can be readily obtained.

Theorem 12.3.1 *For each $\boldsymbol{\kappa}$, the mean behaviour of the corresponding solution of the coupled system* (12.3.6) *with the initial condition* (12.3.9) *is determined by the following system of deterministic differential equations*

$$\frac{d\boldsymbol{\mu}(t)}{dt} = \tilde{A}(t,\boldsymbol{\kappa})\boldsymbol{\mu}(t) \tag{12.3.13a}$$

with the initial condition

$$\boldsymbol{\mu}(0) = \bar{\boldsymbol{\xi}}^0. \tag{12.3.13b}$$

For the covariance matrix of the process $\boldsymbol{\xi}$, we have the following result.

Theorem 12.3.2 *For each $\boldsymbol{\kappa}$, let $\boldsymbol{\xi}(\cdot \mid \boldsymbol{\kappa})$ be the solution of the coupled system* (12.3.6) *with the initial condition* (12.3.9). *Then, the corresponding covariance matrix $\Psi(\cdot \mid \boldsymbol{\kappa})$ is determined by the following matrix differential equation.*

$$\frac{d\Psi(t)}{dt} = \tilde{A}(t,\boldsymbol{\kappa})\Psi(t) + \Psi(T)\left[\tilde{A}(t,\boldsymbol{\kappa})\right]^\top + \tilde{\Gamma}(t,\boldsymbol{\kappa})\tilde{\Lambda}(t)\left[\tilde{\Gamma}(t,\boldsymbol{\kappa})\right]^\top \tag{12.3.14a}$$

with the initial condition

$$\Psi(0) = \Psi^0 = \begin{bmatrix} P^0 & 0 \\ 0 & 0 \end{bmatrix}, \tag{12.3.14b}$$

where

$$\tilde{\Lambda}(t) = \begin{bmatrix} \Lambda(t) & 0 \\ 0 & \Lambda^0(t) \end{bmatrix}$$

with $\Lambda(t) = diag\ (\lambda_1(t), \ldots, \lambda_m(t))$, and $\Lambda^0(t) = diag\ \left(\lambda_1^0(t), \ldots, \lambda_q^0(t)\right)$. Furthermore, $P^0 \in \mathbb{R}^{n\times n}$ is obviously zero in the case when $\boldsymbol{x}^0$ is a deterministic vector.

Proof. From (12.3.11) and (12.3.12), it follows that

$$\boldsymbol{\xi}(t \mid \boldsymbol{\kappa}) - \bar{\boldsymbol{\xi}}(t \mid \boldsymbol{\kappa}) = \tilde{\Phi}(t,0 \mid \boldsymbol{\kappa})(\boldsymbol{\xi} - \boldsymbol{\xi}) + \int_0^t \tilde{\Phi}(t,\tau \mid \boldsymbol{\kappa})\tilde{\Gamma}(\tau,\boldsymbol{\kappa})d\tilde{\boldsymbol{M}}(\tau), \tag{12.3.15}$$

where the second term on the right hand side, which is a stochastic integral with respect to the martingale $\tilde{\boldsymbol{M}}$, is itself a martingale. Now, for any $\boldsymbol{\varphi} \in \mathbb{R}^{n+k}$, define

$$\boldsymbol{\varphi}^\top \Psi(t \mid \boldsymbol{\kappa})\boldsymbol{\varphi} = \mathcal{E}\left\{\left[\boldsymbol{\varphi}^\top(\boldsymbol{\xi}(t \mid \boldsymbol{\kappa}) - \bar{\boldsymbol{\xi}}(t \mid \boldsymbol{\kappa}))\right]^2\right\}, \tag{12.3.16}$$

where $\Psi(t \mid \boldsymbol{\kappa})$ is an $(n+k) \times (n+k)$ matrix yet to be determined.

From (12.3.15), it follows that

$$\left[\boldsymbol{\varphi}^\top(\boldsymbol{\xi}(t|\boldsymbol{\kappa}) - \bar{\boldsymbol{\xi}}(t|\boldsymbol{\kappa})\right]^2 = \boldsymbol{\varphi}^\top \Phi(t,0|\boldsymbol{\kappa})\left(\boldsymbol{\xi}^0 - \bar{\boldsymbol{\xi}}^0\right) + \int_0^t \boldsymbol{\varphi}^\top \tilde{\Phi}(t,\tau|\boldsymbol{\kappa})\tilde{\Gamma}(\tau,\boldsymbol{\kappa})d\tilde{\boldsymbol{M}}(\tau). \tag{12.3.17}$$

Taking the expectation of both sides and using the quadratic variation of the martingale $\tilde{\boldsymbol{M}}$ given by

$$\mathcal{E}\left\{\int_0^t \boldsymbol{\eta}^\top d\tilde{\boldsymbol{M}}(\tau)\right\} = \int_0^t \boldsymbol{\eta}^\top \tilde{\Lambda}\boldsymbol{\eta}d\tau, \quad \boldsymbol{\eta} \in \mathbb{R}^{m+q},$$

we obtain

$$\begin{aligned}\boldsymbol{\varphi}^\top \Psi(t \mid \boldsymbol{\kappa})\boldsymbol{\varphi} &= \boldsymbol{\varphi}^\top \tilde{\Phi}(t,0 \mid \boldsymbol{\kappa})\Psi^0\left(\tilde{\Phi}(t,0 \mid \boldsymbol{\kappa})\right)^\top \boldsymbol{\varphi} \\ &+ \int_0^t \boldsymbol{\varphi}^\top \tilde{\Phi}(t,\tau \mid \boldsymbol{\kappa})\tilde{\Gamma}(\tau,\boldsymbol{\kappa})\tilde{\Lambda}(\tau)\left(\tilde{\Gamma}(\tau,\boldsymbol{\kappa})\right)^\top\left(\tilde{\Phi}(t,\tau \mid \boldsymbol{\kappa})\right)^\top \boldsymbol{\varphi}d\tau. \end{aligned} \tag{12.3.18}$$

Since (12.3.18) is valid for arbitrary $\boldsymbol{\varphi} \in \mathbb{R}^{n+k}$, it follows that

$$\begin{aligned}\Psi(t \mid \boldsymbol{\kappa}) = {} & \tilde{\Phi}(t,0 \mid \boldsymbol{\kappa})\Psi^0\left(\tilde{\Phi}(t,0 \mid \boldsymbol{\kappa})\right)^\top + \int_0^t \tilde{\Phi}(t,\tau \mid \boldsymbol{\kappa})\tilde{\Gamma}(\tau,\boldsymbol{\kappa})\tilde{\Lambda}(\tau) \cdot \\ & \left(\tilde{\Gamma}(\tau,\boldsymbol{\kappa})\right)^\top\left(\tilde{\Phi}(t,\tau \mid \boldsymbol{\kappa})\right)^\top d\tau. \end{aligned} \tag{12.3.19}$$

With $t = 0$ in the above expression, it follows from (12.3.10b) that

$$\Psi(0 \mid \boldsymbol{\kappa}) = \Psi^0. \tag{12.3.20}$$

Now, by differentiating (12.3.19) and then using (12.3.10), we obtain (12.3.14a). Thus, the proof is complete.

Remark 12.3.1 *From* (12.3.14), *we observe readily that* $\Psi(t \mid \boldsymbol{\kappa})$ *is symmetric. Thus, we only need to solve a system of* $\left((n+k)^2 + (n+k)\right)/2$ *distinct differential equations for the determination of* $\Psi(t \mid \boldsymbol{\kappa})$.

Let $\boldsymbol{z}(t \mid \boldsymbol{\kappa})$ be a vector consisting of $\boldsymbol{\mu}(t \mid \boldsymbol{\kappa})$ and the independent components of $\Psi(t \mid \boldsymbol{\kappa})$. Then, for each $\boldsymbol{\kappa}$, $\boldsymbol{z}(t \mid \boldsymbol{\kappa})$ is determined by the following system of differential equations.

$$\frac{d\boldsymbol{z}(t)}{dt} = \boldsymbol{f}(t, \boldsymbol{z}(t), \boldsymbol{\kappa}) \tag{12.3.21a}$$

with the initial condition

$$\boldsymbol{z}(0) = \boldsymbol{z}^0, \tag{12.3.21b}$$

where $\boldsymbol{f}$ (respectively, $\boldsymbol{z}$) is determined by (12.3.13a) together with (12.3.14a) (respectively, (12.3.13b) together with (12.3.14b)).

12.3.1 Two Stochastic Optimal Feedback Control Problems

In this section, our aim is to formulate two classes of stochastic optimal feedback control problems based on the dynamical system (12.3.1), the observation dynamics (12.3.2) and the proposed control dynamics given by (12.3.3) (which is driven by the measurement process $\boldsymbol{y}$).

To begin, let us assume that the vector $\boldsymbol{\kappa}$ is to be chosen from the set $\mathbb{K}$ defined by

$$\begin{aligned}\mathbb{K} &= \left\{\boldsymbol{\kappa} = [\kappa_1, \ldots, \kappa_{2rk}] \in \mathbb{R}^{2rk} : \tilde{\boldsymbol{\beta}} \leq \boldsymbol{\kappa} < \bar{\boldsymbol{\beta}}\right\} \\ &= \left\{\boldsymbol{\kappa} = [\kappa_1, \ldots, \kappa_{2rk}]^\top \in \mathbb{R}^{2rk} : \tilde{\beta}_i \leq \kappa_i \leq \bar{\beta}_i,\ i = 1, \ldots, 2rk\right\}, \end{aligned} \tag{12.3.22}$$

where $\tilde{\boldsymbol{\beta}}$ and $\bar{\boldsymbol{\beta}}$ are given vectors in $\mathbb{R}^{2rk}$. For each $\boldsymbol{\kappa} \in \mathbb{K}$, let $\Psi(t \mid \boldsymbol{\kappa})$ be partitioned as follows.

$$\Psi(t \mid \boldsymbol{\kappa}) = \begin{bmatrix} \Psi_{11}(t \mid \boldsymbol{\kappa}) & \Psi_{12}(t \mid \boldsymbol{\kappa}) \\ \Psi_{21}(t \mid \boldsymbol{\kappa}) & \Psi_{22}(t \mid \boldsymbol{\kappa}) \end{bmatrix}, \tag{12.3.23}$$

where $\Psi_{11}(t \mid \boldsymbol{\kappa}) \in \mathbb{R}^{n\times n}$, $\Psi_{12}(t \mid \boldsymbol{\kappa}) \in \mathbb{R}^{n\times k}$, $\Psi_{21}(t \mid \boldsymbol{\kappa}) \in \mathbb{R}^{k\times n}$ and $\Psi_{22}(t \mid \boldsymbol{\kappa}) \in \mathbb{R}^{k\times k}$. Note that $\Psi_{11}(t \mid \boldsymbol{\kappa})$ and $\Psi_{22}(t \mid \boldsymbol{\kappa})$ are, respectively, the covariances of the processes $\boldsymbol{x}(t \mid \boldsymbol{\kappa})$ and $\boldsymbol{y}(t \mid \boldsymbol{\kappa})$, i.e.,

$$\boldsymbol{\eta}^\top \Psi_{11}(t \mid \boldsymbol{\kappa})\boldsymbol{\eta} = \mathcal{E}\left\{\boldsymbol{\eta}^\top\left(\boldsymbol{x}(t \mid \boldsymbol{\kappa}) - \bar{\boldsymbol{x}}(t \mid \boldsymbol{\kappa})\right)\right\}^2,\ \boldsymbol{\eta} \in \mathbb{R}^n$$

and

$$\boldsymbol{\nu}^\top \Psi_{22}(t \mid \boldsymbol{\kappa})\boldsymbol{\nu} = \mathcal{E}\left\{\boldsymbol{\nu}^\top\left(\boldsymbol{y}(t \mid \boldsymbol{\kappa}) - \bar{\boldsymbol{y}}(t \mid \boldsymbol{\kappa})\right)\right\}^2,\ \boldsymbol{\nu} \in \mathbb{R}^k,$$

while $\Psi_{12}(t \mid \boldsymbol{\kappa})$ and $\Psi_{21}(t \mid \boldsymbol{\kappa})$ are cross covariances.

With these preparations, the first problem may be stated formally as follows.

Subject to the dynamic system (12.3.1), the initial condition (12.3.7), the observation channel (12.3.2) with the initial condition (12.3.8) and the control system given by (12.3.3), find a constant vector $\boldsymbol{\kappa} \in \mathbb{K}$ such that the cost functional

$$g_0(\boldsymbol{\kappa}) = \mathcal{E}\left\{\int_0^T \operatorname{Tr}\left\{(\boldsymbol{x}(t \mid \boldsymbol{\kappa}) - \bar{\boldsymbol{x}}(t \mid \boldsymbol{\kappa}))\left[\boldsymbol{x}(t \mid \boldsymbol{\kappa}) - \bar{\boldsymbol{x}}(t \mid \boldsymbol{\kappa})\right]^\top\right\} dt\right\} \tag{12.3.24}$$

is minimized over $\mathbb{K}$.

For convenience, let this (stochastic optimal feedback) control problem be referred to as Problem (SP_{2a}). Note that Problem (SP_{2a}) aims to find an optimal vector (and hence feedback matrix) $\boldsymbol{\kappa} \in \mathbb{K}$ such that the resulting system (12.3.6) with the initial condition (12.3.9) is least noisy.

In our second problem, our aim is to find a constant vector (and hence constant feedback matrix) $\boldsymbol{\kappa} \in \mathbb{K}$ such that the mean behaviour of the corresponding dynamical system is closest to a given deterministic trajectory, while the uncertainty of the corresponding dynamical system is within a given acceptable limit. Let the given deterministic trajectory be denoted by $\hat{\boldsymbol{x}}(t)$. Then, the corresponding problem, which is identified as Problem (SP_{2b}), may be stated formally as follows.

Given the system (12.3.1) with the initial condition (12.3.7), the observation channel (12.3.2) with the initial condition (12.3.8) and the proposed control dynamics of the form (12.3.3), find a constant vector $\boldsymbol{\kappa} \in \mathbb{K}$ such that the cost functional

$$g_0(\boldsymbol{\kappa}) = \int_0^T \|\bar{\boldsymbol{x}}(t \mid \boldsymbol{\kappa}) - \hat{\boldsymbol{x}}(t)\|^2 \, dt \tag{12.3.25}$$

is minimized subject to $\boldsymbol{\kappa} \in \mathbb{K}$ and the constraint

$$\mathcal{E}\left\{\int_0^T \operatorname{Tr}\left\{(\boldsymbol{x}(t \mid \boldsymbol{\kappa}) - \bar{\boldsymbol{x}}(t \mid \boldsymbol{\kappa}))\left(\boldsymbol{x}(t \mid \boldsymbol{\kappa}) - \bar{\boldsymbol{x}}(t \mid \boldsymbol{\kappa})\right)^\top\right\} dt\right\} \leq \varepsilon, \tag{12.3.26}$$

where ε is a positive constant corresponding to some acceptable level of uncertainty.

12.3.2 Deterministic Model Transformation

The stochastic optimal feedback control problems as stated above are difficult to solve. However, by virtue of the structure of the dynamical system, the observation channel and the form of the control law, we can show that these problems are, in fact, equivalent to certain deterministic optimal parameter selection problems.

We first define the following deterministic optimal parameter selection problem, denoted as Problem (DP_{2a}).

Subject to the system (12.3.21), find a constant vector $\boldsymbol{\kappa} \in \mathbb{K}$ such that the cost functional

$$g_0(\boldsymbol{\kappa}) = \int_0^T \mathrm{Tr}\left\{\Psi_{11}(t \mid \boldsymbol{\kappa})\right\} dt = \int_0^T \mathrm{Tr}\left\{M\Psi(t \mid \boldsymbol{\kappa})\right\} dt \tag{12.3.27}$$

is minimized over $\mathbb{K}$, where $M \in \mathbb{R}^{(n+k)\times(n+k)}$ is given by

$$M = \begin{bmatrix} I_{n\times n} & 0 \\ 0 & 0 \end{bmatrix}$$

and $I_{n\times n}$ is the identity matrix in $\mathbb{R}^{n\times n}$.

Theorem 12.3.3 *Problem (SP_{2a}) is equivalent to Problem (DP_{2a}).*

Proof. Let $\Psi_{11}(t \mid \boldsymbol{\kappa})$ be as defined for the matrix $\Psi(t \mid \boldsymbol{\kappa})$ given by (12.3.23). Then, it is clear that

$$\begin{aligned}
&\mathcal{E}\left\{\int_0^T \mathrm{Tr}\left\{(\boldsymbol{x}((t \mid \boldsymbol{\kappa}) - \bar{\boldsymbol{x}}(t \mid \boldsymbol{\kappa}))\,(\boldsymbol{x}((t \mid \boldsymbol{\kappa}) - \bar{\boldsymbol{x}}(t \mid \boldsymbol{\kappa}))^\top\right\} dt\right\} \\
&= \mathrm{Tr}\int_0^T \mathcal{E}\left\{(\boldsymbol{x}(t \mid \boldsymbol{\kappa}) - \bar{\boldsymbol{x}}(t \mid \boldsymbol{\kappa}))\,(\boldsymbol{x}(t \mid \boldsymbol{\kappa}) - \bar{\boldsymbol{x}}(t \mid \boldsymbol{\kappa}))^\top\right\} dt \\
&= \int_0^T \mathrm{Tr}\left\{\Psi_{11}(t \mid \boldsymbol{\kappa})\right\} dt \\
&= \int_0^T \mathrm{Tr}\left\{M\Psi(t \mid \boldsymbol{\kappa})\right\} dt.
\end{aligned}$$

The proof is complete.

We now turn our attention to Problem (SP_{2b}) and define the following deterministic optimal parameter selection problem, to be denoted as Problem (DP_{2b}).

Given the system (12.3.21), find a constant feedback vector $\boldsymbol{\kappa} \in \mathbb{K}$ such that the cost functional (12.3.25) is minimized subject to $\boldsymbol{\kappa} \in \mathbb{K}$ and the constraint

$$\int_0^T \mathrm{Tr}\left\{M\Psi(t \mid \boldsymbol{\kappa})\right\} dt \le \varepsilon, \tag{12.3.28}$$

where $\varepsilon > 0$ is an appropriate positive number and M is the $(n+k)\times(n+k)$ matrix introduced in (12.3.27).

Theorem 12.3.4 *The stochastic problem (SP_{2b}) is equivalent to the deterministic optimal parameter selection problem (DP_{2b}).*

Proof. We only need to show that the constraint (12.3.26) is equivalent to (12.3.28). The proof of this equivalence is similar to that given for Theorem 12.3.3.

Remark 12.3.2 *Note that our formulation also holds for the case of time-varying control matrices $K = K(t)$, $\hat{K} = \hat{K}(t)$, $t > 0$. In this case, Problems (DP_{2a}) and (DP_{2b}) corresponding to Problems (SP_{2a}) and (SP_{2b}), as described above, are to be considered as deterministic optimal control problems with controls $K(t)$ and $\hat{K}(t)$ rather than as deterministic optimal parameter selection problems with constant matrices K and $\hat{K}$.*

12.3.3 An Example

This example is taken from [234]. Consider a machine maintenance problem, where there are two types of maintenance action. The first type is continuous (minor) maintenance. It works to slow down natural degradation of the machine. The second type is overhaul (major) maintenance. It is carried out at certain discrete time points so as to significantly improve the condition of the machine. For this machine maintenance problem, its condition is modeled as an impulsive stochastic differential equation over the time horizon. The objective is to choose the continuous maintenance rate and the overhaul maintenance times such that the total cost of operating and maintaining the machine is minimized subject to constraints on the state and output of the machine satisfying minimum acceptable levels with high probability.

Let $x(t)$ denote the state of the machine at time t, and let $y(t)$ denote the total output produced by the machine up to time t. The state and output of the machine are, respectively, governed by the following stochastic differential equations:

$$dx(t) = (u(t) - k_1)x(t)dt + k_2 dw(t) \tag{12.3.29}$$

$$dy(t) = k_3 x(t)dt, \tag{12.3.30}$$

where $u(t)$ denotes the continuous maintenance rate; $w(t)$ denotes the standard Brownian motion with mean 0 and covariance given by

$$Cov\{w(t_1), w(t_2)\} = \min\{t_1, t_2\},$$

and k_1, k_2 and k_3 are the given constants. These constants represent, respectively, the natural degradation rate of the machine, the propensity for random fluctuations in the condition of the machine and the extent to which the production is being influenced by the state of the machine. It is assumed that the continuous maintenance rate is subject to the following boundedness constraints:

$$0 \leq u(t) \leq ak_1,\ t \geq 0, \tag{12.3.31}$$

where $a \in (0, 1)$ is a given constant. The initial state of the machine and the initial production level are, respectively, given by

$$x(0) = x^0 + \delta_0, \tag{12.3.32}$$
$$y(0) = 0 \tag{12.3.33}$$

where δ_0 is a normal random variable with mean 0 and variance k_4. The machine is regarded to be operating in an almost perfect condition when $x(t) \approx x^*$.

For each $i = 1, \ldots, N+1$, let τ_i denote the time of the i-th overhaul, where N is the number of times the machine being overhauled and τ_{N+1} is referred to as the final time (i.e., the time at which the machine is replaced). To ensure that overhauls do not happen too frequently, the following constraints are imposed.

$$\tau_i - \tau_{i-1} \geq \rho, \quad i = 1, \ldots, N+1, \tag{12.3.34}$$

where $\rho > 0$ denotes the minimum duration between any two consecutive overhauls. To ensure that the time, τ_{N+1}, for the machine being replaced is greater than or equal to $t_{\min}$, the following condition is imposed.

$$\tau_{N+1} \geq t_{\min}. \tag{12.3.35}$$

We consider the situation where the time required for each overhaul is negligible when compared with the length of the time horizon. Thus, the state of the machine improves instantaneously at each overhaul time. On the other hand, the output level stays the same. This phenomenon is modeled by the following jump conditions:

$$x\left(\tau_i^+\right) = k_5 x(\tau_i^-) + \delta_i, \quad i = 1, \ldots, N \tag{12.3.36}$$
$$y\left(\tau_i^+\right) = y(\tau_i^-), \quad i = 1, \ldots, N, \tag{12.3.37}$$

where k_5 is a positive constant and δ_i is a normal random variable with mean 0 and variance k_6. We assume throughout that the Brownian motion $w(t)$ and the random variables δ_i, $i = 0, \ldots, N$, are mutually statistically independent.

There are two operational requirements that are required to be satisfied. First, the state of the machine state is required to stay above a minimum acceptable level with a high probability. Thus, we impose the following probabilistic state constraint:

$$\Pr\{x(t) \geq x_{\min}\} \geq p_1, \quad t \in [0, \tau_{N+1}], \tag{12.3.38}$$

where $x_{\min}$ is the minimum acceptable level of the state of the machine and p_1 is a given probability level. Second, the accumulated output level over the entire time horizon is required to be greater than or equal to a specified minimum level with a high probability. This requirement can be modeled as given below:

$$\Pr\{y(\tau_{N+1}) \geq y_{\min}\} \geq p_2, \tag{12.3.39}$$

where $y_{\min}$ denotes the minimum output level and p_2 is a given probability level. Note that constraint (12.3.39) is only imposed at the final time, while constraint (12.3.38) is imposed at each time over the time horizon.

Let $\boldsymbol{\tau} = [\tau_1, \ldots, \tau_{N+1}]^\top$ denote the vector of overhaul times. Furthermore, let $\varUpsilon$ be the set defined by

$$\varUpsilon = \left\{\boldsymbol{\tau} \in \mathbb{R}^{N+1} : \tau_i - \tau_{i-1} \geq \rho,\ \ i = 1, \ldots, N+1;\ \tau_{N+1} \geq t_{\min}\right\}. \tag{12.3.40}$$

A vector $\boldsymbol{\tau} \in \varUpsilon$ is called an admissible overhaul time vector. For simplicity, we assume that the continuous maintenance rate is constant between consecutive overhauls (note that it is easy to extend the approach to the case where the continuous maintenance rate takes several different constant levels between overhauls). For a given $\boldsymbol{\tau} \in \varUpsilon$, let $u : [0, \infty) \to \mathbb{R}$ be a piecewise constant function that takes a constant value on each of the intervals $[\tau_{i-1}, \tau_i)$, $i = 1, \ldots, N+1$. If such a u satisfies (12.3.31), then it is called an admissible control.

Let $\mathcal{U}(\boldsymbol{\tau})$ be the class of all admissible controls corresponding to $\boldsymbol{\tau} \in \varUpsilon$. Then, any element $(\boldsymbol{\tau}, u) \in \varUpsilon \times \mathcal{U}(\boldsymbol{\tau})$ is called an admissible pair. Any admissible pair satisfying constraints (12.3.38) and (12.3.39) is called a feasible pair.

Our goal is to choose a feasible pair such that the following cost functional is minimized.

$$\begin{aligned} g_0(\boldsymbol{\tau}, u) = & \int_0^{\tau_{N+1}} \mathcal{E}\left\{ \underbrace{\mathcal{L}_1(x(t))}_{\text{Operating cost}} + \underbrace{\mathcal{L}_2(u(t))}_{\text{continuous maintenance cost}} \right\} dt \\ & + \sum_{i=1}^{N} \mathcal{E}\left\{ \underbrace{\Psi_1(x(\tau_i^-))}_{\text{Overhaul cost}} \right\} - \mathcal{E}\left\{ \underbrace{\Psi_2(x(\tau_{N+1}))}_{\text{Salvage cost}} \right\}, \end{aligned} \tag{12.3.41}$$

where $\mathcal{E}$ denotes the mathematical expectation. This cost functional consists of four components: (1) the operating cost, (2) the continuous maintenance cost, (3) the overhaul cost and (4) the salvage value.

We assume that the following conditions are satisfied.

Assumption 12.3.1 *$\mathcal{L}_1 : \mathbb{R} \to \mathbb{R}$ and $\Psi_2 : \mathbb{R} \to \mathbb{R}$ are quadratic.*

Assumption 12.3.2 *$\mathcal{L}_2 : \mathbb{R} \to \mathbb{R}$ is continuously differentiable with respect to each of its arguments.*

Assumption 12.3.3 *$\Psi_1 : \mathbb{R} \to \mathbb{R}$ is linear.*

The problem may now be stated formally as the following stochastic optimal control problem.

Problem (EP_0). Given the system of stochastic differential equations (12.3.29) and (12.3.30) with the initial conditions (12.3.32) and (12.3.33) and the jump conditions (12.3.36) and (12.3.37), find an admissible pair $(\boldsymbol{\tau}, u) \in \Upsilon \times \mathcal{U}(\boldsymbol{\tau})$ such that the cost functional $g_0(\boldsymbol{\tau}, u)$ defined by (12.3.41) is minimized subject to the constraints (12.3.39) and (12.3.40).

Problem (EP_0) is a stochastic impulsive optimal control problem with probabilistic state constraints. It is solved using the approach proposed in this section. First, it is transformed into a new deterministic optimal control problem. Define

$$\mu_x(t) = \mathcal{E}[x(t)], \quad \mu_y(t) = \mathcal{E}[y(t)] \tag{12.3.42}$$

$$\sigma_{xx}(t) = Var[x(t)], \quad \sigma_{yy}(t) = Var[y(t)], \tag{12.3.43}$$

$$\sigma_{xy}(t) = \sigma_{yx}(t) = Cov\left\{x(t), y(t)\right\}. \tag{12.3.44}$$

Let $\Phi : \mathbb{R} \times \mathbb{R} \to \mathbb{R}^{2\times 2}$ denote the principal solution matrix of the following homogeneous system:

$$\frac{\partial \Phi(t,s)}{\partial t} = \begin{bmatrix} u(t) - k_1 & 0 \\ k_3 & 0 \end{bmatrix} \Phi(t,s), \quad t > s, \tag{12.3.45}$$

$$\Phi(s,s) = I, \tag{12.3.46}$$

where

$$\Phi(t,s) = \begin{bmatrix} \phi_{11}(t,s) & \phi_{12}(t,s) \\ \phi_{21}(t,s) & \phi_{22}(t,s) \end{bmatrix}. \tag{12.3.47}$$

Then, it is known that for each $i = 1, \ldots, N+1$, the solution of the stochastic impulsive system (12.3.29) and (12.3.30) on $[\tau_{i-1}, \tau_i)$ can be expressed as follows:

$$\begin{bmatrix} x(t) \\ y(t) \end{bmatrix} = \Phi(t, \tau_{i-1}) \begin{bmatrix} x\left(\tau_{i-1}^+\right) \\ y\left(\tau_{i-1}^+\right) \end{bmatrix} + \int_{\tau_{i-1}}^{t} \Phi(t,s) \begin{bmatrix} k_2 \\ 0 \end{bmatrix} dw(s). \tag{12.3.48}$$

This can be written as

$$x(t) = \phi_{11}\left(t, \tau_{i-1}\right) x\left(\tau_{i-1}^+\right) + \phi_{12}\left(t, \tau_{i-1}\right) y\left(\tau_{i-1}^+\right) + \int_{\tau_{i-1}}^{t} k_2 \phi_{11}(t,s) dw(s) \tag{12.3.49}$$

and

$$y(t) = \phi_{21}(t, \tau_{i-1}) x\left(\tau_{i-1}^+\right) + \phi_{22}(t, \tau_{i-1}) y\left(\tau_{i-1}^+\right) + \int_{\tau_{i-1}}^{t} k_2 \phi_{21}(t,s) dw(s). \tag{12.3.50}$$

Taking the expectation of $x(t)$ and $y(t)$ gives

$$\mu_x(t) = \phi_{11}(t, \tau_{i-1})\mu_x\left(\tau_{i-1}^+\right) + \phi_{12}(t, \tau_{i-1})\mu_y\left(\tau_{i-1}^+\right) \tag{12.3.51}$$

$$\mu_y(t) = \phi_{21}(t, \tau_{i-1})\mu_x\left(\tau_{i-1}^+\right) + \phi_{22}(t, \tau_{i-1})\mu_y\left(\tau_{i-1}^+\right). \tag{12.3.52}$$

By differentiating (12.3.51) and (12.3.52) with respect to t, we obtain

$$\begin{aligned}\frac{d\mu_x(t)}{dt} &= (u(t) - k_1)\left(\phi_{11}(t, \tau_{i-1})\mu_x\left(\tau_{i-1}^+\right) + \phi_{12}(t, \tau_{i-1})\mu_y\left(\tau_{i-1}^+\right)\right)\\ &= (u(t) - k_1)\mu_x(t) \end{aligned}\tag{12.3.53}$$

and

$$\begin{aligned}\frac{d\mu_y(t)}{dt} &= k_3\left(\phi_{11}(t, \tau_{i-1})\mu_x(\tau_{i-1}^+) + \phi_{12}(t, \tau_{i-1})\mu_y(\tau_{i-1}^+)\right)\\ &= k_3\mu_x(t). \end{aligned}\tag{12.3.54}$$

Now, their variances can be calculated as given below:

$$\begin{aligned}\sigma_{xx}(t) &= \phi_{11}^2(t, \tau_{i-1})\sigma_{xx}\left(\tau_{i-1}^+\right) + \phi_{12}^2(t, \tau_{i-1})\sigma_{yy}\left(\tau_{i-1}^+\right)\\ &\quad + 2\phi_{11}(t, \tau_{i-1})\phi_{12}(t, \tau_{i-1})\sigma_{xy}\left(\tau_{i-1}^+\right) + \int_{\tau_{i-1}}^{t} k_2^2\phi_{11}^2(t, s)ds \end{aligned}\tag{12.3.55}$$

and

$$\begin{aligned}\sigma_{yy}(t) &= \phi_{21}^2(t, \tau_{i-1})\sigma_{xx}\left(\tau_{i-1}^+\right) + \phi_{22}^2(t, \tau_{i-1})\sigma_{yy}\left(\tau_{i-1}^+\right)\\ &\quad + 2\phi_{21}(t, \tau_{i-1})\phi_{22}(t, \tau_{i-1})\sigma_{xy}\left(\tau_{i-1}^+\right) + \int_{\tau_{i-1}}^{t} k_2^2\phi_{21}^2(t, s)ds. \end{aligned}\tag{12.3.56}$$

Furthermore, the covariance is

$$\begin{aligned}\sigma_{xy}(t) &= \sigma_{yx}(t)\\ &= \phi_{11}(t, \tau_{i-1})\phi_{21}(t, \tau_{i-1})\sigma_{xx}\left(\tau_{i-1}^+\right)\\ &\quad + \phi_{11}(t, \tau_{i-1})\phi_{22}(t, \tau_{i-1}) + \phi_{12}(t, \tau_{i-1})\phi_{21}(t, \tau_{i-1})\sigma_{xy}\left(\tau_{i-1}^+\right)\\ &\quad + \phi_{12}(t, \tau_{i-1})\phi_{22}(t, \tau_{i-1})\sigma_{yy}\left(\tau_{i-1}^+\right) + \int_{\tau_{i-1}}^{t} k_2^2\phi_{11}(t, s)\phi_{21}(t, s)ds. \end{aligned}\tag{12.3.57}$$

Differentiating (12.3.55)–(12.3.57) with respect to time, we obtain

$$\begin{aligned}\frac{d\sigma_{xx}(t)}{dt} &= 2(u(t) - k_1)\left(\phi_{11}^2(t, \tau_{i-1})\sigma_{xx}\left(\tau_{i-1}^+\right) + \phi_{12}^2(t, \tau_{i-1})\sigma_{yy}\left(\tau_{i-1}^+\right)\right)\\ &\quad + 4(u(t) - k_1)\phi_{11}(t, \tau_{i-1})\phi_{12}(t, \tau_{i-1})\sigma_{xy}\left(\tau_{i-1}^+\right)\end{aligned}$$

$$
\begin{aligned}
&+ 2(u(t) - k_1) \int_{\tau_{i-1}}^{t} k_2^2 \phi_{11}^2(t,s) ds \\
&= 2(u(t) - k_1)\sigma_{xx}(t) + k_2^2 \qquad (12.3.58) \\
&\frac{d\sigma_{yy}(t)}{dt} \\
&= 2k_3\phi_{11}(t,\tau_{i-1})\phi_{21}(t,\tau_{i-1})\sigma_{xx}\left(\tau_{i-1}^{+}\right) \\
&+ 2k_3\phi_{12}(t,\tau_{i-1})\phi_{22}(t,\tau_{i-1})\sigma_{yy}\left(\tau_{i-1}^{+}\right) \\
&+ 2k_3\left(\phi_{11}(t,\tau_{i-1})\phi_{22}(t,\tau_{i-1}) + \phi_{12}(t,\tau_{i-1})\phi_{21}(t,\tau_{i-1})\right)\sigma_{xy}\left(\tau_{i-1}^{+}\right) \\
&+ 2k_3 \int_{\tau_{i-1}}^{t} k_2^2 \phi_{11}(t,s)\phi_{21}(t,s) ds \\
&= 2k_3\sigma_{xy}(t) \qquad (12.3.59)
\end{aligned}
$$

$$
\begin{aligned}
\frac{d\sigma_{xy}(t)}{dt} &= \frac{d\sigma_{yx}(t)}{dt} \\
&= (u(t) - k_1)\phi_{11}(t,\tau_{i-1})\phi_{21}(t,\tau_{i-1})\sigma_{xx}\left(\tau_{i-1}^{+}\right) \\
&+ (u(t) - k_1)\phi_{11}(t,\tau_{i-1})\phi_{22}(t,\tau_{i-1}) \\
&+ \phi_{12}(t,\tau_{i-1})\phi_{21}(t,\tau_{i-1})\sigma_{xy}\left(\tau_{i-1}^{+}\right) \\
&+ (u(t) - k_1)\phi_{12}(t,\tau_{i-1})\phi_{22}(t,\tau_{i-1})\sigma_{yy}\left(\tau_{i-1}^{+}\right) \\
&+ (u(t) - k_1) \int_{\tau_{i-1}}^{t} k_2^2 \phi_{11}(t,s)\phi_{21}(t,s) ds \\
&+ k_3\left(\phi_{11}^2(t,\tau_{i-1})\sigma_{xx}\left(\tau_{i-1}^{+}\right) + \phi_{21}^2(t,\tau_{i-1})\sigma_{yy}\left(\tau_{i-1}^{+}\right)\right) \\
&+ k_3\left(2\phi_{11}(t,\tau_{i-1})\phi_{12}(t,\tau_{i-1})\sigma_{xy}\left(\tau_{i-1}^{+}\right) + \int_{\tau_{i-1}}^{t} k_2^2 \phi_{11}^2(t,s) ds\right) \\
&= (u(t) - k_1)\sigma_{xy}(t) + k_3\sigma_{xx}(t). \qquad (12.3.60)
\end{aligned}
$$

The mean, variance and covariance of the initial conditions (12.3.32) and (12.3.33) are

$$\mu_x(0) = x^*, \quad \mu_y(0) = 0 \qquad (12.3.61)$$

$$\sigma_{xx}(0) = k_4, \quad \sigma_{yy}(0) = 0, \quad \sigma_{xy}(0) = \sigma_{yx}(0) = 0. \qquad (12.3.62)$$

At the overhaul times $t = \tau_i$, $i = 1, \ldots, N$, the mean, variance and covariance of the state jump conditions (12.3.36) and (12.3.37) are

$$\mu_x\left(\tau_i^{+}\right) = k_5\mu_x(\tau_i^{-}),\ \mu_y\left(\tau_i^{+}\right) = \mu_y(\tau_i^{-}), \qquad (12.3.63)$$

$$\sigma_{xx}\left(\tau_i^{+}\right) = k_5^2\sigma_{xx}(\tau_i^{-}) + k_6, \quad \sigma_{yy}\left(\tau_i^{+}\right) = \sigma_{yy}(\tau_i^{-}) \qquad (12.3.64)$$

$$\sigma_{xy}\left(\tau_i^{+}\right) = \sigma_{yx}\left(\tau_i^{+}\right) = k_5\sigma_{xy}(\tau_i^{-}). \qquad (12.3.65)$$

Since the state equations (12.3.29) and (12.3.30) and the jump conditions (12.3.36) and (12.3.37) are linear, $x(t)$ and $y(t)$ are mixtures of normally distributed random variables. Thus, the probabilistic state constraints (12.3.39) and (12.3.40) can be written as follows:

$$\int_{x_{\min}}^{\infty} \frac{1}{(2\pi\sigma_{xx}(t))^2} \exp\left\{\frac{-(\eta-\mu_x(t))^2}{2\sigma_{xx}(t)}\right\} d\eta \geq p_1, \quad t \in [0, \tau_{N+1}] \quad (12.3.66)$$

$$\int_{x_{\min}}^{\infty} \frac{1}{(2\pi\sigma_{yy}(\tau_{N+1}))^2} \exp\left\{\frac{-(\eta-\mu_y(\tau_{N+1}))^2}{2\sigma_{yy}(\tau_{N+1})}\right\} d\eta \geq p_2. \quad (12.3.67)$$

Constraint (12.3.66) is a continuous inequality constraint in terms of the new state variables μ_x and σ_{xx}, and constraint (12.3.67) is a terminal state constraint involving the new state variables μ_y and σ_{yy}.

As the functions $\mathcal{L}_1(\cdot)$ and $\Psi_2(\cdot)$ appearing in the cost functional (12.3.41) are quadratic, we can express $\mathcal{E}[x(t)]$ and $\mathcal{E}[\Psi_2(x(\tau_{N+1}))]$ in terms of the new state variables by replacing $\mathcal{E}[x(t)]$ and $\mathcal{E}[(x(t))^2]$ with $\mu_x(t)$ and $\sigma_{xx}(t) + \mu_x^2(t)$, respectively. We denote the resulting functions by $\widetilde{\mathcal{L}}_1(\mu_x(t), \sigma_{xx}(t))$ and $\widetilde{\Psi}_2(\mu_x(\tau_{N+1}), \sigma_{xx}(\tau_{N+1}))$, respectively. Similarly, since Ψ_1 is linear, we have

$$\mathcal{E}[\Psi_1(x(\tau_i^-))] = \Psi_1(\mu_x(\tau_i^-)).$$

Thus, the cost functional (12.3.41) can be written as

$$g_0(\boldsymbol{\tau}, u) = \int_0^{\tau_{N+1}} \left\{\widetilde{\mathcal{L}}_1(\mu_x(t), \sigma_{xx}(t)) + \widetilde{\mathcal{L}}_2(u(t))\right\} dt$$
$$+ \sum_{i=1}^{N} \Psi_1(\mu_x(\tau_i^-)) - \widetilde{\Psi}_2(\mu_x(\tau_{N+1}), \sigma_{xx}(\tau_{N+1})). \quad (12.3.68)$$

We are now able to state the transformed problem as follows.

Problem (PE_1). Given the dynamic system (12.3.53) and (12.3.54) and (12.3.58)–(12.3.60) with the initial conditions (12.3.61) and (12.3.62) and the jump conditions (12.3.63)–(12.3.65), find an admissible pair $(\boldsymbol{\tau}, u) \in \Upsilon \times \mathcal{U}(\boldsymbol{\tau})$ such that the cost function (12.3.68) is minimized subject to constraints (12.3.66) and (12.3.67).

In Problem (PE_1), the state of the machine experiences N instantaneous jumps during the time horizon. The times at which these jumps occur are actually decision variables to be optimized. Now, by applying the time scaling transformation (see Section 7.4.2), the time scale $t \in [0, \tau_{N+1}]$ is mapped into the new time scale $s \in [0, N+1]$ such that the variable jump points are mapped into fixed jump points. This mapping is realized by the following differential equation:

$$\frac{dt(s)}{ds} = \widetilde{v}(s) = \sum_{i=1}^{N+1} v_i \chi_{[i-1,i)}(s) \quad (12.3.69)$$

$$t(0) = 0, \quad (12.3.70)$$

where $v_i = \tau_i - \tau_{i-1} \geq \rho$ for each $i = 1, \ldots, N+1$, $v_1 + \cdots + v_{N+1} \geq t_{\min}$, and $\chi_{[i-1,i)}(s)$ is the indicator function of $[i-1, i)$.

Note that v_i denotes the time duration between the $(i-1)$th and ith jump times. We collect the duration parameters into a vector $\boldsymbol{v} = [v_1, \ldots, v_{N+!}]^\top \in \mathbb{R}^{N+1}$. Define

$$\mathcal{V} = \left\{ \boldsymbol{v} \in \mathbb{R}^{N+1} : v_i \geq \rho,\ \ i = 1, \ldots, N+1;\ \ v_1 + \cdots + v_{N+1} \geq t_{\min} \right\}. \tag{12.3.71}$$

A vector $\boldsymbol{v} \in \mathcal{V}$ is called an admissible duration vector. From (12.3.69) and (12.3.70), we have, for each $i = 1, \ldots, N+1$,

$$t(i) = t(0) + \int_0^i \widetilde{v}(s) ds = t(0) + v_1 + \cdots + v_i = \tau_i. \tag{12.3.72}$$

This equation shows the relationship between the variable jump points $t = \tau_i$, $i = 1, \ldots, N+1$, and the fixed jump points $s = i$, $i = 1, \ldots, N+1$.

Define $\widetilde{u}(s) = u(t(s))$. Recall that the continuous maintenance rate is constant between consecutive overhauls. Thus, the admissible controls are restricted to piecewise constant functions that assume constant values between consecutive jump times. As a result, $\widetilde{u}(s)$ can be expressed as

$$\widetilde{u}(s) = \sum_{i=1}^{N+1} h_i \chi_{[i-1,i)}(s), \tag{12.3.73}$$

where h_i, $i = 1, \ldots, N+1$, are control heights to be optimized. In view of (12.3.31), these control heights must satisfy the following constraints:

$$0 \leq h_i \leq ak_1,\ \ i = 1, \ldots, N+1. \tag{12.3.74}$$

Let $\boldsymbol{h} = [h_1, \ldots, h_{N+1}]^\top \in \mathbb{R}^{N+1}$. Furthermore, define

$$\mathcal{H} = \left\{ \boldsymbol{h} \in \mathbb{R}^{N+1} : 0 \leq h_i \leq ak_1,\ i = 1, \ldots, N+1 \right\}. \tag{12.3.75}$$

A vector $\boldsymbol{h} \in \mathcal{H}$ is called an admissible control parameter vector. Furthermore, a pair $(\boldsymbol{h}, \boldsymbol{v}) \in \mathcal{H} \times \mathcal{V}$ is called an admissible pair.

We assume throughout that the control switches coincide with the overhaul times. Note that it is straightforward to consider the case in which the control can switch value between consecutive jump times, as well as at the jump times themselves. However, the notation will be more involved. Let

$$\widetilde{\mu}_x(s) = \mu_x(t(s)),\quad \widetilde{\mu}_y(s) = \mu_y(t(s), \tag{12.3.76}$$

$$\widetilde{\sigma}_{xx}(s) = \sigma_{xx}(t(s)),\, \widetilde{\sigma}_{yy}(s) = \sigma_{yy}(t(s)),\, \widetilde{\sigma}_{xy}(s) = \widetilde{\sigma}_{yx}(s) = \sigma_{xy}(t(s)). \tag{12.3.77}$$

Then, the dynamics (12.3.53) and (12.3.54) and (12.3.58)–(12.3.60) are transformed into

$$\frac{d\widetilde{\mu}_x(s)}{ds} = \widetilde{v}(s)(\widetilde{u}(s) - k_1)\widetilde{\mu}_x(s) \tag{12.3.78}$$

$$\frac{d\widetilde{\mu}_y(s)}{ds} = k_3\widetilde{v}(s)\widetilde{\mu}_x(s) \tag{12.3.79}$$

$$\frac{d\widetilde{\sigma}_{xx}(s)}{ds} = \widetilde{v}(s)\left(2(\widetilde{u}(s) - k_1)\widetilde{\sigma}_{xx}(s) + k_2^2\right) \tag{12.3.80}$$

$$\frac{d\widetilde{\sigma}_{yy}(s)}{ds} = 2k_3\widetilde{v}(s)\widetilde{\sigma}_{xy}(s) \tag{12.3.81}$$

$$\frac{d\widetilde{\sigma}_{xy}(s)}{ds} = \frac{d\widetilde{\sigma}_{yx}(s)}{ds} = \widetilde{v}(s)(\widetilde{u}(s) - k_1)\widetilde{\sigma}_{xy}(s) + k_3\widetilde{\sigma}_{xx}(s). \tag{12.3.82}$$

Furthermore, the initial conditions are

$$\widetilde{\mu}_x(0) = x^*, \quad \widetilde{\mu}_y(0) = 0 \tag{12.3.83}$$

$$\widetilde{\sigma}_{xx}(0) = k_4, \ \widetilde{\sigma}_{yy}(0) = 0, \quad \widetilde{\sigma}_{xy}(0) = \widetilde{\sigma}_{yx}(0) = 0. \tag{12.3.84}$$

At the overhaul times $t = \tau_i$, $i = 1, \ldots, N$, the new state jump conditions are

$$\widetilde{\mu}_x\left(i^+\right) = k_5\widetilde{\mu}_x(i^-), \ \widetilde{\mu}_y\left(i^+\right) = \widetilde{\mu}_y(i^-), \tag{12.3.85}$$

$$\widetilde{\sigma}_{xx}\left(i^+\right) = k_5^2\widetilde{\sigma}_{xx}(i^-) + k_6, \quad \widetilde{\sigma}_{yy}\left(i^+\right) = \widetilde{\sigma}_{yy}(i^-) \tag{12.3.86}$$

$$\widetilde{\sigma}_{xy}\left(i^+\right) = \widetilde{\sigma}_{yx}\left(i^+\right) = k_5\widetilde{\sigma}_{xy}(i^-). \tag{12.3.87}$$

The probabilistic state constraints become

$$\int_{x_{\min}}^{\infty} \frac{1}{\left(2\pi\widetilde{\sigma}_{xx}(s)\right)^{1/2}} \exp\left\{\frac{-(\eta - \widetilde{\mu}_x(s))^2}{2\widetilde{\sigma}_{xx}(s)}\right\} d\eta \geq p_1, \quad s \in [0, N+1] \tag{12.3.88}$$

$$\int_{x_{\min}}^{\infty} \frac{1}{\left(2\pi\widetilde{\sigma}_{yy}(N+1)\right)^{1/2}} \exp\left\{\frac{-(\eta - \widetilde{\mu}_y(N+1))^2}{2\widetilde{\sigma}_{yy}(N+1)}\right\} d\eta \geq p_2. \tag{12.3.89}$$

An admissible pair $(\boldsymbol{h}, \boldsymbol{v}) \in \mathcal{H} \times \mathcal{V}$ is said to be feasible if it satisfies the constraints (12.3.88) and (12.3.89).

After applying the time scaling transformation, Problem (PE_1) becomes Problem (PE_2) defined below.

Problem (PE_2). Given the dynamic system (12.3.78)–(12.3.82) with the initial conditions (12.3.83)–(12.3.84) and the jump conditions (12.3.85)–(12.3.87), find a pair $(\boldsymbol{h}, \boldsymbol{v}) \in \mathcal{H} \times \mathcal{V}$ such that the cost functional

$$\widetilde{g}_0(\boldsymbol{h},\boldsymbol{v}) = \int_0^{N+1} \widetilde{v}(s)\left\{\widetilde{\mathcal{L}}_1(\widetilde{\mu}_x(s),\widetilde{\sigma}_{xx}(s)) + \widetilde{\mathcal{L}}_2(\widetilde{u}(s))\right\} ds + \sum_{i=1}^{N} \Psi_1(\widetilde{\mu}_x(i^-)) - \widetilde{\Psi}_2(\widetilde{\mu}_x(N+1),\widetilde{\sigma}_{xx}(N+1)) \quad (12.3.90)$$

is minimized subject to constraints (12.3.88) and (12.3.89).

Problem (PE_2) is an impulsive optimal parameter selection problem with state constraints. We can apply the constraint transcription technique introduced in Section 4.2 to the continuous inequality constraints (12.3.88) to obtain respective approximate canonical inequality constraints. Furthermore, the gradient formulas for the cost functional and the approximate canonical inequality constraint functionals can be obtained readily from Theorem 7.2.2. Thus, the resulting approximate optimal control problem can be solved as nonlinear programming problem. See Section 7.2.2.

For illustration, we consider the stochastic machine maintenance problem for a brand-new machine costing \$10,000. The manager in charge of the machine plans to replace the machine after 20 overhauls (major maintenance). Meanwhile, the workers in the factory will perform continuous maintenance on the machine (minor maintenance) to ensure that it is kept in good working order. The model parameters are given by

$$\begin{aligned} &k_1 = 1.35\times 10^{-2},\ k_2 = 10^{-3},\ k_3 = 2.5,\ k_4 = 10^{-4},\ k_5 = 1.18\\ &k_6 = 10^{-4},\quad a = 0.1,\quad x^* = 1.0,\ p_1 = 0.8,\quad p_2 = 0.8\\ &x_{\min} = 0.1,\quad y_{\min} = 500,\ \rho = 15.0,\ t_{\min} = 400 \end{aligned} .$$

The explicit forms for the functions in the cost functional are given as follows:

$$\mathcal{L}_1(x(t)) = 2.5x^2(t) - 20x(t) + 40,\ \ \mathcal{L}_2(u(t)) = \frac{40}{k_1}u(t)$$

$$\Psi_1(x(\tau_i^-)) = 1000 - 500x(\tau_i^-),\ \ \Psi_2(x(\tau_{N+1})) = \frac{1}{5}x(\tau_{N+1}) \times 1000.$$

Note that $N = 20$ is the number of overhaul times, and \$10,000 is the original capital cost of the machine.

Table 12.3.1: Optimal jump times for the example

i	τ_i	i	τ_i	i	τ_i	i	τ_i	i	τ_i	i	τ_i	i	τ_i
1	15	4	60	7	105	10	150	13	195	16	240	19	285
2	30	5	75	8	120	11	165	14	210	17	255	20	300
3	45	6	90	9	135	12	180	15	225	18	270	21	400

The optimal value of the cost functional obtained is $\widetilde{g}_0 = 11{,}602.7281$. The optimal jump times (the overhaul times) and the optimal terminal time (replacement time) are given in Table 12.3.1, while the optimal continuous maintenance rates (minor maintenance) are shown in Table 12.3.2.

Note that we are assuming the continuous maintenance rate takes a constant value between consecutive jump points. The optimal trajectories of the state variables, $\mu_x(t), \mu_y(t), \sigma_{xx}(t), \sigma_{xx}(t), \sigma_{yy}(t), \sigma_{xy}(t)$, are shown in Figure 12.3.1a–e, respectively.

Figure 12.3.1a shows the mean of the machine state over the duration of 400 time periods. The mean starts off at 1 and gradually decreases with time. However, with each overhaul, the mean of the state of the machine is restored to a higher value, close to where it was at the previous overhaul. Figure 12.3.1b shows the mean of the accumulated output, which gradually increases over time. Figure 12.3.1c shows the variance of the state of the machine. The variance changes with each overhaul performed and then gradually decreases after the last overhaul. The variance of the output is shown in Figure 12.3.1e, while Figure 12.3.1d shows the covariance of the output with the state of the machine.

To examine the performance of our optimal maintenance policy over a range of scenarios, 500 sample paths are simulated for the machine state (see Figure 12.3.1f) and machine output (see Figure 12.3.1g), respectively. The paths simulated are similar in shape to the paths of mean values as shown earlier.

Table 12.3.2: Optimal continuous maintenance rate for the example

Interval	$u(t)$	Interval	$u(t)$	Interval	$u(t)$
1	1.35×10^{-3}	8	4.6386×10^{-31}	15	6.9333×10^{-33}
2	1.35×10^{-3}	9	4.2867×10^{-31}	16	0
3	1.35×10^{-3}	10	0	17	0
4	1.35×10^{-3}	11	4.0135×10^{-31}	18	0
5	1.35×10^{-3}	12	0	19	0
6	1.35×10^{-3}	13	4.1610×10^{-31}	20	0
7	1.35×10^{-3}	14	7.7037×10^{-31}	21	0

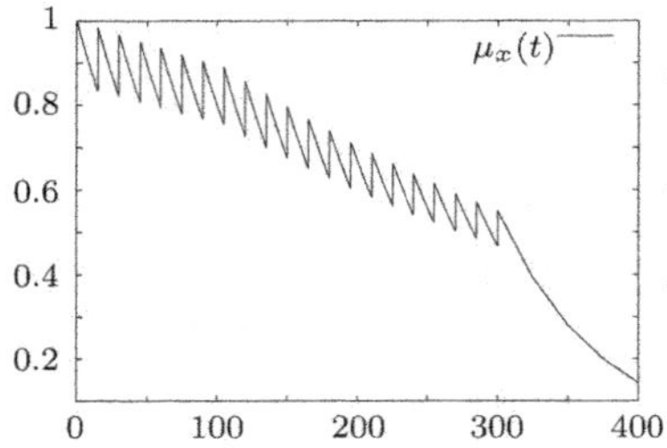

(a) $\mu_x(t)$

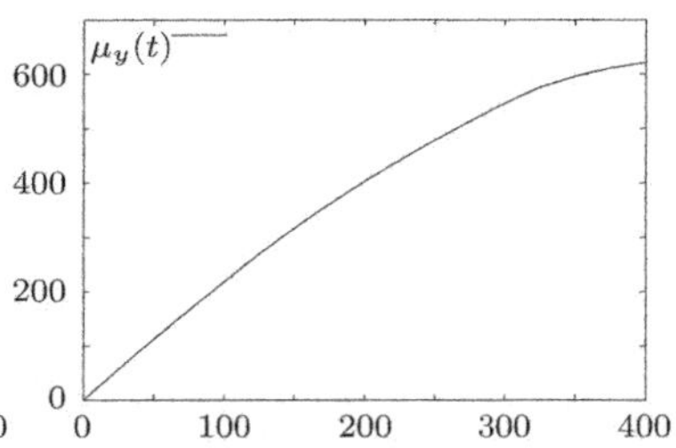

(b) $\mu_y(t)$

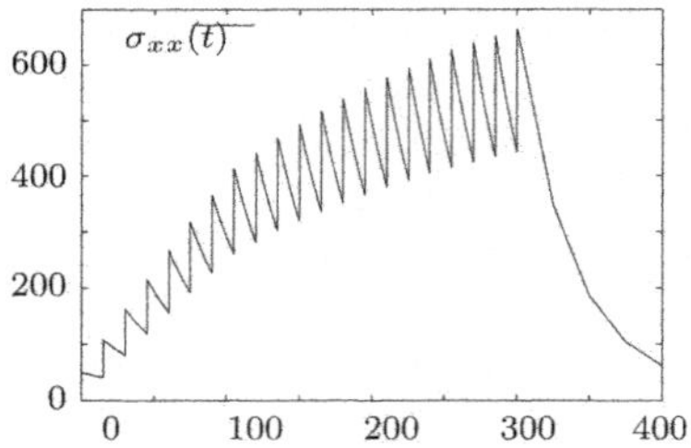

(c) $\sigma_{xx}(t)$

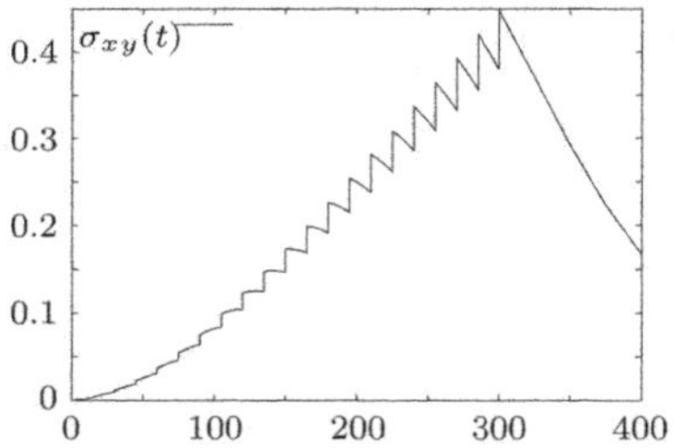

(d) $\sigma_{xy}(t)$

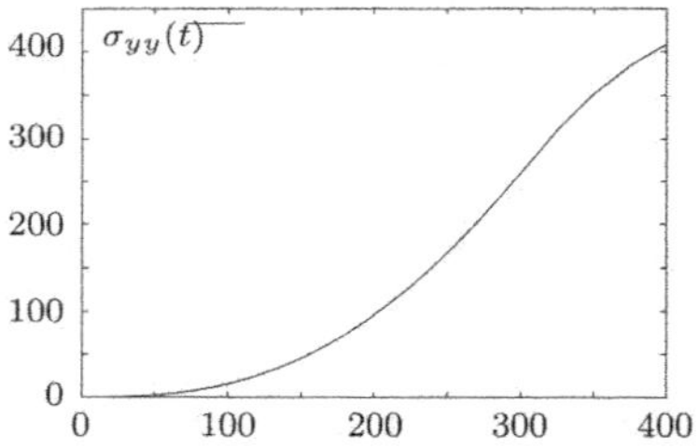

(e) $\sigma_{yy}(t)$

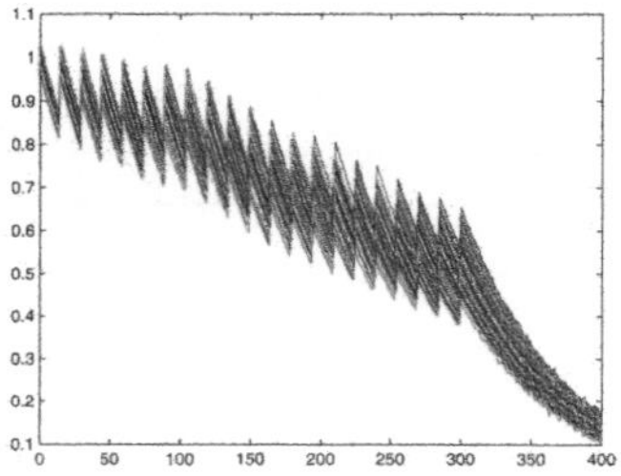

(f) Simulation of $x(t)$ with 500 sample paths.

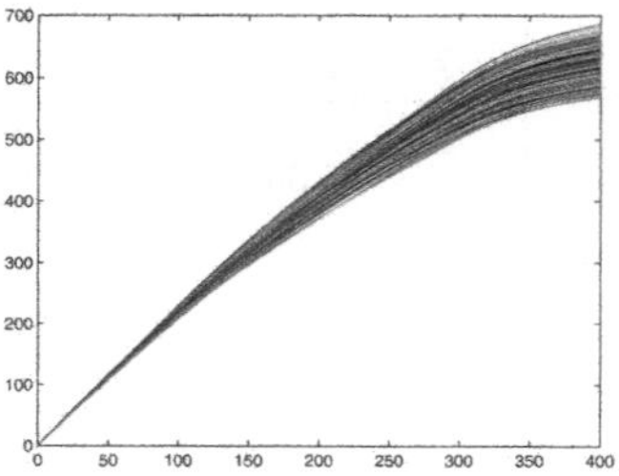

(g) Simulation of $y(t)$ with 500 sample paths.

Fig. 12.3.1: The optimal trajectories of the state variables and simulation with 500 sample paths. (**a**) $\mu_x(t)$. (**b**) $\mu_y(t)$. (**c**) $\sigma_{xx}(t)$. (**d**) $\sigma_{xy}(t)$. (**e**) $\sigma_{yy}(t)$. (**f**) Simulation of $x(t)$ with 500 sample paths. (**g**) Simulation of $y(t)$ with 500 sample paths

12.4 Exercises

12.4.1 *Consider the optimal parameter and optimal control problem (SP) with the control taking the form as given below:*

$$\boldsymbol{u} = K\boldsymbol{x},$$

where K is an $n \times n$ matrix to be determined such that the cost functional (12.2.5) *is minimized. Obtain the corresponding deterministic optimal parameter selection problem.*

12.4.2 *Consider the optimal parameter and optimal control problem* (SP). *However, the probabilistic state constraints* (12.2.4) *are replaced by the following constraints*

$$\alpha_i \leq \mathcal{E}\left[\xi_i(t) - \overline{\xi_i}(t)\right] \leq \beta_i, \quad \textit{for all } t \in [0, T], \quad i = 1, \ldots, N,$$

where α_i, β_i, $i = 1, \ldots, n$, are the given real constants and $\overline{\xi_i}(t)$, $i = 1, \ldots, N$, are the specified desired state trajectories. Obtain the corresponding deterministic optimal parameter and optimal control problem.

12.4.3 *Consider the example of Section 12.2.2. Let the probabilistic constraint* (12.2.23) *be replaced by the following constraint*

$$\alpha \leq \mathcal{E}\left[\xi(t) - \overline{\xi}(t)\right] \leq \beta, \quad \textit{for all } t \in [0, T],$$

where α and β are the given real constants and $\overline{\xi}(t)$ is the specified desired state trajectory. Obtain the corresponding deterministic optimal control problem.

12.4.4 *Consider Problem* (SP_{2b}) *but with the constraint* (12.2.26) *being replaced by appropriate probabilistic constraints of the form given by* (12.2.4). *Derive the corresponding deterministic optimal control problem.*

12.4.5 *Consider the example of Section 12.3.3. Let the probabilistic constraints* (12.3.38) *and* (12.3.39) *be replaced by the constraints of the form given by* (12.3.25). *Derive the corresponding deterministic optimal control problem.*

12.4.6 *Consider the example of Section 12.3.3. Suppose that the control is a piecewise constant function between every pair of overhaul times, where the heights and switching times of the piecewise constant control are decision variables. Derive the corresponding deterministic optimal control problem.*

Appendix A.1
Elements of Mathematical Analysis

A.1.1 Introduction

In this Section, some results in measure theory and functional analysis are presented without proofs. The main references are [3, 4, 40, 51, 90, 91, 198, 206, 216, 250, 253].

A.1.2 Sequences

Let X be a given set. It is called a *finite set* if it is either empty or a finite sequence. Similarly, the set X is called a *countable set* if it is either empty or a sequence.

Let x_n be a sequence. The *limit superior* of the sequence x_n, denoted by $\limsup_{n\to\infty} x_n$ or $\overline{\lim}_{n\to\infty} x_n$, is defined by

$$\overline{\lim_{n\to\infty}}\, x_n = \inf_n \sup_{k\geq n} x_k.$$

Similarly, the *limit inferior* of the sequence x_n, denoted by $\liminf_{n\to\infty} x_n$ or $\underline{\lim}_{n\to\infty} x_n$, is defined by

$$\underline{\lim_{n\to\infty}}\, x_n = \sup_n \inf_{k\geq n} x_k.$$

Let x_n be a sequence. Then, $K \in \mathbb{R}\cup\{+\infty\}$ (respectively, $K \in \mathbb{R}\cup\{-\infty\}$) is the *limit superior* (respectively, *limit superior*) of the sequence $\{x_n\}$ if and only if the following two conditions are satisfied:

K. L. Teo et al., *Applied and Computational Optimal Control*, Springer Optimization and Its Applications 171,
https://doi.org/10.1007/978-3-030-69913-0

1. There exists a subsequence $\{x_{n(\ell)}\}$ of the sequence $\{x_n\}$ such that $\lim_{\ell\to\infty} x_{n(\ell)} = K$.
2. If $\{x_{n(\ell)}\}$ is any subsequence of the sequence $\{x_n\}$ such that $\lim_{\ell\to\infty} x_{n(\ell)} = A$, then

$$A \le K \quad (\text{respectively, } A \ge K).$$

Note that a sequence $\{x_n\}$ can only have one limit superior (respectively, limit inferior). A sequence $\{x_n\}$ in $\mathbb{R} \cup \{\pm\infty\}$ has the limit A, denoted by $\lim\limits_{n\to\infty} x_n = A$, if and only if

$$\overline{\lim_{n\to\infty}}\, x_n = \underline{\lim_{n\to\infty}}\, x_n = A.$$

A.1.3 Linear Vector Spaces

Let X be a set, which is a non-empty collection of elements. We define two operations, called *addition* and *scalar multiplication*, on X. The addition operation satisfies:

1. $x + y \in X$ for all x, $y \in X$;
2. $x + (y + z) = (x + y) + z$ for all x, y, $z \in X$;
3. there exists a unique element, 0, in X such that $0 + x = x + 0 = x$ for all $x \in X$;
4. for every $x \in X$ there exists a unique element, denoted by $-x$, such that $x + (-x) = 0$;
5. $x + y = y + x$ for all x, $y \in X$.
   The scalar multiplication operation satisfies:
6. $\alpha x \in X$ for any $\alpha \in \mathbb{R}$, and $x \in X$;
7. $\alpha(x + y) = \alpha x + \alpha y$ for any $\alpha \in \mathbb{R}$, and x, $y \in X$;
8. $(\alpha + \beta)x = \alpha x + \beta x$ for any $\alpha, \beta \in \mathbb{R}$, and $x \in X$;
9. $\alpha(\beta x) = (\alpha\beta)x$ for any α, $\beta \in \mathbb{R}$, and $x \in X$;
10. there exists a unique element, 1, in X such that$1x = x$ for any $x \in X$.

The space X together with the two operations defined above is called a *(linear) vector space*.

Let A be a non-empty subset of X. A is called a *convex* set if

$$\lambda x + (1 - \lambda)y \in A, \quad \text{for all } x, \; y \in X \quad \text{and} \quad \lambda \in [0, 1].$$

It is easy to show that the set A is convex if and only if

$$\sum_{i=1}^{n} \lambda_i x_i \in A$$

whenever

$$x_i \in A,\ \lambda_i \geq 0,\ \text{and} \sum_{i=1}^{n} \lambda_i = 1.$$

Note that the intersection of any number of convex sets is a convex set. However, the union of two convex sets is, in general, *not* a convex set.

Let $x_1, \ldots, x_m$ be m vectors in a vector space X. A *linear combination* of these m vectors is defined by

$$\alpha_1 x_1 + \cdots + \alpha_m x_m,$$

where α_i, $i = 1, \ldots, m$, are real numbers.

Let $x_1, \ldots, x_m$ be m vectors in a vector space X. The set of these vectors in X is called *linearly dependent* if there exist real numbers $\alpha_1, \ldots, \alpha_m$, not all zero, such that

$$\alpha_1 x_1 + \cdots + \alpha_m x_m = 0. \tag{A.1.1}$$

If this collection of vectors is not linearly dependent, then it is called *linearly independent*. In this case, the only solution to (A.1.1) is

$$\alpha_i = 0, \quad \text{for all } i = 1, \ldots, m.$$

Let M be a non-empty subset of a vector space X. If $\alpha x + \beta y \in M$ for all real numbers α, β and for all $x,\ y \in M$, then M is called a *subspace* of X. The whole space X is itself a subspace of X. If M is a subspace of X such that $M \neq X$, then M is called a *proper subspace of* X.

Let S be a subset of a vector space X. Then, the set $[S]$ is said to be the *subspace generated by* S if it consists of all those vectors in X that can be expressed as linear combinations of vectors in S. Let $x_1, \ldots, x_m$ be m vectors in X. Then,

$$\alpha_1 x_1 + \cdots + \alpha_m x_m$$

is called a *convex combination* of $x_1, \ldots, x_m$, if $\alpha_i \geq 0$ for $i = 1, \ldots, m$, and $\sum_{i=1}^{m} \alpha_i = 1$.

Let $S \equiv \{x_1, \ldots, x_m\}$ be a set of linearly independent vectors in a vector space X. If X is generated by S, then S is called a *basis* for the vector space X. In this case, the vector space X is said to be of *(finite) m-dimension*. If a vector space X is not finite dimensional, it is called *infinite dimensional*. In a finite dimensional vector space, any two bases must contain the same number of linearly independent vectors.

A.1.4 Metric Spaces

Let X be a non-empty set. A topology $\mathcal{T}$ on X is a family of subsets of X such that

1. $X \in \mathcal{T}$, the empty set $\varnothing \in \mathcal{T}$;
2. the union of any number of members of $\mathcal{T}$ is in $\mathcal{T}$; and
3. the intersection of a finite number of members of $\mathcal{T}$ is in $\mathcal{T}$.

The set endowed with the topology $\mathcal{T}$ is called a *topological space* and is written as $(X, \mathcal{T})$. Members of $\mathcal{T}$ are called *open sets.* A set $B \subset X$ is said to be *closed* if its complement $X \setminus B$ is open.

Let $\mathcal{T}_1$ and $\mathcal{T}_2$ be two topologies on X. $\mathcal{T}_1$ is said to be *stronger* than $\mathcal{T}_2$ (or $\mathcal{T}_2$ *weaker* than $\mathcal{T}_1$) if $\mathcal{T}_1 \supset \mathcal{T}_2$.

Let $(X, \mathcal{T})$ be a topological space, and let A be a non-empty subset of X. The family $\mathcal{T}_A = \{A \cap B : B \in \mathcal{T}\}$ is a topology on A and is called the *relative topology* on A induced by the topology $\mathcal{T}$ on X.

Let $(X, \mathcal{T})$ be a topological space, $A \subset X$, and $C \equiv \{G_i\}$ a subfamily of $\mathcal{T}$ such that $A \subset \cup G_i$. Then, C is called an *open covering* of A. If every open covering of X has a finite subfamily $\{G_1, \ldots, G_n\} \subset C$ such that $X = \cup_{i=1}^N G_i$, then the topological space $(X, \mathcal{T})$ is called a *compact space.* A subset A of a topological space is said to be compact if it is compact as a subset of X. Equivalently, A is call compact if every open covering of A contains a finite subfamily that covers A.

A family of closed sets is said to possess the *finite intersection property* if the intersection of any finite number of sets in the family is non-empty.

A topological space is compact if and only if any family of closed sets with the finite intersection property has non-empty intersection.

A point $x \in X$ is said to be an *interior* point of a set $A \subset X$ if there exists an open set G in X such that $x \in G \subset A$. The *interior* $\mathring{A}$ of A is the set which consists of all the interior points of the set A. A *neighbourhood* of a point $x \in X$ is a set $V \subset X$ such that x is an interior point of V. A point $x \in X$ is said to be an *accumulation point* of a set $A \subset X$ if every neighbourhood of $x \in X$ contains points of A other than x. If $A \subset X$ is a closed set, then it contains all its accumulation points. The union of a set B and its accumulation points is called the *closure* of B and is written as $\overline{B}$. A set $A \subset X$ is said to be *dense* in a set $E \subset X$ if $\overline{A} \supset E$. A set A is said to be *nowhere dense* if the interior of its closure is empty. If X contains a countable subset that is dense in X, then it is called *separable.* The *boundary* ∂A of a set A is the set of all accumulation points of both A and $X \setminus A$. Thus, $\partial A = \overline{A} \cap \overline{(X \setminus A)}$.

A family B of subsets of X is a *base* for a topology $\mathcal{T}$ on X if B is a subfamily of $\mathcal{T}$ and, for each $x \in X$ and each neighbourhood U of x, there is a member V of B such that $x \in V \subset U$. A family $\mathcal{F}$ of subsets of X is a *subbase* for a topology $\mathcal{T}$ on X if the family of finite intersections of members of $\mathcal{F}$ is a base for $\mathcal{T}$.

A sequence $\{x^n\} \subset X$ is said to *converge* to a point $x \in X$, denoted by $x^n \to x$, if each neighbourhood of x contains all but a finite number of elements of the sequence.

A topological space X is called *Hausdorff* if it satisfies the separation axiom: if any x, $y \in X$ are such that $x \neq y$, then x and y have disjoint neighbourhoods.

A compact subset of a Hausdorff (topological) space is closed. A closed subset of a compact set is compact.

We now turn our attention to a special topological space, which is known as the *metric space* as follows.

Let X be a vector space and let ρ be a function from $X \times X$ into $\mathbb{R}$ such that the following axioms are satisfied:

(M1) $\rho(x, y) \geq 0$ for all x, $y \in X$;
(M2) $\rho(x, y) = 0$ if and only if $x = y$;
(M3) $\rho(x, y) = \rho(y, x)$ (symmetry) for all x, $y \in X$;
(M4) $\rho(x, y) \leq \rho(x, z) + \rho(z, y)$ (triangle inequality) for all x, y, $z \in X$.

A vector space X equipped with such a metric ρ is called a *metric space*. It is written as (X, ρ).

A set $\Theta \subset X$ is said to be *open* if for every $x \in \Theta$ there exists a $\delta > 0$ such that

$$\{y \in X : \rho(x, y) < \delta\} \subset \Theta,$$

where $\{y \in X : \rho(x, y) < \delta\}$ is an open ball of radius δ with center x, or a δ-neighbourhood of x. The whole set X and the empty set $\emptyset$ are both open sets. A metric space (X, ρ) is a topological space with its open sets generated by the metric ρ. Let D be a subset of X, and let x be a point in X. If, for any $\delta > 0$, there exists a point $y \in D$ such that $\rho(x, y) < \delta$, then x is said to be a point in the closure of the set D. Let $\bar{D}$ denote the closure of the set D. Clearly, $D \subset \bar{D}$. If $A \subset B$, then $\bar{A} \subset B$. Furthermore,

$$\overline{A \cup B} = \bar{A} \cup \bar{B}$$

and

$$\overline{A \cap B} \subset \bar{A} \cap \bar{B}.$$

If $\bar{A} = A$, then the set A is said to be closed. The closure $\bar{B}$ of any set B is a closed set. The whole set X and the empty set $\emptyset$ are closed sets. The union of any two closed sets is a closed set. Although the intersection of any collection (countable or uncountable) of closed sets is closed, the union of a countable collection of closed sets needs not be closed. Similarly, the intersection of a countable collection of open sets needs not be open.

Let A be a subset in X. The complement $\tilde{A}$ of A is defined by

$$\tilde{A} = \{x \in X : x \notin A\}.$$

The complement of an open set is a closed set. Similarly, the complement of a closed set is an open set.

A metric space (X, ρ) is called *separable* if there exists a subset D of X that contains only a countable number of points in X such that $\bar{D} = X$.

We can define different metrics on the same vector space X. Let ρ and σ be two different functions from $X \times X$ into $\mathbb{R}$ such that the properties (M1)–(M3) are satisfied. Then, (X, ρ) and (X, σ) are two different metric spaces.

In a metric space (X, ρ), a sequence $\{x_n\}_{n=1}^{\infty} \subset X$ is called a *Cauchy sequence* if

$$\rho(x_{n+p}, x_n) \to 0 \text{ as } n \to \infty$$

for any integer $p \geq 1$. A metric space (X, ρ) is said to be *complete* if every Cauchy sequence has a limiting point in X.

Let (X, ρ) be a metric space. For any two distinct points x_1, x_2 in X, there exist two real numbers $\delta_1 > 0$ and $\delta_2 > 0$ such that

$$\{y \in X : \rho(x_1, y) < \delta_1\} \cap \{y \in X : \rho(x_2, y) < \delta_2\} = \emptyset.$$

This implies that the metric space (X, ρ) is a Hausdorff space. Therefore, a convergent sequence can have only one limiting point.

Let (X, ρ) be a metric space, and let $A \subset X$. If, for any sequence $\{x_n\}_{n=1}^{\infty}$ in A, there exist a subsequence $\{x_{n(\ell)}\}_{\ell=1}^{\infty}$ and a point $x \in X$ such that

$$\rho(x_{n(\ell)}, x) \to 0 \text{ as } \ell \to \infty,$$

then A is said to be *conditionally sequentially compact.* The set A is said to be *sequentially compact* if it is conditionally sequentially compact and the limiting point x remains in the set A.

Let (X, ρ) be a metric space. If for each $\varepsilon > 0$ there exists a finite collection $\{G_{i,\varepsilon}\}_{i=1}^{n}$ of open balls of radius ε such that $X = \cup_{i=1}^{n} G_{i,\varepsilon}$, then the metric space is called *totally bounded.*

The open ball of radius $\varepsilon > 0$ with center at the point $x^0 \in X$ is the set

$$K_{\varepsilon}(x^0) = \{x \in X : \rho(x, x^0) < \varepsilon\}.$$

The family of sets:

$$\mathcal{B} = \{K_{1/n}(x) : n = 1, 2, \ldots \quad \text{and } x \in X\}$$

forms a basis for X, i.e., for each $x \in X$ and each neighbourhood U of x, there is a member V of $\mathcal{B}$ such that $x \in V \subset U$.

A set A in $\mathbb{R}^n$ is compact if and only if it is closed and bounded. Thus, a closed subset of a compact set A in $\mathbb{R}^n$ is compact.

A metric space is compact if and only if it is complete and totally bounded. If a metric space is compact, then it is separable.

Let (X, ρ) be a metric space and let S be a subset of X. We can restrict the metric ρ to S. Equipped with this induced metric ρ, S becomes a metric space, and is written as (S, ρ). In this case, we call (S, ρ) a subspace of (X, ρ).

Let (S, ρ) be a subspace of the metric space (X, ρ). If $E \subset X$, $\bar{E} \cap S$ is the closure of E relative to S, where $\bar{E}$ denotes the closure of E in X. A set $A \subset S$ is closed relative to S if and only if there exists a closed set F in X such that $A = S \cap F$. A set $A \subset S$ is open relative to S if and only if there exists an open set G in X such that $A = S \cap G$.

Every subspace of a separable metric space is separable. Let (X, ρ) be a metric space. If a subset A of X is complete, then it is closed. On the other hand, a closed subset of a complete metric space is itself complete.

A.1.5 Continuous Functions

Let (X, ρ) and (Y, σ) be two metric spaces, and let f be a function from X into Y. The function f is said to be *continuous* at $x_0 \in X$ if, for every $\varepsilon > 0$ there exists a $\delta \equiv \delta(\varepsilon, x_0) > 0$, such that

$$\sigma(f(x), f(x_0)) < \varepsilon \quad \text{whenever } \rho(x, x_0) < \delta.$$

It is called *uniformly continuous* if δ does not depend on x_0. The function f is said to be *continuous* if it is continuous at every point in X. Let $C(X, Y)$ be the set of all such continuous functions. If the metric space X is compact, then f is uniformly continuous. Furthermore, if $Y = \mathbb{R}$, then f attains both its maximum and minimum.

Let f be a function from an interval $I \subset \mathbb{R}$ to $\mathbb{R}$. Then, the function f is said to be *continuous* at $x_0 \in I$ if, for every $\varepsilon > 0$, there exists a $\delta \equiv \delta(\varepsilon, x_0)$, with $\delta > 0$, such that

$$|f(x) - f(x_0)| < \varepsilon \quad \text{whenever } x \in I \text{ and } |x - x_0| < \delta,$$

where $|y|$ denotes the absolute value of y. The concept of uniform continuity for this special case is to be understood similarly.

Let (X, ρ), (Y, σ), and (Z, η) be three metric spaces. Let f be a continuous function from X into Y, and g a continuous function from Y into Z. Then, the composite function $g \circ f : X \to Z$ is also continuous, where

$$g \circ f(x) \equiv g(f(x)).$$

Let f be a continuous function from an interval $I \subset \mathbb{R}$ to $\mathbb{R}$. Then, for any real number α, the set

$$B = \{x \in I : f(x) = \alpha\}$$

is a compact subset of I. A continuous function is uniformly continuous on any compact subset A of I. It also attains both its maximum and minimum on A. The set $f(A)$ defined by

$$f(A) = \{\ f(x) : x \in A\ \}$$

is compact.

Let $X = (X, \rho)$ be a metric space. A real-valued function f defined on X is said to be *lower semicontinuous* at $x^0 \in X$ if for every real number α such that $f(x^0) > \alpha$, there is a neighbourhood V of x^0 such that $f(x) > \alpha$ for all $x \in V$. *Upper semicontinuity* is defined by reversing the inequalities. We say that f is *lower semicontinuous* if it is lower semicontinuous at every $x \in X$.

Let f be an upper (respectively, lower) semicontinuous real-valued function on a compact space X. Then, f is bounded from above (respectively, below) and assumes its maximum (respectively, minimum) in X.

Theorem A.1.1 (Dini Theorem). *Let $\{f^n\}$ be a sequence of upper semicontinuous real-valued functions on a compact space X, and suppose that for each $x \in X$ the sequence $\{f^n(x)\}$ decreases monotonically to zero. Then, $\{f^n\}$ converges to zero uniformly.*

A.1.6 Normed Spaces

Let X be a vector space and let $\|\cdot\|$ be a function from X into $[0, \infty)$ such that the following properties are satisfied:

(N1) $\|y\| \geq 0$ for all $y \in X$, and $\|y\| = 0$ if and only if $y = 0$.
(N2) $\|x + y\| \leq \|x\| + \|y\|$ for each x, $y \in X$ (triangle inequality).
(N3) $\|\alpha y\| = |\alpha|\ \|y\|$ for all real numbers α.

The function $\|\cdot\|$ is called the *norm*, and $X = (X, \|\cdot\|)$ is called a *normed (linear vector) space.*

Let $X = (X, \|\cdot\|)$ be a normed space, and let ρ be the metric induced by the norm $\|\cdot\|$ as follows:

$$\rho(x, y) \equiv \|x - y\|\ .$$

Then (X, ρ) is a metric space. A *Banach space* is a complete normed space with respect to the metric induced by its norm.

Let X be a vector space, and let $\langle \cdot, \cdot \rangle$ be a function from $X \times X$ into $\mathbb{R}$ such that the following conditions are satisfied:

(I1) $\langle x, x \rangle \geq 0$ for all $x \in X$ and $\langle x, x \rangle = 0$ if and only if $x = 0$;
(I2) $\langle x, y \rangle = \langle y, x \rangle$ for all x, $y \in X$;
(I3) $\langle \lambda x + \beta y, z \rangle = \lambda \langle x, z \rangle + \beta \langle y, z \rangle$ for any real numbers λ, β and for all x, y, $z \in X$.

Such a function $\langle \cdot, \cdot \rangle$ is called the *inner product.* If we let $\|x\| = (\langle x, x \rangle)^{1/2}$, then $(X, \|\cdot\|)$ becomes a normed space. A *Hilbert space* is a complete normed space, where the norm is induced by the inner product.

Let $\boldsymbol{x} = [x_1, \ldots, x_n]^\top$ be a vector in $\mathbb{R}^n$, where the superscript $^\top$ denotes the transpose. The usual Euclidean norm $|\boldsymbol{x}|$ is defined by:

$$|\boldsymbol{x}| = \left(\sum_{i=1}^{n} (x_i)^2 \right)^{1/2} .$$

In particular, if $x \in \mathbb{R}$, then $|x|$ is simply the absolute value of x.

Let $C(I, \mathbb{R}^n)$ be the space of all continuous functions from $I \equiv [a, b] \subset \mathbb{R}$ to $\mathbb{R}^n$. The space $C(I, \mathbb{R}^n)$ is a vector space, and becomes a Banach space when it is equipped with the *sup norm* defined by

$$\|\boldsymbol{f}\|_{C(I,\mathbb{R}^n)} \equiv \sup_{t \in I} |\boldsymbol{f}(t)|,$$

where $\boldsymbol{f} \equiv [f_1, \ldots, f_n]^\top$, and $|\boldsymbol{f}(t)| = \left(\sum_{i=1}^{n} (f_i(t))^2 \right)^{1/2}$.

A set $A \subset C(I, \mathbb{R}^n)$ is said to be *equicontinuous* if, for any $\varepsilon > 0$, there exists a $\delta = \delta(\varepsilon) > 0$ such that for all $\boldsymbol{f} \in A$

$$|\boldsymbol{f}(t') - \boldsymbol{f}(t)| < \varepsilon$$

whenever t', $t \in I$ are such that $|t' - t| < \delta$.

Theorem A.1.2 (*Arzelà-Ascoli*). *Let $I = [a, b] \subset \mathbb{R}$. A set $A \subset C(I, \mathbb{R}^n)$ is conditionally sequentially compact if and only if A is bounded as well as equicontinuous.*

Let $I = [a, b] \subset \mathbb{R}$ and $\boldsymbol{f} \equiv [f_1, \ldots, f_n]^\top \in C(I, \mathbb{R}^n)$. The function $\boldsymbol{f}$ is said to be *absolutely continuous* on I if for any given $\varepsilon > 0$ there exists a $\delta > 0$ such that

$$\sum_{k=1}^{m} |\boldsymbol{f}(t_k) - \boldsymbol{f}(t'_k)| < \varepsilon,$$

for every finite collection $\{(t_k, t'_k)\}$ of non-overlapping intervals satisfying

$$\sum_{k=1}^{m} |t_k - t'_k| < \delta.$$

Let $AC(I, \mathbb{R}^n)$ be the class of all such absolutely continuous functions.

Let $C^m(I, \mathbb{R}^n)$ be the space of all m-time continuously differentiable functions from $I \equiv [a, b] \subset \mathbb{R}$ to $\mathbb{R}^n$. It is a vector space and becomes a Banach space when it is equipped with the *sup norm* defined by

$$\|\boldsymbol{f}\|_{C^m(I,\mathbb{R}^n)} = \max_{o \le i \le m} \left\| \boldsymbol{f}^{(i)} \right\|_{C(I,\mathbb{R}^n)},$$

where $\boldsymbol{f}^{(i)}$ denotes the i-time derivative of the function $\boldsymbol{f}$.

A function f from $\mathbb{R}^n$ to $\mathbb{R}$ is called *convex* if

$$f(\lambda \boldsymbol{y} + (1-\lambda)\boldsymbol{z}) \leq \lambda f(\boldsymbol{y}) + (1-\lambda) f(\boldsymbol{z}) \tag{A.1.2}$$

for all $\boldsymbol{y}$, $\boldsymbol{z} \in \mathbb{R}^n$, and for all $\lambda \in [0,1]$. If a function $f : \mathbb{R}^n \to \mathbb{R}$ is convex and continuously differentiable, then

$$f(\boldsymbol{z}) - f(\boldsymbol{y}) \geq \frac{\partial f(\boldsymbol{y})}{\partial \boldsymbol{x}}(\boldsymbol{z} - \boldsymbol{y}), \tag{A.1.3}$$

where

$$\frac{\partial f(\boldsymbol{y})}{\partial \boldsymbol{x}} \equiv [\partial f(\boldsymbol{x})/\partial x_1, \ldots, \partial f(\boldsymbol{x})/\partial x_n]|_{\boldsymbol{x}=\boldsymbol{y}} \tag{A.1.4}$$

is called the *gradient (vector)* of f at $\boldsymbol{x} = \boldsymbol{y}$.

For *strict convexity* of the function f, we only need to replace the inequality in the conditions (A.1.2) and (A.1.3) by strict inequality.

If $f : \mathbb{R}^n \to \mathbb{R}$ is twice continuously differentiable, the *Hessian matrix* of the function f at $\boldsymbol{x}^0$ is a real $n \times n$ matrix with its ij-th element defined by

$$H(\boldsymbol{x}^0)_{ij} = \left[\frac{\partial^2 f(\boldsymbol{x})}{\partial x_i \partial x_j}\right]_{\boldsymbol{x}=\boldsymbol{x}^0}. \tag{A.1.5}$$

A.1.7 Linear Functionals and Dual Spaces

Let $X \equiv (X, \|\cdot\|_X)$ and $Y \equiv (Y, \|\cdot\|_Y)$ be normed spaces and let $f : X \to Y$ be a linear mapping. If there exists a constant M such that

$$\|f(x)\|_Y \leq M \ \|x\|_X \quad \text{for all} \quad x \in X, \tag{A.1.6}$$

then f is said to be bounded, and

$$\|f\| \equiv \sup\{\|f(x)\|_Y : \|x\|_X \leq 1\}$$

is called the *norm*, or the *uniform norm*, of f. The function f is bounded if and only if it is continuous.

Let $\mathcal{L}(X,Y)$ be the set of all continuous linear mappings from X to Y. Then, $\mathcal{L}(X,Y)$ is a normed space with respect to the uniform norm. If Y is a Banach space, then $\mathcal{L}(X,Y)$ is also a Banach space. In particular, let X be a normed space and let $Y = \mathbb{R}$. Then the set of all continuous linear mappings from X to Y becomes the set of all continuous linear functionals on X. This space $\mathcal{L}(X,\mathbb{R})$, which is denoted by X^*, is called the *dual space* of X. The set X' of all linear functionals (not necessarily continuous) on X is called the *algebraic dual* of X. Clearly $X^* \subset X'$. The dual of X^*, also known as the *second dual* of X, is denoted by X^{**}.

For each $x^* \in X^*$, we can define a continuous linear functional with values

$$\langle x^*, x \rangle \equiv x^*(x) \quad \text{at} \quad x \in X.$$

For a fixed $x \in X$, it is clear that the bilinear form also defines a continuous linear functional on X^*, and we can write it as:

$$x^*(x) \equiv J_x(x^*).$$

The correspondence $x \to J_x$ from X to X^{**} is called the *canonical mapping*. Define

$$X_0^{**} \equiv \{x^{**} \in X^{**} : x^{**} = J_x, \quad x \in X\}.$$

The canonical mapping $x \to J_x$ from X onto X_0^{**} is one-to-one and *norm preserving*, (i.e., $\|x\| = \|J_x\|$). Hence, we may regard X as a subset of X^{**}. If, under the canonical mapping, $X = X^{**}$, then X is called *reflexive*. If X is reflexive, then so is X^*.

Let X be a Banach space and let X^* be its dual. The norm topology of X is called the *strong topology*. Apart from this topology, elements of X^* can also be used to generate another topology for X which is called the *weak topology*. A base for the weak topology consists of all neighbourhoods of the form

$$N(x_0, F^*, \varepsilon) \equiv \{y \in X : |x^*(y - x_0)| < \varepsilon, \quad x^* \in F^*\},$$

where $x_0 \in X$, F^* is any finite subset of X^* and $\varepsilon > 0$.

Similarly, we can introduce two topologies on the dual space X^*: (1) the norm (strong) topology; and (2) the weak topology (i.e., the topology induced by X^{**} on X^*). In addition, since $X \subset X^{**}$ under the canonical mapping, we can introduce another topology which is induced by X on X^*. This topology is called the weak* topology, and it is weaker than the weak topology. A base for the weak* topology consists of all neighbourhoods of the form

$$N(x_0^*, F, \varepsilon) \equiv \{x^* \in X^* : |x^*(x) - x_0^*(x)| < \varepsilon, \quad x \in F\},$$

where $x_0^* \in X^*$ and F is finite subset of X and $\varepsilon > 0$. If X is reflexive, then the weak topology for X^* and the weak* topology for X^* are equivalent.

Definition A.1.1. A sequence $\{x^n\}$ in a normed space X is said to converge to $\bar{x}$ in the weak topology (denoted by $x^n \xrightarrow{w} \bar{x}$) if

$$\lim_{n \to \infty} |x^*(x^n - \bar{x})| = 0, \qquad \text{for every} \quad x^* \in X^*.$$

Note that every weakly convergent sequence is bounded.

Definition A.1.2. A subset F of a normed space X is said to be weakly closed in X, if, for every sequence $\{x^n\} \subset F$ such that $\{x^n\}$ converges to $\bar{x}$ in the weak topology, then $\bar{x} \in F$.

Since a strongly convergent sequence is weakly convergent, a weakly closed set is strongly closed. However, the converse is not necessarily true. But we have the following theorem.

Theorem A.1.3 (Mazur). *A convex subset of a normed space X is weakly closed if and only if it is strongly closed.*

Theorem A.1.4 (Banach-Saks-Mazur). *Let X be a normed space and let $\{x^n\}$ be a sequence in X converging weakly to $\bar{x}$. Then there exists a sequence of finite convex combinations of $\{x^n\}$ that converges strongly to $\bar{x}$.*

Definition A.1.3. A subset F of a normed space X is said to be conditionally weakly sequentially compact in X if every sequence $\{x^n\}$ in F contains a subsequence that converges weakly to a point $\bar{x}$ in X.

Definition A.1.4. A subset F of a normed space X is said to be weakly sequentially compact in X if it is conditionally weakly sequentially compact in X and weakly closed (that is, the limit does not leave F).

Theorem A.1.5. *A subset F of a normed space X is weakly sequentially compact if and only if it is norm (strongly) bounded and weakly closed.*

Theorem A.1.6 (Eberlein-Šmulian). *A subset of a Banach space X is weakly compact if and only if it is weakly sequentially compact.*

Definition A.1.5. Let X^* be the dual space of a Banach space X. A sequence $\{x^n\}$ in X^* is said to converge to $\bar{x} \in X^*$ in the weak* topology (denoted by $x^n \xrightarrow{w^*} \bar{x}$) if

$$\lim_{n \to \infty} |x^n(x) - \bar{x}(x)| = 0, \quad \text{for every} \quad x \in X.$$

Definition A.1.6. A subset F^* of X^* is said to be weak* closed in X^*, if, for every sequence $\{x^n\} \subset F^*$ such that $\{x^n\}$ converges to $\bar{x}$ in the weak* topology, then $\bar{x} \in F^*$.

Definition A.1.7. A subset F^* of X^* is said to be conditionally weak* sequentially compact in X^*, if every sequence $\{x^n\} \subset F^*$ contains a subsequence that converges to a point $\bar{x} \in X^*$ in the weak* topology.

Definition A.1.8. A subset F^* of X^* is said to be weak* sequentially compact in X^*, if it is conditionally weak* sequentially compact and weak* closed.

Remark A.1.1. In a reflexive Banach space X, the weak topology for X^* and the weak* topology for X^* are equivalent.

Theorem A.1.7 (Alaoglu). *A subset of X^* is weak* sequentially compact (i.e., sequentially compact in the weak* topology) if and only if it is norm (strongly) bounded and weak* closed.*

Remark A.1.2. As a direct consequence of Theorem A.1.7, we note that any closed ball

$$K_r = \{x^* \in X^* : \|x\|_{X^*} \leq r\}$$

is weak* sequentially compact in X^*.

A.1.8 Elements in Measure Theory

Let X be a set, and let A and B be two subsets of X. The *symmetric difference* of these two sets is defined by

$$A \Delta B = \{A \setminus B\} \cup \{B \setminus A\}, \tag{A.1.7}$$

where

$$A \setminus B = \{x \in X : x \in A \quad \text{and} \quad x \notin B\} \tag{A.1.8}$$

and $B \setminus A$ is defined similarly.

A class $\mathcal{D}$ of subsets of X is called a *ring* if it is closed under finite unions and set differences. If $\mathcal{D}$ is closed with respect to complements and also contains X, it is called an *algebra*. If an algebra is closed under countable unions, it is called a *σ-algebra*.

Let X be a set, $\mathcal{D}$ a *σ-algebra*, and $\bar{\mu}$ a function from $\mathcal{D}$ to $\mathbb{R}^+ \cup \{+\infty\}$, where $\mathbb{R}^+ \equiv \{x \in \mathbb{R} : x \geq 0\}$, such that the following two conditions are satisfied:

1. $\bar{\mu}(\emptyset) = 0$;
2. If $S_1, S_2, \ldots \in S$ is a sequence of disjoint sets, then

$$\bar{\mu}\left(\bigcup_{i=1}^{\infty} S_i\right) = \sum_{i=1}^{\infty} \bar{\mu}(S_i). \tag{A.1.9}$$

Define

$$\mu^*(E) = \inf \sum_{i=1}^{\infty} \bar{\mu}(A_i), \tag{A.1.10}$$

where the infimum is taken with respect to all possible sequences $\{A_i\}$ from $\mathcal{D}$ such that $E \subset \cup_{i=1}^{\infty} A_i$.

A set E is called μ^* measurable if, for every $A \subset X$,

$$\mu^*(A) = \mu^*(A \cap E) + \mu^*(A \cap (X \setminus E)). \tag{A.1.11}$$

The set function μ^* is called an *outer (Lebesgue) measure*. It satisfies:

1. μ^* is increasing (i.e., if $A \subset B$, $\mu^*(A) \leq \mu^*(B)$);
2. $\mu^*(\emptyset) = 0$;
3. μ^* is countably subadditive (i.e., $\mu^*(\bigcup_{I=1}^{\infty} A_i) \leq \sum_{i=1}^{\infty} \mu^*(A_i)$);
4. μ^* extends $\bar{\mu}$ (i.e., if $A \in \mathcal{D}$, $\mu^*(A) = \bar{\mu}(A)$).

The μ^* measurable sets form a σ-algebra $\mathcal{F}$ containing $\mathcal{D}$, and μ^* restricted to $\mathcal{F}$ is a *(Lebesgue) measure* μ. Also, $(X, \mathcal{F})$ is a *(Lebesgue) measurable space*, and $(X, \mathcal{F}, \mu)$ is a *(Lebesgue) measure space*. The measure space $(X, \mathcal{F}, \mu)$ is complete (i.e., any subset in X with zero μ^* measure, and hence zero μ measure, is in $\mathcal{F}$). The measure μ is said to be finite if $\mu(X) < \infty$.

A function $f : X \to \mathbb{R} \cup \{\pm\infty\}$ is said to be *(Lebesgue) measurable* if for any real number α the set $\{x \in X : f(x) < \alpha\}$ is measurable. The function f is called a *simple function* if there is a finite, disjoint class $\{A_1, \ldots, A_n\}$ of measurable sets and a finite set $\{a_1, \ldots, a_n\}$ of real numbers such that

$$f(x) = \sum_{i=1}^{n} a_i \chi_{A_i}(x), \tag{A.1.12}$$

where χ_{A_i} is the *characteristic (indicator) function* of A_i defined by

$$\chi_{A_i}(x) = \begin{cases} 1 \text{ if } x \in A_i \\ 0 \text{ otherwise.} \end{cases} \tag{A.1.13}$$

Every non-negative measurable function is the limit of a monotonically increasing sequence of non-negative simple functions.

Consider a measure space (X, S, μ). Let $\mathcal{B}$ be the smallest σ-algebra generated by all open sets in X. It is also the smallest σ-algebra that contains all closed sets in X. Elements of $\mathcal{B}$ are called *Borel sets.* Let $\hat{\mu}$ denote the measure μ restricted to $\mathcal{B}$. It is called a *Borel measure.* All open sets and closed sets are Borel sets. A function $f : X \to \mathbb{R} \cup \{\pm\infty\}$ is said to be *Borel measurable* if for any real number α the set

$$\{x \in X : f(x) < \alpha\} \tag{A.1.14}$$

is a Borel set. If f is measurable and B is a Borel set,

$$f^{-1}(B) \equiv \{x \in X : f(x) \in B\} \tag{A.1.15}$$

is a measurable set. Every Borel measurable function is measurable. If f is Borel measurable, and B is a Borel set, then $f^{-1}(B)$ is a Borel set. Every lower (respectively, upper) semicontinuous function is Borel measurable, and hence measurable. There exist measurable functions that are not Borel measurable. In fact, suppose $f(t) = g(t)$ almost everywhere in $[a, b]$ (in the sense of Borel), and f is Borel measurable. It is not necessarily true that g is also Borel measurable, since the Borel measure space $(X, \mathcal{B}, \hat{\mu})$ is not complete. That is, not all sets with zero $\hat{\mu}$ measure are in $\mathcal{B}$.

Let (X, ρ), (Y, σ), and (Z, η) be three metric spaces. Let f be a Borel measurable function from X into Y, and g a measurable function from Y into Z. Then, the composite function $g \circ f : X \to Z$ is measurable, where

$$g \circ f(x) \equiv g(f(x)).$$

Let I be a measurable set and ψ a non-negative simple function

$$\psi(x) = \sum_{i=1}^{n} c_i \chi_{E_i}(x), \tag{A.1.16}$$

where c_i, $i = 1, \ldots, n$, are real non-negative numbers, and the measurable subsets $\{E_i\}_{i=1}^n \subset E$ are disjoint and satisfy $\bigcup_{i=1}^n E_i = I$. Also, let χ_{E_i} be the indicator function of E_i. We define the (Lebesgue) integral of ψ over I by

$$\int_I \psi(x)\, dx = \sum_{i=1}^n c_i \mu(E_i), \tag{A.1.17}$$

where $\mu(E_i)$ denotes the Lebesgue measure of E_i.

Let f be a non-negative measurable function from I to $\mathbb{R} \cup \{+\infty\}$. Then, the (Lebesgue) integral of the function f is defined by

$$\int_I f(x)\, dx = \sup \int_I \psi(x)\, dx, \tag{A.1.18}$$

where the supremum is taken with respect to all non-negative simple functions ψ with $0 \le \psi(x) \le f(x)$ for all $x \in I$. We say f is *integrable* if

$$\int_I f(t)\, dt < \infty.$$

For any measurable function $f : I \to \mathbb{R} \cup \{\pm\infty\}$, we can write $f = f^{\,+} - f^{\,-}$, where $f^{\,+} \equiv \max\{f, 0\}$ and $f^{\,-} \equiv \max\{-f, 0\}$. The function f is said to be integrable if

$$\int_I f^{\,+}(t)\, dt < \infty \quad \text{and} \quad \int_I f^{\,-}(t)\, dt < \infty.$$

The integral of f is

$$\int_I f(t)\, dt = \int_I f^{\,+}(t) - \int_I f^{\,-}(t)\, dt.$$

Theorem A.1.8 (Fatou's Lemma). *If $\{f_n\}_{n=1}^\infty$ is a sequence of non-negative measurable functions on I, then*

$$\int_I \varliminf_{n\to\infty} f_n(t)\, dt \le \varliminf_{n\to\infty} \int_I f_n(t)\, dt.$$

Theorem A.1.9 (The Monotone Convergence Theorem). *If $\{f_n\}_{n=1}^\infty$ is an increasing sequence of non-negative measurable functions on I such that $f_n \to f$ pointwise in I, then f is measurable and*

$$\int_I f(t)\, dt = \lim_{n\to\infty} \int_I f_n(t)\, dt.$$

A property is said to hold *almost everywhere (a.e.)* if it holds everywhere except on a set of measure zero in the sense of Lebesgue. Two (Lebesgue) measurable functions f and g from $I \subset \mathbb{R}$ to $\mathbb{R}$ are said to be *equivalent* if $f(t) = g(t)$ a.e. on I.

A sequence $\{f_n\}_{n=1}^{\infty}$ of measurable functions from $I \subset \mathbb{R}$ to $\mathbb{R}$ is said to converge *almost everywhere (a.e.)* to a function f (written as $f_n \to f$ *a.e.* on I) if there exists a set $A \subset I$ such that $\mu(A) = 0$ and

$$f_n(t) \to f(t)$$

for each $t \in I \setminus A$, where μ denotes the Lebesgue measure. In this situation, the function f is automatically a measurable function from I to $\mathbb{R}$.

Theorem A.1.10 (The Lebesgue Dominated Convergence Theorem). *Let $\{f_n\}_{n=1}^{\infty}$ be a sequence of measurable functions on I. If $f_n \to f$ a.e. and there exists an integrable function g on I such that $|f_n(t)| \le g(x)$ for almost every $x \in I$, then*

$$\int_I f(t)\,dt = \lim_{n\to\infty} \int_I f_n(t)\,dt.$$

Theorem A.1.11 (Luzin's Theorem). *Let $I \subset R$ be such that $\mu(I) < \infty$, and let f be a measurable function from I to R^n. Then, for any $\varepsilon > 0$, there exists a closed set $I_\varepsilon \subset I$ such that $\mu(I \setminus I_\varepsilon) < \varepsilon$ and the function f is continuous on I_ε.*

A.1.9 The L_p Spaces

Let $I = (a, b) \subset \mathbb{R}$, and $1 \le p < \infty$. Two functions are said to be *equivalent* if they are equal almost everywhere. Let $L_p(I, \mathbb{R}^n)$ be the class of all measurable functions from I to $\mathbb{R}^n$ such that

$$\int_I |\boldsymbol{f}(t)|^p\,dt < \infty,$$

where $\boldsymbol{f} \equiv [f_1, \ldots, f_n]^\top$, and $|\mathbf{f}(t)| = \left(\sum_{i=1}^n (f_i(t))^2\right)^{1/2}$. If we do not distinguish between equivalent functions in $L_p(I, \mathbb{R}^n)$, for $1 \le p < \infty$, then it is a Banach space with respect to the norm

$$\|\boldsymbol{f}\|_p = \left(\int_I |\boldsymbol{f}(t)|^p\,dt\right)^{1/p}. \tag{A.1.19}$$

Theorem A.1.12. *If $\boldsymbol{f} \in L_1(I, \mathbb{R}^n)$ and $\mathbf{g}$ is defined by*

$$\boldsymbol{g}(t) = \boldsymbol{g}(a) + \int_a^t \boldsymbol{f}(\tau)\,d\tau, \quad \textit{for } t \in I, \tag{A.1.20}$$

then $\boldsymbol{g} \in AC(I, \mathbb{R}^n)$ and $d\boldsymbol{g}(t)/dt = \boldsymbol{f}(t)$ a.e. on I.

A measurable function $\boldsymbol{f}$ from I to $\mathbb{R}^n$ is said to be *essentially bounded* if there exists a positive number $K < \infty$ such that the set

$$S = \{t \in I : |\boldsymbol{f}(t)| > K\} \tag{A.1.21}$$

has (Lebesgue) measure zero.

Let $L_\infty(I, \mathbb{R}^n)$ denote the space of all such essentially bounded measurable functions. The smallest number K for which (A.1.21) is valid is called the *essential supremum* of $|\boldsymbol{f}(t)|$ over $t \in I$ and is written as:

$$\|\boldsymbol{f}\|_\infty = \operatorname{ess\,sup}\{|\boldsymbol{f}(t)| : t \in I\}. \tag{A.1.22}$$

$L_\infty(I, \mathbb{R}^n)$ is a Banach space with respect to the norm $\|\cdot\|_\infty$, where we identify functions that are equivalent.

For a given $1 \le p \le \infty$, a sequence $\{\boldsymbol{f}^n\}_{n=1}^\infty$ in $L_p(I, \mathbb{R}^n)$ is said to converge to a function $\boldsymbol{f} \in L_p(I, \mathbb{R}^n)$ if $\|\boldsymbol{f}^n - \boldsymbol{f}\|_p \to 0$ as $n \to \infty$.

Let q be a number such that $1/p + 1/q = 1$. Clearly, if $p = 1$, then $q = \infty$, while if $p = \infty$, then $q = 1$.

Theorem A.1.13. *Let* $1 \le p \le \infty$ *and* $q = p/(p-1)$. *Then,*

(a) (Hölder's inequality) for $\boldsymbol{f} \in L_p(I, R^n)$ *and* $\boldsymbol{g} \in L_q(I, R^n)$,

$$\left| \int_I (\boldsymbol{f}(t))^\top \boldsymbol{g}(t)\, dt \right| \le \left(\int_I |\boldsymbol{f}(t)|^p \, dt \right)^{1/p} \left(\int_I |\boldsymbol{g}(t)|^q \, dt \right)^{1/q} ; \tag{A.1.23}$$

and

(b) (Minkowski's inequality) for $\boldsymbol{f}$ *and* $\boldsymbol{g} \in L_q(I, R^n)$,

$$\left(\int_I |\boldsymbol{f}(t) + \boldsymbol{g}(t)|^p \, dt \right)^{1/p} \le \left(\int_I |\boldsymbol{f}(t)|^p \, dt \right)^{1/p} + \left(\int_I |\boldsymbol{g}(t)|^p \, dt \right)^{1/p} . \tag{A.1.24}$$

Remark A.1.3. It is well-known that if I is a finite interval,

(a) $\|\boldsymbol{f}\|_\infty = \lim_{p \to \infty} \left(\int_I |\boldsymbol{f}(t)|^p \, dt \right)^{1/p}$; and
(b) $L_1(I, \mathbb{R}^n) \supset L_2(I, \mathbb{R}^n) \supset \ldots L_\infty(I, \mathbb{R}^n)$.

Let $\boldsymbol{f} : I \equiv (a, b) \to U \subset \mathbb{R}^n$ be a measurable function. If $\theta \in I$ is such that the condition

$$\lim_{\mu(J) \to 0} \frac{\mu(\boldsymbol{f}^{-1}(V) \cap J)}{\mu(J)} = 1 \tag{A.1.25}$$

is satisfied for every neighbourhood $V \subset U$ of $\boldsymbol{f}(\theta)$, where

$$\boldsymbol{f}^{-1}(V) = \{t \in I : \boldsymbol{f}(t) \in V\},$$

J denotes an arbitrary interval that contains θ, and μ denotes the Lebesgue measure, then θ is called a *regular point* for the function $\boldsymbol{f}$.

Let $\theta \in I$ be a continuity point of $\boldsymbol{f}$. Then, J can be made sufficiently small such that $J \subset \boldsymbol{f}^{-1}(V)$, and hence condition (A.1.25) is satisfied. Thus,

θ is a regular point for the function $\boldsymbol{f}$. If $\boldsymbol{f}$ is piecewise continuous, then all but a finite number of points in I are regular points. In fact, almost all points in I are regular points for a general measurable function.

The following well-known result is an important tool in deriving pointwise necessary conditions for optimality in optimal control theory. It is also an important theorem in the convergence analysis of computational algorithms.

Theorem A.1.14. *Let I be an interval of $\mathbb{R}$, $\boldsymbol{f} \in L_1(I, \mathbb{R}^n)$, $t \in I$ a regular point of $\boldsymbol{f}$, and $\{I_k\}$ a decreasing sequence of subintervals of I such that $t \in I_k$ for all k and $\lim_{k\to\infty} \mu(I_k) = 0$. Then,*

$$\lim_{k\to\infty} \frac{1}{\mu(I_k)} \int_{I_k} \boldsymbol{f}(\tau)\, d\tau = \boldsymbol{f}(t).$$

Let F be a set in $L_p(I, \mathbb{R}^n)$, for $1 \le p < \infty$. A sequence $\{\boldsymbol{f}^{(n)}\} \subset F$ is said to *converge weakly* to a function $\hat{\boldsymbol{f}} \in L_p(I, \mathbb{R}^n)$ (written as $\boldsymbol{f}^{(n)} \xrightarrow{w} \hat{\boldsymbol{f}}$) if

$$\lim_{n\to\infty} \int_I \left(\boldsymbol{f}^{(n)}(t)\right)^\top \boldsymbol{g}(t) d\tau = \int_I \left(\hat{\boldsymbol{f}}(t)\right)^\top \boldsymbol{g}(t) dt$$

for every $\boldsymbol{g} \in L_q(I, \mathbb{R}^n)$, where $1/p + 1/q = 1$ and the superscript $^\top$ denotes the transpose. The function $\hat{\boldsymbol{f}}$ is called the *weak limit* of the sequence $\{\boldsymbol{f}^{(n)}\}$.

A set $F \subset L_p(I, \mathbb{R}^n)$, for $1 \le p < \infty$, is said to be *weakly closed* if the limit of every weakly convergent sequence $\{\boldsymbol{f}^{(n)}\} \subset F$ is in F.

A set $F \subset L_p(I, \mathbb{R}^n)$, for $1 \le p < \infty$, is said to be *conditionally weakly sequentially compact* if every sequence $\{\boldsymbol{f}^{(n)}\} \subset F$ contains a subsequence that converges weakly to a function $\hat{\boldsymbol{f}} \in L_p(I, \mathbb{R}^n)$. The set F is said to be *weakly sequentially compact* if it is conditionally weakly sequentially compact and weakly closed.

Let F be s set in $L_\infty(I, \mathbb{R}^n)$. A sequence $\{\boldsymbol{f}^{(n)}\} \subset F$ is said to converge to a function $\hat{\boldsymbol{f}} \in L_\infty(I, \mathbb{R}^n)$ in the *weak* topology* of $L_\infty(I, \mathbb{R}^n)$ (written as $\boldsymbol{f}^{(n)} \xrightarrow{w^*} \hat{\boldsymbol{f}}$) if

$$\lim_{n\to\infty} \int_I \left(\boldsymbol{f}^{(n)}(t)\right)^\top \boldsymbol{g}(t) dt = \int_I \left(\hat{\boldsymbol{f}}(t)\right)^\top \boldsymbol{g}(t) dt$$

for every $\boldsymbol{g} \in L_1(I, \mathbb{R}^n)$. The function $\hat{\boldsymbol{f}}$ is called the *weak * limit* of the sequence $\{\boldsymbol{f}^{(n)}\}$.

A set $F \subset L_\infty(I, \mathbb{R}^n)$ is said to be weak * closed if the limit of every weak* convergent sequence $\{\boldsymbol{f}^{(n)}\} \subset F$ is in F.

A set $F \subset L_\infty(I, \mathbb{R}^n)$ is said to be *conditionally sequentially compact in the weak * topologyof* $L_\infty(I, \mathbb{R}^n)$ if every sequence $\{\boldsymbol{f}^{(n)}\} \subset F$ contains a subsequence that converges to a function $\hat{\boldsymbol{f}} \in L_\infty(I, \mathbb{R}^n)$ in the weak* topology. The set F is called *sequentially compact in the weak * topology of* $L_\infty(I, \mathbb{R}^n)$ if it is conditionally sequentially compact in the weak* topology of $L_\infty(I, \mathbb{R}^n)$ and weak* closed.

The following well-known result is extremely important in proving existence of optimal controls, and in analysing the convergence properties of computational algorithms for linear optimal control problems.

Theorem A.1.15. *Let U be a compact and convex subset of $\mathbb{R}^n$. Then, the set*

$$\mathcal{U} \equiv \{\boldsymbol{u} \in L_\infty(I, \mathbb{R}^n) : \boldsymbol{u}(t) \in U, \text{ a.e. on } I\}$$

is sequentially compact in the weak topology of $L_\infty(I, \mathbb{R}^n)$.*

Definition A.1.9. Let $\mathcal{L} : [0, T] \to U$, where U is a compact and convex subset of $\mathbb{R}^r$. Then, $\mathcal{L}$ is said to be weakly sequentially lower semicontinuous if for any sequence $\{\boldsymbol{u}^n\} \subset L_2([0, T], \mathbb{R}^r)$ such that $\boldsymbol{u}^n \to \boldsymbol{u}$ in the weak topology of $L_2([0, T], \mathbb{R}^r)$, where $\boldsymbol{u} \in L_2([0, T], \mathbb{R}^r)$, then

$$\mathcal{L}(\boldsymbol{u}) \leq \underline{\lim}_{n \to \infty} \mathcal{L}(\boldsymbol{u}^n).$$

Remark A.1.4. The set $\mathcal{U}$ defined by

$$\mathcal{U} = \{\boldsymbol{u} \in L_2([0, T], \mathbb{R}^r) : \|\boldsymbol{u}\| \leq \rho\}$$

is weakly sequentially compact, where $\|\cdot\|$ denotes the L_2-norm in $L_2([0, T], \mathbb{R}^r)$ and ρ is a given positive constant.

Theorem A.1.16. *Let I be an open bounded subset of $\mathbb{R}$, U a compact subset of $\mathbb{R}^r$, and $\boldsymbol{f}$ a function from $I \times U$ to $\mathbb{R}^n$. If $\boldsymbol{f}(t, \cdot)$ is continuous on U for almost all $t \in I$, and $\boldsymbol{f}(\cdot, \boldsymbol{u})$ is measurable on I for every $\boldsymbol{u} \in U$, then for any $\varepsilon > 0$, there exists a closed set $I_\varepsilon \subset I$ such that $\mu(I \setminus I_\varepsilon) < \varepsilon$ and $\boldsymbol{f}$ is a continuous function on $I_\varepsilon \times U$.*

Theorem A.1.17. *Let I be an interval in $\mathbb{R}$ such that $\mu(I) < \infty$, and let $\boldsymbol{f} \in L_p(I, \mathbb{R}^n)$ for all $p \in [1, \infty)$. If there exists a constant K such that $\|\boldsymbol{f}\|_p \leq K$ for all such p, then $\boldsymbol{f} \in L_\infty(I, \mathbb{R}^n)$ and $\|\boldsymbol{f}\|_\infty \leq K$.*

Theorem A.1.18. *Let I be an open bounded subset of $\mathbb{R}$, Ω a compact and convex subset of $\mathbb{R}^n$, f a continuous function defined on $I \times \Omega$ such that $f(t, \cdot)$ is convex on Ω for each $t \in I$, and $\{\mathbf{y}^k\}$ a sequence of measurable functions defined on I with values in Ω. If $\mathbf{y}^k \xrightarrow{w^*} \mathbf{y}^0$ in $L_\infty(I, \mathbb{R}^n)$, then*

$$\int_I f\left(t, \boldsymbol{y}^0(t)\right) dt \leq \underline{\lim}_{k \to \infty} \int_I f\left(t, \boldsymbol{y}^k(t)\right) dt.$$

A measurable function f from $[0, \infty]$ to $\mathbb{R}$ is said to belong to L_1^{loc} if

$$\int_0^t |f(\tau)|\, d\tau < \infty, \quad \text{for any } t < \infty.$$

Theorem A.1.19. *(**Gronwall-Bellman Lemma**) Suppose that*

$$f(t) \leq \alpha(t) + \int_0^t K(\tau) f(\tau)\, d\tau,$$

where α is a continuous and bounded function on $[0, \infty)$ such that $\alpha(t) \geq 0$ for all $t \in [0, \infty)$. If $K \in L_1^{\text{loc}}$ such that $K(t) \geq 0$ a.e. on $[0, \infty)$, and $f(t) \geq 0$ for all $t \in [0, \infty)$, then

$$f(t) \leq \alpha(t) + \int_0^t \exp\left[\int_s^t K(\tau)\, d\tau\right] K(s)\alpha(s)\, ds$$

for $t \in [0, \infty)$.

A.1.10 Multivalued Functions

For results on existence of optimal controls, we are required to make use of the concept of multivalued functions, also known as set-valued functions. We refer the reader to [3, 4, 40] for details. In this section, we shall prove a selection theorem involving lower and upper semicontinuous functions.

Let $\mathcal{K}$ be the set of all non-empty compact subsets of $\mathbb{R}$. For $x \in \mathbb{R},\ A \in \mathcal{K}$, the distance $\rho(x, A)$ of x from A is defined by

$$\rho(x, A) = \inf\ \{\ |x - a| : a \in A\}. \tag{A.1.26}$$

We now define, for $A, B \in \mathcal{K}$,

$$2\rho_H(A, B) = \sup\{\rho\,(a, B) : a \in A\} + \sup\{(b, A) : b \in B\}\,. \tag{A.1.27}$$

Here, ρ_H is called the *Hausdorff metric* on $\mathcal{K}$. Let I be an open bounded set in $\mathbb{R}$. Let F be a multivalued function defined on I such that $F(x) \in \mathcal{K}$. We assume that the set-valued function F is continuous with respect to the Hausdorff metric. That is, if $\bar{x} \in I$ is arbitrary but fixed, then for any $\varepsilon > 0$ there exists a $\delta > 0$ such that

$$\rho_H\left(F\left(x\right), F\left(\bar{x}\right)\right) < \varepsilon$$

whenever

$$x \in I_\delta \equiv \{x \in I : |x - \bar{x}| < \delta\}\,.$$

Lemma A.1.1. *Let I and U be compact subsets in $\mathbb{R}$ and let g be a continuous function from $I \times U$ into $\mathbb{R}$. Furthermore, let $F\left(\cdot\right)$ be a continuous multivalued function defined on I with respect to the Hausdorff metric such that $F\left(x\right)$ is a non-empty compact subset of U for each $x \in I$. Define*

$$r(x) = \inf \{g(x,u) : u \in F(x)\}. \tag{A.1.28}$$

Then, r is an upper semicontinuous function on I.

Proof. Since $F(\bar{x})$ is compact, there exists a $\bar{u} \in F(\bar{x})$ such that

$$r(\bar{x}) = g(\bar{x}, \bar{u}) = \inf_{u \in F(\bar{x})} g(\bar{x}, u). \tag{A.1.29}$$

By the continuity of the function g on $I \times U$, we see that $g(\cdot, \bar{u})$ is continuous at $\bar{x}$. Thus, for any $\varepsilon > 0$ there exists a $\delta_1 > 0$ such that

$$|g(\bar{x}, \bar{u}) - g(x, \bar{u})| < \varepsilon \tag{A.1.30}$$

for all $x \in I_{\delta_1}(\bar{x})$, where

$$I_{\delta_1}(\bar{x}) = \{x \in I : |x - \bar{x}| < \delta_1\}. \tag{A.1.31}$$

In particular, it is clear that

$$r(\bar{x}) = g(\bar{x}, \bar{u}) > g(x, \bar{u}) - \varepsilon \tag{A.1.32}$$

for all $x \in I_{\delta_1}(\bar{x})$.

Now, for any $x \in I_{\delta_1}(\bar{x})$, there are two cases to be considered:

(i) $\bar{u} \in F(x)$; and
(ii) $\bar{u} \notin F(x)$.

For case (i), it follows that

$$g(x, \bar{u}) \geq \inf_{u \in F(x)} g(x, u). \tag{A.1.33}$$

For case (ii), there exists a $u_0 \in F(x)$ such that

$$|\bar{u} - u_0| = \rho(\bar{u}, F(x)) \leq \rho_H(F(\bar{x}), F(x)). \tag{A.1.34}$$

By the continuity of the set-valued function $F(\cdot)$ on I with respect to the Hausdorff metric, it follows that for any $\delta_2 > 0$ there exists a $\delta_3 > 0$ such that

$$\rho_H(F(\bar{x}), F(x)) < \delta_2 \tag{A.1.35}$$

whenever $|\bar{x} - x| < \delta_3$. Thus, by (A.1.34),

$$|\bar{u} - u_0| < \delta_2 \quad \text{whenever} \quad |\bar{x} - x| < \delta_3. \tag{A.1.36}$$

Since the function $g(\cdot, \cdot)$ is continuous on $I \times U$, where $I \times U$ is compact, it is clear that it is uniformly continuous on $I \times U$. Thus, for any $\varepsilon > 0$ as defined in (A.1.30), we can choose a $\delta_2 > 0$ and hence $\delta_3 > 0$ such that

$$|g(x, \bar{u}) - g(x, u_0)| < \varepsilon \tag{A.1.37}$$

whenever $|\bar{u} - u_0| < \delta_2$. This, in turn, implies that

$$g(x, \bar{u}) > g(x, u_0) - \varepsilon \tag{A.1.38}$$

whenever $|\bar{u} - u_0| < \delta_2$. Thus,

$$g(x, \bar{u}) > g(x, u_0) - \varepsilon \geq \inf_{u \in F(x)} g(x, u) - \varepsilon \tag{A.1.39}$$

whenever $|\bar{x} - x| < \delta_3$. Combining (A.1.32), (A.1.33) and (A.1.39) yields

$$r(\bar{x}) \geq \inf_{u \in F(x)} g(x, u) - 2\varepsilon = r(x) - 2\varepsilon$$

for all $x \in I_\delta(\bar{x})$, where $\delta = \min\{\delta_1, \delta_3\}$. This completes the proof.

Lemma A.1.2. *Let I, U, g, and $F(\cdot)$ be as defined in Lemma A.1.1. Let $\{x_k\} \subset I$ be such that $x_k \to \hat{x}$, and $u(x_k) \to \hat{u}$, both as $k \to \infty$. If $u(x_k) \in F(x_k)$ for all $k \geq 1$, then $\hat{u} \in F(\hat{x})$.*

Proof. Since $x_k \to \hat{x}$ as $k \to \infty$, and $F(\cdot)$ is continuous on I with respect to the Hausdorff metric, it follows that for any $\varepsilon > 0$ there exists an integer $N > 0$ such that

$$\rho_H(F(x_k), F(\hat{x})) < \varepsilon$$

for all $k > N$. This, in turn, implies that

$$F(x_k) \subset F^\varepsilon(\hat{x}) \tag{A.1.40}$$

for all $k > N$, where $F^\varepsilon(\hat{x})$ is the closed ε-neighbourhood of $F(\hat{x})$ defined by

$$F^\varepsilon(\hat{x}) = \{v \in U : \rho(v, F(\hat{x})) \leq \varepsilon\}.$$

Since $u(x_k) \in F(x_k)$ for all $k \geq 1$, it is clear from (A.1.40) that $u(x_k) \in F^\varepsilon(\hat{x})$ for all $k > N$. Thus, by the facts that $u(x_k) \to \hat{u}$ as $k \to \infty$, and $F^\varepsilon(\hat{x})$ is closed, we have

$$\hat{u} \in F^\varepsilon(\hat{x}).$$

Since this relation is true for all $\varepsilon > 0$, and $F(\hat{x})$ is closed, it follows that $\hat{u} \in F(\hat{x})$. Thus, the proof is complete.

Theorem A.1.20. *Let I, U, g, and $F(\cdot)$ be as defined in Lemma A.1.1. Define*

$$g(x, F(x)) = \{g(x, u) : u \in F(x)\}, \tag{A.1.41}$$

and

$$r(x) = \inf \{g(x, u) : u \in F(x)\}, \tag{A.1.42}$$

for all $x \in I$. Then, there exists a lower semicontinuous function $u(x)$ with values in $F(x)$ such that

$$r(x) = g(x, u(x)) \tag{A.1.43}$$

for all $x \in I$.

Proof. By Lemma A.1.1, r is an upper semicontinuous function on I. Furthermore,

$$r(x) \in g(x, F(x)) \tag{A.1.44}$$

for all $x \in I$. We choose the smallest $u \in F(x)$ for which

$$r(x) = g(x, u) \in g(x, F(x)). \tag{A.1.45}$$

Since $g(x, \cdot)$ is continuous in U and $F(x)$ is compact, the set

$$\{u \in F(x) : g(x, u) = r(x)\} \tag{A.1.46}$$

is compact. Thus, the choice of such a function u is possible.

For the function u so chosen to be lower semicontinuous, it suffices to show that for any real number α, the set

$$A = \{x \in I : u(x) \leq \alpha\} \tag{A.1.47}$$

is closed. Suppose this is false. Then, there exists a sequence $\{x_k\}$ in $A \subset I$ such that

$$r(x_k) = g(x_k, u(x_k)) \in g(x_k, F(x_k)), \tag{A.1.48}$$

$$x_k \to \hat{x} \in I, \tag{A.1.49}$$

$$u(\hat{x}) > \alpha. \tag{A.1.50}$$

By the definition of the set A given in (A.1.47), there exists an $\varepsilon > 0$ such that

$$u(x_k) \leq u(\hat{x}) - \varepsilon. \tag{A.1.51}$$

Since $x \in F(x) \subset U$ for all $x \in I$, and U is compact, there exists a constant $K > 0$ such that

$$|u(x)| \leq K \quad \text{for all} \quad x \in I.$$

In particular,

$$|u(x_k)| \leq K \tag{A.1.52}$$

for all positive integers k. Thus, there exists a subsequence of the sequence $\{x_k\}$, again denoted by the original sequence, such that

$$u(x_k) \to \hat{u}. \tag{A.1.53}$$

Since $u(x_k) \in F(x_k)$, F is continuous on I with respect to the Hausdorff metric, and $F(\hat{x})$ is closed, it follows from Lemma A.1.2 that $\hat{u} \in F(\hat{x})$. Thus, by (A.1.51) and (A.1.53), we have

$$\hat{u} \leq u(\hat{x}) - \varepsilon. \tag{A.1.54}$$

Since r is an upper semicontinuous function on the closed interval I, it is bounded. In particular, there exists a constant $K_1 > 0$ such that $|r(x_k)| \leq K_1$ for all positive integers k. Thus, there exists a further subsequence, which is again denoted by the original sequence, such that

$$r(x_k) \to \hat{r}, \quad \text{and} \tag{A.1.55}$$

$$\hat{r} \leq r(\hat{x}). \tag{A.1.56}$$

We note that

$$r(x_k) = g(x_k, u(x_k)) \in g(x_k, F(x_k)) = \{g(x_k, u) : u \in F(x_k)\}, \tag{A.1.57}$$

F is continuous on I, g is continuous on $I \times U$ which is compact, $F(\hat{x})$ is closed, and hence $g(\hat{x}, F(\hat{x}))$ is closed. Thus, by (A.1.55), (A.1.49), (A.1.53), and (A.1.57), we have

$$\hat{r} = g(\hat{x}, \hat{u}) \in g(\hat{x}, F(\hat{x})). \tag{A.1.58}$$

However, in view of the definition of $r(\hat{x})$, it is clear that

$$r(\hat{x}) \leq \hat{r}. \tag{A.1.59}$$

Combining (A.1.56) and (A.1.58), we have

$$r(\hat{x}) = \hat{r}. \tag{A.1.60}$$

Therefore,

$$r(\hat{x}) = g(\hat{x}, \hat{u}). \tag{A.1.61}$$

However, by (A.1.54), we see that $u(\hat{x})$ is not the smallest value of u such that

$$r(\hat{x}) = g(\hat{x}, u).$$

This contradicts the definition of $u(x)$. Thus, the set A defined by (A.1.47) must be closed, and hence the function u is lower semicontinuous on I. The proof is complete.

A.1.11 Bounded Variation

By a partition of the interval $[a, b]$, we mean a finite set of points $t_i \in [a, b]$, $i = 0, 1, \ldots, m$, such that

$$a = t_0 < t_1 < t_2 < \cdots < t_m = b.$$

A function f defined on $[a, b]$ is said to be of *bounded variation* if there is a constant K so that for any partition of $[a, b]$

$$\sum_{i=1}^{m} |f(t_i) - f(t_{i-1})| \leq K.$$

The *total variation* of f, denoted by $\overset{b}{\underset{a}{\vee}} f(t)$, is defined by

$$\overset{b}{\underset{a}{\vee}} f(t) = \sup \sum_{i=1}^{m} |f(t_i) - f(t_{i-1})|,$$

where the supremum is taken with respect to all partitions of $[a, b]$. The total variation of a constant function is zero and the total variation of a monotonic function is the absolute value of the difference between the function values at the end points a and b.

The space $BV[a, b]$ is defined as the space of all functions of bounded variation on $[a, b]$ together with the norm defined by

$$\|f\| = |f(a)| \; + \; \overset{b}{\underset{a}{\vee}} f(t).$$

Suppose $f \in BV[a, b]$. Then, f is differentiable a.e. on $[a, b]$. If $f : [a, b] \to \mathbb{R}$ is absolutely continuous, then it is of bounded variation.

Theorem A.1.21. *If $f \in BV[a, b]$, then f is absolutely continuous if and only if*

$$\int_a^b \left| \frac{df(t)}{dt} \right| dt = \overset{b}{\underset{a}{\vee}} f(t).$$

If f is monotone, then $f \in BV[a, b]$ and $\overset{b}{\underset{a}{\vee}} f(t) = |f(b) - f(a)|$.

If $f \in BV[a, b]$, then the jump of f at t is defined as:

$$\begin{cases} |f(t) - f(t-0)| + |f(t+0) - f(t)| & \text{if } a < t < b \\ |f(a+0) - f(a)| & \text{if } \quad t = a \\ |f(b) - f(b-0)| & \text{if } \quad t = b \end{cases}$$

We now consider a function $\boldsymbol{f} \equiv [f_1, \ldots, f_n]^\top : [a, b] \to \mathbb{R}^n$, where $[a, b]$ is a finite closed interval in $\mathbb{R}$. The full variation of $\boldsymbol{f}$ is defined as:

$$\overset{b}{\underset{a}{\vee}} \boldsymbol{f}(t) = \sum_{i=1}^{n} \overset{b}{\underset{a}{\vee}} f_i(t).$$

Let $BV([a, b], \mathbb{R}^n)$ be the space of all functions $\boldsymbol{f} : [a, b] \to \mathbb{R}^n$ which are of bounded variation on $[a, b]$.

Theorem A.1.22. *If $\boldsymbol{f} \in BV([a,b],\mathbb{R}^n)$, then $\boldsymbol{f}(t+0) \equiv \lim_{s\downarrow t+0} \boldsymbol{f}(s)$, the limit from the right at t, exists if $a \le t < b$; and $\boldsymbol{f}(t-0) \equiv \lim_{s\uparrow t-0} \boldsymbol{f}(s)$, the limit from the left at t, exists if $a < t \le b$.*

In order that $\boldsymbol{f}$ shall approach a limit in $\mathbb{R}^n$ as s approaches t from the right (respectively, from the left), the following condition is necessary and sufficient: To each $\varepsilon > 0$ there corresponds a $\delta > 0$ such that

$$|\boldsymbol{f}(\tau) - \boldsymbol{f}(t)| < \varepsilon$$

if $s < \tau < t + \delta$ (respectively, $t - \delta < \tau < s$).

Theorem A.1.23. *If $\boldsymbol{f} \in BV([a,b],\mathbb{R}^n)$, the set of points of discontinuity of $\boldsymbol{f}$ is countable.*

Let $\mathcal{E}$ be a family of functions in $BV([a,b],\mathbb{R}^n)$. It is said to be *equibounded with equibounded total variation* if there exist constants $K_1 > 0$, $K_2 > 0$ such that

$$|\boldsymbol{f}(t)| \le K_1 \quad \text{and} \quad \bigvee_a^b \boldsymbol{f}(t) \le K_2, \quad \text{for all} \quad \boldsymbol{f} \in \mathcal{E}.$$

Theorem A.1.24 (Helly). *Let $\mathcal{E}$ be a family of functions in $BV([a,b],\mathbb{R}^n)$ which is equibounded with equibounded total variation. Then, any sequence $\{\boldsymbol{f}^{(n)}\}$ of elements in $\mathcal{E}$ contains a subsequence $\{\boldsymbol{f}^{(n(k))}\}$ which converges pointwise everywhere on $[a,b]$ toward a function $\boldsymbol{f}^{(0)} \in BV([a,b],\mathbb{R}^n)$ with*

$$\bigvee_a^b \boldsymbol{f}^{(0)}(t) \le \varliminf_{n\to\infty} \bigvee_a^b \boldsymbol{f}^{n(k)}(t).$$

Appendix A.2
Global Optimization via Filled Function Approach

Consider the following optimization problem defined by

$$\min_{\boldsymbol{x} \in X} f(x), \tag{A.2.1}$$

where $X \subset R^n$ is a closed bounded domain containing all global minimizers of $f(\boldsymbol{x})$ in its interior. It is assumed that $f(\boldsymbol{x})$ has only a finite number of local minimizers. Let this optimization problem be referred to as Problem (B).

We note that the solution obtained from solving Problem (B) using a gradient-based method is unlikely to be a global minimizer. We will introduce a filled function [57, 285, 286] and then use it to obtain a global minimizer for Problem (B) by incorporating the filled function with a gradient-based method.

We suppose that $\boldsymbol{x}^{1,*}$ is a local minimizer of Problem (B). Since $f(\boldsymbol{x})$ are differentiable with respect to $\boldsymbol{x}$ and the set X is bounded and closed, there exists a constant $M > 0$ such that for all $\boldsymbol{x}^1$ and $\boldsymbol{x}^2 \in X$, the following condition is satisfied

$$f\left(\boldsymbol{x}^1\right) - f\left(\boldsymbol{x}^2\right) \leq M \left|\boldsymbol{x}^1 - \boldsymbol{x}^2\right|, \tag{A.2.2}$$

where $|\cdot|$ denotes the standard Euclidean norm in $\mathbb{R}^n$. To escape from the local minimizer $\boldsymbol{x}^*$, we construct the function

$$\begin{aligned} p(\boldsymbol{x}, \boldsymbol{x}^*, \rho, \mu) =& f(\boldsymbol{x}^*) - \min\left\{f(\boldsymbol{x}^*), f(\boldsymbol{x})\right\} - \rho \left|\boldsymbol{x} - \boldsymbol{x}^*\right|^2 \\ &+ \mu \left[\max\left\{0, f(\boldsymbol{x}) - f(\boldsymbol{x}^*)\right\}\right]^2, \end{aligned} \tag{A.2.3}$$

where $\boldsymbol{x} \in X$, while μ and ρ are parameters, which are such that $\rho > 0$ and $0 \leq \mu < \rho/M^2$. To proceed further, we need some definitions.

K. L. Teo et al., *Applied and Computational Optimal Control*, Springer Optimization and Its Applications 171,
https://doi.org/10.1007/978-3-030-69913-0

Definition A.2.1. The basin of $f(\boldsymbol{x})$ at an isolated minimizer $\boldsymbol{x}^*$ is a connected domain, denoted as X^*, which contains $\boldsymbol{x}^*$ and within X^* the steepest descent trajectory of $f(\boldsymbol{x})$ converges to $\boldsymbol{x}^*$ from any initial point.

Definition A.2.2. The basin of $f(\boldsymbol{x})$ at an isolated minimizer $\boldsymbol{x}_1^*$ is said to be lower than another basin of $f(\boldsymbol{x})$ at an isolated minimizer $\boldsymbol{x}_2^*$ if and only if $f(\boldsymbol{x}_1^*) < f(\boldsymbol{x}_2^*)$.

Definition A.2.3. The hill of $f(\boldsymbol{x})$ at an isolated minimizer $\boldsymbol{x}^*$ is the basin of $-f(\boldsymbol{x})$ at its isolated minimizer x^*.

In the following, we will show that $p(\boldsymbol{x}, \boldsymbol{x}^*, \rho, \mu)$ satisfies the following properties.

Property A.2.1. $\boldsymbol{x}^*$ is a maximizer of $p(\boldsymbol{x}, \boldsymbol{x}^*, \rho, \mu)$ and the whole basin X^* of $f(\boldsymbol{x})$ at $\boldsymbol{x}^*$ becomes part of a hill of the function $p(\boldsymbol{x}, \boldsymbol{x}^*, \rho, \mu)$.

Property A.2.2. $p(\boldsymbol{x}, \boldsymbol{x}^*, \rho, \mu)$ has no minimizers or saddle points in any basin of $f(\boldsymbol{x})$ higher than X^*.

Property A.2.3. Let $\boldsymbol{x}^{1,*}$ be an isolated minimizer of $f(\boldsymbol{x})$ in X, and let X_1^* be the basin as defined in Definition A.2.1. If $f(\boldsymbol{x})$ has a basin X_2^* at $\boldsymbol{x}^{2,*}$ that is lower than X_1^* at $\boldsymbol{x}^{1,*}$, then there is a point $\boldsymbol{x}' \in X_2^*$ that minimizes $p\left(\boldsymbol{x}, \boldsymbol{x}^{1,*}, \rho, \mu\right)$ on the line through $\boldsymbol{x}^{1,*}$ and $\boldsymbol{x}'$, for every $\boldsymbol{x}'$ in some neighbourhood of $\boldsymbol{x}^{2,*}$.

Property A.2.1 will be established in the next two theorems.

Theorem A.2.1. *Assume that $\boldsymbol{x}^*$ is a local minimizer of $f(\boldsymbol{x})$. If $\rho > 0$ and $0 \le \mu < \rho/M^2$, then $\boldsymbol{x}^{1,*}$ is a strict local maximizer of $p(\boldsymbol{x}, \boldsymbol{x}^*, \rho, \mu)$.*

Proof. Since $\boldsymbol{x}^*$ is a local minimizer of $f(\boldsymbol{x})$, there exists a neighbourhood $\mathcal{N}(\boldsymbol{x}^*, \delta)$ with radius $\delta > 0$ and center $\boldsymbol{x}^*$ such that for any $\boldsymbol{x} \in \mathcal{N}(\boldsymbol{x}^*, \delta)$, we have

$$f(\boldsymbol{x}) \ge f(\boldsymbol{x}^*). \tag{A.2.4}$$

Now, for any $\boldsymbol{x} \in \mathcal{N}(\boldsymbol{x}^*, \delta)$,we obtain

$$\begin{aligned} &p(\boldsymbol{x}, \boldsymbol{x}^*, \rho, \mu) - p(\boldsymbol{x}^*, \boldsymbol{x}^*, \rho, \mu) \\ &\quad = \mu \left[f(\boldsymbol{x}) - f(\boldsymbol{x}^*)\right]^2 - \rho|\boldsymbol{x} - \boldsymbol{x}^*|^2 < 0. \end{aligned} \tag{A.2.5}$$

Thus, $\boldsymbol{x}^*$ is a local maximizer of $p(\boldsymbol{x}, \boldsymbol{x}^*, \rho, \mu)$.

Theorem A.2.2. *Assume that $\boldsymbol{x}^*$ is a local minimizer of $f(\boldsymbol{x})$. Suppose that $\boldsymbol{x}^1$ and $\boldsymbol{x}^2$ are two points such that*

$$\left|\boldsymbol{x}^1 - \boldsymbol{x}^*\right| < \left|\boldsymbol{x}^2 - \boldsymbol{x}^*\right| \tag{A.2.6}$$

and

$$f\left(\boldsymbol{x}^*\right) \leq f\left(\boldsymbol{x}^1\right) \leq f\left(\boldsymbol{x}^2\right). \tag{A.2.7}$$

If $\rho > 0$ *and* $0 \leq \mu < \min\left\{\rho/M^2, \rho/MM_1\right\}$, *where*

$$M_1 \geq \max_{0\leq\alpha\leq 1}\left|\frac{\partial f\left(\boldsymbol{x}^1+\alpha\left(\boldsymbol{x}^2-\boldsymbol{x}^1\right)\right)}{\partial \boldsymbol{x}}\right| \frac{\left|\boldsymbol{x}^2-\boldsymbol{x}^1\right|}{\left|\boldsymbol{x}^2-\boldsymbol{x}^{1,*}\right|-\left|\boldsymbol{x}^1-\boldsymbol{x}^{1,*}\right|}, \tag{A.2.8}$$

then

$$p\left(\boldsymbol{x}^2, \boldsymbol{x}^*, \rho, \mu\right) < p(\boldsymbol{x}^*, \boldsymbol{x}^*, \rho, \mu) < 0 = p(\boldsymbol{x}^*, \boldsymbol{x}^*, \rho, \mu). \tag{A.2.9}$$

Proof. Note that

$$\begin{aligned}
&p\left(\boldsymbol{x}^2, \boldsymbol{x}^*, \rho, \mu\right) - p\left(\boldsymbol{x}^1, \boldsymbol{x}^*, \rho, \mu\right) \\
&\quad = \mu\left\{\left[f\left(\boldsymbol{x}^2\right)-f(\boldsymbol{x}^*)\right]^2-\left[f\left(\boldsymbol{x}^1\right)-f(\boldsymbol{x}^*)\right]^2\right\} \\
&\qquad -\rho\left(\left|\boldsymbol{x}^2-\boldsymbol{x}^*\right|^2-\left|\boldsymbol{x}^1-\boldsymbol{x}^*\right|^2\right) \\
&\quad = \left(\left|\boldsymbol{x}^2-\boldsymbol{x}^*\right|^2-\left|\boldsymbol{x}^1-\boldsymbol{x}^*\right|^2\right)\cdot \\
&\qquad \left\{-\rho+\mu\frac{\left[f\left(\boldsymbol{x}^2\right)-f(\boldsymbol{x}^*)\right]^2-\left[f\left(\boldsymbol{x}^1\right)-f(\boldsymbol{x}^*)\right]^2}{\left|\boldsymbol{x}^2-\boldsymbol{x}^*\right|^2-\left|\boldsymbol{x}^1-\boldsymbol{x}^*\right|^2}\right\} \\
&\quad = \left(\left|\boldsymbol{x}^2-\boldsymbol{x}^*\right|^2-\left|\boldsymbol{x}^1-\boldsymbol{x}^*\right|^2\right)\cdot \\
&\qquad \left\{-\rho+\mu\frac{\left[f\left(\boldsymbol{x}^2\right)+f\left(\boldsymbol{x}^1\right)-2f(\boldsymbol{x}^*)\right]\left[f\left(\boldsymbol{x}^2\right)-f\left(\boldsymbol{x}^1\right)\right]}{\left(\left|\boldsymbol{x}^2-\boldsymbol{x}^*\right|+\left|\boldsymbol{x}^1-\boldsymbol{x}^*\right|\right)\left(\left|\boldsymbol{x}^2-\boldsymbol{x}^*\right|-\left|\boldsymbol{x}^1-\boldsymbol{x}^*\right|\right)}\right\}.
\end{aligned} \tag{A.2.10}$$

From (A.2.1), we obtain

$$\begin{aligned}
&f\left(\boldsymbol{x}^2\right)+f\left(\boldsymbol{x}^1\right)-2f(\boldsymbol{x}^*) = f\left(\boldsymbol{x}^2\right)-f(\boldsymbol{x}^*)+f\left(\boldsymbol{x}^1\right)-f(\boldsymbol{x}^*) \\
&\quad \leq M\left|\boldsymbol{x}^2-\boldsymbol{x}^*\right|+M\left|\boldsymbol{x}^1-\boldsymbol{x}^*\right| = M\left[\left|\boldsymbol{x}^2-\boldsymbol{x}^*\right|+\left|\boldsymbol{x}^1-\boldsymbol{x}^*\right|\right].
\end{aligned} \tag{A.2.11}$$

Combining (A.2.10), (A.2.11) and then using the mean value theorem, we obtain

$$\begin{aligned}
&p\left(\boldsymbol{x}^2, \boldsymbol{x}^*, \rho, \mu\right) - p\left(\boldsymbol{x}^1, \boldsymbol{x}^*, \rho, \mu\right) \\
&\quad \leq \left(\left|\boldsymbol{x}^2-\boldsymbol{x}^*\right|^2-\left|\boldsymbol{x}^1-\boldsymbol{x}^*\right|^2\right)\left\{-\rho+\mu M\frac{f\left(\boldsymbol{x}^2\right)-f\left(\boldsymbol{x}^1\right)}{\left|\boldsymbol{x}^2-\boldsymbol{x}^*\right|-\left|\boldsymbol{x}^1-\boldsymbol{x}^*\right|}\right\} \\
&\quad \leq \left(\left|\boldsymbol{x}^2-\boldsymbol{x}^*\right|^2-\left|\boldsymbol{x}^1-\boldsymbol{x}^*\right|^2\right)\cdot \\
&\qquad \left\{-\rho+\mu M\left|\nabla f\left(\boldsymbol{x}^1+\alpha\left(\boldsymbol{x}^2-\boldsymbol{x}^1\right)\right)\right|\frac{\left|\boldsymbol{x}^2-\boldsymbol{x}^1\right|}{\left|\boldsymbol{x}^2-\boldsymbol{x}^1\right|}\frac{\left|\boldsymbol{x}^2-\boldsymbol{x}^1\right|}{\left|\boldsymbol{x}^2-\boldsymbol{x}^*\right|-\left|\boldsymbol{x}^1-\boldsymbol{x}^*\right|}\right\},
\end{aligned} \tag{A.2.12}$$

where α is some value in $(0, 1)$, and

$$\nabla f\left(\boldsymbol{x}^1 + \alpha\left(\boldsymbol{x}^2 - \boldsymbol{x}^1\right)\right) = \frac{\partial f\left(\boldsymbol{x}^1 + \alpha\left(\boldsymbol{x}^2 - \boldsymbol{x}^1\right)\right)}{\partial \boldsymbol{x}}$$

.

Thus, by (A.2.8) and the assumption that $0 \leq \mu < \rho/M^2$, we obtain

$$\begin{aligned} &p\left(\boldsymbol{x}^2, \boldsymbol{x}^*, \rho, \mu\right) - p\left(\boldsymbol{x}^1, \boldsymbol{x}^*, \rho, \mu\right) \\ &\quad \leq (|\boldsymbol{x}^2 - \boldsymbol{x}^*|^2 - |\boldsymbol{x}^1 - \boldsymbol{x}^*|^2)(-\rho + \mu M M_1) < 0. \end{aligned} \tag{A.2.13}$$

Therefore, we conclude that

$$p\left(\boldsymbol{x}^2, \boldsymbol{x}^*, \rho, \mu\right) < p\left(\boldsymbol{x}^1, \boldsymbol{x}^*, \rho, \mu\right) < 0 = p(\boldsymbol{x}^*, \boldsymbol{x}^*, \rho, \mu).$$

This completes the proof.

Our next task is to show the validity of Property A.2.2. For this, we need the following lemma.

Lemma A.2.1. *Assume that $\boldsymbol{x}^*$ is a local minimizer of $f(\boldsymbol{x})$. Suppose that $\boldsymbol{x}^1$ is a point such that $f\left(\boldsymbol{x}^1\right) > f(\boldsymbol{x}^*)$. Suppose that $\rho > 0$ and*

$$0 \leq \mu < \min\left\{\rho/M^2, \rho/MM_1\right\}. \tag{A.2.14}$$

Then, there exists a sufficiently small $\varepsilon_1 > 0$, such that whenever $\boldsymbol{d}_1$ is chosen satisfying $0 < |\boldsymbol{d}_1| \leq \varepsilon_1$, it holds that

$$\left|\boldsymbol{x}^1 - \boldsymbol{d}_1 - \boldsymbol{x}^*\right| < \left|\boldsymbol{x}^1 - \boldsymbol{x}^*\right| < \left|\boldsymbol{x}^1 + \boldsymbol{d}_1 - \boldsymbol{x}^*\right|, \tag{A.2.15}$$

$$f\left(\boldsymbol{x}^1 + \boldsymbol{d}_1\right) \geq f(\boldsymbol{x}^*) \tag{A.2.16}$$

and

$$\begin{aligned} p\left(\boldsymbol{x}^1 + \boldsymbol{d}_1, \boldsymbol{x}^*, \rho, \mu\right) &< p\left(\boldsymbol{x}^1, \boldsymbol{x}^*, \rho, \mu\right) \\ &< p\left(\boldsymbol{x}^1 - \boldsymbol{d}_1, \boldsymbol{x}^*, \rho, \mu\right) \\ &< 0 = p(\boldsymbol{x}^*, \boldsymbol{x}^*, \rho, \mu). \end{aligned} \tag{A.2.17}$$

Proof. For a given $\varepsilon_1 > 0$, let

$$\boldsymbol{d}_1 = \frac{\varepsilon_1}{2} \frac{\boldsymbol{x}^1 - \boldsymbol{x}^*}{|\boldsymbol{x}^1 - \boldsymbol{x}^*|}. \tag{A.2.18}$$

Then,

$$0 < |\boldsymbol{d}_1| = \frac{\varepsilon_1}{2} \leq \varepsilon_1. \tag{A.2.19}$$

Clearly, if $\varepsilon_1 > 0$ is sufficiently small, we have

$$\left|\boldsymbol{x}^1+\boldsymbol{d}_1-\boldsymbol{x}^*\right|=(1+\frac{\varepsilon_1}{2|\boldsymbol{x}^1-\boldsymbol{x}^*|})\left|\boldsymbol{x}^1-\boldsymbol{x}^*\right|>\left|\boldsymbol{x}^1-\boldsymbol{x}^*\right|, \tag{A.2.20a}$$

$$\left|\boldsymbol{x}^1-\boldsymbol{d}_1-\boldsymbol{x}^*\right|=(1-\frac{\varepsilon_1}{2|\boldsymbol{x}^1-\boldsymbol{x}^*|})\left|\boldsymbol{x}^1-\boldsymbol{x}^*\right|<\left|\boldsymbol{x}^1-\boldsymbol{x}^*\right|. \tag{A.2.20b}$$

Since $f\left(\boldsymbol{x}^1\right)>f(\boldsymbol{x}^*)$ and $0<|\boldsymbol{d}_1|\leq\varepsilon_1$, it follows that

$$f\left(\boldsymbol{x}^1\pm\boldsymbol{d}_1\right)\geq f(\boldsymbol{x}^*), \tag{A.2.21}$$

if $\varepsilon_1>0$ is chosen sufficiently small. Now, choose ρ and μ such that $\rho>0$ and $0\leq\mu<\min\left\{\rho/M^2,\rho/MM_1\right\}$. Then, by using arguments similar to that given for Theorem A.2.2, we can show that

$$\begin{aligned} p\left(\boldsymbol{x}^1+\boldsymbol{d}_1,\boldsymbol{x}^*,\rho,\mu\right) &< p\left(\boldsymbol{x}^1,\boldsymbol{x}^*,\rho,\mu\right) \\ &< p\left(\boldsymbol{x}^1-\boldsymbol{d}_1,\boldsymbol{x}^*,\rho,\mu\right) \\ &< 0=p(\boldsymbol{x}^*,\boldsymbol{x}^*,\rho,\mu). \end{aligned} \tag{A.2.22}$$

This completes the proof.

This lemma shows that any local minimizer of $p(\boldsymbol{x},\boldsymbol{x}^*,\rho,\mu)$ must be in the set

$$S=\{\boldsymbol{x}:f(\boldsymbol{x})\leq f(\boldsymbol{x}^*)\}. \tag{A.2.23}$$

The next theorem shows that the function satisfies Property A.2.2.

Theorem A.2.3. *Assume that $\boldsymbol{x}^*$ is a local minimizer of $f(\boldsymbol{x})$. If $\rho>0$ and $0\leq\mu<\min\left\{\rho/M^2,\rho/MM_1\right\}$, then any local minimizer or saddle point must belong to the set S.*

Proof. It suffices to show that for any $\boldsymbol{x}$, if $f(\boldsymbol{x})>f(\boldsymbol{x}^*)$, then

$$\left[\frac{\partial p(\boldsymbol{x},\boldsymbol{x}^*,\rho,\mu)}{\partial\boldsymbol{x}}\right]^\top\neq\mathbf{0}. \tag{A.2.24}$$

From (A.2.4), we have

$$\left[\frac{\partial p(\boldsymbol{x},\boldsymbol{x}^*,\rho,\mu)}{\partial\boldsymbol{x}}\right]^\top=-2\rho(\boldsymbol{x}-\boldsymbol{x}^*)+2\mu[f(\boldsymbol{x})-f(\boldsymbol{x}^*)]\left[\frac{\partial f(\boldsymbol{x})}{\partial\boldsymbol{x}}\right]^\top. \tag{A.2.25}$$

If $\left[\frac{\partial f(\boldsymbol{x})}{\partial\boldsymbol{x}}\right]^\top=\mathbf{0}$, we have

$$\left[\frac{\partial p(\boldsymbol{x},\boldsymbol{x}^*,\rho,\mu)}{\partial\boldsymbol{x}}\right]^\top=-2\rho(\boldsymbol{x}-\boldsymbol{x}^*)\neq\mathbf{0}. \tag{A.2.26}$$

Now suppose that $\left[\frac{\partial f(\boldsymbol{x})}{\partial\boldsymbol{x}}\right]^\top\neq\mathbf{0}$. Define

$$\boldsymbol{d} = \frac{\boldsymbol{x} - \boldsymbol{x}^*}{|\boldsymbol{x} - \boldsymbol{x}^*|} - \beta \frac{\left[\frac{\partial f(\boldsymbol{x})}{\partial \boldsymbol{x}}\right]^\top}{\left|\frac{\partial f(\boldsymbol{x})}{\partial \boldsymbol{x}}\right|}, \tag{A.2.27}$$

where $\beta > 0$ is sufficiently small. Then, by taking the inner product of $\boldsymbol{d}$ and $\left[\frac{\partial p(\boldsymbol{x}, \boldsymbol{x}^*, \rho, \mu)}{\partial \boldsymbol{x}}\right]^\top$, we obtain

$$\begin{aligned}
&\boldsymbol{d}^\top \left[\frac{\partial p(\boldsymbol{x}, \boldsymbol{x}^*, \rho, \mu)}{\partial \boldsymbol{x}}\right]^\top \\
&= -2\rho|\boldsymbol{x} - \boldsymbol{x}^*| + 2\rho\beta(\boldsymbol{x} - \boldsymbol{x}^*)^\top \left[\frac{\partial f(\boldsymbol{x})}{\partial \boldsymbol{x}}\right]^\top \Big/ \left|\frac{\partial f(\boldsymbol{x})}{\partial \boldsymbol{x}}\right| \\
&+ 2\mu \left[f(\boldsymbol{x}) - f(\boldsymbol{x}^*)\right] \frac{\partial f(\boldsymbol{x})}{\partial \boldsymbol{x}} \frac{\boldsymbol{x} - \boldsymbol{x}^*}{|\boldsymbol{x} - \boldsymbol{x}^*|} \\
&- 2\mu\beta \left[f(\boldsymbol{x}) - f(\boldsymbol{x}^*)\right] \left|\frac{\partial f(\boldsymbol{x})}{\partial \boldsymbol{x}}\right|.
\end{aligned} \tag{A.2.28}$$

If $(\boldsymbol{x} - \boldsymbol{x}^*)^\top \left[\frac{\partial f(\boldsymbol{x})}{\partial \boldsymbol{x}}\right]^\top \leq 0$, then $\boldsymbol{d}^T \left[\frac{\partial p\left(\boldsymbol{x}, \boldsymbol{x}^{1,*}, \rho, \mu\right)}{\partial \boldsymbol{x}}\right]^\top < 0$. Otherwise, choose $\mu \geq 0$ to be sufficiently small. Since $\beta > 0$ can also be chosen to be sufficiently small, it follows that $\boldsymbol{d}^\top \left[\frac{\partial p(\boldsymbol{x}, \boldsymbol{x}^*, \rho, \mu)}{\partial \boldsymbol{x}}\right]^\top < 0$. Thus, $\left[\frac{\partial p(\boldsymbol{x}, \boldsymbol{x}^*, \rho, \mu)}{\partial \boldsymbol{x}}\right]^\top \neq \boldsymbol{0}$. This completes the proof.

The next theorem shows that the filled function p satisfies Property A.2.3.

Theorem A.2.4. *Assume that $\boldsymbol{x}^{1,*}$ is a local minimizer of $f(\boldsymbol{x})$. If $\boldsymbol{x}^{2,*}$ is another minimizer of $f(\boldsymbol{x})$ and satisfies*

$$f\left(\boldsymbol{x}^{2,*}\right) < f\left(\boldsymbol{x}^{1,*}\right), \tag{A.2.29}$$

then there exists a neighbourhood $\mathcal{N}(\boldsymbol{x}^{2,*}, \delta)$ of $\boldsymbol{x}^{2,*}$ such that $p\left(\boldsymbol{x}, \boldsymbol{x}^{1,*}, \rho, \mu\right)$ has a minimizer $\boldsymbol{x}'$ which is on the line segment connecting $\boldsymbol{x}^{1,*}$ and $\boldsymbol{x}^2$ for every $\boldsymbol{x}^2 \in \mathcal{N}(\boldsymbol{x}^{2,*}, \delta)$ when $0 \leq \mu < \rho/M^2$ and $0 < \rho < \varepsilon_1/D_1$, where

$$0 < \varepsilon_1 < f\left(\boldsymbol{x}^{1,*}\right) - f\left(\boldsymbol{x}^2\right) \tag{A.2.30}$$

and

$$D_1 = \max_{\boldsymbol{x} \in N(\boldsymbol{x}^{2,*}, \delta)} \left|\boldsymbol{x} - \boldsymbol{x}^{1,*}\right|^2. \tag{A.2.31}$$

Furthermore, if there exists no basin lower than B_1^* between B_1^* and B_2^*, where B_1^* and B_2^* are the basins of $f(\boldsymbol{x})$ at $\boldsymbol{x}^{1,*}$ and $\boldsymbol{x}^{2,*}$, respectively, then there exists a $\boldsymbol{x}' \in B_2^*$ such that

$$f(\boldsymbol{x}') \leq f\left(\boldsymbol{x}^{1,*}\right). \tag{A.2.32}$$

Proof. By Theorem A.2.1, there is a neighbourhood $\mathcal{N}(\boldsymbol{x}^{1,*}, \delta_1)$ of $\boldsymbol{x}^{1,*}$ with $\delta_1 > 0$ such that for all $\boldsymbol{x}^1 \in \mathcal{N}(\boldsymbol{x}^{1,*}, \delta_1)$,

$$p\left(\boldsymbol{x}^1, \boldsymbol{x}^{1,*}, \rho, \mu\right) < 0 = p\left(\boldsymbol{x}^{1,*}, \boldsymbol{x}^{1,*}, \rho, \mu\right). \tag{A.2.33}$$

Furthermore, there is a neighbourhood $\mathcal{N}(\boldsymbol{x}^{2,*}, \delta_1)$ of $\boldsymbol{x}^{2,*}$ with $\delta_2 > 0$ such that for all $\boldsymbol{x}^2 \in \mathcal{N}(\boldsymbol{x}^{2,*}, \delta_2)$,

$$0 < \varepsilon_1 < f\left(\boldsymbol{x}^{1,*}\right) - f\left(\boldsymbol{x}^2\right). \tag{A.2.34}$$

Thus, by (A.2.3), it follows from (A.2.34) and (A.2.31) that

$$\begin{aligned} p\left(\boldsymbol{x}^2, \boldsymbol{x}^{1,*}, \rho, \mu\right) &= f\left(\boldsymbol{x}^{1,*}\right) - f\left(\boldsymbol{x}^2\right) - \rho\left|\boldsymbol{x}^2 - \boldsymbol{x}^{1,*}\right|^2 \\ &> \varepsilon_1 - \rho D_1. \end{aligned} \tag{A.2.35}$$

If ρ is chosen such that $\rho < \varepsilon_1 / D_1$, then

$$p\left(\boldsymbol{x}^2, \boldsymbol{x}^{1,*}, \rho, \mu\right) > 0. \tag{A.2.36}$$

Thus, by the continuity of the filled function, there exists a minimizer $\boldsymbol{x}'$ which is on the line segment connecting $\boldsymbol{x}^{1,*}$ and $\boldsymbol{x}^2$ for every $\boldsymbol{x}^2 \in \mathcal{N}(\boldsymbol{x}^{2,*}, \delta_2)$.

Now, we consider the case when there exists no basin lower than B_1^* between B_1^* and B_2^*. Let $\boldsymbol{x}^B$ be the boundary point of B_2^* on the line segment connecting $\boldsymbol{x}^{1,*}$ and a $\boldsymbol{x}^2 \in \mathcal{N}(\boldsymbol{x}^{2,*}, \delta_2)$. Since there exists no basin lower than B_1^* between B_1^* and B_2^*, it is clear that

$$f\left(\boldsymbol{x}^B\right) - f\left(\boldsymbol{x}^{1,*}\right) > 0. \tag{A.2.37}$$

Thus, by the continuity of $f(\boldsymbol{x})$, there are three points $\boldsymbol{x}^{0,-}$, $\boldsymbol{x}^0$ and $\boldsymbol{x}^{0,+}$ on the line segment connecting $\boldsymbol{x}^{1,*}$ and $\boldsymbol{x}^2$ such that

$$f\left(\boldsymbol{x}^0\right) = f\left(\boldsymbol{x}^{1,*}\right) \tag{A.2.38}$$

and

$$f\left(\boldsymbol{x}^B\right) > f\left(\boldsymbol{x}^{0,-}\right) \geq f\left(\boldsymbol{x}^0\right) \geq f\left(\boldsymbol{x}^{0,+}\right) > f\left(\boldsymbol{x}^2\right), \tag{A.2.39}$$

where

$$\boldsymbol{x}^{0,-} = \boldsymbol{x}^0 - \eta\left(\boldsymbol{x}^0 - \boldsymbol{x}^{1,*}\right) \tag{A.2.40a}$$

and

$$\boldsymbol{x}^{0,+} = \boldsymbol{x}^0 + \eta\left(\boldsymbol{x}^0 - \boldsymbol{x}^{1,*}\right), \tag{A.2.40b}$$

where $\eta > 0$ is sufficiently small. Since

$$p\left(\boldsymbol{x}^0, \boldsymbol{x}^{1,*}, \rho, \mu\right) = -\rho\left|\boldsymbol{x}^0 - \boldsymbol{x}^{1,*}\right|^2 < 0 = p\left(\boldsymbol{x}^{1,*}, \boldsymbol{x}^{1,*}, \rho, \mu\right). \tag{A.2.41}$$

We note from (A.2.40b) that

$$\begin{aligned}\left|\boldsymbol{x}^{0,+}-\boldsymbol{x}^{1,*}\right| &= \left|\boldsymbol{x}^{0}+\eta\boldsymbol{x}^{0}-\eta\boldsymbol{x}^{1,*}-\boldsymbol{x}^{1,*}\right| \\ &= \left|\left(\boldsymbol{x}^{0}-\boldsymbol{x}^{1,*}\right)+\eta\left(\boldsymbol{x}^{0}+\boldsymbol{x}^{1,*}\right)\right| \\ &= (1+\eta)\left|\boldsymbol{x}^{0}-\boldsymbol{x}^{1,*}\right| > \left|\boldsymbol{x}^{0}-\boldsymbol{x}^{1,*}\right|. \end{aligned} \tag{A.2.42}$$

By (A.2.38) and (A.2.39), we recall that

$$f\left(\boldsymbol{x}^{0,+}\right) \geq f\left(\boldsymbol{x}^{0}\right) = f\left(\boldsymbol{x}^{0,*}\right). \tag{A.2.43}$$

Thus, it follows from Theorem A.2.2 that

$$p\left(\boldsymbol{x}^{0,+},\boldsymbol{x}^{1,*},\rho,\mu\right) < p\left(\boldsymbol{x}^{0},\boldsymbol{x}^{1,*},\rho,\mu\right) < 0 = p\left(\boldsymbol{x}^{1,*},\boldsymbol{x}^{1,*},\rho,\mu\right). \tag{A.2.44}$$

Next, we note from (A.2.40a) that

$$\begin{aligned}\left|\boldsymbol{x}^{0,-}-\boldsymbol{x}^{1,*}\right| &= \left|\boldsymbol{x}^{0}-\eta\boldsymbol{x}^{0}+\eta\boldsymbol{x}^{1,*}-\boldsymbol{x}^{1,*}\right| \\ &= \left|\left(\boldsymbol{x}^{0}-\boldsymbol{x}^{1,*}\right)-\eta\left(\boldsymbol{x}^{0}+\boldsymbol{x}^{1,*}\right)\right| \\ &= (1-\eta)\left|\boldsymbol{x}^{0}-\boldsymbol{x}^{1,*}\right| < \left|\boldsymbol{x}^{0}-\boldsymbol{x}^{1,*}\right|. \end{aligned} \tag{A.2.45}$$

From (A.2.39), we can write down that

$$f\left(\boldsymbol{x}^{0,-}\right) \geq f\left(\boldsymbol{x}^{0}\right) = f\left(\boldsymbol{x}^{1,*}\right). \tag{A.2.46}$$

Thus, by (A.2.45) and (A.2.46), it follows from Theorem A.2.2 that

$$0 > p\left(\boldsymbol{x}^{0,-},\boldsymbol{x}^{1,*},\rho,\mu\right) > p\left(\boldsymbol{x}^{0},\boldsymbol{x}^{1,*},\rho,\mu\right). \tag{A.2.47}$$

From (A.2.44) and (A.2.47), it is clear that $\boldsymbol{x}^{0}-\boldsymbol{x}^{1,*}$ is a descent direction of $p\left(\boldsymbol{x},\boldsymbol{x}^{1,*},\rho,\mu\right)$ at $\boldsymbol{x}^{0}$. Therefore, there exists a $\boldsymbol{x}'\in B_2^*$ such that $f(\boldsymbol{x}')\leq f\left(\boldsymbol{x}^{1,*}\right)$.

Now, by virtue of Theorems A.2.1–A.2.4, we see that $p(\boldsymbol{x},\boldsymbol{x}^*,\rho,\mu)$ satisfies Properties A.2.1–A.2.3. Thus, it is a filled function. By Theorem A.2.3, any local minimizer $\boldsymbol{x}^*$ of $p(\boldsymbol{x},\boldsymbol{x}^*,\rho,\mu)$ in $\varXi\times\varUpsilon$ satisfies

$$f(\boldsymbol{x}^*) \leq f(\boldsymbol{x}^*). \tag{A.2.48}$$

Therefore, we can escape from the current local minimizer $\boldsymbol{x}^*$ of $f(\boldsymbol{x})$ by searching for a local minimizer of $p(\boldsymbol{x},\boldsymbol{x}^*,\rho,\mu)$.

To implement the computation, we need the gradient $\dfrac{\partial p(\boldsymbol{x},\boldsymbol{x}^*,\rho,\mu)}{\partial\boldsymbol{x}}$ with respect to the parameter $\boldsymbol{x}$. Note that the gradient $\dfrac{\partial p(\boldsymbol{x},\boldsymbol{x}^*,\rho,\mu)}{\partial\boldsymbol{x}}$ is needed only when $f(\boldsymbol{x})>f(\boldsymbol{x}^*)$. We give the following theorem.

Theorem A.2.5. *Suppose that $f(\boldsymbol{x}) > f(\boldsymbol{x}^*)$. Then,*

$$\left[\frac{\partial p(\boldsymbol{x}, \boldsymbol{x}^*, \rho, \mu)}{\partial \boldsymbol{x}}\right]^\top = -2\rho(\boldsymbol{x} - \boldsymbol{x}^*) + 2\mu\,[f(\boldsymbol{x}) - f(\boldsymbol{x}^*)]\left[\frac{\partial f(\boldsymbol{x})}{\partial \boldsymbol{x}}\right]^\top. \tag{A.2.49}$$

Proof. The result is obvious.

Now, we give the following algorithm to search for a $\boldsymbol{x}$ such that

$$f(\boldsymbol{x}) < f(\boldsymbol{x}^*).$$

Algorithm A.2.1

Step 1. *Initialize ρ, μ, $\hat{\mu}$ (where $\hat{\mu} < 1$), μ_l (where μ_l is sufficiently small).*

Step 2. *Start from $x^{1,*}$, construct a search direction according to (A.2.49) and select a search step. We minimize the filled function $p(\boldsymbol{x}, \boldsymbol{x}^{1,*}, \rho, \mu)$ along this search direction with the selected search step. Then, we find a point $\boldsymbol{x}$. Go to Step 3.*

Step 3. *If $f(\boldsymbol{x}) < f\left(\boldsymbol{x}^{1,*}\right)$, stop. Otherwise, go to Step 5.*

Step 4. *Continue the search as described in Step 2. Find a new point $\boldsymbol{x}$. Go to Step 3.*

Step 5. *If $\boldsymbol{x}$ is on the boundary of X and $\mu \geq \mu_l$, set $\mu = \hat{\mu}\mu$ and go to Step 1. Otherwise, go to Step 4.*

Algorithm A.2.1 can be modified by incorporating the filled function (A.2.3) so that we can search for a global minimizer.

Algorithm A.2.2

Step 1. *Choose a $\boldsymbol{x}^0 \in X$, and obtain a local minimizer by a gradient-based optimization algorithm. Let it be denoted as $\boldsymbol{x}^*$.*

Step 2. *Use Algorithm A.2.1 to find another initial point $\boldsymbol{x}$ such that $f(\boldsymbol{x}) < f(\boldsymbol{x}^*)$. If this point cannot be found, then go to Step 4.*

Step 3. *Use $\boldsymbol{x}$ as an initial point, and obtain another local minimizer $\boldsymbol{x}^*$ by the gradient-based optimization algorithm. Go to Step 2.*

Step 4. *$\boldsymbol{x}^*$ is a global minimizer of Problem (B).*

Appendix A.3
Elements of Probability Theory

Let S denote the sample space which is the set of all possible outcomes of an experiment. If S is finite or countably infinite, it is called a discrete sample space. On the other hand, if S is a continuous set, then it is called a continuous sample space. An element from S is called a sample point (i.e., a single outcome). A collection of possible outcomes is referred to as an event. In set theory, it is called a subset of S. $A \subset B$ menas that if event B occurs, then event A must occur. What *probability* does is to assign a weight between 0 and 1 to each outcome of an experiment. This weight represents the likelihood or chance of that outcome occurring. These weights are determined by long run experiment, assumption or some other methods.

The probability of an event A in S is the sum of the weights of all sample points in A and is denoted by $P(A)$. It satisfies the following properties. (1) $0 \leq P(A) \leq 1$; (2) $P(S) = 1$; and (3) $P(\varnothing) = 0$, where $\varnothing$ denotes the empty event. Two events A and B are said to be *mutually exclusive* if $A \cap B = \varnothing$. If $\{A_1, \ldots, A_n\}$ is a set of *mutually exclusive* events, then

$$P(A_1 \cup A_2 \cup \cdots \cup A_n) = \sum_{i=1}^{n} P(A_i).$$

If an experiment is such that each outcome has the same probability, then the outcomes are said to be *equally likely.*

Consider the case for which an experiment can result in any one of N different equally likely outcomes. Let the event A consists of exactly n of these outcomes. Then, the probability of event A is

$$P(A) = \frac{n}{N}.$$

K. L. Teo et al., *Applied and Computational Optimal Control*, Springer Optimization and Its Applications 171,
https://doi.org/10.1007/978-3-030-69913-0

The following probability rules are easy to prove.

1. Let $\bar{A}$ denotes the complement of A (i.e., those outcomes not in A). Then, $P(\bar{A}) = 1 - P(A)$.
2. $P(AB) + P(A\bar{B}) = P(A)$.
3. $P(A \cup B) = P(A) + P(B) - P(AB)$.
4. $P(A \cup B \cup C) = P(A) + P(B) + P(C) - P(AB) - P(AC) - P(BC) + P(ABC)$.

Here, AB denotes the intersection of A and B. This abbreviation is used throughout this Appendix.

Conditional Probability: The conditional probability of event A given that event B has occurred is defined by

$$P(A|B) = \frac{P(AB)}{P(B)} \tag{A.3.1}$$

where A and B are events in the same sample space S.

Example A.3.1. A fair coin is tossed 3 times. What is the probability of getting at least 1 head, h, given that the first toss was a tail, t.

Solution. The sample space S is

$$S = \{hhh,\ hht,\ hth,\ thh,\ tth,\ tht,\ htt,\ ttt\}.$$

Define $A = \{\text{at least one head}\}$ and $B = \{\text{first toss resulted in a tail}\}$. Then,

$$\begin{aligned}
&A = \{hhh,\ hht,\ hth,\ thh,\ tth,\ tht,\ htt\}\\
&B = \{thh,\ tht,\ tth,\ ttt\} \Rightarrow AB = \{thh,\ tht,\ tth\}\\
&\Rightarrow P(A) = \frac{7}{8},\ P(B) = \frac{4}{8},\ \text{and } P(AB) = \frac{3}{8}\\
&\Rightarrow P(A|B) = \frac{P(AB)}{P(B)} = \frac{3/8}{4/8} = \frac{3}{4}.
\end{aligned}$$

Random Variables: A function X whose value is a real number determined by each element in the sample space is called a *random variable* (*r.v.*).

The *probability distribution* of X, written as $P(X = x)$, is called the *probability function* of X. Define

$$P(X = x) = f(x).$$

Convention: Capital letters for random variables and lower case for values of the random variable.

A random variable X is *discrete* if its range forms a discrete (countable) set of real numbers. On the other hand, a random variable X is *continuous* if its range forms a continuous set of real numbers. For a continuous random variable X, the probability of a specified outcome being occurred is 0.

The cumulative distribution $F(x)$ of a discrete random variable X with probability function $f(x)$ is given by

$$F(x) = P(X \leq x) = \sum_{t \leq x} f(t) = \sum_{t \leq x} P(X = t).$$

Independence: 2 events A and B are said to be *independent* if and only if $P(A|B) = P(A)$ (or equivalently, $P(B|A) = P(B)$).

If A and B are independent events, then $P(AB) = P(A)P(B)$. From (A.3.1), we have

$$P(AB) = P(A|B)P(B) = P(B|A)P(A) \tag{A.3.2}$$

It can be generalized to any number of events. For example, consider $A_1, A_2, \ldots, A_n$. Then,

$$P(A_1 A_2 \ldots A_n) = P(A_1)P(A_2|A_1)P(A_3|A_1A_2)\ldots P(A_n|A_1 \ldots A_{n-1}).$$

Theorem A.3.1 (Bayes Theorem). *Suppose that $\{B_1, B_2, \ldots, B_n\}$ is a partition of the sample space S, where $P(B_i) \neq 0$ for $i = 1, \ldots, n$; $\bigcup_{i=1}^{n} B_i = S$; and $B_iB_j = \varnothing$ for $i \neq j$ (i.e., B_i, $i = 1, \ldots, n$, are mutually exclusive). Let A be any event in S such that $P(A) \neq 0$. Then, for any $k = 1, \ldots, n$,*

$$P(B_k|A) = \frac{P(B_kA)}{\sum_{i=1}^{n} P(B_iA)} = \frac{P(B_k)P(A|B_k)}{\sum_{i=1}^{n} P(B_i)P(A|B_i)}. \tag{A.3.3}$$

Continuous random variable: A continuous random variable X has a real value function associated with it which is called *probability density function (p.d.f) $f(x)$*. The p.d.f. $f(x)$ is defined by

$$P(a < X < b) = \int_a^b f(x)dx. \tag{A.3.4}$$

Clearly,

$$P(X = a) = P(a \leq X \leq a) = \int_a^a f(x)dx = 0.$$

Properties: (1) $f(x) \geq 0$ for all $x \in \mathbb{R}$; (2) $\int_{-\infty}^{\infty} f(x)dx = 1$; and (3). $P(a < X < b) = \int_a^b f(x)dx$.

In general, since $P(X = x) = 0$, it holds that $P(X < x) = P(X \leq x)$.

The cumulative distribution $F(x)$ of a continuous random variable is

$$F(x) = P(X \le x) = \int_{-\infty}^{x} f(t)dt \Rightarrow f(x) = \frac{dF(x)}{dx}. \tag{A.3.5}$$

Joint Random Variables: Suppose that there are 2 random variables X and Y on the sample space S. Then, each point in S has a value for X and a value for Y. X and Y are said to be *jointly distributed.*

1. If X and Y are both discrete random variables, then X and Y have a joint probability function

$$f(x, y) = P(X = x,\ Y = y)$$

   with properties: (1) $f(x, y) \ge 0$ for all x, y; (2) $\sum_x \sum_y f(x, y) = 1$; and (3) $P\{(X, Y) \in A\} = \sum\sum_A f(x, y)$ for any region A in the xy plane.

2. If X and Y are continuous random variables, then they have a joint probability density function $f(x, y)$

$$f(x, y) = P(X = x,\ Y = y)$$

   with properties: (1)$f(x, y) \ge 0$ for all x, y; (2) $\int_{-\infty}^{\infty}\int_{-\infty}^{\infty} f(x, y)dxdy = 1$; and (3.) $P\{(X, Y) \in A\} = \int_A \int f(x, y)dxdy$ for any region A in the xy plane.

Joint Cumulative Distribution: The joint cumulative distribution is

$$F(x, y) = P(X \le x,\ Y \le y) = \int_{-\infty}^{\infty}\int_{-\infty}^{\infty} f(t, s)dt\, ds.$$

Marginal Distributions: The marginal distribution of X is defined as:

$$G(x) = \begin{cases} \sum_{\text{all } y} P(X = x,\ Y = y), & \text{if } X \text{ and } Y \text{ discrete} \\ \int_{-\infty}^{\infty} f(x, y)dy, & \text{if } X \text{ and } Y \text{ continuous,} \end{cases}$$

where $G(x)$ is also called the marginal probability distribution function of X. Similarly, the marginal distribution of Y is defined as

$$H(y) = \begin{cases} \sum_{\text{all } x} P(X = x,\ Y = y), & \text{if } X \text{ and } Y \text{ discrete} \\ \int_{-\infty}^{\infty} f(x, y)dx, & \text{if } X \text{ and } Y \text{ continuous,} \end{cases}$$

where $H(y)$ is called the marginal probability distribution function of Y.

Conditional distributions: The distribution of X given $Y = y$ is called the *conditional distribution of* X *given* $Y = y$ and is defined as:

$$f(x|y) \equiv P(X = x|Y = y) = \frac{f(x,y)}{H(y)} = \frac{\text{joint distribution of } X \text{ and } Y}{\text{marginal distribution of } Y}.$$

The conditional distribution of Y given $X = x$ is

$$f(y|x) \equiv P(Y = y|X = x) = \frac{f(x,y)}{G(x)}.$$

Independence: The random variables X and Y, which are jointly distributed, are said to be *independent* if and only if

$$f(x,y) = G(x)H(y).$$

(i.e., Joint distribution = product of the marginal distributions of the random variables X and Y).

Clearly, if X and Y are independent, then

$$f(x|y) = G(x) \text{ and } f(y|x) = H(y).$$

Expectation: The expectation of the random variable X is defined by

$$\mathcal{E}(X) = \begin{cases} \sum_x x f(x), & \text{if } X \text{ is discrete,} \\ \int_{-\infty}^{\infty} x f(x) dx, & \text{if } X \text{ is continuous.} \end{cases}$$

Generalization: Let $g(X)$ be a function of the random variable X. Then, the expectation of $g(X)$ is

$$\mathcal{E}[g(X)] = \begin{cases} \sum_x g(x) f(x), & \text{if } X \text{ is discrete,} \\ \int_{-\infty}^{\infty} g(x) f(x) dx, & \text{if } X \text{ is continuous.} \end{cases}$$

Joint random variables: Let X and Y be jointly distributed with probability function $f(x,y)$. Then, the expectation of $g(X,Y)$ is defined as:

$$\mathcal{E}[g(X,Y)] = \begin{cases} \sum_x \sum_y g(x,y) f(x,y), & \text{if } X, Y \text{ discrete,} \\ \int_{-\infty}^{\infty} \int_{-\infty}^{\infty} g(x,y) f(x,y) dx dy, & \text{if } X, Y \text{ continuous.} \end{cases}$$

Let $g(X,Y) = X$. Then, $\mathcal{E}[g(X,Y)] = \mathcal{E}(X)$, meaning that

$$\mathcal{E}(X) = \begin{cases} \sum_x \sum_y x f(x,y), & \text{if } X, Y \text{ discrete,} \\ \int_{-\infty}^{\infty}\int_{-\infty}^{\infty} x f(x,y) dx dy, & \text{if } X, Y \text{ continuous.} \end{cases}$$

In the case when X and Y are continuous,

$$\begin{aligned} \mathcal{E}(X) &= \int_{-\infty}^{\infty}\int_{-\infty}^{\infty} x f(x,y) dx dy \\ &= \int_{-\infty}^{\infty} x \left\{ \int_{-\infty}^{\infty} f(x,y) dy \right\} dx = \int_{-\infty}^{\infty} x G(x) dx, \end{aligned}$$

where $G(x) \equiv \int_{-\infty}^{\infty} f(x,y) dy$ is the marginal probability distribution function of X.

Also,

$\mathcal{E}(Y) = \int_{-\infty}^{\infty} y\{\text{marginal probability distribution function of } Y\} dy$

Similar conclusions are valid for discrete random variables.

Rules of Expectations:

1. Let b be a constant. Then

$$\mathcal{E}(b) = \int_{-\infty}^{\infty} b f(x) dx = b \int_{-\infty}^{\infty} f(x) dx = b.$$

2. Let a and b are two constants. Then

$$\begin{aligned} \mathcal{E}(aX + b) &= \int_{-\infty}^{\infty} \{ax + b\} f(x) dx \\ &= a \int_{-\infty}^{\infty} x f(x) dx + b \int_{-\infty}^{\infty} f(x) dx \\ &= a\mathcal{E}(X) + b. \end{aligned}$$

3. Let $S(X)$ and $T(X)$ be functions of X. Then

$$\begin{aligned} \mathcal{E}(S(X) \pm T(X)) &= \int_{-\infty}^{\infty} [S(x) \pm T(x)] f(x) dx \\ &= \int_{-\infty}^{\infty} S(x) f(x) dx \pm \int_{-\infty}^{\infty} T(x) f(x) dx \\ &= \mathcal{E}(S(X)) \pm \mathcal{E}(T(X)). \end{aligned}$$

4. Let X and Y be jointly distributed, and let g and h be functions of the random variables X and Y. Then

$$
\begin{aligned}
&\mathcal{E}[g(X,Y)] \pm h(X,Y)] \\
&= \int_{-\infty}^{\infty}\int_{-\infty}^{\infty} \{g(x,y) \pm h(x,y)\} f(x,y)dxdy \\
&= \int_{-\infty}^{\infty}\int_{-\infty}^{\infty} g(x,y)f(x,y)dxdy \pm \int_{-\infty}^{\infty}\int_{-\infty}^{\infty} h(x,y)f(x,y)dxdy \\
&= \mathcal{E}[g(X,Y)] \pm \mathcal{E}[h(X,Y)].
\end{aligned}
$$

As a consequence, we have, by setting $g(X,Y) = X$ and $h(X,Y) = Y$,

$$\mathcal{E}(X \pm Y) = \mathcal{E}(X) \pm \mathcal{E}(Y).$$

5. Let X_i, $i = 1, \ldots, n$, be n random variables. Then

$$\mathcal{E}\left[\sum_{i=1}^{n} X_i\right] = \sum_{i=1}^{n} \mathcal{E}(X_i).$$

6. If X and Y are independent random variables, then

$$\mathcal{E}(XY) = \mathcal{E}(X)\mathcal{E}(Y)$$

.

Remark A.3.1. The condition (6) is necessary but *not* sufficient. That is, $\mathcal{E}(XY) = \mathcal{E}(X)\mathcal{E}(Y)$ does *not* necessarily imply that X and Y are independent.

Moment: The expectation of the kth (k is a positive integer) power of a random variable X is called the kth moment of the random variable X and it is denoted by $\widehat{\mu}_k$. That is, for any $k = 0, 1, 2, \ldots$,

$$\widehat{\mu}_k = \mathcal{E}\left(X^k\right) = \begin{cases} \sum_x x^k f(x), & \text{if } X \text{ is discrete,} \\ \int_{-\infty}^{\infty} x^k f(x)dx, & \text{if } X \text{ is continuous.} \end{cases}$$

Clearly,

$$\mathcal{E}(X^0) = \mathcal{E}(1) = 1.$$

The first moment is called the *mean* of X and is denoted by μ.

$$\widehat{\mu}_1 = \mathcal{E}(X) = \text{ mean of } X \equiv \mu.$$

The kth moment about the mean of the random variable X is defined as:

$$\mathcal{E}\left[(X-\mu)^k\right] \equiv \mu_k = \begin{cases} \sum_x (x-\mu)^k f(x), & \text{if } X \text{ is discrete,} \\ \int_{-\infty}^{\infty} (x-\mu)^k f(x)dx, & \text{if } X \text{ is continuous.} \end{cases}$$

Variance: The second moment about the mean of a random variable X is called the *variance* of X. It is denoted by σ^2. More specifically,

$$\sigma^2 = \mathcal{E}\left[(X-\mu)^2\right] = \begin{cases} \sum_x (x-\mu)^2 f(x), & X \text{ discrete,} \\ \int_{-\infty}^{\infty} (x-\mu)^2 f(x)dx, & X \text{ continuous.} \end{cases}$$

Clearly,

$$Var(X) = \mathcal{E}\left(X^2\right) - [\mathcal{E}(X)]^2, \text{ i.e.,} \sigma^2 = \widehat{\mu}_2 - \mu^2.$$

Bivariate or Joint Moments: Let X and Y be two random variables with a joint probability function $f(x,y)$. Then

$$\mathcal{E}(XY) = \begin{cases} \sum_x \sum_y xyf(x,y), & X,Y \text{ discrete,} \\ \int_{-\infty}^{\infty}\int_{-\infty}^{\infty} xyf(x,y)dxdy, & X,Y \text{ continuous.} \end{cases}$$

The "joint" moment of X and Y about their respective means is the *covariance* of X and Y and is denoted by

$$Cov(X,Y) \equiv C(X,Y) \equiv \sigma_{XY}.$$

That is to say,

$$\sigma_{XY} = \mathcal{E}[(X-\mu_X)(Y-\mu_Y)] = \mathcal{E}(XY) - \mu_X\mu_Y,$$

where $\mathcal{E}(X) \equiv \mu_X$ and $\mathcal{E}(Y) = \mu_Y$.

If X and Y are independent, then $\mathcal{E}(XY) = \mathcal{E}(X)\mathcal{E}(Y)$. Under this situation, it is easy to see that $\sigma_{XY} = 0$. However, $\sigma_{XY} = 0$ does not necessarily mean that X and Y are independent.

Rules of Variance: Let $h(X)$ be a function of the random variable X. Then the variance of $h(X)$ is

$$Var[h(X)] = \mathcal{E}\left[(h(X) - \mathcal{E}(h(X))^2\right] = \mathcal{E}\left[(h(X))^2\right] - [\mathcal{E}(h(X))]^2.$$

Consider $h(X) = aX + b$, where a and b are constants. Then,

$$Var(aX+b) = \mathcal{E}\left[((aX+b) - \mathcal{E}(aX+b))^2\right] = a^2 Var(X).$$

For a constant b, $Var(b) = 0$.

For bivariate case, we have:

1. If a, b are constants, then

$$\begin{aligned} Var(aX+bY) &= \mathcal{E}\left[(aX+by - \mathcal{E}(aX+bY))^2\right] \\ &= a^2Var(X) + b^2Var(Y) + 2abCov(X,Y). \end{aligned}$$

2. If X and Y are independent, then

$$Var(aX + bY) = a^2Var(X) + b^2Var(Y).$$

3. $$\begin{aligned} Cov(aX + b, cY + d) &= \mathcal{E}[(aX + b - \mathcal{E}(aX + b))(cY + d - \mathcal{E}(cY + d))] \\ &= acCov(X, Y). \end{aligned}$$

Binomial Distributions: A binomial experiment is one with properties: (1) The experiment consists of m independent trials; (2) Each trial results in one of 2 possible outcomes called success and failure; and (3) The probability of success does not change from trial to trial and is denoted by p.

Define the random variable X as the number of successes in m trials of a binomial experiment. This random variable X is called a binomial random variable. There are 2^m possible sequences in all. Consider the case of x successes in m trials. Then, there are

$$\binom{m}{x} = \frac{m!}{x!(m-x)!}$$

ways this can occur, each with probability

$$P(x \text{ successes and } (m-x) \text{ failures}) = p^x(1-p)^{m-x} \Rightarrow$$

$$P(X = x) = \binom{m}{x} p^x(1-p)^{m-x}, \quad x = 0, 1, \ldots, m.$$

This is the binomial distribution and is written as:

$$X \sim B(m, p).$$

Note that $P(X = x) \geq 0$ for all x and

$$\sum_{x=0}^{m} P(X = x) = \sum_{x=0}^{m} \binom{m}{x} p^x(1-p)^{m-x}.$$

However,

$$(a + b)^m = \sum_{x=0}^{m} \binom{m}{x} a^x b^{m-x}.$$

Thus,

$$\sum_{x=0}^{m} P(X = x) = (p + 1 - p)^m = 1.$$

Hence,

$$P(X = x) = \binom{m}{x} p^x(1-p)^{m-x}$$

is its probability function. The mean of X is

$$\mathcal{E}(X) = \sum_{x=0}^{m} x \binom{m}{x} p^x (1-p)^{m-x}$$

But

$$\binom{m}{x} = \frac{m!}{x!(m-x)!}.$$

Thus,

$$\begin{aligned}\mathcal{E}(X) &= \sum_{x=1}^{m} \frac{m(m-1)!}{(x-1)!(m-x)!} pp^{x-1}(1-p)^{m-x} \\ &= mp[p-(1-p)]^{m-1} = mp.\end{aligned}$$

The variance of X is

$$Var(X) = \mathcal{E}\left(X^2\right) - [\mathcal{E}(X)]^2,$$

where

$$\mathcal{E}\left(X^2\right) = \sum_{x=0}^{m} x^2 \binom{m}{x} p^x (1-p)^{m-x}.$$

Since $X^2 = X(X-1) + X$, it is clear that

$$\mathcal{E}\left(X^2\right) = \mathcal{E}[X(X-1)] + \mathcal{E}(X).$$

On the other hand,

$$\begin{aligned}\mathcal{E}[X(X-1)] &= \sum_{x=0}^{m} x(x-1) \frac{m!}{(m-x)!x!} p^x (1-p)^{m-x} \\ &= m(m-1)p^2[p-(1-p)]^{m-2} = m(m-1)p^2\end{aligned}$$

Therefore,

$$Var(X) = \mathcal{E}\left(X^2\right) - [\mathcal{E}(X)]^2 = mp(1-p).$$

Example A.3.2. Let X be a binomial variate for which

$$P(X = x) = \binom{m}{x} p^x (1-p)^{m-x}, \quad x = 0, 1, \ldots, m.$$

Then,

$$\begin{aligned}P(X = x+1) &= \binom{m}{x+1} p^{x+1}(1-p)^{m-x-1} \\ &= \frac{m!}{(m-x-1)!(x+1)!} p^{x+1}(1-p)^{m-x-1}\end{aligned}$$

$$= \left\{ \frac{m!}{(m-x)!x!} p^x (1-p)^{m-x} \right\} \frac{(m-x)p}{(x+1)(1-p)}$$
$$= \frac{m-x}{x+1} \frac{p}{1-p} P(X=x).$$

Poisson Distribution: A Poisson experiment is one with properties: (1) Number of successes occurring in a single time interval (or region) is *independent* of those occurring in any other disjoint time interval (or region); (2) The probability of a single success occurring in a very short time interval (or region) is *only* proportional to the length of the time interval (or size of the region); and (3) The probability of more than one success occurring in a very short time interval (or region) is *negligible.*

The number X of successes in a Poisson experiment is called a Poisson random variable, which is written as

$$X \sim P(\lambda).$$

The probability function of X is

$$P(X=x) = \frac{e^{-\lambda}\lambda^x}{x!}, \quad x = 0, 1, 2, \ldots,$$

where λ denotes the mean number of successes in a specified interval. The *mean* of X is

$$\mathcal{E}(X) = \sum_{x=0}^{\infty} x \frac{e^{-\lambda}\lambda^x}{x!} = e^{-\lambda} \sum_{x=1}^{\infty} \frac{\lambda^x}{(x-1)!}$$
$$= \lambda e^{-\lambda} \sum_{x=1}^{\infty} \frac{\lambda^{x-1}}{(x-1)!} = \lambda e^{-\lambda} \sum_{x=0}^{\infty} \frac{\lambda^x}{x!} = \lambda e^{-\lambda} e^{\lambda} = \lambda.$$

The variance of X is

$$Var(X) = \mathcal{E}\left(X^2\right) - [\mathcal{E}(X)]^2 = \mathcal{E}[X(X-1)] + \mathcal{E}(X) - [\mathcal{E}(X)]^2$$
$$= \lambda^2 + \lambda - \lambda^2 = \lambda.$$

The sum of 2 independent Poisson random variables is also a Poisson variable. That is,

$$X_1 \sim P(\lambda_1), \quad X_2 \sim P(\lambda_2), \quad X_1 \text{ and } X_2 \textit{ independent}.$$
$$\Rightarrow X_1 + X_2 \sim P(\lambda_1 + \lambda_2).$$

Let X be a Binomial random variable with m trials, and let p be the probability of success. Let $m \to \infty$ and $p \to 0$. If $mp = \lambda$ remains constant and finite, then

$$B(m, p) \to P(\lambda).$$

Example A.3.3. Experiment shows that the number of mechanical failures per quarter for a certain component used in a loading plant is Poisson distributed. The mean number of failures per quarter is 1.5. Stocks of the component are built up to a fixed number at the beginning of a quarter and not replenished until the beginning of the next quarter. Calculate the least number of spares of this component which should be carried at the beginning of the quarter to ensure that the probability of a demand exceeding this number during the quarter will not exceed 0.10. If stocks are to be replenished only once in *6* months, how many ought now to be carried at the beginning of the period to give the same protection?

Solution.

(a) Let $X \sim P(1.5)$ be the number of mechanical failures/quarter, and let t be the smallest v number of spares. Then, $P(X > t) \leq 0.1$ and $P(X \leq t) > 0.9$. Since $P(X \leq t) = \sum_{x=0}^{t} e^{-1.5}\frac{1.5^x}{x!}$, we have $P(X \leq 2) = 0.8088$, and $P(X \leq 3) = 0.9344$ (hence $P(X > 3) = 0.0666$). This implies 3 spares/quarter

(b) For 6 months, $\lambda = 3$, $X \sim P(3)$. Find t such that $P(X > t) \leq 0.1$.

$$P(X > 5) = 0.0839 \Rightarrow 5 \text{ spares/6 months.}$$

Normal Distribution: Let X be a continuous random variable with its value denoted by x. The random variable X is said to have a normal distribution if its probability density function is

$$f(x) = \frac{1}{\sqrt{2\pi}\sigma} \exp\left\{-\tfrac{1}{2}\left(\tfrac{x-\mu}{\sigma}\right)^2\right\},$$

where $-\infty < x < \infty$, $\mu \equiv \mathcal{E}(X)$ is the mean of X such that $-\infty < \mu < \infty$, $\sigma^2 \equiv Var(X)$ is the variance of X, and σ is standard deviation of X such that $\sigma > 0$. Here, μ and σ are the parameters of the distribution.

Characteristic:

1. $f(-x) = f(x)$;
2. $\frac{df(x)}{dx} = \left(\frac{1}{\sigma\sqrt{2\pi}} \exp\left\{-\tfrac{1}{2}\left(\tfrac{x-\mu}{\sigma}\right)^2\right\}\right)\left(-\frac{x-\mu}{\sigma^2}\right).$

   Then, $\frac{df(x)}{dx} = 0 \Rightarrow x = \mu$;
3. $\frac{d^2f(x)}{dx^2} = -\frac{1}{\sigma^3\sqrt{2\pi}}e^{-\frac{1}{2}\left(\frac{x-\mu}{\sigma}\right)^2} + \left(\frac{1}{\sigma\sqrt{2\pi}} \exp\left\{-\tfrac{1}{2}\left(\tfrac{x-\mu}{\sigma}\right)^2\right\}\right)\left(\frac{x-\mu}{\sigma^2}\right)^2.$

   Then,

$$\left.\frac{d^2f(x)}{dx^2}\right|_{x=\mu} = -\frac{1}{\sigma^3\sqrt{2\pi}} \leq 0.$$

This implies that $x = \mu$ is the point at which the function $f(x)$ attains its maximum, $f(x)$ is symmetric about μ, and the points of inflexion are at $x = \mu + \sigma$ and $x = \mu - \sigma$.

Let X be a random variable with its value denoted by x. If X is normally distributed with mean μ and variance σ^2, written as $X \sim N(\mu, \sigma^2)$. If $\mu = 0$ and $\sigma^2 = 1$, then the corresponding random variable U is called a *standard normal* random variable, written as $U \sim N(0,1)$. Let the value of U be denoted by u. Then, its probability density function $\phi(u)$ is

$$\phi(u) = \frac{1}{\sqrt{2\pi}} \exp\left\{-\tfrac{1}{2}u^2\right\},$$

and its cumulative distribution is

$$\Phi(u) = \int_{-\infty}^{u} \frac{1}{\sqrt{2\pi}} \exp\left\{-\tfrac{1}{2}v^2\right\} d\nu.$$

If $X \sim N(\mu, \sigma^2)$, then

$$P(X < x) = \int_{-\infty}^{x} \frac{1}{\sigma\sqrt{2\pi}} \exp\left\{-\tfrac{1}{2}\left(\tfrac{t-\mu}{\sigma}\right)^2\right\} dt.$$

Return again to the case when $U \sim N(0,1)$. Then,

$$\Phi(u) = P(U < u) = \int_{-\infty}^{x} \frac{1}{\sigma\sqrt{2\pi}} \exp\left\{-\tfrac{1}{2}v^2\right\} d\nu$$

$$\Phi(2) = P(U < 2) = 0.9772; \; \Phi(0) = \frac{1}{2}; \; \Phi(\infty) = 1; \; \Phi(-\infty) = 0.$$

$$\begin{aligned} &\Phi(-u) = P(U < -u) = P(U > u) = 1 - P(U < u) \\ &\Rightarrow \Phi(-u) = 1 - \Phi(u), \end{aligned}$$

and

$$P(U > u) = 1 - P(U < u) = 1 - \Phi(u).$$

Thus,

$$\begin{aligned} P(u_1 < U < u_2) &= P(U < u_2) - P(U < u_1). \\ &= \Phi(u_2) - \Phi(u_1). \end{aligned}$$

Note that

$$\begin{aligned} P(|U| < u) &= P(-u < U < u) = \Phi(u) - (1 - \Phi(u)) \\ &= 2\Phi(u) - 1. \end{aligned}$$

Probabilities for $X \sim N(\mu, \sigma^2)$: We can transform $(X \sim N(\mu, \sigma^2))$ to $(U \sim N(0,1))$. To begin with, we recall

$$P(x_1 < X < x_2) = \int_{x_1}^{x_2} \frac{1}{\sigma\sqrt{2\pi}} \exp\left\{-\tfrac{1}{2}\left(\tfrac{t-\mu}{\sigma}\right)^2\right\} dt.$$

Set

$$\nu = \frac{t-\mu}{\sigma} \Rightarrow t = \sigma\nu + \mu \text{ and } d\nu = \frac{1}{\sigma}dt$$

$$t = x_1 \Rightarrow \nu = \frac{x_1 - \mu}{\sigma} \equiv u_1;\ t = x_2 \Rightarrow \nu = \frac{x2 - \mu}{\sigma} \equiv u_2$$

$$\Rightarrow P(x_1 < X < x_2) = \int_{u_1}^{u_2} \frac{1}{\sqrt{2\pi}} e^{-\frac{1}{2}\nu^2} d\nu = \Phi(u_2) - \Phi(u_1)$$

$$= \Phi\left(\frac{x_2 - \mu}{\sigma}\right) - \Phi\left(\frac{x_1 - \mu}{\sigma}\right).$$

Thus, we conclude that if $X \sim N(\mu, \sigma^2)$, then $U = \dfrac{X-\mu}{\sigma} \sim N(0,1)$. This is called standardization.

Theorem A.3.2. *Suppose that* $X \sim N(\mu, \sigma^2)$. *Then,*

$$aX + b \sim N\left(a\mu + b, a^2\sigma^2\right).$$

Theorem A.3.3. *For each* $i = 1, \ldots, n$, *let* $X_i \sim N(\mu_i, \sigma_i^2)$. *If* X_i, $i = 1, \ldots, n$, *are independent, then*

$$Y \equiv \sum_{i=1}^{n} a_i X_i \sim N\left(\sum_{i=1}^{n} a_i\mu_i, \sum_{i=1}^{n} a_i^2\sigma_i^2\right).$$

Lognormal Random Variables: A random variable Z is lognormal if the random variable $\ln Z$ is normal. Equivalently, if X is normal, then $Z = \exp\{X\}$ is lognormal. This means that the density function for Z has the form

$$p(Z) = \frac{1}{\sqrt{2\pi}\sigma z} \exp\left\{-\frac{1}{2\sigma^2}(\ln Z - \upsilon)^2\right\}.$$

We have

$$\mathcal{E}(Z) = \exp\left\{\frac{\upsilon + \sigma^2}{2}\right\};\ \mathcal{E}(\ln Z) = \upsilon$$

$$Var(Z) = \exp\left\{\upsilon + \sigma^2\right\}\left\{\exp\left\{\sigma^2\right\} - 1\right\} \Rightarrow Var(\ln Z) = \sigma^2.$$

It follows from the summation result for joint normal random variables that products and powers of jointly lognormal variables are again lognormal. For example, if U and V are lognormal, then $Z = U^\alpha V^\beta$ is also lognormal.

Normal approximation to a binomial distribution: Suppose that $X \sim B(m,p)$. Then,

$$P(X = x) = \binom{m}{x} p^x (1-p)^{m-x},\ x = 0, 1, \ldots, m;\ 0 < p < 1.$$
$$\mathcal{E}(X) = m;\ Var(X) = mp(1-p).$$

This means that X is approximately $\sim N\left(m, (mp(1-p))^2\right)$. The approximation is very accurate if m is large and p is close to $\frac{1}{2}$. It is fairly good if m is not large and p is not very close to 0 or 1.

Theorem A.3.4 (Central Limit Theorem). *If $X_1, \ldots, X_n$ are independently identically distributed random variables with mean μ and variance σ^2, then, for large n,*

$$U \equiv \frac{\bar{X} - \mu}{\sigma/\sqrt{n}} \sim \textit{(approximately) } N(0,1), \qquad (*)$$

where $\bar{X} = \left\{\sum_{i=1}^{n} X_i\right\}/n$. In other words, for a large sample size, the sample mean is approximately normally distributed no matter what the distribution of the X_i, $i = 1, 2, \ldots$, are as long as they are independently identically distributed.

In conclusion, if a random sample of size n is selected from a population whose variance σ^2 is known, then the confidence interval for μ (i.e., the mean of the population) is

$$\left(\bar{x} - u_{\alpha/2}\frac{\sigma}{\sqrt{n}},\ \bar{x} + u_{\alpha/2}\frac{\sigma}{\sqrt{n}}\right).$$

On the other hand, if σ is not known, then we shall replace σ by the sample standard deviations, provided that $n \geq 30$.

Let $(\Omega, \mathcal{F}, P)$ denote a complete probability space, where Ω represents the sample space, $\mathcal{F}$ the σ-algebra (Borel algebra) of the subsets of the set Ω, and P the probability measure on the algebra $\mathcal{F}$. Let $\mathcal{F}_t$, $t \geq 0$, be an increasing family of complete subsigma algebras of the σ-algebra $\mathcal{F}$. For any random variable (or equivalently $\mathcal{F}$ measurable function) X, let

$$\mathcal{E}\{X\} = \int_\Omega X(\omega)P(\omega)$$

denote the expected value provided it exists. It does exist if $X \in L_1(\Omega, P)$.

For a subsigma algebra $\mathcal{G} \subset \mathcal{F}$, the conditional expectation of X, relative to $\mathcal{G}$, is denoted by

$$\mathcal{E}\{X \mid \mathcal{G}) = Y$$

. The random variable Y is $\mathcal{G}$ measurable.

Let $\mathcal{G}$ and $\mathcal{G}_1 \subset \mathcal{G}_2 \subset \mathcal{F}$ be any three (complete) subsigma algebras of the sigma algebra $\mathcal{F}$. The conditional expectation is a linear operator in the sense that

1. For $\alpha_1, \alpha_2 \in \mathbb{R}$,
$$\mathcal{E}\{\alpha X_1 + \alpha X_2 \mid \mathcal{G}) = \alpha_1 \mathcal{E}\{X_1 \mid \mathcal{G}) + \alpha_2 \mathcal{E}\{X_2 \mid \mathcal{G}).$$
2. $$\mathcal{E}\left\{\mathcal{E}\{X \mid \mathcal{G}_1) \mid \mathcal{G}_2\right\} = \mathcal{E}\left\{\mathcal{E}\{X \mid \mathcal{G}_2) \mid \mathcal{G}_1\right\} = \mathcal{E}\{X \mid \mathcal{G}_1)\}.$$
3. If Z is a bounded $\mathcal{G}$ measurable random variable with $\mathcal{G} \subset \mathcal{F}$, then
$$\mathcal{E}\{ZX \mid \mathcal{G}) = Z\mathcal{E}\{X \mid \mathcal{G}\},$$
which is a $\mathcal{G}$ measurable random variable.
4. If X is a random variable independent of the sigma algebra $\mathcal{G} \subset F$, then
$$\mathcal{E}\{X \mid \mathcal{G}_1) = \mathcal{E}\{X\};$$
5. For any $\mathcal{F}$ measurable and integrable random variable Z, the process
$$Z_t = \mathcal{E}\{Z \mid \mathcal{F}_t\}, \ \ t \geq 0,$$
is an $\mathcal{F}_t$ martingale in the sense that for any $s \leq t < \infty$,
$$\mathcal{E}\{Z_t \mid \mathcal{F}_s\} = Z_s.$$

We say that a process $W(t)$ is a Wiener process (or alternatively, Brownian motion) if it satisfies the following properties: (1) For any $s < t$, the quantity $W(t) - W(s)$ is a normal random variable with mean zero and variance $t - s$; (2) For any $0 \leq t_1 \leq t_2 \leq t_3 \leq t_4$, the random variables $W(t_2) - W(t_1)$ and $W(t_4) - W(t_3)$ are uncorrelated; and (3) $W(t_0) = 0$ with probability 1.

Theorem A.3.5. *Suppose that the random process X is defined by the Ito process:*

$$dX(t) = \alpha(X(t), t)dt + \beta(X(t), t)dW(t),$$

where W is a standard Wiener process.

Suppose also that the process $Y(t)$ *is defined by*

$$Y(t) = F(X(t), t).$$

Then, $Y(t)$ *satisfies the Ito equation*

$$dY(t) = \left\{ \frac{\partial F}{\partial x}\alpha + \frac{\partial F}{\partial t} + \frac{1}{2}\frac{\partial^2 F}{\partial x^2}(\beta)^2 \right\} dt + \frac{\partial F}{\partial x}\beta dW$$

References

1. Åkesson, J., Arzen, K., Gäfert, M., Bergdahl, T., Tummescheit, H.: Modelling and optimization with Optimica and JModelica.org – languages and tools for solving large-scale dynamic optimization problems. Comput. Chem. Eng. **34**(11), 1737–1749 (2010)
2. Abu-Khalaf, M., Lewis, F.L.: Nearly optimal control laws for nonlinear systems with saturating actuators using a neural network HJB approach. Automatica **41**(5), 779–791 (2005)
3. Ahmed, N.U.: Elements of Finite-Dimensional Systems and Control Theory. Longman Scientific and Technical, Essex (1988)
4. Ahmed, N.U.: Dynamic Systems and Control with Applications. World Scientific, Singapore (2006)
5. Ahmed, N.U., Teo, K.L.: Optimal Control of Distributed Parameter Systems. Elsevier Science, New York (1981)
6. Al-Tamimi, A., Lewis, F., Abu-Khalaf, M.: Discrete-time nonlinear HJB solution using approximate dynamic programming: convergence proof. IEEE Trans. Syst. Man Cybern. B Cybern. **38**, 943–949 (2008)
7. Amerongen, J.V.: Adaptive steering of ships-a model reference approach. Automatica **20**(1), 3–14 (1984)
8. Anderson, B.D.O., Moore, J.B.: Linear Optimal Control. Prentice-Hall, Englewood Cliffs (1971)
9. Anderson, B.D.O., Moore, J.B.: Optimal Control: Linear Quadratic Methods. Dover, New York (2007)
10. Aoki, M.: Introduction to Optimization Techniques: Fundamentals and Applications of Nonlinear Programming. Macmillan, New York (1971)
11. Athans, M., Falb, P.L.: Optimal Control. McGraw-Hill, New York (1966)
12. Baker, S., Shi, P.: Formulation of a tactical logistics decision analysis problem using an optimal control approach. ANZIAM J. **44**(E), 1737–1749 (2002)

K. L. Teo et al., *Applied and Computational Optimal Control*, Springer Optimization and Its Applications 171,
https://doi.org/10.1007/978-3-030-69913-0

13. Banihashemi, N., Kaya, C.Y.: Inexact restoration for Euler discretization of box-constrained optimal control problems. J. Optim. Theory Appl. **156**, 726–760 (2003)
14. Banks, H.T., Burns, J.A.: Hereditary control problem: Numerical methods based on averaging approximations. SIAM J. Control. Optim. **16**, 169–208 (1978)
15. Bartlett, M.: An inverse matrix adjustment arising in discriminant analysis. Ann. Math. Stat. **22**(1), 107–111 (1951)
16. Bashier, E.B.M., Patidar, K.C.: Optimal control of an epidemiological model with multiple time delays. Appl. Math. Comput. **292**, 47–56 (2017)
17. Bech, M., Smitt., L.W.: Analogue Simulation of Ship Manoeuvres. Hydro and Aerodynamics Lab. Report No. Hy-14, Denmark (1969)
18. Bellman, R.: Introduction to the Mathematical Theory of Control Processes, Vol. 1. Academic, New York (1967)
19. Bellman, R.: Introduction to the Mathematical Theory of Control Processes, Vol. 2. Academic, New York (1971)
20. Bellman, R., Dreyfus, R.: Dynamic Programming and Modern Control Theory. Academic, Orlando (1977)
21. Bensoussan, A., Hurst, E., Naslund, B.: Management Application of Modern Control Theory. North Holland, Amsterdam (1974)
22. Bertsekas, D.: Constrained Optimization and Lagrange Multiplier Methods. Academic, New York (1982)
23. Bertsimas, D., Brown, D.: Constrained stochastic LQC: a tractable approach. IEEE Trans. Autom. Control **52**, 1826–1841 (2007)
24. Betts, J.: Practical Methods for Optimal Control and Estimation Using Nonlinear Programming. SIAM Press, Philadelphia (2010)
25. Biegler, L.: An overview of simultaneous strategies for dynamic optimization. Chem. Eng. Process. Process Intensif. **46**(11), 1043–1053 (2007)
26. Birgin, E.G., Martinez, J.M.: Local convergence of an Inexact-Restoration method and numerical experiments. J. Optim. Theory Appl. **127**(2), 229–247 (2005)
27. Blanchard, E., Loxton, L., Rehbock, V.: Dynamic optimization of dual-mode hybrid systems with state-dependent switching conditions. Optim. Methods Softw. **33**(2), 297–310 (2018)
28. Blanchard, E., Loxton, R., Rehbock, V.: A computational algorithm for a class of non-smooth optimal control problems arising in aquaculture operations. Appl. Math. Comput. **219**, 8738–8746 (2013)
29. Boltyanskii, V.: Mathematical Methods of Optimal Control. Holt, Rinehart and Winston, New York (1971)
30. Boyd, S., Vandenberghe, L.: Convex Optimization (2013). http://www.stanford.edu/~boyd/cvxbook/

31. Brooke., D.: The design of a new automatic pilot for the commercial ship. In: First IFAC/IFIP Symposium on Ship Operation Automation, Oslo (1973)
32. Broyden, C.: The convergence of a class of double-rank minimization algorithms. J. Inst. Math. Appl. **6**, 76–90 (1970)
33. Bryson, A., Ho, Y.: Applied Optimal Control. Hemisphere Publishing, Washington DC (1975)
34. Büskens, C.: Optimierungsmethoden und sensitivitätsanalyse für optimale steuerprozesse mit steuer und zustands beschränkungen. Ph.D. thesis, Institut für Numerische und Inentelle Mathematik, Universität Münster (1998)
35. Büskens, C., Maurer, H.: Nonlinear programming methods for real-time control of an industrial robot. J. Optim. Theory Appl. **107**(3), 505–527 (2000)
36. Buskens, C., Maurer, H.: SQP-methods for solving optimal control problems with control and state constraints: adjoint variables, sensitivity analysis and real-time control. J. Comput. Appl. Math. **120**, 85–108 (2000)
37. Butovskiy, A.: Distributed Control Systems. American Elsevier, New York (1969)
38. Caccetta, L., Loosen, I., Rehbock, V.: Computational aspects of the optimal transit path problem. J. Ind. Manage. Optim. **4**, 95–105 (2008)
39. Canuto, C., Hussaini, M., Quarteroni, A., Zang, T.: Spectral Methods in Fluid Dynamics. Springer, New York (1988)
40. Cesari, L.: Optimization: Theory and Applications. Springer, New York (1983)
41. Chai, Q., Yang, C., Teo, K.L., Gui, W.: Time-delay optimal control of an industrial-scale evaporation process sodium aluminate solution. Control Eng. Pract. **20**, 618–628 (2012)
42. Chen, T., Xu, C., Lin, Q., Loxton, R., Teo, K.L.: Water hammer mitigation via PDE constrained optimization. Control Eng. Pract. **45**, 54–63 (2015)
43. Cheng, T.C.E., Teo, K.L.: Further extensions of a student related optimal control problem. Int. J. Math. Model. **9**, 499–506 (1987)
44. Choi, C., Laub, A.: Efficient matrix-valued algorithms for solving stiff Riccati differential equations. IEEE Trans. Autom. Control **35**(7), 770–776 (1990)
45. Chyba, M., Haberkorn, T., Smith, R.N., Choi, S.K.: Design and implementation of time efficient trajectories for autonomous underwater vehicles. Ocean Eng. **35**, 63–76 (2008)
46. Cuthrell, J.E., Biegler, L.: Simultaneous optimization and solution methods for batch reactor control profiles. Comput. Chem. Eng. **13**, 49–62 (1989)

47. Denis-Vidal, L., Jauberthie, C., Joly-Blanchard, G.: Identifiability of a nonlinear delayed-differential aerospace model. IEEE Trans. Autom. Control **51**(1), 154–158 (2006)
48. Dontchev, A.L.: In: Balakrishnan, A.V., Thoma, M. (eds.) Perturbations, Approximations and Sensitivity Analysis of Optimal Control Systems. Lecture Notes in Control and Information Sciences. Springer, Berlin (1983)
49. Dontchev, A.L., Hager, W.W.: The Euler approximation in state constrained optimal control problems. Math. Comput. **70**, 173–203 (2000)
50. Dontchev, A.L., Hager, W.W., Malanowski, K.: Error bound for Euler approximation of a state and control constrained optimal control problem. Numer. Funct. Anal. Optim. **21**(6), 653–682 (2000)
51. Dunford, N., Schwartz, J.T.: Linear Operators, Part 1 and Part 2. Wiley, New York (1958)
52. Elnagar, G., Kazemi, M., Razzaghi, M.: The Pseudospectral Legendre method for discretizing optimal control problems. IEEE Trans. Autom. Control **40**(10), 1793–1796 (1995)
53. Esposito, W., Floudas, C.: Deterministic global optimization in nonlinear optimal control problems. J. Glob. Optim. **17**(1–4), 97–126 (2000)
54. Evtushenko, Y.: Numerical Optimization Techniques. Springer, New York (1985)
55. Feehery, W., Barton, P.: Dynamic optimization with state variable path constraints. Comput. Chem. Eng. **22**(9), 1241–1256 (1998)
56. Feng, Z.G., Teo, K.L., Rehbock, V.: Hybrid method for a general optimal sensor scheduling problem in discrete time. Automatica **44**, 1295–1303 (2008)
57. Feng, Z.G., Teo, K.L., Rehbock, V.: A discrete filled function method for the optimal control of switched systems in discrete time. Optimal Control Appl. Methods **30**(6), 585–593 (2009)
58. Fisher, M.E., Jennings, L.: Discrete-time optimal control problems with general constraints. ACM Trans. Math. Softw. **18**(4), 401–413 (1992)
59. Fleming, W., Rishel, R.: Deterministic and Stochastic Optimal Control. Springer, Berlin (1975)
60. Fletcher, R.: A new approach to variable metric algorithms. Comput. J. **13**(3), 317–322 (1970)
61. Fletcher, R.: Practical Methods of Optimization, 2nd edn. Wiley-Interscience, New York (1987)
62. Fletcher, R., Reeves, C.: Function minimization by conjugate gradients. Comput. J. **7**, 149–154 (1964)
63. Fu, J., Chachuat, B., Mitsos, A.: Local optimization of dynamic programs with guaranteed satisfaction of path constraints. Automatica **62**, 184–192 (2015)
64. Gamkrelidze, R.: Principles of Optimal Control Theory. Plenum Press, New York (1978)

65. Gao, Y., Kostyukova, O., Chong, K.T.: Worst-case optimal control for an electrical drive system with time-delay. Asian J. Control **11**(4), 386–395 (2009)
66. Gerdts, M.: Solving mixed-integer optimal control problems by branch and bound: a case study from automobile test-driving with gear shift. Optimal Control Appl. Methods **26**(1), 1–18 (2005)
67. Gerdts, M.: A variable time transformation method for mixed-integer optimal control problems. Optimal Control Appl. Methods **27**, 169–182 (2006)
68. Gerdts, M.: Global convergence of a nonsmooth Newton method for control-state constrained optimal control problems. SIAM J. Control Optim. **19**(1), 326–350 (2008)
69. Gerdts, M.: Optimal control of ODEs and DAEs. De Gruyter, Berlin (2012)
70. Giang, D., Lenbury, Y., Seidman, T.: Delay effect in models of population growth. J. Math. Anal. Appl. **305**, 631–643 (2005)
71. Gill, P., Murray, W., Wright, M.: Practical Optimization. Academic, London (1981)
72. Goh, B.: Necessary conditions for singular extremals involving multiple control variables. SIAM J. Control **4**(4), 716–731 (1966)
73. Goh, B.: The second variation for the singular Bolza problem. SIAM J. Control **4**(2), 309–325 (1966)
74. Goh, B.: Management and Analysis of Biological Populations. Elsevier, Amsterdam (1980)
75. Goh, C.J., Teo, K.L.: Control parametrization: a unified approach to optimal control problems with general constraints. Automatica **24**(1), 3–18 (1988)
76. Goh, C.J., Teo, K.L.: Alternative algorithms for solving nonlinear function and functional inequalities. Appl. Math. Comput. **41**(2), 159–177 (1991)
77. Goldfarb, D.: A family of variable-metric methods derived by variational means. Math. Comput. **24**, 23–26 (1970)
78. Goldstein, A.: On steepest descent. SIAM J. Control **3**, 147–151 (1965)
79. Gong, Z.H., Loxton, R., Yu, C.J., Teo, K.L.: Dynamic optimization for robust path planning of horizontal oil wells. Appl. Math. Comput. **274**, 711–725 (2016)
80. Gong, Z.H., Teo, K.L., Liu, C.Y., Feng, E.: Horizontal well's path planning: an optimal switching control approach. Appl. Math. Model. **39**, 4022–4032 (2015)
81. Gonzaga, C., Polak, E., Trahan, R.: An improved algorithm for optimization problems with functional inequality constraints. IEEE Trans. Autom. Control **25**(1), 211–246 (1980)
82. Graham, K., Rao, A.: Minimum-time trajectory optimization of low-thrust earth-orbit transfers with eclipsing. J. Spacecr. Rocket. **53**(2), 289–303 (2016)

83. Gruver, W., Sachs, E.: Algorithmic Methods in Optimal Control. Research Notes in Mathematics, vol. 47. Pitman, London (1981)
84. Guinn, T.: Reduction of delayed optimal control problems to nondelayed problems. J. Optim. Theory Appl. **18**(3), 371–377 (1976)
85. Hager, W.W.: Runge-Kutta methods in optimal control and the transformed adjoint system. Numer. Math. **87**, 247–282 (2000)
86. Han, S.: Superlinearly convergent variable metric algorithms for general nonlinear programming problems. Math. Program. **11**(1), 263–282 (1976)
87. Han, S.: A globally convergent method for nonlinear programming. J. Optim. Theory Appl. **22**(3), 297–309 (1977)
88. Hartl, R.F., Sethi, S.P., Vickson, R.G.: A survey of the maximum principles for optimal control problems with state constraints. SIAM Rev. **37**(2), 181–218 (1995)
89. Hausdorff, L.: Gradient Optimization and Nonlinear Control. Wiley, New York (1976)
90. Hermes, H., LaSalle, J.P.: Functional Analysis and Time optimal Control. Academic, New York (1969)
91. Hewitt, E., Stromberg, K.: Real and Abstract Analysis. Springer, New York (1965)
92. Hindmarsh, A.: Large ordinary differential systems and software. IEEE Control Mag. **2**, 24–30 (1982)
93. Ho, C.Y.F., Ling, B.W.K., Liu, Y.Q., Tam, P.K.S., Teo, K.L.: Optimal PWM control of switched-capacitor DC–DC power converters via model transformation and enhancing control techniques. IEEE Trans. Circuits Syst. I **55**, 1382–1391 (2008)
94. Hounslow, M.J., Ryall, R.L., Marshall, V.R.: A discretized population balance for nucleation, growth, and aggregation. AIChE J. **34**(11), 1821–1832 (1988)
95. Howlett, P.: Optimal strategies for the control of a train. Automatica **32**(4), 519–532 (1996)
96. Howlett, P.: The optimal control of a train. Ann. Oper. Res. **98**(1-4), 65–87 (2000)
97. Howlett, P.G., Pudney, P.J., Vu, X.: Local energy minimization in optimal train control. Automatica **45**, 2692–2698 (2009)
98. Huang, C., Wang, S., Teo, K.L.: Solving Hamilton-Jacobi-Bellman equations by a modified method of characteristics. Nonlinear Anal. **40**(1–8), 279–293 (2000)
99. Huang, C., Wang, S., Teo, K.L.: On application of an alternating direction method to Hamilton-Jacobi-Bellman equations. J. Comput. Appl. Math. **166**(1), 153–166 (2004)
100. Hull, D., Speyer, J., Tseng, C.: Maximum-information guidance for homing missiles. J. Guid. Control. Dyn. **8**(4), 494–497 (1985)

101. Huntington, G., Rao, A.: Optimal reconfiguration of spacecraft formations using the Gauss pseudospectral method. J. Guid. Control. Dyn. **31**(3), 689–698 (2008)
102. Hussein, I., Bloch, A.: Optimal control of underactuated nonholonomic mechanical systems. IEEE Trans. Autom. Control **53**(3), 668–682 (2008)
103. Jennings, L.S., Teo, K.L.: A computational algorithm for functional inequality constrained optimization problems. Automatica **26**(2), 371–375 (1990)
104. Jennings, L.S., Fisher, M.E., Teo, K.L., Goh, C.J.: MISER3 optimal control software: theory and user manual-both FORTRAN and MATLAB versions (2004).
105. Jennings, L.S., Wong, K., Teo, K.L.: Optimal control computation to account for eccentric movement. J. Aust. Math. Soc. B **38**(2), 182–193 (1996)
106. Jiang, C., Lin, Q., Yu, C., Teo, K.L., Duan, G.R.: An exact penalty method for free terminal time optimal control problem with continuous inequality constraints. J. Optim. Theory Appl. **154**, 30–53 (2012)
107. Jiang, C., Teo, K.L., Duan, G.: A suboptimal feedback control for nonlinear time-varying systems with continuous inequality constraints. Automatica **48**, 660–665 (2012)
108. Jiang, C., Teo, K.L., Loxton, R., Duan, G.R.: A neighboring extremal solution for an optimal switched impulsive control problem. J. Ind. Manage. Optim. **8**, 591–609 (2012)
109. Kailath, T.: Linear Systems. Prentice-Hall Information and System Science Series. Prentice-Hall, Englewood Cliffs (1980)
110. Kamien, M., Schwartz, N.: Dynamic Optimization: The Calculus of Variations and Optimal Control in Economics and Management. North Holland, Amsterdam (1991)
111. Kaya, C.Y., Noakes, J.L.: Computational method for time-optimal switching control. J. Optim. Theory Appl. **117**(1), 69–92 (2003)
112. Kaya, C.Y., Noakes, J.L.: Leapfrog for optimal control. SIAM J. Numer. Anal. **46**(6), 2795–2817 (2008)
113. Kaya, C.Y.: Inexact restoration for Runge-Kutta discretization of optimal control problems. SIAM J. Numer. Anal. **48**(4), 1492–1517 (2010)
114. Kaya, C.Y.: Markov–Dubins path via optimal control theory. Comput. Optim. Appl. **68**, 719–747 (2017)
115. Kaya, C.Y., Martinez, J.M.: Euler discretization for inexact restoration and optimal control. J. Optim. Theory Appl. **134**, 191–206 (2007)
116. Kaya, C.Y., Maurer, H.: A numerical method for nonconvex multi-objective optimal control problems. Comput. Optim. Appl. **57**, 685–702 (2014)
117. Kaya, C.Y Noakes, J.L.: Computations and time-optimal controls. Optimal Control Appl. Methods **17**, 171–185 (1996)

118. Khmelnitsky, E.: A combinatorial, graph-based solution method for a class of continuous time optimal control problems. Math. Oper. Res. **27**(2), 312–325 (2002)
119. Kogan, K., Khmelnitsky, E.: Scheduling: Control-Based Theory and Polynomial-Time Algorithms. Kluwer Academic, Dordrecht (2000)
120. Lee, C., Leitmann, G.: On a student-related optimal control problem. J. Optim. Theory Appl. **65**(1), 129–138 (1990)
121. Lee, E., Markus, L.: Foundations of Optimal Control Theory. Wiley, New York (1967)
122. Lee, H., Ali, M., Wong, K.: Global optimization for a class of optimal discrete-valued control problems. Dyn. Contin. Discrete Impuls. Syst. B **11**(6), 735–756 (2004)
123. Lee, H.W.J., Teo, K.L., Jennings, L.S.: On optimal control of multi-link vertical planar robot arms systems moving under the effect of gravity. J. Aust. Math. Soc. B **39**(2), 195–213 (1997)
124. Lee, H.W.J., Teo, K.L., Lim, A.E.B.: Sensor scheduling in continuous time. Automatica **37**(12), 2017–2023 (2001)
125. Lee, H.W.J., Teo, K.L., Rehbock, V., Jennings, L.S.: Control parameterization enhancing technique for time optimal control problems. Dyn. Syst. Appl. **6**, 243–262 (1997)
126. Lee, H.W.J., Teo, K.L., Rehbock, V., Jennings, L.S.: Control parameterization enhancing technique for optimal discrete-valued control problems. Automatica **35**(8), 1401–1407 (1999)
127. Lee, W., Rehbock, V., Caccetta, L., Teo, K.L.: Numerical solution of optimal control problems with discrete-valued system parameters. J. Glob. Optim. **23**(3-4), 233–244 (2002)
128. Lee, W., Wang, S., Teo, K.L.: Optimal recharge and driving strategies for a battery-powered electric vehicle. Math. Probl. Eng. **5**(1), 1–32 (1999)
129. Lei, J.: Optimal vibration control of nonlinear systems with multiple time-delays: an application to vehicle suspension. Integr. Ferroelectr. **170**, 10–32 (2016)
130. Lewis, F.: Optimal Control. Wiley, New York (1986)
131. Li, B., Teo, K.L., Duan, G.R.: Optimal control computation for discrete time time-delayed optimal control problem with all-time-step inequality constraints. Int. J. Innov. Comput. Inf. Control **6**(7), 3157–3175 (2010)
132. Li, B., Teo, K.L., Lim, C.C., Duan, G.R.: An optimal PID controller design for nonlinear constrained optimal control problems. Discrete Contin. Dyn. Syst. B **16**, 1101–1117 (2011)
133. Li, B., Teo, K.L., Zhao, G.H., Duan, G.: An efficient computational approach to a class of minmax optimal control problems with applications. ANZIAM J. **51**(2), 162–177 (2009)
134. Li, B., Yu, C., Teo, K.L., Duan, G.R.: An exact penalty function method for continuous inequality constrained optimal control problem. J. Optim. Theory Appl. **151**(2), 260–291 (2011)

135. Li, B., Zhu, Y.G., Sun, Y.F., Aw, G., Teo, K.L.: Deterministic conversion of uncertain manpower planning optimization problem. IEEE Trans. Fuzzy Syst. **26**(5), 2748–2757 (2018)
136. Li, B., Zhu, Y.G., Sun, Y.F., Aw, G., Teo, K.L.: Multi-period portfolio selection problem under uncertain environment with bankruptcy constraint. Appl. Math. Model. **56**, 539–550 (2018)
137. Li, C., Teo, K.L., Li, B., Ma, G.: A constrained optimal PID-like controller design for spacecraft attitude stabilization. Acta Astrnaut. **74**, 131–140 (2011)
138. Li, R., Teo, K.L., Wong, K.H., Duan, G.R.: Control parameterization enhancing transform for optimal control of switched systems. Math. Comput. Model. **43**(11-12), 1393–1403 (2006)
139. Li, Y.G., Gui, W.H., Teo, K.L., Zhu, H.Q., Chai, Q.Q.: Optimal control for zinc solution purification based on interacting CSTR models. J. Process Control **22**, 1878–1889 (2012)
140. Liang, J.: Optimal magnetic attitude control of small spacecraft. Ph.D. thesis, Utah State University (2005)
141. Lim, C., Forsythe., W.: Autopilot for ship control. IEEE Proc. **130**(6), 281–294 (1983)
142. Lin, Q., Loxton, R., Teo, K.L.: Optimal control of nonlinear switched systems: computational methods and applications. J. Oper. Res. Soc. China **1**, 275–311 (2013)
143. Lin, Q., Loxton, R., Teo, K.L., Wu, Y.H.: A new computational method for a class of free terminal time optimal control problems. Pac. J. Optim. **7**(1), 63–81 (2011)
144. Lin, Q., Loxton, R., Teo, K.L., Wu, Y.H.: Optimal control computation for nonlinear systems with state-dependent stopping criteria. Automatica **48**, 2116–2129 (2012)
145. Lin, Q., Loxton, R., Teo, K.L., Wu, Y.H.: Optimal feedback control for dynamic systems with state constraints: an exact penalty approach. Optim. Lett. **8**(4), 1535–1551 (2014)
146. Lin, Q., Loxton, R., Teo, K.L., Wu, Y.H.: Optimal control problems with stopping constraints. J. Glob. Optim. **63**(4), 835–861 (2015)
147. Lin, Q., Loxton, R., Teo, K.L., Wu, Y.H., Yu, C.J.: A new exact penalty method for semi-infinite programming problems. J. Comput. Appl. Math. **261**(1), 271–286 (2014)
148. Lin, Q., Loxton, R.C., Teo, K.L.: The control parameterization method for nonlinear optimal control: a survey. J. Ind. Manage. Optim. **10**(1), 275–309 (2014)
149. Lions, J.: Optimal Control of Systems Governed by Partial Differential Equations. Springer, New York (1971)
150. Liu, C., Gong, Z.: Optimal control of Switched Systems Arising in Fermentation Processes. Springer, Berlin (2014)

151. Liu, C., Loxton, R., Teo, K.L.: Switching time and parameter optimization in nonlinear switched systems with multiple time delays. J. Optim. Theory Appl. **163**, 957–988 (2014)
152. Liu, C., Loxton, R., Lin, Q., Teo, K.L. : Dynamic optimization for switched time-delay systems with state-dependent switched conditions. SIAM J. Control Optim. **56**, 3499–3523 (2018)
153. Liu, C., Loxton, R., Lin, Q., Teo, K.L.: Dynamic optimization for switched time-delay systems with state-dependent switching conditions. SIAM J. Control Optim. **56**(5), 3499–3523 (2018)
154. Liu, C.M., Feng, Z.G., Teo, K.L.: On a class of stochastic impulsive optimal parameter selection problems. Int. J. Innov. Comput. Inf. Control **5**(4), 1043–1054 (2009)
155. Liu, C.Y., Gong, Z., Feng, E., Yin, H.: Optimal switching control of a fed-batch fermentation process. J. Glob. Optim. **52**, 265–280 (2012)
156. Liu, C.Y., Gong, Z., Shen, B., Feng, E.: Modelling and optimal control for a fed-batch fermentation process. Appl. Math. Model. **37**, 695–706 (2013)
157. Liu, C.Y., Gong, Z.H., Lee, H.W.J., Teo, K.L.: Robust bi-objective optimal control of 1,3-propanediol microbial batch production process. J. Process Control **78**, 170–182 (2019)
158. Liu, C.Y., Gong, Z.H., Teo, K.L., Feng, E.: Multi-objective optimization of nonlinear switched time-delay systems in fed-batch process. Appl. Math. Model. **40**, 10,533–10,548 (2016)
159. Liu, C.Y., Gong, Z.H., Teo, K.L., Loxton, R., Feng, E.: Bi-objective dynamic optimization of a nonlinear time-delay system in microbial batch process. Optim. Lett. **12**, 1249–1264 (2018)
160. Liu, C.Y., Gong, Z.H., Teo, K.L., Sun, J., Caccetta, L.: Robust multi-objective optimal switching control arising in 1,3-propanediol microbial fed-batch process. Nonlinear Anal. Hybrid Syst. **25**, 1–20 (2017)
161. Liu, Y., Teo, K.L., Agarwal, R.P.: A general approach to nonlinear multiple control problems with perturbation consideration. Math. Comput. Model. **26**, 49–58 (1997)
162. Liu, Y., Teo, K.L., Jennings, L.S., Wang, S.: On a class of optimal control problems with state jumps. J. Optim. Theory Appl. **98**(1), 65–82 (1998)
163. Löberg, J.: YALMIP : A toolbox for modeling and optimization in Matlab. In: Proc. Int. Symp. CACSD, Taipei, pp. 284–289 (2004)
164. Loxton, R., Lin, Q., Teo, K.L.: Minimizing control variation in nonlinear optimal control. Automatica **49**, 2652–2664 (2013)
165. Loxton, R., Teo, K.L., Rehbock, V.: An optimization approach to state-delay identification. IEEE Trans. Autom. Control **55**, 2113–2119 (2010)
166. Loxton, R., Teo, K.L., Rehbock, V.: Robust suboptimal control of nonlinear systems. Appl. Math. Comput. **217**(14), 6566–6576 (2011)

167. Loxton, R., Teo, K.L., Rehbock, V., Ling, W.K.: Optimal switching instants for a switched-capacitor DA/DC power converter. Automatica **45**, 973–980 (2009)
168. Loxton, R.C., Lin, Q., Teo, K.L., Rehbock, V.: Control parameterization for optimal control problems with continuous inequality constraints: new convergence results. Numer. Algebra Control Optim. **2**(3), 571–599 (2012)
169. Loxton, R.C., Teo, K.L., Rehbock, V.: Optimal control problems with multiple characteristic time points in the objective and constraints. Automatica **44**(11), 2923–2929 (2008)
170. Loxton, R.C., Teo, K.L., Rehbock, V.: Computational method for a class of switched system optimal control problems. IEEE Trans. Autom. Control **54**(10), 2455–2460 (2009)
171. Loxton, R.C., Teo, K.L., Rehbock, V., Yiu, K.F.C.: Optimal control problems with a continuous inequality constraint on the state and the control. Automatica **45**(10), 2250–2257 (2009)
172. Luenberger, D.G., Ye, Y.: Linear and Nonlinear Programming, 3rd edn. Springer, New York (2008)
173. Luus, R.: Optimal control by dynamic programming using systematic reduction in grid size. Int. J. Control **51**(5), 995–1013 (1990)
174. Luus, R.: Piecewise linear continuous optimal control by iterative dynamic programming. Ind. Eng. Chem. Res. **32**(5), 859–865 (1993)
175. Luus, R.: Iterative Dynamic Programming. Chapman & Hall/CRC, Boca Raton (2000)
176. Luus, R., Okongwu, O.: Towards practical optimal control of batch reactors. Chem. Eng. J. **75**(1), 1–9 (1999)
177. Malanowski, K., Maurer, H.: Sensitivity analysis for state constrained optimal control problems. Discrete Contin. Dyn. Syst. **4**(2), 241–272 (1998)
178. Malanowski, K., Buskens, C., Maurer, H.: Convergence of approximations to nonlinear optimal control problems. In: Fiacco, A.V. (ed.) Mathematical Programming with Data Perturbations V. Lecture Notes in Pure and Applied Mathematics, vol. 195, pp. 253–284. Springer, New York (1997)
179. Martez, J.M.: Inexact restoration method with Lagrangian tangent decrease and new merit function for nonlinear. J. Optim. Theory Appl. **111**, 39–58 (2001)
180. Martin, R.B.: Optimal control drug scheduling of cancer chemotherapy. Automatica **28**, 1113–1123 (1992)
181. Martin, R., Teo, K.L.: Optimal Control of Drug Administration in Cancer Chemotherapy. World Scientific, Singapore (1994)
182. Martinez, J.M., Pilotta, E.A.: Inexact restoration algorithm for constrained optimization. J. Optim. Theory Appl. **104**(1), 135–163 (2000)
183. The Mathworks, Inc., Natick, Massachusetts: MATLAB version 8.5.0.197613 (R2015a) (2015)

184. Maurer, H.: On the minimum principle for optimal control problems with state constraints. Tech. Rep. 41, Schriftenreihe des Rechenzentrums der Universität Münster (1979)
185. Maurer, H., Osmolovskii, N.P.: Second order sufficient conditions for time-optimal bang–bang control problems. SIAM J. Control Optim. **42**, 2239–2263 (2004)
186. Maurer, H., Buskens, C., Kim, J.-H.R., Kaya, C.Y.: Optimization methods for the verification of second order sufficient conditions for bang–bang controls. Optimal Control Appl. Methods **26**, 129–156 (2005)
187. McCormick, G.: Nonlinear Programming: Theory, Algorithms and Applications. Wiley, New York (1983)
188. McEneaney, W.: A curse-of-dimensionality-free numerical method for solution of certain HJB PDEs. SIAM J. Control Optim. **46**(4), 1239–1276 (2007)
189. Mehra, R., Davis, R.: A generalized gradient method for optimal control problems with inequality constraints and singular arcs. IEEE Trans. Autom. Control **AC-17**(1), 69–78 (1972)
190. Mehta, T., Egerstedt, M.: Multi-modal control using adaptive motion description languages. Automatica **44**, 1912–1917 (2008)
191. Miele, A., Wang, T.: Dual-properties of sequential gradient-restoration algorithms for optimal control problems. In: Conti, R., De Giorgi, E., Giannessi, F. (eds.) Optimization and Related Fields, pp. 331–357. Springer, New York (1986)
192. Miele, A., Pritchard, R.E., Damoulakis, J.N.: Sequential gradient-restoration algorithm for optimal control problems. J. Optim. Theory Appl. **5**, 235–282 (1970)
193. Miele, A., Wang, T., Basapur, V.K.: Primal and dual formulations of sequential gradient-restoration algorithms for trajectory optimization. Acta Astronaut. **13**, 491–505 (1986)
194. Misra, C., White, E.: Kinetics of crystallization of aluminium trihydroxide from seeded caustic aluminate solutions. Chem. Eng. Prog. Symp. Ser. **67**(110), 53–65 (1971)
195. Mitsos, A.: Global optimization of semi-infinite programs via restriction of the right-hand side. Optimization **60**(10–11), 1291–1308 (2011)
196. Mordukhovich, B.S.: Variational Analysis and Generalized Differentiation: Applications, vol. II. Springer, Berlin (2006)
197. Mu, Y., Zhang, D., Teng, H., Wang, W., Xiu, Z.: Microbial production of 1,3-propanediol by Klebsiella pneumoniae using crude glycerol from biodiesel preparation. Biotechnol. Lett. **28**, 1755–1759 (2008)
198. Neustadt, L.: Optimization: A Theory of Necessary Conditions. Princeton University Press, New York (1976)
199. Nocedal, J., Wright, S.: Numerical Optimization, 2nd edn. Springer, Berlin (2006)

200. Oberle, H.J., Sothmann, B.: Numerical computation of optimal feed rates for a fed-batch fermentation model. J. Optim. Theory Appl. **100**(1), 1–13 (1999)
201. Ořuztöreli, M.: Time-Lag Control Systems. Academic, New York (1966)
202. Parlar, M.: Some extensions of a student related optimal control problem. IMA Bull. **20**, 180–181 (1984)
203. Polak, E.: On the use of consistent approximations in the solution of semi-infinite optimization and optimal control problems. Math. Program. **62**(1), 385–414 (1993)
204. Polak, E., Ribiere, G.: Note sur la convergence de méthodes de directions conjuguées. ESAIM Math. Model. Numer. Anal. **3**(R1), 35–43 (1969)
205. Polik, I., Terlaky., T.: A survey of the S-Lemma. SIAM Rev. **49**(3), 371–418 (2007)
206. Pontryagin, L., Boltyanskii, V., Gamkrelidze, R., Mishchenko, E.: The mathematical Theory of Optimal Processes, vol. 4. Gordon and Breach Science Publishers, Montreux (1986)
207. Powell, M.: A fast algorithm for nonlinearly constrained optimization calculations. In: Watson, G. (ed.) Numerical Analysis. Lecture Notes in Mathematics, vol. 630, pp. 144–157. Springer, Berlin (1978)
208. Powell, W.: Approximate Dynamic Programming: Solving the Curses of Dimensionality. Wiley, New York (2007)
209. Raggett, G., Hempson, P., Jukes, K.: A student-related optimal control problem. Bull. Inst. Math. Appl. **17**, 133–136 (1981)
210. Rao, A.: Trajectory optimization: a survey. In: Waschl, H., Kolmanovsky, I., Steinbuch, M., del Re, L. (eds.) Optimization and Control in Automotive Systems. Lecture Notes in Control and Information Sciences. Springer, Cham (2014)
211. Rao, A., Benson, D., Darby, C., Patterson, M., Francolin, C., Sanders, I., Huntington, G.: Algorithm 902: GPOPS, a matlab software for solving multiple-phase optimal control problems using the gauss pseudospectral method. ACM Trans. Math. Softw. **37**(2), 22:1–22:39 (2010)
212. Reddien, G.: Collocation at Gauss points as a discretization in optimal control. SIAM J. Control Optim. **17**, 298–306 (1979)
213. Rehbock, V., Caccetta, L.: Two defence applications involving discrete valued optimal control. ANZIAM J. **44**, 33–54 (2002)
214. Rehbock, V., Livk, I.: Optimal control of a batch crystallization process. J. Ind. Manage. Optim. **3**(3), 585–596 (2007)
215. Rehbock, V., Teo, K.L., Jennings, L.S., Lee, H.: A survey of the control parameterization and control parameterization enhancing methods for constrained optimal control problems. In: Eberhard, A., Hill, R., Ralph, D., Glover, B. (eds.) Progress in Optimization: Contributions from Australasia, pp. 247–275. Kluwer Academic, Dordrecht (1999)
216. Royden, H.L.: Real Analysis, 2nd edn. MacMillan, New York (1968)

217. Ruby, T., Rehbock, V., Lawrance, W.B.: Optimal control of hybrid power systems. Dyn. Contin. Discrete Impuls. Syst. **10**, 429–439 (2003)
218. Sakawa, A.: Trajectory planning of a free-flying robot by using the optimal control. Optimal Control Appl. Methods **20**, 235–248 (1999)
219. Sakawa, Y., Shindo, Y.: Optimal control of container cranes. Automatica **18**(3), 257–266 (1982)
220. Schittkowski, K.: The nonlinear programming method of Wilson, Han, and Powell with an augmented Lagrangian type line search function, Part 1: convergence analysis. Numer. Math. **38**(1), 83–114 (1982)
221. Schittkowski, K.: On the convergence of a sequential quadratic programming method with an augmented Lagrangian line search function. Optimization **14**(2), 197–216 (1983)
222. Schittkowski, K.: NLPQL: a Fortran subroutine solving constrained nonlinear programming problems. Ann. Oper. Res. **5**(2), 485–500 (1986)
223. Schittkowski, K.: NLPQLP: a Fortran implementation of a sequential quadratic programming algorithm with distributed and non-monotone line search - User's guide, version 2.24. University of Bayreuth, Bayreuth (2007)
224. Schwartz, A.: Homepage of RIOTS. http://www.schwartz-home.com/riots/ (1997)
225. Schwartz, A.: Theory and implementation of numerical methods based on Runge-Kutta integration for solving optimal control problems. Ph.D. thesis, Electrical Engineering and Computer Sciences, University of California at Berkeley (1998)
226. Sethi, S., Thompson, G.: Optimal Control Theory: Applications to Management Science, 2nd edn. Kluwer Academic, Dordrecht (2000)
227. Shanno, D.: Conditioning of quasi-Newton methods for function minimization. Math. Comput. **24**(111), 647–656 (1970)
228. Siburian, A., Rehbock, V.: Numerical procedure for solving a class of singular optimal control problems. Optim. Methods Softw. **19**(3–4), 413–426 (2004)
229. Sirisena, H.: Computation of optimal controls using a piecewise polynomial parameterization. IEEE Trans. Autom. Control **18**(4), 409–411 (1973)
230. Sirisena, H., Chou, F.: Convergence of the control parameterization Ritz method for nonlinear optimal control problems. J. Optim. Theory Appl. **29**(3), 369–382 (1979)
231. Sturm, J.F.: Using SeDuMi 1.02, a MATLAB toolbox for optimization over symmetric cones. Optim. Methods Softw. **12**, 625–633 (1999)
232. Sun, W., Yuan, Y.: Optimization Theory and Methods - Nonlinear Programming. Springer, New York (2006)
233. Sun, Y., Aw, E., Teo, K.L., Zhou, G.: Portfolio optimization using a new probabilistic risk measure. J. Ind. Manage. Optim. **11**(4), 1275–1283 (2015)

234. Sun, Y., Aw, G., Loxton, R., Teo, K.L.: An optimal machine maintenance problem with probabilistic state constraints. Inf. Sci. **281**, 386–398 (2014)
235. Sun, Y., Aw, G., Teo, K.L., Wang, X.: Multi-period portfolio optimization under probabilistic risk measure. Financ. Res. Lett. **18**, 60–66 (2016)
236. Sun, Y.F., Aw, G., Loxton, R., Teo, K.L.: Chance constrained optimization for pension fund portfolios in the presence of default risk. Eur. J. Oper. Res. **256**, 205–214 (2017)
237. Teo, K.L.: Control parametrization enhancing transform to optimal control problems. Nonlinear Anal. Theory, Methods Appl. **63**, e2223–e2236 (2005)
238. Teo, K.L., Womersley, R.S.: A control parameterization algorithm for optimal control problems involving linear systems and linear terminal inequality constraints. Numer. Funct. Anal. Optim. **6**, 291–313 (1983)
239. Teo, K.L., Ahmed, N.U., Fisher, M.F.: Optimal feedback control for linear stochastic systems driven by counting processes. Eng. Optim. **15**(1), 1–16 (1989)
240. Teo, K.L., Clements, D.: A control parametrization algorithm for convex optimal control problems with linear constraints. Numer. Funct. Anal. Optim. **8**(5–6), 515–540 (1985)
241. Teo, K.L., Goh, C.J.: A Simple computational procedure for optimization problems with functional inequality constraints. IEEE Trans. Autom. Control **32**(10), 940–941 (1987)
242. Teo, K.L., Goh, C.J.: On constrained optimization problems with nonsmooth cost functionals. Appl. Math. Optim. **18**(1), 181–190 (1988)
243. Teo, K.L., Goh, C.J.: A unified computational method for several stochastic optimal control problems. Int. Ser. Numer. Math. **86**(2), 467–476 (1988)
244. Teo, K.L., Goh, C.J.: A computational method for combined optimal parameter selection and optimal control problems with general constraints. J. Aust. Math. Soc. B **30**(3), 350–364 (1989)
245. Teo, K.L., Jennings, L.S.: Nonlinear optimal control problems with continuous state inequality constraints. J. Optim. Theory Appl. **63**(1), 1–22 (1989)
246. Teo, K.L., Jennings, L.S.: Optimal control with a cost on changing control. J. Optim. Theory Appl. **68**(2), 335–357 (1991)
247. Teo, K.L., Lim., C.C.: Computational algorithm for functional inequality constrained optimization problems. J. Optim. Theory Appl. **56**(1), 145–156 (1998)
248. Teo, K.L., Wong, K.H.: A computational method for time-lag control problems with control and terminal inequality constraints. Optimal Control Appl. Methods **8**(4), 377–395 (1987)
249. Teo, K.L., Wong, K.H.: Nonlinearly constrained optimal control problems. J. Aust. Math. Soc. B **33**(4), 517–530 (1992)

250. Teo, K.L., Wu, Z.S.: Computational Methods for Optimizing Distributed Systems. Academic, Orlando (1984)
251. Teo, K.L., Ang, B., Wang, M.: Least weight cables: optimal parameter selection approach. Eng. Optim. **9**(4), 249–264 (1986)
252. Teo, K.L., Fischer, M.E., Moore, J.B.: A suboptimal feedback stabilizing controller for a class of nonlinear regulator problems. Appl. Math. Comput. **59**(1), 1–17 (1993)
253. Teo, K.L., Goh, C.J., Wong, K.H.: A Unified Computational Approach to Optimal Control Problems. Longman Scientific and Technical, Essex (1991)
254. Teo, K.L., Jennings, L.S., Lee, H.W.J., Rehbock, V.: The control parameterization enhancing transform for constrained optimal control problems. J. Aust. Math. Soc. B Appl. Math. **40**, 314–335 (1999)
255. Teo, K.L., Jepps, G., Moore, E.J., Hayes, S.: A computational method for free time optimal control problems, with application to maximizing the range of an aircraft-like projectile. J. Aust. Math. Soc. B **28**(3), 393–413 (1987)
256. Teo, K.L., Lee, W.R., Jennings, L.S., Wang, S., Liu, Y.: Numerical solution of an optimal control problem with variable time points in the objective function. ANZIAM J. **43**(4), 463–478 (2002)
257. Teo, K.L., Lim, C.C.: Time optimal control computation with application to ship steering. J. Optim. Theory Appl. **56**, 145–156 (1988)
258. Teo, K.L., Liu, Y., Goh, C.J.: Nonlinearly constrained discrete-time optimal-control problems. Appl. Math. Comput. **38**(3), 227–248 (1990)
259. Teo, K.L., Rehbock, V., Jennings, L.S.: A new computational algorithm for functional inequality constrained optimization problems. Automatica **29**(3), 789–792 (1993)
260. Teo, K.L., Wong, K.H., Clements, D.J.: Optimal control computation for linear time-lag systems with linear terminal constraints. J. Optim. Theory Appl. **44**(3), 509–526 (1984)
261. Teo, K.L., Yang, X.Q., Jennings, L.S.: Computational discretization algorithms for functional inequality constrained optimization. Ann. Oper. Res. **98**(1), 215–234 (2000)
262. Thompson, G.: Optimal maintenance policy and sale date of a machine. Manag. Sci. **14**(9), 543–550 (1968)
263. Uhlig, F.: A recurring theorem about pairs of quadratic forms and extensions: a survey. Linear Algebra Appl. **25**, 219–237 (1979)
264. Rehbock, V., Lim, C.C., Teo, K.L.: A stable constrained optimal model following controller for discrete-time nonlinear systems affine in control. Control Theory Adv. Technol. **10**(4), 793–814 (1994)
265. Varaiya, P.: Notes on Optimization. Van Nostrand Reinhold Notes on System Sciences. Van Nostrand Reinhold, New York (1972)
266. Varaiya, P.: Lecture Notes on Optimization (2013). https://people.eecs.berkeley.edu.cn.edu/~varaiya-optimization.pdf

267. Veliov, V.M.: Error analysis of discrete approximations to bang-bang optimal control problems: the linear case. Control Cybern. **34**(3), 967–982 (2005)
268. Vincent, T.L., Grantham, W.J.: Optimality in Parametric Systems. Wiley, New York (1981)
269. Vossen, G., Rehbock, V., Siburian, A.: Numerical solution methods for singular control with multiple state dependent forms. Optim. Methods Softw. **22**(4), 551–559 (2007)
270. Vossen, G.A., Maurer, H.: On L^1-minimization in optimal control and applications to robots. Optimal Control Appl. Methods **27**, 301–321 (2006)
271. Wang, L.Y., Gui, W.H., Teo, K.L., Loxton, R., Yang, C.H.: Optimal control problems arising in the zinc sulphate electrolyte purification process. J. Glob. Optim. **54**, 307–323 (2012)
272. Wang, L.Y., Gui, W.H., Teo, K.L., Loxton, R.C., Yang, C.H.: Time delayed optimal control problems with multiple characteristic time points: computation and industrial applications. J. Ind. Manage. Optim. **5**(4), 705–718 (2009)
273. Wang, S., Gao, F., Teo, K.L.: An upwind finite-difference method for the approximation of viscosity solutions to Hamilton-Jacobi-Bellman equations. IMA J. Math. Control Inf. **17**(2), 167–178 (2000)
274. Wang, S., Jennings, L.S., Teo, K.L.: Numerical solution of Hamilton-Jacobi-Bellman equations by an upwind finite volume method. J. Glob. Optim. **27**(2–3), 177–192 (2003)
275. Wang, Y., Xiu, N.: Theory and Algorithms for Nonlinear Programming (in Chinese). Shanxi Publisher of Science and Technology, Shanxi, China (2004)
276. Warga, J.: Optimal Control of Differential and Functional Equations. Academic, New York (1972)
277. Wächter, A., Biegler, L.T.: On the implementation of an interior-point filter line-search algorithm for large-scale nonlinear programming. Math. Program. **106**, 25–57 (2006)
278. Wilson, R.: A simplicial algorithm for concave programming. Ph.D. thesis, Harvard University, Cambridge (1963)
279. Wong, K.H.: Convergence analysis of a computational method for time-lag optimal control problems. Int. J. Syst. Sci. **19**(8), 1437–1450 (1988)
280. Wong, K.H., Clements, D.J., Teo, K.L.: Optimal control computation for nonlinear time-lag systems. J. Optim. Theory Appl. **47**(1), 91–107 (1985)
281. Wong, K.H., Jennings, L.S., Benyah, F.: The control parametrization enhancing transform for constrained time-delayed optimal control problems. ANZIAM J. **43**, E154–E185 (2002)
282. Wong, K.H., Jennings, L.S., Teo, K.L.: A class of nonsmooth discrete-time constrained optimal control problems with application to hydrothermal power systems. Cybernet. Syst. **24**, 339–352 (2007)

283. Woon, S.F., Rehbock, V., Loxton, R.C.: Towards global solutions of optimal discrete-valued control problems. Comput. Optim. Appl. **33**(5), 576–594 (2012)
284. Wu, C.Z., Teo, K.L.: Global impulsive optimal control computation. J. Ind. Manage. Optim. **2**(4), 435–450 (2006)
285. Wu, C.Z., Teo, K.L., Rehbock, V.: A filled function method for optimal discrete-valued control problems. J. Glob. Optim. **44**(2), 213–225 (2009)
286. Wu, Z.Y., Zhang, L.S., Teo, K.L., Bai, F.S.: A New Filled Function Method for Global Optimization. J. Optim. Theory Appl. **125**, 181–203 (2005)
287. Wu, C.Z., Teo, K.L., Wu, S.Y.: Min–max optimal control of linear systems with uncertainty and terminal state constraints. Automatica **49**, 1809–1815 (2013)
288. Wu, C.Z., Teo, K.L., Li, R., Zhao, Y.: Optimal control of switched systems with time delay. Appl. Math. Lett. **19**(10), 1062–1067 (2006)
289. Wu, D., Bai, Y.Q., Xie, F.S.: Time-scaling transformation for optimal control problem with time-varying delay. Discrete Contin. Dyn. Syst. S (2019). https://doi.org/10.3934/dcdss.2020098
290. Wu, D., Bai, Y.Q., Yu, C.Y.: A new computational approach for optimal control problems with multiple time-delay. Automatica **101**, 388–395 (2019)
291. Xiao, L., Liu, X.: An effective pseudospectral optimization approach with sparse variable time nodes for maximum production of chemical engineering problems. Can. J. Chem. Eng. **95**, 1313–1322 (2017)
292. Xiu, Z., Song, B., Sun, L., Zeng, A.: Theoretical analysis of effects of metabolic overflow and time delay on the performance and dynamic behavior of a two-stage fermentation process. Biochem. Eng. J. **11**, 101–109 (2002)
293. Xiu, Z., Zeng, A., An, L.: Mathematical modeling of kinetics and research on multiplicity of glycerol bioconversion to 1,3-propanediol. J. Dalian Univ. Technol. **40**, 428–433 (2000)
294. Yang, F., Teo, K.L., Loxton, R., Rehbock, V., Li, B., Yu, C.J., Jennings, L.: Visual miser: an efficient user-friendly visual program for solving optimal control problems. J. Ind. Manage. Optim. **12**(2), 781–810 (2016)
295. Yang, F., Teo, K.L., Loxton R., Rehbock, V., Li, B.,Yu, C.J., Jennings, L.: VISUAL MISER: an efficient user-friendly visual program for solving optimal control problems. J. Ind. Manage. Optim. **12**, 781–810 (2016)
296. Yang, X.Q., Teo, K.L.: Nonlinear Lagrangian functions and applications to semi-infinite programs. Ann. Oper. Res. **103**(1), 235–250 (2001)
297. Yu, C.J., Li, B., Loxton, R., Teo, K.L.: Optimal discrete-valued control computation. J. Glob. Optim. **56**(2), 503–518 (2013)

298. Yu, C.J., Lin, Q., Loxton, R., Teo, K.L., Wang, G.Q.: A hybrid time-scaling transformation for time-delay optimal control problems. J. Optim. Theory Appl. **169**, 876–901 (2016)
299. Yu, C.J., Teo, K.L., Bai, Y.Q.: An exact penalty function method for nonlinear mixed discrete programming problems. Optim. Lett. **7**, 23–38 (2013)
300. Yu, C.J., Teo, K.L., Zhang, L.S., Bai, Y.Q.: A new exact penalty function method for continuous inequality constrained optimization problems. J. Ind. Manage. Optim. **6**(4), 895–910 (2010)
301. Yu, C.J., Teo, K.L., Zhang, L.S., Bai, Y.Q.: On a refinement of the convergence analysis for the new exact penalty function method for continuous inequality constrained optimization problem. J. Ind. Manage. Optim. **8**(2), 485–491 (2012)
302. Yuan, J.L., Zhang, Y.D., Yee, J.X., Xie, J., Teo, K.L., Zhu, X., Feng, E.M., Yin, H.C., Xi, Z.L.: Robust parameter identification using parallel global optimization for a batch nonlinear enzyme-catalytic time-delayed process presenting metabolic discontinuities. Appl. Math. Model. **46**, 554–571 (2017)
303. Zhang, K., Teo, K.L.: A penalty-based method from reconstructing smooth local volatility surface from American options. J. Ind. Manage. Optim. **11**(2), 631–644 (2015)
304. Zhang, K., Teo, K.L., Swartz, M.: A robust numerical scheme for pricing American options under regime switching based on penalty method. Comput. Econ. **43**, 463–483 (2014)
305. Zhang, K., Wang, S., Yang, X.Q., Teo, K.L.: A power penalty approach to numerical solutions of two-asset American options. Numer. Math. Theory Methods Appl. **2**(2), 202–223 (2009)
306. Zhang, K., Wang, S., Yang, X.Q., Teo, K.L.: Numerical performance of penalty method for American option pricing. Optim. Methods Softw. **25**(5), 737–752 (2010)
307. Zhang, K., Yang, X.Q., Teo, K.L.: Augmented Lagrangian method applied to American option pricing. Automatica **42**, 1407–1416 (2006)
308. Zhang, K., Yang, X.Q., Teo, K.L.: A power penalty approach to American option pricing with jump diffusion processes. J. Ind. Manage. Optim. **4**, 783–799 (2008)
309. Zhang, K., Yang, X.Q., Teo, K.L.: Convergence analysis of a monotonic penalty method for American option pricing. J. Math. Anal. Appl. **348**, 915–926 (2008)
310. Zhong, W.F., Lin, Q., Loxton, R., Teo, K.L.: Optimal train control via switched system dynamic optimization. Optim. Methods Softw. (2019). https://doi.org/10.1080/0556788.2019.1604704

311. Zhou, J.Y., Teo, K.L., Zhou, D., Zhao, G.H.: Optimal guidance for lunar module soft landing. Nonlinear Dyn. Syst. Theory **10**(2), 189–201 (2010)
312. Zhou, J.Y., Teo, K.L., Zhou, D., Zhao, G.H.: Nonlinear optimal feedback control for lunar module soft landing. J. Glob. Optim. **52**(2), 211–227 (2012)

MIX
Papier aus verantwortungsvollen Quellen
Paper from responsible sources
FSC® C105338

Printed by Books on Demand, Germany